A Century of

SEPARATION
SCIENCE

A Century of

SEPARATION

SCIENCE

edited by
Haleem J. Issaq
SAIC Frederick
NCI-Frederick Cancer Research
and Development Center
Frederick, Maryland

CRC Press
Taylor & Francis Group
Boca Raton London New York

CRC Press is an imprint of the
Taylor & Francis Group, an **informa** business

CRC Press
Taylor & Francis Group
6000 Broken Sound Parkway NW, Suite 300
Boca Raton, FL 33487-2742

First issued in paperback 2019

ISBN-13: 978-0-8247-0576-3 (hbk)
ISBN-13: 978-0-367-39651-0 (pbk)

Visit the Taylor & Francis Web site at
http://www.taylorandfrancis.com

and the CRC Press Web site at
http://www.crcpress.com

It gives me great pleasure to dedicate this book to the memory of my father Mr. Jeries Sheha-deh Issaq; and to my mother Mrs. Lamia Issaq; my sister Miriam; and my brothers Kareem, Akram, Saleem and Elias, for their continued help and support over the years; and my wife Alberta and son Sameer for their patience and understanding.

Preface

Since its introduction by Tswett in 1901 as column chromatography for the separation of plant pigments, separation science has developed many diversified branches, each of which has its advantages, limitations, and fields of application. This book explores many aspects of separation science. For example, Dr. Leslie Ettre discusses the development of chromatography in the twentieth century, a period that witnessed the emergence, development, and dominance of separation science as an analytical technique. Today, chromatography and electrophoresis are "must" techniques in the analytical laboratory, a topic further explored in Professor Johan Roeraade's chapter, "From Crushed Bricks to Microchips," which delineates the major developments in separation science.

As we examine the advances of separation science in the 1900s, we realize that at the beginning of the twentieth century, the emphasis was on separating milligram quantities in minutes—researchers used either glass columns packed with large particle materials or paper chromatography. Detection was done by observing color bands, and quantitation methods were crude at best. As the century progressed, so did development of many separation techniques in addition to column and paper chromatography: thin layer chromatography, slab gel electrophoresis, gas chromatography, ion chromatography, supercritical fluid chromatography, field flow fractionation, countercurrent chromatography, solid phase extraction, affinity chromatography, and capillary electrophoresis. We also simultaneously witnessed the development of microchip technology for fast separations in milliseconds and the emergence of array capillary and microchip electrophoresis for the analysis of multisamples. By the end of the century, we were able to resolve nanogram quantities, in seconds on millimeter internal diameter (i.d.) columns packed with μm particles, or nanochannels (microchips), using sophisticated equipment with accurate detection instrumentation. So, the trend throughout the twentieth century focused on high throughput, increased efficiency, smaller sample size, less waste, high resolution, and more sensitive detection methods.

We expect that this trend will continue and that separation science will be not just a tool for the organic chemist but will be more closely involved in the biological and biomedical sciences. Thus, more molecular biologists, medicinal chemists, forensic scientists, and others will rely on separation science to resolve important biological problems at the single-gene and single-cell level. Increased use of microfluidics and hyphenated techniques will be employed to respond to these needs. The trend of the past century tells us that the twenty-first century will bring high speed (milliseconds), high throughput (multisamples simultaneously), extremely small sample sizes (nanoliters), and nanometer separation channels equipped with fully automated ultrahigh sensitive detectors (single-molecule level). However, that does not mean that the separation of milligram and gram quantities will disappear. On the contrary, such needs may increase as more and more drugs and products are developed.

This book records some of the advances that took place in the twentieth century, enabling the reader to trace separation science from its earliest development to its emergence as a leader in science research today. I asked each contributor to write his or her section so that it would be easy and enjoyable reading and not a dry statement of scientific events. As a result, the reader has in hand a volume that discusses different aspects of separation science and, I believe, one that will be not only informative but also pleasurable reading.

I would like to thank all the contributors for their time and efforts, and for sharing their knowledge. In addition, special thanks are due to my wife and son for their support and understanding during the many months I spent shepherding this project to completion.

I am especially grateful to Barbara Mathieu and her editorial staff at Marcel Dekker, Inc., for an excellent and professional job. Their attention to details and language requirements is commendable.

Haleem J. Issaq

Contents

Contents

Contributors

Daniel W. Armstrong, Ph.D.   Department of Chemistry, Ames National Laboratory, Iowa State University, Ames, Iowa

Victor G. Berezkin, Dr.Sc.   A. V. Topchiev Institute of Petrochemical Synthesis, Russian Academy of Sciences, Moscow, Russia

Phyllis R. Brown, Ph.D.   Department of Chemistry, University of Rhode Island, Kingston, Rhode Island

Jack Cazes, Ph.D.   Florida Atlantic University, Boca Raton, Florida

Jian Chen, Ph.D.   MDS Proteomics, Inc., Toronto, Ontario, Canada

Andreas Chrambach, Ph.D.   Laboratory of Cellular and Molecular Biophysics, National Institute of Child Health and Human Development, National Institutes of Health, Bethesda, Maryland

Frank David, Ph.D.   Research Institute for Chromatography, Kortrijk, Belgium

Zdenek Deyl, Ph.D., D.Sc.   Department for the Analysis of Physiologically Relevant Compounds, Institute of Physiology, Academy of Sciences of the Czech Republic, and Department of Analytical Chemistry, Institute of Chemical Technology, Prague, Czech Republic

Nebojsa M. Djordjevic, Ph.D.   Core Technology Area, Novartis Pharma AG., Basel, Switzerland

Leslie S. Ettre, Dr. Techn.   Department of Chemical Engineering, Yale University, New Haven, Connecticut

xi

Andrew G. Ewing, Ph.D.   Department of Chemistry, The Pennsylvania State University, University Park, Pennsylvania

Salvatore Fanali, Ph.D.   Istituto di Cromatografia, Consiglio Nazionale Delle Ricerche, Rome, Italy

Daniel Figeys, Ph.D.   MDS Proteomics, Inc., Toronto, Ontario, Canada

Niels H. H. Heegaard, M.D., D.Sc.(Med.)   Department of Autoimmunology, Statens Serum Institut, Copenhagen, Denmark

John V. Hinshaw, Ph.D.   ChromSource, Inc., Franklin, Tennessee

Stellan Hjertén, Ph.D.   Department of Biochemistry, Biomedical Center, Uppsala University, Uppsala, Sweden

Haleem J. Issaq, Ph.D.   SAIC Frederick, NCI-Frederick Cancer Research and Development Center, Frederick, Maryland

Yoichiro Ito, M.D.   Laboratory of Biophysical Chemistry, National Heart, Lung, and Blood Institute, National Institutes of Health, Bethesda, Maryland

Pavel Jandera, Ph.D.   Department of Analytical Chemistry, University of Pardubice, Pardubice, Czech Republic

George M. Janini, Ph.D.   SAIC Frederick, NCI-Frederick Cancer Research and Development Center, Frederick, Maryland

Kiyokatsu Jinno, Ph.D.   Department of Materials Science, Toyohashi University of Technology, Toyohashi, Japan

Jerry W. King, Ph.D.   Food Quality and Safety Research, National Center for Agricultural Utilization Research, Agricultural Research Service/USDA, Peoria, Illinois

Peter T. Kissinger, Ph.D.   Department of Chemistry, Purdue University, West Lafayette, Indiana

Ira S. Krull, Ph.D.   Department of Chemistry, Northeastern University, Boston, Massachusetts

James P. Landers, Ph.D.   Department of Chemistry, University of Virginia, Charlottesville, Virginia

Julie A. Lapos, B.A.   Department of Chemistry, The Pennsylvania State University, University Park, Pennsylvania

Maria T. Matyska, Ph.D.   Department of Chemistry, San José State University, San José, California

Uwe Dieter Neue, Ph.D.   Waters Corporation, Milford, Massachusetts

Janusz Pawliszyn, Ph.D.   Department of Chemistry, University of Waterloo, Waterloo, Ontario, Canada

Joseph J. Pesek, Ph.D.   Department of Chemistry, San José State University, San José, California

Fred E. Regnier, Ph.D.   Department of Chemistry, Purdue University, West Lafayette, Indiana

Pier Giorgio Righetti, Ph.D.   Department of Industrial and Agricultural Biotechnology, University of Verona, Verona, Italy

Johan Roeraade, Ph.D.   Department of Analytical Chemistry, Royal Institute of Technology, Stockholm, Sweden

Pat Sandra, Ph.D.   Department of Organic Chemistry, Ghent University, Ghent, Belgium

Gerhard Schomburg, Ph.D.*   Max Planck Institut für Kohlenforschung, Mülheim a.d. Ruhr, Germany

Volker Schurig, Ph.D.   Department of Chemistry, Institute of Organic Chemistry, University of Tübingen, Tübingen, Germany

Raymond P. W. Scott, D.Sc., F.R.S.C., F.A.I.C.   Consultant, Scientific Detectors Limited, Oxfordshire, England

Joseph Sherma, Ph.D.   Department of Chemistry, Lafayette College, Easton, Pennsylvania

Zak K. Shihabi, Ph.D.   Department of Pathology, Wake Forest University School of Medicine, Winston-Salem, North Carolina

Antoine-Michel Siouffi, Ph.D.   Faculté des Sciences St. Jerome, Université d'Aix-Marseille, Marseille, France

Hamish Small, M.Sc.   HSR, Leland, Michigan

Edward Soczewiński, Ph.D.   Department of Inorganic and Analytical Chemistry, Medical University, Lublin, Poland

Jiaqi Wu, Ph.D.   Convergent Bioscience Ltd., Toronto, Ontario, Canada

---

* Retired.

Xing-Zheng Wu, Ph.D.   Department of Materials Science and Engineering, Fukui University, Fukui-shi, Japan

Edward S. Yeung, Ph.D.   Ames Laboratory–USDOE and Department of Chemistry, Iowa State University, Ames, Iowa

1

Chromatography: The Separation Technique of the Twentieth Century*

Leslie S. Ettre

Yale University, New Haven, Connecticut

An essential condition for all fruitful research is to have at one's disposal a satisfactory technique. "Tout progrès scientific est un progrès de méthode"† as somebody once remarked. Unfortunately the methodology is frequently the weakest aspect of scientific investigations.

M. S. Tswett [1]

I. INTRODUCTION

M. S. Tswett contemplated the possibility of "chromatography" in 1899–1901 while carrying out his first research on the physico-chemical structure of plant chlorophylls, and he reported "on a new category of adsorption analysis" in 1903. Thus chromatography was born with the twentieth century.

Chromatography is based on a flow system containing two phases, mobile and stationary, and the sample components are separated according to differences in their distribution between the two phases. This separation principle is flexible and versatile, permitting one to explore various ways to achieve separation, and indeed, during its 100-year evolution chromatography underwent many steps, each representing new approaches and further broadening the scope and application of the technique. Each step in this evolution followed logically from the previous one. It started as liquid adsorption chromatography, which was followed by partition chromatography. Next its use was extended to the analysis of gases and vaporized samples. The principles

---

* Originally published in *Chromatographia*, 51:7–17, January 2000. Reprinted with permission from Vieweg Publishers, Wiesbaden, Germany.
† "All scientific progress is progress in a method." This statement is attributed to the French philosopher René Descartes (1596–1650), the author among others of the book *Discours de la méthode.*

of separation were also broadened by adding ion exchange and separation by molecular size and, most recently, electrochromatography.

Chromatography also permits great flexibility in the technique itself. The flow of the mobile phase might be controlled by gravity, pressure, capillary action, and electroosmosis; the separation may be carried out in a wide temperature range; and sample size can vary from a few atoms to many kilograms. Also, the shape of the system in which the separation takes place can be varied, using columns of various length and diameter or flat plates. Through all this evolution, chromatography was transformed from an essentially batch technique into an automated, instrumental method.

During its evolution chromatography not only interacted with its environment but greatly affected it, initiating a revolution in industrial and biochemical analysis. Through its continuous growth, chromatography became the most widely used analytical separation technique in chemistry and biochemistry. Thus, it is no exaggeration to call it the technique of the twentieth century.

At the beginning of the new millennium it is proper to survey this evolution, point out how the individual variants evolved, and explain their place in the continuously growing family of chromatography. It is not our aim to present a detailed review of the various chromatographic methods: the main purpose of this discussion is to shed light on their origin and interrelationships.

In the title of this paper we used the word "technique." However, chromatography is more than a simple technique: it is an important part of science encompassing chemistry, physical chemistry, chemical engineering, and biochemistry and cutting through different fields. When introduced, it represented a new paradigm and provided the theory and practice of interactions between two different phases.

We could not understand the beginnings of chromatography without knowing the state of art of chemistry at the time of its inception. Therefore, we start our discussion by presenting the background from which chromatography was born.

II. CHEMISTRY 100 YEARS AGO

100 years ago chemistry—particularly organic chemistry—and its techniques already had become well established. It was a world of some highly respected *gurus* with a flock of students and followers, and their main aim was to learn about reactions, to prepare pure substances, to try to reduce the more complex organic substances into their building blocks, and then to synthesize them from these simple compounds. Investigation of natural substances was still in its infancy and the key words were isolation and purification: isolation from the accompanying material and purification from other similar compounds present together with the main substance in small quantities. This was done by extraction and crystallization, and the proof of somebody's success was the showing of a few crystals of the pure substance. Very large amounts of starting material were needed: Richard Willstätter, the great organic chemist of the first half of the twentieth century, mentioned that in his laboratory in Zurich, there was a large basement room where "the isolation of chlorophyll from large vessels containing the powder of dried stinging nettle started" [2]. But isolation, in itself, was not science: it was done by junior associates. Rather, the reactions carried out with the isolated substance were considered as "science." For these investigations relatively large amounts—at least gram quantities—of the pure substance were needed.

Chromatography eventually changed this situation and from the 1930s onwards changed the way how the investigations of complex natural substances were carried out. However, for volatile sample components the situation still remained essentially unchanged until the mid-century. Justus G. Kirchner [3a] mentioned that in the second part of the 1940s, for the study of the flavoring and aroma materials of orange and grapefruit juices, close to 3000 gallons of the juice had to be processed, and Keene P. Dimick told the story [4] that 6 years of work and processing of 30 tons of strawberries was needed to finally obtain 35 mL of an oil, the essence of the fruit, which then permitted the further investigation of its constituents. This situation finally changed in the 1950s by the introduction of gas chromatography. In other words, the way chemistry approached the investigation of complex materials was fundamentally changed through the work of the pioneers of chromatography, the first of whom was M. S. Tswett.

III. TSWETT AND THE INVENTION OF CHROMATOGRAPHY

Mikhail Semenovich Tswett (1872–1919) was not a chemist: his Swiss doctorate as well as his Russian scientific degrees were in botany. We consider him a Russian although he was born in Italy (from a Russian father and a mother of Italian descent) and grew up in Switzerland, graduating in 1896 from the University of Geneva. In fact, he learned Russian only later, as a teenager, from his father, and his French accent could be recognized even later in his life. After graduation he followed his father to Russia, where he spent the rest of his life, first a few years in St. Petersburg and then, between 1901 and 1916, in Warsaw, in the Russian-occupied part of Poland. Because the Russian system did not accept foreign degrees for academic appointments, he had to earn a Russian Magister's degree and, for this, submit a new thesis on new and original research. Tswett's thesis submitted in 1901 to the University of Kazan' dealt with investigations of the physico-chemical structure of plant chlorophyll, and it represented the start of his research, which eventually led to the development of chromatography.

Tswett was interested in the natural systems represented by the plants in which chlorophyll is present, and his goal was to investigate the pigments as closely as possible to their native state. When trying to separate chlorophyll from the plant material, he realized that only polar solvents can extract it; however, after chlorophyll was isolated from the plant material, it could be easily dissolved in nonpolar solvents. Tswett correctly concluded that the pigments are present in the plants as adsorption complexes and the nonpolar solvents are unable to break the adsorption forces. Subsequent work carried out in Warsaw and reported in a lecture in 1903 (its English translation was published recently [5]) resulted in an embryonic separation method, based on stepwise adsorption precipitation and extraction. Finally, this work led to the technique of chromatography, involving selective adsorption/desorption in a flowing system, with the skillful use of various solvents [6]. (For a detailed discussion of the discovery process of chromatography by Tswett, see Ref. 7.)

Tswett's method represented a radical change in the existing philosophy of how natural substances were investigated: instead of obtaining a single compound in crystal form, he separated all the individual pigments from the plant matrix and from one another and characterized them by their spectroscopic properties. Also, he did not study the chemistry of the individual pigments as done, e.g., by Willstätter, but rather concentrated on their behavior in the plants. This is clear from the title of his *magnum opus* published in 1910 (representing his thesis for a Russian Doctor of Science degree): it dealt with the chromophylls present "in the plant and animal world" [1] and provided a unified treatment discussing not only the individual pigments but also their interaction with the plant material. However, this book—which also discussed the chromatographic separation technique in more details than the original publication [6]— was published only in Russian by a small local publisher and was practically unknown to western scientists.

In his lifetime Tswett's achievements were not appreciated by most of his peers, who criticized both his results on pigment research and his claims that chromatography is a superior separation technique that can provide compounds as pure as crystallization. (For a detailed discussion of the controversies related to Tswett and his achievements, see Robinson [8] and Ettre [9].) In fact, the negative opinion about chromatography as a purification method lingered for a long time. As late as 1929, F. M. Schertz, an American agricultural chemist, stated categorically that [10] "... it is evident that Tswett was never at any time dealing with pure pigments for not once were the substances crystallized." It took 10 more years until this argument was finally confuted by Paul Karrer, the great organic chemist, who in 1939 stated that [11] "... it would be a mistake to believe that a preparation purified by crystallization should be purer than one obtained from chromatographic analysis. In all recent investigations chromatographic purification widely surpassed that of crystallization."

IV. THE REBIRTH OF CHROMATOGRAPHY

In the two decades following the invention of chromatography only a handful of researchers utilized Tswett's technique: of these Charles Dhéré, at the University of Fribourg, in Switzerland [12], and Leroy S. Palmer, at the University of Missouri, in Columbia, MO [13], are particularly noteworthy. However, they represented isolated cases: a change in the philosophy was needed as to how complex natural substances are investigated before chromatography could prevail. As stated by Georg-Maria Schwab [14], "... only after biochemistry, pressed by new problems, demanded methods for the reliable separation of small quantities of similar substances, could chromatography celebrate a rapid and brilliant resurrection." This happened in 1930–1931 in Heidelberg, at the Kaiser Wilhelm Institute for Medical Research, and it was Edgar Lederer, a young postgraduate chemist, who first separated the various xanthophylls present in egg yolk by chromatography [15]. His work was picked up almost instantaneously by various groups in Europe, and within a few years chromatography was used for the investigation of a wide variety of natural substances. By 1937 a textbook by Zechmeister and Cholnoky, two Hungarian scientists, was published in German, the language of chemistry at that time [16]; its success was best demonstrated by the fact that within one year a second edition had to be published. This book greatly contributed to the wide acceptance of the technique. This was the time when finally the criticisms against "chromatographic purity" came to rest and its superiority was accepted.

The chromatography used at this time differed very little from Tswett's technique: it was a simple laboratory method carried out in small glass tubes. It needed considerable skill: the adsorbent had to be prepared and packed into a small tube, the sample solution added to the top of the column and then developed using various solvents. The process was stopped before the first sample component emerged from the column. Next, the contents of the tube, with the separated colored rings, was carefully pushed out and the individual rings separated with a sharp knife. Finally, the compounds present in these separate adsorbent fractions were extracted, the solutions were characterized by spectroscopy or by other means, and the pure compounds were obtained by evaporating the solvent. At this time the term "chromatogram" was actually used to describe the column with the separated rings of the sample components. In the second half of the 1930s a new method started to gain acceptance, the so-called *Durchflusschromatogramm* or "flow-through chromatogram": now the individual sample components did not remain in the column but were washed out of it with the eluent (the mobile phase). This technique is practically identical to the way liquid chromatography is carried out today, with two exceptions. The column was not pressurized, and the column effluent was not monitored continuously to measure the amount of individual compounds present: rather it was collected in small fractions, each of which were investigated separately. In the early 1940s continuous detection to monitor the column effluent by refractive index measurement was added as a possibility [17]. Another innovation was the consecutive use of different solvents or solvent mixtures with increasing developing and eluting power. This in effect may be considered to be a precursor to gradient elution chromatography, which was introduced by Alm et al. in 1952 [18]. Two new techniques were also described by Arne Tiselius in the early 1940s, in addition to the original elution-type analysis: *frontal analysis* [17] and *displacement development* [19].

Classical liquid chromatography based on adsorption-desorption was essentially a nonlinear process where the time of retardation (what we call today the retention time) and the quantitative response depended on the position on the adsorption isotherm. Essentially it was a preparative technique: the aim was to obtain the components present in the sample in pure form, which then may be subject to further chemical or physical manipulations. In addition, as we have already pointed out, the columns were not reusable and the success of separation depended to a great extent on the skill of the operator who prepared the adsorbent and packed the columns.

Classical chromatography was also a slow process: the mobile phase flow was mainly caused by gravity, although in some cases slight pressure was added to the head of the column to enhance the flow. However, even with these fairly simple systems, remarkable results were achieved. The next step in the evolution of chromatography further extended its range and eventually revolutionized the field. This was the introduction of partitioning as the basis of separation.

V. PARTITION CHROMATOGRAPHY

Liquid partition chromatography was developed by A. J. P. Martin and R. L. M. Synge and first described in 1941 [20]. The thoughts that led to the invention of the technique are well documented [3b–c,21,22]. They originally tried to separate monoamino monocarboxylic acids present in wool by counter-current extraction but had great practical difficulties with the technique. Eventually Archer Martin had the brilliant idea to fix one of the solvents and move only the other. In their first work, water was used as the liquid stationary phase (the fixed phase), silica gel as the support of it, and chloroform, containing 0.5% alcohol, as the mobile phase. The columns were developed either in the classical way (having the separated colored zones remain on the column: the bands of acids were made visible by methyl orange added to the column) or as a "flow-through chromatogram," eluting the individual bands and collecting them separately. The amount of amino acid present in each fraction was determined by titration. In addition to water and chloroform, Martin and Synge also used other substances as the two phases and emphasized that the technique is not restricted to protein chemistry.

The major importance of the work of Martin and Synge was that, by introducing partition as the basis of separation, it made chromatography a linear process in which retention characteristics—both qualitative and quantitative—were (naturally, within a certain range) independent of the sample size. Also, when using the flow-through technique, the columns were reusable. Because of the linear nature of the separation process, Martin and Synge were able to present a mathematical treatment of the theory of chromatography, including the theoretical plate concept. This also provided a quantitative way to express the separation power of the column. In addition, their 1941 paper also discussed the possibilities for further improvements in separation such as the use of very small particles and a high-pressure difference along the column.

VI. PAPER CHROMATOGRAPHY

It is interesting to note that in spite of the revolutionary nature of Martin and Synge's invention, liquid-liquid partition chromatography in columns was not picked up immediately by other researchers. However, liquid-liquid partition chromatography became very popular within a few years through the microanalytical technique of paper chromatography, also invented by Martin: in fact, paper chromatography overshadowed column chromatography.

Paper chromatography originated from a desire to extend the application of partition (column) chromatography to the separation of dicarboxylic and basic amino acids. However, for these silica gel was found to be too active as a support. Based on some previous knowledge of the analysis of dyes on paper, Martin tried to use filter paper impregnated with water for this purpose, and it was then logical to use the impregnated paper in a flat form and not as a column packing. After A. H. Gordon, a new member of Martin's team, found a color reaction to reveal the amino acids on the paper, the basics of paper chromatography were essentially completed. The new technique was first outlined in 1943 in a paper of which only a brief abstract is available [23]. Meanwhile Synge left the group and the development of the technique was concluded by Martin with Gordon and R. Consden, another new member of the team, in 1944. The detailed paper [24] reporting on their work also described the possibility of further enhancing the separation by using two different eluents in two directions ("two-dimensional separation") and presented the theory of the technique by introducing the term of the retardation factor (R_F).

The particular advantages of paper chromatography were its simplicity and speed as compared to earlier techniques. Its immediate impact had been characterized by W. J. Whelan, professor emeritus of the University of Miami School of Medicine, Miami, FL, who in 1945–1948 was a graduate student in England; it is worthwhile to quote his narrative in its entirety [25]:

> . . . The technological advance represented (by paper chromatography) was astonishing. Amino acids, which were formerly separated by laborious techniques of organic chemistry and where large quantities of protein hydrolysates were needed, could now be separated in microgram amounts and visualized. The technique soon spread to other natural products such as carbohydrates. . . . (The) technique . . . would allow one within the space of a week to carry out first a test for homogeneity and then a structural analysis of an oligosaccharide, which until then could very well have occupied the three years of a Ph.D. dissertation using Haworth's technique* of exhaustive methylation, hydrolysis, and identification of the methylated monosaccharides.

Indeed, within a few years paper chromatography, and through it partition chromatography, became universally accepted. In fact, in October 1948, the British Biochemical Society held in London a special Symposium on Partition Chromatography and the six papers presented— besides Martin's discussion of the theoretical aspects—all dealt with the application of the technique for the investigation of a wide variety of biologically important compound groups [26].

The achievements of Martin and Synge revolutionized not only chromatography, but also the way biochemical investigations were carried out. In addition, within a decade the application of partition chromatography for the analysis of gaseous and volatile compounds completely changed the way industrial analyses were carried out. Thus, it is not surprising that in 1952, Martin and Synge were honored by the Chemistry Nobel Prize. The announcement of this honor in *Nature* [27] concluded by saying that

---

* Sir Norman Haworth (1883–1950), the great British scientist who received the 1937 Chemistry Nobel Prize for his research in carbohydrates and vitamin C.

... the methods evolved by Martin and Synge are probably unique by virtue of simplicity and elegance of conception and execution, and also by the wide scope of their application. It is likely that (this) invention will be considered by future generations as one of the more important milestones in the development of chemical sciences.

Now, 50 years later we see even more the importance of their work and its impact on the whole field of science.

VII. THIN-LAYER CHROMATOGRAPHY

Thin-layer chromatography (TLC) is a planar chromatographic technique, like paper chromatography. However, while in paper chromatography only cellulose is available as the stationary phase matrix, TLC offers a variety of stationary phases. Also, by carrying out the separation on a porous layer consisting of small particles, the speed of analysis can be improved and the technique can be easily adapted to some automation.

TLC did not represent a new invention but only an extension of existing techniques. The first report is from 1938 by N. A. Izmailov and M. S. Shraiber [28] from the Pharmaceutical Institute in Kharkov, the Soviet Union (today: Ukraine). When investigating complex natural samples that form part of pharmaceutical solutions, they used classical column chromatography according to Tswett but found it too slow. Therefore, they tried to use thin layers of the adsorbent on small glass plates, adding only a drop of the sample solution and developing it by the dropwise addition of the solvent. The obtained chromatograms consisted of concentric circles of the separated substances. Izmailov and Shraiber called the technique *spot chromatography* [3d]. About 10 years later J. E. Meinhard and N. F. Hall at the University of Wisconsin, in Madison, WI, also carried out chromatography on adsorbent-coated glass plates and were the first to use a binder in preparing the adsorbent layer [29]: they called the technique *surface chromatography*. However, real thin-layer chromatography started in 1951 by Justus G. Kirchner [30], then at the laboratories of the U.S. Department of Agriculture in Pasadena, CA, who used glass plates coated with silicic acid (the so-called *chromatostrips*) for the analysis of terpenes in essential oils.

In the following years Kirchner widely demonstrated the usefulness of the technique, and he also described a number of possibilities, such as, e.g., carrying out reactions on the plate or utilizing two-dimensional chromatography, similar to the way already known in paper chromatography [3a]. Yet, in the public opinion, not Kirchner but Egon Stahl is usually considered as the originator of thin-layer chromatography. Stahl published his first paper on *Dünnschichtchromatographie* (thin-layer chromatography) in 1956 [31]; however, the technique won wide interest only after his second paper was published 2 years later [32], which also described the equipment useful for the coating of glass plates with a thin layer of special, fine-grain adsorbent. Such equipment as well as standardized adsorbents soon became commercially available, and this fact greatly helped in the wide acceptance of the technique. In addition, Stahl published in 1962 a very useful and highly popular handbook of TLC [33], which was translated into a number of languages; he was also instrumental in the standardization of the procedure and its materials [3e].

The main advantage of TLC is that it is a simple method that permits the parallel analysis of a number of samples. In the last 25 years the technique has undergone a number of improvements, expanding its scope and permitting automation, particularly of sample introduction and the quantitation of detection [34]. It is a living method, and this is best exemplified by the fact that the fourth edition of a popular handbook was just published in the United States [35]. TLC will continue to have a role in the routine analysis of large number of samples.

VIII. GAS CHROMATOGRAPHY

In their 1941 paper on (liquid-liquid) partition chromatography [20], Martin and Synge predicted that ". . . the mobile phase need not to be a liquid but may be a vapour. . . . Very refined separations of volatile substances should therefore be possible in a column in which permanent gas is made to flow over gel impregnated with a non-volatile solvent." However, this suggestion lay dormant for a decade, and *gas-liquid partition chromatography* (GLPC) was finally developed by Martin with A. T. James [3b,3f]. Their first report on some preliminary results was presented on October 20, 1950, at a meeting of the Biochemical Society; a detailed description of the results was published 16 months later [36]. This paper not only reported on the technique but also further expanded the theory of partition chromatography by considering the compressibility of the gas used as the mobile phase. This original paper only dealt with the separation of fatty acids; however, in two additional papers published within a few months [37,38], the application of GLPC to the separation of basic compounds was also demonstrated, and the possibility of extending it to other types of compounds was also indicated.

It should be noted that gas adsorption chromatography was investigated in the late 1940s by Erika Cremer at the University of Innsbruck in Austria [3g,39]; however, it was not followed up and, thus, had no impact on the further evolution of gas chromatography.

The impact of GLPC on analytical chemistry was tremendous and almost instantaneous. A number of reasons contributed to this instant success. The first was that soon after the development of the technique, Martin elaborated on its theory and practice in a major lecture at the First International Congress on Analytical Chemistry held in Oxford, England, in September 1952, before the widest possible audience, and this lecture was soon also published [40]. Also, Martin had early contact with scientists from major industrial organizations (see, e.g., [3h–l]) and not only demonstrated the technique but also advised them how to extend it for other types of samples and how to simplify it, e.g., by using syringe injection and a thermal-conductivity detector. At that time the new processes in petroleum refining and in the petrochemical industries required improved analytical controls, which were not possible using the old laboratory techniques. Gas chromatography provided the ideal way to solve these problems. Finally, we should also mention the fact that soon a very good textbook written by A. I. M. Keulemans [41]— mainly based on the accumulated knowledge gathered from the Shell laboratories—became available, from which newcomers could learn the theory and practice of the technique. This book was published in two English editions and was also translated into other languages.

Gas-liquid partition chromatography was the right method introduced at just the right time, providing a simple and sensitive method for the analysis of volatile compounds. In the first years its applications were mostly concentrated on the analysis of hydrocarbons and fixed gases; however, soon it was utilized for almost every type of organic compound. In fact, the method had such obvious advantages that it was even used for the analysis of many types of nonvolatile compounds, which were converted to volatile derivatives for this purpose. Steroids and amino acids are classical examples of this: there was a period during which their analysis by GLPC was widely reported.

The introduction and the success of gas chromatography also had another important effect on the evolution of modern analytical chemistry. While classical liquid chromatography was a technique based on manual dexterity, gas chromatography could not be carried out in simple glass tubes: it needed *instrumentation*, developed by collaboration between chemists, engineers, and physicists. Thus, its development also helped to create a new industry, the scientific instrument industry, which, from its modest beginnings in the early 1950s grew to the present multibillion dollar giant [42,43].

The evolution of gas chromatography went through a number of stages. It started as a relatively simple operation, with packed columns, isothermal temperature control, and thermal-conductivity detection. Within a short time the system incorporated temperature programming [44,45], ionization detectors [46–49], and open-tubular (capillary) columns [3m,50], and the usable temperature range was also expanded. Also, its theory was further extended by describing the influence of diffusion, gas velocity, particle size, tube diameter, and the resistance to mass transfer between the two phases by Van Deemter et al. [3n,51] and by Golay [50]. These achievements made gas chromatography within two decades the most widely used analytical separation technique, surpassed only relatively recently by modern, high-performance liquid chromatography.

IX. ION-EXCHANGE CHROMATOGRAPHY*

Ion-exchange chromatography was first described in 1938 by Taylor and Urey for the separation of lithium and potassium isotopes, using inorganic zeolites as the stationary phase [53]. Synthetic ion-exchange resins were also introduced at this time and were used for the first time by Olof Samuelson for the separation of interfering anions and cations [54,55].

Ion-exchange chromatography found an important application during World War II for the separation of rare earths, as part of the Manhattan Project. In this work carried out at Iowa State College (today: Iowa State University) in Ames, Iowa, and at Oak Ridge National Laboratories, in Oak Ridge, TN, the possibility of the expansion of liquid chromatography to a truly *production-scale operation* was first demonstrated. The "chromatographic plant" established at Ames consisted of a dozen columns, and over 1000 liters of mobile phase per column were consumed in one operation. As a conclusion of this work a number of rare earth oxides—which up to then were hardly available even in milligram quantities—were produced in quantities of hundreds of grams. Due to its nature, this work was classified during the war and reported only in 1947–1949. Even today it is fascinating to read the summary of these activities [56,57]. After the war, production on an even larger scale was established, using 12 10-ft.-long, 30-inch-diameter columns in series. In this system thousands of pounds of high-purity yttrium and many kilograms of the other rare earths were produced [58].

It should be mentioned that the Oak Ridge group was the first to apply the plate theory introduced in 1941 by Martin and Synge to describe the efficiency of the ion-exchange separation process [59].

While the use of ion-exchange chromatography for the separation of rare earths is fairly well documented, it is little known that the technique also had an important role in the research on transuranium elements. In fact, a number of transuranium elements could be detected only after ion-exchange separation of various irradiation products in very small amounts [3o,60]. Probably one of the most remarkable achievements of separation science was represented by the separation of element 101 (mendelevium) from elements 99 (einsteinium) and 100 (fermium), involving only 17 atoms [61].

It may be mentioned here that the experience gained in the investigation of fission products at Oak Ridge was then turned toward the solution of problems in biochemistry: here Waldo E. Cohn's pioneering work in 1949–1950 on the separation of nucleic acid constituents by ion-exchange chromatography is particularly noteworthy [62,63].

---

* The evolution of ion-exchange chromatography can be followed through the extraordinary collection edited by Harold F. Walton, presenting reprints of 48 benchmark papers [52].

Another important impact of ion-exchange chromatography on the overall evolution of chromatography is related to the work of Stanford Moore and William H. Stein at the Rockefeller Institute for Medical Research (today: Rockefeller University) in New York City, on the *analysis of amino acids* [3p]. After first using starch columns at the end of the 1940s, they soon switched to the new sulfonated polystyrene ion-exchange resins [64,65]. Within a little over a decade the time needed for a complete analysis could be reduced from 2 weeks to 6 hours. At that time, together with Darrel H. Spackman, Moore and Stein also developed the *Amino Acid Analyzer*, an automated instrument [66,67] further reducing the analysis time. This instrument truly revolutionized biochemical analysis and permitted the study of the structure of proteins and enzymes. Moore and Stein were involved in the investigation of ribonuclease, for which they eventually received the 1972 Nobel Prize in Chemistry, together with Christian B. Anfinsen.

X. SIZE-EXCLUSION CHROMATOGRAPHY

Another variant of chromatography that became a very important tool in biochemistry was *gel filtration*, in which the macromolecules are separated by differences in their molecular size, on a hydrophilic gel consisting of dextran cross-linked with epichlorohydrin. This material and the technique were developed by Per Flodin and Jerker O. Porath [3q–r], as a cooperative venture between the Department of Biochemistry of Uppsala University and Pharmacia AB, the biotechnology company, which introduced the gel under the trade name Sephadex in 1959 [68,69]. On the other hand, the use of hydrophobic polystyrene gels by J. C. Moore of the Dow Chemical Co. led to the development of *gel-permeation chromatography* (GPC) [70], enabling the determination of the molecular weight distribution of high-molecular-weight synthetic polymers. In the mid-1960s special instruments were introduced for such measurements, and these may be considered as early dedicated liquid chromatographs. Eventually the two techniques merged under the name *size-exclusion chromatography* (SEC) and continued to be an important branch of liquid chromatography.

XI. HIGH-PERFORMANCE LIQUID CHROMATOGRAPHY

While GLPC already involved fairly sophisticated instrumentation with reusable columns, classical liquid chromatography in the 1950s still relied on manual methods, and the reproducibility of the determination depended mainly on the analyst's skill. Encouraged by the success of the amino acid analyzer, attempts were made to develop more or less automated instruments based on liquid chromatography: the "Steroid Analyzer" developed at the U.S. National Institutes of Health by the group of Erich Heftmann is an example [71,72]. However, this was also a single-purpose system, with limited applications.

By today's standards classical liquid chromatography was an inherently slow technique, and the dedicated instruments did not change in this. We have already mentioned that a full determination on the Amino Acid Analyzer took 6 hours; similarly, the analysis of an adrenal extract with the Steroid Analyzer needed 8.5 hours. By extending the well-developed theory of gas chromatography to liquid chromatography (LC), it became apparent that LC was limited by the slowness of diffusion in the liquid mobile phase, about three orders of magnitude slower than in a gas. Thus, if the technique was to be improved, then this slowness had to be overcome by other means, notably by using uniform, small particles with a short diffusion path and relatively high velocities (as compared to classical LC), requiring high pressures. Based on their laboratory experience, this was already predicted by Martin and Synge in their fundamental

1941 paper [20]. Theoretical papers published in the early 1960s also resulted in the same conclusions but in a more sophisticated way, using exact relationships [73].

By the mid-1960s the field was ripe for a major leap forward: the development of a liquid chromatograph, an integrated system based on the model of a gas chromatograph. Such a system would employ high pressures, sample injection at such pressures, accurate flow controls, and continuously monitoring detectors without band spreading and would permit the analysis of a wide range of sample types with high efficiencies. The first modern liquid chromatograph along these lines was developed by Csaba Horváth at Yale University in 1965 [3s], with particular regard to the investigation of biological substances. After a brief report [74] it was described in more detail during the discussion at the 1966 International Chromatography Symposium [75]. Because at that time small particles with uniform size distribution and other desired properties were simply not available. Horváth developed a pellicular column packing in which a thin (a few μm) porous layer of the sorbent was coated on impermeable glass beads of about 40 μm diameter and narrow particle size distribution. These particles were then packed into columns of 1 mm inner diameter* [76]. This was then followed by the work of Jack Kirkland at DuPont [3t], also developing pellicular material, which soon became commercially available [77]. Parallel to this work, Lloyd R. Snyder also carried out systematic investigations on improvements of adsorption chromatography [3u,78,79].

This new liquid chromatography received a special name based on a suggestion of Horváth: it is now universally called *high-performance liquid chromatography* and is usually identified by its acronym, HPLC.

At the beginning it was not clear what branch of liquid chromatography would prevail. However, soon *reversed-phase chromatography* (RPC), using an essentially nonpolar stationary phase with polar mobile phases, became the predominant technique. The advantages of RPC had been shown earlier [80], but its use was hindered by the lack of suitable stationary phases and by the lack of understanding of the underlying physicochemical phenomena.† Now Horváth established its theory [82], while Kirkland pioneered the development of the corresponding ''bonded'' stationary phases consisting of long-chain hydrocarbon or other organic moieties covalently bound on silica particles [83,84]. These were soon available commercially from DuPont.

All these developments occurred in the lifetime of the present generation, representing an exciting time in the evolution of chromatography. Some of the key players recorded their personal stories [3s–v,85,86], and these make fascinating reading.

In earlier discussions we pointed out the impact of textbooks on the growth of the various chromatographic techniques, such as the book of Zechmeister and Cholnoky on classical liquid chromatography [16], the handbook of Stahl on TLC [33], and Keulemans' compilation on gas chromatography [41]. The situation was also similar to HPLC, where its rapid evolution was aided by the excellent textbook written by Jack Kirkland and Lloyd Snyder, the first edition of which was published in 1971 [87]. In subsequent decades, this book and its revised editions helped thousands of chemists learn the intricacies of the technique.

The evolution of HPLC is still continuing, particularly in the life sciences and biotechnology area. Today it has surpassed gas chromatography as the most widely used analytical technique. A good quantitative measure of this is the yearly sales volume of the respective instruments,

---

* Much later columns with such inner diameter were reinvented by others under the name of ''microbore columns.''

† In fact, a publication of the I.U.P.A.C. from 1972 [81] indicated that reversed-phase chromatography is ''a technique of only historical interest.''

accessories, and materials: according to a recent estimate [88], the annual sales revenues for HPLC instruments are more than double ($2.2 billion) that for GC instruments ($1 billion) and still show a healthy growth.

Finally, we should mention that in the last decade the use of liquid chromatography has extended beyond the analytical laboratory into the field of biopharmaceutical processing, the production of pure substances (see, e.g., [89,90]). A recent summary of this evolution by E. N. Lightfoot, a leading chemical engineering scientist, is very interesting reading [91].

XII. SUPERCRITICAL-FLUID CHROMATOGRAPHY

Supercritical-fluid chromatography (SFC) is considered to be halfway between gas chromatography and liquid chromatography. In supercritical fluids the diffusion is much faster than in liquids, although still slower than in gases; however, separation can be carried out at lower temperatures, permitting the analysis of heat-sensitive, nonvolatile compounds, which cannot be handled by GC.

The first documented suggestion of chromatography with mobile phases in supercritical state was made by James E. Lovelock, in 1958 [92]. Actual investigations with dense gases for the transport of nonvolatile substances through a chromatographic column were first reported in the 1960s by three independent groups. The first was Ernst Klesper, a visiting scientist from Germany at Johns Hopkins University, in Baltimore, MD, working in the group of A. H. Corwin: he demonstrated the separation of nickel etioporphyrin isomers using ''high-pressure gas chromatography,'' with supercritical chlorofluoromethanes as the mobile phase* [93]. This was followed by in-depth studies of GC with very high pressures by M. N. Myers and J. C. Giddings at the University of Utah [94,95]. The most thorough investigations were carried out at the Shell laboratories in Amsterdam, The Netherlands, studying SFC with different mobile and stationary phases [96,97]. However, these investigations were fairly isolated: at that time instrumentation was not yet sufficiently advanced, and within a few years the advantages of HPLC overshadowed the possible advantages of SFC. Nevertheless, interest in SFC was renewed at the beginning of the 1980s. Dennis R. Gere and associates at Hewlett-Packard modified a commercial HPLC instrument for supercritical fluid chromatography and generally used 10 cm $\times$ 4.6 mm i.d. columns packed with 3 μm particles of a bonded-phase material [98]. As the mobile phase they used CO_2 or, in the case of polar solutes insoluble in it, CO_2 to which small volumes of a polar modifier were added. About the same time the groups of Milos Novotny at Indiana University, in Bloomington, IN, and Milton L. Lee at Brigham Young University, in Provo, UT, successfully introduced SFC with open-tubular (capillary) columns, using both CO_2 and n-pentane as the mobile phase [99]. In these works both the flame-ionization detector, the universal GC detector, and the UV detector of HPLC, now modified to permit operation under high pressures, were used.

In the second part of the 1980s supercritical-fluid chromatography went through a sudden upsurge. In addition to some major instrument companies offering chromatographs for supercritical-fluid operation, a few smaller companies specializing in the development and marketing of SFC instruments were also founded. At that time enthusiastic supporters of the technique even

---

* It is practically unknown that at that time the Perkin-Elmer Corporation built for the Baltimore group on special order a ''supercritical chromatograph.'' I remember well the instrument, the construction of which was the responsibility of Stanley Norem, one of my colleagues at that time; however, I have no information about the fate of the unit.

predicted that it would revolutionize the way analytical separations are carried out. However, this did not materialize and only a limited number of applications were found where SFC offered some advantages as compared to GC and HPLC. Because of the vanishing interest, most of the "big players" slowly withdrew from the field, and the specialized companies that flourished at the end of the 1980s have ceased to exist. Recent evaluation of the future of SFC has been fairly skeptical [100,101].

Another employment of supercritical fluids is *supercritical-fluid extraction* (SFE). The use of supercritical CO_2 as an efficient solvent is, of course, not new: analytical as well as industrial processes have been based on its use. During the period of the upsurge of SFC, it was proposed that the combination of SFE with SFC could yield an essentially integrated process, using the same fluid as both the solvent and the mobile phase. Unfortunately such combined systems were found to be too complicated, with many pitfalls; however, the use of supercritical CO_2 as a solvent for the preparation of samples for chromatographic analysis remained one possible alternative.

XIII. CAPILLARY ELECTROCHROMATOGRAPHY

Electrophoresis as a separation technique was introduced in the 1920s by Arne Tiselius, who was awarded the 1948 Chemistry Nobel Prize for his pioneering work [102]. Electrophoresis, particularly slab (thin-layer) electrophoresis, continues to remain an important method in biochemical analysis.

In 1979 Dandeneau and Zerenner introduced fused-silica capillary tubing for the preparation of open-tubular columns in gas chromatography [103,104]. These columns not only revolutionized GC, but also made it possible to scale down the other separation techniques. In this respect the pioneering work of James Jorgenson on the development of *capillary zone electrophoresis* (CZE) should be emphasized [105].

Although usually lumped together with the chromatographic methods, CZE is not chromatography: in this technique there is no mobile phase flow and the charged molecules migrate to the respective electrodes with different velocities, the magnitude of which depends on the potential difference as well as the shape and size of the molecules. It is true that CZE is a separation method of very high efficiency, however, the initial great expectations soon dissipated, mainly because—as expressed by Georges Guiochon [101]—"the method has proven exceedingly difficult to reduce to a quantitative analytical technique"; also, its reproducibility was found to be much poorer than that of HPLC.

While CZE did not fulfill its original promise, it eventually led to *capillary electrochromatography* (CEC), the newest member of the chromatography family. CEC represents a combination of capillary electrophoresis and high-performance liquid chromatography. As characterized by Csaba Horváth, one of the leading scientists in this field [106], CEC "employs capillary columns packed with a suitable stationary phase having fixed charges at the surface, and electroosmotic flow of the eluent generated by applying a high electric field. . . . It is a *bona fide* chromatographic technique for the separation of both neutral and charged sample components."

In science, it is often the case that new developments are based on ideas which were put forward long ago but which were not practicable at that time because other segments of science and technology were lagging in overall development. This is also true about CEC. Actually, the concept of using electro-osmosis to generate flow of the mobile phase in liquid chromatography was first proposed 25 years ago by Victor Pretorius in a lecture presented at the opening session of the 9th International Symposium on Advances in Chromatography (November 4–7, 1974, Houston, TX) [107]. I chaired the session and remember well the presentation, which not only covered the theory of electro-osmosis but also presented some crude experimental data on thin layers as well as on packed and open-tube columns of 1-mm diameter. However, this brilliant suggestion did not have any immediate follow-up. Pretorius, himself, realized this situation and in a personal retrospection expressed his hopes that "sometime, someone . . . will further exploit this crude, but, I believe, potentially worthwhile idea" [3w]. Finally, his hope was fulfilled.

At the present time CEC is creating a lot of excitement. However, it is too early to make any prediction concerning its future.

XIV. CHROMATOGRAPHY: AN APPRAISAL

We have reviewed a century of development: this survey was, however, brief by its nature, and aimed only at a few interesting points, mainly related to the beginnings of the individual techniques and the improvements achieved relative to earlier methods. Our main aim was to show how Tswett's original idea grew beyond the inventor's wildest dreams, becoming the most widely used laboratory separation technique.

During the twentieth century chromatography changed the way in which complex natural substances are investigated, and it has became an indispensable analytical tool in all areas of science and technology. Chromatography also greatly aided biochemistry and life sciences: as expressed by Synge in his Nobel Lecture [22], it helped to merge chemistry with biology. The impact of chromatography on biochemistry and biotechnology has became even more pronounced in the last two decades, and this evolution will be even more important in the years to come. One thing is certain: chromatography is still continuously growing and its fields of application are widening. Future chromatographers will encounter new and exciting developments, and their efforts will be highly rewarded by new discoveries.

REFERENCES

1. MS Tswett. Khromofilly v Rastitel'nom i Zhivotnom Mire (Chromophylls in the Plant and Animal World). Warsaw: Karbasnikov Publishers, 1910.
2. R Willstätter. Aus meinem Leben: Von Arbeit, Musse und Freuden. 2nd ed. Weinheim: Verlag Chemie, 1973, p. 156.
3. LS Ettre, A Zlatkis, eds. 75 Years of Chromatography—A Historical Dialogue. Amsterdam: Elsevier, 1979: (a) JG Kirchner, pp. 201–298; (b) AJP Martin, pp. 285–296; (c) RLM Synge, pp. 447–452; (d) MS Shraiber, pp. 413–417; (e) E Stahl, pp. 425–435; (f) AT James, pp. 167–172; (g) E Cremer, pp. 21–30; (h) ER Adlard, pp. 1–10; (i) H Boer, pp. 11–19; (j) DH Desty, pp. 31–42; (k) GR Primavesi, pp. 339–343; (l) NH Ray, pp. 345–350; (m) MJE Golay, pp. 109–114; (n) JJ Van Deemter, pp. 461–465; (o) GT Seaborg, G Higgins, pp. 405–412; (p) S Moore, WH Stein, pp. 297–308; (q) P Flodin, pp. 67–74; (r) JO Porath, pp. 323–331; (s) C Horváth, pp. 151–158; (t) JJ Kirkland, pp. 209–217; (u) LR Snyder, pp. 419–424; (v) JFK Huber, pp. 159–166; (w) V Pretorius, pp. 333–338.
4. KP Dimick. LC·GC North Am 8:782–786, 1990.
5. VG Berezkin, compiler, Chromatographic Adsorption Analysis: Selected Works of M. S. Tswett. New York: Ellis Horwood, 1990, pp. 9–19.
6. M Tswett. Ber Dtsch Botan Ges 24:316–326, 384–393, 1906.
7. LS Ettre, KI Sakodynskii. Chromatographia 35:223–231, 329–338, 1993.
8. T Robinson. Chymia 6:146–160, 1960.
9. LS Ettre. In: C Horváth, ed., High Performance Liquid Chromatography—Advances and Perspectives. Vol. I. New York: Academic Press, 1980, pp. 1–74.

10. FM Schertz. Plant Physiol 4:337–348, 1929.

11. P Karrer. Helv Chim Acta 22:1149–1150, 1939.

12. V Meyer, LS Ettre. J Chromatogr 600:3–15, 1992.

13. LS Ettre, RL Wixom. Chromatographia 37:659–668, 1993.

14. GM Schwab, K Jockers. Angew Chem 50:546–553, 1937.

15. E Lederer. J Chromatogr 73:361–366, 1972.

16. L Zechmeister, L Cholnoky. Die Chromatographische Adsorptionsmethode. Vienna: Springer Verlag, 1937; 2nd ed. 1938. English translation published by Chapman & Hall, London, in 1941.

17. A Tiselius. Ark Kem Mineral Geol 14B(22):1–5, 1940.

18. RS Alm, RJP Williams, A Tiselius. Acta Chem Scand 6:826–836, 1952.

19. A Tiselius. Ark Kem Mineral Geol 16A(18):1–11, 1943.

20. AJP Martin, RLM Synge. Biochem J 35:1358–1368, 1941.

21. AJP Martin. In: Nobel Lectures: Chemistry 1942–1962. Amsterdam: Elsevier, 1964, pp. 355–373.

22. RLM Synge. In: Nobel Lectures: Chemistry 1942–1962. Amsterdam: Elsevier, 1964, pp. 374–389.

23. AH Gordon, AJP Martin, RLM Synge. Biochem J 37:xiii only, 1943.

24. R Consden, AH Gordon, AJP Martin. Biochem J 38:224–232, 1944.

25. WJ Whelan. FASEB J 9:287–288, 1995.

26. RT Williams, RLM Synge, eds. Partition Chromatography. Biochemical Society Symposia No. 3, October 30, 1948, Cambridge: Cambridge University Press, 1951.

27. Anon. Nature (London) 170:826 only, 1952.

28. NA Izmailov, MS Shraiber. Farmatsiya 3:1–7, 1936. For English translation of the paper see: J Planar Chromatogr 8:402–405, 1995.

29. JE Meinhard, NF Hall. Anal Chem 21:185–188, 1949.

30. JG Kirchner, JM Miller, G Keller. Anal Chem 23:420–425, 1951.

31. E Stahl. Pharmazie 11:633–637, 1956.

32. E Stahl. Chem Ztg 82:323–328, 1958.

33. E Stahl, ed. Dünnschichtchromatographie. Ein Laboratoriumshandbuch. Berlin: Springer Verlag, 1962. English translation was published in 1965.

34. H Kalász, LS Ettre, M Báthori. LC·GC North Am 15:1044–1050, 1997.

35. B Fried, J Sherma. Thin-Layer Chromatography, 4th ed. New York: Marcel Dekker, Inc., 1999.

36. AT James, AJP Martin. Biochem J 50:679–690, 1952.

37. AT James, AJP Martin, GH Smith. Biochem J 52:238–242, 1952.

38. AT James. Biochem J 52:242–247, 1952.

39. LS Ettre. Am Lab 4(10):10–16, October 1972.

40. AT James, AJP Martin. Analyst (London) 77:915–932, 1952.

41. AIM Keulemans. Gas Chromatography, 2nd ed. New York: Reinhold, 1957, 1959.

42. LS Ettre. J Chromatogr Sci 15:90–110, 1977.

43. LS Ettre. LC·GC North Am 8:716–724, 1990.

44. J Griffith, D James, CSG Phillips. Analyst (London) 77:897–903, 1952.

45. S Dal Nogare, CE Bennett. Anal Chem 30:1157–1158, 1958.

46. IG McWilliam, RA Dewar. Nature (London) 181:760 only, 1958.

47. IG McWilliam, RA Dewar. In: DH Desty, ed., Gas Chromatography 1958 (Amsterdam Symposium). London: Butterworths, 1958, pp. 142–152.

48. JE Lovelock. J Chromatogr 1:35–46, 1958.

49. JE Lovelock. Nature (London) 182:1663–1664, 1958.

50. MJE Golay. In: DH Desty, ed., Gas Chromatography 1958 (Amsterdam Symp.). London: Butterworths, 1958, pp. 36–55.

51. JJ Van Deemter, FJ Zuiderweg, A Klinkenberg. Chem Eng Sci 5:271–289, 1956.

52. HF Walton, ed. Ion-Exchange Chromatography. Stroudsburg, PA: Dowden, Hutchinson & Ross, Inc., 1976.

53. TI Taylor, HC Urey. J Chem Phys 6:429–438, 1938; reprinted in Ref. 52, pp. 113–122.

54. O Samuelson. Z Anal Chem 116:328–334, 1939.

55. O Samuelson. Svensk Kem Tidskr 51:195–206, 1930; reprinted in Ref. 52, pp. 10–24.

56. FH Spedding. Disc Faraday Soc 7:214–231, 1949.

57. ER Tompkins. Disc Faraday Soc 7:232–237, 1949.

58. LS Ettre. LC·GC North Am 17:1104–1109, 1999.

59. SW Mayer, ER Tompkins. J Am Chem Soc 69:2866–2874, 1947; reprinted in Ref. 52, pp. 28–36.

60. RM Diamond, K Street, Jr, GT Seaborg. J Am Chem Soc 76:1461–1469, 1954; reprinted in Ref. 52, pp. 97–105.

61. A Ghiorso, BG Harvey, GR Choppin, SG Thompson, GT Seaborg. Phys Rev 98:1518–1519, 1955; reprinted in Ref. 52, pp. 106–107.

62. WE Cohn. Science 109:377–378, 1949; reprinted in Ref. 52, pp. 295–296.

63. WE Cohn. J Am Chem Soc 72:1471–1478, 1950; reprinted in Ref. 52, pp. 297–304.

64. S Moore, WH Stein. J Biol Chem 192:663–681, 1951; reprinted in Ref. 52, pp. 329–347.

65. S Moore, WH Stein. J Biol Chem 211:893–906, 907–913, 1954.

66. S Moore, DH Sparkman, WH Stein. Anal Chem 30:1185–1190, 1958.

67. SH Sparkman, WH Stein, S Moore. Anal Chem 30:1190–1206, 1958; partly reprinted in Ref. 52, pp. 348–350.

68. J Porath, P Flodin. Nature (London) 183:1657–1659, 1959.

69. JC Janson. Chromatographia 23:361–369, 1987.

70. JC Moore. J Polymer Sci A2:835–843, 1964.

71. FO Anderson, LR Crisp, GC Riggle, GG Vurek, E Heftmann, DF Johnson, D Francois, TD Perrine. Anal Chem 33:1606–1610, 1961.

72. D Francois, DF Johnson, E Heftmann. Anal Chem 35:2019–2022, 1963.

73. JC Giddings. Anal Chem 35:2215–2216, 1963.

74. C Horváth, SR Lipsky. Nature (London) 211:748–749, 1966.

75. IV Mortimer, reporter. In: AB Littlewood, ed., Gas Chromatography 1966 (Rome Symp.). London: Institute of Petroleum, 1967, pp. 414–418.

76. C Horváth, BA Preiss, SR Lipsky. Anal Chem 39:1422–1428, 1967; reprinted in Ref. 52, pp. 305–311.

77. JJ Kirkland. Anal Chem 41:218–220, 1969.

78. LR Snyder. Anal Chem 33:698–704, 705–709, 1967.

79. LR Snyder. Principles of Adsorption Chromatography. New York: Marcel Dekker, Inc., 1960.

80. GA Howard, AJP Martin. Biochem J 46:532–538, 1950.

81. Recommendations on Nomenclature for Chromatography. Information Bulletin, Appendices on Tentative Nomenclatures, Symbols, Units and Standards, No. 15, I.U.P.A.C. Secretariat, Oxford, February 1972.

82. C Horváth, W Melander, I Molnár. J Chromatogr 125:129–156, 1976.

83. JJ Kirkland. J Chromatogr Sci 9:206–214, 1971.

84. JJ Kirkland, JJ Di Stefano. J Chromatogr Sci 8:309–314, 1970.

85. LR Snyder. J Chem Ed 74:37–44, 1997.

86. BL Karger. J Chem Ed 74:45–48, 1997.

87. JJ Kirkland, LR Snyder. Modern Practice of Liquid Chromatography. New York: Wiley-Interscience, 1971.

88. B Erickson. Anal Chem News Features 71:271A–276A, 1999.

89. G Guiochon, SG Shirazi, AM Katti. Fundamentals of Non-linear Chromatography. New York: Academic Press, 1994.

90. FG Lote, A Rosenfeld, QS Yuan, TW Root, EN Lightfoot. J Chromatogr A 796:3–14, 1998.

91. EN Lightfoot. Am Lab 31(12):13–23, June 1999.

92. JE Lovelock, notarized suggestion from 1958. Cited in: ML Lee, KE Markides, eds., Analytical Supercritical Fluid Chromatography and Extraction. Provo, UT: Chromatography Conference, Inc., 1990, p. 7.

93. E Klesper, AH Corwin, DA Turner. J Org Chem 27:700–701, 1962.

94. MN Myers, JC Giddings. Separ Sci 1:761–776, 1966.

95. MN Myers, JC Giddings. Anal Chem 37:1453–1457, 1965.

96. ST Sie, W Van Beersum, GWA Rijnders. Separ Sci 1:459–490, 1966.

97. ST Sie, GWA Rijnders. Separ Sci 2:699–727, 729–753, 755–777, 1967.

98. DR Gere, R Board, D McManigill. Anal Chem 54:736–740, 1982.

99. M Novotny, SR Springston, PA Peaden, JC Fjeldsted, ML Lee. Anal Chem 53:407A–414A, 1981.

100. HM McNair. Am Lab 29(12):10–13, June 1997.

101. G Guiochon. Am Lab 30(18):14–15, Sept. 1998.

102. AWK Tiselius. In: Nobel Lectures: Chemistry 1942–1962. Amsterdam: Elsevier, 1964, pp. 195–215.

103. RD Dandeneau, EH Zerenner. HRC/CC 2:351–356, 1979.

104. RD Dandeneau, EH Zerenner. LC·GC North Am 8:908–912, 1990.

105. JW Jorgenson, KD Lukas. Anal Chem 53:1298–1302, 1981.

106. C Horváth. LC·GC Int 10:622–623, 1990.

107. V Pretorius, BJ Hopkins, JD Schieke. J Chromatogr 99:23–30, 1974.

2

Mikhail Semenovich Tswett: The Father of Modern Chromatography

Haleem J. Issaq

SAIC Frederick, NCI-Frederick Cancer Research and Development Center, Frederick, Maryland

Victor G. Berezkin

A. V. Topchiev Institute of Petrochemical Synthesis, Russian Academy of Sciences, Moscow, Russia

Chromatography provides us, as it were, with a new "vision" and insight, which enables us to obtain important information on the chemical composition of the world around us. It was developed in the twentieth century and became a new and effective tool for scientific research, operation of chemical processes in industry, and for control of human health.

In the early 1900s Mikhail Semenovich Tswett discovered chromatography as a new phenomenon and a new technique for investigating complex mixtures. "No other discovery has exerted as great an influence and widened the field of investigation of the organic chemist as much as Tswett's chromatographic adsorption analysis. Research in the field of vitamins, hormones, carotinoids and numerous other natural compounds could never have progressed if it had not been for this new method, which has also disclosed the enormous variety of closely related compounds in nature" said organic chemist and Nobel prize winner P. Karrer, expressing his appreciation of the importance of chromatography at an IUPAC Congress in 1947. Many interesting books and papers are devoted to the unique biography of M. S. Tswett [1–5]. With these data in mind, we present here a record of his achievements.

M. S. Tswett was born on May 14, 1872, in Asti, Italy. His father, Semen Nikolaevich Tswett, a Russian, was a high-ranking official in the finance ministry. His mother, Maria de Dorozza, was an Italian, who died soon after his birth.

Tswett spent his childhood and youth in Switzerland. In 1891 he entered the Physicomathematical Department at Geneva's University, and in 1896 he finished his doctoral dissertation, entitled "Investigation of cell's physiology." In 1896 Tswett moved to Russia, where he defended the Master's degree thesis "Physicochemical study of chlorophyll particle: Experimental and critical study" in 1901 in order to qualify to work at a university. He dedicated the thesis to the memory of his father, who died not long before Tswett finished the thesis. The dedication read: "To the memory of my father Semen Nikolaevich Tswett, thinker and figure."

In quest of work, Tswett moved to Warsaw in 1902. At first he worked as a temporary laboratory assistant at Warsaw University. Then he won a position as an assistant in the Department of Anatomy and Physiology of Plants. At the same time he was given the title of private assistant professor and granted permission to lecture.

At the meeting of the Biological Section of the Warsaw Society of Natural Scientists, March 3, 1903, Tswett for the first time spoke of his chromatographic method. In a presentation entitled "On a New Category of Adsorption Phenomena and on Their Application to Biological Analysis," which was published in the same year in the "Proceedings of the Warsaw Society of Natural Scientists," he described his method. Therefore, the year 1903 is considered to be the year in which chromatography was born.

In 1906 Tswett published in *Berichte der Deutschen Botanischen Gesellschaft* (Transactions of the German Botanical Society) two important papers in which he described the chromatographic method and its application to practical problems.

In 1907 Tswett married Helena Trusevich, who worked at the library of Warsaw University and helped him in his scientific work. In 1910 Tswett presented his doctoral thesis at Warsaw University. "The investigations of Mr. Tswett are bringing about real revolution in the theory of photosynthetic pigments, and he is now among the outstanding scientists studying this question," said his opponent Professor Ivanovsky [2, p. 100].

In 1911 the Petersburg Academy of Sciences (Russian Academy of Sciences) awarded Tswett the N. N. Akhmatov Grand Prize of 1000 rubles for his monograph "Chromophylls in Plants and Animals." At the time, 1000 rubles was a large amount of money. Commenting on this event, academician Famintsyn noted that the most valuable result of the new adsorption method developed by Professor Tswett was that it made possible the analysis of chlorophylls in plant extracts with a much greater degree of precision than other methods.

Unfortunately, not all scientists of Tswett's time were able to appreciate the importance of chromatography. Further development of the chromatographic method began only in 1931 after the publication of a work by R. Kuhn, A. Winterstein, and E. Lederer, "Zur Kenntnis der Xanthophylle" [6].

During the years 1912–1914 Tswett's health deteriorated, which prevented him from being as active and hard-working as he would have liked. In 1917 he was appointed Professor at the University of Yurev (now Tartu, Estonia). In the same year the university was moved to the city of Voronezh. On June 26, 1919, Tswett, who had been suffering from heart disease, died in his hospital bed. He was buried in Voronezh.

In conclusion, let us quote Professor I. I. Bevad's opinion on Tswett as a teacher and a person: "M. S. Tswett, who holds a Doctor's degree in botany, is a teacher of botany, and has for a number of years read lectures and conducted laboratory work with students. He has proved himself to be an excellent instructor. He has put the teaching of botany on a strictly scientific basis, and does this with great competence. He gets on well with his students; thanks to his tactfulness there has never been even the slightest conflict between them. He is a man of high moral standards: he is wholly devoted to his work, without a thought for personal advantages. He is modest, and despite his fame as a scientist, a diligent, kind and generous person, all in all, a man of lofty views" [2, p. 110].

M. S. Tswett's importance to the discovery of chromatography is acknowledged worldwide. Unfortunately, there is no detailed, logical analysis of his contribution to chromatography development, with the exception of a short communication [7].

M. S. Tswett's creative heritage was not completely understood or used by the next generation of scientists. Hence, many important chromatography variants and the solutions he suggested had to be discovered anew. In this context let us quote well-known scientist P. Langevin: "Nothing could replace direct access to the pioneering classical works."

I. FRONTAL AND DISPLACEMENT CHROMATOGRAPHY

"Frontal chromatography is a procedure in which the sample (liquid or gas) is fed continuously into the chromatographic bed. In frontal chromatography no additional mobile phase is used" [8]. In the study [9, p. 25] Tswett wrote: "When the petroleum ether solution of chlorophyll is passed through an adsorbent column, usually a narrow glass tube packed with calcium carbonate, the pigments are distributed in accordance with their position in the adsorption series, and they form distinctly colored zones. This can be explained by the fact that these pigments with more pronounced adsorption properties displace substances that are less firmly bound. Subsequent washing with a pure solvent leads to an even more complete separation of the pigments. Like high beams in a spectrum, the various components of the pigment mixture are distributed in the calcium carbonate column, which permits their quantitative and qualitative assay. The preparation obtained as a result of this procedure was termed a "chromatogram" and the method suggested "chromatographic."

The above passage describes the frontal technique. Since the adsorption of individual components to be separated differs sharply, in this specific case displacement (or quasi-displacement, more precisely) variation also occurs, with the most adsorbed component on the given sorbent being a displacer. Thus, Tswett established frontal and displacement chromatography, although most chromatographers (see, e.g., Refs. 10,11) believe that Tiselius suggested frontal [12] and displacement [13] techniques in 1940–1943.

Note that Tswett used frontal displacement technique for sample injection with the purpose of sample preparation for subsequent separation. Application of frontal and displacement chromatography as a preliminary stage allowed for (a) concentration of target components and (b) preliminary separation. The technique improved separation and increased the sensitivity of determination.

II. ELUTION CHROMATOGRAPHY

Elution chromatography is a procedure in which the mobile phase is continuously passed through or along the chromatographic bed and the sample is fed into the system as a finite plug [8].

In the studies by M. S. Tswett, the solution of the mixture to be analyzed was fed onto a column filled with dry adsorbent before chromatographic separation. As a result, the process of frontal displacement separation and concentration occurred on the adsorbent layer. The zones of individual components, containing admixtures of other compounds, formed a general zone of short width of the test sample. Then the general zone was separated into individual zones by elution chromatography in which a flow of pure solvent passes through the column. In the book "Chlorophyll in Plants and Animals," Tswett described the technique that he used as follows [9, p. 67]:

When the desired quantity of solution has passed through the column, the chromatogram obtained is subjected to final "development" by passing pure solvent through it. Carbon disulfide and benzene were shown to be the most suitable developers for chlorophyll α and α′ bands in benzene, these pigments can be washed out of the column and isolated separately as benzene solutions by passing benzene through the column for a sufficiently long time.

As soon as the adsorption bands become sufficiently separated in the course of "development," solvent inflow is stopped. If this was carbon disulfide or benzene, it is then displaced from the adsorbent surface by passing petroleum ether and then air (or illuminating gas) through the column to displace liquid from the capillary space.

This procedure leads to some segmentation of the column, and the latter can, therefore, be forced out of the instrument as a snow-white cylinder colored with motley bands. The cylinder is laid on glossy paper and individual bands immediately cut off with a scalpel. The pigments can then be extracted with ethanol, ether, or petroleum ether with the addition of ethanol. If two colored bands are in contact or separated by a curved plane, their isolation in the pure state is difficult to achieve. In such a case, the boundary layers of both bands should be sacrificed.

Earlier Tswett described the technique of selection of separate eluent fractions from a chromatographic column: "The extracts are filtered under a pressure of 250–300 mm and as full suction of the water pump, when large adsorption funnels are used. After filtration of a certain amount of the test liquid, pure solvent is passed through the column, after which the individual adsorption zones slightly broaden and become maximally differentiated. Non-adsorbing substances are totally removed, whereas substances that bind with the adsorbent slowly migrate downward as rings that can be collected at the lower opening of the tube and separated from each other" [8, p. 30]. Thus, Tswett was the first to suggest and use eluent (liquid) chromatography, which is currently the most commonly used technique.

III. THE PHENOMENON OF CHROMATOGRAPHIC SEPARATION

M. S. Tswett considered chromatography to be new physicochemical phenomenon comprising formation and movement of chromatographic zones along adsorbent layer in a column [9, p. 57]:

> If now we pass a pure solution through the adsorption column colored in its upper part, the following will occur: the first layer of the liquid, on passing through the successive layers of colored adsorbent, will dissolve part of the adsorbed, according to the already mentioned isotherm for the adsorption equilibrium. The concentration of the solution will gradually increase until it approaches the limiting value a/v* of equilibrium adsorption

---

* The values used deal with the known Freundlich adsorption equation used by Tswett in this study:

$$\frac{x}{m} = \beta(a/v)^{1/P} \qquad \text{or} \qquad C_a = \beta(C_s)^{1/P},$$

where

$$C_a = \frac{x}{m}$$

is the concentration of adsorbed compound, C_s is the compound concentration in the solvent (mobile phase), m is the adsorbent mass, x is the amount of the adsorbed compound, a is the amount of the compound in the solution, V is the volume of solution, β and 1/p are the constants of the Freundlich equation, X is the amount of the adsorbed compound being in equilibrium with its initial concentration.

concentration X/m dominating in the adsorbate. Since this maximal concentration is smaller than a/v, the liquid will filter further until it leaves the region of adsorptive concentration X/m, i.e., down to the lower boundary of the adsorbate, where it will precipitate the dissolved pigment. The fate of the second and subsequent layers of the pure solvent through the adsorbate will be the same.

Therefore, the pigment will gradually be washed out at the upper boundary of the colored adsorbent layer and be adsorbed at the low boundary of the colored adsorbent layer and will slowly move with the flow of pure solvent. The width of the adsorbent zone occupied by the adsorbate will gradually increase owing to the gradual drop of adsorptive concentration in it. Indeed, each layer of the adsorbate, as it passes through the adsorbent, depletes all the other layers, including the layer the adsorptive concentration of which is maximal, because the increase of concentration in solution (C_s) can occur only at the expense of decreasing C_a in the washed layers. The rate at which the adsorbate of a definite substrate will travel apparently depends on the values of adsorption coefficients β and $1/p$; it decreases with the former and increases with the latter. But since different substances generally have different constants, passage of a solvent through an adsorbent column saturated in the upper part with several substances, the substances will (notwithstanding the displacement phenomena) move at various rates and thus gradually be separated into independent adsorption zones.

These predictions deduced from the basic law of adsorption equilibrium are confirmed in the observation of the filtration through a lime carbonate column of carbon disulfide or benzene solutions of separate or mixed chlorophyll pigments. In the latter case (i.e., with the mixture) there occurs, according to the law of substitution established by us, separation of the pigments into layers, and the subsequent passage of pure solvent completes the process of separation, thus causing the uneven movement of separate substances as independent zones. The xanthophyll alpha zone moves especially rapidly. The zones of xanthophyll alpha prime and alpha double prime, which were initially mixed or touching become separated by this movement. Hence, it can be concluded that generally the distribution of substances by the rate of their adsorptive migration coincides with the series of adsorptive substitution.

This rule, not yet theoretically substantiated, certainly needs more extensive empirical support.

The above quantitative explanation of chromatographic separation is close to the modern interpretation of chromatographic phenomenon using the Kreig machine (see, e.g., Ref. 13).

Tswett's interpretation of the "adsorption concept" satisfies the term "chromatographic phenomenon" in essence [15].

The adsorption phenomena observed during filtration through powder are especially instructive. The liquid flowing from the lower end of the funnel is first colorless then yellow. The yellow color results from carotene, as shown by spectroscopic investigation, and from its solubility behavior in the system 80% alcohol + petroleum ether. The adsorbed pigments are retained in the upper layers of the adsorbent, the mass of the adsorbent being not uniformly colored, but segregated into several differently colored zones which correspond to various constituent pigments and follow one another in a definite order. This remarkable phenomenon which, besides being of theoretical interest, is of great practical importance, and its investigation in detail will be described below [9, p. 48].

IV. THE MOBILE PHASE IN CHROMATOGRAPHY

The mobile phase is an important and simple factor in separation optimization. Tswett described the role of the mobile phase as follows [9, p. 54]:

> If pure petroleum ether is passed through the colored column, as mentioned above, the separate layers extend slightly more and the boundaries become more distinct. The lower yellow boundary, which during filtration slightly mixed with the above layers, is completely separated as it goes down, the separation being into two tiers of not altogether identical colors. The upper layer becomes colorless; the third layer, blue-green that, acquires a greenish-blue color.
>
> Instead of petroleum ether, carbon disulfide can be passed through the preparation; complete separation and final differentiation then occur more rapidly. All layers in this case are considerably extended, and a yellow band is distinctly outlined at the boundary between the upper colorless layer and the yellow-green layer. At the bottom of the preparation two yellow layers are separated by a narrow colorless band.
>
> A flow of benzene causes even more spreading of the layers. The lower yellow layers in this case slowly move down, the upper of them being divided into two. Even with the small height of the adsorption column and the rather prolonged flow of benzene, the yellow layers can be completely washed out of the instrument, and each of them collected separately as a benzene solution. If during the downward movement of the yellow layers the direction of the benzene flow is changed, the yellow layers also move upwards, and, after passing through the upper layers, are washed from the preparation. A similar separation of the pigments is observed when benzene or carbon disulfide solutions of chlorophyll are passed through the adsorbent column. Benzene solutions yield lower yellow layers that are very pale and diffused.

Selection of various mobile phases for separation optimization is one effective modern method for the improvement of chromatographic methods.

V. GRADIENT ELUTION

"Gradient elution is the procedure in which the composition of the mobile phase is changing continuously or stepwise with time during the elution process" [8].

In the first study on chromatography, M. S. Tswett noted the possibility of changing of one eluent for another when separating the components of the same sample. He was the first to realize gradient elution, where he used stepwise change of the mobile phase (naphtha) for another mixed phase (naphtha, containing alcohol admixture) [9, p. 17]:

> Several milliliters of the resulting solution were passed through an insulin column. The out flowing liquid was first colorless and then yellow, containing much fatty matter. Pure naphtha was passed through the column for a long time until the out flowing liquid stopped being yellow, but gave a transparent fatty spot on silk paper after drying out. Then, naphtha with the addition of ethanol was allowed to flow through the filter. A light yellow solution began to flow out, and the dry residue from this turned out to be a waxy substance exhibiting double refraction, swelling in water and becoming liquid in a concentrated aqueous solution of resorcinol; that is, showing properties characteristic of lecithin.

VI. THE ROLE OF ADSORBENT (STATIONARY PHASE) IN CHROMATOGRAPHY

When developing a chromatographic technique, M. S. Tswett used more than 100 solid compounds in his experiments. Tswett was the first to intuitively use many adsorbents for separation that are now in common use in chromatography experiments, e.g., silicon (silicium) dioxide (SiO_2), aluminum oxide (Al_2O_3), silicon (Si), coal, glass wool, and carbohydrates (saccharose, galactase, insulin, dextrin, starch, cellulose filter paper). He tested not only inorganic adsorbents but organic materials as well.

Tswett was the first to pay attention to band broadening of chromatographic zones. To decrease this negative phenomenon, he recommended the use of fine-grained sorbents (Sakodynskii had noted this peculiarity of M. S. Tswett's experiments [16]). This conforms well with the modern practice of chromatography.

In one of his first studies, Tswett wrote [9, p. 29]: "The fineness of the powder is especially important, since band broadening is obtained when crude grainy material is used—this can be explained by the adsorption being coupled with diffusion in wide capillaries. Of all the adsorbents, I recommend the use of precipitated calcium carbonate, which ensured a high quality of chromatogram. Sucrose, known to be one of the most chemically inert substances, can also be quite easily dispersed to the required degree."

It is worth noting that Tswett worked with short chromatographic columns for analytical separation. The columns were 20–30 mm long and 2–3 mm in diameter. Such short columns are in common use in modern liquid chromatography [17].

VII. CONCLUSION

Performance analysis of M. S. Tswett's studies show that he was not only a founder of chromatography, but a developer of its scientific concept and basic variants. When considering Tswett's contribution to science, it is necessary to note that he not only chronicled chromatographic techniques, but also performed seminal research that became the basis of chromatographic science.

Thomas Edison, among others, noted that societies are slow to accept innovative ideas. No doubt the field of chromatography would have evolved much more quickly had M. S. Tswett's ideas been appreciated in his own time.

ACKNOWLEDGMENT

This project has been funded in whole or in part with federal funds from the National Cancer Institute, National Institutes of Health, under Contract No. NO1-CO-56000.

By acceptance of this article, the publisher or recipient acknowledges the right of the U.S. Government to retain a nonexclusive, royalty-free license and to any copyright covering the article. The content of this publication does not necessarily reflect the views or policies of the Department of Health and Human Services, nor does mention of trade names, commercial products, or organizations imply endorsement by the U.S. Government.

REFERENCES

1. AA Richter, TA Krasnosel'skaya. The role of M. S. Tswett in the development of the method of chromatographic adsorption analysis and a biographical sketch of Tswett. In: AA Richter, TA Krasnosel'skaya, eds. MS Tswett. Chromatographic Adsorption Analysis. Moscow: USSR Academy of Sciences Publishing House, 1946, p. 215 (in Russian).
2. EM Senchenkova. Mikhail Semenovich Tswett. Moscow: Nauka, 1973 (in Russian).
3. KI Sakodynskii, KV Chmutov. Chromatographia 5:471, 1972.
4. KI Sakodynskii. Michael Tswett. Life and Work. Milan: Viappiani, 1982.
5. LS Ettre, C Horvath. Anal Chem 47:422A, 1975.
6. R Kuhn, A Winterstein, E Lederer. Zur kenntnis der xanthophylle. Z Physiol Chem Bd 197, S. 141, 1931.
7. VG Berezkin. In: VG Berezkin, compiler; MR Masson, transl., ed., MS Tswett. Chromatographic Adsorption Analysis. Selected Words. Chichester: Ellis Horwood, 1990, p. 103.
8. LS Ettre. Nomenclature for chromatography. Pure Appl Chem 65:819, 1993.
9. MS Tswett. Chromatographic Adsorption Analysis. Selected works. VG Berezkin, compiler; MR Masson, transl., ed. Chichester: Ellis Horwood, 1990.
10. G Schay. In: KV Chmutov, KI Sakodynskii, eds., Advances in Chromatography. Moscow: Nauka, 1972, p. 179 (in Russian).
11. S Claesson. Arkiv Kemi Mineral Geol 14B:133, 1940.
12. A Tiselius. Arkiv Kemi Mineral Geol 14B:5, 1940.
13. A Tiselius. Arkiv Kemi Mineral Geol 16A:11, 1943.
14. G Guiochon, CL Guillemin. Quantitative Gas Chromatography. Amsterdam: Elsevier, 1988.
15. VG Berezkin. Industrial Laboratory, Diagnostic of Materials 65:2, 1999 (in Russian).
16. KI Sakodynskii. In: KV Chmutov, KI Sakodynskii, eds., Advances in Chromatography. Moscow: Nauka, 1972, p. 9 (in Russian).
17. D Ishii, ed. Introduction to Microscale High-Performance Liquid Chromatography. Weinheim: VCH Verlagsgesellschaft, 1988.

3

From Crushed Bricks to Microchips

Johan Roeraade

Royal Institute of Technology, Stockholm, Sweden

I. THE EARLY DAYS

I can imagine that the title of this chapter could be somewhat confusing for some readers. Nearly all chromatographers are aware of the ongoing efforts to employ microchips in separation science, but only the older chromatographers will remember the days when crushed bricks were used as a support material for gas chromatography (GC) stationary phases. Indeed, this was how it also started for me in the early 1960s at the Dutch State Mines (DSM) in the Netherlands, where I worked one semester in the frame of my chemical engineering studies. Fire brick was crushed and sieved to yield a narrow size fraction of particles, typically 60–80 mesh, which were impregnated with 10% or more of stationary phases like stopcock grease (Apiezon grease), squalane, or other high boiling products from the petroleum industry. The impregnated material was packed into U-shaped glass tubes, which were mounted in the standard workhorse, the legendary Fractovap Gas Chromatograph from Perkin Elmer, equipped with a hot wire detector. The results that I obtained with this setup made an incredible impression on me. Earlier I had carried out some fractionation work on petroleum distillates with Podbielniak precision distillation equipment, one of the best laboratory stills available at that time. I was very proud of the outcome of that work. However, results from the gas chromatograph were so much better and were obtained in such a short time. I had to run the samples over and over again to believe it, and I could not get enough of looking at the movement of the recorder pen! It became the start of a lifetime devotion to separation science.

Since the majority of our analytes were nonpolar hydrocarbons, we did not really need to deactivate the firebrick at that stage. In those days, most applications, as well as the main developments in gas chromatography, were driven by the petroleum industry. The huge potential of the method as a general separation technique for all types of volatile products was of course realized very early, and an increasing number of polar stationary phases such as Carbowax, high boiling esters, and liquids with other functional groups rapidly became popular. Also, the excellent support material, based on presieved diatomaceous earth (marketed by Johns Manville under the trade name Chromosorb), started to replace the home-made pulverized brick. The commercial material could be obtained in various treated forms, such as reflux-calcined and/or silylated, which markedly improved peak symmetry for polar analytes.

However, the real fundamental breakthrough came with the open tubular column. During my stay at DSM, some instruments were being equipped with adapters for Golay columns (the trade name for the open tubular column, invented by the brilliant scientist Marcel Golay), but unfortunately I left DSM before I had seen the phenomenal performance of these columns. However, in the mid-1960s I got my next chance when I became involved in a research project at the Agricultural University of Sweden dealing with the formation of aroma compounds of apples during postharvest storage. This was a fortunate turn, since the area of flavor research was going to become one of the most important fields supporting the development of high-resolution GC. The assignment at the University was challenging since it involved the entire analytical chain, i.e., sampling of the volatiles on active charcoal, release of the collected material, and a qualitative and quantitative analysis of the trapped concentrate. The instrumentation I used was an early model of Wilkens (a company later taken over by Varian) and Golay columns from Perkin Elmer, who held a worldwide patent on open tubular (capillary) columns. Gas chromatography, and particularly capillary GC, was still a rather unfamiliar technique for many people. I still remember the concern of the local sales representative of Wilkens, who wondered how in earth I would get apples through such a long narrow tube. Fortunately I did not follow his various suggestions. The capillary column performed outstandingly on my aroma concentrates, yielding chromatograms with more than 200 peaks and many more unresolved ones, which demonstrated the enormous complexity of the natural aroma product.

A considerable amount of fundamental research was carried out during the early 1960s. One of the great innovators who should be mentioned was Dennis Desty. His "Capillary Chromatography: Trials, Tribulations and Triumphs" [1] has remained one of the more enjoyable articles to read. Also, the first books on the subject, from pioneers like Leslie Ettre [2] and Rudolph Kaiser [3], started to appear. In the frame of my own work, I became tempted to try making my own capillary columns, partly because the Perkin Elmer columns were quite expensive, but mostly out of pure curiosity. Stainless steel tubing was ordered from the Pennsylvania-based company Handy & Harman, and I followed a column-making recipe from the literature. The procedure was very simple: a 10% solution of Apiezon L in hexane was pushed through the tube with pressurized air, which was left on overnight to dry the liquid film. Unbelievably, my first coated tube outperformed my expensive commercial column. However, repeated experiments gave inconsistent results, which showed that there was a real need for improvements.

Two issues were clearly of central importance: reducing adsorption of polar analytes and finding a better way to prepare a reproducible, smooth, uniform film of stationary phase. Several workers came with suggestions regarding the first issue, but no major improvements were made. Compounds like amines or alcohols tailed rather badly, if at all eluted from the column. The best results were obtained with self-deactivating polar stationary phases like Carbowax, sometimes with additives as drastic as sodium hydroxide. One of the ideas I found very useful was the addition of long-chain quaternary ammonium salts, as suggested by Metcalfe and Martin [4] (their material became commercially available under the name Gas-Quat). Indeed, adsorption of compounds like amines was significantly reduced, but results were inconsistent; for example, a new batch of stainless steel could show inferior behavior.

It became obvious that column materials other than stainless steel had to be utilized, and glass was clearly the most interesting material. This was possible due to Dennis Desty's marvelous invention of the glass drawing machine [5], which could produce coiled capillary tubes of various dimensions and long lengths. Figure 1 is a picture of his original setup.

This machine became commercially available from Hupe & Busch in Germany. I had started

Figure 1 Photograph of the original glass capillary drawing machine of Desty et al. (From Ref. 5. Copyright © 1960 American Chemical Society.)

to work with research on aroma compounds of tobacco for a new employer, and I was very happy to be able to get one of these machines. This opened a new world. I made hundreds of columns with various inner diameters and lengths and experimented with different glasses like Pyrex, soda lime, and Duran glass. Very soon the extremely smooth inner wall of the glass capillary proved to be a major obstacle to getting a uniform coating of stationary phase. This was due to factors initially not well understood. Substantial work to improve the situation was carried out by Kurt Grob, one of the true pioneers of glass capillary GC. He precoated the glass tubes with a very thin carbon layer, produced by pyrolysis of methylene chloride [6]. This layer had better wetting characteristics than the bare glass wall. He also made other intermediate coatings by in situ polymerization technologies as well as direct chemical modification of the column wall [6].

Another successful approach was that of Tesarik and Novotny in Czechoslovakia, who had developed etching procedures to obtain a roughened surface [7]. This was a sound approach, since a roughened surface allows easier spreading of liquids due to a reduction in the apparent contact angle. Other groups, e.g., Bruner and Cartoni [8], had reported on methods of surface corrosion using bases in liquid form. We adopted Tesarik and Novotny's idea of etching the glass wall with HCl gas, which was particularly successful when using soft glass. The procedure resulted in a very fine layer of sodium chloride crystals, which proved to be an efficient anchor for the thin film of the stationary phase.

However, a real breakthrough in capillary GC was the invention of the static coating method described by Bouche and Verzele [9]. Instead of pushing a plug of diluted solution of stationary phase through the column, which frequently led to droplet formation and irregular film thickness along the column, the static method produced consistent results. The original setup used by Bouche and Verzele is shown in Fig. 2.

The difficulties with the dynamic coating method were clarified much later by Lee and coworkers, who found that the initial film of coating solution was inherently unstable due to

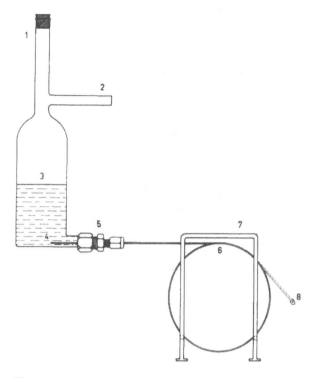

Figure 2   Schematic of the filling procedure by the static coating method according to Bouche and Verzele. The column (6) is first filled with a diluted solution of stationary phase (4) and then sealed on one end (8). Thereafter the open end of the column is connected to a vacuum pump to remove the solvent. (From Ref. 9. Reproduced by permission of Preston Publications, a Division of Preston Industries, Inc.)

Rayleigh effects [10], which invariably leads to a distorted layer of stationary phase. Our homemade glass capillary columns became a key tool in our flavor research, particularly when we combined them with mass spectrometry, sniff outlets, and selective detectors. However, adsorption of highly polar trace components was still a major problem, and the search for improvements continued to be a major issue for at least another two decades.

II.   GLASS CAPILLARY GAS CHROMATOGRAPHY

The situation in the early 1970s was rather diverse. There was considerable activity in the field of glass capillary GC, and a number of laboratories fabricated and developed their own glass capillary columns. Also, capillary gas chromatography–mass spectroscopy (GC-MS) became established. However, the majority of users still continued to work with packed columns. To some extent this was due to a lack of suitable instrumentation. Injectors and detectors were not optimized for work with glass capillary columns. Although capillary GC was a hot topic at conferences, as in the series "Advances in Chromatography" of Al Zlatkis, the most important initiative was made by Rudolph Kaiser, who organized the First International Symposium on Glass Capillary Chromatography in Hindelang, a small village in the Bavarian Alps. This meeting, which drew fewer than a hundred participants from different backgrounds, was an immense success due to a beautiful blend of an informal, relaxed atmosphere, excellent (sometimes controversial and quite vigorous) scientific discussions, mixed with a lot of fun. I remember when Denis Desty discussed the results of his high-speed GC obtained in the early 1960s on a very short capillary column [11] (see Fig. 3).

Desty had been concerned about extra column band broadening due to insufficient speed of injection and therefore had to use the stroke of a hammer to accelerate the movement of the syringe plunger! A representative for a syringe manufacturer who saw an unexpected business opportunity expressed the sincere hope that all chromatographers would start hitting their syringes with a hammer. These kinds of jokes were typical elements of the discussions, well chaired and beautifully fueled by Rudi Kaiser. The entire Hindelang meeting proceedings were published in *Chromatographia* [12], and the event triggered an intense wave of research in the field. Three more meetings were held in Hindelang in 1977, 1979, and 1981, after which the meeting outgrew the local facilities and was moved to Riva del Garda, where it is still going strong.

It is not possible in a short review to give justice to all the persons who have contributed in an essential way to the development of capillary GC. ''Small'' as well as large inventions have been vitally important. However, some work should perhaps be highlighted. Gerhard Schomburg should be remembered for many efforts, particularly for development of two-dimensional techniques [13–16]. Likewise, the Eindhoven group (Carel Cramers, Jaques Rijks, and coworkers) made numerous contributions to the field; particularly important was their work on narrow-bore, high-speed GC [17–20]. Also, the whisker columns of Victor Pretorius should be mentioned [21,22], which in our hands was the only method for keeping a stable layer of very polar stationary phases. Kurt Grob, who had already made many outstanding contributions, made the important discovery of the solvent effect [23–25]. Apparently, as he told me, this was by pure accident. He was operating his injector in the splitless mode and was ready to raise the temperature of the oven to elute his high boiling trace components trapped on the stationary phase of the column entrance. However, at the start of his experiment he had forgotten to switch on the oven heater to bring the initial column temperature above the temperature of the boiling point of the solvent. (At that time, this was the general practice in order not to evoke a condensation of the solvent inside the capillary column, which could dissolve and destroy the stationary phase layer). To his surprise, small sharp peaks appeared directly after the large solvent peak. After some more experiments, he realized that the small sharp peaks were not anomalies, but impurities of the solvent, which normally were not observed due to their low capacity ratios and the extended

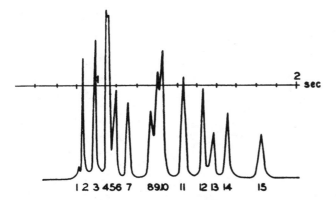

Figure 3   Gas chromatogram of the first ultra-high-speed analysis (2 s) of a volatile hydrocarbon mixture performed by Desty. (From Ref. 11. Copyright © 1962, with permission from Academic Press, New York.)

injection time. The condensation of the solvent, which formed a film in the beginning of the column, acted as a temporary trap for the volatile impurities, which were auto-focused during the evaporation of the solvent film. In fact, the intriguing principle had already been observed on packed columns [26] by another pioneer, David Deans from ICI in the United Kingdom, who also was the original inventor of valveless column switching [27]. The use of the solvent effect opened an important route towards trace analysis. Figure 4 shows early results of Grob.

The preparation of improved stationary phases was a further important development. At the first Hindelang conference the need for purified and stable stationary phases was emphasized, advice that was followed by companies like Ohio Valley, which specialized in making purified silicones (the OV phases). A breakthrough was made by Madani et al. [28–30], which was followed by Blomberg and coworkers [31–33], who developed the cross-linked stationary phase concept. This improved the thermostability of the stationary phase layer while it was permanently fixed to the wall. Thus, direct injections of liquids onto the column (on purpose or accidental) did not rearrange or harm the stationary phase. Today, practically every classical liquid stationary phase for capillary columns has been replaced by a cross-linked alternative.

In spite of these efforts, the development of improved column deactivation methods was still a major issue until the detrimental effect of metal impurities in glass was revealed. In this context, the work of Welsch [34] was of major importance. He combined a leaching procedure, which removed the metal ions, with a high-temperature silanization, which resulted in columns with a new level of inertness. One of the most important breakthroughs in glass capillary GC came from Dandeneau and Zerenner, who invented the flexible fused silica column [35] (Fig. 5). Their presentation at the third Hindelang conference in 1979 produced a shock wave at the meeting. Although several established glass capillary scientists were initially sceptical, I am quite sure that it was clear to everybody that the fused silica capillary technology was going to evoke a paradigm shift, which it did. Today, glass capillary columns have been fully replaced by fused silica columns. Practically all columns are now produced by a few commercial companies.

It is interesting to note that the main reason for the success of the fused silica columns is convenience. The glass capillary columns of the early 1980s performed at least as well as the fused silica ones, and they were cheaper as well. However, handling of the glass columns

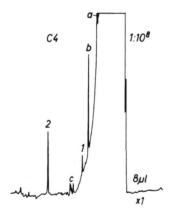

Figure 4 Trace analysis of volatile components in pentane using the solvent effect and a septum leak. Compounds 1 and 2 are benzene and toluene, respectively. Concentration level: 10 ppb. (From Ref. 23. With permission from K. Grob and G. Grob.)

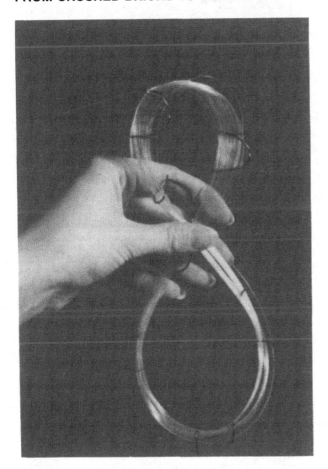

Figure 5 Photograph of the first flexible fused silica capillary column, with an outer coating of silicone rubber, as reported by Dandeneau and Zerenner in 1979. (From Ref. 35.)

(straightening column ends as well as their inherent vulnerability) was too much of a problem. With the advent of fused silica column, capillary GC became one of the most important (arguably the most important) analytical techniques.

In the beginning, the thin-walled columns were a bit unfamiliar to handle and small mistakes were often made. I remember when we mounted our first fused silica column in the detector end of the GC. As every chromatographer knows, it is important to assure that the column end is positioned very close to the flame tip in order to avoid dead volumes and/or exposure of the sample to unnecessary surface areas. For glass capillaries with an OD of ca. 1 mm this was no problem, but with the fused silica columns we had problems igniting the FID until we found that the FID jet was partially blocked by the column itself or sometimes by fused silica pieces that broke off the column end. Another reported problem was the occurrence of distorted peaks with a Christmas tree–like appearance [36], but this proved to be only related to inadequate oven heating control of certain gas chromatographs. This was initially hyped as a general problem, but the steam went out of this issue rather fast. The situation was well summarized by René de Nijs of Chrompack during a lecture at a GC meeting in Stockholm in 1985 in a single crushing comment: ''We do not see the Christmas tree peaks in our chromatograms, so we must be

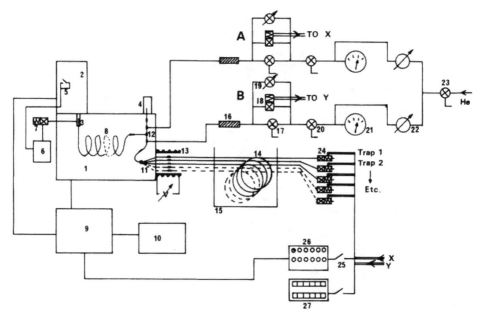

(a)

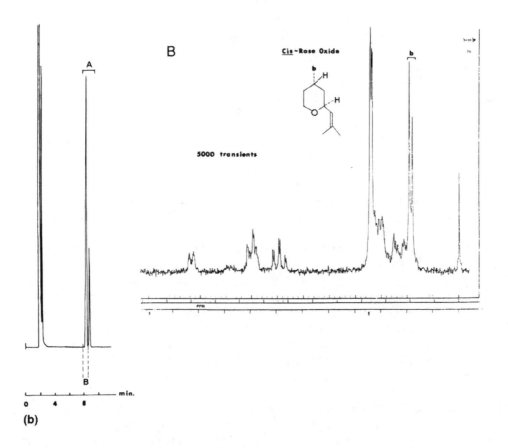

(b)

doing something wrong.'' However, this is all history. Today, fused silica GC columns are used routinely, and excellent results are obtained all over the world.

Looking back at our own work during the 1970s, an important part of my efforts were focused on the fundamentals of column preparation [37], since good columns were of central importance for our flavor research. However, we had severe difficulties in identifying some of the interesting unknown trace components, because of a lack of reference mass spectra. Mass spectrometric information is usually not sufficient for structure elucidation of unknown components, although some mass spectroscopists may have a different opinion. The usual method of revealing the structure of unknown components is by obtaining complementary spectral information, particularly using nuclear magnetic resonance (NMR). Much of our work was therefore dealing with isolation of the relevant trace components in sufficient (mg) quantities.

Extensive fractionation schemes were developed, which included some exotic chromatographic tools, such as a gigantic preparative silica liquid chromatography (LC) column [38–40]. Also, preparative GC (with packed columns) was an indispensable technology. However, many of the refined fractions that we obtained remained complex mixtures. This prompted us to develop a technique for preparative collection using high-resolution glass capillary columns. The low loadability of these columns was a problem, but the use of a fully automated computer-controlled set up allowed us to make a massive number of runs for obtaining mg amounts of material. Indeed, this was a breakthrough, since clean NMR spectra could now be obtained from trace components with closely related structures [41]. Figure 6 shows a schematic of our first setup and some results obtained.

The work was continued at the Royal Institute of Technology in the mid-1980s, where we improved the computer software, which allowed us to collect up to 32 pure trace components from a complex mixture [42] in sufficient amounts for ^{13}C-NMR.

III. TRACE ANALYSIS

During the 1980s many of the developments in GC were fueled by the explosive growth of environmental chemistry. Analysis of residues from persistent chemicals (chlorinated pesticides, PCBs, dioxins, etc.) became an important assignment for chromatographers. Therefore, new methodologies for trace analysis, like improved preconcentration and trapping systems, were a first-order issue. K. Grob, Jr., who carried out numerous investigations of column trapping and refocusing mechanisms, made major contributions. His findings are summarized in detail in his books on the subject [43–45], which describe pitfalls and provide a wealth of practical guidelines. On-column methods for preconcentration of diluted samples using a length of empty precolumn (coined by Grob as the ''retention gap'') proved to be one of the more powerful strategies in trace analysis [46]. This setup exploits the solvent effect to reconcentrate volatile components from large volume injections or, for less volatile components, makes use of the difference in phase ratio between the empty tube and the coated column, which will lead to a solute focusing at the column inlet when a temperature program is applied. By combining these

Figure 6 (a) Schematic of our preparative glass capillary GC setup with automated repetitive sample injection. Sample collection is computer controlled, where narrow fractions corresponding to single peaks are directed into cooled traps. (b) Left side: Glass capillary chromatogram of *cis*- and *trans*-rose oxide. Fraction B (*cis*-rose oxide) was collected for NMR spectroscopy (shown on the right side). (From Ref. 41. With permission from J. Roeraade and C.R. Enzell.)

mechanisms, it was possible to reconcentrate both volatile and less volatile components from a sample into a narrow band.

The development of more versatile and better sampling systems was an important goal for gas chromatographers. The original split injection system [47] and the splitless method [48,49] had a number of deficiencies and limitations. The design of injector liners proved to be of crucial importance. Schomburg [50] and others reported that the commonly employed empty glass liner could lead to large variations of reproducibility. Numerous constructions were suggested, which aimed at improved quantitative sample transfer. Examples of these are the falling needle concept of van den Berg et al. [51], the cold on-column injector [1,50,52,53], the direct injection device of Schomburg [50], and the Programmed Temperature Vaporizer (PTV) from Vogt et al. [54,55] as well as the at-column injector from our group [56]. Our at-column design, which allows direct large volume injections onto narrow-bore columns, is shown in Fig. 7.

It also became clear from the studies of K. Grob, Jr. and others [50,57,58] that sample fractionation and discrimination could occur in the syringe needle when inserted in a hot injector. This was caused by an uncontrolled boiling from the syringe needle, particularly noticeable when dealing with samples containing compounds with a wide range of boiling points. The problem was elegantly solved by Hewlett Packard, who constructed an automated injector that performed a very fast (100 ms) injection. Thus, there was no time for the syringe needle to become hot, and discrimination effects were avoided.

Our own work was also directed towards trace analysis, including fundamental investigations and expanded use of the solvent effect. Already in 1979, during a live demonstration at the third Hindelang symposium, I performed a 50 µL direct injection of sample in an empty

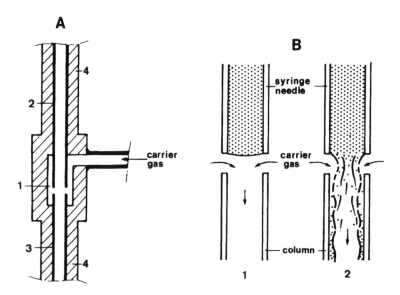

Figure 7 The principle of at-column injection. The syringe needle containing the sample is brought close to the column entrance (B, left). The needle end and the column entrance are in a small cavity, which is pressurized with carrier gas (A). When the sample is dispensed, the uniform surrounding flow and pressure of the carrier gas will lead to a perfect transfer of the sample into the column (B, right). In this way, ultra-low loss sample transfer to narrow bore columns is possible. (From Ref. 56. Copyright © 1990, with permission from Wiley-VCH.)

precolumn using a syringe equipped with a very thin glass needle. The large plug of solvent is spread out into a quasi-film by the flowing carrier gas. This solvent film, which could be as long as 10–50 m, acts as a temporary trap for volatile trace components. However, partial escape of some of the analytes (depending on differences in polarity and volatility between the solutes and the solvent) cannot be avoided. By utilizing different solvent barriers, a quantitative trapping and solute focusing was accomplished [26,59,60]. The principle is a unique form of enrichment, since the sample acts as its own "stationary" phase. When the "stationary" phase has evaporated, the solutes are present in the form of a concentrate in a narrow band. On the basis of this mechanism, we designed a system for continuous operation, which allows quantitative concentration of ultra-trace components present in solvents or extracts [61]. This system is quite versatile and is also suitable for quantitative concentration of trace components from gases, dynamic headspace, etc. [62,63]. Figure 8 shows an example of its application.

Another trend in the field of trace analysis during the 1980s, which culminated in the 1990s, was the use of coupled techniques. GC-MS can be considered as one of the first, and definitely one of the most successful examples of hyphenation. However, by combining several separation modes (GC, LC, SFC, 2D-GC, etc.), detectors (MS, MS-MS, AED, ECD, etc.) as well as sample work-up procedures (solid phase extraction, purge & trap, size exclusion, etc.) into an integrated line, more powerful systems can be obtained for trace analysis of dedicated components. Basically, combining LC and GC is a marvelous idea because of the fact that the separation mechanisms of the methods are so different. From a general perspective, LC separations are mainly based on polarity, while GC mainly separates according to volatility. The combination has therefore been quite successful [64–67]. Very early work was done by R. Majors [65], and the result of one of his experiments is shown in Fig. 9.

When the issue of hyphenation is brought up, Udo Brinkman must be mentioned. He and his group in Amsterdam have convincingly shown the analytical power of combining instrumental techniques (e.g., Refs. 68–70). His recent book on the subject [71] demonstrates some of the

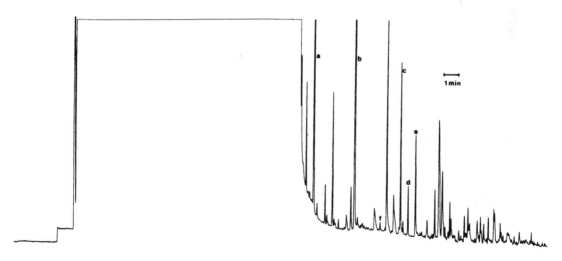

Figure 8 Chromatogram of volatile trace components (ppt levels) from 1 liter of drinking water. The water was extracted with 5 mL of pentane. The extract was concentrated by continuous chromatographic evaporation in order to avoid losses of volatile components. (From Ref. 63. Copyright © 1989, with permission from Wiley-VCH.)

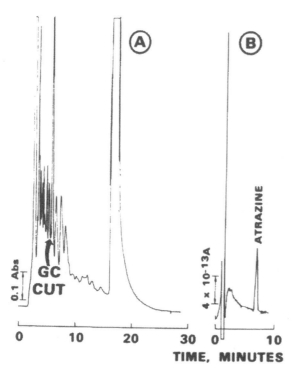

Figure 9 Chromatograms of atrazine in sorghum using combined LC-GC: (A) HPLC chromatogram; (B) gas chromatogram of a fraction from the HPLC eluate containing atrazine. (From Ref. 65. Reproduced by permission of Preston Publications, a Division of Preston Industries, Inc.)

ongoing trends in this area, and it is clear that the field will continue to develop in the future, not the least in the field of bio-analysis.

We also accomplished several new developments in the field of trace analysis. First, we combined an automated liquid-liquid extraction (based on flow injection) with capillary GC, equipped with an empty precolumn for large volume injection. Thus, trace volatile compounds on a ppt level in wastewater could be automatically monitored [72]. Additionally, we developed a technology for preparing ultra-thick film open tubular traps, with a film thickness up to 100 μm (Fig. 10). Such traps had an extremely low phase ratio (β = 1.5 or less), which resulted in

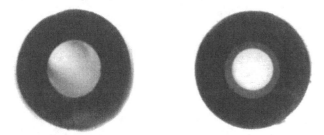

Figure 10 Photograph (right) of the cross section of an open tubular thick film trap, where the film of stationary phase (thickness ~ 80 μm) can be clearly observed. The cross section of the inner diameter of the tube is ~ 0.7 mm (original tube cross section shown on left side). (From Ref. 73. With permission from S. Blomberg and J. Roeraade.)

a strong retention even for comparatively volatile substances [73]. Quantitative concentration of solutes from gases (headspace, air) as well as from water were shown to be possible [74]. The interesting feature of these traps is that the retention mechanism is sorption, with predictable breakthrough volumes, in contrast to adsorption, which is governed by nonlinear adsorption isotherms.

The SPME technology developed by Pawliszyn [75–77] is another real success story of contemporary GC. The method utilizes the same sorption mechanism as our thick film traps. However, instead of using an open tube, the SPME concept employs a fused silica needle, which is coated on the outside with a silicone rubber. Thus, the needle can simply be dipped into a sample, and after a selected time the needle is inserted into a hot injector, using a syringe-like setup. Despite a number of shortcomings, numerous workers have used this method, and the number of applications is still growing [78]. Again, this shows that simplicity and convenience are of major importance for the widespread acceptance of new technology.

IV. THIN LAYER CHROMATOGRAPHY

Thin layer chromatography is one of the oldest forms of chromatography. In a sense, it is a refined version of Tswett's original paper chromatography, and it was used extensively in the early days. However, when high-performance liquid chromatography (HPLC) became established, TLC rapidly lost its importance. Nowadays, TLC is mainly employed as a preoptimization method of HPLC. In some laboratories, TLC is still in use as a complementary analytical tool, e.g., for rapid quality control or screening, and the technique has been instrumentalized to a large extent. However, TLC has some unique features compared to HPLC [79]: (1) TLC has a multilane capacity; (2) TLC plates provide the format for true two-dimensional separations; (3) components not eluted from the "column" remain visible and can be quantified; (4) plates containing original samples can be stored and archived after the separation has been carried out. The latter can be important, for example, in forensic applications.

Our interest in TLC was triggered by the work of Tyihak and coworkers from Hungary [80,81], who described a method in which the mobile phase was forced through the sorbent layer of the TLC plate by means of a pump. This was achieved by keeping the TLC plate in a sandwich-like arrangement, where the sorbent layer was covered with a thin Mylar foil. The outside of the Mylar foil was pressurized with water in order to provide the necessary counter-force to avoid an overflow of mobile phase. The technique was called "overpressured TLC." Although Tyihac et al. used only moderate eluent pressure, due to instrumental limitations, they showed that some of the inherent limitations of classical TLC could be eliminated. The maximum distance of the solvent front was extended, and it was even possible to elute compounds from the TLC plate since the mobile phase flow was not limited by the action of capillary force.

We constructed a system based on the principles of the Hungarian workers, but with the possibility of applying mobile phase pressures up to 300 bar [82]. This became a huge device since the largest plates required counterpressures of up to 200 tons. Figure 11 shows a photograph of our apparatus. This device allowed us to run the TLC plates with very high mobile phase velocities. The development time for a 20 cm plate was only a few minutes. We also evaluated plates, coated with 3 μm spherical particles, with the hope that these could replace HPLC columns (we are grateful to Macherey–Nagel, who made such plates for us). Unfortunately, the performance of these plates (or, rather, the flatbed HPLC columns) was not as good as we had predicted; we found this to be partly due to the low mechanical stability of the silica material [83]. Thus, our studies were discontinued after some time. However, I believe that it would be worthwhile to continue research on the subject. New, mechanically stable column

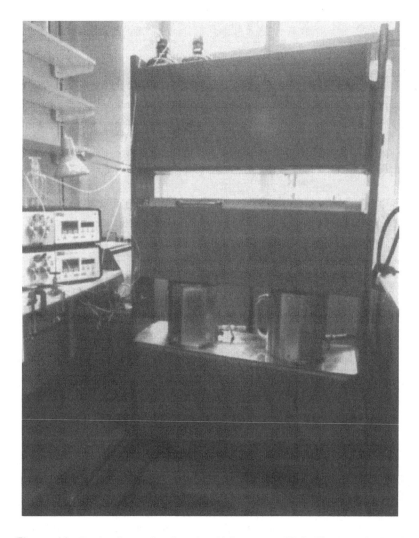

Figure 11 Device for performing ultra-high-pressure TLC. The large hydraulic press can deliver a counter pressure of 200 tons. Mobile phase is forced through the TLC plate with an HPLC pump. Counterpressure between the upper surface of the hydraulic press and the TLC plate is accomplished with a mylar foil cushion, which is pressurized with water. (S. Flodberg and J. Roeraade, unpublished.)

packing materials have become available as well as array detection systems, which could be used as an end-on-plate detector. Moreover, the urgent need for high-throughput parallel analysis systems (e.g., in combinatorial chemistry) would justify renewed efforts in forced flow TLC.

V. CAPILLARY ELECTROPHORESIS

The development of capillary electrophoresis (CE) is another fascinating, more recent story. Pioneering work of Virtanen [84], Everaerts et al. [85], and Hjertén [86] pointed towards miniaturization as the future direction of electrophoresis, but it is the work of Jim Jorgensson in the early 1980s [87] that is usually regarded as the start of modern capillary electrophoresis. Several researchers saw the potential of the technique and started to work in the area, but the progress reports were a bit scattered and felt somewhat untangible.

It was not until the initiative of Barry Karger, who organized the First International Symposium on High Performance Capillary Electrophoresis in 1989 in Boston, that everything started to fall into place. Many chromatographers (including the author) were drawn to this event out of curiosity without knowing too much about the state of the art, and it was an eye opener to many of us. There was a kind of euphoric atmosphere, and the attendence at the lectures was massive (and I am convinced that this was not due to bad weather in Boston). In fact, the flavor of the event reminded me of the first Hindelang symposium.

It was obvious that CE had an enormous potential in the field of bioanalysis, where there was an urgent need for new high-resolution separation techniques. However, in contrast to chromatography, any interaction of the analytes other than the transport medium can lead to detrimental adsorption. This was, and still is a major issue in CE, the importance of which was perhaps not immediately recognized by some chromatographers. Many suggestions to solve this problem came up very quickly in subsequent HPCE meetings, such as the use of increased salt concentrations [88], the use of surfactant (e.g., cationic or zwitterionic) additives [89–93], various column wall treatments, etc. [94–97]. In our own laboratory, we tried to utilize some of these methods, but initially we were not too successful. Therefore we searched for new concepts and started testing perfluorinated surfactants as buffer additives. We had a series of these components on the shelf, which I had used in the early 1980s as stationary phase modifiers in GC. To our surprise, excellent results were obtained, particularly with the cationic surfactant additive (with a quarternary ammonium headgroup). Using some model proteins with high pI, we were able to generate more than 500,000 plates while running the system at pH 7 with a weak phosphate buffer [98]. The reason for the outstanding performance was the unique properties of the fluorocarbon chain. The cationic group adheres to the negatively charged fused silica group by electrostatic interaction and a double layer is created in the same way as occurs with cetylpyridinium ammonium bromide (CTAB). Thus, the surface charge is changed from negative to positive, which leads to repelling of positively charged analytes. With the fluorosurfactants, a very dense admicellar layer is created due to the extreme hydrophobic properties of the perfluoro chain. Figure 12 shows an electropherogram of some of our first results.

Interestingly, perfluoro groups show poor interaction with hydrocarbon chains [99]. Therefore, interactions with the fluorosurfactant layer are mainly of electrostatic nature and are not due to hydrophobic interaction. We exploited this effect to separate various complex protein mixtures [100,101], and by mixing cationic and anionic fluorosurfactants we succeeded in separating both basic and acidic proteins in the same run at pH 7, using a weak buffer [102]. Moreover, a "charge tuning" at constant pH was possible by altering the proportions of the different fluorosurfactant additives.

Today, capillary electrophoresis is a well-established technique, and many of the initial problems have been solved. It would hardly be possible (or adequate) to review the enormous development of this area in the frame of this chapter. However, the real success story of CE must be mentioned: DNA sequencing. Already at the first High Performance Capillary Electrophoresis meeting (HPCE), I remember some people (e.g., Barry Karger) discussing the possibility of using gel-based CE for DNA sequencing. They had the vision that a large number of gel-filled columns in parallel could be used to provide a throughput capacity, necessary for sequencing entire genomes. Also, the human genome was discussed, but as usual many people were sceptical. However, it became the goal of many talented and determined people, and very rapidly the first array-based systems were constructed. Today it is extremely satisfying for a separation scientist to conclude that parallel capillary electrophoresis was of crucial importance for sequencing the human genome, completed in 2000.

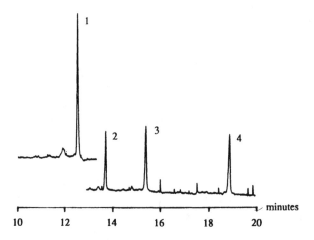

Figure 12 Electropherogram of some basic model proteins (1 = myoglobin, 2 = ribonuclease, 3 = cytochrome c, 4 = lysozyme) using an untreated fused silica column. Running buffer: 0.01 M phosphate, pH = 7, additive: cationic fluorosurfactant FC-134, 50 μg/mL. (From Ref. 98. Copyright © 1991, with permission from Elsevier Science.)

VI. MICROCHIPS

My involvement in microchips dates back to 1981, when I happened to see a report about a miniature gas chromatograph on a silicon wafer [103]. The work had been performed by a group at Stanford University [104]. A schematic of their chromatography is shown in Fig. 13.

I became intrigued, and after some time I decided to visit this group. At that time, a company (IC-Sensors) had started to manufacture the chip gas chromatograph. I was overwhelmingly impressed by the technology and visualized numerous other possibilities, particularly after reading Kurt Petersen's classic article "Silicon as a Mechanical Material" [105]. Unfortunately, the chip gas chromatograph was not too much of a success story at that time, partly because of overoptimistic performance expectations, but I think mainly because the time was not ripe for this device. I still find it difficult to understand why chromatographers did not jump at this technology!

My chance to do some real work with chips came in 1986–87, when I found out that a research group at Uppsala University in Sweden, directed by Bertil Hök, had started work in micro mechanics. This group had the necessary facilities to manufacture three-dimensional structures in silicon. Although their initial facilities were quite limited, we started to outline some microfluidic projects, and lots of "revolutionary" ideas were conceived. This was a wonderful time.

However, many people were sceptical about the real benefits of the miniaturized devices, and development costs were high. Thus, it was difficult to obtain adequate funding at that time. Our first work was therefore theoretical, dealing with simulations and predictions of chip-based CE separations [106]. Soon, however, we started making different channels in silicon. For electrokinetic fluid propulsion, we had to oxidize the inside of the chip to obtain an insulating layer in order to avoid a voltage breakdown. Figure 14 shows our first setup, where we tested electro-osmotic flow in a silicon microchannel.

In the early 1990s, several groups started to work with microchips. The efforts of Michael Widmer and his group at Ciba-Geigy should be mentioned in particular. Andreas Manz from

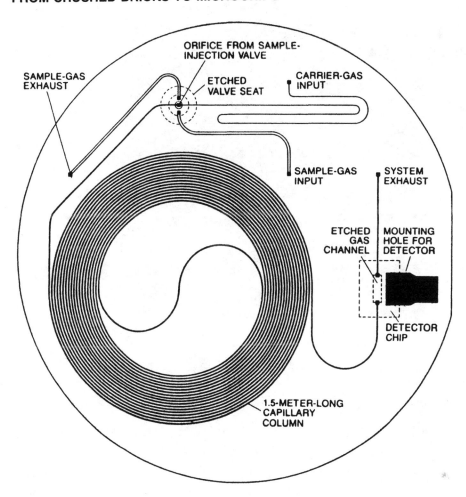

Figure 13 Schematic of the Stanford silicon wafer gas chromatograph: diameter = 5 cm. (From Ref. 104. Copyright © 1979 IEEE.)

this group did pioneering work and showed the first electrophoretic separations with glass chips [107]. Soon other researchers were engaged in the area, for example, Jed Harisson [108], Michael Ramsey [109], and Richard Mathies [110], among others. Our subsequent work with microchips has been mainly in the area of microvials and nanoscale chemistry [111–113], fundamental questions of sample handling [114,115], and mass spectrometry [116].

Today the interest in microchips is enormous. Several conference series are specifically devoted to the subject (e.g., the microTAS series), and at all major conferences on separation science a substantial part of the programs deals with microchips—and the field is still growing.

Although this book is about the history of chromatography, I cannot resist giving some of my personal thoughts about two questions: What is the future role of microchips? and Can we foresee any fundamentally new breakthroughs in separation science? Regarding the first question, I believe that chip technology will be of crucial importance for providing us with the analytical tools of tomorrow. But I am not convinced that chips will replace all classical separation columns. For biochemical applications, it seems obvious that polymer-based chips will replace glass or silicon devices for reasons of better chemical compatibility and of cost/perfor-

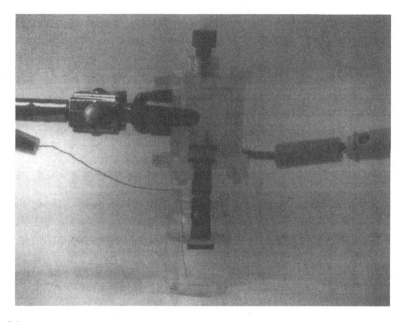

(a)

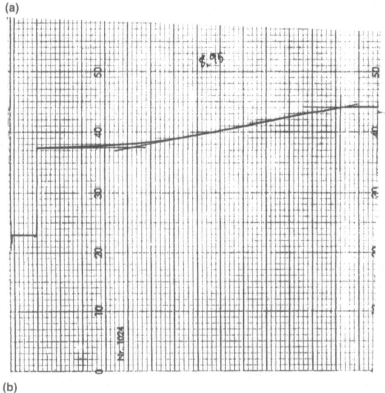

(b)

Figure 14 (a) Experimental setup used for testing electro-osmotic flow in a silicon chip–based micro-channel. The ends of the channel were connected to salt solutions. The solutions had different ionic strengths. (b) After voltage was applied, the change in resistance over the channel vs. time was monitored. (M. Jansson and J. Roeraade, 1989, unpublished.)

mance. It can also be foreseen that polymer-based microstructures will find extensive use in other applications.

Prediction of discoveries and fundamentally new principles is, by definition, rather uncertain. However, I believe that the truly novel things are hidden in the nanoworld. Separation science deals with selective mass transport, molecular properties, and interactions. The basic principles involved are well known, and results are fairly predictable. If separations could be performed in discrete channels reduced to submicrometer dimensions, the forces needed to accomplish selective mass transport would scale very differently when comparing with macro systems. Some forces, which are negligible in the macroworld, will become dominant in the nanoworld. Current systems such as HPLC are usually based on multipath (packed bed) concepts, where selective effects of the nanoworld are blurred by dispersion caused by differential mass transport. Single path LC (open tubular LC) has usually been performed with systems in which the cross-sectional dimensions of the column has been too large to accomplish an efficient reduction of the Taylor dispersion.

To conclude, I believe that we still have many exciting developments ahead of us. There are always people who claim that separation science is mature and that fundamentally new developments can no longer be expected in this field. They have always proven to be wrong. I strongly believe in a flourishing, fruitful, and fascinating upcoming century of separation science.

ACKNOWLEDGMENTS

I would like to thank the following persons for their essential contributions to my past research work: Curt Enzell, Sverker Blomberg, Göran Flodström, Mårten Stjernström, Gunnar Hagman, Åsa Emmer, Peter Lindberg, Erik Litborn, Bertil Hök, Ulf Lindberg, Anders Hanning, Johan Sjödahl, Andreas Woldegiorgis, Mario Curcio, Theres Redeby, and Johan Pettersson. I would also like to thank the Swedish Natural Science Foundation, The Swedish Board for Technical Development, The Swedish Council for Engineering Sciences, and The Swedish Foundation for Strategic Research for their financial support.

REFERENCES

1. DH Desty. In: JC Giddings, RA Keller, eds. *Capillary Columns: Trials, Tribulations and Triumphs.* New York: Marcel Dekker, Inc., 1965, p. 199.
2. LS Ettre. Open Tubular Columns in Gas Chromatography. New York: Plenum Press, 1965.
3. R Kaiser. Chromatographie in der Gasphase, II, "Kapillar-Chromatographie." 2d ed. Mannheim: Bibliographisches Institut, 1966.
4. LD Metcalfe, RJ Martin. Anal Chem 39:1204, 1967.
5. DH Desty, JN Haresnape, BHF Whyman. Anal Chem 32:302, 1960.
6. K Grob. Helv Chim Acta 51:718, 1968.
7. K Tesařík, M Novotny. In: HG Struppe, ed. Gaschromatographie 1968. Berlin: Akademie Verlag GmbH, 1968, p. 575.
8. FA Bruner, GP Cartoni. Anal Chem 36:1522, 1964.
9. J Bouche, M Verzele. J Gas Chromatogr 6:501, 1968.
10. KD Bartle, CL Woolley, KE Markides, ML Lee, RS Hansen. HRC&CC 10:128, 1987.
11. DH Desty, A Goldup, T Swanton. In: N Brenner et al., eds. Gas Chromatography. New York: Academic Press, 1962, p. 105.
12. Chromatographia 8:421–530, 1975.
13. G Schomburg, E Ziegler. Chromatographia 5:96, 1972.

14. G Schomburg, F Weeke. In: SG Perry, ER Adlard, eds. Gas Chromatography 1972 (Montreux Symposium). Barking: Applied Science Publishers, 1973, pp. 285–294.

15. G Schomburg, H Husmann, F Weeke. In: A Zlatkis, LS Ettre, eds. Advances in Chromatography 1974 (Houston Symposium). Houston: University of Houston, 1974, pp. 63–79.

16. G Schomburg, H Husmann, F Weeke. J Chromatogr 112:205, 1975.

17. CA Cramers, JA Rijks, CPM Schutjes. Chromatographia 14:439, 1981.

18. CPM Schutjes, PA Leclercq, JA Rijks. J Chromatogr 289:163, 1984.

19. CA Cramers, CE van Tilburg, CPM Schutjes, JA Rijks, GA Rutten, R de Nijs. J Chromatogr 279: 83, 1983.

20. CPM Schutjes, EA Vermeer, JA Rijks. J Chromatogr 253:1, 1982.

21. JD Schieke, NR Comins, V Pretorius. J Chromatogr 112:97, 1975.

22. JD Schieke, NR Comins, V Pretorius. Chromatographia 8:354, 1975.

23. K Grob, G Grob. Chromatographia 5:3, 1972.

24. K Grob, K Grob, Jr. J Chromatogr 94:53, 1974.

25. K Grob, K Grob, Jr. HRC&CC 1:57, 1978.

26. DR Deans. Anal Chem 43:2026, 1971.

27. DR Deans. Chromatographia 1:18, 1968.

28. C Madani, EM Chambaz, M Rigaud, P Chebroux, J Breton. Chromatographia 10:466, 1977.

29. C Madani, EM Chambaz. Chromatogr Symp Ser 1:175, 1979.

30. C Madani, EM Chambaz. Chromatographia 11:725, 1978.

31. L Blomberg, T Waennman. J Chromatogr 168:159, 1979.

32. L Blomberg, J Buijten, J Gawdzik, T Waennman. Chromatographia 11:521, 1978.

33. L Blomberg, T Waennman. J Chromatogr 168:81, 1979.

34. T Welsch, W Engelwald, C Klaucke. Chromatographia 10:22, 1977.

35. RD Dandeneau, EH Zerenner. HRC&CC 2:351, 1979.

36. F Munari, S Trestianu. Proc. Fifth Int. Symp. Capillary Chromatogr., Riva del Garda, Italy. Amsterdam: Elsevier, 1983, p. 327.

37. J Roeraade. Chromatographia 8:511, 1975.

38. CR Enzell, A Rosengren, I Wahlberg. Tobacco Sci 13:127, 1969.

39. RA Appleton, CR Enzell, B Kimland. Beitr Tabakforsch 5:266, 1970.

40. B Kimland, RA Appleton, AJ Aasen, J Roeraade, CR Enzell. Phytochemistry 11:309, 1972.

41. J Roeraade, CR Enzell. HRC&CC 2:123, 1979.

42. J Roeraade, S Blomberg, HDJ Pietersma. J Chromatogr 356:271, 1986.

43. K Grob. In: Split and Splitless Injection of Capillary Gas Chromatography. 3rd rev. ed. Heidelberg, Germany: Huethig, 1993, p. 547.

44. K Grob. In: On-Column Injection in Capillary Gas Chromatography: Basic Technique, Retention Gaps, Solvent Effects. Heidelberg: Huethig Verlag, p. 591.

45. K Grob, Jr. In: Advanced On-Column Injection. Heidelberg: Hüthig, 1986.

46. K Grob, Jr. J Chromatogr 213:3, 1981.

47. DH Desty, A Goldup, BAF Whymann. J Inst Petroleum 45:287, 1959.

48. K Grob, G Grob. J Chromatogr Sci 7:584, 1969.

49. K Grob, G Grob. J Chromatogr Sci 7:587, 1969.

50. G Schomburg, H Behlau, R Dielmann, F Weeke, H Husmann. J Chromatogr 142:87, 1977.

51. PMJ van den Berg, T Cox. Chromatographia 5:301, 1972.

52. A Zlatkis, JQ Walker. J Gas Chromatogr 1:9, 1963.

53. K Grob, K Grob, Jr. J Chromatogr 151:311, 1978.

54. W Vogt, K Jacob, HW Obwexer. J Chromatogr 174:437, 1979.

55. W Vogt, K Jacob, A-B Ohnesorge, HW Obwexer. J Chromatogr 186:197, 1979.

56. G Hagman, J Roeraade. HRC 13:461, 1990.

57. K Grob, Jr., HP Neukom. HRC&CC 2:15, 1979.

58. G Schomburg, R Dielmann, F Weeke, H Husmann. J Chromatogr 122:55, 1976.

59. J Roeraade, S Blomberg. Chromatographia 17:387, 1983.

60. V Pretorius, K Lawson, W Bertsch. HRC&CC 6:419, 1983.
61. S Blomberg, J Roeraade. Chromatographia 25:21, 1988.
62. J Roeraade, S Blomberg. In: P Sandra, ed. Proc. Ninth Int. Symp. Capillary Chromatography (Monterey). Heidelberg: Hüthig, 1988, p. 614.
63. J Roeraade, S Blomberg. HRC 12:138, 1989.
64. HJ Cortes, CD Pfeiffer, GL Jewett, BE Richter. J Microcol Sep 1:28, 1989.
65. RE Majors. J Chromatogr Sci 18:571, 1980.
66. F Munari, K Grob. J Chromatogr Sci 28:61, 1990.
67. K Grob. On-Line Coupled LC-GC. Heidelberg: Hüthig, 1991.
68. EC Goosens, D de Jong, JHM van den Berg, GJ de Jong, UAT Brinkman. J Chromatogr 552:489, 1991.
69. AJH Louter, CA van Beekvelt, P Cid Montanes, J Slobodnik, JJ Vreuls, UAT Brinkman. J Chromatogr 725:67, 1996.
70. A Farjam, JJ Vreuls, WJGM Cuppen, GJ de Jong, UAT Brinkman. Anal Chem 63:2481, 1991.
71. UAT Brinkman, ed. Hyphenation: Hype and Fascination. Amsterdam: Elsevier, 1999.
72. J Roeraade. J Chromatogr 330:263, 1985.
73. S Blomberg, J Roeraade. HRC&CC 11:457, 1988.
74. S Blomberg, J Roeraade. J Chromatogr 394:443, 1987.
75. RP Belardi, J Pawliszyn. Water Pollut Res J Can 24:179, 1989.
76. CL Arthur, J Pawliszyn. Anal Chem 62:2145, 1990.
77. DW Potter, J Pawliszyn. J Chromatogr 625:247, 1992.
78. H Kataoka, HL Lord, J Pawliszyn. J Chromatogr A 880:35, 2000.
79. F Geiss. Fundamentals of Thin Layer Chromatography (Planar Chromatography). Heidelberg: Hüthig, 1987.
80. E Tyihak, E Mincsovics. J Chromatogr 174:75, 1979.
81. E Tyihak, E Mincsovics, F Körmendi. Hung Sci Instr 55:33, 1983.
82. G Flodberg, J Roeraade. J Planar Chromatogr 8:10, 1995.
83. G Flodberg, J Roeraade. J Planar Chromatogr 6:252, 1993.
84. R Virtanen. Acta Polytech Scand 123:1, 1974.
85. AJP Martin, FM Everaerts. Anal Chim Acta 38:223, 1967.
86. S Hjertén. Arkiv Kemi 13:151, 1958.
87. JW Jorgenson, KD Lukacs. Anal Chem 53:1298, 1981.
88. JS Green, JW Jorgenson. J Chromatogr 478:63, 1989.
89. M-D Zhu, R Rodriguez, D Hansen, T Wehr. J Chromatogr 516:123, 1990.
90. G Mandrup. J Chromatogr 604:267, 1992.
91. HK Kristensen, SH Hansen. J Chromatogr 628:309, 1993.
92. JE Wiktorowicz, JC Colburn. Electrophoresis 11:769, 1990.
93. K Tsuji, RJ Little. J Chromatogr 594:317, 1992.
94. S Hjertén. J Chromatogr 347:191, 1985.
95. GJM Bruin, JP Chang, RH Kuhlman, K Zegers, JC Kraak, H Poppe. J Chromatogr 471:429, 1989.
96. KA Cobb, V Dolnik, M Novotny. Anal Chem 62:2478, 1990.
97. JK Towns, J Boa, FE Regnier. J Chromatogr 599:227, 1992.
98. Å Emmer, M Jansson, J Roeraade. J Chromatogr 547:544, 1991.
99. E Kissa. Fluorinated Surfactants—Synthesis, Properties, Applications. New York: Marcel Dekker, 1994.
100. J Sjödahl, Å Emmer, B Karlstam, J Vincent, J Roeraade. J Chromatogr B 705:231, 1998.
101. Å Emmer, M Jansson, J Roeraade. J Chromatogr A 672:231, 1994.
102. EL Hult, Å Emmer, J Roeraade. J Chromatogr A 757:255, 1997.
103. SC Terry. A gas chromatography system fabricated on a silicon wafer using integrated circuit technology. Ph.D. dissertation, Dept. of Electrical Engineering, Stanford University, Stanford, CA, 1975.
104. SC Terry, JH Jerman, JB Angell. IEEE Trans Electron Devices ED-26:1880, 1979.

105. KE Petersen. Silicon as a mechanical material. Proc IEEE 70:420, 1982.
106. M Jansson, Å Emmer, J Roeraade. HRC&CC 12:797, 1989.
107. DJ Harrison, A Manz, Z Fan, H Lüdi, HM Widmer. Anal Chem 64:1926, 1992.
108. DJ Harrison, K Fluri, K Seiler, Z Fan, CS Effenhauser, A Manz. Science 261:895, 1993.
109. SC Jacobson, R Hergenröder, LB Koutny, RJ Warmack, JM Ramsey. Anal Chem 66:1107, 1994.
110. AT Woolley, RA Mathies. Anal Chem 67:3676, 1995.
111. M Jansson, Å Emmer, J Roeraade, B Hök, U Lindberg. J Chromatogr 626:310, 1992.
112. E Litborn, Å Emmer, J Roeraade. Anal Chim Acta 401:11, 1999.
113. E Litborn, M Stjernström, J Roeraade. Anal Chem 70:4847, 1998.
114. E Litborn, Å Emmer, J Roeraade. Electrophoresis 21:91, 2000.
115. E Litborn, J Roeraade. J Chromatogr B 745:137, 2000.
116. S Jespersen, WMA Niessen, UR Tjaden, J van der Greef, E Litborn, U Lindberg, J Roeraade. Rapid Commun Mass Spectrom 8:581, 1994.

4

Thin Layer Chromatography

Joseph Sherma

Lafayette College, Easton, Pennsylvania

The term ''thin layer chromatography'' will be used throughout this chapter rather than ''planar chromatography,'' a general classification that has been used recently to denote chromatography modes having a flat stationary phase. Paper chromatography is a type of planar chromatography, but the term has been used incorrectly to include electrochromatography or electrophoresis because this method does not involve the elements necessary for designation as a chromatographic method as defined by Strain in 1942 [1], i.e., substances in a narrow initial zone are caused to undergo differential migration on a porous, sorptive stationary phase having a selective resistive action by the nonselective driving force of a liquid or gaseous mobile phase.

I. EARLY HISTORY OF THIN LAYER CHROMATOGRAPHY (1889 TO EARLY 1960s)

The history of thin layer chromatography (TLC) has been traced back to experiments performed by the Dutch biologist Beyerinck in 1889 [2]. These predate the early work on column chromatography reported by Tswett in 1906 [3] but are predated by the paper chromatography work of Runge, Schoenbein, and Goppelsroeder [4] during the period 1834–1888. Beyerinck allowed a drop of a mixture of hydrochloric and sulfuric acids to diffuse through a thin layer of gelatin on a glass plate. The hydrochloric acid traveled faster than the sulfuric acid and formed a ring around the latter. The hydrochloric acid zone was made visible by reaction with silver nitrate and the sulfuric acid with barium chloride. Tswett did not study TLC but discussed adsorption of compounds on strips of paper during capillary analysis [5]. Consden, Martin, and Gordon reintroduced paper chromatography in 1944 [6], and it grew into a universally used microanalytical method during the next 10 years. These authors also defined the term R_f, the ratio of the distance moved by the sample zone to that of the mobile phase front, in this paper.

Nine years later, Wijsman used the same techniques to show that there were two enzymes in malt diastase, and he proved that only one of them split off maltose from soluble starch. Wijsman was also the first to use a fluorescent indicator to detect chromatographic zones. He incorporated fluorescent bacteria from seawater in a gelatin layer containing starch and allowed the amylase mixture to diffuse in the layer. A fluorescent band appeared only where the beta-amylase reacted with the starch. This method was sufficiently sensitive to detect about 40 pg of maltose, a value rarely surpassed in TLC detection methods today. Wijsman's contributions were reviewed by Kirchner [7].

Izmailov and Schraiber [8], in 1938 at the Institute of Experimental Pharmacy of the University of Kharkov, used a horizontal 2 mm layer of aluminum oxide without binder on a glass microscope slide for circular chromatography of alcoholic plant extracts such as belladona, reported in the Russian Pharmacopeia VII. The use of a thin layer, or open column, may have been suggested by the wider adoption of Tswett's column chromatography method by the late 1930s. Izmailov and Schraiber placed one drop of the pharmaceutical sample solution on the adsorbent and developed it into concentric zones by dropwise addition of methanol on the spot to produce circular zones of substances that were seen under an ultraviolet (UV) lamp. This method has been called spread-layer chromatography in some sources. The authors pointed out the usefulness of drop chromatography for testing adsorbents and solvents for application in column chromatography, compared to which it was faster and economical in terms of adsorbent, solvent, and sample usage.

In 1940, Lapp and Erali [9] spread a loose layer of aluminum oxide 8 cm long on a glass slide that was supported on an inclined aluminum sheet. The latter was cooled at its upper end and heated at the lower end. The sample to be separated was placed at the top of the adsorbent and was gradually washed down with the developing solvent.

Crowe, in 1941 [10], used a technique similar to that of Izmailov and Schraiber by placing adsorbent in the cups of spot plates. After selecting the adsorbent and solvent, a thin, wedge-shaped loose layer of adsorbent was formed in a tilted petri dish. A drop of the solution was allowed to flow onto the adsorbent and was then developed dropwise. Like Izmailov and Schraiber, Crowe used his method to select suitable solvents for column chromatography.

In 1942, von Bekesy [11] used an adsorbent slurry to fill a channel between two glass plates separated by cork gaskets. This apparatus was then used in the same manner as a chromatographic column. Von Bekesy prepared a micro column by covering a shallow channel in a glass plate with another glass plate. Agar was used to hold the two plates together, and the column was formed by filling with an adsorbent slurry.

Brown [12] demonstrated circular paper chromatography in 1939 by placing filter paper between two glass plates. The upper plate contained a small hole for sample application and dropwise addition of solvent. Because of the mild adsorption characteristics of paper, Brown proposed the use of a thin layer of alumina between the sheets of paper. T. L. Williams, in 1947 [13], eliminated the paper and used only the adsorbent between the glass plates.

Meinhard and Hall, in 1949 [14], were the first to use a binder to hold the adsorbent on the support and produce a layer without cracks. They bound a mixture of aluminum oxide and Celite with starch to microscope slides. These prepared slides were used to perform circular drop chromatography of inorganic ions with slow application of solvent by means of a special developing pipet. This method was termed surface chromatography.

During the period 1945–1954, Kirchner and his associates at the U.S. Department of Agriculture were working on the isolation and identification of the flavoring components (terpenes) present in orange and grapefruit juice and required a sensitive microchromatographic method for their purification and identification. Paper chromatography was tried, but it was quickly evident that paper was too limited in its adsorption capabilities to be of any great value. The use of silica gel–impregnated paper was studied in 1950, but its capacity was also limited and the preparation was rather tedious. The work of Meinhard and Hall was published at this time, and Kirchner conceived the idea that it would be possible to modify the drop chromatographic technique to combine the advantages of column and paper chromatography. As a result, they carried out TLC on silica gel "chromatostrips" essentially as it is practiced today. Thin layers were developed in a closed tank in a manner analogous to the ascending paper chromatography technique; different adsorbents, binders, and detection reagents were studied, and a zinc silicate and zinc cadmium sulfate mixture was incorporated in the layer as a fluorescence indicator; larger plates and two-dimensional development were used to allow greater resolution of complex mixtures; and quantitative TLC by the elution method was introduced for the determination of biphenyl in citrus fruits [15]. These studies on the discovery, modification, and application of TLC were published by Kirchner and coworkers starting in 1951 [16] through 1957 in a series of papers that were the basis of modern TLC.

Instead of narrow glass strips, Reitsema [17], in 1954, used larger glass plates, first reported by Kirchner, on which several samples could be chromatographed side by side or two-dimensional chromatograms could be developed. He originated the use of the term "chromatoplate."

Stahl published his first paper on TLC and was the first to use the name "thin layer chromatography" in 1956 [18]. He used a fine-grained silica gel (1–5 μm) without binder to prepare his layers but returned in 1958 [19] to the use of gypsum binder, first reported by Kirchner, to prepare "kieselgel G" layers ("silica gel according to Stahl for TLC"). Stahl was respon-

Figure 1 (Left to right) C. Horvath, E. Stahl, H. Kalasz, H. Issaq, and K. Macek. Horvath and Kalasz were pioneers in the development of gas chromatography and forced-flow TLC, respectively, while Macek wrote many important books and papers on both paper and thin layer chromatography.

sible for the standardization of materials through creation of a basic TLC kit, procedures, and nomenclature; design of a convenient, practical spreader for preparing layers about 250 μm thick; and applications of the new equipment first with lipophilic compounds (e.g., steroids) and then hydrophilic ones (sugars). These advances, along with (1) publicity and availability of commercial products from Merck (introduced standardized silica gel, aluminum oxide, and kieselguhr "according to Stahl" in 1958) and Desaga (supplied a basic TLC kit), (2) indexing by *Chemical Abstracts* of papers under the separate term "thin layer chromatography," thereby making it easier to locate information on the method, and (3) publication of Stahl's first TLC laboratory manual in 1962 with contributions by about 25 TLC specialists [20] and books by Bobbitt [21], Randerath [22], Truter [23], and Kirchner [24], were responsible for the growth of popularity of TLC in the 1950s and 1960s until it replaced paper chromatography in most applications because of better efficiency and resolution, versatility, and sample capacity. A second, expanded handbook by Stahl published in 1967 with over 1000 pages reflected the greatly increased use of TLC in this period. Stahl has provided an introspective view of his contributions to the field over a period of 25 years [25,26]. Figure 1 is a photograph of Stahl taken in Budapest at the second annual American-Hungarian Chromatography Symposium in 1982.

Camag began selling basic TLC equipment such as plate coaters, developing tanks, sample application devices, and UV lamps in 1961 and eventually became one of the leading companies in the development of instruments for modern, quantitative high-performance (HP) TLC.

II. LATER ADVANCES IN TLC (MID-1960s TO 2000)

The availability of commercially precoated layers that could be used in place of hand-coated plates was an important advance in TLC because of their improved convenience and uniformity. No longer did plates have to be prepared in the laboratory using pouring and spreading methods or a commercial plate-coating device (Fig. 2). Around 1965, layers with organic binders were introduced and were of higher quality and more durable than gypsum-bound layers; however, a significant number of G plates are still sold today according to plate manufacturing company sources. Plastic-backed precoated sheets developed by Eastman Kodak, which combined the handling advantages of paper chromatography and the separation quality of TLC, were described in 1965 by Przybylowicz et al. [27], and Baker-Flex sheets, prepared for J. T. Baker by Macherey-Nagel, became available at about the same time. Other companies introduced precoated analytical and preparative glass-, aluminum-, or polyester-backed TLC layers composed of silica gel, cellulose, aluminum oxide, polyamide, and silica gel mixed with ion exchangers. Silica gel 60 plates with a pore size of 60 Å have probably been the most widely used. Soft sheets consisting of silica gel or silicic acid impregnated into a glass microfiber support were a hybrid between chromatography paper and a precoated thin layer; these were sold by Gelman starting in the late 1960s as instant thin layer chromatography (ITLC) media. The use of smaller sorbent particles to improve resolution was suggested by Thoma in 1968 [28]. A 5 μm particle diameter silica gel for the method, termed high-performance thin layer chromatography (HPTLC), was developed by Kaiser, Halpaap, and Ripphahn in the period 1975–1977 [29], and commercial plates with these smaller particles and a narrower size distribution were first produced by Merck in this period. HPTLC involved application of small volumes (200 nL) and 3–5 cm development distances, leading to smaller zone diameters, better sensitivity and resolution, faster analysis time, and the ability to apply more samples per plate. HPTLC is becoming more widely used each year compared to TLC, and the method is especially popular for qualitative and instrumental quantitative analysis in Europe and Asia. A particle size of 3–5 μm appears to be the practical

Figure 2 Hand-operated TLC plate coater (1961). (Photograph supplied by D. Jaenchen, Camag, Muttenz, Switzerland.)

limit for capillary flow HPTLC, but smaller particles may be usable with forced flow development, e.g., in overpressured layer chromatography (OPLC). The first C-18–bonded plate was produced by Whatman in the early 1980s [30], and precoated plates are now available with most of the sorbents available in HPLC columns and others, including hydrophobic, reversed-phase C-2, C-8, C-18, and phenyl-modified silica gel and hydrophilic, multimodal amino- and cyano-derivatized silica gel plates [31]. Laboratory-made biphasic plates with a lower 2 cm ion exchange zone for direct desalting of biological samples were reported in 1968 [32]. In 1978, Issaq (Figs. 1 and 3) started making biphasic TLC plates with two adjacent stationary phases for two-dimensional TLC with different selectivities in each direction by dipping a precoated silica gel plate up to half its height in a silylating agent and drying three times [31]. The first commercial preadsorbent plates with a lower kieselguhr concentration zone and upper silica gel layer were made by Beesley (Fig. 3) in 1971 at his company Quantum Industries, which was founded in 1966 and was taken over by Whatman in 1977. Quantum was also the first to make laned or channeled plates about 2 years later and to offer a TLC kit for screening abused drugs in urine (Quantum toxicology system). Normal and reversed-phase plates with kieselguhr or wide-pore silica gel preadsorbent or concentration zones are now commercially available, as are chiral layers for separation of enantiomers [33]. Laboratory-prepared Sephadex layers were reported in 1970 for the size exclusion TLC of proteins for molecular weight determination [34]. The first book on stationary phases by Macek and Hais was published in 1965 [35], and an important paper by Halpaap on the standardization of commercially produced plates was published in 1968 [36].

Figure 3 Photograph taken in Budapest of (left to right) T. Beesley, J. Sherma, and H. Issaq.

A method called "thin film chromatography" was introduced by Cremer in 1968 [37]. It involved a vacuum-evaporated 1 mm layer of indium oxide on a glass plate developed within 100 s for a distance of 1–2 cm with a thin film (~4 mm) of mobile phase. The method was not successful because of low capacity (10 nL application volume) and difficulty in detecting zones.

Theoretical studies of TLC make wide use of the important terms R_f and R_m, the latter being first defined as $\log(1/R_f-1)$ by Bate-Smith and Westall in 1950 [38]. Although the elutropic series of solvents listed according to increasing strength or polarity for adsorbents such as alumina was introduced around 1940 and Stahl described the concept of considering the properties of the three components of the TLC system—the sample, layer, and mobile phase—in order to help predict successful combinations, layer and mobile phase selection has been mostly based on trial and error guided by literature reports of similar separations. Soczewinski mathematically characterized solvent composition effects in certain types of adsorption systems using R_m values in 1969 [39]. Snyder described a systematic theory of adsorption chromatography in 1971 [40] involving the concepts of resolution, selectivity, and theoretical plates, and several years later he introduced the selectivity triangle by use of which mobile phases with optimum selectivity could be chosen using solvents with proton donor, proton acceptor, and dipolar properties. The theoretical aspects of TLC were presented in detail in 1972 by Geiss in a book [41] that explained to the reader the fundamentals influencing TLC during development with different kinds of layers and mobile phases in various kinds of chambers, such as equilibration of the layer and solvent demixing to produce multiple fronts, and how to adopt practical experimental parameters that lead to improved resolution and reproducibility. This book for the first time allowed practitioners to convert from an empirical approach to one based on understanding and control of the underlying processes occurring during TLC. Geiss also presented an equation for transfer of separation conditions from a layer to a column. Kaiser introduced in 1977 [42] a unique parameter for TLC, the separation number, or SN, defined as the number of substances completely separated (resolution = 1) between R_f values of 0 and 1 by a homogeneous mobile phase (no gradients in the development direction). Guiochon and Siouffi [43] studied the relationship of SN to theoretical plate height (HETP). More detailed systematic strategies, many computer-based, for mobile phase optimization in TLC began to appear in the literature around 1980; the PRISMA method, introduced by Nyiredy and coworkers [44], has been among the most widely used of these.

In addition to advances in layer materials and understanding of chromatographic theory as described above, progress in TLC has been fueled primarily by development of new techniques and instrumentation. Various types of chambers were introduced in addition to the original large-volume rectangular glass tanks (N-tanks) used for vertical development, including horizontal chambers (Desaga and Camag) [45] that allow the development of double the number of samples from opposite sides of the plate to the middle (Fig. 4) and twin-trough (Camag) chambers. Sandwich development, in which the layer is covered with a closely spaced counter plate, is another early type of development. These methods provided better control over the effects of saturation with the gas phase inside the chamber. The horizontal KS-Vario chamber [46] allowed testing and optimization of various mobile phases and vapor-saturation conditions. The short-bed continuous-development (SB-CD) chamber, which effectively increased the separation distance, was sold by Regis starting in 1979. Chambers for circular (U-chamber) and anticircular TLC (suggested by Kaiser) were designed by Camag in the 1976–1978 period to improve the

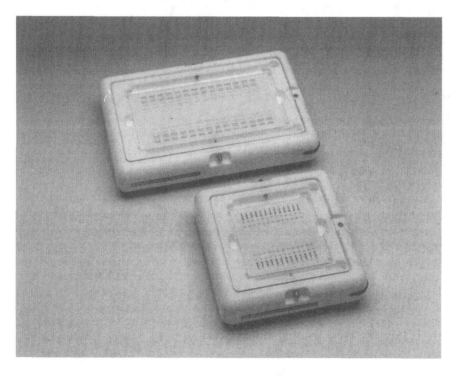

Figure 4   Horizontal developing chamber. (Photograph supplied by D. Jaenchen, Camag, Muttenz, Switzerland.)

resolution of zones with low or high R_f values, respectively. High-pressure planar liquid chromatography (HPPLC) is the latest version of rapid circular chromatography; solvent is pumped under high pressure continuously through a layer covered with a glass plate while samples are injected at constant intervals [47]. OPLC was described by Tyihak et al. in 1979 [48]; this method, which was proposed to be more reproducible than capillary development and provides constant flow rate and efficiency over the length of the layer, is the TLC method most closely related in procedure to HPLC but has not been widely used up to this time despite the availability of instruments from a number of manufacturers. Continuous development was suggested in 1961 by Brenner and Niederwieser [45] to increase effective separation distance and improve resolution. Repeated development with the same solvent in the same direction over the same distance was studied by Thoma in 1968, while programmed multiple development (PMD) with a single mobile phase over increasing distances was first reported by Perry et al. in the mid-1970s [49]. This was followed in 1984 by Burger's introduction of automated multiple development (AMD) [50], in which the layer is developed over 10–25 stages in the same direction for increasingly longer distances (3–10 cm) with a stepwise mobile phase gradient of decreasing strength (Fig. 5). AMD is among the most efficient HPTLC methods and has been used widely combined with densitometry for quantification of complex mixtures, such as residues of pesti-

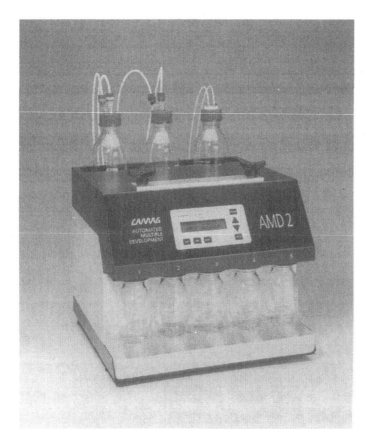

Figure 5 Automated multiple development (AMD) system 2. (Photograph supplied by D. Jaenchen, Camag, Muttenz, Switzerland.)

cides and explosives in contaminated water and soil samples [51]. According to Jaenchen [52], AMD is the method to resort to when one-step TLC development does not give the desired resolution; however, he suggests that it cannot be expected that AMD will become the standard, routine procedure in TLC, and that the routine developing step is the weakest link today in the TLC procedure.

Sample and standard solutions were first spotted by use of hand-held glass capillary tubes such as Drummond Microcaps or syringes. Over time, various partly and fully automated instruments were designed and sold commercially to apply solutions in the form of round spots or bands in nL to μL volumes. Among the first spotting instruments was the one reported by Getz in 1971 [53] and commercialized by Kontes, consisting of a series of tubes filled with equal volumes of sample and standard solutions that emptied through capillary needles by gravity flow onto the plate origins with drying by a stream of air or nitrogen. This type of manual, simultaneous applicator was followed by automated simultaneous applicators, such as the Camag automatic sample applicator (1973) with nine syringes providing automatic spot-size control in the 5–50 μL volume range regulated by measurement of electrical conductivity and the Analytical Instrument Specialties (AIS) spotter that applied solutions in 1–2 mm spots from 10 or 19 syringes that were discharged at a controlled rate by (1) a motor-driven bar with drying by heat directed at the tips; (2) manual, sequential applicators using a single dispensing device; and (3) currently by instruments that automatically apply selectable, variable volumes of sample and standard solutions in sequence in the form of bands or spots to precisely chosen locations of the layer (Fig. 6). The development of small-volume, sequential automated applicators was critical in allowing the enhanced properties of HPTLC plates to be exploited fully.

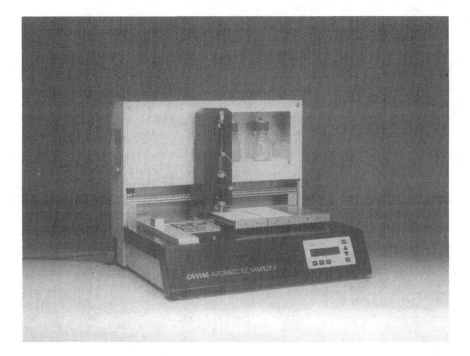

Figure 6 Automatic TLC Sampler 4. (Photograph supplied by D. Jaenchen, Camag, Muttenz, Switzerland.)

Quantitative analysis in TLC was first performed by comparing visually the sizes and intensities of standard zones versus unknown sample zones developed on the same plate. The next approach was scraping of sample and standard zones from the plate followed by collection in a vacuum-tube apparatus (Desaga), elution with a solvent, and measurement by solution spectrometry. Camag produced an instrument (the Eluchrom) for automated elution of separated zones directly without scraping off of the layer to obtain compound solutions for quantification (Fig. 7). Commercial TLC densitometers became available in the 1960s, among the earliest of which were the Zeiss [54], Kontes [55], Camag-Turner [56], Schoeffel [57], and Vitatron TLD 100 [58] scanners. At first, the visible or UV absorption or fluorescence of sample and standard zones was scanned in situ by reflection or transmission, and quantification was based on measurement of areas by manual triangulation of the peaks produced by an analog recorder or peak area measurement by a computing integrator. Figure 8 shows Camag's first monochromator-equipped densitometer (TLC/HPTLC Scanner I), produced in 1979. Software for computer control, developed by Ebel and coworkers at Wurzburg University, was available for this instrument at this early date but was seldom applied. Modern densitometers today, such as those from Camag, Shimadzu, and Desaga, provide scanning and data recording and processing that are totally automated by a computer. Unlike the theoretical linear Beer's law correlation between absorbance and concentration in solution spectrophotometry, densitometry scan area and zone concentration were found to be nonlinearly related. Many studies of the theory of quantitative TLC were carried out in order to find the optimal mathematical relationship, among the earliest

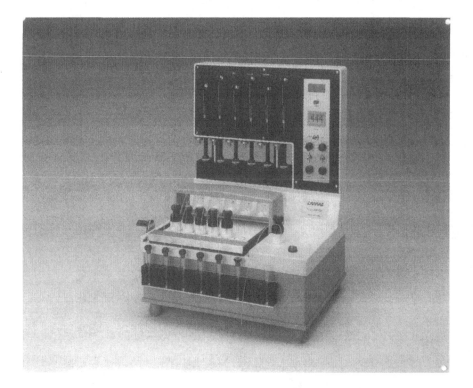

Figure 7 Eluchrom instrument (1972) for automatic elution of separated zones directly without scraping off the layer. (Photograph supplied by D. Jaenchen, Camag, Muttenz, Switzerland.)

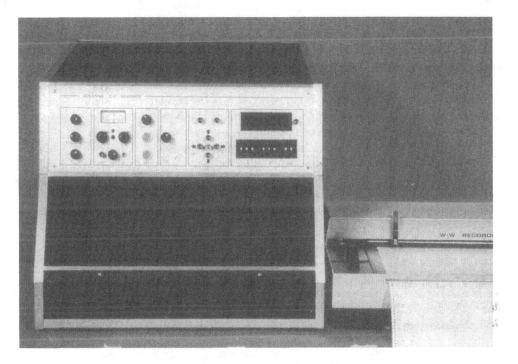

Figure 8 TLC/HPTLC Scanner I. (Photograph supplied by D. Jaenchen, Camag, Muttenz, Switzerland.)

of which were those by Goldman and Goodall [59] and Pollak and Boulton [60]. Digital cameras were first used successfully for quantitative evaluation of chromatograms in 1984 [61], and Analtech produced the first commercial video densitometer in 1988. Video technology is widely used today, along with photography and computer scanning, for chromatogram documentation and is being increasingly applied for quantification. Advantages of video densitometry include speed and simple instrument operation, while disadvantages include application only to colored, fluorescent, or fluorescence quenched zones and the absence of compound specificity because measurement is made from a totally illuminated layer rather than by use of a beam of radiation with a variable wavelength range selected by a monochromator. Books by Shellard [62] and Touchstone [63] were important in furthering the early adoption of quantitative TLC.

Radioactive zones were detected and quantified by TLC using autoradiography, liquid scintillation counting, and in situ scanning using instruments with gas-flow proportional counters beginning in the 1960s. A review paper by Prydz [64] provided impetus to wider application of thin layer radiochromatography. Preparative layer chromatography (PLC) was performed from the earliest days of TLC using basic equipment with a thicker layer and manual application of larger volumes of sample usually in the form of a band across the origin of the layer. Hopf [65] introduced a centrifugal apparatus to accelerate the mobile phase (called the Chromatofuge) in 1947 for separation of substances on a 100 mg scale. In 1972, Heftmann and coworkers [66] modified this apparatus to make it more suitable for preparative separations. Today, automated sample application with various instruments and forced-flow overpressured or rotational development are used often for PLC.

To increase reproducibility, quantitative accuracy, and sample throughput of TLC, mechanized and automated instruments have been introduced for many of the procedural steps, e.g., for a linear ascending development and for dip and spray application of detection reagents [67]. The TAS oven was an early instrument for transfer of volatile sample components to a TLC plate [68]. By the 1980s, TLC had become highly instrumentalized, with Camag and Desaga among the most important companies supplying instruments for TLC. Jaenchen and Issaq in 1988 [69] reviewed advances in TLC high-performance, multimodal stationary phases and sample application, development, and quantification devices for the preceding decade. Topics covered included two-phase plates with adjacent layers of silica gel (normal phase) and C-18–bonded silica gel (reversed phase); multimodal amino- and cyano-modified silica gel plates that can be used to separate hydrophilic or charged compounds; instruments for applying round spots or bands ranging from nanoliter to microliter volumes; horizontal sandwich, AMD, and overpressured TLC chambers; and linear and zig-zag slit-scanning reflection densitometers for performing quantitative analysis by measurement of sample and standard zones that are colored, UV-absorbing, or fluorescent, either naturally or after post-chromatographic derivatization. Jaenchen and Issaq also discussed applications of a very powerful multidimensional analytical method for complex mixtures achieved by combining reversed-phase gradient elution HPLC, spraying of column fractions onto a normal phase silica gel HPTLC layer, gradient elution AMD, and identification of separated zones by multiwavelength densitometric scanning.

The use of a flame ionization detector for detection of separated zones on narrow quartz plates coated with silica gel began in the early 1970s [70]; today, adsorbent-coated quartz rods are used (Iatroscan TLC-FID). Issaq [71] was the first to use multidimensional TLC/MS; this was carried out by elution of separated zones with methanol from a layer using the Eluchrom instrument, and the eluents were then introduced into a mass spectrometer by using a specially built Pyrex inlet probe. Other MS interfaces were reported in the mid-1980s, and DRIFT interface and laser Raman techniques began to be reported in the early 1990s. These and other coupled TLC-spectrometric methods, such as TLC-FTIR and TLC-solid phase NMR, allow confirmation of identity of unknown separated zones [72]. The latest coupled method reported is fluorescence line-narrowing spectrometry with poly(ethylenimine)–cellulose TLC [73].

J. T. Baker designed a totally automated quantitative TLC instrument in the late 1960s employing a "chromatape" composed of 15 cm silica gel sections on a moving 10 m Mylar band. The band containing the applied samples moved through a tunnel where separation, drying, derivatization, and measurement were performed. The instrument was tested by, among others, the New York City Police Department for analyzing street drugs [61], but according to personnel employed at Baker at the time, none of the instruments were sold and the project was abandoned. A fully automated robotic TLC system with postrun evaluation using a digital camera was proposed in 1989 [74].

III. PERSONAL INVOLVEMENT IN TLC

The author of this chapter began work in the field of chromatography by earning a Ph.D. degree in analytical chemistry from Rutgers, the State University of New Jersey, in 1958 under the supervision of William Rieman III, a world-famous authority in ion exchange chromatography. My thesis research involved development solubilization chromatography, a new method involving the separation of hydrophobic nonelectrolytes of moderate to high molecular weight by elution through a column of ion exchange resin with aqueous acetic acid or methanol as the mobile phase [75].

In the fall of 1958, I was hired to teach analytical chemistry by Lafayette College. Lafayette has a great tradition in analytical chemistry; the current ACS journal *Analytical Chemistry* made its debut in January 1887 as the *Journal of Analytical Chemistry*, a quarterly publication edited by Edwin Hart, then head of Lafayette's chemistry department.

I began research with undergraduate student collaborators using commercial ion exchange papers (Reeve Angel-Whatman) for the separation of organic compounds (paper solubilization chromatography) [76] and metal ions [77]. David Locke, the first Lafayette student to coauthor a paper with me, went on to earn a Ph.D. in analytical chemistry and develop an outstanding research program in liquid chromatography as professor of chemistry at Queens College, City University of New York. My first papers on TLC involved the use of laboratory-made ion exchange layers in 1965 for separations of phenols (thin layer solubilization chromatography) [78] and metals [79].

I was fortunate to be able to work during many summers when classes were not in session with some of the leading chromatography researchers in the world in order to gain knowledge and experience in TLC and other types of chromatography and to publish papers with them. The experience gained in these outside laboratories led to projects carried out at Lafayette that resulted in many student coauthored papers. With Harold H. Strain at Argonne National Laboratory in 1961, 1962, 1964, 1966, and 1968, I developed methods for the paper electrophoresis of metal ions and for column, paper, and thin layer chromatography [80] of chloroplast pigments. We demonstrated that these pigments are quite labile and that precautions are required during sample preparation and TLC so that they are not decomposed and artifacts formed. Research in ion exchange and impregnated-paper chromatography of metal ions was carried out with James S. Fritz at Iowa State University in 1965 [81]. In 1967 and 1971 at the Syracuse University Research Corporation with Gunter Zweig, research on the TLC of pigments was continued [82], and I began to study pesticide residue analysis by gas chromatography.

My visits to Zweig's laboratory marked the start of an extensive career of writing books and review articles on chromatography and pesticide analysis. I joined Zweig in writing the 1970 biennial review of column liquid, paper, and thin layer chromatography for the Fundamental Reviews issue of *Analytical Chemistry* [83]. I have continued to write this review, which covered only planar chromatography starting in 1976, for publication in each even-numbered year through 2000, either alone or with Zweig or Professor Bernard Fried of Lafayette's Biology Department as a coauthor. I wrote the bienniel review on pesticide analysis for the Applications Reviews issues of *Analytical Chemistry* from 1983 to 1995, first with Zweig and then alone after his death; each review contained a section on pesticide TLC.

My first book was *Paper Chromatography*, written with Zweig and published in 1971 by Academic Press. It was Volume II of "Paper Chromatography and Electrophoresis," Volume I having been written by John R. Whittaker, and was the last book ever written on paper chromatography because this method lost popularity as TLC became more important. These two volumes represented a new edition of the classic *Paper Chromatography and Paper Electrophoresis* by R. J. Block, E. L. Durrum, and Zweig, the second edition of which was published in 1958 by Academic Press. In 1972, Zweig was preparing Volume VI of his celebrated series of books published by Academic Press titled "Analytical Methods for Pesticides and Plant Growth Regulators." I became his coeditor for 10 volumes and edited Volumes XVI and XVII alone, the last appearing in 1989. I wrote chapters reviewing the techniques and applications of TLC in pesticide analysis in Volumes VII and XI, which had significant impact on furthering research in this area. Also in 1972, I began the "Handbook of Chromatography" series with Zweig for CRC Press. We wrote general Volumes I and II and edited 22 more volumes on individual compound classes before his death. I edited five more volumes alone before the series ended in 1994. Each volume contained techniques and data for all areas of chromatography, including TLC.

I worked at the University of Pennsylvania Hospital with Joseph Touchstone during the summers of 1973 and 1974 and during a sabbatical leave from Lafayette in 1979. Touchstone was a prominent researcher in clinical analysis who gave me my first serious training in quantitative TLC-densitometry, and we worked together on developing methods for determination of steroids, amino acids, pesticides, drugs, bile acids, and phospholipids in samples such as blood, urine, saliva, and semen. Our first of many joint papers was published in 1974 [84], and we edited, and wrote several chapters in, an influential book on the techniques, instrumentation, and applications of quantitative TLC in 1979 [85]. We also coauthored many review articles and chapters in other books on quantitative TLC.

By the early 1970s, my research program at Lafayette involved mostly development of quantitative TLC methods using the Kontes Chromaflex fiber optics densitometer. I taught short courses on TLC throughout the eastern United States for Kontes and wrote 25 issues of the Kontes Quant Notes from August 1975 to August 1981, which contained abstracts and articles on densitometric analysis.

During the summers of 1975 and 1976, I worked with Melvin Getz at the USDA-Beltsville laboratory developing methods for the quantitative TLC of pesticide residues [86]. I spent the summer of 1978 working with Thomas Beesley at Whatman, Inc. developing HPLC and TLC methods, especially using channeled, preadsorbent plates in combination with scanning densitometry [87]. I was commissioned by Whatman to write three manuals describing the techniques and applications of TLC on their C-18, preadsorbent, and high-performance layers that were published in 1981 and 1982 and distributed free of charge to thousands of customers. Producing the manuals led me to design numerous research projects at Lafayette to study separations and determinations of important compound classes on these layers (e.g., Ref. 88). We demonstrated in many papers that channeled, preadsorbent TLC and HPTLC plates using manual spotting of sample solutions and a single standard solution in variable volumes between 1 and 25 µL with a Drummond digital microdispenser and nonautomated scanning can provide quantitative analyses with accuracy and precision approaching those obtained with automated instrumental band-wise sample application and automated densitometry. This can be seen by comparing our publications by these two approaches for quantitative analysis of analgesic and decongestant tablets [89,90] and sunscreens in cosmetics [91,92]. We also studied systematic mobile phase optimization for C-18 plates [93]. I traveled to Hungary in May 1981 with Beesley and Issaq to lecture on pesticide TLC at the first annual American-Hungarian Chromatography Symposium (Fig. 3).

While working during the summer of 1984 at the USDA Eastern Regional Research Laboratory, Philadelphia, with Daniel Schwartz, we developed a method for spectrophotometric screening of sulfathiazole (ST) residues in honey involving isolation of the drug on tandem laboratory-made alumina and anion exchange minicolumns. This method was later extended by undergraduates working at Lafayette to include TLC quantification of ST residues in honey [94]. Using the experience gained with Schwartz, my research group at Lafayette became among the first to combine solid phase extraction (SPE) on commercial columns with TLC-densitometry for the concentration, cleanup, and quantification of pesticide residues of various classes in environmental water samples [95]. In addition, I have had an active program in developing densitometric methods for active ingredients in pharmaceutical products, starting with a paper on the determination of caffeine in APC tablets in 1978 [96]. Other important areas of research have been development of methods for determination of active ingredients and impurities in cosmetics and analysis of foods and beverages. My first paper on beverages reported the HPTLC determination of caffeine in sodas [97]. This paper is one of a number in which I presented a new, practical analytical method that can also be used effectively for teaching purposes in college laboratory courses.

Another significant aspect of my research at Lafayette has been a unique collaborative program in analytical chemistry and invertebrate biology with Fried, one of the world's foremost experts in parasitology [98]. This program has resulted in 65 publications coauthored with 76 chemistry, biology, and biochemistry major undergraduates in various peer-reviewed analytical, chromatography, and biological journals since our first joint publication in 1987 [99]. The research has involved mainly studies on the chemical content, i.e., lipids, phospholipids, pigments, sugars, and amino acids, of parasitic flatworms and medically important snails, and we have also analyzed pheremones released by parasites and snails and the chemical constituents of various foods fed to our animals. The major analytical techniques in this research have been qualitative and quantitative TLC and HPTLC. An example of a student coauthored paper resulting from this interdisciplinary research program with Fried is Ref. 100. In addition, Fried and I have collaborated with Bruce Young, a biologist specializing in herpetology, and three biology majors to publish a paper describing the analysis of scent gland lipids from snakes [101], and I have been involved in research with a botany professor, Lorraine Mineo, and a biology major in analyzing *Hedeoma pulegioides* and peppermint oil for pulegone [102] and with a developmental biology professor, Elaine Reynolds, and a biochemistry major to determine phospholipids in fly (*Drosophila*) heads [103]. Development and application of new qualitative and quantitative TLC methods for analysis of a wide range of biological samples has been an important and unique aspect of my undergraduate research program. As of the date of this writing, I have published a total of 171 papers with 128 student coauthors resulting from all aspects of my TLC-based research program, the productivity and quality of which was recognized by my receiving the 1995 ACS Award for Research at an Undergraduate Institution, sponsored by Research Corporation.

I have undertaken a very active writing program in order to increase appreciation for the advantages of TLC; educate readers about techniques, instrumentation, and applications; and promote its use throughout the world. In addition to my manuals and reviews mentioned above, I have written numerous general [104] and applications reviews in the areas of environmental [105], food [106], pharmaceutical [107], and pesticide [108] analysis. Fried and I have written four editions of a book describing the principles, techniques, and applications of TLC with some emphasis on biological applications [109], and we have edited two editions of the most comprehensive treatise of TLC ever published, with chapters on theory, techniques, instrumentation, and applications organized by analyte type [110], and a practical book on TLC analysis organized by sample type [111]. I have served as editor for residues and trace element analysis of the *Journal of AOAC International* since 1981 and edited special sections of that journal on TLC in issue 1 of Volume 82 (1999) and on TLC-densitometry and in issue 6 of Volume 83 (2000). Fried and I guest-edited three special issues on TLC for the *Journal of Liquid Chromatography & Related Technologies* (issues 1 and 10, Volume 22, 1999, and issue 10, Volume 24, 2001). In reviewing the fourth edition of our smaller book, Kissinger wrote the following in *Analytical Chemistry* [112]: "What is there to say about a book appearing in its fourth edition? That alone suggests it is a work of considerable value. Overall, this is a update to previous editions and a credit to the (Dekker) Chromatographic Science Series, of which this is Volume 81." In his paper on the history of TLC, Kreuzig [61] ends by stating that "all the essential knowledge of TLC has been presented in an extensive book by J. Sherma and B. Fried." His reference was to the first (1991) edition of our *Handbook of Thin Layer Chromatography*, which has now been followed by an expanded and updated second edition (1996) [110]. My published books and papers and invited lectures on analytical chemistry and chromatography total 508 at this writing.

IV. EPILOG

This historical review is necessarily selective rather than exhaustive. Readers interested in further detailed information on the history of TLC should consult the book chapter by Pelick et al. [113], which contains a tabulation of historical events from 1938 to 1958 and complete translations of the original papers by Izmailov and Schraiber [8] and Stahl [19], and the book by Wintermeyer [114]. Berezkin also offered a translation of the Izmailov and Schreiber paper [115]. Tswett's historic paper describing the invention of chromatography was translated from the German [116] and critically evaluated [117] by Strain and Sherma.

The *Journal of Liquid Chromatography* was the first analytical journal to publish special issues devoted to TLC; these began in the mid-1970s and were edited by Issaq. The Camag Bibliography Service (CBS), first issued in 1965, contains rather detailed abstracts of TLC books and papers; it is available today in paper and searchable CD-ROM formats and is a valuable, comprehensive resource for keeping up with the latest advances in TLC. The field of TLC received a significant boost with the start of publication in 1998 of the first journal devoted entirely to TLC, the *Journal of Planar Chromatography-Modern TLC*, by Dr. Alfred Huethig Verlag, with Prof. Dr. Sz. Nyiredy as editor-in-chief. Since 1994 the journal has been published by the Research Institute for Medicinal Plants (Budakalasz, Hungary) in cooperation with Springer (Budapest, Hungary).

Colin Poole, a recognized expert in TLC and other chromatographic areas, recently reviewed the advantages of TLC, future prospects for improved separation performance, and advances in coupling TLC with column chromatographic and spectrometric methods. He offered some predictions as to how TLC will be practiced in the future in terms of layers and development, quantification, and coupled and multimodal methods [51].

ACKNOWLEDGMENTS

The author is deeply indebted to Dr. Dieter Jaenchen of Camag, Muttenz, Switzerland, for supplying information on the history of TLC instrumentation for sample application, chromatogram development, and densitometric chromatogram evaluation and figures illustrating Camag instruments, which were invaluable in preparing this chapter. It is recommended that interested readers consult the article written by Dr. Jaenchen comparing general TLC laboratory equipment and TLC instrumentation between 1965 and 1982, which was published in the 50th (October 1982) issue of the CBS.

REFERENCES

1. HH Strain. Chromatographic Adsorption Analysis. New York: Interscience, 1942.
2. MW Beyerinck. Ein einfacher Diffusionsversuch. Z Phys Chem 3:110–112, 1889.
3. HH Strain, J Sherma. Michael Tswett's contributions to sixty years of chromatography. J Chem Educ 44:235–237, 1967.
4. SV Heines. Three who pioneered in chromatography. J Chem Educ 46:315–316, 1969.
5. KI Sakodynskii. The contribution of M.S. Tswett to the development of planar chromatography. J Planar Chromatogr-Mod TLC 5:210–211, 1992.
6. R Consden, AH Gordon, AJP Synge. Qualitative analysis of proteins: a partition chromatographic method using paper. Biochem J 38:224–232, 1944.
7. JG Kirchner. History of thin layer chromatography. In: JC Touchstone, D Rogers, eds. Thin Layer Chromatography—Quantitative Environmental and Clinical Applications. New York: Wiley-Interscience, 1980, pp. 1–6.
8. NA Izmailov, MS Schraiber. Spot chromatographic adsorption analysis and its application in pharmacy, communication I. Farmatsiya 3:1–7, 1936.
9. C Lapp, K Erali. Essais de chromatographie a deux dimensions. Bull Sci Pharmacol 47:49–57, 1940.
10. M O'L Crowe. Micromethod for chromatographic analysis. Anal Chem 13:845–846, 1941.

11. N von Bekesy. Über eine zerlegbare Kuvette zur chromatographischne Adsorption. Biochem Z 312: 100–113, 1947.
12. WG Brown. Micro separations by chromatographic adsorption on blotting paper. Nature 143:377–378, 1939.
13. TL Williams. Introduction to chromatography. Glasgow: Blackie, 1947, p. 36.
14. JE Meinhard, NF Hall. Surface chromatography. Anal Chem 21:185–188, 1949.
15. JG Kirchner, JM Miller, RG Rice. Quantitative determination of diphenyl in citrus fruits and fruit products by means of chromatostrips. J Agric Food Chem 2:1031–1033, 1954.
16. JG Kirchner, JM Miller, GJ Keller. Separation and identification of some terpenes by a new chromatographic technique. Anal Chem 23:420–425, 1951.
17. RH Reitsema. Characterization of essential oils by chromatography. Anal Chem 26:960–963, 1954.
18. E Stahl, G Schroter, G Kraft, R Renz. Dunnschicht-chromatographie: Methode, Einflussfaktoren und einige Anwendungsbeispiele. Pharmazie 11:633–637, 1956.
19. E Stahl. Thin layer chromatography. II. Standardization, visualization, documentation, and application. Chem Ztg 82:323–329, 1958.
20. E Stahl. Dunnschlicht-chromatographie, ein Laboratoriumshandbuch. Berlin: Springer-Verlag, 1962.
21. JM Bobbitt. Thin Layer Chromatography. New York: Reinhold, 1963.
22. K Randerath. Thin Layer Chromatography. New York: Academic Press, 1963.
23. EV Truter. Thin Film Chromatography. New York: Wiley, 1963.
24. JG Kirchner. Thin Layer Chromatography. New York: Interscience, 1967.
25. E Stahl. Twenty five years of thin layer chromatography-an intermediate balance. Angew Chem 95:515–524, 1983.
26. E Stahl. Twenty five years of thin layer chromatography—a report on work with observations and future prospects. J Chromatogr 165:59–73, 1979.
27. EP Przybylowicz, WJ Staudenmayer, ES Perry, AD Baitsholts, TN Tischer. Precoated sheets for thin layer chromatography. J Chromatogr 20:506–513, 1965.
28. JA Thoma. Polar solvents, supports, and separation. In:JC Giddings, RA Keller, eds. Advances in Chromatography. Vol. 6. New York: Marcel Dekker, 1968, pp. 61–117.
29. H Halpaap, J Ripphahn. High performance thin layer chromatography: development, data, and results. In: A Zlatkis, RE Kaiser, eds. HPTLC: High Performance Thin Layer Chromatography. Amsterdam: Elsevier, 1977, pp. 95–127.
30. TE Beesley. Chemically bonded reverse-phase TLC. In: JC Touchstone, ed. Advances in Thin Layer Chromatography. New York: John Wiley & Sons, 1982, pp. 1–11.
31. HJ Issaq. Recent advances in multimodal thin layer chromatography. Trends Anal Chem 9:36–40, 1990.
32. R Kraffczyk, R Helger. Die Dunnschicht-chromatographie der Aminosäuren im Harn auf Zweischichtenplatten. Z Anal Chem 243:536–541, 1968.
33. HY Aboul-Enein, MI El-Awady, CM Heard, PJ Nicholls. Application of thin layer chromatography in enantiomeric chiral analysis—an overview. Biomed Chromatogr 13:531–537, 1999.
34. D Jaworek. Thin layer gel chromatography. Part 2. Molecular weight determination of proteins. Chromatographia 3:414–417, 1970.
35. K Macek, IM Hais. Stationary Phases in Paper and Thin Layer Chromatography. Amsterdam: Elsevier, 1965.
36. H Halpaap. Standardization of thin layer chromatography using prefabricated layers. J Chromatogr 33:144–164, 1968.
37. E Cremer, H Nau. Chromatographie mit ultradunnen Flüssigkeitsfilmen. Naturwissenschaft 55:651, 1968.
38. EC Bate-Smith, RG Westall. Chromatographic behaviour and chemical structure. I. Some naturally occurring phenolic substances. Biochim Biophys Acta 4:427–440, 1950.
39. E Soczewinski. Solvent composition effects in thin layer chromatography systems of the type silica gel-electron donor solvent. Anal Chem 41:179–183, 1969.

40. LR Snyder. Solvent selectivity in adsorption chromatography on alumina. non-donor solvents and solutes. J Chromatogr 63:15–44, 1971.

41. F Geiss. Die Parameter der Dunnschichtchromatographie. Braunschweig: Vieweg, 1972.

42. RE Kaiser. Simplified theory of TLC. In: A Zlatkis, RE Kaiser, eds. High Performance Thin Layer Chromatography. Amsterdam: Elsevier, 1977, pp. 15–38.

43. G Guiochon, A Siouffi. Study of the performance of TLC. II. Band broadening and plate height equation. J Chromatogr Sci 16:470–481, 1978.

44. Sz Nyiredy, B Meier, CAJ Erdelmeier, O Sticher. "PRISMA": geometrical design for solvent optimization in HPLC. J High Resolut Chromatogr 8:186–188, 1985.

45. M Brenner, A Niederwieser. Overrun thin layer chromatography. Experientia 17:237–238, 1961.

46. F Geiss, H Schlitt. A new and versatile thin layer chromatography separation system. Chromarographia 1:392–402, 1968.

47. RE Kaiser. Einführung in die HPPLC. Heidelberg: Huethig, 1987.

48. E Tyihak, E Mincsovics, H Kalasz. New planar liquid chromatographic technique: over-pressured thin layer chromatography. J Chromatogr 174:75–81, 1979.

49. JA Perry, TH Jupille, LJ Glunz. Programmed multiple development. Anal Chem 47:65A–74A, 1975.

50. K Burger. TLC-programmed multiple development, thin layer chromatography with gradient elution in comparison to HPLC. z Anal Chem 318:228–233, 1984.

51. CF Poole. Planar chromatography at the turn of the century. J Chromatogr 856:399–427, 1999.

52. D Jaenchen. Personal communication, May 2000.

53. M Getz. An automatic spotter for quantitative thin layer and paper chromatographic analysis by optical scanning. J Assoc Off Anal Chem 54:982–985, 1971.

54. H Jork. Direct quantitative thin layer chromatographic analysis. Z Anal Chem 236:310–326, 1968.

55. M Beroza, KR Hill, KH Norris. Determination of reflectance of pesticide spots on thin layer chromatograms using fiber optics. Anal Chem 40:1608–1613, 1968.

56. D Jaenchen, G Pataki. Quantitative in situ fluorimetry of thin layer chromatograms. J Chromatogr 33:391–399, 1968.

57. JC Touchstone, SS Levin, T Murawec. Quantitative aspects of spectrodensitometry of thin layer chromatograms. Anal Chem 43:858–863, 1971.

58. W Schmidtmann, L Reshke, L Baumeister, M Koch. Experience of the Vitatron densitometrt TLD 100 for quantitative evaluation of thin layers. Chromatographia 3:163–169, 1970.

59. J Goldman, RR Goodall. Quantitative analysis on thin layer chromatograms. J Chromatogr 32:24–42, 1968.

60. V Pollak, AA Boulton. Optical noise in photometric scanning of thin layer media chromatograms II double beam difference systems. J Chromatogr 45:200–207, 1969.

61. F Kreuzig. The history of planar chromatography. J Planar Chromatogr-Mod TLC 11:322–324, 1998.

62. EJ Shellard, ed. Quantitative Paper and Thin Layer Chromatography. New York: Academic Press, 1968.

63. JC Touchstone, ed. Quantitative Thin Layer Chromatography. New York: Wiley-Interscience, 1973.

64. S Prydz. Summary of the state of the art in radiochromatography. Anal Chem 45:2317–2326, 1973.

65. PP Hopf. Radial chromatography in industry. Ind Eng Chem 22:938–940, 1947.

66. E Heftmann, JM Krochta, DF Farkas, S Schwimmer. The chromatofuge, an apparatus for preparative rapid radial chromatography. J Chromatogr 66:365–369, 1972.

67. F Kreuzig. Derivatization of substances, separated on thin layer plates, by means of an automated spraying device. Chromatographia 13:238–240, 1980.

68. E Stahl. TAS-verfahren and Dunnschicht-chromatographie zur Mikroanalytik der Morphaktine. Naturwissenschaft 55:651–652, 1968.

69. DE Jaenchen, HJ Issaq. Modern thin layer chromatography: advances and perspectives. J Liq Chromatogr 11:1941–1965, 1988.

70. T Cotgreave, A Lynes. Quantitative analysis by thin layer chromatography using a flame ionization detector. J Chromatogr 30:117–129, 1967.

71. HJ Issaq, JA Schroer, EW Barr. A direct on-line thin layer chromatography/mass spectrometry coupling system. Chem Instrument 8:51–53, 1977.

72. T Cserhati, E Forgacs. Hyphenated techniques in thin layer chromatography. J AOAC Int 81:329–332, 1998.

73. SJ Kok, R Evertsen, NH Velthorst, UATh Brinkmen, C Gooijer. On the coupling of fluorescence line-narrowing spectroscopy with poly(ethylenimine)-cellulose thin layer chromatography. Anal Chim Acta 405:1–7, 2000.

74. M Prosek, M Pukl, A Smidovnik, A Medja. Automation of thin layer chromatography with a laboratory robot. J Planar Chromatogr-Mod TLC 2:244–246, 1989.

75. J Sherma, W Rieman III. Solubilization chromatography III ethers, carboxylic acids, and hydrocarbons. Anal Chim Acta 20:357–365, 1959.

76. D Locke, J Sherma. Paper solubilization chromatography I phenols. Anal Chim Acta 25:312–316, 1961.

77. J Sherma, CW Cline. Anion exchange paper chromatography of metal ions. Anal Chim Acta 30:139–147, 1964.

78. J Sherma, LVS Hood. Thin layer solubilization chromatography I phenols. J Chromatogr 17:307–315, 1965.

79. J Sherma. Separation of certain metal ions by thin layer ion exchange chromatography. J Chromatogr 19:458–459, 1965.

80. HH Strain, J Sherma, M Grandolfo. Comparative chromatography of the chloroplast pigments. Anal Biochem 24:54–69, 1968.

81. JS Fritz, J Sherma. Reversed phase paper chromatography of metal ions with phenylbenzohydroxamic acid. J Chromatogr 25:153–160, 1966.

82. J Sherma, G Zweig. Separation of chloroplast leaf pigments by thin layer chromatography on cellulose sheets. J Chromatogr 31:439–445, 1967.

83. G Zweig, RM Moore, J Sherma. Chromatography. Anal Chem 42:349R–362R, 1970.

84. J Sherma, MF Dobbins, JC Touchstone. Quantitative spectrodensitometry of morphine and amphetamine on thin layer chromatograms. J Chromatogr Sci 12:300, 1974.

85. JC Touchstone, J Sherma. Densitometry in Thin Layer Chromatography. New York: Wiley-Interscience, 1979.

86. J Sherma, KE Klopping, M Getz. Determination of pesticide residues by quantitative thin layer densitometry on microscope slides. Am Lab 9(12):66–73, 1977.

87. J Sherma. Determination of pesticides in water by densitometry on preadsorbent TLC plates. Amer Lab 10(11):105–109, 1978.

88. J Sherma, BP Sleckman. Reversed phase TLC of phenols on chemically bonded C-18 layers. J High Resolut Chromatogr Chromatogr Commun 11:557–560, 1981.

89. MB Lippstone, J Sherma. Analysis of tablets containing naproxin and ibuprofen by HPTLC with ultraviolet absorption densitometry. J Planar Chromatogr-Mod TLC 8:427–429, 1995.

90. MB Lippstone, E Grath, J Sherma. Analysis of decongestant pharmaceutical tablets containing pseudoephedrine hydrochloride and guaifenesin by HPTLC with ultraviolet absorption densitometry. J Planar Chromatogr-Mod TLC 9:456–458, 1996.

91. B Musial, J Sherma. Determination of the sunscreen 2-ethylhexyl-p-methoxy cinnamate in cosmetics by reversed phase HPTLC with ultraviolet absorption densitometry on preadsorbent plates. Acta Chromatogr 8:5–12, 1998.

92. E Westgate, J Sherma. Determination of the sunscreen oxybenzone in lotions by reversed phase HPTLC with ultraviolet absorption densitometry. J Liq Chromatogr Relat Technol 23:609–615, 2000.

93. J Sherma, S Charvat. Solvent selection for reversed phase TLC. Am Lab 15(2):138–144, 1983.

94. J Sherma, W Bretschneider, M Dittamo, N DiBiase, D Huh, D Schwartz. Spectrometric and thin layer chromatographic quantification of sulfathiazole residues in honey. J Chromatogr 463:229–233, 1989.

95. J Sherma. Determination of triazine and chlorophenoxy acid herbicides in natural water samples

by solid phase extraction and quantitative thin layer chromatography. J Liq Chromatogr 9:3433–3438, 1986.

96. J Sherma, M. Beim. Determination of caffeine in APC tablets by densitometry on chemically bonded C-18 reversed phase layers. High Resolut Chromatogr Chromatogr Commun 1:309–310, 1978.

97. J Sherma, DA Gray. Quantitation of caffeine in beverages by TLC-densitometry-a student experiment. Am Lab 11(10):21–24, 1979.

98. J Sherma, B Fried. Collaborative research in invertebrate biology and analytical chemistry at Lafayette College. Chromatography 22:17–24, 2001.

99. M Duncan, B Fried, J Sherma. Lipids in fed and starved *Biomphalaria glabrata* (Gastropoda). Comp Biochem Physiol 86A:663–665, 1987.

100. E Muller, L Rosa Brunet, B Fried, J Sherma. Effects of the neutral lipid contents of the liver, ileum and serum during experimental schistosomiasis. Int J Parasitol 31:285–287, 2001.

101. BA Young, BA Frazer, B Fried, M Lee, J Lalor, J Sherma. Determination of cloacal-scent-gland lipids from two sympatric snakes, the eastern diamondback rattlesnake and the Florida cottonmouth. J Planar Chromatogr-Mod TLC 12:166–201, 1999.

102. BP Sleckman, J Sherma, LC Mineo. Determination of pulegone in H. Pulegioides and peppermint oil by thin layer chromatography with densitometry. J Liq Chromatogr 6:1175–1182, 1983.

103. M Nyako, C Marks, J Sherma, ER Reynolds. Tissue-specific and developmental effects of the easily shocked mutation on ethanolamine kinase activity and phospholipid composition in *Drosophila melanogaster*. Biochemical Genetics, submitted for publication, 2000.

104. J Sherma. Thin-layer chromatography-the underappreciated analytical alternative. Inside Lab Manage 4(1):10–11, 2000.

105. J Sherma. Thin layer chromatography. In: RA Meyers, ed. Encyclopedia of Environmental Analysis and Remediation. New York: John Wiley & Sons, 1998, pp. 4805–4813.

106. J Sherma. Thin layer chromatography in food analysis. Food Test Anal 2(2):39–44, 1996.

107. J Sherma. Thin layer chromatography. In: J Swarbrick, JC Boylan, eds. Encyclopedia of Pharmaceutical Technology. Vol. 15. New York: Marcel Dekker, 1997, pp. 81–106.

108. J Sherma. Recent advances in the thin layer chromatography of pesticides. J AOAC Int 82:48–53, 1999.

109. B Fried, J Sherma. Thin Layer Chromatography. 4th ed. New York: Marcel Dekker, 1999.

110. J Sherma, B Fried, ed. Handbook of Thin Layer Chromatography. 2nd ed. New York: Dekker, 1996.

111. B Fried, J Sherma, eds. Practical Thin Layer Chromatography—A Multidisciplinary Approach. Boca Raton, FL: CRC Press, 1996.

112. PT Kissinger. Book review: Thin Layer Chromatography, fourth edition. Anal Chem 71:827A, 1999.

113. N Pelick, HR Bolliger, HK Mangold. The history of thin layer chromatography. In: JC Giddings, RA Keller, eds. Advances in Chromatography. Vol. 3. New York: Marcel Dekker, 1968, pp. 85–118.

114. U Wintermeyer. The Root of Chromatography: Historical Outline of the Beginning to Thin Layer Chromatography. Darmstadt: GIT-Verlag, 1989.

115. V Berezkin. The discovery of thin layer chromatography. J Planar Chromatogr-Mod TLC 8:401–405, 1995.

116. HH Strain, J Sherma. M. Tswett: Adsorption analysis and chromatographic methods; application to the chemistry of the chlorophylls. J Chem Ed 44:238–242, 1967.

117. HH Strain, J Sherma. Michael Tswett's contributions to sixty years of chromatography. J Chem Ed 44:235–237, 1967.

5

From Thin Layer Chromatography to High-Performance Thin Layer Chromatography to Planar Chromatography

Antoine-Michel Siouffi

Université d'Aix-Marseille, Marseille, France

When I was a student I wanted to be an organic chemist. I was fascinated by the beauty of a synthesis and the geometry of organic molecules. At that time education in analytical chemistry was only devoted to chemical equilibria in solutions and identification of functionalities through spot tests as proposed in the textbook from Feigl [1]. I graduated from the School of Engineering Chemistry in Marseille, France, in 1961 and started a Ph.D. thesis. I then joined a lipid chemistry laboratory, where I studied allylic bromination of unsaturated fatty acids. The gap between what I learned and what I was working on was enormous. The reaction of *N*-bromosuccinimide gave rise to a great number of products. At that time organic chemists heavily relied on distillation. It was impossible to do so with fatty esters, and we were not at all sure that isomerization could occur during distillation.

At that time, a French company (Prolabo) displayed a commercially available gas chromatograph. Our institute bought one and I was asked to use it. In the early 1960s lipid analysis relied mainly on saponification and acidity indexes, and the iodine value was the single determination of unsaturation. Starting a new method was exciting.

I. FIRST STEPS IN CHROMATOGRAPHY

The gas chromatography (GC) instrument was enormous (Fig. 1a). The column was placed in a big oven, and we had to push it inside and connect it to a system similar to a submarine valve. The stainless steel columns were 6 m long by 6 mm wide. We had to prepare the stationary phase by physically coating a gum onto a silicoaluminate. Columns were packed by tap filling. We were happy to obtain 1500–2000 plates (Fig. 1b)!

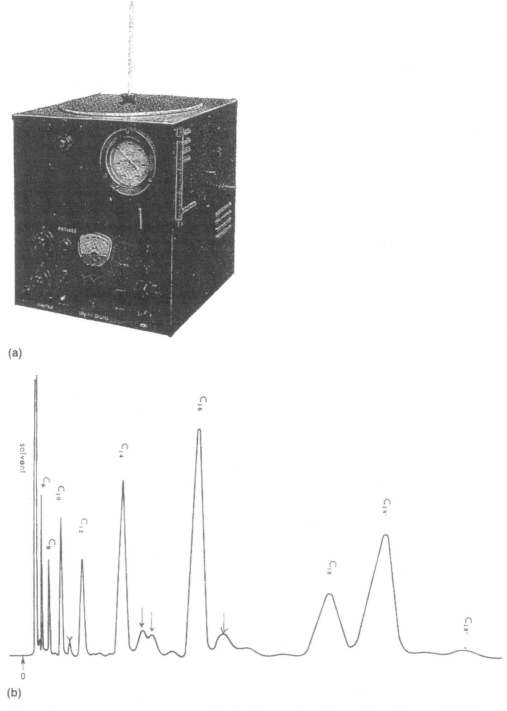

Figure 1 (a) A Prolabo GC instrument, 1961. (b) GC separation of fatty acid methyl esters (FAME) on diethylene glycol succinate stationary phase (DEGS) in 1963. Temperature, 180°C; detection:conductivity.

Conductivity detectors were not very sensitive. Nevertheless, we obtained good separations of stearic, oleic, linoleic, linolenic fatty esters and were able to check the quality of olive oils. We performed separations with standards. Brominated esters were transformed into volatile derivatives, and we were able to determine the transposition rate.

Most analyses were performed by trial and error, and we started reading Keulemans [2] and later bought an excellent book by Giddings [3]. This book was really the one that directed my attention to analytical chemistry. I realized that a sound theoretical basis is mandatory. Separations in lipid chemistry are tedious. When solutes are not volatile, what does one do? Stahl's book [4] on thin layer chromatography (TLC) appeared in 1962, and we started quantitative separations of glycerides. We prepared our own layers by coating glass plates. We still have the racks on which the plates were placed in the oven to be dried.

Preparative work was rather tedious since we had to scratch the plates and extract the solutes of interest from silica. Do not forget that in the 1960s we needed gram amounts of compound to perform nuclear magnetic resonance (NMR) or mass spectrometry (MS). To overcome the drawback of remaining silica we used the old liquid chromatography system with glass columns. The mobile phase slowly percolated by gravity. Properly filling the column was an art, and it was disappointing to watch a circular band becoming twisted. Elution lasted nights and days, so we had the idea to accelerate elution by pressurizing the liquid with an inert gas. The device was constructed and worked pretty well but was limited in pressure by the breaking of the glass column. Our first attempt to perform high-pressure liquid chromatography was in 1966. In 1967 the Philipps and Pye Unicam liquid chromatography instrument, a moving belt system, appeared. The liquid solution from the HPLC column was deposited onto a continuous moving wire. The solvent was removed by applying vacuum and heat while the belt was passing through vacuum locks. The remaining solute was carried to an oxidation chamber, where carbon was transformed in carbon dioxide, which was subsequently reduced to methane, and detection was performed through flame ionization. Obviously the system was well suited for lipid analysis. The 15 km long wire was very fragile due to heating and cooling, and we spent many hours fixing the wire or something else. Some of the results were really exciting, and I still have some separations that would be worth publishing.

Two events changed my scientific life: I was asked to teach a practical course on analytical chemistry and I wanted to do a postdoctoral stint. The Institute of Lipids was interested in applications of chromatography but not in the design of new systems. I had the opportunity to move to Paris to the Ecole Polytechnique.

II. STARTING WORK WITH GEORGES GUIOCHON

I was asked to work on enantiomeric separations. The topic was new in HPLC. At that time HPLC was gaining wide acceptance and performance was improving quickly. Particles were better and better manufactured. Packing a column was not so easy. The laboratory was equipped with a big pump, and we were able to pack our own columns through the slurry packing method. It was quite frightening to turn on the valve when pressure reached 600 bars and let the slurry go. Once a day a 50,000-plate column was achieved with a 50 cm long stainless steel column, and some organic chemists thought it might be possible to achieve enantiomeric enrichment. It failed, of course, but we gained experience in efficient column preparation. To obtain chiral resolution, I proposed using an optically active mobile phase. After a careful literature survey, I synthesized an enantiomerically pure alcohol. Asymmmetric synthesis was not so easy, but we were able to obtain approximately 200 mL of optically pure alcohol. I used my experience in TLC to perform some experiments to check the selectivity. Preliminary results were not so bad, but the alcohol reacted with some parts of metal in the instrument and we had to give up considering this alcohol as a mobile phase for chiral separations.

Georges Guiochon was a distinguished scientist and was a lecturer in many symposia. He went to a congress in England and had a stimulating talk with I. Halasz about TLC. When he returned he asked me about the performance of this technique. Most people did not really use TLC as it was envisioned. It was time when manufacturers gained experience in finer and finer particles. Available HPLC materials in narrow size distribution allowed the achievement of very homogeneous beds in TLC.

III. THE HEPT EQUATION IN TLC

The problem was as follows: some reports advocated the use of fine particles to obtain good resolution, whereas others concluded that they behave less satisfactorily than large ones. In 1979–1980 the Merck Company launched its HPTLC (high-performance thin layer chromatography) plates with layers made of fine particles, and since there were contradictory results some people translated HP as high price instead of high performance. It was thus necessary to consider the parameters governing efficiency in TLC.

In HPLC (in GC as well) the analyst can handle the fluid flow rate and therefore set it to correspond to the optimum in HETP equation from either Van Deemter or Knox. Furthermore, the plate count is generally considered as almost constant in tubular column chromatography. The situation is completely different in TLC, since the plate number for a spot is proportional to the retardation factor, Rf, which leads to the definition of a bed efficiency [5]:

$$N = N'Rf \tag{1}$$

where N is the number of theoretical plates and N' is the bed efficiency. The separation is observed inside the chromatographic bed and not at the outlet of the column, and thus spots that are strongly retained are poorly resolved.

Resolution, Rs, between two spots can be written as follows:

$$Rs = (\sqrt{N'}/4)(\Delta K/K)(1 - Rf)\sqrt{Rf} \tag{2}$$

where $\Delta K = K_2 - K_1$ is the difference in partition coefficient of the two compounds to separate. Rf is the mean value of two closely related spots.

The plot of Rs versus Rf (Fig. 2) demonstrates that maximum resolution is achieved when Rf is 0.33. Nevertheless, the resolution does not vary greatly for Rf between 0.20 and 0.50.

In normal TLC the liquid ascends the plate and the position, z_f, of the solvent front on a thin layer plate at time t is given by the quadratic equation:

$$z_f^2 = kt \tag{3}$$

where k is the velocity constant. We started a study of the flow velocity of the mobile phase [6], in which k is given by

$$k = 2bk_o d_p \gamma / \eta \cos \theta \tag{4}$$

where b is the proportionality constant between the velocity of the solvent front and the velocity of the bulk liquid behind this front, taking into account the fact that capillary rise is faster in narrow capillaries; k_o is the specific permeability (function of the external porosity); γ is the surface tension of the solvent exhibiting η as viscosity; and θ is the wetting angle of the solvent on the adsorbent.

TLC is often plagued by the presence of a gas phase. The "normal" situation is one in which the mobile phase evaporates from the moistened layer and adsorbs on the dry adsorbent.

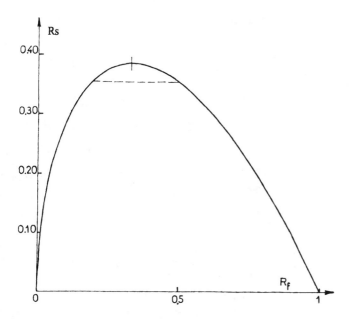

Figure 2 Plot of the resolution between two compounds as a function of the retention of the second compound of the pair. (Reproduced with permission of the *Journal of Chromatographic Science*.)

The general equation is then:

$$z_i^2 = kt[1 + (v' - 2v)/2e\varepsilon_0] \tag{5}$$

where v is the rate of vaporization of solvent from the wet layer, v' is the rate of adsorption of solvent vapor on the dry bed, e is the thickness of the layer, with a porosity of ε_0. The problem was to determine what could actually be achieved with small particles.

We knew from our experience in HPLC that the smaller the particles, the better the efficiency. We performed many experiments with different batches of carefully sized particles (kindly supplied by I. Halasz and H. Engelhardt). By chance we were able to conduct those experiments with solutes exhibiting very high UV absorption (*bis*-chlorophenyl triazenes, whose molecular extinction coefficient ε is $>20,000$). This allowed us to use very small amounts of solute in the linear part of the adsorption isotherm. We observed that diffusion in the mobile phase was the main contributor to band broadening. Performing experiments with old TLC plates together with our home-made plates prompted us to derive a relationship between efficiency and development length.

Working with Georges Guiochon was fascinating. He was (and still is) as quick as an eagle hunting a snake and what he was able to do with a pencil and a paper was really incredible. His group was doing laborious work, and we had the chance to meet some prominent scientists visiting his lab.

From the concept of local plate height introduced by Giddings [7]:

$$H_l = d\sigma^2/dx \tag{6}$$

where $d\sigma^2$ is the differential increase of a Gaussian band traveling a distance dx and the Knox equation in HPLC [8]:

$$h = Av^{1/3} + B/v + Cv \tag{7}$$

where h is the reduced plate height, ν is the reduced velocity of the mobile phase, and A, B, and C are coefficients.

The local plate height can be written as

$$H_l = BD_m/u + (Ad_p^{4/3}/D_m^{1/3})u^{1/3} + (Cdp^2/D_m)u \tag{8}$$

The velocity u of the mobile phase is given by

$$u = \chi d_p/2z \tag{9}$$

where χ is $2bk_o\gamma/\eta\cos\theta$ in the velocity constant or χ is k/dp. (Note that χ is θ in Refs. 6 and 9, but we here use χ to prevent confusion with the wetting angle called θ.) D_m is the diffusion coefficient of the solute in the mobile phase.

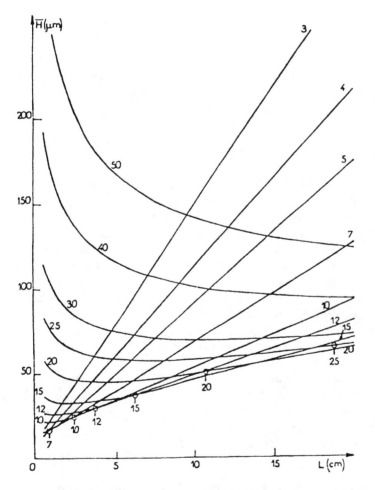

Figure 3 Plot of HETP versus development length for various particle sizes. The number on each curve is particle diameter. Diffusion coefficient: 10^{-5} cm^{2}s^{-1} (Coefficient a, b, and c of Eq. (11) in Ref. 9. Reproduced with permission of the *Journal of Chromatographic Science*.)

Following Eq. (9), it follows that

$$\overline{H} = \frac{B}{\chi d_p}\left[L + z_o\right] + \frac{3}{2}\frac{Ad_p^{5/3}\chi^{1/3}}{(2D_m)^{1/3}}\frac{L^{2/3} - z_o^{2/3}}{L - z_o}$$
$$+ \frac{C\chi d_p^3}{2D_m(L - z_o)}\log\frac{L}{z_o} \tag{10}$$

where z_o is the distance of the sample spot above the solvent level in the developing chamber. For the sake of convenience we may write:

$$\overline{H} = b(L + Z_0) + \frac{a}{L - Z_0}(L^{2/3} - Z_0^{2/3}) + \frac{c}{L - Z_0}\log\frac{L}{Z_0} \tag{11}$$

where b is $BD_m/\chi dp$, a is $3/2Ad_p^{5/3}\chi^{1/3}/(2D_m)^{1/3}$, and c is $C\chi d_p^3/2D_m$. The three-term equation is H = f (development length) instead of H = f(u) in GC or LC.

Plots of average HETP versus development length are displayed in Fig. 3. The plots are similar to the conventional Van Deemter or Knox curves. The c term is usually small. Maximum efficiency with coarse (40 µm) particles is achieved with a 10 cm development length. This is exactly the development distance advocated in the famous book by Stahl [4].

From the plot we derived an explanation of the contradictory results with different particle diameters. The best efficiencies are obtained with small particles when the solute diffusion coef-

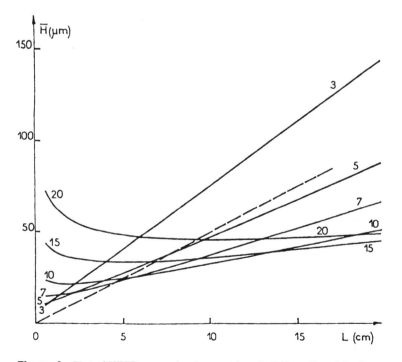

Figure 4 Plot of HETP versus development length. Effect of particle diameter. Rf of the solute = 0.7, Diffusion coefficient: 5.10^{-6} cm²s⁻¹. Coefficients of the HETP equation are as follows: a = $6.3–9.4.10^{-4}$; b = $8.5–2.5.10^{-4}$; c = $2.9–9.8.10^{-4}$. (Reproduced with permission of the *Journal of Chromatographic Science*.)

Table 1 Coefficients of the HETP Equation; Effect of Diffusion Coefficient

$D_m(cm^2/s)$	$a * 10^4$	$b * 10^4$	$c * 10^6$
$3 * 10^{-5}$	4.36	25.4	0.98
$2 * 10^{-5}$	4.99	17.0	1.47
$1 * 10^{-5}$	6.29	8.48	2.95
$5 * 10^{-6}$	7.93	4.24	5.90
$3 * 10^{-6}$	9.40	2.54	9.83

Calculations with: $A = 1$; $C = 0.01$; $\chi = 47.2$ cm^2/s; $z_0 = 0.5$ cm; $R_f = 0.7$; $\gamma = 0.7$; $d_p = 5$ μm; $B = 2\gamma D_m/\chi d_p R_f$

ficient is small and the development length is short. Conversely, large (10–15 μm) particles are more efficient when the diffusion coefficients of the solutes are large. Layers made with 20 μm particles can exhibit a better efficiency than layers made with 5 μm particles with a careful choice of development length.

In the HETP equation the first term is a diffusion term. Diffusion occurs in both mobile and stationary phases. One cannot neglect diffusion in stationary phase. We can write

$$B = 2(\gamma_m D_m + (1 - Rf/Rf)\gamma_s D_s) \tag{12}$$

where the subscripts s and m refer to stationary and mobile phase and γ_m and γ_s are tortuosity factors. When $\gamma_m D_m$ and $\gamma_s D_s$ are similar, the B term becomes inversely proportional to Rf and the spot diameter is independent of Rf.

Figure 4 demonstrates that when very fine particles are utilized, the H vs. L curve is linear. H depends only on diffusion. Tables 1 and 2 display the dependence of solute diffusion coefficient and particle diameter on coefficients of the HETP equation.

IV. PROPER USE OF TLC

From the above HETP equation it becomes easy to select the experimental conditions necessary to achieve a separation in conventional TLC. The plate number necessary to achieve a separation is given by:

Table 2 Coefficients of the HETP Equation; Effect of Particle Size

$d_p(μm)$	$a * 10^4$	$b * 10^4$	$c * 10^6$
3	3.38	7.06	1.27
5	7.93	4.24	5.90
7	13.9	3.03	16.2
10	25.2	2.12	47.2
15	49.5	1.41	153.3
20	79.9	1.06	378.0

Calculations with: $A = 1$; $C = 0.01$; $\chi = 47.2$ cm^2/s; $z_0 = 0.5$ cm; $R_f = 0.7$; $\gamma = 0.7$; $D_m = 5 * 10^{-6}$ cm^2/s; $B = 2\gamma D_m/\chi d_p R_f$

$$N_{ne} = 16Rs^2(\alpha/\alpha - 1)^2(1 + k/k)^2 \tag{13}$$

When dealing with TLC it becomes:

$$N_{ne} = 16Rs^2(\alpha/\alpha - 1)^2(1/(1 - Rf)^2 \tag{14}$$

If we consider an average plate length:

$$H = Rf(L/N_{ne}) \tag{15}$$

it is easy [10] to determine the necessary length from the intercept between the line from the above relationship, $L = N_{ne}H/Rf$, and the HETP curve.

Figure 5 displays such plots. With an Rf of 0.7 and Dm around $5.10-6$ cm/s^2, it is possible to obtain 1400 plates with a layer of 5 μm particles or a layer of 15 μm particles. It is obvious from the figure that it may be that no separation is possible if the required plate number is higher than can be reached on a plate (2800, for example). Some practical examples of this are given in Ref. 11. For example, tocopherols can be separated on a layer of 5 μm particles of silica with a 2.5 cm development length.

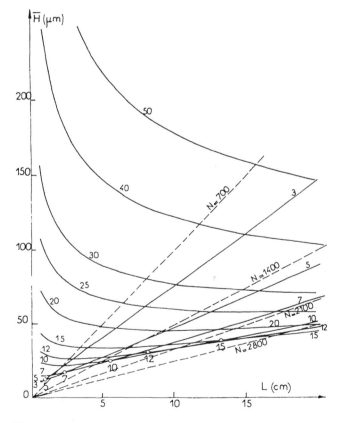

Figure 5 Example of the selection of the experimental conditions to achieve a separation. Dotted lines have a slope Rf/N corresponding to the number of theoretical plates given on the line. Intersections with H versus development length give the proper conditions. Calculations with A = 1; C = 0.01; $D_m = 5 * 10^{-6}$ cm^2/s; $R_f = 0.7$; $\gamma = 0.70$; $\chi = 47$ cm/s; $z_0 = 0.2$ cm; $\gamma_m D_m = \gamma_s D_s$. (Reproduced with permission of the *Journal of Chromatographic Science*.)

Time of analysis is given by:

$$t_R = N_{ne}^2 H^2 / Rf^2 k \tag{16}$$

where H is the HETP and k the velocity constant, which can be written when the c term is negligible:

$$t_R = 1/k[a/Rf/N - b]^{3/2} \tag{17}$$

with the condition Rf/N > b.

A plot of the variation of the analysis time with the relative retention of two solutes to separate is given in Fig. 6.

On a TLC plate there is a maximum number of plates that can be achieved. When development time increases, migration distance and spot variance increase with $\sqrt{t}$, and the chromatogram only expands without increase in resolution. Visual detection is biased by the fact that maximum concentration is only visible. We must remember that a TLC plate is made of fine

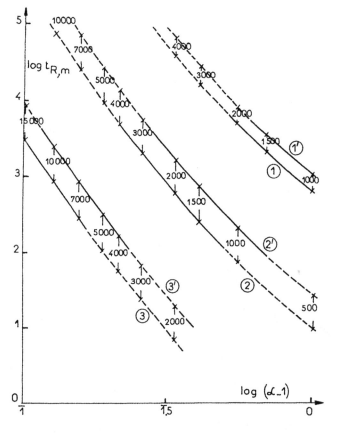

Figure 6 Plot of the variation of analysis time with the relative retention of the two solutes to separate. Plot of ln tr versus ln($\alpha - 1$). Curves 1,2,3 $\gamma_s D_s = 0$; curves 1',2',3' $\gamma_m D_m = \gamma_s D_s$. Resolution Rs = 1, coefficients in Ref. 10. The dotted parts of the times correspond to unrealistic experimental conditions. (Reproduced with permission of the *Journal of Chromatographic Science*.)

particles, and the eye does not detect small amounts of solutes adsorbed on particles. For a discussion, see the paper from Kubelka and Munk [12].

What the analyst is looking for is obviously the resolution, which has a great influence on the analysis time. In the two cases, when axial diffusion in the stationary phase is negligible compared to axial diffusion in the mobile phase, or proceeds at a comparable rate in both phases, the minimum analysis time is given by:

$$t_{R,m} = 65\ 536 \left(\frac{3A\gamma}{2}\right)^{3/2} \frac{D_m}{\chi^2} R^6 \frac{(1 + k')^9}{k'^6} \left(\frac{\alpha}{\alpha - 1}\right)^6 \tag{18}$$

We can see the dependence of the analysis time upon resolution (sixth power!). It takes more than ten times to increase the resolution from 1 to 1.5.

Our assumption was that the spot variance at the start of the development was negligible. It is far from the reality in 1980. Fortunately, manufacturers were aware of this, and two major developments appeared: zone concentration plates from Merck and the Linomat sample application device from Camag. In zone concentration plates there are two layers, one of which is 1 cm wide and made of kieselguhr, which is not retentive. The adjacent layer is made of real chromatographic support. You can spot on the kieselguhr layer without great difficulty. When the plate is immersed in the tank the mobile phase moves up the sample to the gap between the layers. Here the sample is "shrunk" to a very small line. The variance is thus reduced. In the Linomat, a flow of nitrogen prevents the spreading of the spot.

V. MULTIDIMENSIONAL TLC

The capacity of a separation system can be defined as the number of compounds that can be separated from each other with a resolution of unity. Giddings [13] has given mathematical expressions for the peak capacity, n, of chromatographic separations. An approximation has been proposed by Grushka [14].

Assuming that the number of theoretical plates is constant,

$$n = 1 + (\sqrt{N}/4) \ln(1 + k_z) \tag{19}$$

k_z is the last eluted solute. In HPLC you can increase the efficiency (N) by increasing the column length. Unfortunately, the procedure is limited by the dramatic increase in pressure drop. Moreover, connecting one column exhibiting N plates to another similar column yields 2N plates. On the other hand, TLC has a unique ability: two developments of the sample mixture can be carried out successively in two perpendicular directions, using two different mechanisms or two solvents that provide different selectivities. The sample is placed at a corner of the plate and two developments are carried out successively, parallel to the two sides of the plate. If the same mechanism is used for both developments, the correlation coefficient between both systems is obviously 1 and the spots are aligned along a line at 45°. The goal is thus to obtain the maximum scatter in order for the spots to spread across the whole surface of the plate. In a very rough approximation we can expect N × N plates. This is not true, since the spot variance must be taken into account. Two-dimensional TLC has been used since the early work on paper chromatography [15], but the performance of such a technique was not considered. The spot capacity in one dimension [16] is calculated from the differential equation:

$$dn = dz/4\sigma \tag{20}$$

where z is the distance along the plate and 4σ is the width of a spot. This equation is the same as the one used in HPLC, but one must take into account the variation of the plate number with development length and with retention. When small particles are utilized the broadening is only dependent on diffusion, which is the same assumption that plate count is constant in HPLC:

$$\sigma^2 = \sigma_i^2 + 2\gamma Dt \tag{21}$$

where γ is the tortuosity of the packing and σ_i^2 is the spot variance at the injection point. From integration, we see that:

$$n = \frac{L}{4\sqrt{\sigma_i^2 + \dfrac{2\gamma DL^2}{k}}} \tag{22}$$

Taking into account the variation of the plate number with development length and retention:

$$\sigma^2 = \sigma_i^2 + zH \tag{23}$$

where H is the plate height given by Eq. (10).

$$dn = dz/4\sqrt{\sigma_i^2 + Hz} \tag{24}$$

Spot capacity is:

$$n = \frac{\sqrt{\sigma_i^2 + LH} - \sqrt{\sigma_i^2}}{2H} \tag{25}$$

If H is not independent of Rf, it is necessary to perform a numerical calculation. Assuming a negligible spot variance at the injection point, 16–22 peaks can be recorded on HPTLC plates with 5 μm particles and a development length of 2 or 5 cm, respectively. With TLC plates (10 μm particles) 25 peaks can be recorded with a 10 cm development length. This means that the recording device has the capability to do so. Since the spot variance cannot be neglected, Fig. 7 displays the variation in spot capacity with different spot variance.

In the calculation of spot capacity in two-dimensional TLC [17,18], we must take into account that all spots spread during the two successive developments unequally in the direction of development and in the perpendicular direction. Furthermore, during the second development the spots also spread laterally, so they must be separated with a resolution higher than unity at the beginning of the second development. Let n_1 and n_2 be the spot capacities obtained in one-dimensional TLC along the two different development directions and ^2n the spot capacity in two-dimensional TLC; ^2n is less than $n_1 * n_2$ and should be written $n_1' * n_2'$. Very approximately, when zones of area $\approx a * b$ are confined to a rectangular space of area $Lx * Ly$, the peak capacity is $Lx/a * Ly/b$.

The spot variance after the second development becomes

$$\sigma_2^2 = \sigma_i^2 + zH_1 + 2\gamma D_2 t_2 \tag{26}$$

where H_1 is the plate height in first direction and t_2 is time of the second development.

When σ_i^2 is negligible and the plate characteristics are similar:

$$^2n = \frac{L}{4H^2}\left(\sqrt{2\gamma \cdot \frac{D}{u} + H} - \sqrt{\frac{2\gamma D}{u}}\right)^2 \tag{27}$$

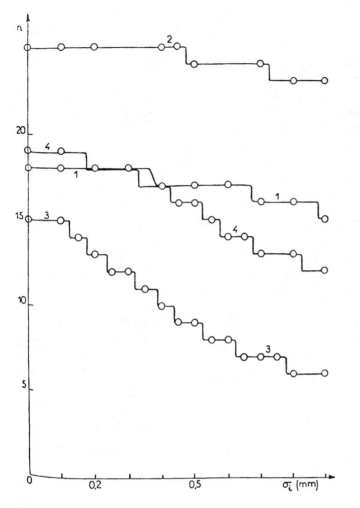

Figure 7 Variation of spot capacity with the standard deviation of the sample spot. (From Ref. 16 with permission of the *Journal of Chromatography*.)

Table 3 Spot Capacity in Two-Dimensional TLC: Influence of Development Length (L) and Particle Size (d_p)

L(cm)	$d_p = 5\ \mu m$			$d_p = 10\ \mu m$		
	n_1'	n_2'	^2n	n_1'	n_2'	^2n
2	12	12	144	11	10	110
3	14	14	196	13	13	169
5	15	15	225	16	15	240
7	15	15	225	17	17	289

Calculations with $\sigma_i = 0.01$ cm; $A = 1$; $C = 0.01$; $\gamma = 0.70$; $\chi_1 = 120$ cm/s; $\chi_2 = 60$ cm/s; $D_{m1} = 5 * 10^{-6}$ cm2/s; $D_{m2} = 2 * 10^{-6}$ cm2/s; $z_0 = 0.2$ cm

Table 3 shows the maximum spot capacity that can be achieved. We can see that the numbers (roughly between 100 and 400) are far beyond what can be achieved with the best columns available in HPLC.

The acceptance of two-dimensional planar chromatography is very much dependent on a reliable method for in situ detection. Conventional slit scanning densitometers are designed for spots distributed along a single lane. Special software is required to map the whole surface of the layer. Electronic scanning provides images of the surface and is gaining attraction due to fast data acquisition, absence of moving parts, and software capabilities. Further developments are expected.

VI. COOPERATIVE WORK

As was pointed out above, mobile phase velocity in conventional TLC with capillary rise is continually changing. It would be highly valuable to be able to master the mobile phase velocity. Furthermore, Rf values are often unreliable due to the vapor phase above the solvent layer, which varies with experimental conditions.

For this reason, devices have been designed, such as the DS chamber from Soczewinski et al. [19], but capillary forces were still the driving force of the mobile phase. Then appeared the forced flow planar chromatography techniques. The real breakthrough was the embedding of the plate. In this method, pressure is applied onto a membrane covering the horizontal layer (Fig. 8). A pump delivers the mobile phase to the side of the layer that is not pressurized, which

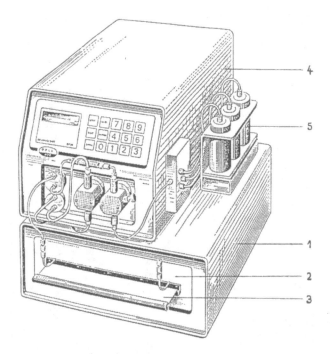

Figure 8 Pressurized ultramicrochamber of linear type, personal OPLC BS-50 system. 1: separation chamber; 2: holding unit; 3: layer cassette; 4: liquid delivery system; 5: eluent reservoirs. (Reproduced with permission from OPLC NIT, Budapest, Hungary.)

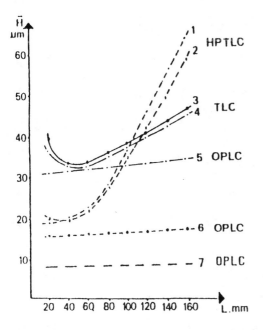

Figure 9 Plot of HETP versus development length in different TLC/HPTLC and OPLC systems. Differences in curves represent different types of development (saturated or unsaturated chambers) and different OPLC instruments.

is separated from the pressurized layer by a small trough. When the pressure of the liquid is higher than the pressure on the layer the liquid flows through, a transparent covering device permits one to monitor the eluent and observe the separated spots, their shapes, and possible deformation of the fronts.

The system can be used as a conventional TLC plate with spot deposition before development or as a flat square capillary column that can be equilibrated before injection through an injection port. Tyihak and Mincsovics [20–22] extensively published about the technique that they called overpressurized layer chromatography (OPLC). We had the same idea [23], but we used a rather different approach since we were mainly interested in two-dimensional separations. We combined our efforts and achieved a fruitful collaboration with the Hungarians [24,25]. TLC became a fully instrumentalized technique. Figure 9 illustrates the variation of the plate height as a function of the solvent front migration. H is practically constant along the plate in OPLC, which means that the theoretical plate number increases linearly with development distance. OPLC can be used off-line, with the analyst deciding when to stop the development, or on-line, in which case the excess mobile phase leaves the chamber continuously and an injection port permits the sample deposition when the ''flat column'' is equilibrated. This is of paramount importance when mobile phase additives (cyclodextrins, metals, etc.) are utilized. Conventional TLC is thus replaced by planar chromatography (PC).

In Rome at the 22nd International Symposium on Chromatography, I was awarded the Halasz Medal. I am fully aware of the debt I owe to my educators in chromatography and the skillfulness of my collaborators. During these years I have learned that achievements only come by combining forces and with experience, and I enjoyed the friendship of my colleagues.

VII. SOME FEATURES OF TLC AND HPLC

TLC advantages	TLC drawbacks
Large number of samples simultaneously analyzed in one single run	Limited efficiencies
	Reproducibility often difficult to achieve
Detection free of time constraints	Off-line detection
Possible multiple detection	Presence of a gas phase
Preservation of the chromatogram	Interfacing with MS not satisfactory
Two-dimensional capability	
Simultaneous sample cleanup and separation	
Hyphenation with Raman or Fourier transform infrared spectroscopy	

HPLC advantages	HPLC drawbacks
Very high efficiencies	One sample analyzed per run
Reproducibility	Impossibility of checking all solutes
Full automation	
MS detection	

VIII. A LOOK AT THE FUTURE

Personnel costs are higher in TLC than in HPLC. In the future we can expect a fully instrumentalized instrument to appear. The dream of the analyst is to be able to use a TLC plate like a compact disc—you put it inside a slot and wait for the answer.

In the last few years there has been an incredible shrinking of mass spectrometers. A detection device of this type will certainly appear. Is it a matter of "plumbing"? Some hyphenations between OPLC and MS have been developed, but the attempts did not surpass the prototypes.

REFERENCES

1. F Feigl. Qualitative Analyse mit Hilfe von Tupfelreaktionen. Leipzig: Akad. Verlag., 1931.
2. AIM Keulemans. Gas Chromatography. New York: Reinhold Pub., 1959.
3. JC Giddings. Dynamics of Chromatography. New York: Marcel Dekker, 1965.
4. E Stahl. Dunnsicht Chromatographie. Berlin: Springer-Verlag, 1965.
5. LR Snyder. Principles of Adsorption Chromatography. New York: Marcel Dekker, 1965.
6. G Guiochon, A-M Siouffi. J Chromatogr Sci 16:598, 1978.
7. JC Giddings. In: E Heftmann, ed. Theory of Chromatography. New York: Van Nostrand, 1975.
8. JH Knox. Practical High Performance Liquid Chromatography. London: Heyden, 1976.
9. G Guiochon, A-M Siouffi. J Chromatogr Sci 16:470, 1978.
10. G Guiochon, F Bressolle, A-M Siouffi. J Chromatogr Sci 17:368, 1979.
11. A-M Siouffi, F Bressolle, G Guiochon. J Chromatogr 209:129, 1981.
12. P Kubelka, F Munk. Z Tech Physik 12:593, 1931.
13. JC Giddings. Anal Chem 39:1027, 1967.

14. E Grushka. Anal Chem 42:181, 1970.
15. R Consden, AH Gordon, AJP Martin. Biochem J 38:244, 1944.
16. G Guiochon, A-M Siouffi. J Chromatogr 245:1, 1982.
17. G Guiochon, M-F Gonnord, A-M Siouffi, M Zakaria. J Chromatogr 250:1, 1982.
18. G Guiochon, LA Beaver, MF Gonnord, A-M Siouffi, M Zakaria. J Chromatogr 255:415, 1983.
19. TH Dzido, E Soczewinski. J Chromatogr 516:461, 1990.
20. E Tyihak, E Mincsovics, H Kalasz. J Chromatogr 174:75, 1979.
21. E Tyihak, E Mincsovics. J Planar Chromatogr 1:6, 1988.
22. E Mincsovics, K Ferenczi-Fodor, E Tyihak. In: J Sherma, HB Fried, eds. Handbook of Thin Layer Chromatography. 2nd ed. New York: Marcel Dekker, 1996, pp. 171–203.
23. A-M Siouffi, M Righezza, J Kantasubrata. In: RE Kaiser, ed. Instrumental High Performance Thin Layer Chromatography, Proceedings of the 2nd Intern. Symposium, Wurzburg 1985. Bad-Durkheim: Institute for Chromatography, 1985, pp. 201.
24. E Mincsovics, E Tyihak, A-M Siouffi. In: E Tyihak, ed. Proc. Intern. Symp. on TLC with special Emphasis on OPLC, Szeged, Hungary. Budapest: Labor MIM, 1986, p. 251.
25. E Mincsovics, E Tyihak, A-M Siouffi. J Planar Chromatogr 1:141, 1988.

6

The Way It Was

Raymond P. W. Scott
Scientific Detectors Limited, Oxfordshire, England

Professional success results from the coalescence of a number of social and technical elements. These elements will obviously include good professional training, a capacity for innovation, and a properly organized and well-equipped environment in which to practice. In addition, success is usually accompanied by a challenge, arising early in one's career, that is not only intellectually stimulating but that leads to concepts, systems, or devices that have long-term and wide-ranging practical use. Serendipity also plays her part—one must also be at the right place at the right time when opportunity arrives. I am, with others in this book, one of the fortunate few who has been favored with these "elements" of professional success.

Gas chromatography erupted onto the laboratory scene in the early 1950s under almost ideal conditions for growth. Internationally, there was a large reservoir of well-trained scientists available, all with excellent experimental skills (at that time universities in Europe placed a strong emphasis on a scientist being experimentally adept), and science (the word meaning a study of measurement) was still considered an experimental subject. I and my peers graduated from university, not merely with a sound knowledge of chemistry, physics, and mathematics, but also proficient in the use of the lathe and in the working of glass and familiar with electronic circuit design and construction. The major advances made in chromatography during the 1950s and 1960s (with one or two notable exceptions) came largely from those with strong experimental skills. Archer Martin, the inventor of gas chromatography, was the most accomplished experimental chemist I have had the privilege and pleasure of knowing.

The 1950s and 1960s was the golden age of research, both in academia and in industry. The people of the western world firmly believed that their future personal safety, health, and happiness lay in the successful progress of scientific research. Indeed, this was true; for that matter, is true today. However, in those days a significant proportion of the population had been schooled in basic science concepts and many appreciated the importance of much of the research that was being proposed. As a result, they were prepared to make available large sums of money (both private and public) to fund scientific research. Into this politically and economically fertile soil were planted the first seeds of modern chromatography.

In 1951, at the age of 27, I took the position of physical chemist at the research department of Benzole Producers in Watford, England. My wife and I and our two young sons took possession of our first house (a small townhouse) in Watford Hertfordshire. Benzole Producers was situated on what was then the Watford Bypass, a 10-minute bus ride from home and a few miles up the road from the research laboratories of the Medical Research Council at Mill Hill, the significance of which was yet to become apparent. Benzole Producers extracted aromatic hydrocarbons from coal gas produced by the pyrolysis of coal,the accompanying product, coke, being used in the production of iron and for domestic heating. The addition of aromatic hydrocarbons to petroleum fractions, as an alternative to tetraethyl lead, raised the octane rating of the fuel to a level suitable for use with high compression-ratio internal combustion engines. At that time, the only method of analyzing hydrocarbon mixtures involved a very complex and time-consuming distillation process. Among my other designated responsibilities, I was to develop a more efficient method for mixed hydrocarbon analysis.

The British Medical Research Council adopted in the 1950s a very enlightened attitude towards the research they funded. All authors were expected to publish full details of their work and, when appropriate, describe their research at seminars to which outside scientists were invited. Very little was patented; in fact, the patent process was considered by some a restrictive practice, which, in fact, it is. For nearly a decade this freedom of disclosure, initiated by the Medical Research Council and strongly supported by the inventor of gas chromatography himself, was adopted by all who joined in the development of chromatography. The consequent free flow of information contributed in no small way to the rapid advancement of the technique, especially in the early days of its development. Sometime in 1951–52, A. T. James and A. J. P. Martin gave a seminar on Vapor Phase Chromatography (as gas chromatography was known then), and since the venue was only a few miles up the road and seemed to be pertinent to my work on hydrocarbon analysis, I attended.

Both Martin and James always gave excellent lectures, and Martin's description of the apparatus and its operation was clear and concise. However, we were absolutely stunned by the separations Martin showed toward the end of his lecture, and one in particular, that of a benzole mixture, showed that in a half hour Martin had discovered more about benzole than our research department had in over a decade. I obviously needed to build a gas chromatograph and build it quickly—I found myself visiting their laboratory the next day.

I was made exceedingly welcome and shown their apparatus, which was described to me in detail. Their laboratory was relatively small, and a little way down the corridor worked a young biophysicist, quiet and reserved, named Jim Lovelock, who would, in due course (among his many other achievements) invent two of the most sensitive gas chromatographic detectors available today. The problem for me (and it was a difficult problem) was the construction of the detector. Martin's detector, the gas density balance, consisted of a Wheatstone network of capillary tubes drilled out of a solid block of high conductivity copper. Using six copper blocks and filling all but the last with broken drills (an unfortunate characteristic of high-conductivity copper), we eventually succeeded in constructing a working density balance. I would need at least six chromatographs for the analytical work I had in mind, and the difficulties associated with constructing six gas density balances provoked me to search desperately for an alternative method of detection.

Initially I attempted to exploit the change in carrier gas viscosity during the elution of a peak as a means of sensing the solute vapor. To emphasize the viscosity change I employed hydrogen as the carrier gas, and for convenience the waste hydrogen was burnt at a small jet. The system was shown to be a very inefficient detector, but the elution of a peak became clearly visible by the elongation of the waste hydrogen flame accompanied by an increase in luminosity.

The jet was then situated inside an old cocoa tin, a thermocouple was placed 3 mm above the flame and connected to a simple battery-operated "backing-off" circuit, and another detector became available for use in gas chromatography (GC). The detector, reported in *Nature* [1], had a sensitivity of between 1×10^{-7} and 1×10^{-6} g/mL and a linear dynamic range of about three orders of magnitude. It was, in fact, the precursor of the flame ionization detector. This detector allowed the simple and rapid fabrication of the six gas chromatographs that were completed and in operation within 2 months. Benzole Producers could now monitor and control the quality of their products more accurately and reliably than ever before.

About the same time, Dennis Desty at British Petroleum was also having difficulties constructing the Martin gas density balance and (although eventually successful) learning of my detector, visited me at Watford. This was to be the beginning of a long (and for him a lifetime of) professional association and personal friendship. Subsequently, I returned the visit and was shown (incidentally) a large laboratory with dozens of low-temperature decommissioned distillation units, having been replaced by gas chromatographs. The low-temperature distillation unit was probably the first of many commercial instrument casualties that resulted from the invention of gas chromatography.

Dennis and I traveled together to the first meeting of the Gas Chromatography Discussion Group (initiated and organized by Dennis) and held in the laboratories of ICI at Ardeer in Scotland. About 80–100 attended, 20–30 of whom constituted most of those actively working in gas chromatography at that time. The Discussion Group was to play an important part in the development of the technique for nearly a decade. Initially, the technique expanded in two directions: first into the field of biochemistry, which was driven by the activities and publications of the inventors James and Martin, second into industry (usually by direct contact) driven initially by Desty and his coworkers in BP Sunbury and Ted Adlard, Bill Whitham, and their coworkers at Shell Thornton, then to Shell Amsterdam, A. I. M. Keulemans, Van der Craats, Van Deemter, and their colleagues and then to Shell Emeriville in the United States. From these centers details of the new techniques spread rapidly to other industries and to the universities. In fact, at the beginning the universities were rather slow to become involved with the technique, and, with the exception of Purnell at Cambridge, Phillips at Oxford, Pollock and Harvey at Bristol, and one or two others, the majority of the early development work was carried out in industrial laboratories.

Even at the time of the Ardeer meeting, no commercial gas chromatography instrument was available. The first instrument with real commercial possibilities was one constructed in the Shell laboratories by Ted Adlard and his colleagues. It was over 6 feet tall, mounted on an impressive plinth, surrounded by vacuum pumps, and painted a glorious British racing green. Ted and his friends referred to it as the Green Goddess, which also happened to be the name of a popular cocktail of that period. It used the katharometer as a detector (introduced by Ray of ICI) and showed a distinct similarity to the chromatograph eventually produced commercially by Griffin and George in the mid-1950s. The instrument companies appeared rather hesitant to become involved with gas chromatography, and, in fact, very early in the 1950s Martin demonstrated his chromatograph to the owner of a well-established American instrument company and suggested that it might be manufactured. The device, though novel and interesting, was deemed to have no commercial future.

Gas chromatography seemed to attract the attention of scientists of widely differing skills and disciplines. Physicists, chemists, engineers, and mathematicians all contributed to the development of GC, with almost limitless enthusiasm, diligence, and effort. This symbiosis was highly productive. In the 1950s, largely due to the original attitude of James and Martin, there was a camaraderie between the workers in the field that was unique in the history of analytical instrument development. There was a continuous interchange of information between those involved, and the content of most technical reports was well known by all long before actual publication. The first symposium on GC, held in London in 1956, was entitled "Vapor Phase Chromatography," the original name used by James and Martin. At that meeting a number of fundamental papers were presented that were to have a long-term impact on the technique. The available detectors were discussed, such as the gas density balance and the katharometer, and two new detectors were introduced, the flame thermocouple detector [2] and the β-ray detector [3], the latter being the first ionization detector. Probably the most significant contribution to the meeting was the paper by Keulemans and Kwantes [4], which provided experimental support for the rate theory developed by Van Deemter et al. [5]. The rate theory was the long-awaited and essential piece of the chromatography puzzle that would provide an understanding of the processes that caused band dispersion in a chromatography column. This understanding would initially lead to improved GC columns, but, more importantly, it would confirm the predictions of Martin and Synge [6] on how efficient liquid chromatography (LC) columns could be made.

The simple thermocouple detection system with its higher sensitivity allowed the operation of long, high-efficiency packed columns, and in 1957 I was the first to demonstrate the partial separation of *m*- and *p*-xylene [4] on a packed column 25 feet long having 12,000 theoretical plates [7]. The column was packed as a straight length by standing on a garage roof, with my coworkers on ladders tapping the column at various points along its length during the packing process. After packing, the column was folded to fit into a vapor jacket for thermostatting purposes. The separation, shown in Fig. 1, was achieved on a dispersive stationary phase with no polar selectivity and thus relied almost entirely on the low dispersion of the high-efficiency column to achieve a separation. The advent of the argon detector developed by Lovelock [8] provided even greater sensitivity and permitted further development of high-efficiency packed columns. At the 1958 Amsterdam symposium, Gas Chromatography 1958, I described a 50-foot packed column having an efficiency of 30,000 theoretical plates, operated at an inlet pressure of 200 psi, that would separate all the isomeric heptanes and thirteen of the isomeric octanes [9]. It would also provide baseline separation of the isomeric xylenes. This column was packed in 10-foot straight lengths, each bent into a U shape and then connected by low-dispersion connecting capillaries. My company could now also monitor and control the quality of another of its prod-

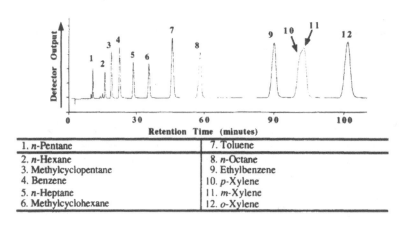

1. *n*-Pentane	7. Toluene
2. *n*-Hexane	8. *n*-Octane
3. Methylcyclopentane	9. Ethylbenzene
4. Benzene	10. *p*-Xylene
5. *n*-Heptane	11. *m*-Xylene
6. Methylcyclohexane	12. *o*-Xylene

Figure 1   Separation achieved using thermocouple detection system.

ucts, the xylenes. In 1959 we moved into a large ranch-style house and I bought my first new car.

The contents of the Gas Chromatography 1958 symposium probably represented the climax of GC development, and although a plethora of symposia would appear over the following half century, none would have the excitement and novel technical content of this one. The theory of GC was extended by Littlewood [10] and Boheman and Purnell [11], and the capillary column, which was eventually to revolutionize GC, was introduced by Golay [12]. Golay's theory would have important ramifications in LC, not so much with respect to LC capillary columns, but in controlling the dispersion that occurs in connecting tubes. McWilliams and Dewer described the flame ionization detector [13], which was to become the ''workhorse'' detector for all future gas chromatographs. Grant described the emissivity detector [14], and temperature programming was introduced by Harrison and coworkers [15], another technique that would be incorporated in the design of all future chromatographs. At the same meeting I also described the first success-ful *moving bed* continuous preparative gas liquid chromatography system [16] and eventually used it to continuously extract pure benzene from coal gas. It would appear that only now is this concept being actively developed for large-scale liquid chromatography separations.

Unfortunately, the great technical success of the 1958 symposium was somewhat marred by the first patent being taken out on the chromatographic system itself. Chromatography had lost its innocence. I suppose as a result of the large sums of money involved in the design, manufacture, and sales of chromatographs, this was inevitable. Up to that time, although patent applications had been made for certain detectors and detector sensors, advances in the basic separating system had been free for all, and any improvements that came out of research were made available to everyone. The patent was that of the open tubular or capillary column, and even the validity of the patent was questioned by many. Some thought that the suggestion by Martin in his introduction to the 1956 symposium [17] anticipated the open tubular column. Martin's comment was, ''we shall have columns only two tenths of a millimeter in diameter and these will carry, I believe, advantages of their own that I have no time to go into.'' Indeed this does seem to portend the introduction of the capillary column. However, it was argued that Martin was referring to packed columns, but the practicability of packing columns of significant length that are only 200 μm in diameter is not considered viable even today. In fact, this particu-lar patent angered the field to such an extent that although capillary columns were examined in research and development laboratories and their properties measured, with the exception of the hydrocarbon industry, capillary columns were scarcely used for general GC analysis for over a decade after the patent was field.

Many attendees returned from the Amsterdam symposium inspired to exploit the various concepts and devices that had been described and to develop them even further. My first interest was to employ my long high-efficiency columns to monitor the composition of benzole formed during the coking cycle of a coke oven [18]. This involved erecting the chromatograph on the top of an experimental coke oven with a continuous sampling device connected to the oven gas exit. The coke oven had a coke capacity of about 400 cubic feet, and the surface of the oven top was, on average, about 80°C. The carbonizing time was 14 hours, and samples were taken every 30 minutes—a long, tedious, and uncomfortable job. There is no romance in watching the sun set and rise from the top of a coke oven. The chromatogram of two samples of benzole taken early and late in the coking sample are shown in Fig. 2. The period of high aromatic hydrocarbon production was unambiguously identified. As a result of Golay's description of the performance of open tubular columns, I examined the effect of different gases—hydrogen, helium, nitrogen, argon, and carbon dioxide—on peak dispersion in a capillary column. I also

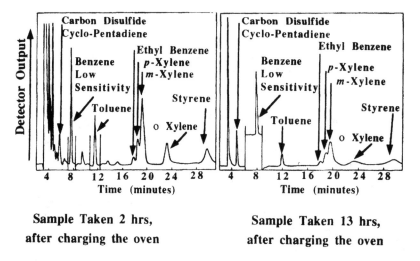

Figure 2 Chromatogram of benzole formed at beginning and end of a coke oven cycle.

introduced nylon open tubular columns and used them to provide very high efficiencies [19] and very fast separations [20].

In 1960, the third of the triad of symposia that were considered to contain the essential technical foundations of GC was held in Edinburgh in 1960. This meeting, although highly successful, lacked the verve of its predecessor. Desty and his coworkers at British Petroleum [21] and Ongkiehong and his group at Shell Amsterdam [22] described the properties of the flame ionization detector in great detail, and Lovelock [23] described the electron capture detector. I and my coworkers reported the separation of a five-component mixture in 5 seconds using short lengths of nylon capillary column [20], and, at the other extreme, we demonstrated the use of 1000 feet of nylon capillary tubing to produce one million theoretical plates [19]. The majority of the papers presented were largely extensions and verifications of the ideas put forward in the previous symposia together with many new applications. Although, as the future would show, there would be further developments in the technique, the impression at the end of the meeting was that the development of analytical GC was more or less complete. This impression may have been the catalyst that initiated the renaissance of LC in the middle and late 1960s. There would be many more symposia on GC, but they would deal, for the most part, with applications, relatively minor modifications to the procedure and apparatus, or the extrapolation of work already reported in the first triad of synopsis. As the years passed, it became increasingly difficult to determine whether the GC symposia were organized merely for the sake of having symposia or for the proper purpose of disseminating knowledge.

During the 1950s my coworkers and I were extremely active and published regularly and frequently in chromatography and analytical journals. In 1958 I was elected Fellow of the Royal Society of Chemistry, and in 1960 I was made a Doctor of Science by the University of London. I spent a short time in 1962 in charge of research and development at W. G. Pye and then joined Tony James and accepted the position of Principle Scientist in the Physical Chemistry Department at the Food Research Laboratories of Unilever Ltd. at Colworth House, Sharnbrook, Bedfordshire. At Unilever I had a relatively large group involved not only in separation science but also general physical chemistry and the various forms of spectroscopy. There was, at the time, a strong interest in associating the gas chromatograph with the mass spectrometer. The gas chromatographer regarded the mass spectrometer as another type of detector, whereas the mass spectroscopist regarded the gas chromatograph as a complicated form of sampling system. Our mass spectroscopist, Bill Kelley, with Ted Adlard of Shell, together specified the type of fast scanning mass spectrometer that could be used with a gas chromatograph. A.E.I., the manufacturer of the high-resolution mass spectrometer MS2, built the specified fast scanning MS 12 according to Ted and Bills specifications, and Ted received the first model that was produced and Bill the second. Having established the viability of the GC/MS tandem system, many other manufacturers developed similar fast scanning mas spectrometers to complete with the MS 12.

The low-resolution mass spectrum produced at that time, however, was often inadequate for absolute solute identification, and the support of IR spectra was frequently needed to identify specific groups present in the molecule. As a consequence we developed a GC/IR/MS system that operated on a stop-start procedure [24]. As each solute was eluted from the column it passed into a chamber, the carrier gas arrested, the infrared spectrum taken, and simultaneously a sample was drawn into a mass spectrometer and the mass spectrum taken. The carrier gas was then turned on again and the next peak eluted. The procedure of stopping and starting the carrier gas flow was programmed to eliminate peak dispersion. The apparatus was demonstrated at the Sixth International Symposium at Rome where it ran continuously throughout the entire meeting. An example of the chromatogram and the spectra obtained is shown in Fig. 3.

At the end of the symposium we turned off the carrier gas halfway through the chromatogram. We capped the ends of the column and carried it in a truck over the Alps back to our laboratories at Colworth House. There we reassembled the apparatus, heated the column to its normal temperature, started the gas flow, and completed the development of the mixture. We found that despite its extreme treatment, the column had lost only about one third of its theoreti-

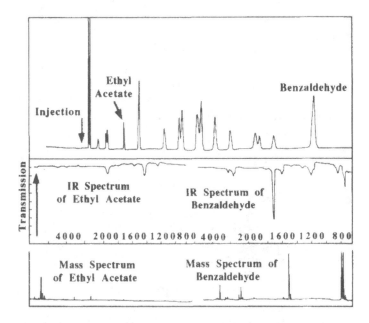

Figure 3   Chromatogram of a 17-component mixture separated by intermittent development for infrared and mass spectrometric peak scanning.

cal plates. Longitudinal diffusion was not a serious contributor to peak dispersion providing the solute diffusivity was small enough.

By 1965 the renaissance of liquid chromatography was well underway. The factors that invoked this renaissance are, even today, obscure and not easily identifiable. Perhaps, besides the contemporary glamour that surrounded GC in 1960, the recognition that it could only separate volatile materials played a part. There were certainly more involatile materials associated with separation problems than there were volatile ones, particularly in the biochemical and biotechnology fields. Perhaps it was felt that the development of GC was virtually played out (which in fact it was) and there were more exciting discoveries to be made in LC. It was to be seen, however, that although progress in GC had been fast and relatively easy, progress in LC was to be slow and arduous.

My first excursion into liquid chromatography was to examine the extra column dispersion effects that took place so that a chromatographic system itself could be designed that would not impair the efficiency of the column by extra column effects [25]. I then worked with Tony James and John Ravenhill to develop an LC transport detector [26]. The original was crude in form but functional, and it established the viability of the transport concept for LC detection and other uses. The great advantage of the transport detector was its immunity to the nature of the solvents used in the mobile phase, providing, of course, that they were volatile. This opened up the technique to a wide range of distribution systems that hitherto could not be used because of the effect of the solvent on the detector. Later Graham Lawrence and I [27] developed the concept into a far more effective detection device. We employed a stainless steel wire as the transport medium, and the column eluent passed over the wire leaving a coating from which the solvent was evaporated. The solute layer was then oxidized in an enclosed atmosphere of air, and the carbon dioxide so formed was entrained into a hydrogen stream, which was passed over a heated catalyst converting the carbon dioxide to methane. The methane was detected by means of a flame ionization detector. The modification increased the sensitivity of the detector by more than an order of magnitude. During this period I also examined the possibility of chemical amplifiers or molecular multipliers to improve detector sensitivity [28]. The solute was oxidized to carbon dioxide and water, and the carbon dioxide alternately passed first over heated carbon, where each molecule of carbon dioxide produced two molecules of carbon monoxide, and second over heated copper oxide, which produced two molecules of carbon dioxide. Thus, each time the combustion gases were passed over the carbon/copper oxide pair, the amount of carbon dioxide was doubled. Consequently, 8 pairs of reactors were shown to provide an amplification of 64. This technique was shown to work well, but due to the complexity of the apparatus it did not arouse commercial interest.

The professional encouragement and the financial support that was provided by Unilever Ltd. at Colworth House and at other of its research centers contributed significantly to the national achievements in science at that time. The shareholders placed their confidence in the scientists to use their resources prudently, and the scientists, in turn, respected the confidence placed in them and were careful to use the revenue as wisely as possible. This situation was true for many other major industries in Britain and, for that matter, in the free world. Unfortunately, this mutual trust between scientists and shareholders and between scientists and politicians was to be slowly eroded over the coming years, leading to an inevitable reduction in research investment.

At this stage further improvement in chromatographic performance would only result from a better understanding of the factors that controlled the selectivity of the phase system and the extent of peak dispersion. The solutes must be moved apart and the dispersion of the solute bands kept to a minimum to provide maximum resolution. For several years my colleagues and

I examined the fundamental theory involved in solute retention and band dispersion in order to identify the optimum conditions that would provide fast separations or high resolution. Unfortunately, this work would only come to fruition after I had left Unilever and joined Hoffman La Roche in Nutley, New Jersey.

In 1969 my two sons were ensconced in university and my wife and I decided we could indulge in what we chose to call "a middle-aged adventure"—we would emigrate to the United States. I accepted a very attractive position at Hoffman-La Roche, Inc. (a pharmaceutical company in New Jersey) to be in charge of their physical chemistry department. My wife and I bought a small farm in Ridgefield, Connecticut, and enjoyed a real country life with horses, sheep, goats, and numerous other farm animals. The commute, however, was 75 miles each way. This was, for me, to be the start of a highly productive period in the development of various aspects of separation science.

Initially, I developed a number of solvent protocols to improve selectivity and extend sample polarity range, but it soon became apparent that a LC/MS interface was needed to help identify the multitude of solutes separated on the LC column. We started work on a wire transport system, similar to that developed by Lawrence and me at Unilever [29]. It was envisioned that the interface would allow electron impact spectra to be obtained that would provide ion fragments to facilitate interpretation of the spectra. A diagram of the transport LC/MS interface is shown in Fig. 4. The eluent from the column passed over a moving stainless steel wire, coating it with a film of column eluent in the conventional manner. The wire then passed through two orifices into two chambers joined in series. Each chamber was connected to a vacuum pump that reduced the pressure in the first to a few μm and in the second to about 10^{-5} mm of mercury. During

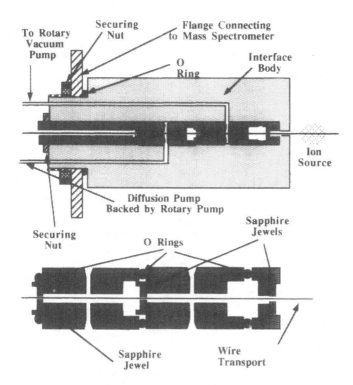

Figure 4   Wire transport interface.

this process the solvent evaporated from the wire, leaving a coating of the solute on the wire surface. A current of a few milliamperes was passed continuously through the wire, which only became hot when it entered the high vacuum of the ion source where it could no longer lose heat to its surrounding. The temperature of the wire rose rapidly to about 350°C in the ion source, vaporizing the sample directly into the electron beam. The wire left the ion source in the same manner as it entered, via a second pair of differentially pumped chambers.

The main body of the interface was constructed of stainless steel and was fitted to side flanges of a Finnigan quadrupole mass spectrometer, such that the interfaces are re-entrant to the ion source and terminated a few millimeters from the electron beam. The two chambers were separated and terminated by ruby jewels 0.1 in. I.D. and 0.018 in. think. The jewels in the left-hand interface where the sample is introduced had apertures 0.010 in. I.D. The jewels in the right-hand interface where the wire transport leaves the mass spectrometer to the winding spool had apertures 0.007 in. I.D. The larger diameter apertures on the feed side were employed to reduce ''scuffing'' of the wire and possible solute loss. Ruby jewels were necessary to prevent frictional erosion of the apertures by the stainless steel wire. The first chamber of each interface was connected to a 150 L/min rotary pump, which reduced the pressure in the first chamber to about 0.1 mm of mercury. The second chamber of each interface was connected to an oil diffusion pump backed by a 150 L/min rotary pump. The pressure in the second chambers was reduced to about 5–10 µm of mercury. The entrance and the exit of each interface were fitted with a helium purge that passed over the aperture through which the wire was entering or leaving and ensured that only helium was drawn into the interfaces. In this way background signals from air contaminants were greatly reduced. The purge also allowed the use of methane as a chemical ionization agent if it were required. The pressure in the source was maintained at about 1×10^{-6} mm of mercury. The sensitivity of the device to diazepam, monitoring the eluent by total ion current, was found to be about 4×10^{-6} g/mL, which in the original chromatographic system was equivalent to about 7×10^{-10} g per spectrum. A total ion current chromatogram of a sample of the mother liquor from some vitamin A acetate crystallization is shown in figure 5. The separation was carried out employing incremental gradient elution using 12 different solvents.

It is seen that a good separation is obtained and the integrity of the separation is maintained after passing through the interface. It is also seen that the system is entirely independent of the

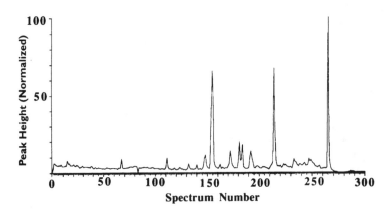

Figure 5   Total ion current chromatogram of a sample of mother liquor from a vitamin A acetate crystallization.

solvent used in the separation, which ranged from the very dispersive *n* paraffins to highly polar aliphatic alcohols and included chlorinated hydrocarbons, nitroparaffins, esters, and ketones as intermediate solvents. This interface was eventually produced commercially using a continuous plastic belt as opposed to a stainless steel wire, which, in my opinion, was not nearly as effective.

The 1970s produced profound changes in the research environment, particularly in separation science. The mutual confidence that had existed between the research scientist and the funding agency or shareholder began to wear thin. This was caused partly by profligate spending on the part of some scientists and partly by a tightening economy and a drive for increased profits. In addition, a form of political correctness crept into research funding. Grant applications appeared to require the presence of certain "buzz words" for success, for example, *pollution*, *environmental damage*, *carcinogens*, *global warming*, *toxicity*, etc., would illicit sympathetic response, but terms such as *basic theory*, *fundamental studies*, *precise measurement*, *accuracy*, etc., did not appear to evince anything like the same enthusiastic support. This tended to divert research from long-term investigations and fundamental experimentation to short-term, troubleshooting, "pop" research, often with strong public appeal but with limited depth and long-term value.

Despite this trend, many of us continued fundamental work on separation techniques generally. In the search for higher efficiencies and shorter analysis times in LC, the use of small-bore columns in LC appeared to be an attractive approach. Kucera and I explored the use of microbore columns in a wide range of different applications [30]. An example of a separation obtained from a microbore column is shown in Fig. 6.

The column was 2 m long, 1 mm I.D., packed with a reverse phase having a particle diameter of 10 μm. The inlet pressure was 6000 p.s.i., the mobile phase was an acetonitrile/water mixture, and the chromatogram was developed isocratically. The column efficiency was about 250,000 theoretical plates, and it should be noted that the retention time is over 40 hours. Unfortunately, very high efficiencies can only be achieved by employing very high pressures or accepting long retention times. In the particular example given, the maximum operating pressure of the sample valve was 6000 p.s.i., which is usually the limiting factor in the use of high inlet pressures. A similar type of column, 10 meters long and packed in a similar manner, produced an efficiency of three quarters of a million plates but with commensurably longer elution times.

The factors that controlled retention and selectivity became an important study in the 1970s and 1980s. Pioneering work by Purnell and his group [31] demonstrated that in GC, providing

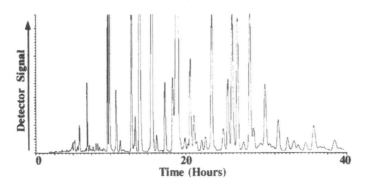

Figure 6   Chromatogram of an aromatic hydrocarbon extract from coal obtained from a high-efficiency small-bore column.

there were no strong interactions between the components of the stationary phase (i.e., the components did not associate, or if they did so only very weakly), the retention volume of a solute was proportional to the volume fraction of either one of the stationary phase components. This could be explained simply by assuming that the probability of interaction was proportional to the volume fraction of the solvent (in much the same way as the partial pressure of a gas determines its probability of collision). This simple and rational explanation caused great controversy, as many reputations were built on the assumption that the logarithm of the corrected retention volume was linearly related to the solvent concentration (although there is no simple explanation as to why that should be). The elementary relationship between volume fraction of solvent and distribution coefficient was confirmed by Katz, Ogan, and myself [32] for nonassociating solvents in liquid–liquid systems. However, the behavior of methanol–water mixtures (and to some extent tetrahydrofuran–water and acetonitrile–water mixtures) was used in an attempt to disprove the theory, as the relationship broke down for strongly associating solvents. Lochmüller, Katz, and I [33] subsequently showed that a mixture of methanol water contained *three* distinct components, not two, water unassociated with methanol, methanol unassociated with water, and the methanol–water associate, so the simple relationship for a binary liquid could not be expected to hold. They also showed that where dispersive interactions dominated between the mobile phase and the solute, retention was exclusively controlled by the volume fraction of methanol *unassociated* with water. The controversy has continued into the twenty-first century. Although the concept that strong association between the components of the mobile phase will significantly influence the nature and strength of solute-phase interactions is now beginning to be generally accepted, the simple relationship predicted and demonstrated by Purnell and confirmed by my colleagues and me in LC is still not universally accepted. However, the evidence is mounting up.

By the end of the 1970s the return on research investment throughout industry had become commercially unacceptable, and this seemed particularly true for the pharmaceutical industry. As I saw the situation, scientists had become reckless with the money entrusted to them. It was not just the excessive expense accounts and the unnecessary attendance at expensive technical meetings, but the purchase of irrelevantly costly equipment that provided services that were either unnecessary or could be obtained from external sources at a fraction of the price. The spending tide seemed inexorable and any attempt to swim against it resulted in one being designated as a cheapskate or accused of not fully supporting the projects. It was time to look elsewhere. In 1980 I joined the Perkin Elmer Corporation as Director of Applied Research at Norwalk Connecticut. A short time after I left Hoffman-La Roche, the company reduced its staff in Nutley by about 20% and not long after by another 20%. Much of the chemical service was replaced by outsourcing. I was not the only one to be concerned about return on research investment.

My years at Perkin Elmer were interesting and good fun. Although research and development had to be directed towards new products, they were exclusively involved with separation science. It was fortunate that shortly after my joining the company, Mike Moore, a man for whom I had a great respect and with whom I enjoyed working, was promoted to CEO. We had a very productive period together. My department was involved in the development of sampling systems, low-dispersion connecting tubing, multifunctional detectors, atomic adsorption spectrometer interfaces, and many other instruments and devices. One particularly interesting project was the successful development of an all quartz/glass argon detector [34] that functioned with no radioactive source and with no compromise on original specifications. The multifunctional detector [35] was combined with a sample valve, column, and low-dispersion connecting tubing, which constituted a complete liquid chromatograph without a pump and was sold for $5,000.

In my opinion, this is the way analytical instrumentation will go in the future (low cost but still precise and accurate), but the device failed with the sales forces. If the commission is the same, salesmen will always prefer to sell one $25,000 instrument as an alternative to selling five $5,000 instruments. However, in due course, it will be the market that dictates the product, not the sales force. Early in the 1980s I learned that Mike Moore was going to leave the company, and as I was in my early sixties I decided to retire and enjoy life with my family and friends. Nevertheless, I still wanted to be involved in separation science but with greater freedom.

Since my retirement I have held visiting chairs in the Chemistry Department at Georgetown and the Chemistry Department at Birkbeck College, University of London. I have worked with Tom Beesley at Advanced Separation Technology, Whippany, New Jersey, Chris Little at Scientific Detectors Ltd., Banbury, Oxford, written books and other contributions for John Wiley and Sons (UK and USA), Marcel Dekker, Academic Press, and Permagon Press. On average, for the last 10 years I have written one book per year, and at present I am publishing technical papers at a rate of about two a year.

It is now the year 2000, and I am retired and my wife and I are still in the United States. We have had a wonderful 30 years of fun, excitement, and success in a great country, with exceptionally beautiful countryside and full of very friendly people. We now have our home in the USA and a holiday home in the UK and enjoy the best of both worlds. In addition, I can still be active in separation science research. This year my wife and I celebrated our fifty-fourth wedding anniversary, and we have two great sons, two super daughters-in-law, and four splendid grandchildren. My wife and I are 76 years old, and inevitably the quality of life deteriorates as one ages. Nevertheless, up to this day we have had a wonderfull time together, great companionship and lots of laughs. We pray the good times will continue to roll.

REFERENCES

1. RPW Scott. Nature 176:793, 1955.
2. RPW Scott. In: DH Desty, ed. Vapour Phase Chromatography. London: Butterworths Scientific Publications, 1957, p. 131.
3. H Boer. In: DH Desty, ed. Vapour Phase Chromatography. London: Butterworths Scientific Publications, 1957, p. 169.
4. AIM Keulemeans, A Kwantes. In: DH Desty, ed. Vapour Phase Chromatography. London: Butterworths Scientific Publications, 1957 p. 15.
5. JJ Van Deemter, FJ Zuiderweg, A Klinkenberg. Chem Eng Sci 5:271, 1956.
6. AJP Martin and RL Synge. Biochem J 35:1358, 1941.
7. RPW Scott, JD Cheshire. Nature (Lond) 180:702, 1957.
8. JE Lovelock. In: RPW Scott, ed. Gas Chromatography 1960. London: Butterworths Scientific Publications Ltd., 1960, p. 9.
9. RPW Scott. In: DH Desty, ed. Gas Chromatography 1958. London: Butterworths Scientific Publications, 1957, p. 197.
10. AB Littlewood. In DH Desty, ed. Gas Chromatography 1958. London: Butterworths Scientific Publications, 1957, p. 23.
11. J Bohemen, JH Purnell. In: DH Desty, ed. Gas Chromatography 1958. London: Butterworths Scientific Publications, 1957, p. 6.
12. MJE Golay. In: DH Desty, ed. Gas Chromatography 1958. London: Butterworths Scientific Publications, 1957, p. 36.
13. IG McWilliams, RA Dewer. In: DH Desty, ed. Gas Chromatography 1958. London: Butterworths Scientific Publications, 1957, p. 142.
14. DW Grant. In: DH Desty, ed. Gas Chromatography 1958. London: Butterworths Scientific Publications, 1957, p. 153.

15. GF Harrison, P Knight, RP Kelley, MT Heath. In: DH Desty, ed. Gas Chromatography 1958. London: Butterworths Scientific Publications, 1957, p. 216.

16. RPW Scott. In: DH Desty, ed. Gas Chromatography 1958. London: Butterworths Scientific Publications, 1957, p. 287.

17. AJP Martin. In: DH Desty, ed. Vapour Phase Chromatography. London: Butterworths Scientific Publications, 1957, p. 1.

18. RPW Scott, GW Girling, NB Coupe. J Appl Chem 11:335, 1961.

19. RPW Scott, GSF Hazedean. In: RPW Scott, ed. Gas Chromatography 1960. London: Butterworths Scientific Publications, 1957, p. 144.

20. RPW Scott, CA Cummings. In: RPW Scott, ed. Gas Chromatography 1960. London: Butterworths Scientific Publications, 1957, p. 117.

21. DH Desty, A Goldup, CJ Geach. In: RPW Scott, ed. Gas Chromatography 1960. London: Butterworths, London, 1958, p. 156.

22. L Ongkiehong. In: RPW Scott, ed. Gas Chromatography 1960. London: Butterworths, 1958, p. 9.

23. JE Lovelock. In: RPW Scott, ed. Gas Chromatography 1960. London: Butterworths, 1958, p. 9.

24. RPW Scott, IA Fowliss, D Welti, T Wilkens. In: AB Littlewood, ed. Gas Chromatography 1966, London: The Institute of Petroleum, 1966, p. 318.

25. RPW Scott, DWJ Blackburn, T Wilkins. J Gas Chromatogr April:183, 1967.

26. AT James, JR Ravenhill, RPW Scott. Chem Ind 746, 1964.

27. RPW Scott, JF Lawrence. J Chromatogr Sci 8:65, 1970.

28. AJP Martin, RPW Scott, T Wilkins. Chromatographia 2:85, 1969.

29. RPW Scott, CG Scott, M Munroe, J Hess, Jr. The Poisoned Patient: The Role of the Laboratory. New York: Elsevier, 1974, p. 395.

30. RPW Scott, P Kucera. J Chromatogr 169:51, 1979.

31. M McCann, JH Purnell, CA Wellington. Proceedings of the Faraday Symposium, Chemical Society, 1980, p. 83.

32. ED Katz, K Ogan, RPW Scott. J Chromatogr 352:67, 1986.

33. ED Katz, CH Lochmüller, RPW Scott. Anal Chem 81(4):349, 1989.

34. SA Beres, CD Halfmann, ED Katz, RPW Scott. Analyst 112:91, 1987.

35. GJ Schmidt, RPW Scott. Analyst 110:757, 1985.

7

Gas Chromatography: A Personal Retrospective

John V. Hinshaw

ChromSource, Inc., Franklin, Tennessee

I. INTRODUCTION

The field of chromatography celebrates its centennial multiple times during this first decade of the third millennium. In 1901 M. S. Tswett completed his Master's thesis on plant physiology in St. Petersburg, Russia, a period that he later described as the source of his discovery of chromatography. Shortly thereafter he took a position at Warsaw University, and in 1903 he published his first paper on chromatography followed by additional articles in 1906 and culminating in the 1910 publication of his doctoral thesis, "Chromophylls in the Plant and Animal World." Thus, the next 10 years will provide ample opportunity to commemorate the beginnings of chromatography and the achievements of the hundreds of scientists who developed Tswett's method over the past century into today's widespread chromatographic techniques.

It is impossible to recognize all who played a role in these developments in a single chapter, so I present a personal retrospective of one scientist's experiences with gas chromatography (GC), beginning 70 years after Tswett's initial work, but only 20 years after the initial reduction to practice in 1951 of partition gas chromatography by James and Martin [1]. During the first two decades of partition gas chromatography—the 1950s and 1960s—chromatographers created and refined most of the theoretical and practical groundwork for GC as we now perform it at the beginning of the twenty-first century. Packed and open-tubular columns, split and on-column inlet systems, temperature and pressure/flow programming, and a wide variety of detectors can all trace their beginnings back to this time [2]. Yet the instruments of the first two decades— some of which are still in use—fell far short of the capabilities that we now take for granted. Significant advances in sensitivity, speed, resolution, accuracy, repeatability, selective detection, automation, and the range of compatible samples are the hallmarks of modern gas chromatography. Advances in chromatographic research, coupled with material, mechanical, electrical, and software engineering developments, have engendered these changes over the years. The statement has sometimes been made that nothing new has been achieved in gas chromatography in recent years: the truth is that everything is being done better than ever before. Far from suffering of old age at its centenary, chromatography continues to develop at a rapid pace.

II. EARLY INVOLVEMENT

I first become involved with the practical aspects of gas chromatography as an undergraduate student pursuing a Bachelor of Science degree in organic chemistry. I had commenced a student internship with Burroughs-Wellcome (now Glaxo-Wellcome) in Research Triangle Park, North Carolina, in August 1971. After several months of effort synthesizing candidate substances for anti-inflammatory testing, by January 1972 my supervisor and I had obtained an apparently pure, white crystalline material that produced the correct NMR and IR spectra. Its melting point, however, was too broad, so she suggested using GC to investigate potential impurities. I had used an isothermal packed column GC briefly the year before, as part of an instrumentation course in college, so I proudly proclaimed that I knew how to use the instrument presented for this analysis. I demonstrated my competence by dissolving the crystals in hexane, setting up the instrument, and deftly injecting a microliter of solution.

At this point the analytical lab manager and my supervisor evidently concluded that I really did know what I was doing, and they left me to finish the job. I set my coffee cup on top of the instrument and settled down to wait for the chromatogram to be eluted. As the peaks began to appear on the chart recorder, they seemed too broad and were coming out too slowly. I remembered that my instrumentation instructor had mentioned that temperature programming the oven could speed up the analysis and sharpen the peaks. So, I stopped the run and, briefly consulting the manual, inserted some little color-coded pegs into the oven-programming panel on the front of the instrument.

Restarting the run, I found that additional peaks were eluted, and they were considerably sharper than before. The presence of more than one major peak meant that I had additional work to do, but I didn't mind. At the moment the programmed run ended, while deep in thought about just how I would now proceed with the synthesis, I was very surprised to see the top of the instrument rapidly move up, to release heat from the oven. Naturally, I wasn't quick enough to catch my coffee cup, which spilled directly into the flame ionization detector (FID) and created a plume of steam accompanied by the aroma of roasting coffee. Later that day, with the lab manager's assistance, I learned how to clean and rebuild a FID: my first practical experience with GC instrumentation.

In 1972, the science and practice of gas chromatography was highly developed. A trained gas chromatographer of that time would be quite comfortable with the benchtop instrumentation of the year 2000, except for the proliferation of computers into all aspects of the analytical laboratory. They would easily recognize the GC oven, inlets, columns and detectors, syringes, vials, septa, fittings, flow meters, and gas purifiers. They would have a harder time with electronic gas-control modules, but keypad and display operation might be intuitively accessible if they had a fair amount of experience with the hand-held calculators of their time. The operation of a networked PC data system would impose a steeper learning curve, but the principles of data acquisition, peak identification, and quantitation wouldn't have changed much.

What yesterday's chromatographer would find significantly changed at the start of the new century is the detailed practice of gas chromatography: refinements in methods, instrumentation, and techniques. The role reversal of open-tubular and packed columns, the myriad of specialized sampling and injection techniques, the proliferation of benchtop mass spectrometric detectors, and the miniaturization of fieldable, self-sustained instruments, to name a few significant developments, reflect the contributions and combined experience of hundreds of researchers, engineers, and experimenters.

III. CHIRAL SEPARATIONS

After my initial experience with gas chromatography as a work-study intern, I completed the Bachelor's of Science Degree in Organic Chemistry in 1973 and went to work at a well-known specialty chemical manufacturer. My group's job was to synthesize chemical candidates for cancer screening under a contract with the then-called National Cancer Institute (NCI), now part of the National Institutes of Health (NIH). We would receive small batches of substances originally obtained from laboratories around the country that had passed an initial anticarcinoma activity test: more material was required for further testing. The materials had an identity, a structure, a synthetic route assigned by its originator, and a set of physical characteristics including spectra, chromatograms, and other data. These pieces of information were not always consistent, and the specified synthesis did not always produce an identifiably exact replica of the original, which was rarely a pure substance in the first place. Thus, we would have to devise alternate synthetic routes and demonstrate the identity of the material we produced. More often than not, we would employ small-scale preparative liquid chromatography (LC) in 0.5–1.5 m gravity-fed columns, collecting fractions and testing each by thin-layer chromatography (TLC) for the presence of one or more substances. We relied heavily on TLC, along with infrared (IR), nuclear magnetic resonance (NMR), and mass spectrometry (MS) to characterize our synthetic materials. We had two GCs at our disposal, only one of which was equipped with a temperature-programmable oven and FID—the other was a small isothermal unit with only a thermal conductivity detector (TCD). The analytical lab would run our unknown samples on two different packed columns—polar and nonpolar—and we would receive zero, one, or two chromatograms in return, depending on how successful the lab technician was in getting our materials to be eluted from either of the columns and detected. In truth, however, GC was a minor player in these endeavors. We had no open-tubular (capillary) column instrumentation, no GC–MS system, and no chromatographic data-handling computers, or, for that matter, any laboratory computers at all.

That job had its risks and hazards. Chemical vapors were constantly in the air, and explosions were fairly common. After several harrowing incidents, one in particular involving the ignition in the lab next to mine of a mixture of air, diethyl ether, hydrogen, and diborane with spectacular results, I applied to graduate school in order to return to a less hazardous existence behind ivied walls.

In the fall of 1975 I matriculated in the Graduate School Chemistry Department at Duke University. I had been admitted on the strengths of my organic chemistry skills, but I soon switched to analytical chemistry and joined C. H. Lochmüller's group there. His research interests included chiral chromatographic separations, and I found a place synthesizing and characterizing a series of liquid crystalline carbonyl-*bis*(amino acid ester) homologs that exhibited high GC selectivities for stereoisomeric compounds volatile enough to be eluted.

The significance of chiral separations of amino acids, as well as of stereoisomeric drugs such as ephedrines and epinephrines, was well understood by the 1970s, and researchers in the field, notably Gil-Av and coworkers [3,4], had demonstrated effective chiral separations of the trifluoroactyl (TFA) derivatives of amino acid esters on various derivatized amino acid and dipeptide stationary phases, as well as on carbonyl-*bis*(amino acid ester) compounds. The latter were found to possess liquid crystalline characteristics over defined temperature ranges [5], in which state they exhibited greatly enhanced separating power for enantiomeric pairs of amino acid and chiral amine derivatives. One immediate goal in synthesizing these materials was to extend their liquid crystalline—and therefore gas chromatographic—temperature ranges; another goal was to better understand the ways in which chiral compounds interact. An activity–structure relationship existed that yielded high separation factors ($\alpha > 2.5$) for even-numbered members of a homologous solute series, but almost no separation for odd-numbered members [6–8]. This line of investigation was eventually abandoned due primarily to the limited and low temperature ranges of the liquid crystalline materials, but the structure-activity relationships implied that a three-point or three-dimensional solute interaction with the stationary phase was responsible for the unusually strong correlation of carbon chain number and separation factor.

In order to characterize these compounds by GC, I learned how to make columns, both packed and capillary. Coating supports and filling packed columns were well developed practices, but making glass capillary columns was a different story. We had a commercial glass capillary drawing machine (Fig. 1), based on a design by Desty [9], that had seen better times, and one of the first tasks that fell to me was to reengineer it. I rebuilt the motor-and-pulley drives that fed the stock tubing into and out of the high-temperature oven, made a new oven with improved heating, and modified the capillary tubing coiling device that finished the process. I drew hundreds of meters of borosilicate capillary tubing with the rejuvenated machine, which required nearly constant attention for useful results; sometimes I had to sit with it until late at night carefully adding graphite lubricant to the coiling tube to ensure that the coils didn't become tangled. Success was measured by the lack of broken capillary tubing on the floor. The glass tubing coils were rigid in the radial direction, although they flexed somewhat along the longitudinal axis. In order to connect the tubing to the GC inlet and detector, the column installer had to straighten the ends of the tubing by heating it to the softening point and allowing it to straighten out under gravitational influence. This required a steady hand, and many columns finished the installation process tens of centimeters shorter than they started.

Glass capillary column tubing often required surface pretreatment prior to coating with a stationary phase. Not only were there impurities at the column inner surface that interacted unfavorably with many solutes, but the stationary phase did not always form a smooth stable coating. In particular, the low molecular weight material I used did not form a stable film on the column wall, and initial attempts gave columns with poor efficiency. A number of stabilization and deactivation techniques had been described up to that time, which involved etching the column wall or depositing inert material on it, as well as chemical deactivation treatments. I turned to the work of the Grobs [10], which employed barium carbonate crystal deposition

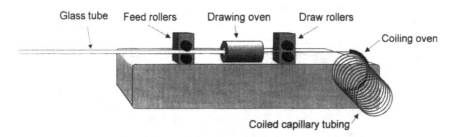

Figure 1 Diagram of a glass-capillary drawing machine. A pair of rollers feeds a glass tube—approximately 1 m long, 7 mm outer diameter, and 2 mm inner diameter—at a controlled rate into a furnace at about 800–1200°C, depending on the softening point of the glass. A second pair of rollers draws the capillary tube out of the furnace at a higher rate, the ratio of which to the feed rate was called the drawing ratio. The capillary tube proceeds into a resistively heated semicircular tube that softens the capillary enough to bend it into a spiral form, but not enough to distort its cross-sectional shape.

on the column wall for stationary phase stabilization, and followed it with surface pretreatment using a nonextractable layer of Carbowax 20-M [11]. Coating the stationary phase onto the inner wall by pushing a solution through with a drop of mercury, the dynamic coating method [12] gave sufficiently high efficiencies. I also tried static coating [13], but found it too slow for my purposes. With some work I became fairly proficient at producing capillary columns that usually exhibited better than 50% coating efficiencies, and which were stable long enough for a complete evaluation. These types of machinations were typical of capillary column preparation procedures at the time.

The very limited temperature range of the liquid crystalline phases severely restricted the solutes that could be separated and eluted successfully. Several amino acid–substituted polysiloxanes were investigated at the time for their potentially higher thermal stability, as well as improved capillary column coating behavior, compared to the lower molecular weight carbonyl-*bis*(amino acid ester) and dipeptide phases. Bayer et al. [14] in particular reported a valine-substituted polysiloxane that later became commercialized under the trade name "Chirasil-Val." Another approach attached chiral amino acid side chains as the carbonyl-*bis*(amino acid ester) to a 3-aminopropyl methyl polysiloxane backbone in a peptide-like arrangement [15]. A chiral separation on one such phase appears in Fig. 2.

At present, chiral GC separations are carried out primarily on a variety of derivatized α-, β-, or γ-cyclodextrin stationary phases [16–18]. The cyclodextrins are crystalline, toroid-shaped cyclic polyglucosides that exhibit enantiomeric selectivity for a wide variety of chiral compounds. The sugar-hydroxyl groups are chemically accessible for alkylation or trifluoroacetyla-

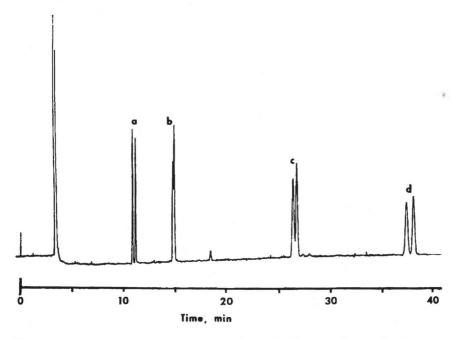

Figure 2 Separation of enantiomers on a chiral polysiloxane stationary phase. Sample: the series $C_6H_5(CH_2)_nCH(CH_3)NHCOCF_2CF_3$; a, $n = 0$; b, $n = 1$; c, $n = 2$; d, $n = 3$. Column: 40 m long × 0.23 mm i.d. glass column coated with a 32% peptidized 3-aminopropyl-dimethylsilicone copolymer. Temperature: 140°C. Carrier gas: helium. Split injection with flame ionization detection. (Reprinted courtesy of Elsevier Science Publishers.)

tion to produce a wide range of selectivity. For capillary column applications, the cyclodextrins are suspended in a moderately polar polysiloxane stationary phase, which establishes a stable coating and lends additional selectivity to the separation. In a recent application, enantioselective GC separations of sulfur-containing fruit lactones were coupled with olfactory evaluation [19].

IV. FUSED-SILICA COLUMNS

In June 1979, word of the invention of flexible fused-silica columns reached the graduate school. Earlier, Desty [20] and others had realized that fused silica or natural quartz could present a more inert surface than borosilicate glass, but they attempted to use silica tubing with relatively thick walls, which was more friable than the glass capillaries, and the concept did not proceed further. At the third Hindelang Symposium, R. D. Dandeneau and E. H. Zerenner presented their work with flexible, thin-walled fused-silica tubing [21]. Not only was the tubing flexible and thereby much easier to work with than glass tubing, but the surface was more inert than borosilicate glass due to the lack of nonsilica components. The fused silica was not completely inert, however, and the surface chemistry presented a challenge, which was to find deactivation and coating techniques suitable to the new environment. Our group in graduate school acquired a sample of 25 meters of fused-silica tubing in July of that year, but I did not have the opportunity to use it before I graduated in August and left the university for a research job with Varian Instruments in Walnut Creek, California.

The opportunity to work with fused silica was not long in coming: my first assignment was to devise deactivation and coating techniques for nonpolar fused-silica GC columns. To do so I followed in the footsteps of earlier workers, employing high-temperature dehydration, acid washing [22,23], silane surface deactivation [24], and static coating techniques [13] that resulted in a series of suitable highly efficient polydimethylsiloxane columns. This being said, one of the more significant early problems with fused-silica columns was their tendency to break spontaneously. The silica surface is prone to microflaws, as is any glass. But fused-silica tubing is inherently straight: coiling it into a 15 cm radius introduces stresses that help propagate minor flaws into significant defects. In production, immediately after drawing, the tubing is protected by the application of one or more layers of a protective coating, which prevents subsequent handling and exposure to air from degrading the surface. The earliest fused silica columns were coated externally with silicone polymers that had a low temperature limit. Some time elapsed after the tubing was in commercial production until the external coating technology matured, however, and until then it was a common occurrence to open the GC oven door and find that a single column had fractured into a dozen or more pieces.

In theory, fused-silica columns had the potential for much higher operating temperatures than borosilicate glass columns. The nonsilica components in glass columns—mainly boron and aluminum oxides—could be removed from the inner surface by leaching and other deactivation techniques, but oxides entrained inside the column wall could migrate back up to the inner surface [25], increasingly so at elevated temperatures, and cause the reappearance of polar-solute activity and stationary phase decomposition. The lack of such impurities in fused-silica glass would suppress stationary phase decomposition and improve column lifetime. Initially, however, fused-silica column upper temperature limits were restricted by the thermal stability of the silicone external coating. A replacement for silicone coatings, polyimide, provided a temperature limit boost, but at temperatures much above 280°C the early polyimide coatings would start to decompose and become brittle; they would slough off the column to leave exposed fused silica surfaces. Polyimide coating techniques gradually improved and permitted higher temperatures, eventually reaching the limit of about 400–425°C for modern polyimide-coated fused-silica columns. Aluminum column coatings are an alternative to polyimide and provide an extended temperature limit up to the practical top end of GC—around 500°C, above which pyrolysis begins to take over.

A. High-Temperature GC

Gas chromatography has always been closely allied with the petroleum industry. Analysis of high molecular weight oils, tars, waxes, and residues is a natural extension that drives GC systems toward higher temperatures. Other high-temperature applications include the analysis of naturally occurring fats and oils. The availability of high-purity column material that could withstand elevated temperatures was essential to the development of high–temperature GC, and with such tubing on hand, the stationary phase, not the tubing, limits the upper operating temperature. Development of high-temperature stationary phases occurred in parallel with the extension of fused-silica tubing upper temperature limits, and in the 1980s increasingly higher limit columns became available.

In 1983 I accepted a position at Perkin-Elmer, where I was occupied at first with GC applications. Soon, however, I became involved in the development of programmed-temperature inlet systems that possessed much less mass discrimination against high molecular weight compounds than conventional split/splitless inlets; high-temperature columns became essential for characterizing inlet performance.

High-temperature stationary phases rely on three essential characteristics. First, the phase must possess an elevated average molecular weight and a very low concentration of smaller oligimers, so that its vapor pressure remains as low as possible at the higher temperatures, thereby limiting column bleed and stationary phase losses. Cross-linking between polymer units in the stationary phase had already been applied to stationary phase chemistry on glass columns [26], and its use for high-temperature fused silica columns was a natural extension. In addition, the phase chemistry itself had to be stable at elevated temperatures. Polysiloxanes with methyl and phenyl substitution and polysilphenylene-siloxane copolymers are two stationary phase chemistries widely used for high-temperature applications. Polyglycol phases, such as Carbowax, are not suitable. Second, the stationary phase should be bonded to the column inner surface for further stability [27], and this rapidly became a standard feature for fused-silica columns based on silicone chemistry. Third, the stationary phase must be free of impurities and chemical residues from the manufacturing process, which could otherwise catalyze stationary phase decomposition. Free radical–catalyzed cross-linking reactions that use peroxides or azo compounds [28] leave behind relatively neutral and volatile byproducts, which are easily eliminated from the column during its initial conditioning.

The availability of high-temperature fused silica columns makes possible separations such as that shown in Fig. 3 [29]. Here, a sample of a polyethylene was separated with hydrogen carrier gas at high linear velocity on a short, thin film polydimethylsiloxane column. Peaks can be seen up to C_{100} and higher. Today, such high-temperature separations are used routinely for simulated distillation and hydrocarbon distribution analyses as well as in quality control for edible fats and oils. Recent applications of high-temperature GC include the analysis of crude extracts of propolis [30] and nonionic surfactants [31].

B. Programmed Temperature Inlet Systems

Conventional heated injection systems do not always quantitatively transfer sample components from the syringe into the column. A number of side effects can occur, including the loss or gain of heavier components relative to lighter ones, which is termed mass discrimination; thermally

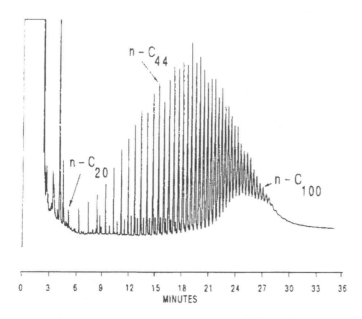

Figure 3   Separation of Polywax 1000 polyethylene on an 8 m long, 0.25 mm i.d., 0.1 μm film dimethyl-polysiloxane column. Column program: 50°C initial at 15°C/min to 430°C. Carrier gas: hydrogen at 25 psig, initial average carrier gas velocity 222 cm/s (calculated). Inlet: PTV split mode, 50°C initial to 440°C with ballistic heating. Injection: 0.5 μL of a 1 mg/mL solution, split 20:1. FID: 450°C, attenuation ×8. (Courtesy of Preston Publications.)

catalyzed decomposition of sensitive compounds such as pesticides or triglycerides; the appearance of decomposition products in the chromatogram; and poor run-to-run reproducibility. Many of these side effects have been attributed to the rapid sample vaporization that occurs in a hot inlet system and associated problems with sample transport inside the inlet on the way to the column entrance. For a comprehensive treatment of this subject, the reader should refer to the book by Grob [32].

Programmed-temperature (PT) injection, in which the inlet temperature increases postinjection in order to volatize sample, divides the sample injection process into two steps: syringe-to-inlet transfer and inlet-to-column transfer. When carried out at temperatures close to or below the solvent boiling point, PT injection avoids flash sample vaporization, and many of the concomitant side effects are suppressed or disappear entirely. The technique was originally described by Vogt et al. [33], and was subsequently developed by Poy et al. [34], Schomburg et al. [35], and others. Cold on-column injection, in which the entire sample enters the column prior to vaporization, also alleviates many flash-vaporization side effects, but it is not always suitable for materials containing high-boiling residue that can accumulate at the beginning of the column and affect retention. Programmed-temperature split injection gives the added flexibility of an adjustable split ratio, making possible the injection of relatively concentrated samples that would otherwise have to be diluted for on-column injection. In PT split or splitless injection, sample residue accumulates primarily in the removable inlet liner, thereby reducing the magnitude of this particular problem.

In addition, the separation of the heated inlet zone from the column oven itself makes it possible to perform cold injection into a relatively hot oven, as shown in Fig. 4. In this case,

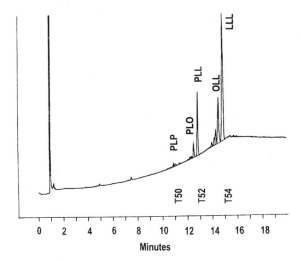

Figure 4 Separation of safflower oil triglycerides on a 25 m long × 0.25 mm i.d. × 0.12 μm film high-temperature 65% (phenyl) methylsiloxane column. Column program: 300°C initial at 4°C/min to 360°C. Carrier gas: hydrogen at 18 psig constant pressure. Inlet: PTV split mode, 50°C initial, ballistic heating *to* 400°C. FID: 450°C. T50, T52, T54 refer to the total number of carbons in the triglyceride side chains consisting of P = palmitin, O = olein, L = linolein. (Courtesy of Hüthig.)

the sample contains no peaks of interest that elute below 330°C. The total analysis time per sample is reduced considerably by not having to cool the oven down to the solvent boiling point prior to injection; only the inlet zone itself need be cooled. For labile compounds such as triglycerides, programmed-temperature injection suppresses thermal breakdown in the inlet and gives quantitative results similar to cold on-column injection [36,37]. Care must be taken when using programmed-temperature splitless injection to allow sufficient splitless sampling time for all sample components to transfer into the column; for most high-boiling mixtures, PT split injection is preferred [38]. The utilization of PT split injection in milk-fat triglycerides analysis by a modified European Union (E.U.) standard procedure has recently been reported [39,40].

Conventional hot splitless injection is used often for trace-level analysis, but it accommodates only a limited amount of injected sample. An excess amount of solvent entering the column leads to peak distortion from solvent flooding [32], and too much solvent in the inlet can produce mass discrimination against later-eluting components [41]. With PT splitless injection, the ability to change the inlet temperature postinjection also makes it possible to preseparate solvent from sample. The inlet is maintained close to the solvent boiling point in the split mode during and immediately after injection; solvent evaporation from the packed inlet liner occurs preferentially over semi-volatile solute vaporization. After the bulk of solvent has been eliminated through the split vent, the inlet is switched to the splitless mode—all flow from the inlet liner now passes into the column—and the solutes remaining in the inlet liner are vaporized by heating the inlet. This type of solvent-purged injection was described by Vogt et al. [42] originally and has been developed by many researchers since [43,44]. It permits the injection of sample volumes on the order of 20–150 μL but is limited by the partial or complete evaporation of solutes with volatilities close to the solvent. Optimization of liner dimensions [44] and packings [45] gives improved performance with larger injected volumes and with polar analytes. The related vapor-overflow technique relies on solvent expansion-driven flow to remove solvent from the inlet liner instead of carrier-gas purging [46]. PT large-volume injection has been applied recently to environmental trace analyses [47] and to the analysis of in situ methylated organic acids and phenols [48].

C. Selectivity Tuning

In addition to the advantages over glass columns detailed above, fused silica columns are much more easily interconnected. Their inherently straight nature, their strength, and their flexibility make them ideal for coupled column experimentation using either low dead-volume fittings or more recently press-fit glass connectors. In the mid-1980s we experimented with coupled pairs of fused silica columns configured so that the midpoint pressure at the junction of the columns was controlled independently of the inlet pressure. The effect of varying the carrier flow through the coupled columns was to modulate the observed overall selectivity of the column pair. If a polar and nonpolar column are coupled, for example, the longer the portion of a peak's complete retention time that it spends in the polar column, the more polar in nature the resulting composite separation will be. The selectivity of the column pair towards solutes traversing both columns is related to the average linear velocities or the unretained peak times by the following simple relationships:

$$\phi_2 = \frac{\bar{u}_1}{\bar{u}_1 + \bar{u}_2} = \frac{L_1\, t_{M2}}{L_2\, t_{M1} + L_1\, t_{M2}}$$

$$\phi_1 = 1 - \phi_2 \tag{1}$$

where ϕ_n represents the apparent contribution of column n to the overall polarity of the coupled column system. Earlier, researchers had investigated the characteristics of mixed stationary phases [49] and coupled packed columns [50]. Subsequently a great many workers reported on such systems; an extensive list of papers up to 1986 appears in Ref. 51. Figure 5 shows a simplified diagram of a coupled ''tunable'' dual column system in which excess carrier gas from the first column is relieved to vent, when necessary, to maintain the set midpoint pressure. Figure 6 illustrates a series of overlaid chromatograms obtained by incrementally varying the

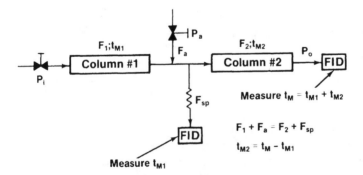

Figure 5 Diagram of a coupled ''tunable'' dual column system. P_i = inlet pressure, P_a = midpoint pressure, P_o = outlet pressure; F_n = flow in column n, F_a = input midpoint flow, F_{sp} = outlet split flow. (Courtesy of Vieweg Publishing.)

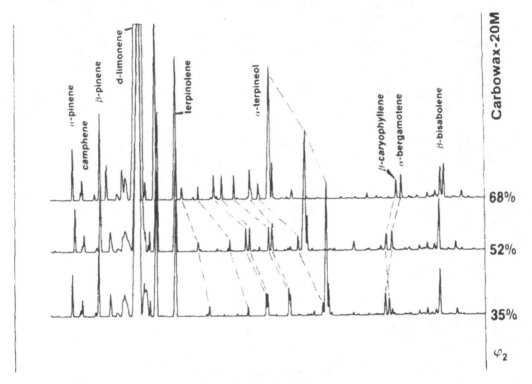

Figure 6 The effect of selectivity tuning on the temperature-programmed separation of lime oil components. Column dimensions: 25 m long × 0.25 mm i.d. × 0.25 μm film. Stationary phase: column 1, methylsilicone; column 2, Carbowax 20M. Temperature program: 50°C initial, at 8°C/min to 230°C. Carrier gas: helium. ϕ_2 refers to the relative fraction of time spent in the Carbowax column at the initial temperature, as determined by the measured unretained peak times for the first column (t_{M1}) and the coupled column pair (t_M); see Fig. 5 and Eq. (1). (Courtesy of Vieweg Publishing.)

relative linear velocities in a coupled methylsilicone and Carbowax 20M column pair for a natural oil. As the overall polarity of the system increases, the more polar peaks migrate towards longer retention times relative to the less polar peaks. This technique, sometimes termed ''multichromatography'' [51,52], is distinguished from multidimensional chromatography by the continuous nature of the elution process, that is, the lack of a trap and release stage. The principle of combining column polarities can also be applied to columns with mixed stationary phases. Recently, column manufacturers have pursued mixed-phase ''designer'' GC columns by devising stationary phases with mixed functionality that yield separations customized for specific applications.

   Developments in coupled columns since 1986 comprise a virtual explosion in work on ''comprehensive'' multidimensional GC [53] and in tunable GC systems [54]. The pioneering work of John Phillips in these areas was recognized recently in a special edition of *The Journal of High Resolution Chromatography* [55]. Recent research into the use of computer programmable pneumatics has applied tunable column couples to high-speed separations and time-of-flight mass spectrometry (TOFMS) [56].

V. PRESENT AND FUTURE

I have described the development and present status of gas chromatography for a few areas in which I have had some experience. This is not a comprehensive record, but it may reveal how an idea that traces it origins back to 1901 has grown and transmuted into a myriad of different but related techniques under the influence of the need to determine significant chemical substances. Chromatography, and in particular gas chromatography, will continue to grow and change over the ensuing decades. In GC, there is potential for the practical realization of even higher speeds and lower limits of detection, although these two improvements may not sit together in the same implementation. For enhanced resolution, gas chromatographers will increasingly turn to multidimensional techniques in combination with spectrometric detectors in order to unravel progressively more complex samples and their matrices. A new impetus towards miniaturization may arise as portable devices acquire more sophisticated wireless networked interfaces that make them part of the developing information web that we know today as the Internet.

REFERENCES

1.   AT James, AJP Martin. Biochem J 50:679–690, 1952.
2.   LS Ettre. The development of gas chromatography. J Chromatogr 112:1–26, 1975.
3.   E Gil-Av, B Fiebush, R Charles-Sigler. Tetrahedron Lett 1009, 1966.
4.   E Gil-Av, B Feibush. Gas chromatography with optically active stationary phases. Resolution of primary amines. J Gas Chromatogr 5:257–260, 1967.
5.   CH Lochmüller, RW Souter. Direct gas chromatographic resolution of enantiomers on optically active mesophases. I. J Chromatogr 87:243–245, 1973.
6.   CH Lochmüller, RW Souter. Direct gas chromatographic resolution of enantiomers on optically active mesophases. II. J Chromatogr 88:41–54, 1974.
7.   CH Lochmüller, JV Hinshaw, Jr. Direct gas chromatographic resolution of enantiomers on optically active mesophases. III. J Chromatogr 171:407–410, 1979.
8.   CH Lochmüller, JV Hinshaw, Jr. Direct gas chromatographic resolution of enantiomers on optically active mesophases. IV. J Chromatogr 178:411–417, 1979.
9.   DH Desty, JN Haresnape, BHF Whyman. Construction of long lengths of coiled glass capillary. Anal Chem 32:302–304, 1960.
10.   K Grob, G Grob. A new, generally applicable procedure for the preparation of glass capillary columns. J Chromatogr 125:471–485, 1976.
11.   DA Cronin. The preparation of stable glass capillary columns coated with Carbowax 20M. J Chromatogr 97:262–266, 1974.
12.   C Schomburg, H Hussman, F Weeke. Preparation, performance and special applications of glass capillary columns. J Chromatogr 99:63–79, 1974.
13.   GAFM Rutten, JA Rijks. Technique for static coating of glass capillary columns. J High Resol Chromatogr 1:279–280, 1978.
14.   H Frank, GJ Nicholson, E Bayer. Angew Chem 90:396–398, 1978.
15.   CH Lochmüller, JV Hinshaw, Jr. Direct gas chromatographic resolution of enantiomers on optically active mesophases. V. J Chromatogr 202:363–368, 1980.
16.   V Schurig. Enantiomer analysis by complexation gas chromatography. Scope: merits and limitations. J Chromatogr 441:135–153, 1988.
17.   Y Li, H Jin, D Armstrong. 2,6-Di-O-pentyl-3-O-trifluoroacetyl cyclodextrin liquid stationary phases for capillary gas chromatographic separation of enantiomers. J Chromatogr 509:303–324, 1990.
18.   J Krupĉík, I Špánik, P Oswald, I Skaĉáni, FI Onuska, P Sandra. Evaluation of enantioselective interactions in direct separation of enantiomers by gas chromatography. J High Resol Chromatogr 21:197–199, 1998.

19. T Beck, A Mosandl. γ(δ)-Thionolactones—enantioselective capillary GC and sensory characteristics of enantiomers. J High Resol Chromatogr 22:89–92, 1999.

20. DH Desty. The origin, development and potentialities of glass capillary columns. Chromatographia 8:452–455, 1975.

21. RD Dandeneau, EH Zerenner. An investigation of glasses for capillary chromatography. J High Resol Chromatogr 2:351–356, 1979.

22. H Borowitsky, G Schomburg. Separation and identification of polynuclear aromatic compounds in coal tar by using glass capillary chromatography including combined gas chromatography-mass spectrometry. J Chromatogr 170:99–124, 1979.

23. MW Ogden, HM McNair. Hydrothermal treatment of fused silica capillary columns. J High Resol Chromatogr 8:326–331, 1985.

24. K Grob, G Grob, K Grob, Jr. Deactivation of glass capillary columns by silylation. Part I. J High Resol Chromatogr 2:31–35, 1979.

25. BW Wright, ML Lee, SW Graham, LV Phillips, DM Hercules. Glass surface analytical studies in the preparation of open tubular columns for gas chromatography. J Chromatogr 199:355–369, 1980.

26. W Jennings, K Yabumoto, RH Wohleb. Manufacture and use of the glass open tubular column. J Chromatogr Sci 12:344–350, 1974.

27. K Grob, G Grob. Immobilization of vinylated OV-17 and OV-1701 based on combined surface bonding and crosslinking. J High Resol Chromatogr 6:153–155, 1983.

28. BW Wright, PA Peaden, ML Lee, T Stark. Free-radical cross-linking in the preparation of non-extractable stationary phases for capillary gas chromatography. J Chromatogr 248:17–34, 1982.

29. JV Hinshaw. Modern Inlet Systems for Capillary Gas Chromatography. J Chromatogr Sci 25:49–55, 1987.

30. A d-S Pereira, AC Pinto, JN Cardoso, FB de-A Neto, MF de-S Ramos, GM Dellamora-Ortiz, and E.P. Dos Santos. Application of high temperature high resolution gas chromatography to crude extracts of propolis. J High Resol Chromatogr 21:396–400, 1998.

31. C Asmussen and H-J Stan. Determination of non-ionic surfactants of the alcohol polyethoxylate type by means of high temperature gas chromatography and atomic emission detection. J High Resol Chromatogr 21:597–604, 1998.

32. K Grob. Split and Splitless Injection. Heidelberg: Hüthig, 1993.

33. W Vogt, K Jacob, HW Obwexer. Sampling method in capillary column GLC allowing injections of up to 250 µL. J Chromatogr 174:437–439, 1979.

34. F Poy, S Visani, F Terrosi. Automatic injection in high resolution GC: a programmed temperature vaporizer as a general purpose injection system. J Chromatogr 217:81–90, 1981.

35. G Schomburg, H Husmann, H Behlau, F Schulz. Cold sample injection with either split or splitless mode of temperature-programmable sample transfer. Design and testing of a new, electrically heated construction for universal application of different modes of sampling. J Chromatogr 279:251–258, 1983.

36. JV Hinshaw, Jr, W Seferovic. Programmed-temperature split-splitless injection of triglycerides: Comparison to cold on-column injection. J High Resol Chromatogr 9:69–72, 1986.

37. JV Hinshaw, Jr, W Seferovic. Analysis of triglycerides by capillary gas chromatography with programmed-temperature injection. J High Resol Chromatogr 9:731–736, 1986.

38. JV Hinshaw, LS Ettre. Aspects of high-temperature capillary gas chromatography. J High Resol Chromatogr 12:251–254, 1989.

39. S Banfi, M Bergna, M Povolo, G Contarini. Programmable temperature vaporizer (PTV) applied to the triglycerides analysis of milk fat. J High Resol Chromatogr 22:93–96, 1999.

40. M Povolo, E Bonfitto, G Contarini, PM Toppino. Study on the performance of three different capillary gas chromatographic analyses in the evaluation of milk fat purity. J High Resol Chromatogr 22:97–102, 1999.

41. JV Hinshaw. The effects of inlet liner configuration and septum purge flow rate on discrimination in splitless injection. J High Resol Chromatogr 16:247–253, 1993.

42. W Vogt, K Jacob, A-B Ohnesorge, HW Obwexer. Capillary GC injection system for large sample volumes. J Chromatogr 186:187–192, 1979.

43. K Grob, Z Li. PTV splitless injection of sample volumes up to 20 µL. J High Resol Chromatogr 11:626–632, 1988.

44. HGJ Mol, H-G Janssen, CA Cramers, UAT Brinkman. Large volume sample introduction using temperature programmable injectors: implications of liner diameter. J High Resol Chromatogr 18: 19–27, 1995.

45. JGJ Mol, PJM Hendriks, H-G Janssen, CA Cramers, UAT Brinkman. Large volume injection in capillary GC using PTV injectors: comparison of inertness of packing materials. J High Resol Chromatogr 18:124–128, 1995.

46. K Grob, C Siegrist. Determination of mineral oil on Jute bags by 20–50 µL splitless injection onto a 3 m capillary column. J High Resol Chromatogr 17:674–675, 1994.

47. HG-J Mol, M Althuizen, H-G Janssen, CA Cramers. Environmental applications of large volume injection in capillary GC using PTV injectors. J High Resol Chromatogr 19:69–79, 1996.

48. A Zapf, H-J Stan. GC analysis of organic acids and phenols using on-line methylation with trimethyl-sulfonium hydroxide and PTV solvent split large volume injection. J High Resol Chromatogr 22: 83–88, 1999.

49. RS Porter, RL Hinkins, L Tornheim, JF Johnson. Computer optimization of mixed liquid phases for gas chromatography. Anal Chem 36:260–262, 1964.

50. GP Hildebrand, CN Reilley. Use of combination columns in gas liquid chromatography. Anal Chem 36:47–58, 1964.

51. JV Hinshaw, Jr, LS Ettre. Selectivity tuning of serially-connected open-tubular columns in gas chromatography. Part I: Fundamental relationships. Chromatographia 21:561–572, 1986.

52. JV Hinshaw, Jr, LS Ettre. Selectivity tuning of serially-connected open-tubular columns in gas chromatography. Part II: Implementation and applications. Chromatographia 21:669–680, 1986.

53. JB Phillips, J Xu. Comprehensive multi-dimensional gas chromatography. J Chromatogr A 703:327–334, 1995.

54. H Smith, R Sacks. Pressure-tunable GC columns with electronic pressure control. Anal Chem 69: 5159–5164, 1997.

55. P Marriott, Guest Editor. J High Resol Chromatogr 23:165–266, 2000.

56. R Sacks, C Coutant, T Veriotti, A Grall. Pressure-tunable dual-column ensembles for high-speed GC and GC/MS. J High Resol Chromatogr 23:225–234, 2000.

8

The Evolution of Capillary Column Gas Chromatography: A Historical Overview

Leslie S. Ettre

Yale University, New Haven, Connecticut

The present is the future of the past and the past of the future.

I. INTRODUCTION

In the first years of gas chromatography one of the (now defunct) instrument companies liked to use a slogan in its advertisements, saying that "The column is the heart of the gas chromatograph." Indeed, this is true: after all, separation takes place in the column and even the best instrument cannot improve it.

At the beginning the separation columns used in gas chromatography (GC) consisted of 1–4 m long tubes with 1/4 in. (6.4 mm) outside diameter, packed with relatively inert support particles coated with a (liquid) stationary phase. Gas chromatography with such columns already represented orders-of-magnitude improvements as compared to other techniques previously available (e.g., low-temperature distillation). However, when researchers started to look more seriously into the theory of chromatographic separation, it soon became clear that the performance of the packed columns is far from the theoretical possibilities. As a conclusion much research was carried out with the aim to improve column performance. The invention of open-tubular (capillary) columns should be considered in this context.

The invention of capillary columns can be attributed to a single person—Marcel J. E. Golay. However, their continuous improvement is the result of scores of chromatographers around the world. Today, when capillary columns superseded the packed columns in almost every application and when they are routinely available, the activities of these pioneers are mostly forgotten.

The purpose of this report is to briefly go through the fascinating story of the evolution of capillary column gas chromatography. I will elaborate on the key problems the pioneers were facing and point out individual achievements; readers who are interested in more details are advised to consult the literature [1–5]. Since Golay carried out his investigations at the Perkin-

115

Elmer Corporation, with which I have also been associated for over 30 years, I will always mention our activities in the individual fields. The chromatograms used as illustration were selected from our work.

II. NOMENCLATURE

In this discussion we are going to use the two interchangeble expressions ''open-tubular'' and ''capillary.'' It should be realized that the latter term is ambiguous and sometimes misleading. After all, one can also have a packed column with a diameter in the ''capillary'' dimension; at the same time open-tubular columns may have a diameter outside the ''capillary'' range. Golay emphasized this ambiguity in a lecture presented at the 1960 Edinburgh Symposium [6], saying that ''it is not the smallness, it is the 'openness' of the open-tubular columns which permits us to realize a two orders of magnitude efficiency over the packed column.'' He continued by explaining that ''the 24-inch diameter gas pipe with an oil-coated inner wall, stretching from Texas to Maine'' may serve as a very good open-tubular column for high-boiling hydrocarbons; however, one certainly could not speak about this pipeline as a ''capillary'' column.

In spite of this clear understanding, popular usage defied its true meaning, and today the term ''capillary column'' is used almost universally. This is one of the misnomers that one can often find in science: after all, ''chromatography'' does not represent only the analysis of colored substances and a ''symposium'' is certainly much more than a drinking party, although the original meaning of these Greek words would suggest otherwise.

III. INVENTION

To start, we have to go back over 40 years to the heyday of gas chromatography. The first American gas chromatographs were introduced in the spring of 1955 by the Burrell Corporation and The Perkin-Elmer Corporation. About that time, Marcel J. E. Golay (Fig. 1) joined Perkin-Elmer as a consultant after a 25-year distinguished career at the U.S. Signal Corps Engineering Laboratories in Fort Monmouth, NJ. He was originally trained as an electrical engineer and mathematician at the Federal Technical University of Zurich, Switzerland, at that time the world's most prestigious technical school, and he received his Ph.D. at the University of Chicago in nuclear physics. Golay had a very broad range of interest and has worked in a number of fields. His connection with Perkin-Elmer was mainly due to his involvement in the development of an IR detector, originally conceived as an aircraft detecting device, and of a multislit IR spectrometer.

When Golay joined Perkin-Elmer everybody was excited by the versatility and the incredible separation power of gas chromatography, and inevitably he became involved in various discussions on the new technique, for him a totally unknown field. Encouraged by his new associates he tried to consider—first theoretically—the separation process taking place in the chromatographic (packed) column. Let us not forget that at that time, very little information was available on these questions: the plate theory of A. J. P. Martin only discussed the overall performance of the column, and the detailed treatise of Van Deemter et al. [7] was not yet available. Golay was particularly interested in the question of how close actual column performance is to the theoretical possibilities. In order to simplify the system, he constructed in his mind a model consisting of a bundle of capillary tubes, each corresponding to a passage through

Figure 1   M. J. E. Golay, around 1960. (Author's collection.)

the column packing. These ideal capillaries would not be restricted by the geometry of the packing or the randomness of the passages through it, which are beyond our control. Therefore, they should behave close to the theoretical possibilities. Golay's considerations were outlined in a number of internal reports from which the one dated September 5, 1956, was most important [8]. In this he suggested some experiments with a capillary tube, 0.5–1 mm in diameter, wetted with a suitable stationary phase (corresponding to one of these passages), to determine the correctness of his assumptions.

These experiments were carried out in the fall of 1956. A serious handicap was represented by the large volume of the then existing (thermal-conductivity) detectors and of the injection block. Therefore, Golay designed a micro thermal-conductivity detector and used this for the investigations. The results were indeed striking: Fig. 2 presents one of Golay's early chromatograms obtained on a 25 ft × 0.010 in (0.25 mm) i.d. column. Note the extremely fast separation, unsurpassed in the 42 years since.

During the next 16 months Golay carried out a very intensive theoretical study, in addition to some experimental work. An interim report was presented at the GC symposium organized by the Instrument Society of America and held in August 1957 in East Lansing, MI [9]. Golay's final report including the full theory of open-tubular (capillary) columns—which is still valid today, 42 years later—was presented at the next international GC symposium held in May 1958 in Amsterdam, The Netherlands [10].

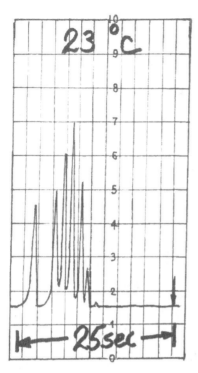

Figure 2 One of Golay's early chromatograms on capillary columns obtained in the winter of 1958–59. Column: 25 ft × 0.010 in. (0.25 mm) i.d. stainless-steel tube, coated with 2,5-hexanedione as the stationary phase. Column temperature: 23°C. Micro thermal-conductivity detector; Sanborn high-speed recording galvanometer. Sample: Phillips 37 hydrocarbon mixture; peaks (from right to left): 2-methylpropane, *n*-butane, 2-methylpropene + butene-1, butene-2 *trans*, butene-2 *cis*, butadiene-1,3. (Author's collection.)

IV. REALIZATION

Golay's treatise submitted to the Amsterdam Symposium with its 93 equations was impressive enough in itself. Still, in the original form published in the preprints of the lectures, it probably would have had little impact: it sounded too theoretical. However, in his actual presentation he showed two chromatograms obtained just a couple of days earlier by Richard D. Condon, his young associate at Perkin-Elmer (Fig. 3), showing the separation of C_8 hydrocarbons and of the xylene isomers, on a 150 ft × 0.010 in. (0.25 mm) i.d. stainless-steel column, coated with diisodecyl phthalate. Thirty years later, Dennis H. Desty (Fig. 4), the organizer of the Symposium (who in the subsequent years had a major role in the general use of capillary columns), still remembered the excitement caused by these chromatograms, demonstrating the exceptional separation power of capillary columns, up to then impossible to achieve on even the best packed column [11]:

> I well remember the gasp of astonishment from the audience at this fantastic performance that was to change the whole technology of gas chromatography over the next decade. We were all enraptured by the elegant simplicity of Marcel's concept and I could not wait to dash off to my laboratory to start experiments with the wonderful new tool.

Figure 3   Richard D. Condon (left) and Leslie S. Ettre. Photo was made at the International GC Symposium, Hamburg, June 1962. (Author's collection.)

As mentioned earlier, Golay had to build a special micro thermal-conductivity detector for his investigations because the existing detectors had a much too large volume and not enough sensitivity for the small sample sizes and low carrier gas flow rates needed. Fortunately, at the same Amsterdam Symposium, I. G. McWilliam and R. A. Dewar described in detail the flame-ionization detector (FID) [12], and within a short time James E. Lovelock modified his argon-ionization detector, making it suitable for capillary column work [13]. Thus soon, all the necessary basic ingredients were available for the practical use of capillary columns.

Figure 4   Dennis H. Desty, in the mid-1960s. (Author's collection.)

Most likely, the first to utilize a capillary column–FID system was Desty at British Petro-leum Co. Ltd. According to his personal recollections [14], they put together a crude set-up within a few days of returning from Amsterdam: it consisted of an FID and a 250-ft-long stain-less-steel tube coated with squalane, using the dynamic coating technique described by G. Dijks-tra and J. De Goey at the Amsterdam Symposium [15]; it also included a split injection system. Then within a short time his group, consisting of B. H. F. Whyman, A. Goldup, and W. T. Swanton, constructed a complete apparatus for operation up to 250°C and explored the separa-tion of a wide variety of samples using columns made of stainless-steel and copper tubes. Figure 5 is a photograph of their system.

Desty first reported on this system at a symposium held October 9–11, 1958, in Leipzig, East Germany [16]. Rudolf E. Kaiser, in his personal recollections [17], emphasized the impact of this presentation:

> I remember in great detail the atmosphere of total silence in an overcrowded lecture hall when (Desty) showed his over-100,000 theoretical plate gasoline run, using a 100-m metal capillary column and when he discussed the potentialities of glass capillary columns. At that instant, we were more-or-less converted to continuing our work with capillary columns.

In England the first major presentation by Desty on the potentialities of capillary columns for the solution of analytical problems in the petroleum industry was at the meeting of the Gas Chromatography Discussion Group held on April 10, 1959, in London [18].

Figure 5 The first capillary column–FID system constructed by the group of D. H. Desty at British Petroleum Co., in the fall of 1958. (From Ref. 14. Courtesy Elsevier Science Publisher.)

Figure 6 James E. Lovelock (left) and Albert Zlatkis at the University of Houston in 1960. (Author's collection.)

At Perkin-Elmer, prototypes of the FID were also constructed soon after the Amsterdam Symposium. I remember that when I joined the company in the fall of 1958, Dick Condon was already obtaining one excellent chromatogram after the other on a working prototype of a gas chromatograph with open-tubular columns and an FID. This instrument was then introduced at the Pittsburgh Conference, in March 1959, where Condon had a major presentation describing this system and illustrating the wide-range applications of capillary columns [19].

Parallel to but independent of our work, Albert Zlatkis at the University of Houston [20] and S. D. Lipsky at Yale University Medical School [21,22]—both with the help of Lovelock— explored the use of capillary columns (Fig. 6) and reported on their results in the first months of 1959. Due to its historical interest, the very first chromatogram obtained by Zlatkis on November 16, 1958, is shown in Fig. 7: besides Zlatkis and Lovelock it has the signatures of M. C.

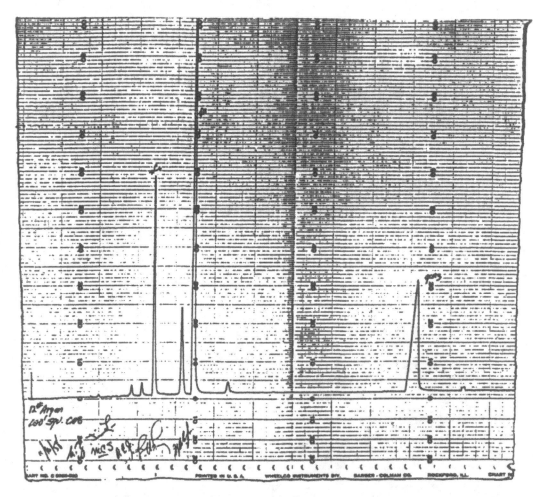

Figure 7   The first chromatogram obtained at the University of Houston on November 16, 1958. A 30.5-m-long stainless-steel capillary column coated with squalane was used with a micro argon-ionization detector. The three large peaks on the left-hand side correspond to *n*-hexane, benzene, and toluene. Identified signatures are (from the left): A. Zlatkis, J. E. Lovelock, M. C. Simmonds, R. E. Johnson, and H. Lilly. (Author's collection.)

Simmonds of Shell Development Co. and R. E. Johnson and H. Lilly of the now-defunct Barber-Colman Co., who were invited to witness the occasion.

V. METAL CAPILLARY COLUMNS

The very first "capillary column" investigated by Golay was actually an (uncoated) 10 m × 3 mm i.d. Teflon tube; however, he soon switched to glass and then to stainless-steel tubes of two internal diameters: 0.010 in. (0.25 mm) and 0.020 in. (0.51 mm). Such tubes could be easily obtained: as mentioned by Golay during the discussion of his Amsterdam paper, "you buy them by weight." Capillary columns made of stainless steel (and, to a lesser extent, copper) have been in general use for well over a decade.

The early stainless-steel columns (see Fig. 1) were made of thick-wall tubing (about 1.6 mm), and these were heavy and bulky. However, by 1962 thin-wall tubing (wall thickness: 0.12–0.15 mm) became available. This tubing was much more flexible and had a much improved quality.

Recently some statements could be found in the literature about the "terrible state of affairs" with metal capillary columns, implying that they had a very poor performance and a very short life time [24]. Such statements are gross exaggerations. It is important to state that properly coated metal capillary columns—with both nonpolar and polar stationary phases—have been successfully used for the analysis of a wide variety of samples, and this is well documented in the scores of publications from this period. I just want to show here two chromatograms from our own work, both from 1962: the fast analysis of a fatty acid methyl ester mixture (Fig. 8) and the analysis of a phenolic mixture (Fig. 9; Table 1).

With respect to the lifetime of the columns, it is sufficient to quote the statement of Halász during a discussion at the 1961 Lansing Symposium [27]: "With a copper column coated with squalane, we worked for about 7 months, eight to ten, sometimes for 24 hours daily." Naturally, metal capillary columns had some inherent problems, but there were ways to overcome most of these limitations. Three such problems may be mentioned here.

A. Activity of the Tubing

Metal is not inert. If a sample consisting of paraffins or cycloparaffins was analyzed, or the column was coated with a polar phase, there was no serious problem: paraffins are not polar, while the polar groups of the stationary phase blocked the tube's active sites. However, in the case of a nonpolar or semi-polar phase and a polar sample, the active sites of the tubing could be detrimental, causing severe peak tailing or even sample loss through permanent adsorption or decomposition.

For the reduction of the activity of the inner tube surface, Warren Averill proposed in 1961 the addition of a small amount (about 1–2%) of a surface-active agent to the stationary phase [28]. A typical such additive is Atpet 80, which is chemically sorbitan monooleate:

$$CH_3(CH_2)_7CH{=}CH(CH_2)_7COOC_6H_8(OH)_5$$

The polar hydroxy groups at one end of the molecules are permanently adsorbed by the tube wall, deactivating its active sites. At the same time, the long hydrocarbon chain will remain free, resulting in a velvet-like structure, spanning the interface between the tube wall and the coated stationary phase film. As pointed out by F. Farré-Rius et al. [29], another effect of such additives may be that they reduce surface tension and thus facilitate the spreading of the station-

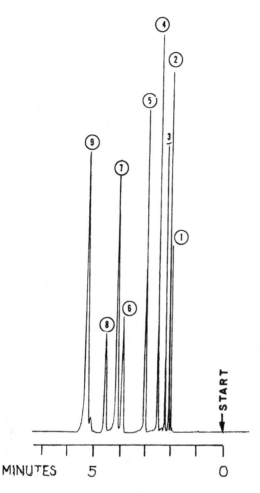

Figure 8 Fast analysis of a mixture of the methyl esters of C_8–C_{18} fatty acids. Column: 150 ft × 0.020 in. (0.51 mm) i.d. stainless-steel capillary, coated with butanediol succinate (BDS). Carrier gas: helium, inlet pressure: 10 psig. Column temperature: 200°C. Sample volume: 1μL, split 1:25. Flame-ionization detector. Peaks: methyl esters of 1 = caprylic, 2 = capric, 3 = lauric, 4 = myristic, 5 = palmitic, 6 = stearic, 7 = oleic, 8 = linoleic, 9 = linolenic acids. (From Ref. 25. Courtesy The Perkin-Elmer Corporation.)

ary phase. At Perkin-Elmer we routinely used such additives in the preparation of stainless-steel capillary columns.

It should be noted that the use of such additives is temperature limited: above 175–180°C they will soon bleed off. However, this fact did not represent a serious limitation, because at that time capillary gas chromatography was generally carried out below 200°C. Even in 1974, G. Schomburg gave the upper temperature limit for most (glass!) open-tubular columns as at or below 200°C [30].

B. Residual Coating

An interesting and little known problem with the metal tubing was that, as a result of the manufacturing process, it had a residual liquid coating on its inside surface, which could actually be

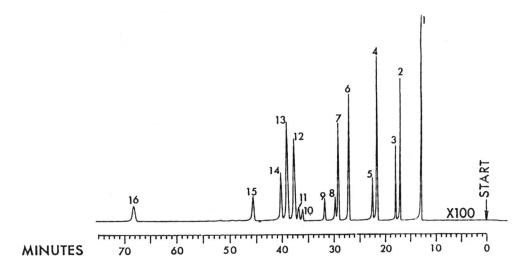

Figure 9 Analysis of a phenolic mixture. Column: 150 ft × 0.010 in. (0.25 mm) i.d. stainless-steel capillary, coated with di-*n*-decyl phthalate, with Atpet 80 additive. Carrier gas: helium, inlet pressure: 50 psig. Column temperature: 120°C. Split injection. Flame-ionization detector. For peak identification, see Table 1. (From Ref. 26. Courtesy The Perkin-Elmer Corporation.)

relatively significant. Porcaro, in 1963, indicated the presence of such a film consisting mostly of polyisobutylene [31], and its thickness could be computed to be as much as 0.32 μm. Because of this the metal tubing had to be cleaned prior to coating with the stationary phase. This cleaning procedure may have been fairly elaborate: for example, O. L. Hollis recommended successive washing with five solvents prior to coating [32]. It is possible that some of the complaints about the poor performance of self-made metal capillary columns may be traced to improper or no prior cleaning of the tubing.

C. Too-Thick Liquid Phase Film

It is well known that for very high column performance, very thin stationary phase coating is desired: column efficiency (the value of the HETP) is directly related to the square of the film

Table 1 Identification of the Peaks in Fig. 9

Peak no.	Compound	Peak no.	Compound
1	Phenol	9	2,4,6-Trimethylphenol
2	*o*-Cresol	10	2,3-Xylenol
3	2,6-Xylenol	11	2,4-Xylenol
4	*p*-Cresol	12	*p*-Ethylphenol
5	*m*-Cresol	13	*m*-Ethylphenol
6	*o*-Ethylphenol	14	3,5-Xylenol
7	2,4-Xylenol	15	3,4-Xylenol
8	2,5-Xylenol	16	3-Methyl-5-ethylphenol

thickness. In spite of this, however, the early metal capillary columns had a too thick coating. For example, Desty's first published chromatogram [33] was obtained on a 0.25-mm i.d. copper column with a 1-μm stationary phase film, and one can find in the early literature information on columns with a film thickness up to 2μm. In general there were two reasons for this situation.

The first was connected to the condition of the inner surface of the metal capillary tubing: it was not smooth. Thus, in the case of a too-thin coating some uncoated metal pieces might have protruded through the coating. Obviously the presence of such uncoated, hot metal in the open channel of the column would represent a disaster; therefore, one wanted to be sure that the metal surface is completely covered by the stationary phase coating. This problem was particularly evident with the early, thick-walled stainless-steel tubing: the thin-walled tubing already had a much improved quality.

The second reason for the too-thick film was connected to the coating technique used at that time. As discussed below, the dynamic coating technique of G. Dijkstra and J. De Goey [15] was used by practically everybody. Using this technique the film thickness depends on the concentration of the coating solution and the velocity with which it is pushed through the tubing. Desty, in his first investigations, used a 10% solution with a slow coating velocity, and almost everyone adopted these conditions, which generally resulted in a film thickness of 0.5–1μm. Such columns were fine for the analysis of low-boiling compounds; however, in the case of wide boiling-range samples, the analysis time was too long and column efficiency for the later eluting peaks suffered. Recognizing this problem, around 1964 we modified the conditions of the coating, halving the concentration of the coating solution and adjusting the coating velocity. As a result of this, our standard stainless-steel (and later, glass) capillary columns had a film thickness of about 0.25 μm. At the same time we also prepared columns with a film thickness of about 0.9–1.0 μm, permitting the analysis of low-boiling compounds without the need for cryogenic cooling at the start of the temperature program. (We identified these columns as 4×, i.e., having four times the standard thickness.)

It should be noted that today capillary columns with a wide variety of stationary phase film thickness are produced. These serve different purposes, e.g., the analysis of low-boiling compounds or increasing the sample capacity of the columns. However, the standard 0.25-mm i.d. columns used for the majority of analyses still have a film thickness of about 0.25 μm or less.

D. Nickel Capillary Columns

Because of the limitations of stainless steel (and copper), capillary tubes of other metals were also tried. Gold was one possibility, but it obviously was much too expensive [18]. Another possibility was to use nickel tubing and treat it by multiple solvent washing and acidic etching. A procedure was described in 1974 by the group of Zlatkis [34], who also showed chromatograms that were quite comparable to chromatograms obtained on glass columns. In spite of these good results, however, nickel tubing has never became popular: the etching procedure was too cumbersome and complicated.

VI. CAPILLARY COLUMNS MADE OF PLASTIC TUBING

In spite of the convenience of metal (stainless-steel) columns, it was obvious from the beginning that a more inert tube material would be needed. Plastic tubing was proposed by R. P. W. Scott [3], but there were obvious disadvantages with such materials: temperature limitations, poor coatability, and short life due to plasticizer migration. Thus, except for some early work, such columns never gained ground. However, one interesting work should be mentioned here that demonstrated the ultimate possibilities of open-tubular columns: the one-mile-long column made of Nylon tubing and coated with *n*-hexadecane, which was described in 1959 by Zlatkis and Kaufman [36]. The i.d. of the column was 1.7 mm, and nitrogen was used as the carrier gas with an inlet pressure of 2.3 atm (gauge) and a flow rate of 250 mL/min, which corresponded to an average linear velocity of 79.5 cm/s. According to the paper, one million theoretical plates were achieved, corresponding to an HETP of 1.61 mm, less than twice the tube radius. This was an excellent result, particularly since the carrier gas velocity was obviously much higher than optimum. It may be interesting to note that the gas hold-up time was about 36 minutes, and a peak with a retention ratio (*k*) of two emerged about 10 minutes after sample introduction.

VII. THE ERA OF GLASS CAPILLARY COLUMNS

An obvious choice of the tube material would have been glass, and this possibility was explored by Golay in his early work: in fact, in the publicity photo made around 1960 (Fig. 1) he was holding a glass capillary tube in his hands! At that time, a number of chromatographers tried to prepare capillary columns made of glass. However, this was not as simple as it sounds.

Usually the "era of glass" is considered to start with the development of an ingenious device to prepare glass capillary tubes by Desty and his coworkers in 1960 [37]; a similar device was also described at that time in France by A. Kreyenbuhl [38]. In 1960–61 Desty and his associates published a number of papers in which they used glass capillary columns, the most famous being the analysis of a Ponca Crude petroleum sample on a 263-m-long, 0.14 mm i.d. column, in 3½ hours [39]. Then, in the second part of the 1960s, Desty's machine became commercially available from Hupe & Busch (Karlsruhe, Germany) and a few years later also from Shimadzu (Kyoto, Japan). With these machines capillary tubes of various lengths and diameters could be prepared using both soda-lime and borosilicate (Pyrex) glass tubes. The capillary tubes produced in these machines had a thick wall, and their final form was that of rigid coils. Typical dimensions were: coil diameter, 13–15 cm, and column internal diameter, 0.23–0.27 mm, with a wall thickness of about 0.20–0.25 mm [40]. Naturally, glass capillary columns were fragile; in the hands of skilled operators, however, very little damage was done. Thus, one may ask, why did glass tubing replace metal only in the early 1970s, 10 years after the description of the glass drawing machine?

The problem resulted from the poor coatability and short life of glass capillary columns prepared in this period. The situation was well characterized by Halász in his (already quoted) remark at the 1961 Lansing Symposium [27]. While emphasizing the long life and good performance of metal columns, he continued by saying that "with glass capillary columns . . . coated with squalane, we were unable to work longer than 2 or 3 days. On glass columns coated with squalane, you can see with your eyes after two days that your film is not in one place."

It was years before the reason for this problem was understood: the strong cohesive forces of liquids on the glass surface. These forces are characterized by the surface tension, which in turn can be characterized by the contact angle of a drop on the solid surface: the higher the contact angle, the poorer the spreading of the liquid. The extent of this phenomenon was first investigated in 1962 by Farré-Rius and coworkers, who measured the contact angles of liquid phases on various potential column tube materials [29]. Later, J. Simon and L. Szepesy connected the problem to the changes occurring on the inner surface of the glass tube during and after drawing [41]. During drawing the surface is dehydrated and an oxide-type surface is formed. However, when cooled down in a normal atmosphere (i.e., containing water vapor) it will sorb water and in a few minutes is transformed into a hydrate form. This surface is autophobic against most organic compounds and cannot support organic layers. In order to be able to coat it with a stable film, the surface must either be kept in the oxide form, e.g., by continuous purging with a dry gas during and after drawing before coating, or chemically modified into a form whose critical surface tension is high enough to support a layer of the stationary phase.

As pointed out by Halász in 1961, squalane was a particularly bad substance to be coated on a glass capillary columns, and this experience had been repeated by many chromatographers using not only untreated but even treated glass tubing. It is, therefore, particularly interesting that Desty and his associates published a number of excellent chromatograms obtained on glass capillary columns coated with squalane. I recently asked his two former (now retired) associates, Dr. A. Goldup and W. T. Swanton, whether they used any "tricks" when coating their columns. Their uniform answer was that this was not the case. However, they usually coated the capillary tubing immediately after drawing, and thus it is possible that the inside surface was not yet hydrated. They recalled that if the capillary tubing had been stored for some time before coating, the result was often unsatisfactory.

The breakthrough in modifying the inner surface of glass capillaries to permit the preparation of a stable coating came in 1965–68 through the work of K. Grob [42–44], describing a way to deposit a carbon layer, and M. Novotný and K. Tesařik [45,46], who etched the internal surface of the tube with dry HCl or HF. In the following decade scores of different methods were developed for the treatment of the inside surface of the glass capillary tube. Such treatment was needed not only to improve the coatability, but also to make the tube more inert. People often do not realize that glass is not inert: metal or other ions in its composition could be detrimental and had to be eliminated. A particular problem was boron, present in fairly high concentration (13% as B_2O_3) in borosilicate glass [47].

It would be practically impossible to even mention the many methods developed to improve the glass capillary tubing: interested readers are referred to the detailed review of Novotný and Zlatkis dealing with the early period [48], the monograph of W. G. Jennings [49], as well as to a few selected books and review articles [30,50–55], which in turn provide scores of further references.

The decade of the 1970s was a most exciting period in the evolution of capillary columns. This was the time when these columns really started to became everybody's tool, and chromatographers even created a special acronym to describe their field, calling it $(GC)^2$ for Glass Capillary Gas Chromatography. Columns with different lengths and diameters and a wide variety of film thicknesses started to be available, and this activity then continued after the introduction of the columns made of fused-silica tubing and having a bonded (immobilized) stationary phase coating (see below). In other words this was the time when the full utilization of the variables of open-tubular columns became possible [56–58]. One of the most impressive chromatograms utilizing the full possibilities of GC is shown in Fig. 10: it was obtained on a glass capillary column having a coated film thickness of less than 0.1 µm and demonstrates the analysis of amino acid derivatives from a ribonuclease hydrolyzate [59].

The increased importance of glass capillary columns was also demonstrated by the organization of the International Symposia on Glass Capillary Columns in 1975. These symposia became a regular event: they were held in the first decade in Hindelang, Bavaria, and then from 1983 on mainly in Riva del Garda, Italy, and periodically also in the United States and Japan. These symposia are still held annually, but now in an enlarged form, encompassing the whole field of capillary column chromatography including liquid chromatography, supercritical-fluid chromatography, and electrochromatography in addition to gas chromatography. Also, this was the period when supply houses specializing in the manufacturing of glass (and after 1980 fused-silica) capillary columns started to supply them for chromatographers.

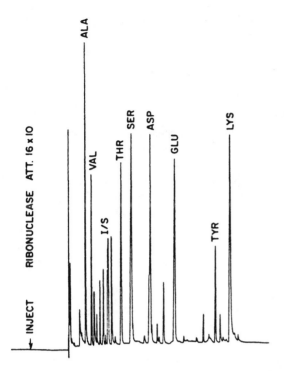

Figure 10 Analysis of the amino acids in ribonuclease hydrolyzate, in the form of the *n*-propyl, *N*-acetyl derivatives. Column: 50 m × 0.27 mm i.d. glass capillary, coated with a 1:1 mixture of Carbowax 20M and Silar CP. Film thickness: <0.1 μm. Carrier gas: helium with a flow rate of 0.75 mL/min. Column temperature programmed at 8°C/min from 110 to 190°C and then ballasted to 250°C. Split injection. Thermionic detector sensitive for nitrogen-containing compounds. Peaks: ALA = alanine, VAL = valine, I/S = norleucine (internal standard), THR = threonine, SER = serine, ASP = aspartic acid, GLO = glutamic acid, TYR = tyrosine, LYS = lysine. The retention time of the last peak is less than 30 minutes. (From Ref. 59. Courtesy Preston Publications.)

VIII. COATING THE CAPILLARY TUBES

Coating of capillary tubes—whether they are made of metal, plastic, glass, or fused silica—has always been a delicate procedure requiring considerable skill. In his original work, Golay used the so-called *static procedure*. The column tube was completely filled with a dilute solution of the stationary phase, one end closed, and then the (metal) tube was slowly drawn through an oven, open end first. However, this technique was too cumbersome and required special apparatus; also in the original form it could not be used for precoiled (e.g., glass) columns. The method was revised by Ilkova and Mistryukov [60], who were threading the coiled glass column filled with the coating solution through an oven, as if it were a gigantic screw. However, this was too complicated.

Because of the difficulties in adopting the static coating method to common practice, practically everybody used the *dynamic procedure* originally described by G. Dijkstra and J. de Goey [15]. In this, a plug of the stationary phase solution is slowly forced through the tubing with the aid of a dry inert gas, wetting in this way the inside wall of the tube with the solution. Subsequently the solvent is evaporated by blowing dry gas through the column for a few hours.

This technique has been discussed in detail in the literature [61,62], and if carried out skillfully it resulted in columns with good performance and long life. A major advantage of the dynamic method is that it does not need any complicated setup; however, its shortcoming is that the thickness of the coated film depends very much on the coating conditions (concentration of the coating solution and the velocity of its travel through the tube), and it cannot be readily established but only estimated.

In 1968 a modified form of the static coating technique was described by J. Bouche and M. Verzele [63]. In this the tube is fully filled with the stationary phase solution, one of its ends is closed, and then the solvent is slowly evaporated through the open end under reduced pressure at a temperature below the solvent's boiling point. The major advantage of this technique is that from the concentration of the coating solution the thickness of the coated film can be readily established. A detailed discussion of the intricacies of this method was given by Merle d'Aubigne et al. [50]. Today this technique is used universally by the column supply houses for column coating.

IX. FUSED-SILICA CAPILLARY COLUMNS

Back in 1960, when the glass drawing machine was developed by Desty's group, its use to make capillary columns from quartz was also mentioned [64]. At that time no further work was done along this line, because the burner would have had to be modified to facilitate the needed much higher temperatures. Later, however, Desty went back to this possibility and built a modification of the capillary drawing machine for this application. In a paper presented at the 1975 Hindelang Symposium, he briefly described this modified capillary drawing machine, showing also a photo of it [65], but he had no data on actual column manufacturing. It should be noted that the system developed by Desty would have produced rigid, thick-walled quartz (fused-silica) columns, and his major problem at that time was to find suitable platinum tubes that could be heated to the needed 1250–1350°C for the formation of the coiled capillary tubes.

Recognizing that quartz is more inert than glass, Grob in the mid-1970s was also considering the possibility of using it as the column tube material. However, he was confident that the methods developed by him to modify the inner surface of the glass tubes were satisfactory and therefore saw no need to change to a new tube material [66].

At the Third Hindelang Symposium (April 29–May 3, 1979) R. D. Dandeneau and E. H. Zerenner of Hewlett-Packard presented a major paper [67], describing the production and use of *thin-walled, flexible* fused-silica columns; this tubing was an adaptation of the production of fiber optics already carried out at Hewlett-Packard's Palo Alto, California, facilities. I was present and remember well the excitement caused by this presentation, which came as a complete surprise to everybody.

Dandeneau and Zerenner also realized that cracks can develop in the thin wall of the tubing, eventually leading to breakage. In order to prevent this, the outside of the tubing was coated immediately after drawing with silicone rubber, but soon this coating was changed to polyimide [68,69].

There is no question that the introduction of fused-silica columns changed not only capillary gas chromatography, but the whole field of separation techniques. Within a few months column supply houses already started to offer fused-silica capillary columns to the users, and very soon these made the glass columns obsolete. I remember an advertisement by J&W—one of the major suppliers of capillary columns—from this period, in which a glass capillary column was placed around the neck of a toy dinosaur, indicating that by then, both were fossils. A good comparison of the characteristics of glass and fused-silica capillary columns was given soon after their introduction in a small monograph by Jennings [70].

It may be interesting to note that thin-walled capillary tubing can also be made of soft glass, and this tubing will also be flexible, like the thin-walled fused-silica tubing [71]. However, the initially strong tubing rapidly becomes weak and brittle. This phenomenon can be only partially overcome by the application of a protective outer coating. Furthermore, such glass tubes still require extensive treatment to control the inner surface activity and to improve coatability.

Above I used the words "quartz" and "fused silica" interchangeably. In practice, however, the first term is always used for the natural material and the latter for the synthetic product prepared from silicon tetrachloride. The difference between the two is the amount of impurities present: natural quartz may contain trace amounts of metals up to a total concentration of about 50–60 ppm, while the total amount of metallic impurities in the synthetic fused silica is in the order of 0.08–0.5 ppm [72].

Following the presentation of Dandeneau and Zerenner, other scientists also investigated in detail the various questions associated with fused-silica capillary columns. In this regard I must particularly emphasize the activities of S. R. Lipsky [72,73].

Today, fused-silica capillary columns are used universally in gas chromatography. They are manufactured by a number of companies providing columns with specified parameters and performance, illustrating also their application fields. These supply houses also provide columns that are "tailor-made" for special applications. From the point of the users, this represents a significant advance because they no longer have to worry about the selection of the proper phase, column parameters, and column preparation. Columns are ordered by their code name, and the manufacturer has already taken care to optimize the column parameters that are automatically specified by the part number. In many cases the user does not even know the chemical composition of the stationary phase; even the analytical conditions for a certain application may be indicated by the column supplier. Naturally this helps the user; on the other hand, its obvious disadvantage is that the chromatographer carrying out the analyses is not involved in the intricacies of the column.

However, there are still many unresolved problems that call for further studies and improvements. A good example for this is the recent investigation of J. E. Cahill and D. H. Tracy [74], who have shown that at elevated temperatures, helium (used as the carrier gas) will permeate through the wall of thin-walled fused-silica capillary tubing, even with the normal outer polyimide coating, causing peak broadening and changes in the gas hold-up time. This effect may be particularly pronounced when programming the column to high temperatures.

X. IMMOBILIZED AND BONDED STATIONARY PHASES

One cannot finish a discussion of the continuous improvements in capillary gas chromatography without mentioning one additional subject: improvements in the stationary phases.

In the first two decades of the evolution of GC, most of the stationary phases used in the columns were taken "from the shelf": chemicals readily available in the laboratory were utilized. Except for a few, these were fairly low boiling, low molecular weight compounds, with significant vapor pressures even at moderate temperatures. This fact restricted the temperature range of GC, and experience has shown that the upper temperature limit of capillary columns was actually lower than that of packed columns containing the same stationary phase. In fact, by the middle of the 1970s one rarely found a capillary chromatogram in which the column was heated above 200°C.

In the 1970s the situation changed drastically with the introduction of silicone (polysiloxane) phases, custom-made for GC and particularly for capillary column use. These were high molecular weight polymers with molecular weights in the thousands or tens of thousands range, with a few even exceeding 100,000. In contrast, squalane, one of the most common phases of the 1960s, has a molecular weight of 423 and the average molecular weight of Carbowax 1540 poly(ethylene glycol)—another popular phase of the period—is 1540. These new phases can be coated well from their solution on the inner surface of the glass (and also fused-silica) tubing; they provide excellent chemical and thermal stability, with low bleeding, permitting the extension of the column upper temperature limit. A good summary of the characteristics of the new silicone phases was given by Blomberg [75,76] and Haken [77].

An obvious requirement in capillary column preparation is that the stationary phase should be soluble; after all, the inner tube wall is coated using a solution of the phase. This condition represents a limitation on the molecular weight of the substance to be used as the stationary phase, because in general, the higher the molecular weight, the more difficult it is to dissolve the substance. Therefore, methods had to be found to overcome this limitation. This was accomplished by an additional step: a secondary polymerization in the column, resulting in a coated stationary phase film with very high molecular weight. We call the products of this process *immobilized phases*. Additionally, a chemical bond may also be formed between the stationary phase molecules and the surface silanol groups on the inside surface of the fused-silica tubing: in this way, the so-called *bonded phases* are created.

Immobilization is a result of cross-linking of the primary polymer molecules, initiated by a variety of free radical initiators such as, e.g., organic peroxides or azo compounds, or by gamma irradiation. In the stationary phases used for chemical bonding, the molecules are terminated by hydroxyl groups. These are coated onto the inner wall of the column tubing, and then the column is temperature programmed to an elevated temperature where condensation reactions occur between the surface silanols of the fused-silica surface and the terminal OH groups of the phase molecules.

The basic work on immobilization and cross-linking was carried out in the decade between 1976 and 1986 by a number of groups such as those of C. Madani in France, L. Blomberg in Sweden, K. Grob in Switzerland, P. Sandra in Belgium, G. Schomburg in Germany, V. Pretorius in South Africa, as well as S. R. Lipsky and M. L. Lee in the United States. A very good summary of the questions associated with this "revolution in column technology" (as paraphrased by K. Grob [78]) was provided by E. F. Barry, giving also the pertinent references [79]. From the mid-1980s on, these techniques have been part of routine column manufacturing technology.

The temperature of capillary columns containing such immobilized and bonded phases can be programmed up to 400–420°C [80–83]. In the case of fused-silica columns intended for programming over 380°C, the polyimide outer coating protecting the tube was replaced by a very thin aluminum layer.

A good illustration for the possibilities of using such bonded-phase columns is Fig. 11, showing the analysis of the triglycerides in peanut oil [81]. The T-numbers indicate the total number of carbon atoms on the three side chains in the particular peak cluster; in addition, the individual side chains corresponding to the peaks are also identified. This column had a very thin bonded phase, and its temperature was programmed up to 380°C. It should be noted that triglycerides are high molecular weight compounds—e.g., the molecular weight of the triglyceride having two stearic and one oleic acid side chain is 890!—and these compounds will easily decompose during injection due to the catalytic effect of the syringe needle, which will rapidly warm up in a hot sample inlet system. Therefore, the so-called PTV (programmed-temperature vaporizer) inlet system (see below) was used here.

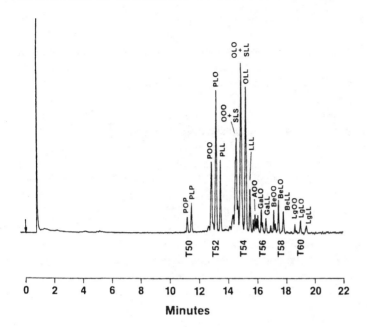

Figure 11 Separation of the triglycerides present in a peanut oil. Column: 25 m × 0.25 mm i.d. fused-silica capillary containing bonded methyl phenyl (65%) silicone stationary phase with a film thickness of 0.1 μm. Carrier gas: hydrogen, inlet pressure: 25 psig. Column temperature programmed from 300 to 360°C at 4°C/min. PTV injector, heated ballistically from 50 to 400°C. Sample: 0.5 μL of a solution in 2,2,4-trimethylpentane, split 1:50. Flame-ionization detector. The T-numbers give the total number of carbon atoms in the triglyceride's side chains with are identified by the following code letters: P = palmitic acid (C16:0), S = stearic acid (C18:0), O = oleic acid (C18:1), L = linoleic acid (C18:2), A = arachidic acid (C20:0), Ga = gadolenic acid (C20:1), Be = behenic acid (C22:0), Lg = ligniceric acid (C24:0). (From Ref. 82. Courtesy the *Journal of High-Resolution Chromatography*.)

Besides providing capillary columns that can be safely used at higher temperatures, immobilization and/or chemical bonding of the stationary phase result in two additional advantages. Such a phase can tolerate the injection of large volumes of a solvent without dissolution in it. In addition, it is also possible in this way to prepare stable capillary columns with a wide range of film thickness, even with a very thick film [84,85]. Without this treatment, such columns would soon lose most of their original coating through bleeding.

XI. P.L.O.T. COLUMNS

In porous-layer open-tubular (P.L.O.T.) columns, the inside surface area of the column tubing is increased by the deposition of a porous layer of either an inert support to be coated with the stationary phase or an adsorbent acting in itself as the stationary phase. This porous layer may also be formed by the chemical treatment of the inside surface of the (glass) capillary tubing. The idea of such columns was already indicated in Golay's 1958 Amsterdam paper [10]; 2 years later, at the 1960 Edinburgh Symposium, he then further elaborated his proposal for preparing columns that—as stated by him—would "constitute a nearly ideal column" [6].

Adsorption-type P.L.O.T. columns were first described in 1962 by Mohnke and Saffert [86], who prepared an active porous layer by the chemical treatment of the inside wall of the glass tubing. Their columns were successfully used for the separation of hydrogen isotopes and their nuclear spin isomers. This work was soon followed by J. E. Purcell of our group at Perkin-Elmer, who deposited both silica gel and molecular sieve particles on the inside wall of a metal capillary with 0.020 in. (0.51 mm) i.d. [87]. Such adsorption-type P.L.O.T. columns were reintroduced in the latter part of the 1980s, and since then they have become commercially available containing various active layers.

Golay's proposal to prepare partition-type P.L.O.T. columns was realized in 1962–63 by C. Horváth at the University of Frankfurt, Germany, who developed the technology of simultaneous deposition of support particles and their coating by the stationary phase with help of the static coating method [88,89]. We continued his work at Perkin-Elmer and extensively studied the theory and practice of support-coated open-tubular (S.C.O.T.) columns with various stationary phase loading, using a variety of phases. Such columns also became commercially available and have been successfully used for a variety of applications [90]. One example is shown in Fig. 12, illustrating the analysis of C_{24}–C_{27} fatty acid methyl esters on a S.C.O.T. column with a low stationary phase loading and programmed up to 260°C [91].

The major advantages of S.C.O.T. columns are their increased sample capacity and the possibility to use higher flow rates. Recently these columns were incorrectly characterized as

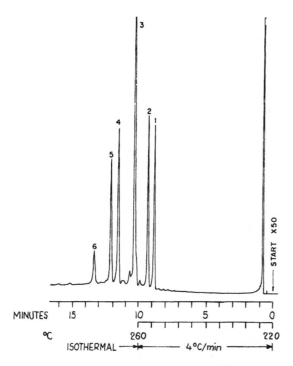

Figure 12 Analysis of the methyl esters of C_{24}–C_{27} fatty acids. Column: 50 ft × 0.020 in. (0.51 mm) i.d., S.C.O.T., prepared with SE-30 methylsilicone stationary phase. Phase ratio: $\beta = 360$. Carrier gas: helium, flow rate: 5 mL/min. Column temperature: programmed, as given. Sample size: 1 µL, split 1:30. Flame-ionization detector. Peaks: methyl esters of 1 = 22-methyl tricosanoic acid, 2 = tetracosanoic acid (lignoceric acid, C24:0), 3 = 22-methyl tetracosanoic acid, 4 = 24-methyl pentacosanoic acid, 5 = hexacosanoic acid (cerotic acid, C26:0), 6 = 24-methyl hexacosanoic acid. (From Ref. 91. Courtesy The Perkin-Elmer Corporation.)

a compromise between packed and open-tubular columns. This is not true: Golay, in his original discussions of 1958 and 1960 [6,10], specifically proposed these columns as an *improvement* to the smooth-wall (''wall-coated'') open-tubular columns. Today the S.C.O.T. columns have been mostly replaced by large-diameter (0.53 mm i.d.) capillary columns prepared with a thick liquid phase film. In fact, such columns have an inherently poorer performance than one would obtain when using properly prepared S.C.O.T. columns.

XII. INSTRUMENTATION

Because of the low carrier gas flow rates and the necessary small sample volumes, the instrumentation in which the capillary column is used can easily reduce the column's apparent performance. The particularly critical factors are the connections between the column to the injector and the detector, the method of heating the column, the detector, and the sample introduction system.

A. Connecting Lines

Particularly in earlier GC systems, the ''dead'' volumes of the connecting lines caused peak broadening, reducing column efficiency. This was soon improved but still not entirely satisfactory in instruments developed in the 1960s and 1970s, resulting in a slight tailing of the peaks characteristic of many chromatograms from this period. This was finally eliminated in later systems, bringing the column's end directly into the injector and the detector. With the glass columns, however, connection created another problem: these were coiled while one would have needed a short straight column end. Because of this the column ends had to be straightened and some way found for easy connections. Various ways were described in the literature to solve this problem (see, e.g., Refs. 40,50,92,93).

B. Detectors

All detectors have a finite volume and, thus, may cause peak broadening. To reduce this effect, make-up gas is frequently added to the column effluent, particularly in the case of concentration-sensitive detectors. Another problem that may arise is limitation in the detector response time for sharp peaks. These problems existed mainly with instruments designed for packed columns and adapted for capillary column use. In present-day instruments optimized for capillary column operation, one does not have to worry about such questions, except perhaps in very high-speed analysis with narrow-bore columns (i.d.$\leq$ 0.10 mm) where the detector's response time may not be fast enough for the extremely narrow peaks they can generate.

C. Column Heating

Column heating represented a limitation in the case of the early thick-walled stainless-steel columns, which had a high thermal mass. It took a long time until such a column equilibrated at an elevated temperature in the instrument's oven, and if temperature programmed, only low program rates could be used, because otherwise the column's temperature could not follow the nominal program. Thin-walled stainless-steel and then glass and fused-silica columns, together with improved construction of the instrument's oven (air thermostat), greatly improved the situation, and today even fast program rates can be reproducibly used.

A unique heating system was used in the Model 226 gas chromatograph of Perkin-Elmer introduced in 1962. In this instrument, the use of the bulky oven was eliminated: the thin-walled stainless-steel capillary column was in a flat spiral form between two aluminum disks, and this was pressed together with a heater of similar configuration, just like placing one pancake over the other. This system provided excellent heat transfer and permitted fast programming (Fig. 13).

Recently other column heating systems have been developed, and this trend, using alternate approaches to the classical oven heating (''air thermostat''), is continuing. A good review of the newer methods was given recently by Hinshaw [94].

D. Inlet Systems

Sample introduction into capillary columns has always been problematic since the sample capacity of such columns is generally lower than one can introduce by direct means. Desty, in his very first paper on the possibilities of capillary GC presented in October 1958 in Leipzig [16], already emphasized this problem and proposed split sampling:

> With capillary columns, where plate capacities are of the order of a thousand times lower than of packed columns and efficiencies ten or more times higher, the introduction of extremely small samples very sharply becomes of paramount importance. We have achieved this by a dynamic division, by far the largest proportion of the sample escaping through the external adjustable needle valve.

Early split systems represented a simple tee design; however, it was soon found out that such systems do not always split the sample ''linearly.'' Therefore, their design was soon changed following the principle of two concentric tubes, with the capillary column taking a linear segment of the gas flow including the sample vapor (Fig. 14A). Naturally it is important that prior to splitting the sample vapor and the carrier gas be homogeneously mixed [95,96]. Therefore, a tortuous path was provided for the carrier gas and the sample vapor molecules (Fig. 14B). Later some filling—e.g., silanized glass wool—in the mixing chamber provided this tortuous path and a removable glass (quartz) liner was also applied for easy cleaning (Fig.

Figure 13   Column and heater of the Model 226 gas chromatograph of Perkin-Elmer introduced in 1962.

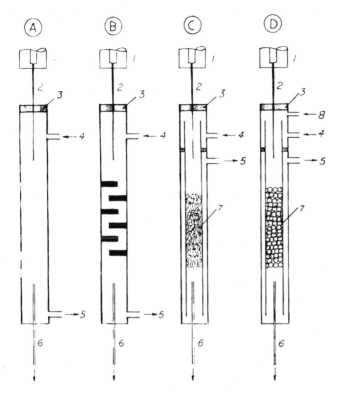

Figure 14   Functional schematics of various injector designs for capillary GC. A = Empty evaporation chamber; B = mixing chamber containing baffles; C = glass (quartz) liner containing silanized glass wool; D = glass (quartz) liner containing silanized glass beads, with septum purge. 1 = Syringe, 2 = syringe needle, 3 = septum, 4 = carrier gas inlet, 5 = carrier gas outlet (split vent), 6 = capillary column, 7 = glass (quartz) liner, 8 = septum purge line.

14C). In 1972 German and Horning pointed out that if the sample consisted of high-boiling compounds present in a relatively low-boiling solvent, there is a possibility of aerosol formation in the vaporization chamber, which then would not mix homogeneously with the carrier gas. To break the aerosol, they suggested to use a kind of precolumn in the injector, by having the liner filled with column packing or support particles [97]. This design was generally adapted for split-type injectors using small, silanized glass beads as the liner packing, providing the tortuous path for the sample to ensure vaporization and mixing with the carrier gas before splitting (Fig. 14D) [40]. An additional practical advantage of such a system is that it prevents undesirable material—such as nonvolatile components—from reaching the column.

Other differences in split design are related to the way the inlet is heated. In simpler systems the injector is kept hot and the liquid sample is injected into this hot zone for instantaneous evaporation. A disadvantage of such systems is that the temperature of the syringe needle increases very rapidly in the injector before withdrawal, and this hot metallic surface may catalyze decomposition of some of the sample components. In order to prevent this, the sample may be injected into a (relatively) cold injector, the needle quickly withdrawn, and then the temperature of the injector increased very rapidly to ensure complete evaporation. These are the so-called programmed-temperature vaporizers (PTV) [81,82,98–100].

The split ratio—i.e., the ratio of the sample amount entering the column vs. its amount discarded through the split vent—is set by the relative flow rates through the column and the vent. In turn, the vent flow rate is controlled by a restrictor, a flow controller, or needle valve.

Above we used the expression ''split linearity.'' This term refers to three basic criteria:

1. Every sample component is split in the same ratio, without any loss.

2. When analyzing mixtures with different concentrations, the individual peak areas are proportional to the concentration.

3. When changing the analytical conditions (temperature, column flow rate, split ratio, etc.), the relative sizes of the individual peaks in the chromatogram remain constant.

To the best of my knowledge, these basic criteria were first formulated in 1961 by Ettre and Averill [101].

While split injection is the most frequently used sample introduction system with capillary columns, sometimes the vapors of the whole sample are introduced into the column (''splitless injection''). In other cases a different kind of ''split'' technique is used: the division is due to boiling point differences. In such systems after sample introduction the low-boiling sample components (the solvent) are first evaporated by gradual heating of the injector and discarded through the vent; then, by increasing the injector temperature, the rest of the sample is evaporated and its vapor conducted into the column. Such a technique may be applied with very dilute solutions, when large sample volumes need to be injected in order to be able to detect the components present in very low concentrations (LVI: ''large-volume injection'') [102].

Above we mentioned the principal designs of the capillary column inlet systems. There are many variants, and one can find a bewildering number of expressions describing them. The literature dealing with the capillary column inlet systems is vast, but there are a few good summaries and review-type publications that provide information on the various systems and their relative merits [102–110]. In addition, a number of books specializing in the problems associated with capillary column inlet systems have been published [111–114].

XIII. FUTURE DEVELOPMENTS

Today fused-silica open-tubular (capillary) columns are used almost universally in gas chromatography. A number of supply houses offer a wide variety of columns with reliable performance. However, this does not mean that the evolution of capillary columns is finished. Continuous development work is carried out by the supply houses, instrument companies, and a number of research groups. Instrument companies further improve the systems in which the columns are used for separation; the supply houses optimize their columns for new applications and further improve their stability; and various research groups are mainly working in two areas. The first usually is characterized by one word: *miniaturization*. There is a definite trend to drastically reduce the size of the instruments and the dimensions of the columns, not only to use less space in the laboratory, but also to further the second aim—to *increase the speed of separation*. This implies the use of short capillary columns with small diameter, use of hydrogen as the carrier gas, very fast temperature programs, and instrumental systems that can scope with the resulting very sharp peaks, having a width of less than one second.

Another interesting innovation is the selection of the column tube material. Reading the literature extolling the importance of fused silica, one would believe that it is the ultimate column tube material. However, this is not necessarily so, at least not in every application. Recent developments in surface treatment technology have made it possible to coat the inside of stainless-steel tubing with a robust layer of silicon, which is then oxidized to form a silica surface, accepting deactivation and bonded stationary phase deposition. Such columns flex without disturbing the inner surface layer, and their physical strength, robustness, and thermal stability are superior to fused silica. Thus, in certain applications (e.g., portable instruments or high-temperature operation where the outer coating of the fused-silica tubing is no longer stable) such tubing may be preferable.

Capillary gas chromatography is a living science: every day still brings something new and exciting, benefiting the users, the chromatographers, who are utilizing the technique in their daily work for the analysis of the widest variety of samples.

ACKNOWLEDGMENTS

This chapter is based on an article entitled ''The Evolution of Capillary Columns for Gas Chromatography,'' published in the January 2001 issue of *LC·GC North America*. The permission of the publisher to utilize the text of this publication is most appreciated.

Particular thanks are due to Drs. A. Goldup and W. T. Swanton (formerly at British Petroleum Co., now retired), who were kind enough to share their recollections of the early period of glass capillary gas chromatography.

REFERENCES

1. LS Ettre. The evolution of open-tubular columns. In: WG Jennings, ed. Applications of Glass Capillary Gas Chromatography. New York: Marcel Dekker, 1981, pp. 1–47.
2. LS Ettre. Capillary columns—from London to London in 25 years. Chromatographia 16:18–25 1982; 17:117, 1983.
3. LS Ettre. Open-tubular columns: Evolution, present status and future. Anal Chem 57:1419A–1438A, 1985.
4. LS Ettre. MJE Golay and the invention of open-tubular (capillary) columns. J High Resolut Chromatogr 10:221–230, 1987.
5. LS Ettre. Open-tubular columns: past, present and future. Chromatographia 34:613–528, 1992.
6. MJE Golay. Brief report on gas chromatographic theory. In: RPW Scott, ed. Gas Chromatography 1960 (Edinburgh Symposium). London: Butterworths, 1960, pp. 139–143.
7. JJ van Deemter, FJ Zuiderweg, A Klinkenberg. Longitudinal diffusion and resistance to mass transfer as a cause of non-ideality in chromatography. Chem Eng Sci 5:271–289, 1956.
8. MJE Golay. Discussion of the results with three experimental gas-liquid chromatographic columns. Engineering Report No. 523, The Perkin-Elmer Corp., Norwalk, CT, September 5, 1956.
9. MJE Golay. Theory and practice of gas-liquid partition chromatography with coated capillaries. In: VJ Coates, HJ Noebels, IS Fagerson, eds. Gas Chromatography (1957 Lansing Symposium). New York: Academic Press, 1958, pp. 1–13.
10. MJE Golay. Theory of chromatography in open and coated tubular columns with round and rectangular cross-sections. In: DH Desty, ed. Gas Chromatography 1958 (Amsterdam Symposium). London: Butterworths, 1958, pp. 36–55.
11. DH Desty. 85th birthday of MJE Golay. Chromatographia 23:319 only. 1987.
12. IG McWilliam, RA Dewar. Flame ionization detector for gas chromatography. In: DH Desty, ed. Gas Chromatography 1958 (Amsterdam Symposium). London: Butterworths, 1958, pp. 142–152.
13. JE Lovelock. Detector for use with capillary tube columns in gas chromatography. Nature (London) 182:1663–1664, 1958.
14. DH Desty. Personal recollections. In: LS Ettre, A Zlatkis, eds. 75 Years of Chromatography—A Historical Dialogue. Amsterdam: Elsevier, 1979, pp. 31–42.

15. G Dijkstra, J De Goey. The use of coated capillaries as columns for gas chromatography. In: DH Desty, ed. Gas Chromatography 1958 (Amsterdam Symposium). London: Butterworths, 1958, pp. 56–68.

16. DH Desty. The potentialities of coated capillary columns for gas chromatography. In HP Angelé, ed. Gas-Chromatographie 1958. Berlin: Akademie Verlag, 1959, pp. 176–184.

17. RE Kaiser. Personal recollections. In: LS Ettre, A Zlatkis, eds. 75 Years of Chromatography—A Historical Dialogue. Amsterdam: Elsevier, 1979, pp. 187–192.

18. DH Desty, A Goldup, BHF Whyman. The potentialities of coated capillary columns for gas chromatography in the petroleum industry. J Inst Petrol 45:287–298, 1959.

19. RD Condon. Design considerations of a gas chromatography system employing high efficiency Golay columns. Anal Chem 31:1717–1722, 1959.

20. A Zlatkis, JE Lovelock. Gas chromatography of hydrocarbons using capillary columns and ionization detectors. Anal Chem 31:620–621, 1959.

21. SR Lipsky, RA Landowne, JE Lovelock. Separation of lipides by gas-liquid chromatography. Anal Chem 31:852–856, 1959.

22. SR Lipsky, JE Lovelock, RA Landowne. Use of high-efficiency capillary columns for the separation of certain cis-trans isomers of long-chain fatty acid esters by gas chromatography. J Am Chem Soc 81:1010 only, 1959.

23. A Zlatkis. Personal recollections. In: LS Ettre, A Zlatkis, eds. 75 Years of Chromatography—A Historical Dialogue. Amsterdam: Elsevier, 1979, pp. 473–482.

24. R Stevenson. Fused-silica capillaries: the key enabling technology for analytical chemistry. Am Lab 30(5):30–34, February 1998.

25. LS Ettre, W Averill, FJ Kabot. Gas Chromatographic Analysis of Fatty Acids. GC-AP-001. Norwalk, CT: Perkin-Elmer, 1962.

26. W Averill. Analysis of Phenols. GC-DS-001. Norwalk, CT: Perkin-Elmer, 1962.

27. I Halász. Discussion remark. In: N Brenner, JE Callen, MD Weiss, eds. Gas Chromatography (1961 Lansing Symposium). New York: Academic Press, 1962, p. 560 only.

28. W Averill. Columns with minimum liquid phase concentration for use in gas-liquid chromatography. In: N Brenner, JE Callen, MD Weiss, eds. Gas Chromatography (1961 Lansing Symposium). New York: Academic Press, 1962, pp. 1–6.

29. F Farré-Rius, J Henniker, G Guiochon. Wetting phenomena in gas chromatography capillary columns. Nature (London) 196:63–64, 1962.

30. G Schomburg, H Husmann, F Wecke. Preparation, performance and special applications of glass capillary columns. J Chromatogr 99:63–79, 1974.

31. PJ Porcaro. Observations on the use of 'empty' copper tubular capillary columns. J Gas Chromatogr 1 (6):17–19, 1963.

32. OL Hollis. Gas-liquid chromatographic analysis of trace impurities in styrene using capillary columns. Anal Chem 33:352–355, 1961.

33. DH Desty, A Goldup, WT Swanton. Separation of m-xylene and p-xylene by gas chromatography. Nature (London) 183:107–108, 1959.

34. W Bertsch, F Shunbo, RC Chang, A Zlatkis. Preparation of high-resolution nickel open-tubular columns. Chromatographia 7:128–134, 1974.

35. RPW Scott. Nylon capillary column for use in gas-liquid chromatography. Nature (London) 183: 1753–1754, 1959.

36. A Zlatkis, HP Kaufman. Use of coated tubing as columns for gas chromatography. Nature (London) 184:2010 only, 1959.

37. DH Desty, JN Haresnape, BHF Whyman. Construction of long lengths of coiled glass capillary. Anal Chem 32:302–304, 1960.

38. A Kreyenbuhl. Construction of glass capillary columns for gas-liquid chromatography. Bull Soc Chim France 2125–2127, 1960.

39. DH Desty, A Goldup, WT Swanton. Performance of coated capillary columns. In: N Brenner, JE Callen, MD Weiss, eds. Gas Chromatography (1961 Lansing Symposium). New York: Academic Press, 1962, pp. 105–135.

40. MJ Hartigan, LS Ettre. Questions related to gas chromatographic systems with glass open-tubular columns. J Chromatogr 119:187–206, 1976.

41. J Simon, L Szepesy. Manufacture of glass open-tubular columns by dehydration of the surface. J Chromatogr 119:495–504, 1976.

42. K Grob. Polar coating of glass capillaries for gas chromatography. Helv Chim Acta 48:1362–1370, 1965.

43. K Grob. Glass capillaries with thin liquid films. In: AB Littlewood, ed. Gas Chromatography 1966 (Rome Symposium). London: Institute of Petroleum, 1967, pp. 113–114.

44. K Grob. Glass capillaries for gas chromatography. Improved preparation and testing of stable liquid separation films. Helv Chim Acta 57:718–737, 1968.

45. K Tesařik, M Novotný. Surface treatments of glass capillaries for gas chromatography. In: HG Struppe, ed. Gas-Chromatographie 1968 (Berlin Symposium). Berlin: Akademie Verlag, 1968, pp. 575–584.

46. M Novotný, K Tesařik. Surface treatment of glass capillaries for gas chromatography. Chromatographia 1:332–333, 1968.

47. ML Lee, FJ Yang, KB Bartle. Open-Tubular Column Gas Chromatography: Theory and Practice. New York: Wiley, 1984, p. 53.

48. M Novotný, A Zlatkis. Glass capillary columns and their significance in biochemical research. Chromatogr Rev 14:1–44, 1971.

49. WG Jennings. Gas Chromatography with Glass Capillary Columns. New York: Academic Press, 1978(1st ed.), 1980 (2nd ed.).

50. J Merle d'Aubigne, C Landault, G Guiochon. Methods of preparation of capillary columns used in gas chromatography. True capillary columns. Chromatographia 4:309–326, 1971.

51. K Grob. Twenty years of glass capillary columns. An empirical model for their preparation and properties. J High Resolut Chromatogr 2:599–604, 1979.

52. G Alexander. Preparation of glass capillary columns. Chromatographia 13:651–660, 1980.

53. WG Jennings, ed. Applications of Glass Capillary Gas Chromatography. New York: Marcel Dekker, 1981.

54. W Bertsch, WG Jennings, RE Kaiser, eds. Recent Advances in Capillary Gas Chromatography. Heidelberg:Huethig Verlag, 1981.

55. K Grob, Jr. Making and Manipulating Capillary Columns for Gas Chromatography. Heidelberg: Huethig Verlag, 1986.

56. LS Ettre. Performance of open-tubular columns as function of tube diameter and liquid phase film thickness. Chromatographia 18:477–488, 1984.

57. LS Ettre. Variation of tube diameter and film thickness of capillary columns. J High Resolut Chromatogr 8:497–503, 1985.

58. LS Ettre. The full utilization of the variables of open-tubular columns. In: F Bruner, ed. The Science of Chromatography. Amsterdam: Elsevier, 1985, pp. 87–109.

59. RF Adams, FL Vandemark, GL Schmidt. Ultramicro GC determination of amino acids using glass open-tubular columns and a nitrogen-sensitive detector. J Chromatogr Sci 15:63–68, 1977.

60. EL Ilkova, EA Mistryukov. A simple versatile method for coating of glass capillary columns. J Chromatogr Sci 9:569–570, 1971.

61. R Kaiser Chromatography in the Gas Phase, Vol. II: Capillary Gas Chromatography. Original German edition: Mannheim: Bibliographisches Institut, 1961; English translation: London: Butterworths, 1963.

62. LS Ettre. Open Tubular Columns in Gas Chromatography. New York:Plenum Press, 1965.

63. J Bouche, M Verzele. A static coating procedure for glass capillary columns. J Gas Chromatogr 6:501–505, 1968.

64. DH Desty, JN Haresnape, BHF Whyman (assigned to British Petroleum Co.). Construction of long lengths of coiled glass capillary. British Patent No. 899,909 (applied: April 9, 1959; issued:June 27, 1962).

65. DH Desty. The origination, development and potentialities of glass capillary columns. Chromatographia 8:452–455, 1975.

66. K Grob, G Grob. A new method for the manufacturing of glass capillary columns. Wissenschaftl Zeitschr der Karl-Marx-Univ, Math-Naturwiss Reihe 26(4):379–384, 1977.

67. RD Dandeneau, EH Zerenner. An investigation of glasses for capillary chromatography. J High Resolut Chromatogr 2:351–356, 1979.

68. RD Dandeneau, EH Zerenner. The invention of fused-silica column: an industrial perspective. LC·GC North Am 8:908–912, 1990.

69. RD Dandeneau, quoted in Ref. 24.

70. WG Jennings. Comparison of Fused-Silica and Other Glass Columns in Gas Chromatography. Heidelberg: Huethig Verlag, 1981.

71. KL Ogan, C Reese, RPW Scott. Strong, flexible soft-glass capillary columns: a practical alternative to fused silica. J Chromatogr Sci 20:425–428, 1982.

72. SR Lipsky, WJ McMurray, M Hernandez, JE Purcell, KA Billeb. Fused-silica glass capillary columns for gas chromatographic analysis. J Chromatogr Sci 18:1–9, 1980.

73. SR Lipsky, WJ McMurray. Role of surface groups in affecting the chromatographic performance of certain types of fused-silica glass capillary columns. J Chromatogr 217:3–17, 1981.

74. JE Cahill, DH Tracy. Effects of permeation of helium through the walls of fused-silica capillary GC columns. J High Resolut Chromatogr 21:531–538, 1998.

75. L Blomberg, Current aspects of the stationary phase in gas chromatography. J High Resolut Chromatogr 5:520–535, 1982.

76. L Blomberg, Contemporary capillary columns for gas chromatography. J High Resolut Chromatogr 7:282–284, 1984.

77. JK Haken. Developments in polysiloxane stationary phases in gas chromatography. J Chromatogr 300:1–77, 1984.

78. K Grob, G Grob. New approach to capillary columns for gas chromatography: condensation of hydroxyl-terminated stationary phases. J Chromatogr 347:351–356, 1985.

79. EF Barry. Columns: packed and capillary—column selection in gas chromatography. In: RL Grob, ed. Modern Practice of Gas Chromatography, 3rd ed. New York: Wiley, 1995, pp. 198–203.

80. SR Lipsky, ML Duffy. High-temperature gas chromatography:the development of new aluminum-clad flexible fused-silica glass capillary columns coated with thermostable non-polar phases. J High Resolut Chromatogr 9:376–382, 725–730, 1986.

81. JV Hinshaw, W Seferovic. Programmed-temperature split-splitless injection of triglycerides: comparison to cold on-column injection. J High Resolut Chromatogr 9:69–72, 1986.

82. JV Hinshaw, W Seferovic. Analysis of triglycerides by capillary gas chromatography with programmed-temperature injection. J High Resolut Chromatogr 9:731–736, 1986.

83. JV Hinshaw, LS Ettre. Aspects of high-temperature capillary gas chromatography. J High Resolut Chromatogr 12:251–254, 1989.

84. LS Ettre. Open-tubular columns prepared with very-thick liquid-phase film. I. Theoretical basis. Chromatographia 17:553–559, 1983.

85. LS Ettre, GL McClure, JD Williams. Open-tubular columns prepared with very-thick liquid-phase film. II. Investigation of column efficiency. Chromatographia 17:560–569, 706, 1983.

86. M Mohnke, W Saffert. Adsorption chromatography of hydrogen isotopes with capillary columns. In: M van Swaay, ed. Gas Chromatography 1962 (Hamburg Symposium). London: Butterworths, 1962, pp. 216–224.

87. JE Purcell. Separation of permanent gases by open-tubular adsorption columns. Nature (London) 201:1321–1322, 1964.

88. C Horváth. Separation columns with porous layers for gas chromatography. Inaugural dissertation, University of Frankfurt/Main, 1963.

89. I Halász, C Horváth. Open tube columns with impregnated thin-layer support for gas chromatography. Anal Chem 35:499–505, 1963.

90. LS Ettre, JE Purcell. Porous-layer open-tubular columns—theory, practice and applications. In: JC Giddings, RA Keller, eds. Advances in Chromatography, Vol 10. New York: Marcel Dekker, 1974, pp. 1–97.

91. LS Ettre, K Billeb, JE Purcell. Applications of S.C.O.T. columns with various liquid phase loading. 18th Pittsburgh Conference, Pittsburgh, PA, March 6–10, 1967.

92. K Grob, G Grob. Techniques of capillary gas chromatography: possibilities of the full utilization of high-performance columns. I. Direct sample injection. Chromatographia 5:3–12, 1972.

93. K Grob, J Jaeggi. Techniques of capillary gas chromatography: possibilities of the full utilization of high-performance columns. II. Manipulation and operation. Chromatographia 5:382–391, 1972.

94. IV Hinshaw. GC ovens: a "hot" topic. LC·GC North Am 18:1142–1147, 2000.

95. I Halász, W Schneider. Quantitative gas chromatographic analysis of hydrocarbons with capillary column and flame-ionization detector. Anal Chem 33:978–982, 1961.

96. I Halász, W Schneider. Quantitative gas chromatographic analysis of hydrocarbons with capillary column and flame-ionization detector. In: N Brenner, JE Callen, MD Weiss, eds. Gas Chromatography (1961 Lansing Symposium). New York: Academic Press, 1962, pp. 287–306.

97. AL German, EC Horning. Thermostable open-tubular capillary columns for the high resolution gas chromatography of human urinary steroids. J Chromatogr Sci 11:76–82, 1973.

98. W Vogt, K Jacob, AB Ohnesorge, HW Obwexer. Capillary gas chromatographic injection system for large sample volumes. J Chromatogr 186:197–205, 1979.

99. F Poy, S Visani, F Terrosi. Automatic injection in high-resolution gas chromatography: a programmed-temperature vaporizer as a general-purpose injection system. J Chromatogr 217:81–90, 1981.

100. JV Hinshaw. Programmed-temperature inlets. LC·GC North Am 10:748–752, 1992.

101. LS Ettre, W Averill. Investigation of the linearity of a stream splitter for capillary gas chromatography. Anal Chem 33:680–684, 1961.

102. JV Hinshaw. Large-volume injection in capillary GC: problems and solutions. LC·GC 7:26–31, 1989.

103. RD Condon, LS Ettre. Liquid sample processing. In: J Krugers, ed. Instrumentation in Gas Chromatography. Eindhoven: Centrex Publishing Co., 1968, pp. 87–105.

104. G Schomburg, H Behlau, R Dielmann, F Weeke, H Husmann. Sampling techniques in capillary gas chromatography. J Chromatogr 142:87–102, 1977.

105. JV Hinshaw, LS Ettre. Introduction to Open-Tubular Column Gas Chromatography. Cleveland: Advanstar, 1994, pp. 127–148.

106. JV Hinshaw, FJ Yang. Solute focusing technique for on-column injection in capillary GC. J High Resolut Chromatogr 6:554–559, 1983.

107. JV Hinshaw. Modern inlet systems for capillary gas chromatography. J Chromatogr Sci 25:49–55, 1987.

108. JV Hinshaw. Inlets. LC·GC North Am 5:954–960, 1987.

109. JV Hinshaw. Splitless injection: principles, optimization and problem solving. LC·GC North Am 9:622–627, 1991.

110. JV Hinshaw. Splitless injection: corrections and further information. LC·GC North Am 10:88–92, 1992.

111. P Sandra, ed. Sample Introduction in Capillary Gas Chromatography. Heidelberg: Huething Verlag, 1985.

112. K Grob. Split and Splitless Injection in Capillary Gas Chromatography. Heidelberg: Huethig Verlag, 1986.

113. K Grob. On-Column Injection in Capillary Gas Chromatography. Heidelberg: Huethig Verlag, 1987 (1st ed.), 1991 (2nd ed.).

114. A Tipler, LS Ettre. The PreVent System and Its Applications in Open-Tubular Column Gas Chromatography. Norwalk, CT: Perkin-Elmer Corp., 1997.

9

Forty Years in Gas Chromatography

Victor G. Berezkin

A. V. Topchiev Institute of Petrochemical Synthesis,
Russian Academy of Sciences, Moscow, Russia

James and Martin's discovery [1] of partition gas-liquid chromatography stimulated the development of gas chromatography not only in Russia but all over the world. Chromatography has been progressing rapidly since the study [1] was published. This new high-performance technique for the analysis of multicomponent volatile mixtures is of prime importance in the fields of chemistry, chemical industry, medicine, pharmaceutical industry, gas and petroleum industry, etc. Unfortunately, this contribution of chromatography to the world economic development went unnoticed by many people.

My Ph.D. thesis, dealing with radiation chemistry of hydrocarbons, was performed at the A. V. Topchiev Institute of Petrochemical Synthesis of the USSR Academy of Science under the supervision of Professor and Doctor of Physicomathematical Sciences L. S. Polak. The study was only possible to perform due to the use of gas chromatography, because a complex mixture of radiolysis products is formed as a result of γ-radiation of even an individual n-alkane. I have been working in the field of gas chromatography since the late 1950s. At that time neither gas chromatographs nor packed chromatography columns were commercially available, so I had to be a designer, a gas chromatograph manufacturer, a developer of chromatographic techniques, a manufacturer of chromatographic columns, and a chromatographer analyzing hydrocarbon samples, all at the same time. I was like a person who cannot swim but had to swim. I swam out to the river entitled "chromatography." This "swimming" was very difficult for me, but it made possible my introduction to chromatography and an understanding of its peculiarities. My participation in a joint monograph devoted to hydrocarbon radiolysis was one of the main projects of that period [2].

In late 1961 I spent 4 months learning chromatography at the Chromatography Laboratory of the Czechoslovak Academy of Sciences (Brno) headed by the well-known Czech scientist Doctor of Sciences J. Janak. For the first time I learned about the general formulation and "technology" of chromatographic science. I am very grateful to Professor Janak and to all my

145

Table 1 Classification of Chromatographic Variants by Phase State of Phases Participating in Chromatographic Process

	Stationary phase			
	Monophase		Binary phase	
Mobile phase	Liquid	Solid	Gas-solid	Liquid-solid
Gas	Variant not realized	Gas-[solid]	Variant not realized	Gas–[liquid-solid]
Liquid	Liquid-[liquid]	Liquid-[solid]	Liquid–[gas-solid]	Liquid–[liquid-solid]

Czech colleagues for their kind attitude and sincere help in helping me understand the scientific basis of chromatography.

Before I got my Ph.D. studying radiolysis of hydrocarbons, I was appointed Head of Chromatography at the Laboratory at the Institute of Petrochemical Synthesis, USSR Academy of Sciences, by its Director, A. V. Topchiev. I have held this position for nearly 40 years. In 1968 I defended my thesis for a doctoral degree, and in 1973 I was awarded the title of Professor.

O. Balsak correctly noted that "a question mark is the key for a science." To my mind, when considering the aphorism in detail, it should be noted that a scientific study usually involves two questions: "Why?" and "How?" Although most researchers respond to the questions at the same time, to me the development of the fundamentals of chromatography (i.e., the response to the question "Why?") has been of most importance.

The main trends of my studies and the results obtained over the last 40 years are discussed below.

I. ADSORPTION PHENOMENA IN GAS-LIQUID CHROMATOGRAPHY AND THEIR INFLUENCE ON RETENTION

In the early 1960s most researchers believed that in gas-liquid chromatography (GLC) the stationary phase was comprised of the liquid phase only (see, e.g., Refs. 3, 4). This is a conventional point of view reflected in the title of one of the main chromatography variants ("gas-liquid" chromatography). It was shown, first, in widespread classification of gas chromatography (Table 1 and Fig. 1). Second, the equation for net retention volume V_{Ni} for the i-solute contained a term that takes into account the solute's dissolution and its partition in the system's stationary liquid phase (SLP)/mobile phase (gas):

$$V_{Ni} = K_{li}V_l \tag{1}$$

 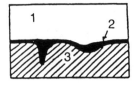

Figure 1 Scheme of polyphase sorbent in capillary GLC.

where K_{li} is the partition coefficient for the i-solute (reflecting the solute absorption by SLP) and V_l is the SLP volume in a column. But "pure" GLC where retention and other chromatographic characteristics are determined by SLP properties only (i.e., GLC without solid support) has not yet been realized, and the term GLC does not take into account the polyphase nature of a real sorbent (see Fig. 1).

There are at least two interactions (SLP–gas phase and solid support–SLP) in the real sorbent. These interactions are able to adsorb and retain the compounds to be chromatographed. Gas-liquid chromatography is the only variant in real chromatography (see Table 1).

In 1968 I was the first to suggest the additive trinomial equation [Eq. (2)] for the net retention volume for i-solute V_{Ni} [5]. This equation is still actively used by many researchers. Within a year the outstanding chromatographers Conder, Locke, and Purnell [6] independently obtained and published the same equation:

$$V_{Ni} = K_{li}V_l + K_{gli}S_{gl} + K_{li}K_{lsi}S_{ls} \tag{2}$$

where K_{gli} is the distribution (adsorption) coefficient for the i-solute in the system (interface of SLP-gas)/(gas); K_{lsi} is the distribution (adsorption) coefficient for the i-solute in the system (interface of solid support–SLP)/(SLP); S_{gl} is the interface of SLP/gas in a column; S_{ls} is the interface of solid support–SLP in a column.

Our study was published in a Russian scientific journal [5]. This seems to be a plausible explanation (taking into consideration language limitation) for the fact that most chromatographers, with few exceptions (see, e.g., Ref. 7), usually quote the study by Conder and coworkers [6] in the context of Eq. (2) but not our paper [5]. We also suggested [5] a new general additive polynomial equation for retention volume which takes into account the contribution of more than three phases being in the real sorbent. In this equation, V_{Ni} of an actual polyphase sorbent is the sum of the partial retention volumes due to absorption and adsorption, respectively, of the chromatographed substance by the individual sorbent phase [5], which is capable of both processes—adsorption or absorption. Martin's equation [8], Keller and Steward's equation [9], Kiselev's equation [10], Karger's equation [11], and others can each be obtained as a particular case of the general polynomial equation.

Detailed experimental examination of Eq. (2) conformed well to numerous experimental data by various researchers for various chromatographic systems. Analysis of Eq. (2) allowed us to elaborate on new physicochemical gas chromatographic applications in real chromatographic systems, namely:

1. accurate determination of the distribution (adsorption) constant K_{lsi} for solute adsorption in chromatographic system: SLP/(SLP-solid support);
2. accurate determination of the distribution (adsorption) constant K_{gli} for solute adsorption in chromatographic system: gas/(gas-SLP);
3. accurate determination of the partition constant K_{li} (absorption) for solute adsorption on chromatographic system: gas/SLP.

Also, Eq. (2) made possible the explanation of all basic chromatographic regularities from a common point of view and one theoretical standpoint.

A detailed consideration of the theory and experimental data of real gas-liquid chromatography has been presented in a monograph [12]. Unfortunately, in the monograph little attention was paid to adsorption phenomena in capillary gas-liquid chromatography. Most chromatographers believe that adsorption on a fused silica surface can be neglected due to its low adsorption activity. As is evident from Fig. 2 [13], the adsorption activity of inner surface of the fused

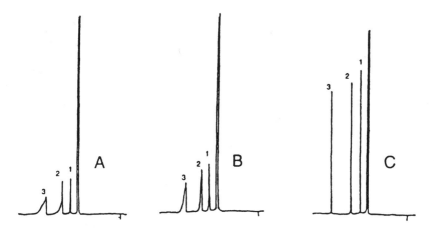

Figure 2 Adsorption activity of internal surface of fused silica capillaries by separation of C_{11}–C_{13} n-alkanes. Compounds: 1—n-$C_{11}H_{24}$; 2—n-$C_{12}H_{26}$; 3—n-$C_{13}H_{28}$. Experimental conditions: fused silica capillary column 25 m length $\times$ 0.2 mm i.d., 70°C, carrier gas = He. Treatment of internal capillary surface: (A) Hydration of the surface (H_2O, 200°C, 3 hours); (B) dehydration of the surface (carrier gas—He, 300°C); (C) internal surface was modified by a very thin film of polar SLP (PEG-20M).

silica capillaries cannot be neglected (Fig. 2). Also, a solute's adsorption on the interface gas (mobile phase)/SLP can be measurable (Fig. 2). Hence the concept of polyphase chromatography and solute adsorption on interfaces should be applied to capillary gas–[liquid-solid] chromatography as well [13].

Relative retention (relative retention time r_i, retention index I_i) is currently the main focus of analytical chromatography, not absolute retention (retention time, net retention volume). Therefore, developing the theory of polyphase equilibrium chromatography I deduced the following equations for relative retention [12,14]:

$$r_i = \frac{K_{\ell i}}{K_{\ell st}} + \lambda_r \cdot \frac{1}{V_\ell} = r_{io} + \lambda_r \cdot \frac{1}{V_\ell} \tag{3}$$

$$r_{io} = \frac{K_l}{K_{lst}} \tag{4}$$

$$r_i = r_{io} + \lambda_k \frac{1}{k_{st}} \tag{5}$$

$$I_i = I_{oi} + \lambda_l \frac{1}{k_{st}} \tag{6}$$

$$I_{oi} = 100n + \frac{100 \log \left(\dfrac{K_{li}}{K_{\ell n}} \right)}{\log \left(\dfrac{K_{l(n+1)}}{K_{\ell n}} \right)} \tag{7}$$

where K_{li} and K_{lst} are the partition constants in the system SLP/gas for the i-solute and standard (st), respectively; K_{ln}, $K_{l(n+1)}$ are the partition constants for n-alkanes containing (n) and ($n+1$) carbon atoms, respectively; λ_r, λ_k, and λ_l are the constants of Eqs. (3), (5), and (6), respectively.

As follows from the above equations, the relative retention is linearly dependent on reciprocal SLP volume in a column or on the proportional value [e.g., on the reciprocal retention factor of standard (k_{st})]. A compound whose adsorption effects can be neglected is used as a standard for Eqs. (3), (5), and (6). Numerous examples of good agreement between the equations and experimental data for various solutes, stationary liquid phases, types of columns (packed and capillary), and experimental conditions have been published in a monograph [12] and a review [13].

The following results were obtained using the concept of polyphase chromatographic system and solute retention in gas–[liquid-solid] chromatography taking into account not only the solute's absorption by SLP but the solute's adsorption by interfaces as well. First, it was shown theoretically and experimentally that relative retentions $(r_i$ and $I_i)$ that are measured experimentally are not true constants for the given SLP and the solute. Second, new invariant values of relative retention r_{io} and I_{oi} were suggested. These values depend only on the partition constants for the i-solute and standard in a chromatographic system: SLP/gas. Third, when adsorption effects are neglected, the developed retention theory is transformed into a classic one in which the relative retention is a ratio of the partition constants.

As a whole, note that the developed theory and its experimental examination led to a new, more detailed understanding of the chromatographic process and more precise definition of retention nature. These allowed an explanation of all general regularities of absolute and relative retention in packed and capillary columns in gas–[liquid-solid] chromatography. Adsorption phenomena in gas-liquid chromatography play an important role in the fundamental theory of modern gas chromatography.

II. INFLUENCE OF CARRIER GAS ON THE RELATIVE RETENTION AND SEPARATION SELECTIVITY IN CLASSICAL CAPILLARY GLC

According to conventional thinking, a carrier gas has no influence on retention and separation selectivity. For example, chromatographers C. F. Poole and S. A. Schuette [15] noted: ''The mobile phase in gas chromatography is generally considered to be inert in the sense that it doesn't react chemically with the sample or stationary phase nor does it influence the sorption-desorption or partitioning processes that occur within the column. Thus the choice of carrier gas doesn't influence the selectivity. It can influence resolution through its effect on column efficiency which arises from differences in solute diffusion rates.''

Taking into consideration the high efficiency of capillary columns, it should be expected that even small variations in relative retention (or in separation selectivity) could lead to a new qualitative result (variation in elution order; resolution, or lack of it, of a given pair of compounds to be chromatographed).

From the 1950s to the 1970s a group of famous English physicochemists and chromatographers [16–18] developed the theory of absolute retention volume dependence on carrier gas nature and column pressure. They did not consider carrier gas influence on relative retention, but relative retention values are commonly used in analytical chromatography. Therefore, in the 1990s I deduced the following equations for relative retention (relative retention r_{ij} or separation α_{ij}, retention index I_i for i and j solutes):

$$\alpha_{ij}P_{av}) = \alpha_{ij}(0) + b_{\alpha i}P_{av} \tag{8}$$

$$I_i = I_i(0) - b_{Ii}P_{av} \tag{9}$$

$$P_{av} = P_o \, J_3^4 \tag{10}$$

$$J_3^4 = \frac{3}{4} \frac{[(P_i/P_o)^4 - 1]}{[(P_i/P_o)^3 - 1]} \tag{11}$$

where P_{av} is column pressure averaged by residence time of solute [18]; P_i and P_o are inlet and outlet pressures of carrier gas, respectively; $\alpha_{ij}(0)$ is separation factor at "zero" pressure, $\alpha_{ij}(0) = \lim \alpha_{ij}(P_{av})$ at $P_{av} \rightarrow 0$; $I_i(0)$ is retention index at "zero" pressure, $I_i(0) = \lim I_i(P_{av})$ at $P_{av} \rightarrow 0$; $b_{\alpha i}$ and b_{Ii} are constants depending on the chromatographic system, the nature of the solute, and the experimental conditions.

We performed the experimental evaluation of Eqs. (8) and (9) over the range of average column pressures $P_{av} < 5$ atm (i.e., at conditions of classic gas-liquid chromatography). It was shown that the equations conformed well to experimental data for chromatographic systems at different experimental conditions (SLP, carrier gas, solutes, and temperature) [19]. Limiting (invariant) relative retention $\alpha_{ij}(0)$ and $I_i(0)$ [see Eqs. (8) and (9)] are considered to be true chromatographic constants of solutes of a given system at a given temperature.

Hence, in our studies we showed that relative retention values $\alpha_{ij}(P_{av})$ and (P_{av}) are not chromatographic constants of the solute. They are dependent on the pressure and nature of carrier gas. "Light" carrier gases influence the retention factor in the following order: helium < hydrogen < nitrogen < carbon dioxide. This order correlates with the order of gases, which is analogous with the order of increasing nonideality.

Note that the limiting relative retention values $\alpha_{ij}(0)$ and $I_i(0)$ do not practically depend on the nature of "light" carrier gas. Therefore, these values are true invariant values.

The regularities for the retention dependence of the nature and the pressure of carrier gas considered theoretically and experimentally are also true for the retention factor $k_i(P_{av})$. We considered the retention factor to be one relative retention value.

Peak resolution R_s is a quantitative separation characteristic. It was shown [19] that R_s is linearly dependent on the pressure and nature of the carrier gas:

$$R_s(P_{av}) = R_s(0) + b_{\alpha ij}P_{av} \tag{12}$$

where $b_{\alpha ij}$ is constant, $R_s(0) = \lim R_s(P_{av})$ at $P_{av} \rightarrow 0$.

Two chromatograms of a pesticide mixture (dicofol + methoxychlor) obtained on a column coated with polar SLP, Carbowax 20 M, using helium and carbon dioxide are shown in Fig. 3. As can be seen, with carbon dioxide as a carrier gas, methoxychlor elutes first whereas dicofol elutes last. When helium is used as the carrier gas, dicofol elutes before methoxychlore. Hence, the carrier gas is an important factor in the separation process.

More pronounced variation in relative retention can be observed for carrier gases that shift the acidic-basic equilibrium of some solute groups to form ions when using polar SLP (acidic-basic chromatography). This leads to an increase in the total partition constant of the solutes in the chromatographic system, SLP/(carrier gas), and therefore increases the retention time for solutes of this type. Variation in retention factor with the nature and temperature of the carrier gas is listed in Table 2. As follows from the data, with an acidic carrier gas, retention of primary amines increases significantly, that of secondary amines increases to a lesser extent, and that of tertiary amines shows a slight increase.

The resultant theory [21] leads to the following equation for the retention time t_R of a basic solute:

$$t_R = t_m \left[\left(\frac{K_2}{\beta} + 1 \right) + \alpha_\beta \frac{[H_2O_{(g)}]^3[CO_{2(g)}]}{[HCO_3^-]} \right] \tag{13}$$

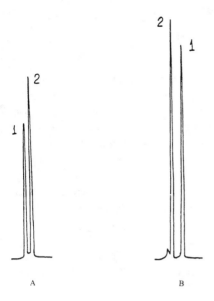

Figure 3 Chromatogram of pesticides [dicofol (1) and methoxychlor (2)] using helium (A) and carbon dioxide (B) as carrier gas. Experimental conditions: capillary column 12 m × 0.32 mm coated with Carbowax 20M (film thickness 0.25 µm), 260°C.

where t_m is hold-up time, $K_2 = [B_{(g)}]/[B_{(l)}]$, $[B_{(g)}]$ and $[B_{(l)}]$ are concentrations of basic solute in the gas (g) and liquid (l) phases, respectively, β is phase ratio, and α_β is the constant depending on the characteristics of equilibrium processes occurring in the given system.

Note that the observed dependence correlates qualitatively with Eq. (13). Linear dependence of the basic solute retention on the concentration $[CO_{2(g)}]$ in the gas phase was confirmed quantitatively [21]. We called this new gas chromatography variant acidic-basic chromatography [21].

As a result of the above-mentioned studies, the conventional concept of the role of the carrier gas in capillary GLC changed, and therefore the fundamentals of chromatography changed as well.

Table 2 Influence of Acidic Humid Carrier Gas (CO_2) on Amine Retention at Various Temperatures on a Capillary Column Coated with Polar SLP (Polyethylene glycol)

	Retention factor, k			
	He + H_2O,	$CO_2 + H_2O$		
Solute	70°C	70°C	90°C	110°C
Propanol	0.817	0.811	0.380	0.202
Dibutylamine	1.299	1.694	0.690	0.334
n-Butanol	1.697	1.752	0.741	0.369
Tributylamine	2.694	2.690	1.177	0.583
n-Hexanol	7.112	6.99	2.70	1.185
Octylamine	5.373	—	5.01	1.521
Diaminopropane	6.440	—	—	3.20

III. WATER-CONTAINING INORGANIC AND ORGANIC SOLUTIONS AS SUPERSELECTIVE PHASES FOR STEAM CHROMATOGRAPHY

In the early 1950s and 1960s some examples of water use as a superselective phase were published (see, e.g., Ref. 22). Due to the high volatility of water its practical application was not possible. Water-containing inorganic phases [23] (and recently, water-containing organic phases [24]) as stable analogs of the water phase were suggested to overcome this disadvantage (high volatility of water). The general experimental scheme using the above-mentioned phases is as follows: The mobile phase (pure superheated steam or inert gas containing water vapors) moves along a sorbent layer. The sorbent used is a solid support coated with a hydrophilic (inorganic or organic) nonvolatile compound, which is highly soluble in water. Water interacts with the nonvolatile compound and forms a water solution under conditions of chromatography. The process can be shown as follows:

$$H_2O(vapor) + Me(salt) \Leftrightarrow MeX - H_2O(water\ solution\text{-}SLP) \qquad (14)$$

At constant temperature (T) the salt concentration in the water solution (C_{MeX}) is dependent on partial pressure of water vapors in the mobile phase P_{H_2O}:

$$P_{H_2O} = \phi(C_{MeX}, T) \qquad (15)$$

At a given temperature the composition of the water-salt solution (SLP) is determined by the water content in the mobile phase. Therefore, the partial pressure of water vapors is considered to be an additional factor that influences the SLP content and, therefore, its selectivity.

We originally suggested the application of water-salt solutions as SLP in the 1980s. The compounds are superselective for organic compounds of the RX type (R is a hydrocarbon radical, X is a polar group: hydroxyl, carboxyl, amino). As an example, the elution order of light C_1–C_6 alcohols on a column with water-containing SLP is: hexanol < pentanol < butanol < propanol < ethanol < methanol. The nature of an electrolyte and its concentration in the water-containing SLP has a significant influence on column selectivity and the separation of the compounds to be chromatographed. Note that the column temperature with water-containing SLP can considerably exceed 100°C. For example, water-salt solutions can be used at 150°C. This allows the analysis of alcohols with boiling temperatures up to 350°C.

We obtained good separation of C_2–C_8 organic acids with water solutions of phosphoric acid and separated amines with potassium hydrate (for review, see Ref. 25).

Application of water solutions of complexing salts as SLP has an additional advantage for the separation of high-boiling olefins C_{10}–C_{13} [25]. The packed columns with water-salt SLP have the same efficiency as analogous columns with SE-30.

Some of the main trends in the practical application for water solutions as SLP in steam chromatography are: (a) superselective separation of polar compounds; (b) separation of polar and nonpolar compounds (e.g., determination of methanol admixtures in petrol, trace determination in alcoholic drinks; (c) preparative production of pure compounds; and (d) physicochemical applications (e.g., determination of complexing constants).

IV. INVERSE GAS CHROMATOGRAPHY OF POLYMERS

With the use of inverse gas chromatography, nonvolatile compounds (e.g., polymers) can be studied. In this type of GC, contrary to the classical variant, the system under investigation is the stationary phase, on which known volatile test compounds (known compounds) and the stationary phase (the unknown compound under investigation) have changed (inverse) places. The term "inverse gas chromatography" was suggested in 1966 independently by Davis and coworkers [26] and Berezkin [27].

E. Nesterov and Y. S. Lipatov described the first studies of polymers by inverse chromatography as follows [28]:

> V. G. Berezkin and coworkers pioneered the studies of change of phase in polymers by inverse gas chromatography [Ref. 29 cited here in original]. It was shown that over the range of change of phase for polyethylene and polypropylene maximum is observed on the curve for the retention time and on the peak width dependence of time. The maximum is caused by the changes in partition constants and diffusion coefficients of low molecular weight compounds as a result of change of phase. In this case, a new stationary phase forms instead of the initial stationary phase. Physicochemical characteristics of the new phase differ from those of the initial one. It was shown that 'prehistory' of sample (technique for stationary phase preparation) has a significant influence on chromatographic characteristics over the range of change of phase. In this study V. G. Berezkin suggested the application of inverse chromatography for estimation of polymer's degree of crystallinity.

Later, Guillet and coworkers [30] made a great contribution to the method's development. The development of inverse chromatography has been presented in a review [31].

In conclusion, let us note that the development of this simple and effective method for polymer investigation has not been accomplished. It seems reasonable to estimate the possibilities of the new variants of the method like inverse fluid (supercritical) chromatography, inverse liquid chromatography and inverse gas chromatography with vapor organic eluents. This would allow study of swelling processes and phase changes in swollen polymers.

V. DEVELOPMENT OF CHROMATOGRAPHIC FIELDS OF APPLICATION

Like any branch of science, chromatography generates new areas as a result of its development. The new fields are characterized by special theory, techniques (or special apparatus), and practical application.

In my opinion, the dissemination of published data including its systematization is an important stage in the development of a new field. Hence, the publication of a monograph (especially the first monograph in the given field) stimulates the further development of the field.

I have actively participated in writing monographs; all monographs I participated in as an author were written on my initiative, and I also developed the schemes of a number of books. Most of the monographs were pioneering ones. Other works include books on reaction chromatography [27], analysis of trace impurities [32], polymer study [33], gas-liquid-solid chromatography [12] (still incorrectly referred to as gas-liquid chromatography), capillary gas-solid chromatography [34], and other fields [35–40].

Recently, I have turned my attention to the development of the definition of chromatography and showed the limitation of this definition as a separation method. In my opinion, chromatography should be considered another branch of science [41].

I have been involved in the preparation of scientific specialists. Forty-five Ph.D. theses were defended under my leadership, and I was a scientific consultant for 3 theses for a Doctor of Sciences degree.

As a whole, I authored 350 scientific papers. I was a member of the advisory board of the

Journal of Chromatography, the *Journal of High Resolution Chromatography*, and *Zhurnal Analiticheskoi Khimii* (Russ J Anal Chem).

In 1996 I received the State Prize of the Russian Federation for the development of new principles for selectivity operation and creation of new sorbents.

REFERENCES

1. AT James, AJP Martin. Biochem J 50:679, 1952.
2. Radioliz uglevodorodov. Pod red. AV Topchieva i LS Polaka. Moskva, Izd. AN SSSR, 1962 [Radiolysis of Hydrocarbons. AV Topchiev, LS Polak, eds. Moscow: AN SSSR Publisher, 1962] (in Russian).
3. S Dal Nogare, RS Juvet. Gas-Liquid Chromatography. New York: Wiley Interscience, 1962.
4. AA Zhukhovitskii, NM Turkel'taub. Gazovaya Khromatografiya. Moskva: Gostoptekhizdat, 1962 [AA Zhukhovitskii, NM Turkel'taub. Gas Chromatography. Moscow: Gostoptekhizdat, 1962] (in Russian).
5. VG Berezkin, VP Pakhomov, VS Tatarinskii, VM Fateeva. Doklady AN SSSR 180:1135, 1968 (in Russian).
6. JR Conder, DC Locke, JH Purnell. J Phys Chem 73:700, 1969.
7. A Voelkel, J Janik. J Chromatogr A 693:315, 1995.
8. RL Martin. Anal Chem 35:116, 1963.
9. RA Keller, GH Steward. Anal Chem 34:1834, 1962.
10. AV Kiselev, YI Yashin. Gas-Adsorption Chromatography. New York: Plenum Press, 1969.
11. BL Karger, A Hartkopf, H Postmanter. J Chromatogr Sci 7:315, 1967.
12. VG Berezkin. Gas-Liquid-Solid Chromatography. New York: Marcel Dekker, 1991.
13. VG Berezkin. In: PhR Brown, E Grushka, eds. Advances in Chromatography. Vol. 40. New York: Marcel Dekker, 2000, p. 599.
14. VG Berezkin, NS Nikitina, VM Fateeva. Doklady AN SSSR 209:1131, 1973 (in Russian).
15. CF Poole, SA Schuette. Contemporary Practice of Chromatography. Amsterdam: Elsevier, 1984.
16. DH Everett, CTH Stoddart. Trans Faraday Soc 57:746, 1961.
17. AJB Cruichshank, ML Windsor, CL Young. Proc Roy Soc 295A:271, 1966.
18. JR Conder, CL Young. Physicochemical Measurements by Gas Chromatography. Chichester: J. Wiley, 1978.
19. VG Berezkin. Neftekhimiya 37:366, 1997 (in Russian).
20. VG Berezkin. Zh Fiz khimii 74:521, 2000 (in Russian).
21. VG Berezkin, VR Alishoev, IV Malyukova, VV Kuznetsov. J Microcolumn Separation 11:331, 1999.
22. LN Phyfer, HR Plummer. Anal Chem 38:331, 1966.
23. VG Berezkin, VR Alishoev, EN Viktorova, VS Gavrichev, VM Fateeva. J High Resolution Chromatogr 6:42, 1983.
24. VG Berezkin, EYu Sorokina, AI Sokolov, AP Arzamastsev, LK Golova. Russian Chemical Bulletin, 2000, No. 6, 1077.
25. VG Berezkin. Russian Chem Bull N10:1831, 1999 (in Russian).
26. TC Davis, JC Petersen, WE Haines. Anal Chem 38:241, 1966.
27. VG Berezkin. Analytical Reaction Gas Chromatography. New York: Plenum Press, 1968 (Russian Edition published in 1966).
28. ZA Nesterov, YuS Lipatov. Obrashennaya gazovaya khromatografiya v termodinamike polimerov. Kiev: Naukova Dumka, 1976 [Inverse gas chromatography in polymer thermodynamics. Kiev: Naukova Dumka, 1976] (in Russian).
29. VR Alishoev, VG Berezkin, YuV Mel'nikova. Zh Fiz Khimii 39:200, 1965 (in Russian).
30. JM Braun, JE Guillet. Advances Polymer Sci 21B:108, 1976.
31. VG Berezkin. Zh Fiz Khimii 70:775, 1996 (in Russian).

32. VG Berezkin, VS Tatarinsky. Gas-Chromatographic Analysis of Trace Impurities. New York: Consultants Bureau, 1973.
33. VG Berezkin, VR Alishoev, IB Nemirovskaya. Gas Chromatography of Polymers. Amsterdam: Elsevier, 1977.
34. VG Berezkin, J de Zeeuw. Capillary Gas Adsorption Chromatography. Heidelberg: Huethig, 1996.
35. VG Berezkin, YuS Drugov. Gas Chromatography in Air Pollution Analysis. Leipzig: Geest & Portig/ Amsterdam: Elsevier, 1991.
36. VG Berezkin, VP Pakhomov, KI Sakodynsky. Solid Supports in Gas Chromatography. Bellefonte: Supelco Inc., 1980.
37. VG Berezkin, VS Gavrichev, LN Kolomiets, AA Korolev, VN Lipavsky, NS Nikitina, VS Tatarinskii. Gazovaya Khromatografiya v Neftekhimii. Moskva: Nauka, 1975 [Gas Chromatography in Petrochemistry. Moscow: Nauka, 1975] (in Russian).
38. VG Berezkin, VD Loshilova, AG Pankov, VD Yagodovsky. Khromato-raspredelitelnyi metod (Chromato-Distribution Method). Moskva: Nauka, 1976 (in Russian).
39. VN Lipavsky, VG Berezkin. Avtomaticheskie Gazovye Potokovye Khromatografy. Moskva: Khimiya, 1982 [On-line gas chromatographs. Moscow: Khimiya, 1982] (in Russian).
40. VG Berezkin, G Stoev, O Georgiev. Applied Capillary Gas Chromatography. Sofia: Nauka i izkustvo, 1986.
41. VG Berezkin. Industrial Laboratory. Diagnostics Materials 65:2, 1999 (in Russian).

10

Liquid Crystal Stationary Phases in Gas Chromatography: A Historic Perspective

George M. Janini

SAIC Frederick, NCI-Frederick Cancer Research and Development Center, Frederick, Maryland

This chapter is not meant to be a review of the literature on the use of liquid crystals as stationary phases in gas chromatography (GC), rather it tells the story of the development of the subject from my prospective as I followed it over a period of 30 years. The history of this subject spans about 40 years. My personal knowledge and involvement started in 1970, and I was an active participant for a period of 20 years. In 1990 my research interest shifted from chromatography to capillary electrophoresis, but I kept myself up to date with the literature. The 1990s witnessed a surge in interest in this field on the part of Russian and Chinese researchers, concomitant with a slow-down in United States and western Europe. Unfortunately, many valuable contributions in the Russian and Chinese literature will not be covered in this review.

I. THE PERIOD 1963–1980

The first time I heard about liquid crystals and actually observed how a cholesteric liquid crystal changes upon heating from solid to a colorful turbid liquid and upon further heating to clear liquid was in 1970 in the laboratory of Professor Daniel Martire at Georgetown University. At the time I was a student working on a Ph.D. in physical chemistry under Professor Martire's supervision. My thesis project dealt with the thermodynamics of normal alkane solutions and weak complexes. Although I never had a chance to work with liquid crystals in Professor Mart-

By acceptance of this article, the publisher or recipient acknowledges the right of the U.S. Government to retain a nonexclusive, royalty-free license and to any copyright covering the article. The content of this publication does not necessarily reflect the views or policies of the Department of Health and Human Services, nor does mention of trade names, commercial products, or organizations imply endorsement by the U.S. Government.

ire's laboratory, I was very much intrigued by these exotic materials, and Professor Martire's fascination and love of liquid crystals was contagious. One fellow student, Walter L. Zielinski, Jr., who, like me, did not work with liquid crystals, was employed full time at the National Bureau of Standards (NBS), now known as the National Institute of Standards and Technology (NIST). Walter used to come in the evening to work on his thesis, and we became, and still are, good friends. One of the difficult problems that he was grappling with at NBS was the separation of *m*- and *p*-divinylbenzene that was present in the commercial monomer that had to be purified prior to use as a cross-linker in the preparation of poly(styrene/divinylbenzene) copolymers. One day he was venting his frustration about the difficulties of his assignment when Professor Martire suggested that the desired separation might be achieved by gas chromatography with a liquid crystal stationary phase. Sure enough, the isomers were baseline separated with 4-4'-dihydroxyazoxybenzene (a nematic liquid crystal available in the laboratory). The event was a cause for celebration in the laboratory because this was the first time, at least in Professor Martire's physical chemistry laboratory, a liquid crystal was used to solve a real practical separation problem.

The first use of liquid crystals as stationary phases in GC was reported a few years earlier almost simultaneously by Kelker [2] and by Dewar and Schroeder [3]. The subject was first reviewed in 1968 by Kelker and Jon Schivizhoffen [4]. Most separations reported in this period were with low-temperature nematic liquid crystals for the separations of predominantly isomeric substituted benzenes. Compounds capable of existing in the liquid crystalline state were known for almost a century prior to their first use as a separation medium in GC. The liquid crystalline (mesomorphic) state was first recognized in 1888 by Reinitzer [5], who synthesized cholesteryl benzoate and found that on heating a solid sample it melted at 150°C to give a turbid liquid, which transformed sharply to a clear amorphous isotropic liquid at 179°C. Lehman [6] subsequently showed that the turbid liquid was birefringent and hence anisotropic and that other cholesteryl esters exhibited similar behavior. Up to that time only crystalline solids were known to have these properties. Lehman also showed that this ordered fluid state was thermodynamically stable in a temperature range between the solid and isotropic liquid states, and he classified compounds that have these properties as liquid crystals. The liquid crystal state was also termed mesomorphic because it exhibited properties in between those of the three-dimensionally ordered crystalline solids and the three-dimensionally disordered isotropic liquids. Since this first discovery, an increasing number of liquid crystals (mesogens) have become available. A list of liquid crystals was first compiled by Kast [7], and a more comprehensive list of over 5000 compounds was published by Demus, and Zaschke [8]. To date, the number of available liquid crystals is much bigger as more monomeric liquid crystals have been developed for display devices and polymeric liquid crystals were introduced [9,10].

My first contribution to the field of liquid crystal stationary phases in GC was conducted at the Frederick Cancer Research and Development Center (FCRDC), Frederick, MD, in the summer of 1974 and was published in *Analytical Chemistry* in 1975 [11]. FCRDC began operation under the authority of the National Cancer Institute (NCI) in the summer of 1972, right after then President Nixon announced the closure of the former Biological Defense Research Laboratories at Fort Detrick, Frederick, MD, and ordered the conversion of the laboratories into a leading center for cancer research. My friend Walter was recruited from NBS to join FCRDC soon after its official start of business as the Head of the Synthesis and Analysis Laboratory (CSAL). After graduation from Georgetown I accepted a faculty position overseas to start in September 1974, but I was free for the summer of 1974. This is when Walter offered me an intern summer research appointment at CSAL, which I enthusiastically accepted because I was eager to establish a collaboration program with a reputable research laboratory while working overseas. When I arrived at the laboratory, Walter Zielinski was grappling with another separation problem, which I immediately got involved in, namely, polycyclic aromatic hydrocarbons (PAH) in biological and environmental situations.

At that time, PAH had been extensively studied, principally because of the exhibited carcinogenic [12,13] and mutagenic [14] properties of members of this class of compounds. The possible adverse effect of these chemicals on human health was, and still is, a matter of international concern. Continuing efforts have been made to elucidate the role of the various components and impurities in PAH in order to assist in the development of preventive measures and to provide a basis for assessing hazards. PAH have been detected in such diverse sources as soot, carbon black, coal tar, mineral shale, crude oils, rubber tires, etc. More significant is the observed occurrence of PAH in environmental situations of concern to public health (e.g., tobacco smoke, smoked food, and automobile exhaust).

Accurate and reliable methods for the separation and quantitative determination of these compounds in occupational, environmental, and biological samples have been a target of many researchers. Several methods have been published and extensively reviewed by Sawicki [15], Schaad [16], and Hutzinger et al. [17]. These methods include column chromatography [18], thin layer and paper chromatography [17], high-pressure liquid chromatography [19], UV spectroscopy [20], and many other less versatile techniques [17]. However, the separation of many of the PAH positional isomers has been found to be difficult, if not impossible. Furthermore, these methods do not, in general, provide satisfactory analytical accuracy because of the inherent resolution limitations. GC has played an important role in the analysis of PAH [16,21–26]. Applications of capillary GC have been reported in the analysis of PAH and related compounds in cigarette smoke [27,28], air, and automobile exhaust [29,30]. Although the GC technique has been used with varying degrees of success, in which columns for specific narrow ranges of PAH compounds have been developed [21–30], no liquid phase has been reported that wholly meets the specifications of the UICC/IARC Joint Working Group [31,32]. This group has specified that an acceptable method should be capable of separating at least benz[a]anthracene, benzo[a]-pyrene, benzo[e]pyrene, benzo[ghi]perylene, pyrene, benzo[k]fluoranthene, and coronene. One packed column [25] and another capillary column [27] achieved partial success; however, a mixture of anthracene and phenanthrene, a mixture of benz[a]anthracene, triphenylene, and chrysene, or a mixture of benz[a]pyrene, benzo[e]pyrene, and perylene and benzo[k]fluoranthene has not been adequately separated on these or other GC columns.

After digesting this background, I immediately thought of GC with a liquid crystal stationary phase as the only hope for resolving these isomers. The problem was that PAH requires high temperature for their GC analysis because of their low volatility, and all the liquid crystals that have been used in GC were either of low molecular weight and high volatility and/or did not exhibit liquid crystalline properties at high temperatures. By searching the chemical catalogs I stumbled upon a high-temperature nematic liquid crystal that had been there for many years but was completely overlooked by chromatographers. This liquid crystal, N,N'-bis(p-methoxy-benzylidine)-α,α'-bi-p-toluidine), which we named BMBT, was unusual in its high molecular weight and broad nematic temperature range, 181°C (solid-nematic) and 320°C (nematic-isotropic). I thought that such properties would make it appropriate for the separation of rigid PAH isomers, and the experimental results were not disappointing. The separations reported [11], especially the separation of benzo[a]pyrene from benzo[e]pyrene, shown in Figure 1, have never been achieved by any GC or other chromatographic technique. The results were enthusiastically met by the analytical scientific community, and the paper received the 1977 Applied Analysis Chemistry Award of the Pittsburgh Conference on Analytical Chemistry and Applied Spectroscopy.

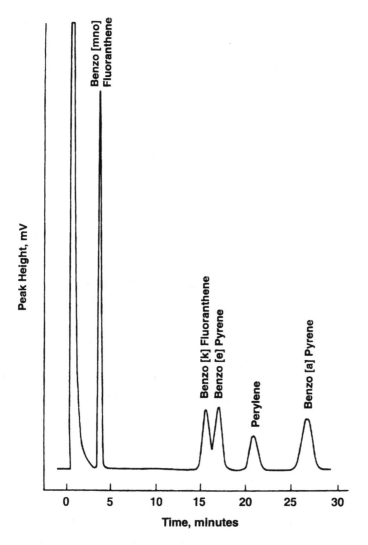

Figure 1   Chromatogram of selected pentacyclic aromatic hydrocarbons with BMBT. Column: 4-ft ×
0.125-in. o.d. stainless steel. Packing: 2.5% (w/w). Conditions: oven 260°C; injector, 260°C; flow rate,
40 mL/min. Concentration: About 0.2 μg of each component. (From Ref. 11.)

This initial success later led to the deployment of BMBT for new GC separations of underivatized steroid epimers [33], polychlorinated biphenyls [34], methoxy benzanthracenes [35],
methyl naphthalenes [36], drug isomers [37], azaheterocyclic compounds [38], and PAH from
combustion effluents [39]. Thus, a single liquid crystal (BMBT) provided improved resolution
of several classes of compounds not attainable by conventional GC with polymeric phases.
Despite its immediate success, BMBT exhibited measurable column bleed when operated close
to its higher temperature operating limit.

    Considering the need for stable, low-volatility liquid crystal stationary phases, the author
and coworkers undertook the synthesis and chromatographic evaluation of several other high
molecular weight homologs of BMBT [40–46]. The first to be synthesized was the butoxy

homolog of BMBT, abbreviated BBBT [40]. The primary objective was to prepare a liquid crystalline material having separation capabilities comparable to BMBT, but with substantially lowered column bleed levels at column operating temperatures required for GC analysis of high molecular weight PAH. BBBT was found to fully satisfy the first requirement, with the added advantage of diminished column bleed. Comparison of the longevity of BBBT versus BMBT columns operated for prolonged periods at elevated temperatures showed the superiority of the former (see Fig. 1 of Ref. 40). Later it was observed [41] that the analysis of PAH having five, six, and seven rings on either BMBT or BBBT columns at temperature at which bleed levels are not excessive (estimated at 220°C for BMBT and 245°C for BBBT) results in long solute retention times and broad peaks. In order to extend the usefulness of liquid crystals as GC phases for the analysis of five- to seven-ring PAH, the author and coworkers synthesized and GC evaluated two thermally stable liquid crystals, namely the hexyloxy (BhxBT) and phenyl (BPhBT) homologs [41]. In particular, BPhBT exhibited a substantially diminished column bleed at an operating temperature of 275°C while maintaining the separation capabilities of BMBT. The low bleed levels and high efficiency characteristics observed for BPhBT enabled its application as liquid phase in GC–mass spectrometry (MS) systems [41]. Isothermal BPhBT column operation at high temperatures (270–290°C) has permitted the analysis of high molecular weight PAH, providing, for the first time, baseline separation of some of the six- and seven-ring PAH geometric isomers, bile acid isomers [42], and benzo[a]pyrene in cigarette smoke [43]. Both BBBT and BPhBT are commercially available from commercial suppliers like Alltech, Deerfield, IL. Our laboratory also conducted physicochemical studies with liquid crystal stationary phase in an effort to understand and explain their chromatographic behavior [47–52]. This together with other GC studies of the same topic will be discussed later in a separate section.

More details about the material discussed so far are given in two extensive reviews published by us [53] and by Witkiewicz [54].

II. WHAT ARE LIQUID CRYSTALS AND HOW DO THEY AFFECT SHAPE SELECTIVITY?

To help in answering these questions, we present an overview of what liquid crystals are and how separations are achieved with liquid crystal stationary phases.

A. Classification by Chemical Structure and Molecular Order

Many organic compounds and some organometallic compounds transform on melting into mesomorphic phases (thermodynamically stable anisotropic liquids known as thermotropic liquid crystals). The term thermotropic serves to distinguish such liquid crystal systems from lyotropic liquid crystals, which are formed by the action of appropriate amounts of specific solvents on certain compounds, generally pure materials or mixtures of amphiphiles (compounds containing two distinct regions, one hydrophobic and the other hydrophilic, e.g., soaps, detergents, and polypeptides). Lyotropic liquid crystals have not been shown to be useful as stationary phases in chromatography and will not be discussed here. Compounds that exhibit thermotropic liquid crystalline properties are nonamphiphilic. They are of a variety of chemical types, but all are markedly elongated with fairly rigid rodlike or lathlike common geometric features such as:

$$Y - R_1 - X - R_2 - Y'$$

where R_1 and R_2 are aromatic systems with one or more aromatic rings, X is a central group capable of electronic conjugation with the aromatic systems, and Y and Y′ are *para*-terminal

Table 1 Common Central Moieties and Terminal Substituents

Common central moieties	Common terminal substituents
$(-CH=CH-)_n$	H–
$(-O-CH_2-)_n$	R–
$-C\equiv C-$	RO–
$-N\equiv C-$	CN–
$-N=N-$ $\downarrow$ O	Cl–
	Br–
$-CH=N-$ $\downarrow$ O	F–
$-C-O-$ ‖ O	O ‖ $R-C-O-$
$(-CH_2-CH_2-)_n$	$R-O-(CH_2)_n-O-$
$-O-CH_2-CH_2-O-$	$(CH_3)_2-N-$
$-CH=N-N=CH-$	

substituents. Table 1 lists some common central moieties and terminal substituents found in most known liquid crystals.

Rod-like molecules tend to pack in the solid crystal lattice with their long axes parallel. The crystal lattice could be layered with a three-dimensional arrangement of molecules, which either have their centers of gravity lined in planes and their long axes parallel, tilted, or orthogonal to the planes or are interdigitated with the only constraint being that their long axes lay parallel. On melting, thermotropic liquid crystals (abbreviated as liquid crystal throughout the chapter) pass through one or more intermediate states of order in a stepwise manner over a range of temperatures. This multiplicity of phases is a manifestation of residual order on the molecular level. In contrast to isotropic melts, they appear as turbid birefringent fluids with associated viscosities ranging from that of a syrupy paste to that of a free-flowing liquid. Depending on their microscopic molecular order such melts are classed as (a) smectic phases, which were first observed with soaps under the polarizing microscope, (b) nematic phases that owe their name to their microscopic threadlike appearance, and (c) chiral nematic phases with the designation ''cholesteric'' derived from cholesterol, derivatives of which were the first known liquid crystals [5,6]. The smectic phase is further subdivided into at least eight distinguishable states, smectic A to smectic H. The molecules in smectic liquid crystals are ordered in layers with their long axes parallel and their centers of gravity in planes orthogonal to, or at a specific angle with, the preferred direction of the long axes of the molecules. The end-to-end intermolecular interactions are weak, and therefore the layers can readily slip over one another. As a result of this two-dimensional order, smectic liquid crystals are viscous. In the nematic mesophase the molecules merely exhibit a time and specially averaged parallel orientation of their long axes within relatively small clusters and are consequently much less viscous than smectic phases. The molecules in the chiral nematic mesophase are arranged in layers. Within each layer the molecular arrangement resembles that of the nematic mesophase, but the direction of the long axes of the molecules in each layer is systematically displaced with respect to the direction of the long axes in the adjacent layers, with the overall displacement tracing a helical path in a direction normal to the molecular planes.

Figure 2 gives generalized structural models for smectics A, B, and C, and Figure 3 gives generalized models for a nematic and a cholesteric phase. A compound may give a nematic, cholesteric, or one or more smectic phases. It may also give one or more smectic phases in addition to either a nematic or a cholesteric phase. The nematic mesophase has the lowest order of all liquid crystal classes, and hence, for a compound that exhibits a nematic and other mesophases, the nematic always immediately precedes the transition to the isotropic state. This was believed to be the case until reentrant liquid crystal phases were observed, where the same phase (e.g., a nematic) may appear at temperatures both higher and lower than a smectic phase. The nematic phase appearing at the lower temperature is called the reentrant phase. There are very few examples in literature where reentrant phases were used in chromatography [55–57].

Discotic (disc-like) liquid crystals, is another type of liquid crystal, were first observed by Chandrasekhar et al. [58]. Examples of discotic liquid crystals are those formed by flat or disc-shaped molecules such as substituted triphenylenes [59]. Discotic liquid crystals exhibit narrower temperature ranges compared to the other types of liquid crystals, nevertheless they were used as a separation media in chromatography [60].

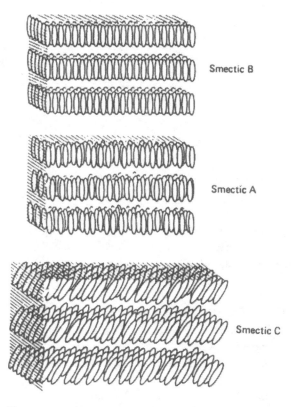

Smectic B

Smectic A

Smectic C

Figure 2 Idealized representation of smectic mesophases.

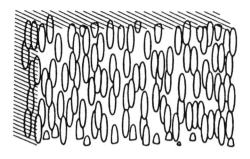

a

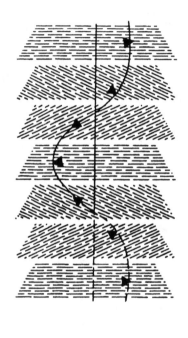

b

Figure 3 Idealized representation of (a) a nematic and (b) a cholesteric mesophase.

Metal-containing liquid crystals (metallomesogens), another interesting variety of liquid crystals, have seen some use as stationary phases in GC. These exist as monomers but can also be introduced into polymer systems. They are unique in that they combine shape selectivity with ligand exchange properties [61].

B. Chemical Nature and Properties of Liquid Crystals

The acceptance of liquid crystals as stationary phases in GC depends very much on the availability of chemically and thermally stable compounds of wide mesomorphic temperature ranges and high mesomorphic-isotropic transition temperatures. The thermal stability of the mesophase has generally been characterized in terms of the mesomorphic-isotropic transition temperature (the clearing point). Comparisons of the clearing points of analogous chemical structures with systematic variations have revealed considerable knowledge of the effect of molecular structure on liquid crystal properties. By changing the molecular structure, the molecular dimensions, shape, polarizability, and permanent dipole moment vary, resulting in systematic modifications of the long-range macroscopic properties of the condensed mesophase.

In summary, the greater the anisotropy of the molecular polarizability, the higher the clearing point and the greater the thermal stability of the mesophase. Additional permanent dipole moments will serve to shift the clearing point to higher temperatures. Accordingly, extension of the conjugated central linkage (e.g., replacing of $-CH=N-$ by $-CH=N-N=CH-$) enhances the thermal stability of the mesophase through increased molecular anisotropy of polarizability. For the same reason extension of the aromatic system (e.g., replacing phenyl by biphenyl) also enhances the thermal stability of the mesophase. On the other hand, lateral branching in the central rigid core destabilizes the anisotropic melt because of steric hindrance to optimum parallel arrangement of the molecules. The disruptive effect of the lateral substituents increases as the branching point becomes closer to the central linkage. Terminal substituents generally enhance the anisotropy of polarizability and raise the clearing temperature provided that they increase the molecular length-to-breadth ratio.

For more comprehensive accounts of the subject discussed in this section, the reader is referred to recent books by Gray [62] and Chandrasekhar [63].

C. Chromatographic Selectivity and Order in the Nematic State

It has generally been well established that the nematic phase is superior to smectic and cholesteric phases for GC separations because of the appreciably greater column efficiency that can be achieved with it [4,53,54]. The nematic state, which is considered to be the simplest of all liquid crystal classes, is characterized by long-range orientational order of the major axes of neighboring molecules with the preferred orientation of the long axes varying continuously with position. The degree of order in the nematic state is given by the *order parameter* (S), which is a measure of the fraction of molecules aligned with their long axes parallel to the preferred direction (the optical director). The order parameter is defined by the expression:

$$S = \frac{1}{2} <3 \cos^2 \theta \ 1>$$

where θ denotes the angle between the major molecular axis and the optical director and the brackets denote average valve. For perfect alignment, S = 1 (unattainable in the absence of external constraints because of the normal thermal motion of the molecules), while for orientational disorder, S = 0 (such as in the isotropic liquid state). The normal range of S valves encountered in the nematic phase is $0.8 \geq S \geq 0.4$, with the upper limit being found at the lowest temperature to which the nematic phase can be cooled without crystallization or transition to a smectic phase, and the lower limit at or near the nematic-isotropic transition temperature.

The Maier-Saupe theory [64] (the most widely quoted theory of the nematic state) is particularly successful in accounting for the orientational properties of nematic mesophases and in explaining the order-disorder transition. According to this theory, S is predicted to be a universal function of reduced temperature ($\tilde{T}$) defined as:

$$\tilde{T} = \frac{T}{T \ (N-I)}$$

where T is the experimental temperature and T (N–I) is the nematic-isotropic transition temperature. S($\tilde{T}$) versus $\tilde{T}$, for all nematogens, should fall on a single universal curve, decreasing with increasing temperature and vanishing at T (N -I).

The selective affinity of nematic mesophases toward linear rodlike molecules has been repeatedly demonstrated in the separation of close-boiling *meta-* and *para*-disubstituted benzenes by GC using liquid crystalline stationary phases [4,53,54]. The more rodlike *para* isomer is invariably retained longer. The remarkable shape selectivity exhibited by such phases is closely related to the parallel molecular alignment of the mesophase molecules that exercise preferential solubility for rigid linear molecules and steric discrimination against bulky molecules. The unusual solvent properties of such compounds should be greater, the larger the degree of order of the molecules in the mesophase. This, in turn, should be greater, the stronger the anisotropic forces responsible for order in the mesophase. According to the Maier-Saupe theory [64], stronger anisotropic forces are reflected in higher mesomorphic-isotropic transition temperatures, and for a single compound anisotropic forces are strongest at the lower temperature end of the mesomorphic region.

A direct relationship between selectivity of nematic stationary phases and the degree of order in nematic solvents has long been sought. Kelker [4] compared the relative retention of a *para*-xylene/*meta*-xylene mixture on four GC nematic stationary phases and observed that the selectivity is higher with liquid crystals of wider nematic ranges. When Schroeder [65] attempted a similar correlation between maximum selectivity and nematic range with data on eight GC nematic phases, expectations were apparent. Although this empirical approach is partially successful, it is not theoretically justified, since the nematic range depends on the crystal-nematic transition temperature, and this, in turn, does not show regular trends for mesomorphic compounds. Alternatively, the author theorized that a direct correlation should exist between the relative retention of two close-boiling solutes ($\alpha_{1,2}$) and the order parameter of the nematic phase [53]. Since according to the Maier-Saupe theory S is a universal function of reduced temperature, then $\alpha_{1,2}$ of two close-boiling solutes should also be a universal function of reduced temperature. This hypothesis was tested with relative retention data for the two isomeric solutes anthracene and phenanthrene at different temperatures in the nematic range of five different liquid crystals, and the results were in qualitative agreement with the theoretical prediction of the theory [53]. Accordingly, when comparing the separation efficiency of liquid crystals with closely related chemical structures, selectivity improves the higher the nematic-isotropic transition temperature, and optimum selectivity is ultimately achieved at the lowest temperature to which the nematic phase could be cooled.

D. Chromatographic Behavior of Liquid Crystal Stationary Phases

Because of their unique selectivity towards rigid solute isomers, liquid crystalline stationary phases were considered at one time to be a very promising class of materials that give GC separations very different from those that can be obtained with any other stationary phase. In contrast to conventional polymeric phases, little attention has been paid to the investigation of column efficiency in relation to the chromatographic properties of liquid crystal stationary phases. This is, perhaps, because the remarkable selectivity of these phases for the particular solutes studied results in large separation factor (α) values for pairs of the most troublesome isomers. This is especially significant since large α values result in considerable improvement in resolution, even with the use of mediocre packed columns of 300–600 theoretical plates per foot. However, liquid crystal stationary phases proved in practice to have a number of drawbacks. Most fail to wet inert GC supports such as Chromosorbs, which results in very uneven coatings that give rise in turn to poor efficiency with packed-column systems. The use of capillary columns did not result in any substantial improvement as the number of theoretical plates of a capillary liquid crystal column was reportedly about one-fifth that which can be obtained by a conventional stationary phase [66,67]. Many of the compounds also tend to be unacceptably volatile at as little as 10°C above their clearing points, and many have somewhat narrow useful mesomorphic temperature ranges.

Laub and coworkers [66,67] attempted to overcome some of the above difficulties by blending, for example, BBBT with a methylphenylsiloxane, with only little success. For example, while the efficiency of the mixed solvents was improved substantially over that of pure BBBT, and while the volatility of the mesomorphic compounds was also reduced, its transition temperature was not thereby altered. Furthermore, when columns were filled with blends of liquid crystal and conventional phases (e.g., BBBT + SE-30), their efficiency varied with the composition of the mixture and with the method of column preparation [68].

III. 1980 TO THE PRESENT

The early 1980s witnessed a major advance in the subject of this review, namely, the introduction of polymeric liquid crystals and their subsequent use as stationary phases in GC. In this section we present the history of the development of polymeric liquid crystal stationary phases and provide examples of their application in GC and superentical fluid chromatography.

A.   Why the Need for Polymeric Phases?

As mentioned earlier, the use of high-temperature monomeric liquid crystal stationary phases for the analysis of high-boiling solute classes such as polycyclic aromatic hydrocarbons was initiated in our laboratory [40–53]. At the time of our initial activity only 16 liquid crystals with mesomorphic-isotropic transition temperatures higher than 200°C were known (Ref. 53, Table 4). In a 1982 review [54], Wilkiewicz listed about 30 liquid crystals that fit this criterion, and only a few more were added in the late 1980s and 1990s [69–72]. For a liquid crystal to be ideal for use in gas chromatography at elevated temperatures, it needs to have as high a mesomorphic-isotropic transition temperature as possible and as low a melting point as possible. It needs to be thermally and chemically stable at temperatures as high as 300°C. Although some of the monomeric liquid crystals fit this requirement, the utility of such low molecular weight phases (compared to conventional polymeric phases) is limited by column bleed at elevated temperatures and restricted useful mesomorphic temperature ranges. Furthermore, the efficiency of liquid crystal columns is moderate, and fabrication of open-tubular columns with these materials has also proven in practice to be difficult [45,66–68]. The need for more efficient materials have prompted researchers to look for alternatives.

B.   How Polymeric Liquid Crystal Stationary Phases
Were Introduced

The first person to think of liquid crystal polymers as an alternative was Richard Laub, who at the time was in transition from Ohio State University, Columbus, OH to San Diego State University, San Diego, CA. Grafted polymers, in which low-temperature monomeric liquid crystals are attached with flexible spacers, like the teeth of a comb, running the length of a polymer backbone, had just been introduced by Finkelmann and coworkers [73]. Richard Laub was somehow alerted to this seminal work and cleverly thought of grafting high-temperature liquid crystals to backbone polymers and testing them as stationary phases for GC [74]. A sample of one of the products was also independently evaluated by Lee and coworkers [75]. The first time I heard about polymeric liquid crystal stationary phases was in the summer of 1982 when I met Richard Laub and he showed me his preliminary results. I was very much impressed with the potential of these novel materials and immediately agreed to collaborate with him. His earlier materials suffered from comparatively meager column efficiency and poor thermal stability arising from incomplete substitution of all available hydrogens on the starting poly(methyl siloxane), as well as polymer cross-linking. Moreover, the batch-to-batch reproducibility of the synthetic procedure, as well as yields, were poor.

In September 1983 I took a one-year sabbatical leave and joined Laub in his laboratory at San Diego State University. It proved to be a very fruitful collaboration as we combined Laub's experience in the manufacture of glass capillary columns (fused-silica capillaries were not available) and their application for high-resolution separations with my experience in the synthesis and characterization of liquid crystals and their use as stationary phases in GC.

C. Development of MEPSIL Phases

Our first objective was to clarify and then simplify the synthesis scheme, and then to synthesize and characterize a variety of MEPSIL products. Our first combined paper [76] presented the synthesis and GC application of biphenyl- and terphenyl-based nematic systems. In addition, we provided for the first time a detailed description of each synthetic step, including equivocal aspects that we have come across during the course of our studies, in order to encourage continued investigation of these materials by others. This, to me, represented another milestone, second in importance only to our seminal paper, in which we had introduced high-temperature monomeric liquid crystals as GC stationary phases [11].

The general reaction scheme for the preparation of MEPSILs is given in Figure 4, and the products reported in this paper are presented in Table 2. MEPSIL IV.B (Table 2) was selected for GC application since it provided the broadest nematic range and the highest column efficiency of

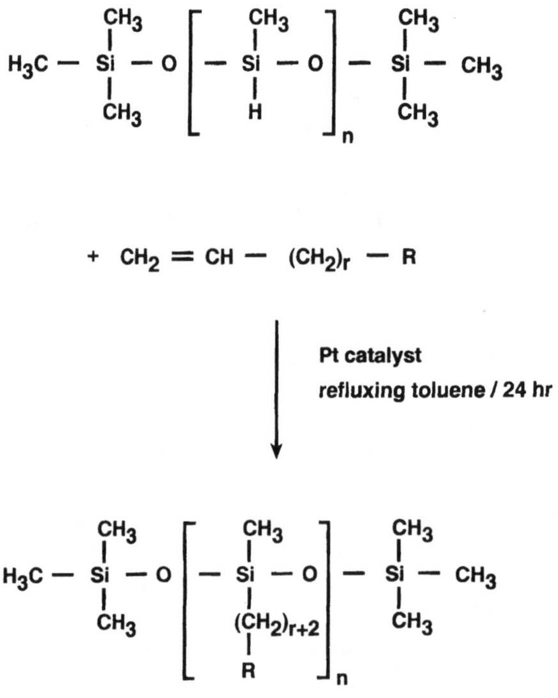

Figure 4 Reaction scheme for synthesis of MEPSIL polymers.

Table 2 Transition Temperatures[a] of Parent 4-(Allyloxy)benzoyl Ester Monomers II.A–II.E and Resultant MEPSIL Polymers IV.A–IV.G

monomer R group	R =	polymer designation	transition temp/°C monomer	polymer
	$\left[H_3C-\overset{O}{\underset{}{Si}}-(CH_2)_3-O-\bigcirc-CO_2R \right]_{n\,=\,36}$			
II.A	—⬡—OCH₃	IV.A	k 88 i	g 15 n 61 i
II.B	—⬡—⬡—OCH₃	IV.B	k 147 n 249 i	k 139 n 319 i
II.C	—⬡—⬡—⬡—OCH₃	IV.C	k > 350	k 200 n 360 (dec) i
II.D	—⬡—⬡ (fused)	IV.D	k 139 i	k 95 n 125 i
II.E	—⬡—⬡—⬡ (fused)	IV.E	k > 350	k 200 n 360 (dec) i
II.B	—⬡—⬡—OCH₃	IV.F[b]	k 147 n 249 i	k 91 n 317 i
II.B	—⬡—⬡—OCH₃	IV.G[c]	k 147 n 249 i	k 87 s 124 n 170 i

[a] DSC, GLC, microscopy: g, glassy; k, crystalline; s, smectic; n, nematic; i, isotropic.
[b] n = 38.
[c] Methylsiloxane/dimethylsiloxane number ratio of 10/17.

all MEPSILs. To illustrate the practical utility of MEPSIL IV.B we presented several applications, including Figure 5, which gives the programmed-temperature separation of a synthetic mixture of 17 of the three- to five-ring PAH encountered most frequently in environmental samples. MEPSIL IV.B provided excellent column efficiency (see Fig. 5), which we ascribed mainly to the unique flexibility of the polysiloxane backbone and its concomitant low surface tension. Glass-capillary columns containing MEPSIL IV.B yielded about 2450 plates/m, N/m (chrysene solute; 230°C), which compares favorably with about 3000 N/m observed with conventional columns of identical dimensions containing SE-30. In contrast, a BBBT column yielded at best only 880 N/m. Following this initial development period, several other MEPSILs were synthesized by us [77] and others [75,78–82]. Of these, two MEPSIL phases have wide mesomorphic temperature ranges as well as high mesomorphic-isotropic transition temperatures and therefore stand out as being potentially the most useful in analytical GC. They are the nematic MEPSIL (IV.B, Table 2) and the smectic MEPSIL (Fig. 1 of Ref. 79).

The latter was commercially available from Lee Scientific, Salt Lake City, UT, and later from Dionex, Lee Scientific Division, Salt Lake City, UT, and was named SB Smectic. We have conducted a comparison of the aforementioned columns and observed that the SB Smectic column was more selective towards polycyclic aromatic hydrocarbons, but that the nematic Mepsil (IV.B) was more useful at higher temperatures because of its higher mesomorphic-to-isotropic transition temperature [83]. Also, while the MEPSIL (IV.B, Table 2) is a homopolymer, SB Smectic is a copolymer. This may account for the batchwise properties of SB Smectic, particularly its transition temperature, which varies by upward of 30°C [84]. One interesting application of MEPSIL columns is the programmed-temperature separation of Aroclors 1254 and 1260 [85]. The separation of Aroclor 1254 is shown in Figure 6. The Ballschmiter-Zell

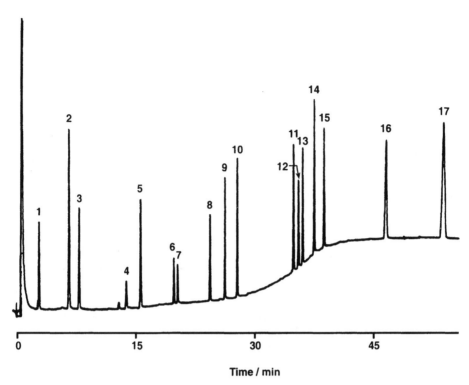

Figure 5 Programmed-temperature separation of common PAH with MEPSIL IV.B. Column temperature: 140° C for 3 min; 140–280°C at 4°C min^{-1}. Column: Pyrex glass (15 m × 0.25 mm i.d.); linear flow velocity 40 cm s^{-1}; solutes: fluorene (1); phenanthrene (2); anthracene (3); fluoranthene (4); pyrene (5); 1,2-benzofluorene (6); 2,3-benzofluorene (7); triphenylene (8); benz[a]anthracene (9); chrysene (10); benzo[b]fluoranthene (11); benzo[k]fluoranthene (12); benzo[e]pyrene (13); perylene (14); benzo[a]pyrene (15); 1,2,3,4-dibenzanthracene (16); benzo[ghi]perylene (17). (From Ref. 76.)

number and substituting pattern for each standard are given in the figure, and the numbered peaks were identified with standards. Since there are 209 discrete PCB compounds (called congeners) comprised of 1–10 chlorine atoms attached to the biphenyl molecule, commercial materials as well as environmental samples that contain PCBs generally give quite complex GC patterns. Moreover, single-run analysis of such mixtures is not possible at present with any one stationary phase. MEPSIL solvents hence will complement, rather than supplant, the use of conventional polysiloxanes for separations of these types. Even so, there is no question that MEPSILs yield a distinctly different elution pattern for PCBs than do the aforementioned materials, and they can therefore be used to validate the identification of PCB congeners in complex mixtures. This was demonstrated by Issaq et al. [86], who used multidimensional GC for the separation of Aroclor congeners. The separation of co-eluting congener peaks from the first column (low-polarity polysiloxane) was achieved when cuts of those peaks were transferred onto a SB smectic column. Although not frequently utilized, the potential advantages of MEPSIL phases in multiple-column switching systems must also be regarded as substantial.

The melting temperatures (solid-mesomorphic) of the most useful of the MEPSIL solvents range from 90 to 140°C. This precludes their application to low-temperature separations of volatile solutes such as benzenes and naphthalenes as they elute with or very close to the solvent

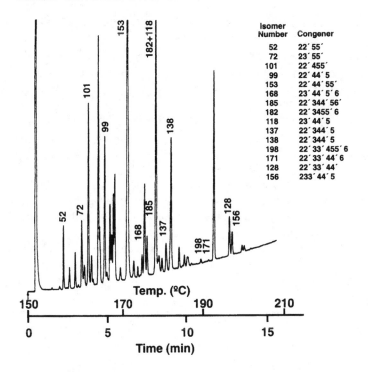

Isomer Number	Congener
52	22′ 55′
72	23′ 55′
101	22′ 455′
99	22′ 44′ 5
153	22′ 44′ 55′
168	23′ 44′ 5′ 6
185	22′ 344′ 56′
182	22′ 3455′ 6
118	23′ 44′ 5
137	22′ 344′ 5
138	22′ 344′ 5
198	22′ 33′ 455′ 6
171	22′ 33′ 44′ 6
128	22′ 33′ 44′
156	233′ 44′ 5

Figure 6 Programmed-temperature separation of Aroclor 1254 with MEPSIL IV.B. Temperature: 150°C for 1 min; 150–210°C at 4°C min^{-1}. Column: fused silica (15 m × 0.25 mm i.d.); linear flow velocity 40 cm s^{-1}. Numbered peaks identified by injection of standards. (From Ref. 85.)

peak. We, therefore, attempted to widen the mesomorphic range of MEPSIL IV.B of Table 2 by blending it with SE-30 poly(dimethylsiloxane) [85]. This resulted in an improvement in column performance at temperatures below MEPSIL's melting point, and the blended phase gave a measurable adjusted retention time for naphthalene. Blending also improved column efficiency, and blended-phase columns exhibited in excess of 3200 plates per meter at elevated temperatures. This was sufficient to provide excellent separations as, for example, the resolution of 9 of 10 methylbenzo[a]pyrene solutes at 275°C, which is impossible to achieve with neat polysiloxane columns [85].

In retrospect, we can safely state that, although other types of polymeric liquid crystal materials were introduced [10,71,87,88], MEPSIL phases were the most widely used. The polysiloxane backbone used in MEPSILs is the most flexible of all polymer backbones, which perhaps explain why polysiloxane is the backbone of most GC stationary phases. Because of the flexibility of the backbone, MEPSILs do not crystallize on cooling, rather they solidify into a glassy state that preserves their mesogenic character. Most important for chromatographers is that glassy solids perfectly wet solid surfaces of capillary walls, making uniform coatings and resulting in efficient columns.

Polyacrylate is another backbone that yields mesomorphic polymers. These polymers are obtained from the corresponding mesogenic acrylate monomer by radical polymerization with 2,2′-azoisobutyronitrile (AIBN) initiator. The first use of polyacrylate liquid crystals as stationary phases in GC was reported in 1985 [81] followed by other reports from the same group [82,89]. One of the most useful polymers reported by the group has a solid-nematic transition temperature of 85°C and a nematic-isotropic transition temperature of 291°C. The separations reported by this group using these phases are equal to those reported with MEPSILs. Of particular interest is the resolution of isomeric insect phermones such as E, E-, Z,Z-, Z,E-, and E,Z-, 12-tetradecadienyl acetates. There is no particular advantage of polyacrylates over MEPSILs, as they are subject to the same drawback that governs MEPSILs, namely, decomposition at the side-chain ester linkage at elevated temperatures. Also, chromatograms reported with a packed column show peak tailing, giving indication of poor column efficiency [81]. Chromatographers were not enthused about polyacrylate stationary phases as no other cited literature reported their use. Most recently, one member of the group that introduced mesomorphic polyacrylates reverted to MEPSILs and reported a new naphthalene-containing MEPSIL [90].

MEPSILs have been coated in narrow-bore open tubular columns and successfully used in supercritical fluid chromatography (SFC), even without cross-linking the stationary phase after coating the column. Higher selectivity is obtained with SFC compared to GC with the same MEPSIL because SFC is operated at a lower temperature [91,92].

In a series of reports [93–97], Lee and coworkers described the synthesis and GC and SFC application of a series of cyanobiphenyl-substituted poly(methylsiloxanes). Some members of this series of phases exhibited retention characteristics that are similar, but not equal, to SB Smectic [97]. They exhibit liquid crystal properties, but their shape selectivity is poor compared to SB Smectic, indicating that they are less ordered [93]. On the other hand, these phases form very efficient columns and were shown to have polar and shape-selective characteristics that are very useful for the separation of fat-soluble vitamins [96] and PCB congeners [97].

D. Chiral MEPSILs

As is clear from the above-cited literature, most polymeric liquid crystals stationary phases are nematic and some are smectic. Cholesteric polymers, in general, are scarce [10,73,98], and they were rarely used as stationary phases in GC [99]. This is because when a cholesteric moiety is attached to a polymer backbone, the resulting polymer is invariably smectic [76,77,100]. The main reason for the selection of a cholesteric liquid crystal stationary phase would be for the separation of optical isomers. This, however, is much more easily achieved with nonmesogenic optical selectors, such as cyclodextrins. Even smectic liquid crystals can be used for the separation of optical isomers provided that the molecules forming the smectic phase have chiral centers [101–103]. Also, chiral monomers that are not mesogenic can be attached to backbone polymers, giving chiral (yet nonmesogenic) stationary phases. The best known compounds of this type are those containing amino acid residues, which were first synthesized by Bayer and Frank [104] and were used for the GC separation of optical isomers [105].

E. Polymeric Versus Monomeric Liquid Crystals

The prevailing consensus among chromatographers is that polymeric liquid crystals make better stationary phases. They argue that polymeric materials are less volatile because of their high molecular weights and that polymers wet the capillary wall or the solid support, resulting in uniform coatings and, hence, more efficient columns. In contrast, most monomeric materials are crystalline, and,hence, they do not coat uniformly, resulting in less efficient columns. Proponents of monomeric liquid crystal stationary phases contend that monomers are single homogeneous compounds while polymers are heterogeneous mixtures of compounds of different backbone lengths and different degrees of side-chain substitution. Thus, while monomers can be synthesized reproducibly, polymers, in contrast, show batch-to-batch variations. Moreover, some monomers, such as isothiocyanates, not only exhibit wide mesomorphic ranges, but they also are thermally and chemically stable, have low viscosities, and show good chromatographic behavior [87].

IV. GC STUDIES OF THE PHYSICOCHEMICAL PROPERTIES OF LIQUID CRYSTALS AND THERMODYNAMICS OF SOLUTION IN LIQUID CRYSTAL STATIONARY PHASES

GC is a particularly attractive method for investigating the thermodynamics of solutions of solutes in liquid-crystalline solvents. The advantages of the GC method include simplicity, accuracy, speed, and the ability to work at infinite dilution, where the solute molecules do not disrupt the long-range order of the mesophase. Furthermore, solutes do not have to be pure, as impurities are removed in the GC process. The method has been successfully applied to the investigation of the thermodynamics of solutions of solutes in nematic [1,2,47,106–108], smectic [109], and cholesteric phases [110,111]. As is evident from the cited literature, Martire and coworkers are the most active in this research area. They have proposed [106] and, subsequently, refined [110,112] an infinite dilution model for the molecular interpretation of the thermodynamics of solution of rigid solutes in liquid crystal solvents. The results of this theory can be qualitatively summarized as follows. Upon transfer of two rigid isomers from the disordered gaseous state (similar for both isomers) to the mesophase, the more rod-like isomer sacrifices more rotational freedom, but at the same time its more favorable geometry allows stronger enthalpic solute-solvent interactions. The entropy loss is overcome by the enthalpy gain, resulting in better solubility (longer GC retention) for the more rod-like solute. More recently, Chao and Martire derived an equation based on the lattice fluid model to describe the chromatographic selectivity enhancement experienced by geometric rigid isomers in liquid crystal solvents [113,114]. The derived equation is applicable in both GC and SFC.

　　GC studies also provide information on mesophase stability [115], melting [116], mesomorphic-mesomorphic [40,41], and mesomorphic-isotropic [2–4] transition behavior. Measurements of several physical properties that depend on the degree of order in the mesophase such as heat capacity [117], IR spectra [118], and NMR linewidth [119] are in qualitative agreement with the GC results on transition behavior. Supercooling is another phenomenon, related to liquid crystal transitions, that have been studied by GC. Upon cooling a liquid crystal melt may stay in the mesomorphic state at temperatures below the normal melting point and is said to be supercooled. This phenomenon has been observed with packed [116] and capillary [120] columns. Studies described in this paragraph where GC is conducted to provide information on the stationary phase and not the injected solutes are termed inverse chromatography. This technique is increasingly being used to probe monomeric [121] as well as polymeric [122,123] liquid crystals.

V. LIQUID CRYSTALS IN HPLC

Liquid crystals have been used as stationary phases in HPLC, albeit with much less frequency than crystals physically coated on solid support [124,125], yet this route is not popular because of the difficulties of retaining the phase in the column as it is continuously dissolving in the running solvent. A much more promising approach is the chemical bonding of liquid crystal moieties to the solid support [126–130]. Whether or not liquid crystal properties are preserved when mesogenic monomers are chemically attached to the surface of silica support is still a subject of contention. However, what is clear from the above-cited literature is that these phases have better planarity and shape-selectivity capabilities compared to commercially available polymeric octadecylsilica (ODS). Readers who are interested in the use of liquid crystals in HPLC can consult other chapters in this book written by J. J. Pesek (Chapter 22), K. Jinno (Chapter 15), and A. M. Siouffi (Chapter 5), who are the leaders in this budding line of research.

ACKNOWLEDGMENT

This project has been funded in whole or in part with federal funds from the National Cancer Institute, National Institutes of Health, under Contract No. NO1-CO-56000.

REFERENCES

1. WL Zielinksi, Jr, DH Freeman, DE Martire, LC Chow. Anal Chem 42:176, 1970.
2. H Kelker. Z Anal Chem 198:254, 1963.
3. MJS Dewar, JP Schroeder. J Am Chem Soc 86:5235, 1964.
4. H Kelker, E Von Schivizhoffen. Advances in Chromatography, Vol. 6. New York: Marcel Dekker, 1968, p. 247.
5. F Reinitzer. Monatsh Chem 9:421, 1888.
6. OZ Lehmann. Phys Chem 4:462, 1889; 56:750, 1906.
7. W Kast. In Landolt-Börnstein, 6th ed., Vol.II, Part 2a. Berlin: Springer-Verlag, York, 1960, p. 266.
8. D Demus, H Zaschke. Flüssige-Kristallen in Tabellen, Vol. II. VEB. Leipzig: Deutscher Verlag für Groundstoffindustrie, 1974.
9. D Demus, ed. Handbook of Liquid Crystals, Vols. II A and II B: Low Molecular Weight Liquid Crystals. New York: John Wiley and Sons, 1998.
10. D Demus, ed. Handbook of Liquid Crystals, Vol. III: High Molecular Weight Liquid Crystals. New York: John Wiley and Sons, 1998.
11. GM Janini, K Johnston, WL Zielinski, Jr. Anal Chem 47:670, 1975.
12. P Shubik. Proc Natl Acad Sci USA 69:1052, 1972.
13. HW Gerade. Toxicology and Biochemistry of Aromatic Hydrocarbon. Amsterdam: Elsevier, 1960.
14. EC Miller, JA Miller. In: A Hallaender, ed. Chemical Mutagens. Vol. 1. New York: Plenum Press, 1971, p. 105.
15. E Sawicki. Chemist-Analyst 53:24, 56, 88, 1964.
16. R Schaad. Chromatogr Rev 13:61, 1970.
17. O Hutzinger, S Safe, M Zande. Analabs Inc. Res. Notes, Vol. 13,No. 3, 1973.
18. E Sawicki, RC Corey, AE Dooley, JB Gisclard, JL Monkman, RE Neligan, LA Ripperto. Health Lab Sci 1:31, 1970.
19. NF Ives, L Guiffrida. J Ass Offic Anal Chem 55:757, 1972.
20. E Clar. Spectrochim Acta 4:116, 1950.
21. V Cantoti, GY Cartoni, A Liberti, AG Torri. J Chromatogr 17:60, 1965.
22. L DeMaio, M Corn. Anal Chem 38:131, 1966.
23. K Bhatia. Anal Chem 43:609, 1971.
24. J Frycka. J Chromatogr 65:341, 432, 1972.
25. DA Lane, HK Moe, M Katz. Anal Chem 45:1776, 1973.
26. A Zane. J Chromatogr 38:130, 1968.
27. N Carugno, S Rossi. J Gas Chromatogr 5:103, 1967.
28. K Grob. Chem Ind (London), 248, 1973.
29. G Grimmer, A Hildebrandt, H Bohnke. Erdöl Kohle 25:442, 531, 1972.
30. G Grimmer. Erdöl Kohle 25:339, 1972.
31. UICC Technical Report Series, Vol. 4, 1970.

32. IARC Internal Techn. Rep., No. 71/002, 1971.
33. WL Zielinski, Jr, K Johnston, GM Muschik. Anal Chem 48:907, 1976.
34. Analabs Tech. Bull., North Haven, CT, 1975.
35. JC Wiley, Jr, CS Menon, DL Fisher, JE Engel. Tetrahedron Lett 33:2811, 1975.
36. S Wasik, S Chesler. J Chromatogr 122:451, 1976.
37. M Hall, DNB Mallen. J Chromatogr 118:268, 1976.
38. M Pailer, V Hlozek. J Chromatogr 128:163, 1976.
39. PE Strup, RD Giammer, TB Stanford, PW Jones. In: R Freudenthal, PW Jones, eds. Carcinogenesis-A Comprehensive Survey. Vol. 1. New York: Raven Press, 1976, pp. 241–251.
40. GM Janini, GM Muschik, WL Zielinski, Jr. Anal Chem 48:809, 1976.
41. GM Janini, GM Muschik, JA Schroer, WL Zielinski, Jr. Anal Chem 48:1879, 1976.
42. GM Janini, GM Muschik, W Manning, WL Zielinski, Jr. J Chromatogr 193:444, 1980.
43. GM Janini, B Shaikh, WL Zielinski, Jr. J Chromatogr 132:136, 1977.
44. GM Janini, GM Muschik, CM Hanlon. Mol Cryst Liq Cryst 53:15, 1979.
45. GM Janini, RI Sato, GM Muschik. Anal Chem 54:2417, 1980.
46. GM Janini, GM Muschik. Mol Cryst Liq Cryst 87:281, 1982.
47. GM Janini, MT Ubeid. J Chromatogr 236:329, 1982.
48. GM Janini, MT Ubeid. J Chromatogr 248:217, 1982.
49. GM Janini, GH Hovakeemian, A Katrib, N Fitzpatrick, S Attari. Mol Cryst Liq Cryst 90:271, 1983.
50. GM Janini, NT Filfil. J Chromatogr 469:43, 1989.
51. GM Janini, AM Al-Ghoul, GH Hovakeemian. Mol Cryst Liq Cryst 172:69, 1989.
52. GM Janini, NT Filfil, GM Muschik. Liq Cryst 7:454, 1990.
53. GM Janini. Advances in Chromatography, Vol. 17. New York: Marcel Dekker, 1979, p. 231.
54. Z Witkiewicz. J Chromatogr 251:311, 1982.
55. ZP Vetrova, NT Karabanov, LA Ivanova, OB Akopova, GT Maidatshenko, YI Yashin. Zh Obshch Khim 57:646, 1987.
56. S Sakagami. J Chromatogr 246:121, 1982.
57. G Kraus, Tran thi Hong Van, W Weissflog. Z Chem 22:448, 1982.
58. S Chandrasekhar, BK Sadashiva, KA Suresh. Pramana 9:471, 1977.
59. Nguyen Huu Tinh, H Gasparoux, C Destrade. Mol Cryst Liq Cryst 68:101, 1981.
60. Z Witkiewicz, B Goca. J Chromatogr 402:73, 1987.
61. C-Y Liu, C-C Hu, J-L Chen, K-T Liu. Anal Chim Acta 385:51, 1999.
62. GW Gray, ed., Thermotropic Liquid Crystals. Critical Reports on Applied chemistry, Vol. 22. New York: John Wiley & Sons, 1987.
63. S Chandrasekhar. Liquid Crystals. Cambridge: Cambridge University Press, 1992.
64. W Maier, A Saupe. Z Naturforsch 13A:564, 1969; 14A:882, 1959; 15A:287, 1960.
65. JP Schroeder. In: GM Gray, PA Winsor, eds. Liquid Crystals and Plastic Crystals. Vol. 1. Chichester, England: Ellis Horwood, 1974, p. 356.
66. RJ Laub. In: DA Young, ed. Chromatography, Equilibria and Kinetics. Proc. Faraday Soc. Symp. (No. 15), Royal Society of Chemistry, 1981, p. 179.
67. RJ Laub, WL Roberts, CA Smith. HRC CC J High Resolut Chromatogr Chromatogr Commun 3: 355, 1980.
68. T Kreczmer, A Gutorska. Chem Anal (Warsaw) 30:419, 1985.
69. F Ammar-Khodja, S Guermouche, MH Guermouche, P Berdague, JP Bayle. Chromatographia 50: 338, 1999.
70. M Baniceru, S Radu, C Sarpe-Tudoran. Chromatographia 48:427, 1998.
71. WJ Leigh, MS Workentin. In: D Demus, ed. Handbook of Liquid Crystals. Vol. I. New York: John Wiley and Sons, 1998, pp. 839–895.
72. A Kraus, G Kraus, R Kubinec, I Ostrovsky, L Sojak. Chem Anal (Warsaw) 42:497, 1997.
73. H Finkelmann. Adv Polymer Sci 60/61:99, 1984.
74. H Finkelmann, RJ Laub, WL Roberts, CA Smith. Paper Presented at the Sixth International Symposium on Polynuclear Aromatic Hydrocarbons, Columbus, OH, October 1981.

75. RC Kong, ML Lee, Y Tominaga, R Pratap, M Iwao, RN Castle. Anal Chem 54:1802, 1982.
76. MA Apfel, H Finkelmann, GM Janini, RJ Laub, B-H Lühmann, A Price, WL Roberts, TJ Shaw, CA Smith. Anal Chem 57:651, 1985.
77. GM Janini, RJ Laub, TJ Shaw. Makromol Chem Rapid Commun 6:57, 1985.
78. BA Jones, JS Bradshaw, M Nishioka, ML Lee. J Org Chem 49:4947, 1984.
79. KE Markides, M Nishioka, BJ Tarbet, JS Bradshaw, ML Lee. Anal Chem 57:1296, 1985.
80. M Nishioka, BA Jones, BJ Tarbet, JS Bradshaw, ML Lee. J Chromatogr 357:79, 1986.
81. S Rokushika, KP Naikwadi, AL Jadhav, H Hatano. HRC CC J High Resolut Chromatogr Chromatogr Commun 8:480, 1985.
82. AL Jadhav, KP Naikwadi, S Rokushika, H Hatano, M Ohshima. HRC CC J High Resolut Chromatogr Chromatogr Commun 10:77, 1987.
83. GM Janini, GM Muschik, SD Fox. Chromatographia 27:436, 1989.
84. KE Markides, HC Chang, CM Schregenberger, BJ Tarbet, JS Bradshaw, ML Lee. HRC CC J High Resolut Chromatogr Chromatogr Commun 8:516–520, 1985.
85. GM Janini, GM Muschik, HJ Issaq, RJ Laub. Anal Chem 60:1119, 1988.
86. HJ Issaq, SD Fox, GM Muschik. J Chromatographic Sci 27:172, 1989.
87. Z Witkiewicz. J Chromatogr 466:37, 1989.
88. GM Janini, RJ Laub, JH Purnell, OS Tyagi. In: CB McArdle, ed. Side Chain Liquid Crystal Polymers. London: Blackie, New York: Chapman and Hill, 1989, pp. 395–441.
89. KP Naikwadi, AL Jadhav, S Rokushika, H Hatano, M Ohshima. Makromol Chem 187:1407, 1986.
90. KP Naikwadi, PP Wadgaonka. J Chromatogr A 811:97, 1998.
91. HC Chang, KE Markedis, JS Bradshaw, ML Lee. J Microcol (Sept. 1):131, 1989.
92. HC Chang, KE Markedis, JS Bradshaw, ML Lee. J Chromatogr Sci 26:280, 1988.
93. A Malik, I Ostrovsky, SR Sumpter, SL Reese, S Morgan, BE Rossiter, JS Bradshaw, ML Lee. J Microcol (Sept. 4):529, 1992.
94. BE Rossiter, SR Reese, S Morgan, A Malik, JS Bradshaw, ML Lee. J Microcol (Sept. 4):521, 1992.
95. YS Shen, W Li, A Malik, SL Reese, BE Rossiter, ML Lee. J Microcol (Sept. 7):411, 1995.
96. Y Shen, JS Bradshaw, ML Lee. Chromatographia 43:53, 1996.
97. BR Hillery, JE Jirard, MM Schantz, SA Wise, A Malik, ML Lee. J Microcol (Sept. 7):221, 1995.
98. SK Aggarwal, JS Bradshaw, M Eguchi, S Parry, BE Rossiter, KE Markides, ML Lee. Tetrahedon 43:451, 1987.
99. JS Bradshaw, SK Aggarwal, CA Rouse, BJ Tarbet, KE Markides, ML Lee. J Chromatogr 405:169, 1987.
100. NW Adams, JS Bradshaw, J-M Bayona, KE Markides, ML Lee. Mol Crys Liq Crys 147:43, 1987.
101. CH Lochmüller, JV Hinshaw. J Chromatogr 171:407, 1979.
102. CH Lochmüller, JV Hinshaw. J Chromatogr 178:411, 1979.
103. CH Lochmüller, JV Hinshaw. J Chromatogr 202:363, 1980.
104. E Bayer, H Frank. German Patent 3005024, August 1981.
105. B Koppenhoefer, E Bayer. Chromatographia 19:123, 1984.
106. LC Chow, DE Martire. J Phys Chem 75:2005, 1971.
107. LC Chow, DE Martire. Mol Crys Liq Crys 14:293, 1971.
108. DE Martire, A Nikolic, KL Vasanth. J Chromatogr 178:401, 1979.
109. DE Martire, PA Blasco, PE Carone, LC Chow, H Vicini. J Phys Chem 72:3489, 1968.
110. JM Schnur, DE Martire. Mol Crys Liq Crys 26:213, 1974.
111. AA Jekmavorian, P Barrett, AC Watterson, EF Barry. J Chromatogr 107:317, 1975.
112. DE Martire. Mol Crys Liq Crys 28:63, 1974.
113. Y Chao, DE Martire. J Phys Chem 96:3505, 1992.
114. Y Chao, DE Martire. Anal Chem 64:1246, 1992.
115. DE Martire, GA Oweimreen, GI Agren, SG Ryan, HT Peterson. J Chem Phys 64:1456, 1976.
116. S Wasik, S Chesler. J Chromatogr 122:451, 1976.
117. H Arnold, H Sackman. Z Phys Chem 213:137, 1959.
118. W Maier, G Englert. Z Electrochem 64:689, 1960.

119. KH Weber. Disc Faraday Soc 25:7, 1958.
120. E Grushka, JF Solsky. Anal Chem 45:1836, 1973.
121. GJ Price, IM Shillcock. Can J Chem 73:1883, 1995.
122. M Romansky, JE Guillet. Polymer 35:584, 1994.
123. GJ Price, IM Shillcock. Polymer 34:85, 1993.
124. PJ Taylor, PL Sherman. J Liq Chromatogr 3:21, 1980.
125. AA Arastkova, VP Vetrova, YI Yoshin. J Chromatogr 365:27, 1986.
126. O Ferroukhi, S Guermouche, MH Guermouche, P Berdague, JP Balye, E Lafontaine. Chromatographia 48:823, 1998.
127. I Terrien, MF Achard, G Felix, F Hardouin. J Chromatorg A 810:19, 1998.
128. JJ Pesek, T Cash. Chromatographia 27:559, 1989.
129. JJ Pesek, AM Siouffi. Anal Chem 61:1928, 1989.
130. Y Saito, K Jinno, JJ Pesek, Y-L Chen, G Luehr, J Archer, JC Fitzer, WR Biggs. Chromatographia 38:295, 1994.

11

Quantitative Retention–Eluent Composition Relationships in Partition and Adsorption Chromatography

Edward Soczewiński

Medical University, Lublin, Poland

I. THE BEGINNINGS

My first contact with chemistry occurred in September 1942. It was a difficult year. In March my family was expelled from our small house, Maidanek Street 15 in the eastern suburb of Lublin; the Germans formed there a provisional smaller ghetto for the Jews. We and our neighbors stayed for a fortnight in the Lublin churches while Jewish teams cleaned the flats left by the Jews; then we got an empty flat in the former ghetto, still smelling of sulfur dioxide. The Jewish teams were burning in the streets the furniture, books, and other belongings of the Jews who were earlier mostly transported to the death camps in Treblinka, Belzec, and Sobibor in the East; the flames reached the second floor of the buildings.

The notorious "Konzentrationslager der Waffen SS Lublin," with its gas chamber and crematorium, was named by the Polish population Maidanek camp from the street we lived in. In autumn 1942 the provisional ghetto in our suburb was liquidated. We were allowed to return to our devastated house in February 1943.

In June 1942 I finished the 7-class elementary school. The Nazis closed all universities and high schools; Poles were intended to have only primitive education. However, there was a chemical school for technicians—Chemische Fachschule—and I started chemical education. In July 1944 the German army retreated before the advancing Red Army, and the front stabilized until January 1945 on the Vistula river, some 30 miles from Lublin. The chemical school was extended to 6 years. In 1948 I was granted the maturity certificate with the degree of chemical engineer.

In the autumn of 1944, still during the war, the University of Lublin (UMCS) was organized "from zero," the first University in postwar Poland, Maria Curie-Sklodowska, the double Nobel Prize winner (1903, 1911), was chosen as its patron. In the autumn of 1948 I started to study chemistry. I liked most the lectures on physical chemistry by Professor Andrzej Waksmundzki, who graduated in 1936 from the Yagellonian University in Cracow and in 1939 received the doctor of chemistry degree as a coworker of professor Bogdan Kamienski. Professor Waksmundzki, as an underground soldier, spent the years 1942–1945 in concentration camps in Auschwitz, Gross Rosen, and Mauthausen. After the liberation in May 1945 by the U.S. Army, he returned to Poland and in 1945 was nominated head of the newly organized Department of Physical Chemistry of the University and in the next year head of Department of Inorganic Chemistry of the Pharmaceutical Faculty. In 1952, after graduating from the University, I was employed in the Department of Inorganic Chemistry, but after one year was delegated by the newly formed (1950) Medical Academy to the Department of Physical Chemistry of UMCS to teach physical chemistry to the students of the Pharmaceutical Faculty.

The topic of my study for the M.Sc. Degree (master of chemistry), directed by Professor Waksmundzki, was concerned with the separation of alkyl pyridines by countercurrent distribution (CCD). The experiments were quite tedious; 20 test tubes attached to a horizontal axis were rotated to attain liquid-liquid equilibrium and then the lower phases were transferred to the next tubes using a pipette. After the leading lower phase reached the end of the series, the pyridine content in the 20 tubes was determined by titration and the results plotted as concentration vs. tube number diagram. The introduction of my M.Sc. Thesis was based on a chapter of Craig and Craig in the book *Techniques of Organic Chemistry* edited by Weissberger [1].

This experience directed my attention for many years (until 1968) to separations in liquid-liquid systems. My first publications were reviews on CCD [2,3]. My contact with chromatography was rather accidental: helping a student of chemistry, Richard Aksanowski (1955), and having no idea of the then-applied paper chromatographic aqueous one-phase systems (e.g., the Partridge system, *n*-butanol–acetic acid–water 4:1:5, the organic water–saturated phase, or phenol saturated with water), I advised him, on the basis of my experience with CCD, to impregnate paper strips with aqueous buffer solution and then, after partial drying (to regain porous structure of cellulose), to develop the strip with nonpolar or weakly polar water-immiscible solvents. In this way, a true two-phase liquid-liquid partition system was formed. Its advantage was that in most cases the R_F values were strictly related to the equilibrium distribution coefficients ($D = C_{org}/C_w$; $r = V_{org}/V_w$):

$$R_F = \frac{Dr}{Dr + 1} \tag{1}$$

We called this technique the "moist buffered paper" method.

My first experimental publication was based on repetition of Aksanowski's experiments [4]. The R_F values plotted against the pH of buffer solution formed regular, S-shaped curves, which reflected the ionisation equilibria of quinolines (Fig. 1).

Substitution of an equation for $D = f(pH)$ relationship (k^0 = partition coefficient of unionized base, C_{org}/C_w; K_a = acidic ionization constant of base):

$$D = \frac{k^0}{1 + \frac{[H^+]}{K_a}} \tag{2}$$

into Eq. (1) resulted in Eq. (3) for R_F vs. pH relationship:

$$R_F = \frac{k^0 r}{k^0 r + 1 + \frac{[H^+]}{K_a}} \tag{3}$$

For acids, a symmetrical equation was derived:

$$R_F = \frac{k^0 r}{k^0 r + 1 + \frac{K_a}{[H^+]}} \tag{3a}$$

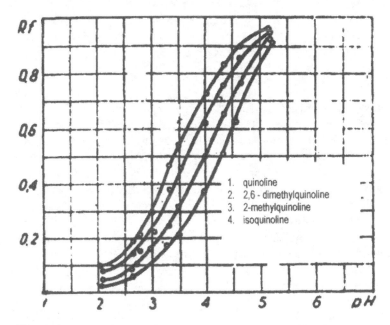

Figure 1 R_F vs. pH plots of quinoline, 2,6-dimethyl quinoline, 2-methyl quinoline, and isoquinoline for the system: paper + aqueous buffer solution/petroleum ether. (From Ref. 4.)

R_F vs. pH relationships for bases and acids (Fig. 2) were in accordance with experimental curves [5–7].

The liquid-liquid partition mechanism in moist buffered paper chromatography made it possible to apply this technique for the estimation of selectivity of extraction and choice of optimal solvent/pH conditions for preparative separation of alkaloids from plant extracts by CCD [8–12]. The theory of buffered paper chromatography, with experimental verification, was the subject of my Ph.D. Dissertation (1960, adviser—Prof. Dr. Andrzej Waksmundzki).

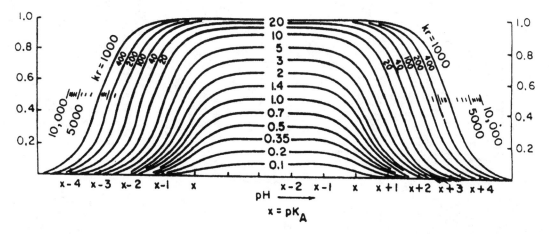

Figure 2   Theoretical R_F vs. pH curves for bases and acids. (From Refs. 5, 22.)

Already in my dissertation I mentioned that retention in liquid-liquid chromatography (LLC) can also be controlled by variation of composition of one of the phases, citing the equation derived by Kemula and Buchowski [13] for partition coefficient K:

$$\log K = X_1 \log K_1 + X_2 \log K_2 \tag{4}$$

where X is the mole fraction and the indices 1,2 denote the two components of the binary phase immiscible with the other phase (e.g., heptane (1) + benzene (2)/water).

On October 3, 1960, I passed the examinations (physical chemistry, philosophy) for the doctor of chemistry degree and defended my thesis (UMCS); the next day I went to Stockholm on a 3-month scholarship at the Institute of Chemistry and Biochemistry, University of Stockholm. This short stay and cooperation with Prof. Dr. Carl Axel Wachtmeister, who specialized in the extraction of phenolic acids from plants, their isolation by chromatography, elucidation of structures, and verification by synthesis, proved to have a very positive effect on my further scientific career: my main topic of investigations, which can be termed "quantitative retention–solvent composition relationships," started with a paper published with Prof. Wachtmeister [14] in which Eq. (4) expressed as chromatographic retention parameter, $R_M[R_M = \log (1 - R_F)/R_F = \log k; k =$ retention factor], was derived:

$$R_M = X_1 R_{M1} + X_2 R_{M2} \tag{5}$$

or

$$\log k = X_1 \log k_1 + X_2 \log k_2 \tag{5a}$$

and verified for the system di-*n*-butyl ether/water + dimethyl sulfoxide (Fig. 3). Later it was found that owing to compensation of nonideality effects, mole fractions can be substituted by volume fractions φ of component solvents 1,2:

$$\log k = \varphi_1 \log k_1 + \varphi_2 \log k_2 \tag{5b}$$

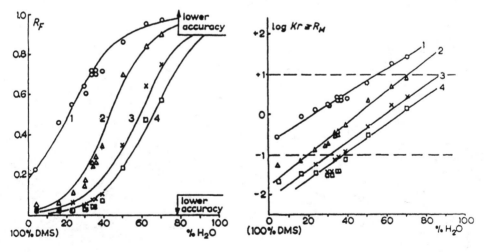

Figure 3 R_F and R_M values of dihydroxydibenzofurans plotted against % H_2O for the system: paper + dimethylsulfoxide + water/di-*n*-butyl ether. (From Ref. 14.)

An analogous semilogarithmic equation was reported by Snyder [15] to hold for very popular high-performance liquid chromatography (HPLC) systems of the type octadecyl silica/water + methanol. Snyder's equation has the more convenient form [it can be deduced from Eq. (5b), substituting φ_1 by $(1 - \varphi_2)$]:

$$\log k = \log k_w - S \, \varphi_{mod} \tag{6}$$

where k_w is the retention factor for pure water and φ_{mod} the volume fraction of modifier (e.g., methanol). For $\varphi_{mod} = 1$ (pure modifier), $S = \log k_w - \log k_{mod}$, i.e., $S = \log(k_w/k_{mod})$—the logarithm of hypothetical partition coefficient of solute between water and modifier (actually miscible).

Boyce and Milborrow [16] applied Eq. (5) to the determination of high partition coefficients by thin-layer chromatography (TLC) on silica impregnated with paraffin oil and developed with mixtures of water and acetone (modifier). The measurable range of R_F values was obtained only for $\varphi_{mod} > 0.4$, but the R_M value for pure water was obtained by linear extrapolation to $\varphi_{mod} = 0$. The extrapolated R_M value for pure water can be considered as hydrophobicity parameter, an essential property that characterizes the rate of penetration of drugs through aqueous/lipid membranes in living organisms and is thus important for the biological activity of substances. Earlier Hansch and Fujita (see Ref. 17) in their QSAR concept (quantitative structure–(biological) activity relationships), applied as the measure of hydrophobic properties the partition coefficient between n-octanol and water (P; log P is an additive property relative to component groups of the solute molecule, like R_M). However, determination of static P values is tedious, and chromatographic R_{Mw}(log k_w) values can be determined very simply in TLC for numerous compounds in several experiments, so that determination of hydrophobic properties by chromatographic methods (TLC, HPLC) became a routine procedure in QSAR investigations [18–20]. Anyway, good correlations between log P and log k_w values are frequently reported.

After some further studies on liquid-liquid equilibria for binary phases [21] in 1963, I received the higher university degree of docent, necessary in Poland for obtaining a professorship.

II. HEAD OF DEPARTMENT: INVESTIGATION OF LLC PARTITION SYSTEMS

In 1964 Professor Waksmundzki did not get approval from the Ministry of education to be chairman of two departments, and I was appointed chairman of the Department of Inorganic Chemistry in the Pharmaceutical Faculty. Close cooperation with Prof. Waksmundzki continued for many years.

To be responsible for the scientific careers of my coworkers was a difficult task; the department had virtually no scientific equipment, which remained the case until the 1980s. The Medical Faculty predominated in the Medical Academy, and the constant argument of the clinics was that they needed equipment to save the lives of patients. Fortunately, the physicochemical approach to chromatography I learned from Prof. Waksmundzki was a great asset; we specialized in the theory of optimization of chromatographic analysis, first in partition and then also in adsorption systems, including molecular models, effects of molecular structure on retention, and selectivity vs. eluent composition relationships. These could be investigated by "cheap" techniques—paper and later thin-layer chromatography—so that gradually my coworkers received doctoral and docent degrees and our investigations were noticed in the international literature, especially when we started publishing in international journals. I summed up my investigations on buffered paper chromatography and binary eluents in partition chromatography in a review published at the invitation of the editors of *Advances in Chromatography* [22].

Similar types of retention–solvent composition relationships were confirmed for gas-liquid chromatographic systems in the Ph.D. thesis of Suprynowicz [23]; however, for liquid-solid adsorption TLC on silica, strong deviations were observed [24].

Deviations from linearity were also observed for certain liquid-liquid partition systems, especially those composed of a nonpolar diluent and a polar (but immiscible with water) solvent—e.g., heptane + tributyl phosphate/water. By analogy to the partition of metal ions in analogous systems, where the law of mass action can be applied to formation of solvates of metal ions and log D is linearly dependent on the log concentration of extractant, with the slope of the plot related to the composition of the extracted complex, I tried to plot the retention vs. solvent composition relationships in the log vs. log coordinate system, and in fact linear plots were then obtained. It was thus concluded that two types of retention vs. eluent composition relationships are frequent [25]:

$$\log - \log: R_M = \text{const.} - m \log C_{mod} \tag{7}$$

$$\text{semilogarithmic: } R_M = \text{const.} - S\,C_{mod} \qquad\qquad \text{equivalent to Eq. (6)}$$

depending on the type of system.

Equation (7) was applied to the investigation of solvation and liquid-liquid partition equilibria of numerous organic substances (phenols, quinolines, anilines, alkaloids) for numerous polar solvents (modifiers). In Eq. (7), in an idealized case, the slope m corresponds to the stoichiometry

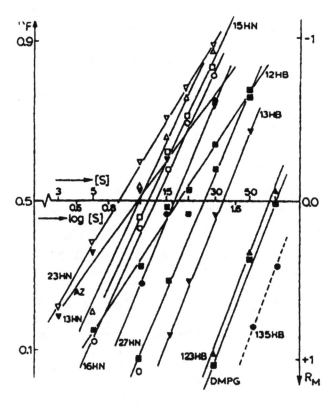

Figure 4 R_M vs. log %S plots of hydroxy derivatives of benzene (B) and naphthalene (N). Stationary phase: paper impregnated with water; mobile phase: cyclohexane + methyl-*n*-amyl ketone (S). (From Ref. 26.)

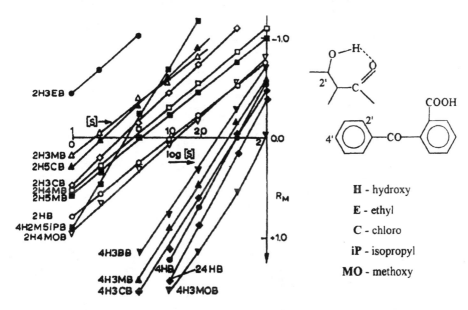

Figure 5 R_M vs. log %S plots of 2-benzoylbenzoic acid derivatives. Stationary phase: aqueous buffer solution of pH = 6.01. Mobile phase; cyclohexane + n-hexanol (S). (From Ref. 27.)

of the solvation complex ZS_m of solute Z with polar solvent S (especially H-bonding the polar groups of the solute); thus, it is determined by the number of polar groups of the solute molecule. In fact, analogies in structure were found to be reflected by parallel R_M vs. log C_{mod} lines and were close to 2 for, e.g., dihydroxy compounds and 1 for monohydroxy compounds. However, some molecular effects are also reflected in the plots: for instance, for the *ortho* position of two polar groups (especially when they can form an internal H-bond), the slope m is <2.0 (e.g., it is about 2.0 for 2,4- and 2,5-dihydroxynaphthalene but 1.5 for 2,3-dihydroxynaphthalene and 1,2-dihydroxybenzene (Fig. 4) [26]). The effect is especially evident for two series of benzoylbenzoic acids (Fig. 5) [27].

Electron donor solvents can form hydrogen bonds with the carboxyl group and hydroxyl group, and in fact the slopes for 4′-hydroxyl derivatives (4′H) are about 2.0; however, the OH group in the 2′ derivatives (2′H) can form an internal H-bond with the carbonyl group, and then the value of m is distinctly lower (about 1.5) so that the R_M vs. log C_{mod} lines form two sets (4′H and 2′H) of differing slopes.

III. INVESTIGATIONS ON LIQUID-SOLID ADSORPTION SYSTEMS

As mentioned earlier, R_M vs. C_{mod} plots for nonaqueous systems of the type silica/nonpolar solvent (diluent) + polar solvent (modifier) are parabolic in shape (Fig.6) [24], similar to partition systems with a solvating modifier [Eq. (7)] plotted in the "wrong" semilogarithmic system of coordinates [Eq. (5b)]. I found that the data replotted in a log-log system were linear

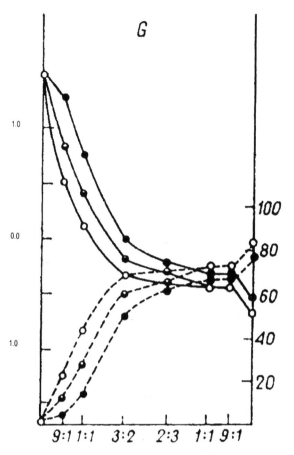

Figure 6 R_F (····) and R_M (—) values of nitroanilines ($\bigcirc$ = 2-; $\ominus$ = 3-; $\bullet$ = 4-) for the system silica/cyclohexane + dioxane plotted against content of dioxane in the eluent. (From Ref. 24.)

(Fig. 7). In consequence, I elaborated a simple model of adsorption from nonaqueous binary solutions (diluent N − polar modifier S), analogous to ion exchange equilibria except that ion pairing is replaced by formation of H-bonds in a competition between solute Z and modifier S for adsorption sites A (Fig. 8).

The competitive exchange equilibrium could be represented for single function solute Z by:

$$A - S + Z \rightleftarrows A - Z + S$$

Treating the system as a whole (composed of N, A, S, Z, AS, AZ) and expressing concentrations of the species in mole fractions and applying the law of mass action to equilibrium, we can write

$$K = \frac{X_{AZ} \cdot X_S}{X_{AS} \cdot X_Z} \tag{8}$$

Thus, the retention factor

$$k = \frac{X_{AZ}}{X_Z} = K \frac{X_{AS}}{X_S} \tag{9}$$

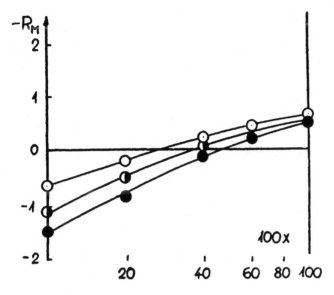

Figure 7 Data of Fig. 6 plotted in the log-log scale ($\bigcirc$ = 2-; $\ominus$ = 3-; $\bullet$ = 4-). (From Ref. 29.)

In logarithmic form, assuming that $X_{AS} \gg X_{AZ}$:

$$\log k = R_M = \text{const.} - \log X_S \qquad (10)$$

When $X_s = 1$, the $\log k = \log k_{mod}$, that is, the constant is equal to the right-hand side of the plot, the $\log k$ value of the solute adsorbed from pure modifier S.

Analogous reasoning for two-point adsorption (two-function solute Z) results in the corresponding formula:

$$\log k = R_M = \text{const.} - 2 \log X_S \qquad (11)$$

Generalizing, for m-point adsorption:

$$\log k = \text{const.} - m \log X_S \qquad (12)$$

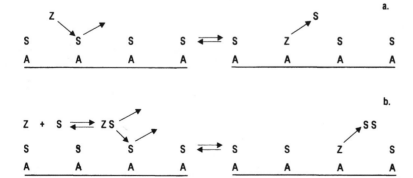

Figure 8 Mechanistic model of competitive adsorption of solute Z on surface silanol groups A covered by monomolecular layer of modifier S; (a) simple displacement, (b) displacement combined with decomposition of solvate ZS.

However, in the first paper from a TLC series on molecular adsorption model, it was mentioned that solvation of the polar solute group by the formation of an H-bond (Fig. 8b) with the modifier molecule can increase the slope m. Thus for the single function solute molecule, its attachment to the surface silanol group entails dissociation of the solvation complex ZS; this can be represented by a different equation:

$$AS + ZS \rightleftarrows AZ + 2S$$

so that (for strong solvation) the slope of a monofunctional solute is m = 2.0. The slope of R_M vs. log X_s plots thus reflects the stoichiometry of adsorption and (competitive) solvation. In real systems the variation of the activity coefficients of the species involved should be taken into account; extensive thermodynamic investigations, especially by Rudzinski, Jaroniec, and Martire, were recently discussed by Rizzi [28].

I sent the first paper from the adsorption series to *Analytical Chemistry*. It was risky—I entered a new field of investigations! The first opinion of the Editor's Referee, Dr. Lloyd R. Snyder, a leading specialist in adsorption chromatography, was critical; however, after exchange of correspondence and modifications of the text, the paper appeared in *Analytical Chemistry* [29]; Snyder's comments were very helpful in making the model more precise. The paper met with interest by readers; I received a record number of reprint requests. It started a new direction of research in our department; Golkiewicz in his Ph.D. dissertation systematically investigated adsorption on silica from binary eluents for numerous solutes and modifiers (Fig. 9) [30], analogous investigations for alumina were published by Wawrzynowicz [31] and for polyamide

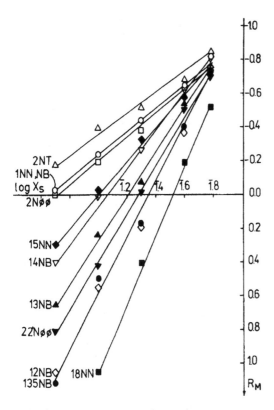

Figure 9 R_M values of aromatic nitro compounds on Florisil plotted against log mole fraction of methyl ethyl ketone in cyclohexane. (From Ref. 30.)

(rather a liquid-liquid system; the polyamide gel forming the stationary phase) by Szumilo [32]. For numerous metal ions Przeszlakowski [33] confirmed Eq. (7) for the so-called liquid ion exchangers (quaternary hydrophobic amines, di-(2-ethylhexyl) orthophosphoric acid, HDEHP, actually ion pairing reagents) and other extractants (e.g., tri-*n*-butyl phosphate, TBP).

Investigations on liquid-liquid partition were likewise continued. The effect of solvation equilibria on distribution coefficients was investigated for numerous solutes and polar water-immiscible solvents (e.g., Refs. 25–27); Kuczynski [34] demonstrated that organic bases can be extracted from strongly acidic solutions as ion pairs with chloride and other anions.

Especially interesting results were obtained by application of liquid cation exchangers (e.g., HDEHP) to extraction of organic bases [35,36]; it was found that the extraction coefficients of, e.g., alkaloids are increased by 1–2 orders of magnitude by the addition of HDEHP to the chloroform phase or other solvents. These publications belong to early reports on the use of ion association (ion pairing) reagents in chromatography of organic electrolytes, later very popular in reversed-phase HPLC.

The simple model of LSC has become popular in the literature [28]; in later publications it was extended to more complex molecular mechanisms [37]. Snyder in his later papers [38–41] considered the model a special case of his more general theory and termed the two approaches as a common Snyder-Soczewinski model; the formation of definite molecular complexes was interpreted as strong localization of solute molecules on the adsorbent surface. Rudzinski, Jaroniec, and Martire published a series of papers in which more detailed thermodynamic interpretations were discussed (e.g., Refs. 42,43); their complex multiparameter equations simplify to Eq. (7) if assumptions accepted in its derivation are introduced. The simple logarithmic equation, Eq. (7), was utilized in many practical optimization procedures (e.g., Refs. 44–49) and for interpretation of experimental results. Its advantages are due to its relation to a real mechanistic model of retention: analogies in molecular structure result in parallel log k vs. log C_{mod} plots. It should be noted that linear plots are convenient for the determination of solvent composition range by interpolation from few experimental points for the choice of optimal solvent composition [28]. Quantitative retention vs. solvent composition relationships are also necessary for the use of computers in optimization procedures (e.g., Drylab [50]) and in theory of gradient elution [41,51]. Equation (12) can also be valid for reversed-phase aqueous systems [49,52,53].

IV. HORIZONTAL SANDWICH CHAMBERS

Determination of numerous retention–eluent composition relationships in traditional TLC chambers required large volumes of solvents; moreover, effects peculiar for TLC (solvent demixing, preadsorption of eluent vapors, formation of a false solvent front) distorted the $R_F = f(k)$ relationships. To eliminate these disadvantages of classical tanks, constructions of horizontal sandwich tanks were devised in our laboratory analogous to well-known chambers of the Brenner-Niederwieser (BN) type.

The first model, the ES (equilibrium sandwich) chamber [54–56], is illustrated in Fig. 10a. The eluent is delivered to the layer from a container with a U-shaped siphon; the solvent is distributed across the plate in a slit between the distributor (strip of glass, 5 × 100 mm, welded to the smaller cover plate with an adjacent orifice for the siphon) and the narrow margin of the plate cleaned of adsorbent, so that an even front of mobile phase is obtained in spite of pointwise delivery of the eluent. The sample can be spotted on-line behind the solvent front at stop-flow (with a marker of $R_F = 1.0$) so that solvent demixing effects are eliminated, and development proceeds under equilibrium conditions, as in column chromatography. Continuous development can be carried out by evaporation of solvent at the other end of the plate; eluent composition can be varied using a pipette instead of the siphon to produce stepwise gradient elution of

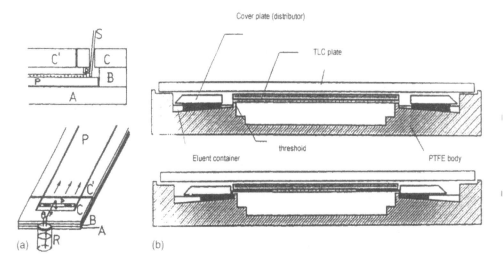

Figure 10 Horizontal chambers. (a) Glass ES chamber: A—base plate; B—distancing frame; C,C′—cover plates; D—distributor; P—TLC plate; R—eluent reservoir; S—capillary siphon. Cross section (expanded vertically) and perspective view. (From J Chromatogr 355, 363, 1986. Courtesy of Elsevier, Amsterdam. (b) Teflon DS chamber cross section. I—filling the two reservoirs (black areas = eluent); II—start of development.

complex samples of wide polarity ranges (e.g., plant extracts). The distributor can also be used to introduce as zones large sample volumes from the edge of the adsorbent layer, under conditions of frontal chromatography, with partial separation and focusing of zones already during sample application. In comparison to the traditional, large-volume chambers, in the ES chamber the required volume of eluent is reduced by a factor of 10 or more and the composition of eluent is strictly controlled.

A still better chamber (DS) was constructed mainly by Dzido [57]; its cross section is illustrated in Fig. 10b. The eluent containers are located slightly below the level of the plate; their bottoms are slanted so that the capillary forces form a vertical meniscus, which during absorption of solvent by the layer is shifted in the direction of the threshold. The cover plates do not touch the TLC plate when the containers are filled with eluent; development is started by moving the cover plates to contact the carrier plate. The eluent forms an even front until the last drop is absorbed by the layer; as in ES chambers the consumption of solvents is very low—e.g., ~2 mL for 100 × 100 × 0.3 mm layers; stepwise gradient elution is also possible as well as zonal application of larger sample volumes from the edge of the layer for micropreparative separations. With the layer on the lower surface of the carrier plate, the adsorbent can be conditioned by pouring a small volume of the eluent on the tray below the plate; the layer can also be conditioned with other solvents to improve separation.

The DS chambers have become popular in Poland, and some have been purchased by western European laboratories.

V. SPECIAL TECHNIQUES—ZONAL MICROPREPARATIVE TLC, STEPWISE GRADIENT ELUTION

As mentioned earlier, the construction of our horizontal sandwich chambers allows for changes of composition of mobile phase in the layer, e.g., by introduction of sample solution or eluent

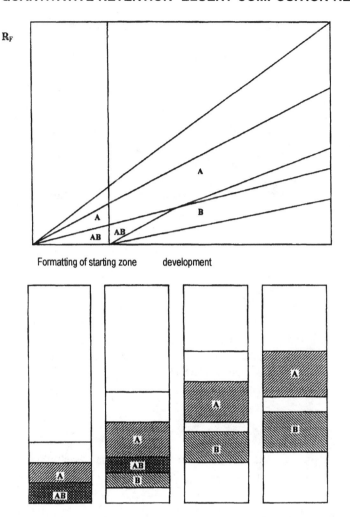

Figure 11   Application of wide starting band from the edge of adsorbent layer in the frontal chromatography mode: separation improved owing to displacement effects. (A and B are hypothetical components of binary mixture A + B.)

with a pipette under the distributor; the front of mobile phase remains even to the absorption of the last drop, then a new portion of eluent containing more of the modifier can be introduced. This feature of the ES and DS chambers was utilized for zonal micropreparative TLC separation and for stepwise gradient development. As mentioned, the introduction of the sample from a container (or pipette) under the distributor of ES or DS chamber to the edge of the layer, in the frontal chromatography mode, leads to partial separation of the sample components in the order of their adsorption affinities. Delivery of the eluent under the distributor then separates the components much more easily because of the stronger displacement effects in the starting zone (Fig. 11). In case of colored solutes, the formation of the starting zone, migration of component zones, and their gradual separation can be observed, recorded, and represented by diagrams [58]. The separation is especially improved by gradient elution owing to stronger focusing of the zones; even wide starting zones are compressed to narrow bands [59]. This advantage is especially favorable in the separation of very complex plant extracts in phytochemical analysis [60,61]. For the theoretical basis of stepwise gradient TLC, see Refs. 62–64.

VI. APPLICATION OF COMPUTERS FOR METHOD DEVELOPMENT

The advances in the application of personal computers greatly simplified and accelerated method development in chromatographic analysis [50,51]; commercial optimization software (e.g., Drylab, Chromsword) became available. In our research, we applied our own programs for special purposes.

The first program that permitted prediction of zone migration under conditions of stepwise gradient TLC [63,64] required the introduction of log k vs. log φ_{mod} relationships for the individual components of the sample (which could be obtained from several isocratic experiments by plotting log k vs. log φ_{mod} diagrams as slope m and ordinate log k_{mod}) and the gradient program φ_{mod} values and the corresponding volumes of the eluent fractions (relative to the hold-up volume of the thin layer) [65]. The computer displays migration paths of the solutes and the final R_F values.

A modified program was applied to optimization of TLC analysis of complex extracts of medicinal plants [61].

In another study [66] a retention data basis (R_{Mmod} and m values) for numerous solutes and 16 binary eluents (combinations of four nonpolar diluents and four polar modifiers) was formed; the program (elaborated by one of the coauthors, M.D.) made it possible to choose optimal combination of the diluent-modifier pair and its composition for a given set of solutes from the database. In toxicological studies where many toxic substances may occur in the sample, it is usual to identify the analyte in preliminary TLC analysis with a universal eluent; from the R_F value of the detected spot, within a narrow ΔR_F window, several compounds are tentatively identified. In the second stage it is necessary to apply a selective eluent for this set of compounds, and the elaborated searching program should help in this task.

VII. CONCLUDING REMARKS

The mechanistic molecular models and equations are not restricted to chromatographic systems; because they are based on bulk partition and adsorption equilibria, they are also valid for batch preparative separations.

The interaction of chromatography with physical chemistry is a reciprocal one: physical chemistry helps us to understand chromatographic processes and to formulate rational method development principles. On the other hand, chromatography can be used as a tool in physicochemical investigations, provided that its mechanism is well understood and any additional superimposed effects are taken into account (especially the validity of V_R vs. k and R_F vs. k equations is essential).

I was the chairman of the Department of Inorganic and Analytical Chemistry, in which up to 20 coworkers were employed, for 34 years (1964–1998). My coworkers contributed much to our common scientific achievements, forming a kind of combined intelligence and integrated experience. Cooperation with them was, with very few exceptions, a pleasure. Gradually independent but closely cooperating research groups were formed under the direction of newly appointed professors. The department, since 1998 headed by Prof. Dr. W. Golkiewicz, undertook new directions as well as developing former topics of research, such as separation of enantiomers, trace analysis of metals using chelating adsorbents, preparative separations, QSAR studies, phytochemical analysis, and investigations on comparative selectivity of various adsorbents.

We owe much to international cooperation—visits to leading western European and U.S. laboratories, participation in international symposia. In the difficult years of the Iron Curtain, with severe currency restrictions, the activity of the foundation ''Scientific Exchange Agreement'' was especially helpful. Among others, in the form of Danube Symposia on Chromatography, their motto was: ''Chromatography separates substances, but unites scientists from various countries'' (needless to say, the Danube flows through western and eastern Europe).

REFERENCES

1.   LC Craig, D Craig. Extraction and Distribution. In: A Weissberger, ed. Technique of Organic Chemistry. Vol 3. New York: Interscience Publishers, 1950, pp. 171–331.
2.   A Waksmundzki, E Soczewiński. Countercurrent extraction in the liquid-liquid systems as a method of separation of mixtures (in Polish). Wiad Chem 9:435–459, 1955.
3.   A Waksmundzki, E Soczewiński. Craig's method of countercurrent distribution (in Polish). In: J Opieńska-Blauth, A Waksmundzki, M Kański, eds. Chromatografia. Warsaw: PWN, 1957, pp. 197–219.
4.   A Waksmundzki, E Soczewiński, R Aksanowski. Chromatographic separation of quinoline bases on buffered paper (in Polish). Chem Anal (Warsaw) 2:459–462, 1957.
5.   A Waksmundzki, E Soczewiński. Relationship of R_F values of weak organic acids and bases on their partition coefficients, dissociation constants and pH of aqueous phase (in Polish). Roczn Chem 32: 863–870, 1958.
6.   A Waksmundzki, E Soczewiński. A paper-chromatographic method for the determination of suitable buffer systems for countercurrent distribution. Nature 184:997, 1959.
7.   A Waksmundzki, E Soczewiński. A simple method of plotting theoretical curves of R_F as a function of pH in buffered paper chromatography. J Chromatogr 3:252–255, 1960.
8.   A Waksmundzki, E Soczewiński. Determination of optimal conditions for Craig's method from paper chromatographic data (in Polish). Roczn Chem 35:1363–1372, 1961.
9.   L Jusiak. Separation of *Chelidonium majus* alkaloids by courtercurrent distribution. Acta Polon Pharm 24:65–70, 1967.
10. M Przyborowska. Separation of alkaloids from the flowers of *Consolida regalis* by countercurrent distribution. Dissert Pharm Pharmacol (Cracow) 18:89–93, 1966.
11. T Wawrzynowicz, A Waksmundzki, E Soczewiński. Partition chromatography and countercurrent distribution of alkaloids of *Fumaria officinalis*. Chromatographia 1:327–331, 1968.
12. E Soczewiński. Determination of optimum solvent systems for countercurrent distribution and column partition chromatography from paper chromatographic data. In: JC Giddings, RA Keller, eds. Advances in Chromatography, Vol. 8. New York: Marcel Dekker, 1969, pp. 91–117.
13. W Kemula, H Buchowski. Partition equilibria in dilute solutions. I Relation between phase composition and partition coefficients (in Polish). Rozniki Chem 29:718–729, 1955.
14. E Soczewiński, CA Wachtmeister. The relation between the composition of certain ternary two-phase solvent systems and R_M values. J Chromatogr 7:311–320, 1962.
15. LR Snyder, JW Dolan, JR Gant. Gradient elution in HPLC. 1. Theoretical basis for reversed-phase systems (review). J Chromatogr 165:3–30, 1979.
16. CBC Boyce, BV Milborrow. A simple assessment of partition data for correlating structure and biological activity using thin-layer chromatography. Nature 208:537–539, 1965.
17. A Leo, C Hansch, D Elkins. Partition coefficients and their uses. Chem Rev 71:525–554, 1971.
18. R Kaliszan. Quantitative Structure-Chromatographic Retention Relationships. New York: Wiley, 1987, pp. 232–290.
19. R Kaliszan. Structure and retention in chromatography. A chemometric approach Amsterdam: Harwood Acad Publ., 1997, pp. 155–202.
20. T Braumann. Determination of hydrophobic parameters by reversed-phase liquid chromatography. J Chromatogr 373:191–225, 1986.
21. E Soczewiński. Distribution coefficients and R_M, R_F values in some multicomponent two-phase systems (in Polish). Chem Anal (Warsaw) 8:337–346, 1963.
22. E Soczewiński. Prediction and control of zone migration rates in ideal liquid-liquid partition chroma-

tography. In: JC Giddings, RA Keller, eds. Advances in Chromatography. Vol. 5. New York: Marcel Dekker, 1968, pp. 3–78.

23. A Waksmundzki, E Soczewiński, Z Suprynowicz. On the relation between the composition of the mixed stationary phase and the retention time in gas-liquid partition chromatography. Coll Czechoslov Chem Comm 27:2000–2006, 1962.

24. A Waksmundzki, JK Różyło. Effect of structure of adsorbents on R_F and R_M values in thin-layer chromatography with binary mobile phase. Chem Anal (Warsaw) 11:101–110, 1966.

25. E Soczewiński, G Matysik. Two types of R_M— composition relationships in liquid-liquid partition chromatography. J Chromatogr 32:458–471, 1968.

26. E Soczewiński, G Matysik. Investigations of the relationship between molecular structure and chromatographic parameters. VIII. Partition of phenols and their derivatives in systems of the type (cyclohexane + polar solvent)-water. J Chromatogr 111:7–19, 1975.

27. M Ciszewska, E Soczewiński. Investigations of the relationship between molecular structure and chromatographic parameters. IX. Extraction of phenolic 2-benzoylbenzoic acids with cyclohexane solutions of n-hexanol and oleic acid. J Chromatogr 111:21–27, 1975.

28. A Rizzi. Retention and selectivity. In: E Katz, R Eksteen, P Schoenmakers, N Miller, eds. Handbook of HPLC. New York: Marcel Dekker, 1998, pp. 1–54.

29. E Soczewiński. Solvent composition effects in thin-layer chromatography systems of the type silica gel-electron donor solvent. Anal Chem 41:179–182, 1969.

30. E Soczewiński, W Gołkiewicz, T Dzido. A simple molecular model of adsorption chromatography. X. Solvent composition effects in TLC of aromatic nitro compounds using magnesium silicate as the adsorbent. Chromatographia 10:221–225, 1977 (and earlier papers).

31. T Wawrzynowicz, TH Dzido. Solvent composition effects in column and thin-layer chromatography of aromatic nitro compounds on alumina. Chromatographia 11:335–340, 1978.

32. H Szumiło, E Soczewiński. Investigations on the mechanism and selectivity of chromatography on thin layers of polyamide. II. Comparison of polyamide and cellulose impregnated with formamide using moderately polar developing solvents. J Chromatogr 94:219–227, 1974.

33. S Przeszlakowski. Paper chromatography of halide and thiocyanate complexes of metals using solutions of liquid ion exchangers as mobile phases. Chromatogr Rev 15:29–91, 1971.

34. J Kuczyński, E Soczewiński. Extraction of organic bases from strongly acidic aqueous solutions. III. The effect of concentration and type of acid on R_M values of organic bases. Chem Anal (Warsaw) 22:247–253, 1977.

35. E Soczewiński, M Rojowska. Chromatography of organic electrolytes on paper impregnated with liquid ion exchangers. J Chromatogr 27:206–213, 1967.

36. E Soczewiński, G Matysik, H Szumiło. Paper chromatography of alkaloids using liquid ion exchangers as developing solvents. Separation Sci 2:25–37, 1967.

37. E Soczewiński. Coadsorption effects in liquid-solid systems of the type silica-heptane + dioxane. I. Theoretical considerations. J Chromatogr 388:91–98, 1987.

38. LR Snyder. Role of the solvent in liquid-solid chromatography—review. Anal Chem 46:1384–1392, 1974.

39. LR Snyder, H Poppe. Mechanism of solute retention in liquid-solid chromatography and the role of the mobile phase in affecting separation. Competition versus "sorption." J Chromatogr 184:363–413, 1980.

40. LR Snyder. Mobile-phase effects in liquid-solid chromatography. In: C Horvath, ed. High Performance Liquid Chromatography—Advances and Perspectives. Vol. 3. New York: Academic Press, 1983, pp. 157–223.

41. LR Snyder, JW Dolan. The linear-solvent—strength model of gradient elution. Advan Chromatogr 38:115–187, 1998.

42. B Borówko, M Jaroniec. Current state in adsorption from multicomponent solutions of nonelectrolytes on solids. Advances Colloid Interfaces Sci 19:137–177, 1983.

43. W Rudziński. Retention in liquid chromatography. In: JA Jönsson, ed. Chromatographic Theory and Basic Principles. New York: Marcel Dekker, 1987, pp. 246–310.

44. M Palamareva, I Kozekov. Theoretical treatment of the adsorptivity of esters of racemic and meso-2,3-dibromobutane-1,4-dioic acids on alumina. J Planar Chromatogr 9:439–444, 1996 (and earlier papers cited therein).

45. T Ando, Y Nakayama, S Hara. Characterization of column packings in normal-phase liquid chromatography II. J Liq Chromatogr 12:739–755, 1989 (and earlier papers cited therein).

46. P Jandera, M Kucerova. Prediction of retention in gradient-elution normal phase high performance liquid chromatography with binary solvent gradients. J Chromatogr A, 759:13–25, 1997.

47. SM Petrovic, E Lončar, M Popsavin, V Popsavin. Thin-layer chromatography of aldopentose and aldohexose derivatives. J Planar Chromatogr 10:406–410, 1997.

48. QS Wang. Optimisation. In: J Sherma, B Fried, eds. Handbook of Thin-Layer Chromatography, 2nd ed. New York: Marcel Dekker, 1996, pp. 81–99.

49. N Sadlej-Sosnowska, I Śledzińska. Validation of chromatographic retention models in reversed-phase high-performance liquid chromatography by fitting experimental data to the relevant equations. J Chromatogr 595:53–61, 1992.

50. LR Snyder, JJ Kirkland, JL Glajch. Practical HPLC Method Development, 2nd ed. New York: Wiley and Sons, 1997, pp. 439–478.

51. P Jandera, J Churáček. Gradient Elution in Column Liquid Chromatography. Amsterdam: Elsevier, 1985, pp. 9–11.

52. F Murakami. Retention behaviour of benzene derivatives on bonded reversed-phase columns. J Chromtogr 178:393–399, 1979.

53. X Geng, FE Regnier. Retention model for proteins in reversed-phase liquid chromatography. J Chromatogr 296:15–30, 1984.

54. E Soczewiński. Simple device for continuous thin-layer chromatography. J Chromatogr 138:443–445, 1977.

55. E Soczewiński, T Wawrzynowicz. A sandwich tank for continuous quasi-column development of precoated high-performance thin-layer chromatography plates. Chromatographia 11:466–468, 1978.

56. E Soczewiński. Equilibrium sandwich TLC chamber for continuous development with a glass distributor. In: RE Kaiser, ed. Planar Chromatography, Vol. 1. Heidelberg: Huethig Verlag, 1986, pp. 79–117.

57. TH Dzido, E Soczewiński. Modification of a horizontal sandwich chamber for thin-layer chromatography. J Chromatogr 516:461–466, 1990.

58. T Wawrzynowicz, E Soczewiński, K Czapińska. Formation and migration of zones in overloaded preparative liquid-solid chromatography. Chromatographia 20:223–227, 1985.

59. E Soczewiński, K Czapińska, T Wawrzynowicz. Migration of zones of test dyes in preparative thin-layer chromatography: stepwise gradient elution. Sep Sci Technol 22:2101–2110, 1987.

60. G Matysik, E Soczewiński, B Polak. Improvement of separation in zonal preparative thin-layer chromatography by gradient elution. Chromatographia 39:497–504, 1994.

61. G Matysik, E Soczewiński. Computer-aided optimization of stepwise gradient TLC of plant extracts. J Planar Chromatogr 9:404–412, 1996.

62. W Golkiewicz. Gradient development in thin-layer chromatography. In: J Sherma, B Fried, eds. Handbook of Thin-Layer Chromatography, 2nd ed. New York: Marcel Dekker, 1996, pp. 149–170.

63. E Soczewiński, W Markowski. Stepwise gradient development in thin-layer chromatography. III. A computer program for the simulation of stepwise gradient elution. J Chromatogr 370:63–73, 1986.

64. W Markowski, E Soczewiński, G Matysik. A microcomputer program for the calculation of R_F values of solutes in stepwise gradient thin-layer chromatography. J Liq Chromatogr 10:1261–1276, 1987.

65. W Markowski, E Soczewiński. Computer aided optimization of stepwise gradient and multiple-development thin-layer chromatography. Chromatographia 36:330–336, 1993.

66. M Matyska, M Dabek, E Soczewiński. Computer-aided optimization of liquid-solid systems in thin-layer chromatography. 3. A computer program for selecting the optimal eluent composition for a given set of solutes from a database. J Planar Chromatogr 3:317–321, 1990.

12

Hollow Sticks with Mud Inside: The Technology of HPLC Columns

Uwe Dieter Neue

Waters Corporation, Milford, Massachusetts

I. INTRODUCTION

If one wants to review the development of a technology, one is often hard-pressed to pinpoint a true beginning. New technologies are not standing still; they are always evolving, often in many different places in the world at the same time. Also, different elements of the technology have to come together to provide the combination of breakthroughs that make it applicable to a broad range of potential users.

For this summary of the evolution of high-performance liquid chromatography (HPLC) column technology, I have selected the time frame of late 1972 to the beginning of 1973 as a starting point. I joined the research laboratory of Istvan Halász at the beginning of 1973, and I remember distinctly that in the middle of the winter, we were still packing 30–50 μm particles, and at the end of the spring, we had moved to 5 and 10 μm particles. Since the step of moving from larger particles to smaller particles presented a giant step in performance, it appears to be quite appropriate to choose this time as the dawn of HPLC technology. Around the same time frame, the preparation of monolayer-type bonded phases with good stability in aqueous mobile phases progressed, which in turn resulted in the breakthrough of reversed-phase chromatography. High-performance columns packed with small particles and reversed-phase chromatography were important elements in the success of high-performance liquid chromatography. Therefore I have selected column packing technology and surface chemistry as the central themes of this chapter.

One has even more difficulties with the selection of material to cover at the end of a review period. Which one of all the newly evolving technologies has the capability to significantly shape the future of HPLC columns? Or even deeper: Is there a future for high-performance columns, or does the power of the mass detector reduce the column to a simple sample clean-up tool, without a significant role in the analysis? Such questions will only be answered in the future.

II. COLUMNS AND PARTICLES

A. The Beginning

Liquid chromatography did not start in 1973. It has its true roots in the investigations of M. S. Tswett, first published in the Proceedings of the Warsaw Society of Natural Scientists in 1903 [1]. However, the investigation of particle size as a performance factor in liquid chromatography did not take place until over half a century later (e.g., Ref. 2, and references therein). It was recognized that the packing of columns with particle sizes of <30 μm should result in an improved separation power due to faster mass transfer in and out of the particles (e.g., Ref. 3). The common column packing technique at the time was a dry packing technique, using vibration and tapping for the formation of the bed. This technique gave good and reproducible results with particles larger than about 30 μm. Unfortunately, this technique failed for particles smaller than 30 μm. This failure had an important consequence: if one could not pack particles of a sufficiently small size, liquid chromatography would be doomed forever to remain a secondary and vastly inferior technique compared to gas chromatography, the dominant high-performance separation technique of the time.

B. The Breakthrough

During 1972, this problem was finally solved, and slurry packing techniques were developed for the packing of particles down to 5 and 10 μm [4,5]. One of the techniques was the "balanced-density" technique, which had been used previously by Kirkland [6] for the packing of larger particles. It used high-density solvents such as tetrachloroethylene and tetrabromoethane to match the density of silica to suspend the particles in a slurry. High pressures, 35 MPa and higher, were used to filter the particles out of the slurry and into the columns. However, the solvents needed for this technique were highly toxic and not suitable for a routine production environment. Therefore, Kirkland [5] developed an alternative: the silica particles were suspended in an aqueous solution of ammonia. The alkaline solution creates a negative charge on the surface of the silica particles, which prevents the agglomeration of the particles and provides a slurry of reasonable stability.

Now, slurry packing had proven its merits. However, much refinement was still necessary. The balanced-density technique used noxious solvents, and the aqueous suspension technique was limited to underivatized silica. Other solutions were necessary. In 1974, Asshauer and Halász reported on the development of an alternative: the slurries were prepared based on high-viscosity solvents [7]. This opened the technology to a wide range of solvents or solvent mixtures. The solvent mixture used could be adjusted to the properties of the packing material and to the needs of a production environment in industry. Now, finally, the routine packing of a range of different particles with different surface chemistries had become possible. High-performance columns with a separation power comparable to gas chromatography had become commercially available, and liquid chromatography was on its path to become the primary analytical separation technique.

It is interesting to note that the thought process behind the use of high-viscosity solvents is highly flawed. Yes, high-viscosity solvents prevent the settling of particles, but the time to pack a column at a given pressure also increases in direct proportion to the viscosity of the slurry solvent [8]. Since both the speed of settling and the speed of packing are proportional to each other, the use of a high-viscosity slurry solvent does not give any advantage and is not necessary. If a high-viscosity solvent is not needed, one can manipulate the packing process to proceed at higher speed by using a low-viscosity solvent. A high speed for the packing process is advantageous for the routine industrial production of large quantities of high-quality HPLC columns. The preparation of high-performance columns was one of the enabling technologies of HPLC, and now it apparently had become routine. But appearances can be deceiving.

C. The Difficulties and the Solutions

Seen superficially, and from the standpoint of the HPLC user, columns packed with high-quality packings had indeed become a commodity. However, for the technologists working on the details of column preparation, the process was a nightmare. Process failures happened often, and often the reasons for the failure were not understood. In addition, and most importantly, the lifetime of a column in the hands of a user appeared to be arbitrary. Sometimes columns failed even before they reached the laboratory of the HPLC user. Something seemed to be wrong with the standard column packing processes. Therefore the question arose if alternatives to standard column packing processes and the standard particle technology could be found. Was it possible to create stable columns that could be produced with a high reproducibility and that could be shipped around the world without deterioration?

1. Radial Compression

A common mode of failure of these high-performance columns packed with irregular particles of a size around 10 μm was the formation of "voids." A perfectly full and apparently well-packed column was left to sit around. When it was tested, it was found to give horrible results: shoulders on the peaks or double peaks. When one now opened the column at the column inlet, one saw a large empty space in the column, often 2–3 cm deep for a 30 cm column. Where had the packing gone?

The first suspicions were that the packing had somehow dissolved or had left the column through badly performing filters at the column outlet. However, careful measurements showed that the amount of packing material in the column before and after "voiding" was identical. No material was lost. The packed bed structure just had collapsed. How could this happen, and—more importantly—what can be done to prevent this from happening?

Similar issues of variable column performance had been encountered previously in the preparation of columns for preparative chromatography. At Waters Associates, Pat McDonald and Carl Rausch were involved in the development of a new instrument for preparative chromatography. At the same time they were searching for a convenient method to deliver preparative columns to customers. Columns prepared by classical packing techniques such as tapping and vibrating also exhibited difficulties in routine preparation. Alternatives were investigated. The idea of delivering a bag of powder to the user and having the user put the bag into a holder was explored. To form the bed, the bag needed to be compacted by applying an external pressure to the bag. During the experimentation with the technique, it became obvious that a well-filled, soft-walled container compressed in the direction perpendicular to the flow provided a good and efficient way to create reproducibly a high-quality and stable chromatographic bed for preparative chromatography. The technique of radial compression was born, and the inventors applied for a patent [9] in 1975.

In 1976, several researchers at Waters were scratching their heads about the problems with column voiding, this time for analytical columns. The task of the team was succinctly put into a simple slogan: "Avoid a void!" Alternatives to standard column preparation techniques were explored, but after a short time the team settled on a continued exploration of radial compression technology [10]. However, there were several fundamental problems. The pressures needed to use the standard 30 cm 10 μm column were quite high; 10 MPa was not unusual. Since the active radial compression technology established for preparative chromatography required a compression pressure in excess of the use pressure, the packing needed to be strong enough to withstand these compression pressures. The standard irregular packings based on 10 μm particles were not strong enough, largely due to the large porosity of the packing.

Two elements were essential to the implementation of radial compression for analytical chromatography. One was the development of high-quality and hard spherical particles, which will be discussed in more detail in the next paragraph. The other was a new thought process on the design of HPLC columns. This thought process had been outlined by Martin and coworkers in a series of papers with the title "Study of the Pertinency of Pressure in Liquid Chromatography" [11–13]. A key element of these papers was the idea that columns should be operated at the minimum of the plate-height vs. velocity curve to minimize the operating pressure. The authors were able to demonstrate that essentially the same column efficiency and analysis time could be obtained using a long 30 cm 10 μm column and a short 10 cm 10 μm column, provided the short column is operated properly. From the standpoint of the column designer, the use of the shorter column reduced the typical operating pressure to less than 1 MPa while maintaining the column efficiency that chromatographers were used to. Consequently, the first radially compressed analytical columns were 10 cm long and were prepared using 10 μm particles. To make them compatible with commonly used flow rates, injection volumes and instrument bandspreading, the column diameter was 8 mm. Shortly after the introduction of the technology, smaller column diameters and smaller particle sizes were added. With this combination of short columns with radial compression technology, high-efficiency analytical columns had been created that no longer suffered from the problem of column voiding.

2. *Spherical Particles*

An important element in the implementation of radial compression for analytical chromatography was the existence of hard spherical particles. Spherical particles with a sufficiently low specific pore volume could be compressed to pressures as high as 20 MPa without breaking. Since the operating pressure of a radially compressed column was limited by the compression pressure, the high strength of the particles chosen for radially compressed analytical columns allowed great freedom in the choice of mobile phase composition and flow rates.

Spherical silica particles specifically designed for HPLC were developed in several laboratories around the world, for example, by K. K. Unger and coworkers [14–17] in Darmstadt, Germany, by I. Halász and I. Sebestian in Saarbrücken, Germany, by J. J. Kirkland at Du Pont de Nemours, United States, and by M. Holdoway at Harwell, United Kingdom. Early commercial versions of high-performance spherical silicas were Lichrospher from Merck KGaA, Darmstadt, Germany, Zorbax-Sil [18] from E. I. Du Pont de Nemours & Co., Wilmington, DE, Spherisorb from Phase Separations Ltd., Deeside, United Kingdom, and Nucleosil from Macherey-Nagel, Düren, Germany. Other manufacturers followed.

While the different spherical silicas had different properties, they shared the common element that they were designed specifically for HPLC. The common pore size was around 10 nm. The manufacturing processes could be adjusted to yield particles in the preferred size range. In addition, the properties of the surfaces could be controlled to give the desired results in subsequent surface modification steps (see below). The same technologies could then be applied to the preparation of other packings for more specialized applications. Manipulation of the pore size of a packing made possible the application of packings for size-exclusion chromatography or for the separation of biomacromolecules like proteins.

In addition, investigators discovered rapidly that spherical silicas had several advantages over irregular silicas of comparable size. They are easier to pack to a consistent performance and often exhibit better stability. A key element that contributes to the improved stability of spherical packings is the packing density, best expressed as the interstitial fraction. While it is easy to prepare high-quality columns from spherical particles with an interstitial fraction close to 0.4, the best column performance for irregular particles requires an interstitial fraction around 0.43–0.45. It is highly likely that this difference contributes to the difference in stability between columns prepared from irregular-shaped packings and spherical packings. Also, some researchers claimed that spherical particles exhibit better permeability [19], but this is complicated due to the fact that the size of an irregular-shaped particle is difficult to define.

A consequence of the ease of column packing with spherical particles was the rapid development of columns with a particle size of 5 μm, which outperformed the older columns based on 10 μm irregular particles. Another consequence was that the good stability of columns based on spherical particles made the application of radial compression unnecessary, at least for the preparation and use of analytical HPLC columns.

3. Today's Technology

The groundwork for most of the technological advances in particle technology had been laid in the early to mid-1970s. Today's column packing technology allows the routine packing of porous particles down to 2 μm [20]. The difficulties today are rather in the preparation of fully porous particles of a small size and with a particle size distribution of sufficient quality. On the other hand, techniques have been developed for the preparation of columns and application of nonporous particles down to 1 μm [21].

Small, nonporous particles were developed and described by Unger and coworkers in the mid-1980s [22]. Commercial versions have been available since the mid-1990s from Micra Scientific, Inc., Northbrook, IL. The key characteristic of nonporous particles is the small surface area, which in turn results in low retention and low loadability compared to fully porous particles. The low retention can be compensated for to some degree by an adjustment of the composition of the mobile phase. About 30% more water is needed in the mobile phase to achieve the same retention as is achieved with fully porous particles. This works well for very hydrophobic analytes, but is a big disadvantage for polar analytes. On the other hand, such packings give significant advantages for analytes with a large molecular weight [8], such as proteins, and early applications [24] demonstrated this principle.

For a long time, porous particles were limited to 3 μm or larger. In 1987, Danielson and Kirkland [25] described the "synthesis and characterization of 2-μm wide-pore silica microspheres as column packings for the reversed-phase liquid chromatography of peptides and proteins." Indeed, the first entry for smaller particle sizes is the separation of large molecular weight compounds such as proteins, and the key advantage of small particles is the speed of the separation. A few isocratic peptide separations were shown in this paper with an analysis time of less than one minute. The authors also demonstrated the rapid separation of synthetic protein mixtures with run times under 2.5 minutes. However, the time had not yet come for high-speed protein separations.

Recently there has been great interest in driving the analysis time for low molecular weight pharmaceuticals down to the one minute time frame. This need has initiated the development of very short columns packed with particles smaller than 3 μm. "Ballistic" gradients executed at high flow rates are used with these short columns to achieve very rapid separations. The philosophy behind such rapid separations is based on the optimization of gradient resolution within a given target analysis time [26]. On the one hand, high flow rates result in lower column plate count. On the other hand, high flow rates expand the gradient and thus extend the separation space in a given time. The compromise between both effects results in a new optimization scheme beyond the thought process established for isocratic separations [11–13]. An example of such a fast separation is shown in Fig. 1.

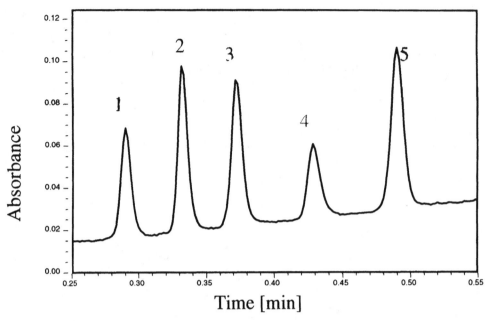

Figure 1 Ultra-fast separation: Separation of five standards with elution of the last peak in 30 seconds. Column: 2 mm × 20 mm XTerra MS C_{18}. Gradient: 5–95%. A: 10 mM NH_4^+ CH_3COO^-, pH 5, 3% MeCN. B: MeCN; flow rate: 1.5 mL/min; temperature: 60°C; instrument: Alliance™ 2690. Compounds: 1, lidocaine, 2, prednisolone, 3, naproxen, 4, amitriptyline, 5, ibuprofen. (Chromatogram courtesy of Y.-F. Cheng and Z. Lu, Waters Corporation.)

4. *Beyond Particles*

Throughout the history of HPLC, attempts have been made to explore structures other than beds packed with particles. The incentive was generally to prepare devices that are mechanically stable, can be prepared with a high reproducibility, and, unlike packed beds, cannot collapse. Examples are stacked membranes [27,28] or aligned fibers [29,30]. A good measure of the capability of a chromatographic device is the ratio of separation performance to the backpressure, for example, the separation impedance [8]. Generally, this ratio was not favorable for these alternative technologies. Recently, the formation and the use of continuous silica rods has been explored [31,32], and an issue of the *Journal of High Resolution Chromatography* [33] has been dedicated to this technology and related endeavours [34]. With proper preparation technologies [32], a structure can be prepared that exhibits low backpressure combined with the separation power of a 5 μm column. The reduction in backpressure claimed by the authors is around 30% [35]. This is the first time that separation devices were developed that exhibit properties superior to packed beds. While such a device is extremely interesting to the technologist, a commercial implementation only became available at the time of this writing. Time will tell if the promise of this technology can be fulfilled.

III. SURFACE CHEMISTRY

A. Silica

Silica has been the backbone of the success of HPLC. Silica is hard, practically incompressible. Some of the standard high-performance packings can be exposed to pressure gradients in excess of 10 MPa/cm before the structure collapses [36]. This high strength is a necessity for the high pressures used in HPLC. However, other materials such as alumina have similar properties. The key advantage of silica is the fact that its surface can be modified without difficulty with a range of agents, some of which provide a surface coating that is stable to hydrolysis. The properties of the resulting packings will be discussed next.

B. Bonded Phases

Before the birth of true high-performance liquid chromatography in 1972/1973, silica had already proved itself in gas chromatography. Simple and reproducible ways for the modification of its surface via esterification with alcohols had been developed previously [37–39]. However, the Si-O-C bond is readily hydrolyzed, which puts some severe limitations on its use in liquid chromatography. Bonded phases based on a Si-N-C-bond were explored as an alternative [40–42]. Improved stability was observed compared to the Si-O-C bond, but a complete stability to hydrolysis is not possible with Si-N-C–bonded phases. While such packings can be used without difficulties in a nonaqueous mobile phase, their use in water-containing solvents is still problematic. However, packings based on a Si-C bond to the silica surface [43,44] are much more stable and can be used without difficulty in an aqueous environment. The use of silanization techniques for the preparation of chromatographic packings extends back to the 1960s [45,46]. Commercial versions of such packings on larger superficially porous particles [47,48] or on fully porous particles [49] had been available since the late 1960s.

C. The Advantage of Reversed-Phase Packings

The silanization of the silica surface provided many different bonded phases. However, the breakthrough was achieved through the broad application of reversed-phase bonded phases. In normal-phase chromatography, the stationary phase is polar and the mobile phase is nonpolar. The chromatographic retention on these phases depends strongly on very small concentrations of water in the mobile phase. On a planet covered up to 71% with water, it is consequently quite difficult to obtain reproducible normal-phase chromatography. On the other hand, in reversed-phase chromatography the mobile phase contains a large proportion of water and the stationary phase is nonpolar. Thus, reversed-phase chromatography is not as finicky as normal phase chromatography, and reproducible results can be obtained by most people without difficulty [50].

As described above, the technology of the preparation of columns packed with small particles became available in 1972. Soon the preparation of reversed-phase packings based on these new high-performance silicas was investigated [51]. In August 1973 Waters introduced the first version of μBondapak C_{18}, which became an immediate commercial success. Charlie Pidacks in the applications laboratories at Waters developed separations for most known drugs using μBondapak C_{18}, which convinced the analytical chemists in the pharmaceutical industry of the merits of HPLC.

Now the chromatographers had the combination of two important developments in their hands: a high-performance column based on the new 10 μm particle size and a stable reversed-phase packing that could be used with aqueous mobile phases. Soon, the development of new applications of the new technique skyrocketed.

D. The Difficulties and the Solutions

1. Reproducibility

Reversed-phase HPLC was an important technology. The usable separation mechanisms can be understood without difficulty [8,50–53]. The retention of an analyte is proportional to its hydrophobic area, and with some experience the chromatographer was able to predict the mobile phase that gives appropriate retention. With a little more experience, he was able to predict the elution order of the compounds. In addition, tools were soon developed to aid in the development of a separation [54]. Nevertheless, a substantial flaw of the technology needed to be addressed: the reproducibility of a separation using different preparations of the bonded phase often left much to be desired.

The underlying details were addressed by several suppliers and the scientific community in the mid-1980s. J. J. Kirkland [55] published his investigations on the properties of reversed-phase packings. Engelhardt and coworkers worked out a stationary phase test that was published in detail in 1990 [56]. At Waters, a similar test was implemented in 1985 that used basic as well as neutral test compounds. This test, described in detail in a later publication [57], demonstrated clearly that the retention and selectivity properties of a reversed-phase bonded phase do not depend exclusively on the bonded ligands, but that surface silanols contribute significantly to the retention, especially for basic analytes. The variation of the bonded-phase properties revealed by this test resulted in a significant effort to improve the reproducibility of packings by superior control of the bonding processes. Figure 2 shows the progress made at Waters during the mid to late 1980's with respect to the batch-to-batch reproducibility of reversed-phase packings. The reproducibility of the relative retention of a basic and a neutral analyte is compared for the time prior to 1985 to a time frame in the 1990s after the improvement of process controls. One can see that the standard deviation was narrowed by roughly a factor of 2 for both the preparation of Bondapak C_{18}, a product of the 1970s, and Nova-Pak C_{18}, a product of the 1980s.

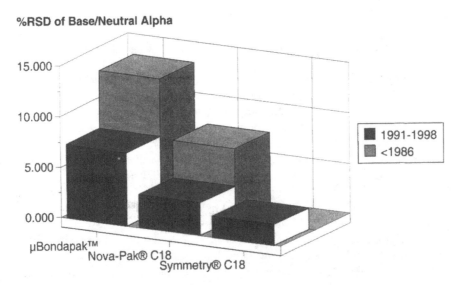

Figure 2 Reproducibility improvement of reversed-phase packings: The reproducibility of the relative retention between a basic analyte (a tricyclic antidepressant) and a neutral analyte (acenaphthene) is compared for two different packings between the times before 1986 and between 1991 and 1998. The improvement in reproducibility is roughly twofold. Also shown is the reproducibility of the same value for Symmetry, a 1994 product. (From Ref. 57.)

2. High-Purity Silica

While manufacturers in the United States were busy improving the control of their processes, a new technology emerged in Japan [58]: high-purity silica. The manufacturing processes of some of the older silicas were based on inorganic raw materials such as silica sols, which contain measurable amounts of other elements such as aluminum or iron. The preparation of a high-purity silica was based on the inexpensive commercial availability of tetraethoxysilane, which in turn was a consequence of the focus on high-purity silicon as a raw material for the semiconductor industry.

From the standpoint of the chromatographer, a high-purity silica packing provides improved peak shapes for analytes with basic functional groups. The improvement can be seen without difficulty using the batch test [57] implemented at Waters or other tests with strongly basic analytes. The improved properties of high-purity silicas changed the landscape of HPLC in the mid-1990s. Waters introduced a line of packings in 1994 based on the high-purity Symmetry® silica [59], and other manufacturers followed in subsequent years. Today the use of packings based on a high-purity silica is the rule rather than the exception.

At Waters, the step forward towards the use of a high-purity silica was accompanied by an additional improvement of process control and consequently further progress in the reproducibility of the packings. As can be seen in Fig. 2, the standard deviation of the relative retention of a base-neutral pair is reduced to 2.2% between different batches of Symmetry C_{18}. Studies by independent researchers [60,61] confirm the excellent reproducibility of the Symmetry C_{18} packing.

3. Incorporated Polar Groups

The use of high-purity silicas for the preparation of bonded phases had a significant impact on the peak shapes of basic analytes, but for very difficult analytes one can still observe peak tailing at neutral pH. However, another approach developed parallel to the success of packings based on high-purity silicas: the incorporation of a polar functional group into the bonded phase close to the surface of the silica. The general idea is that such polar functions prevent the interaction of polar groups of the analyte with surface silanols. As a result, packings with incorporated polar groups yield still further improved peak shapes for all types of analytes.

At Waters, the origin of bonded phases with incorporated polar groups goes back to studies investigating the suppression of the interaction of proteins with the surface of a capillary in capillary electrophoresis. It was known that this interaction can be suppressed by using a high salt concentration, but this increased the current and thus led to excessive heating of the electrolyte. John Petersen at Waters found that one can solve the problem of protein adsorption without an increase in current: adding high concentrations of zwitterions to the electrolyte suppresses the interactions of proteins with the capillary wall just like a high salt concentration, but zwitterions do not conduct current and therefore do not contribute to the heating of the electrolyte. In a sideline to this study, the use of zwitterionic surfactants was investigated as well, and we needed to know the concentration of the surfactant on the surface of the capillary. This was not possible using a capillary; therefore, I proposed to carry out the measurements on the surface of a chromatographic column packed with particles. After the column was coated with the zwitterionic surfactant, we carried out additional measurements looking at the peak shape and the elution volume of proteins and basic analytes on this column. Beautiful nontailing peaks were observed independent of the analyte that we tried. The next idea was to create a reversed-phase bonded phase with an incorporated zwitterionic group. Carsten Niederländer joined the group shortly after these preliminary investigations, and his task was the creation of such a zwitterionic bonded phase. The simplest way to do this was to create first a zwitterionic layer on the surface of the silica, followed by a second derivatization with a C_{18} silane:

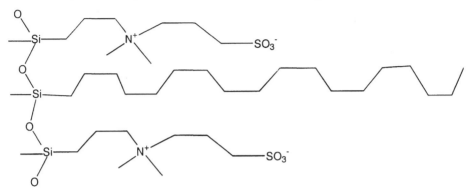

The resulting bonded phase gave the best peak shapes for a wide range of analytes [62,63]. Nevertheless, for reasons of stability and reproducibility, it was preferred to create such a bonded phase in a single step. However, the creation of a long-chain silane with an incorporated zwitterion turned into a nightmare, and John Petersen proposed to substitute the zwitterion with a simpler polar function, such as a carbamate group. It worked equally well, and we applied for a patent [64]. The synthesis of the final product was refined and published by O'Gara et al. [65,66], and the first product based on this idea, SymmetryShield RP$_8$, was introduced in 1996. Figure 3 shows the batch test chromatogram for SymmetryShield RP$_8$ at neutral pH using neutral, acidic, and basic analytes. The very good peak shape for the basic analytes does not allow us to identify them simply by visual inspection of the chromatogram.

Started earlier, but partially overlapping with the developments at Waters, were similar investigations at Supelco by Binyamin Feibush [67] and Tracy Ascah [68,69]. In order to sup-

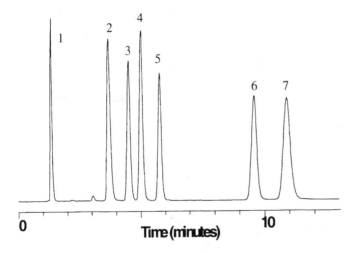

Figure 3 Batch test chromatogram for SymmetryShield RP$_8$ at neutral pH: The test consists of basic, neutral, and weakly acidic analytes at neutral pH. One cannot see a difference in peak shape between neutral and basic analytes. Test conditions: mobile phase: 35% 100 mM potassium phosphate buffer at pH 7.0, 65% methanol. Sample: 1, uracil; 2, propranolol; 3, butylparaben; 4, dipropylphthalate; 5, naphthalene; 6, amitriptyline; 7, acenaphthene. (Chromatogram courtesy of B. A. Alden, Waters Corporation.)

press the interaction of analytes with the acidic silanols, Feibush first created a propylamino bonded phase. Then he derivatized the resulting packing in a second step with a long-chain activated acid to obtain the desired reversed-phase character. The packing gave good peak shapes for bases but suffered from stability problems due to the presence of a large number of amino groups on the surface. In the next generation packing, Supelcosil ABZ, a large portion of the residual amino groups was further derivatized to yield acetamide groups. Nevertheless, even this second derivatization remained incomplete, and a small number of residual amino groups can be detected chromatographically. While residual silanols on classical packings induce bad peak shapes and variable retention for basic analytes, the residual amino groups result in similar effects for acidic analytes. The newest version of the Supelco packings, Discovery RP-AmideC16, is now free from residual amino groups, presumably due to a single-step surface derivatization. The lack of amino groups is not a disadvantage, and good peak shapes for basic analytes are obtained due to the incorporated amide group.

Similar technologies had been investigated at Keystone. The Prism packing contains a urea group incorporated into the packing. This polar functional group also creates a good peak shape for all types of analytes, and no trace of unfavorable side reactions can be found chromatographically.

Another benefit of the packings with an incorporated polar group is their wettability with mobile phase containing no organic modifier. Therefore such packings are useful for the analysis of very polar compounds that require such mobile phases, especially under gradient conditions.

Well-designed packings with incorporated polar groups therefore have some clear advantages over classical reversed-phase packings. The success of these packings recently led to the introduction of a range of similar packings by different manufacturers. Unfortunately, some of the approaches used are similar to the older, multistep surface reactions, whose disadvantages have been demonstrated. For other versions, the nature of the surface modification is not known. However, chromatographers are most successful if they know the properties of a packing and can understand the interaction of their analytes with the surface of a packing.

4. Beyond Silica

The introduction of packings with incorporated polar functions allows the use of mobile phases without organic modifiers and enables us to obtain good peak shapes for practically all analytes independent of the pH of the mobile phase. What else do we need?

A limitation of silica-based bonded phases has always been the pH stability of the packings. Most manufacturers do not recommend the use of their packings at a pH higher than 8, but with a combination of a high-density silica and a thorough surface coating, higher pH values can be reached. However, if we could use a still wider pH range, we would be less constrained in our selection of optimal mobile phases.

Organic packings such as divinylbenzene-based packings allow the use of the entire pH range, from pH 0 to 14. Still, these packings have been plagued since their inception with inferior mass-transfer properties, resulting in broad peaks and low-efficiency columns. Recent improvements in the technology have not yet entirely eliminated this problem. What other alternatives are there?

In the mid-1970s, Klaus Unger and coworkers [70] pursued for a short time an alternative approach to bonded phases based on silica. They synthesized packings from a mixture of tetraethoxysilanes and triethoxysilanes already containing a reversed-phase ligand. In the turmoil of the early times of HPLC, the idea was not pursued further. Zhiping Jiang and coworkers at Waters Corporation reinvestigated this principle in the late 1990s. After some additional work, it turned out that inorganic/organic hybrid packings indeed have a superior high pH stability

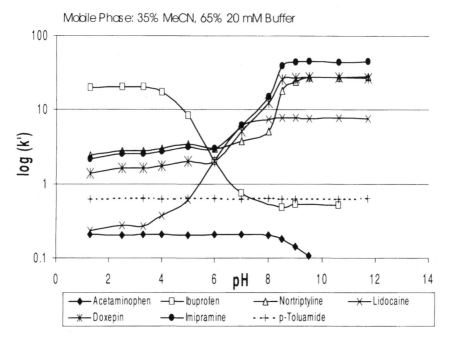

Figure 4   Log k versus pH for acidic, basic, and neutral analytes: The retention of different analytes is shown as a function of the pH of the mobile phase. The retention of basic analytes increases with increasing pH and becomes stable at high pH values. (Data courtesy of Y.-F. Cheng and Z. Lu, Waters Corporation.)

compared to classical silica-based packings [71,72]. On the other hand, the surface derivatization techniques used for classical silica-based packings can still be applied, and the retention properties of the hybrid packings are very similar to those of silica-based packings. Therefore, the procedures used for the development of separations on classical reversed-phase packings can be applied without difficulty, and the chromatographer does not need to learn any new techniques.

In 1999 hybrid packings based on this approach were introduced by Waters Corporation under the name of XTerra®. Two of the packings use bonded phases with an incorporated polar group, as described in the last section; the other two packings use classical trifunctional silanes. The XTerra packings allow a broader exploration of the pH range compared to silica-based packings. Of course, for ionizable analytes changes in the pH of the mobile phase are an important tool in the adjustment of the selectivity of a separation, and therefore the broad pH range is a significant benefit. In addition, the stability of the packing extends sufficiently far into the alkaline pH range, allowing the analysis of basic analytes under conditions where the analyte is not charged. This results in stable retention, independent of small fluctuations in the mobile phase pH (Fig. 4). As a consequence, hybrid packings have expanded the usable range of reversed-phase HPLC.

IV. CONCLUSION

The rapid development of high-performance columns in the early 1970s led to the success of high-performance liquid chromatography. The subsequent years completed the consolidation of the technology, characterized by a slow refinement of the knowledge base. This was true for both the technology of columns and particles and the preparation of chromatographic surfaces. However, the technological development did not stop then, nor has it stopped now. In the last few years, significant improvements have been made, and the nature of the "hollow sticks" as well as the "mud inside" is still changing, even as this sentence is written.

REFERENCES

1. MS Tswett. Proc Warsaw Soc Nat Sci Biol Sec 14, No. 6, 1903.
2. JC Giddings. Dynamics of Chromatography, Part I, Principles and Theory. New York: Marcel Dekker, Inc., 1965.
3. LR Snyder. Anal Chem 39:698, 1967.
4. RE Majors. Anal Chem 44:1722, 1972.
5. JJ Kirkland. J Chromatogr Sci 10:593, 1972.
6. JJ Kirkland. J Chromatogr Sci 9:206, 1971.
7. J Asshauer, I Halász. J Chromatogr Sci 12:139, 1974.
8. UD Neue. HPLC Columns, Theory, Technology and Practice. New York: Wiley-VCH, 1997.
9. PD McDonald, CW Rausch. U.S. Patent 4,211,658 (1981).
10. CW Rausch, Y Tuvin, UD Neue. U.S. Patent 4,228,007 (1980).
11. M Martin, C Eon, G Guiochon. J Chromatogr 99:357, 1974.
12. M Martin, C Eon, G Guiochon. J Chromatogr 108:229, 1975.
13. M Martin, C Eon, G Guiochon. J Chromatogr 110:213, 1975.
14. K Unger. German Patent 2155281 (1973).
15. K Unger, J Schick-Kalb, KF Krebs. J Chromatogr 83:5, 1973.
16. K Unger, R Kern, C Ninou, KF Krebs. J Chromatogr 99:435, 1974.
17. KK Unger. Porous Silica. Amsterdam: Elsevier, 1979.
18. RE Leitch, JJ DeStefano. J Chromatogr Sci 11:105, 1973.
19. R Endele, I Halász, K Unger. J Chromatogr 99:377, 1974.
20. UD Neue, TH Walter, BA Alden, RP Fisk, JL Carmody, JM Grassi, Y-F Cheng, Z Lu, R Crowley, Z Jiang. Am Lab 31(22):36, 1999.
21. KK Unger, G Jilke, JN Kinkel, MTW Hearn. J Chromatogr 359:61, 1986.
22. B Anspach, KK Unger, H Giesche, MTW Hearn. Paper 103 at the 4th International Symposium on HPLC of Proteins, Peptides and Polynucleotides, Baltimore, 1984.
23. TJ Barder, PJ Wohlman, C Thrall, PD DuBois. LC-GC 15:918, 1997.
24. C Horvath. Separation Science and Biotechnology Seminar, Ft. Lauderdale, FL, Jan 21–23, 1987.
25. ND Danielson, JJ Kirkland. Anal Chem 59:2501, 1987.
26. UD Neue, JL Carmody, YF Cheng, Z Lu, CH Phoebe, TE Wheat. Advances Chromatogr 41:93, 2001.
27. JA Gerstner, R Hamilton, SJ Cramer. J Chromatogr 596:173, 1992.
28. DK Roper, EN Lightfoot. J Chromatogr A 702:3, 1995.
29. RF Meyer, PB Champlin, RA Hartwick. J Chromatogr Sci. 21:433, 1983.
30. M Czok, G Guiochon. J Chromatogr 506:303, 1990.
31. H Minakuchi, K Nakanishi, N Soga, N Ishizuka, N Tanaka. Anal Chem 68:3498, 1996.
32. K Cabrera, G Wieland, D Lubda, K Nakanishi, N Soga, H Minakuchi, KK Unger. TrAC 1:50, 1998.
33. J High Resol Chromatogr 23(1), 2000.
34. JMJ Fréchet, F Svec. Anal Chem 64:820, 1992.
35. D Lubda, K Cabrera, H-M Eggenweiler. GIT Spezial—Separation 1/2000:34, 2000.
36. R Groh. Diplomarbeit, Saarbrücken, Universität des Saarlandes, 1975.
37. I Sebastian. Dissertation, J. W. Goethe Universität, Frankfurt, 1969.
38. I Halász, I Sebastian. Angew Chem Intern Ed 8:453, 1969.
39. DF Horgan, Jr, JN Little. J Chromatogr Sci 10:76, 1972.
40. OE Brust. Dissertation, 1972, Saarbrücken, Universität des Saarlandes, 1972.

41. OE Brust, I Sebestian, I Halász. J Chromatogr 83:15, 1973.
42. UD Neue. Diplomarbeit, Saarbrücken, Universität des Saarlandes, 1973.
43. DC Locke, JT Schmermund, B Banner. Anal Chem 44:90, 1972.
44. DC Locke. J Chromatogr Sci 11:120, 1973.
45. EW Abel, FH Pollard, PC Uden, G Nickless. J Chromatogr 22:23, 1966.
46. HNM Steward, SG Perry. J Chromatogr 37:97, 1968.
47. JJ Kirkland, JJ DeStefano. J Chromatogr Sci 8:309, 1970.
48. JJ Kirkland. J Chromatogr Sci 9:206, 1971.
49. Waters Associates, Chromatogr. Notes 2, 1972.
50. K Karch, I Sebestian, I Halász, H Engelhardt. J Chromatogr 122:171, 1976.
51. RB Sleight. J Chromatogr 83:31, 1973.
52. K Karch, I Sebestian, I Halász. J Chromatogr 122:3, 1976.
53. C Horváth, WR Melander, I Molnár. J Chromatogr 125:129, 1976.
54. LR Snyder, JW Dolan, JR Grant. J Chromatogr 165:3, 1979.
55. J Köhler, DB Chase, RD Farlee, AJ Vega, JJ Kirkland. J Chromatogr 352:275, 1986.
56. H Engelhardt, M Jungheim. Chromatographia 29:59, 1990.
57. UD Neue, E Serowik, P Iraneta, BA Alden, TH Walter. J Chromatography A 849:87–100, 1999.
58. Y Ohtsu, Y Shiojima, T Okumura, J-I Koyama, K Nakamura, O Nakata, K Kimata, N Tanaka. J Chromatogr 481:147, 1989.
59. UD Neue, DJ Phillips, TH Walter, M Capparella, B Alden, RP Fisk. LC-GC 12(6):468, 1994.
60. M Kele, G Guiochon. J Chromatogr A 830:41, 1999.
61. M Kele, G Guiochon. J Chromatogr A 830:55, 1999.
62. UD Neue. Presentation at the Dal Nogare Award Ceremony honoring H. Engelhardt, Pittsburgh Conference, 1994.
63. UD Neue, C Niederländer, JS Petersen. Presentation at the 17th International Symposium on Column Liquid Chromatography, Hamburg, Germany, 1993.
64. UD Neue, C Niederländer, JS Petersen. U.S. Patent 5,374,755 (1994).
65. JE O'Gara, BA Alden, TH Walter, JS Petersen, C Niederländer, UD Neue. Anal Chem 67:3801, 1995.
66. JE O'Gara, DP Walsh, BA Alden, P Casellini, TH Walter. Anal Chem 71:2992, 1999.
67. B Feibush. U.S. Patent 5,137,627 (1992).
68. TL Ascah, B Feibush. J Chromatogr 506:357, 1990.
69. TL Ascah, KMR Kallury, CA Szafranski, SD Corman, F Liu. J Liq Chromatogr Rel Technol 19:3049, 1995.
70. KK Unger, N Becker, P Roumeliotis. J Chromatogr 125:115, 1976.
71. UD Neue, TH Walter, BA Alden, RP Fisk, JL Carmody, JM Grassi, Y-F Cheng, Z Lu, R Crowley, Z Jiang. Am Lab 31(22):36, 1999.
72. Y-F Cheng, TH Walter, Z Lu, P Iraneta, BA Alden, C Gendreau, UD Neue, J Grassi, J Carmody, JE O'Gara, RP Fisk. LC-GC 18:1162, 2000.

13

On the Way to a General Theory of Gradient Elution

Pavel Jandera

University of Pardubice, Pardubice, Czech Republic

I. BEGINNINGS OF RESEARCH AT THE UNIVERSITY OF PARDUBICE

In late 1960s, having finished my undergraduate studies in chemistry, I continued my formation as a Ph.D. student of Jaroslav Churáček, who introduced me into the fascinating field of chromatography in his undergraduate lectures, by then the first course on analytical separation in the former Czechoslovakia. He and his small group were practicing planar chromatography and gas chromatography, and Jaroslav suggested I try doing something in the emerging new field of high-performance liquid chromatography (HPLC). First I spent some time developing instrumentation, making possible liquid chromatographic separations on columns packed with relatively coarse silica gel particles. Since then my research has been connected to liquid chromatography. In 1973 I finished my Ph.D. thesis on the effects of working conditions on LC separations and I joined the permanent staff at the Department of Analytical Chemistry of the Institute of Chemical Technology in Pardubice (now the Faculty of Chemical Technology of the University of Pardubice), and I continued my cooperation with Jaroslav, who always showed deep interest in my work and encouraged my efforts and became a good friend.

In 1972 we purchased our first commercial liquid chromatograph (Waters LC 100), but we were still working with relatively inefficient separation columns. In 1977 I was lucky to have the opportunity to spend 3 months in the research group in Saarbrücken of the late Prof. Isztván Halász, an excellent scientist and a very fine man, who was always ready to help. This short period had a great impact on my future scientific development. Not only did I learn how to synthesize chemically bonded alkyl silica stationary phases and how to prepare efficient HPLC columns with fine particle size packing materials, I had the opportunity to learn many new aspects of the theory of chromatography—and of the philosophy of science—during the famous Halász's late Friday afternoon seminars. In his laboratory I became acquainted with very inspiring fields of chromatographic research, including new approaches to improve the efficiency of

HPLC columns and broad application fields. Together with Heinz Engelhardt, I worked out efficient HPLC methods for reversed-phase separations of strong sulfonic acids and important dye intermediates [1]. After my return to Pardubice, our group continued in the development of HPLC and later capillary electrophoresis (CE) separation methods for dyes and their intermediates.

At our department, we constructed a column packing device, and within a short time we were able to prepare our own efficient HPLC columns packed with 5–10 μm particles. We also upgraded our instrumentation to allow high-performance operation.

II. BASIC ASSUMPTIONS FOR DEVELOPING A GENERAL THEORY OF GRADIENT ELUTION

Since the beginning of my involvement in HPLC, I was attracted by the user's approaches to the development of HPLC separation methods. Once the adequate column is selected, proper adjustment of the composition of the mobile phase is the most efficient tool a chromatographer can employ to approach the desired quality of separation of a particular sample mixture. To this end, it is important to understand the effects of the mobile phase composition on both retention and the selectivity of separation. Our investigations of these effects began in 1973.

Many samples can be successfully separated using isocratic elution, i.e., running the separation at a constant composition of the mobile phase, but the resolution of complex samples containing compounds with great differences in the affinities to the stationary phase usually can be much improved by gradient elution techniques, where the composition of the mobile phase is programmed to change with time. The selection of an optimum gradient program is often not straightforward and can be greatly facilitated if we can use a suitable theory for predictive calculations of retention volumes, bandwidths, and resolution of sample compounds in dependence on the parameters characterizing the profile of the gradient. After a few early attempts at calculating retention volumes for a few specific gradient-elution applications, a breakthrough contribution to the theory of gradient elution was published by Lloyd Snyder [2]. In his "linear solvent strength gradient" (LSS) approach, the gradient program is adjusted so that the logarithms of retention factors of sample compounds are changed in a linear manner with time.

Linear concentration gradients of organic solvents in aqueous-organic mobile phases in reversed-phase systems are often close to LSS gradient profiles, but it is difficult to program the gradient profiles to match the LSS requirements in other chromatographic systems, such as in ion-exchange or in normal-phase HPLC. Further, the LSS theory generally is limited to binary gradients. To overcome these limitations, we started working on a more general theory applicable in various HPLC modes, including reversed-phase, normal-phase, and ion-exchange systems, with binary or ternary gradients with different linear and nonlinear profiles.

The approach is based on the solution of the basic differential equation [3]:

$$dV = k \, dV_0 \tag{1}$$

assuming that the retention factor k is constant during the migration of the solute band by an infinitesimally small distance corresponding to a differential fraction of the column hold-up volume, V_0, caused by a differential increase in the volume of the mobile phase that had passed through the column, dV. Equation (1) is integrated in the limits from zero to V_0 and from zero to V_R', to obtain the net elution volume in gradient-elution HPLC, V_R'. To this end, the dependence of k on the volume of the eluate that passed through the column from the start of the gradient should be introduced into Eq. (1). Instead of using the LSS relationship for this purpose, our approach introduces a combination of two separate dependencies to make possible flexible calculations of the elution data for various combinations of HPLC modes and gradient profiles potentially useful for solving practical separation problems:

1. The first dependence is the "retention equation" describing the relationship between the k of sample solutes on the concentration(s) of one or more efficient eluting component(s) in the mobile phase, which reflects the thermodynamics of the distribution process of sample solutes between the stationary phase and the mobile phase in the chromatographic system.

2. The second dependence, the "gradient function," describes the profile of the gradient, i.e., the time program of the composition of a binary or more complex mobile phase from the start of the gradient run.

We started with binary gradients using mobile phases composed of a component with a higher elution strength (strong solvent **B**) and a component with a weaker elution strength (weak solvent **A**). The elution strength of the mobile phase increases and the retention of sample solutes decreases as the concentration of the solvent **B** increases during the chromatographic run. Consequently, adequate resolution can be achieved during a single analysis for early eluted compounds while keeping acceptably short elution times of the last eluted compounds. Further, the peaks in gradient-elution chromatography become narrower, the sensitivity of detection is improved and the analysis is accelerated with respect to isocratic separation.

The most simple gradient function, which is used most often in practice, describes the linear binary gradient profile:

$$\varphi = A + B't = A + B'\frac{V}{F_m} = A + BV = A + (\varphi_G - A)\frac{t}{t_G} = A + (\varphi_G - A)\frac{V}{V_G} \quad (2)$$

Here, A is the initial concentration φ of the solvent **B** in the mobile phase at the start of the gradient, and B or B′ characterizes the steepness (slope) of the gradient per time or per volume unit of the eluate. t_G and V_G are the gradient time and the gradient volume, respectively, necessary to complete the conentration change of **B** from the initial concentration of solvent **B**, $\varphi = A$, to the final concentration φ_G at the end of the gradient.

Sometimes a nonlinear gradient shape can improve the spacing of peaks and resolution in the chromatogram and shorten the analysis time. This occurs often in normal-phase chromatography, for which we introduced a convenient gradient function (Eq. 3) allowing characterization of various curved gradient profiles by a single curvature parameter, κ [4]:

$$\varphi = (A^{(1/\kappa)} + B.V)^{\kappa} \quad (3)$$

where $A = \varphi_0$ is the initial concentration of the strong eluent at the start of the gradient and $B = [\varphi_G^{(1/\kappa)} - A^{(1/\kappa)}]/V_G$ is the steepness (slope) of the gradient. Figure 1 shows a few examples of gradient functions described by Eq. (3). The gradients are linear for $\kappa = 1$, concave for $\kappa > 1$, and convex for $\kappa < 1$.

To select a suitable retention equation to be used for the calculation of gradient-elution retention data, different interaction forces controlling the retention in various chromatographic systems should be taken into account and the problem should be solved separately for the individual HPLC modes. We compared various earlier retention models and suggested equations that allowed the calculation of retention data in combination with various gradient functions [5]. A really useful retention equation should be as simple as possible but should provide reasonable description of a wide variety of specific separation problems in various HPLC modes [6].

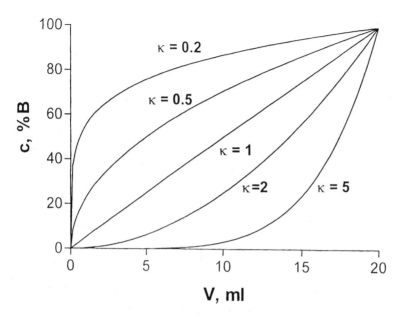

Figure 1 Examples of linear, concave, and convex gradients from 0 to 100% stronger eluent, **B**, in 20 min (at 1 mL/min) described by a gradient function (dependence of the concentration of **B**, φ on the volume of the eluate from the start of the gradient, V) with various values of the gradient shape parameter κ. (From Ref. 13.)

III. RETENTION BEHAVIOR AND GRADIENT ELUTION IN NORMAL-PHASE AND ION-EXCHANGE HPLC

We started with the retention in normal-phase HPLC systems. Based on the Snyder theory of adsorption chromatography [7], we derived a three-parameter retention equation [6,8]:

$$k = \frac{1}{(a + b\varphi)^m} \tag{4}$$

The constant m in Eq. (4) is the ratio of the adsorbent surface area occupied by one adsorbed molecule of the solute to the area occupied by one adsorbed molecule of the strong solvent **B**, $a = (k_A)^{-1/m}$ and $b = [(k_B)^{-1/m} - (k_A)^{-1/m}]$, where k_A is the retention factor of the solute in pure solvent **A** and k_B is k in pure solvent **B**. If the retention in pure solvent **A** is very strong, the term a can be neglected and Eq. (4) simplifes to the following equation [6,8]:

$$k = \frac{1}{(b\varphi)^m} = k_B \varphi^{-m} \tag{5}$$

We compared the validity of Eqs. (4) and (5) for HPLC of various classes of compounds, e.g., nonionic azo dyes [9], some steroids [8], pesticides [10,11], and phenols [12] on columns packed with silica gel, alumina and polar bonded nitrile and amine stationary phases in binary mobile phases containing 2-propanol, dioxane, or dichloromethane as the solvent **B** in *n*-hexane or *n*-heptane as the nonpolar solvent A. In most cases, Eq. (5) describes adequately the retention behavior in binary mobile phases, but the retention of some less retained compounds is better characterised by Eq. (4).

In normal-phase HPLC systems controlled by Eq. (4), the integration of the differential Eq. (1) provides Eq. (6) for the calculation of retention volumes in gradient-elution chromatography with linear gradients following Eq. (2):

$$V_R = \frac{1}{bB} [b\,B(m+1)\,V_0 + (a + Ab)^{(m+1)}]^{1/m+1} - \frac{a+b}{bB} + V_0 \tag{6}$$

With nonlinear gradients following Eq. (3), the integration of Eq. (1) yields Eq. (7) for retention volumes in normal-phase systems controlled by Eq. (5):

$$V_R = \frac{1}{B} [(\kappa m + 1)\,Bk_B V_0 + A^{(\kappa m+1)/\kappa}]^{1/\kappa m+1} - \frac{A^{1/\kappa}}{B} + V_0 \tag{7}$$

For calculation of retention data in gradient-elution HPLC, the constants of the retention equations should be determined in a few isocratic or gradient experiments. We verified the suitability of Eqs. (6) and (7) for predictive calculations of the retention of various classes of compounds on columns packed with bare silica gel and with chemically bonded polar stationary phases using linear and nonlinear gradients with various polar (**B**) and nonpolar (**A**) solvents as mobile phase components [9–13].

We found that Eq. (5) in many cases can adequately describe the retention of ionic compounds in ion-exchange chromatography, where φ is the concentration of a salt in aqueous or aqueous-organic mobile phases (instead of the strong solvent **B** in normal-phase systems). Here, m represents the stoichiometric coefficient of the ion-exchange reaction between the ionized solute and the salt counter-ions in the mobile phase and k_B is proportional to the conventional selectivity constant of ion exchange and to the ion-exchange capacity per volume unit of the ion exchanger. For weakly acidic or weakly basic compounds, k_B is a more or less complex function of the pH of the mobile phase [5]. Hence, Eq. (7) can be used for calculations of elution volumes of ionic solutes in ion-exchange chromatography with linear or nonlinear gradients of increasing concentration of a salt or of a buffer (φ) at a constant pH of the mobile phase. The validity of Eq. (7) was later verified by Baba et al. [14] for anion-exchange chromatography of nucleotides.

IV. RETENTION BEHAVIOR, CALIBRATION OF RETENTION AND SELECTIVITY, AND GRADIENT ELUTION IN REVERSED-PHASE HPLC

In the early years of reversed-phase chromatography on chemically bonded alkyl silica column packing materials, the mechanism of retention was understood by many workers as a specific case of liquid-liquid partition between the bulk mobile phase and the immobilized liquid stationary phase. We employed this model to describe the effect of the concentration of the organic solvent (**B**), φ, in binary aqueous-organic mobile phases on the solute retention factor, k. Using the Hildebrand regular solution theory [15], we derived the following equation [6]:

$$\log k = \log k_w - m\varphi + d\varphi^2 \tag{8}$$

where k_w is the retention factor extrapolated to pure water as the mobile phase, m is a constant proportional to the product of differences between the solubility parameters of the solute and of water and between the solubility parameters of the organic solvent (**B**) and of water, respectively. The constant d is proportional to the second power of the difference between the solubility parameters of the solvent **B** and of water, and its importance increases with decreasing polarity of the organic solvent. A similar approach was later used by Schoenmakers et al. [16] to derive a relationship that is virtually identical to Eq. (8), but with slightly different definition of the coefficients m and d. In practice, experimental log k versus φ plots are usually more curved in tetrahydrofuran-water than in methanol-water mobile phases. We showed that the quadratic term in Eq. (8) often can be neglected over a limited composition range of binary mobile phases, and in that case Eq. (8) is simplified to Eq. (9) [6], suitable for the description of mobile phase effects on the retention in many reversed-phase systems [17]:

$$\log k = \log k_w - m\varphi \tag{9}$$

Retention behavior according to Eq. (8) was observed, e.g., by Karch et al. [18], and equations formally identical with Eq. (8) were later suggested by Snyder et al. [19] and by Schoenmakers et al. [20]. This equation has since been widely used for the prediction of retention and optimization of reversed-phase HPLC separations.

The idea that liquid-liquid partition mechanism controls the retention in reversed-phase systems has gradually been abandoned, as a monolayer of bonded alkyl chains cannot be expected to behave like a bulk liquid. Even though attractive forces between the solute molecules and the stationary phase or with preferentially adsorbed solvent layer can contribute significantly to the retention, the predominating role of mobile phase interactions in the retention mechanism (the solvophobic effect) has become recognized [21,22].

Because of rapidly increasing practical importance of reversed-phase liquid chromatography, we started a closer investigation of various phenomena in this chromatographic mode, especially of the effects of the mobile phases on the separation. During my very pleasant 6-month stay in Georges Guiochon's group at Ecole Polytechnique in Palaiseau in 1980, I cooperated closely with Henri Colin on this topic. We developed a semi-empirical model of retention for reversed-phase systems [23,24]. This model, based on interaction indices as a measure of polarity of sample solutes and of mobile phase components, leads to the description of retention by an equation formally identical with Eq. (8). Using a set of adequately selected standard compounds for the calibration of the retention scale, the retention of sample solutes in different mobile phases can be calculated from their interaction indices.

After my return to Pardubice, I further elaborated and refined this model by introducing two types of indices for each solute: the lipophilic index characterizing general lipophilicity of a solute as an equivalent of the hydrocarbon retention and a polar index as a measure of the contribution to the retention of specific polar interactions of the polar groups in the molecules of solutes with mobile phase components [25,26]. The two indices improved the correlations between the retention and the selectivity data measured in different mobile phases. This approach was also helpful to develop theoretical description of reversed-phase retention behavior in homologous and oligomeric series [27,28].

Integration of differential Eq. (1) after the introduction of the retention Eq. (9) provides Eq. (10) for calculation of elution volumes in reversed-phase HPLC with linear gradients [2,17]:

$$V_R = \frac{1}{mB}\log[2.31\, mBV_0 k_w 10^{-mA} + 1] + V_0 \tag{10}$$

Equation (10) is very similar to the equations derived by Snyder for linear solvent strength gradients and for linear concentration gradients in reversed-phase systems [2,20].

V. PREDICTION OF RETENTION AND OPTIMIZATION OF RESOLUTION IN GRADIENT ELUTION

To predict the resolution and to optimize the separation in gradient-elution chromatography, a reliable estimate of bandwidths is necessary. In gradient-elution HPLC, bandwidths w_g are generally narrower than in isocratic HPLC and are controlled mainly by the actual composition of the mobile phase at the time of the elution of a sample solute. We suggested setting w_g equal to the isocratic bandwidth corresponding to the retention factor k_f at the time of elution of the band maximum [4]:

$$w_g = \frac{4V_0}{\sqrt{N}} (1 + k_f) \tag{11}$$

N in Eq. (11) is the number of theoretical plates of the column determined under isocratic conditions. k_f can be calculated from the appropriate retention equation, for instance, from Eqs. (4), (5), or (9), introducing the isocratic concentration of the strong solvent **B** corresponding to the elution volume determined from the appropriate equation, e.g., from Eqs. (6), (7), or (8), according to the gradient function describing the actual gradient profile. Eq. (11) neglects additional band compression during the migration through the column as a result of a faster migration of the trailing edge of the band in a mobile phase with greater elution strength with respect to a slower migration of the leading edge. However, as other effects contribute to band broadening in gradient elution, the error caused by neglecting it usually is not very significant.

Using Eq. (11) and the appropriate equation for the elution volume applicable to a particular gradient profile in the HPLC system used, the calculation of gradient-elution resolution of sample compounds **1** and **2** with adjacent peaks is straightforward:

$$R_{s,g} = \frac{2(V_{R,2} - V_{R,1})}{w_{g,1} + w_{g,2}} \tag{12}$$

where the indices 1 and 2 relate to the individual solutes.

Using the present theory for prediction of the basic retention characteristics in gradient-elution HPLC, we are in principle able to optimize the gradient profile for a desired separation of various samples in different HPLC separation modes, using constants of the retention equations determined in a few initial isocratic or gradient-elution experiments. Commercial software "DryLab G" was developed by Snyder and coworkers [29] for optimization of linear binary gradients in reversed-phase chromatography, based on the theory of linear-solvent-strength gradients. This approach is based on the optimization of gradient time (i.e., steepness) in the first step, followed by the optimization of gradient range (i.e., the initial and final concentration of the strong solvent) in the second step, and, if necessary, other experimental conditions in the last step.

The effect of the concentration of the strong-eluting component in the mobile phase at the start of the gradient, A, on the resolution and on the time of analysis is equally important as that of the gradient steepness, B. With increasing A, not only do the retention volumes decrease, but the selectivity and even the elution order of some compounds may change (Fig. 2). The two parameters of the gradient profile show synergistic effects on separation. Therefore we have developed two strategies for simultaneous optimization of A and B. If there is a "critical pair" of sample compounds that are most difficult to resolve, we can directly calculate the combination of A and B necessary for a desired resolution and minimize elution time of the most strongly retained compound, using numerical iterative approach [30]. However, in complex sample mixtures the separation of more than two solutes may be difficult, and in this case we need a detailed picture of the resolution of all adjacent peaks independent of the gradient profile.

For this purpose, we shall use different optimization approach. First, we set a desired time for gradient elution, i.e., for the gradient volume, V_G, in which a preset final concentration of the polar solvent, φ_G, should be achieved. The setting of V_G does not affect significantly the

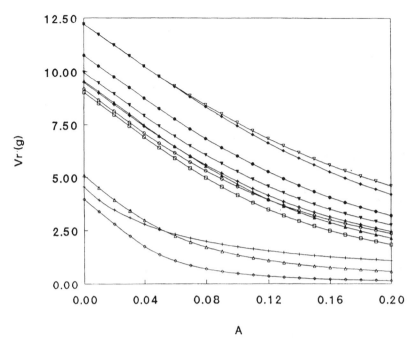

Figure 2 Dependence of the retention volumes, $V_r(g)$, in gradient elution chromatography on the initial concentration of 2-propanol in n-heptane, A (in %vol. 10^{-2}) in gradients with gradient time of 30 min. Column: silica gel Separon SGX, 7.5 μm, 150 mm × 3.3 mm I.D., 40°C, flow rate 1 mL/min. Compounds: phenylurea herbicides. (From Ref. 10.)

results if V_G is large enough. With a preset V_G, the slope B of the gradient and the initial concentration of solvent **B**, A, are linked by Eq. (13):

$$B = \frac{(\varphi_G - A)}{V_G} \tag{13}$$

The elution volume V_R can be calculated as a function of a single parameter A, introducing Eq. (13) into the appropriate equation [e.g., Eq. (6), Eq. (7), or Eq. (10)]. The differences between the V_R of compounds with adjacent peaks or the resolution $R_{s,g}$ can be plotted versus A in the form of a "window diagram" to select optimum initial concentration A and gradient steepness B [from Eq. (13)] [31,32]. The selection of the highest value of A at which the desired resolution (e.g., $R_{s,g} = 1.5$) is achieved for all compounds in the sample mixture in most cases automatically minimizes the time of the analysis, as the elution volumes decrease with increasing A (Fig. 2). The optimization can be repeated for various preset values of of V_G to find really optimal combination of the gradient steepness B and of the initial concentration of the strong solvent, A. Fig. 3 illustrates this optimization approach on the example of reversed-phase separation of 12 phenylurea herbicides. From the window diagram (top), optimum separation is predicted for the gradient from 24 to 100% methanol in water in 73 minutes at 1 mL/min (bottom).

This approach can be applied also to optimization of binary gradients in normal-phase chromatography, as illustrated by an example of the separation of herbicides on a silica gel column (Fig. 4). Here, the window diagram (Fig. 4A) predicts two linear gradients of 2-propanol in n-heptane (from 12% or from 25%) to yield the desired resolution $R_{s,g} \geq 1.25$ for all peaks

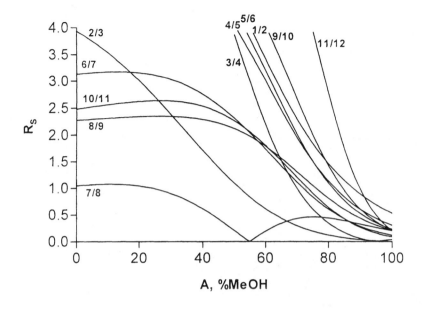

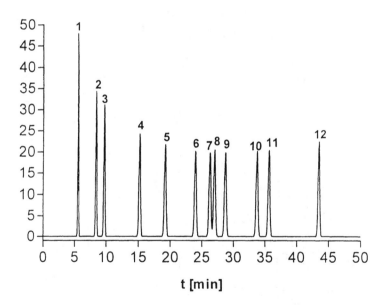

Figure 3 (Top) Resolution window diagram for the gradient elution separation of a mixture of twelve phenylurea herbicides on a Separon SGX C 18, 7.5 μm, 150 mm × 3.3 mm I.D. column, in dependence on the initial concentration of methanol in water at the start of the gradient, A, with optimum gradient volume V_G = 73 ml. Column plate number N = 5000. Sample compounds: hydroxymetoxuron (1), desphenuron (2), phenuron (3), metoxuron (4), monuron (5), monolinuron (6), chlorotoluron (7), metobromuron (8), diuron (9), linuron (10), chlorobromuron (11), neburon (12). (Bottom) Separation of the 12 phenylurea herbicides with optimized binary gradient, 24–100% methanol in water in 73 min, 1 mL/min. (Adapted from Ref. 35.)

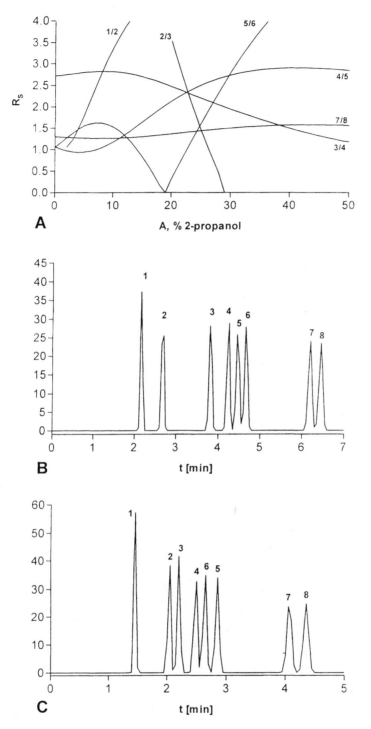

Figure 4 (A) The resolution window diagram for the gradient-elution separation of a mixture of eight phenylurea herbicides on a Separon SGX, 7.5 μm, silica gel column (150 × 3.3 mm I.D.) in dependence on the initial concentration of 2-propanol in *n*-heptane at the start of the gradient, A, with optimum gradient volume $V_G = 10$ mL. Column plate number N = 5000. (B, C) Separation of eight phenylurea herbicides with optimized gradient-elution conditions for two maxima of lowest resolution in Fig. 4A, with gradients from 12 to 38.6% 2-propanol in *n*-heptane in 7 min (B) and from 25 to 37.5% 2-propanol in *n*-heptane in 5 min (C). Flow rate 1 mL/min. Sample compounds: neburon (1), chlorobromuron (2), 3-chloro-4-methylphenylurea (3), desphenuron (4), isoproturon (5), diuron (6), metoxuron (7), deschlorometoxuron (8). (From Ref. 33.)

Linear Gradient

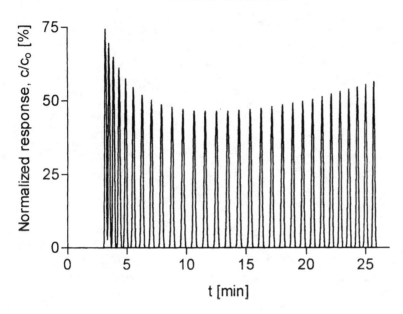

Non-linear Gradient

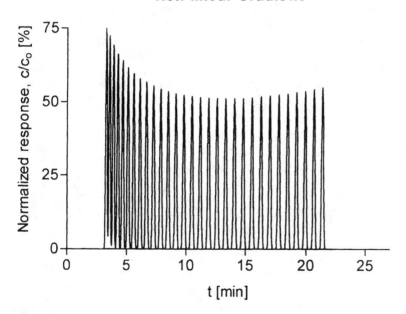

Figure 5 Normal-phase gradient-elution separation of 30 lower oligostyrenes on two Separon SGX C18, 7 μm, silica gel columns in series (150 × 3.3 mm I.D. each), using the optimized linear and the optimized convex gradients of dioxane in *n*-heptane. Flow rate = 1 mL/min. Normalized response relates to the original concentrations of the oligomers in the sample. (From Ref. 13.)

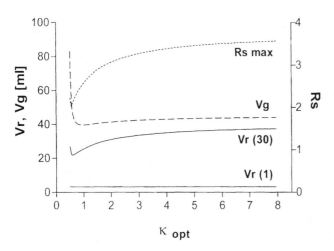

Figure 6 Gradient volume, V_g, retention volumes of the first (1) and last (30) oligostyrene and maximum resolution, $R_{s\,max}$, of the oligomers with 9 and 10 oligostyrene units at the optimized shape parameter, κ_{opt}, in gradient elution separation of 30 lower oligostyrenes on two silica gel columns in series ($V_0 = 2.35$ mL, N = 5000) with nonlinear gradients of dioxane in n-heptane optimized for minimum resolution ≥ 1.5. Flowrate = 1 mL/min. (From Ref. 13.)

in the chromatogram, but the gradient from 25 to 37.5% 2-propanol in 5 minutes (Fig. 4B) provides better band spacing and shorter time of analysis than the gradient from 12 to 38.6% 2-propanol in 7 minutes (Fig. 4C).

Later, our optimization strategy was adapted to enable simultaneous optimization of the initial concentration of the strong solvent, of the gradient steepnes, and of the gradient shape characterized by the curvature parameter κ in gradient function Eq. (3) [13]. The optimization of the gradient shape can significantly improve band spacing and peak capacity for the separation of oligomers, as illustrated by an example of separation of lower oligostyrenes on a silica gel column using optimized linear (Fig. 5, top) and nonlinear (Fig. 5, bottom) gradients of dioxane in n-heptane. Fig. 6 illustrates the effect of the curvature parameter κ on resolution, on the gradient volume, V_G, and on the elution volumes of the first and the thirtieth oligomer.

VI. DESCRIPTION AND OPTIMIZATION OF TERNARY GRADIENTS IN REVERSED-PHASE AND NORMAL-PHASE HPLC

Having developed the theory of binary gradient elution, we turned our attention to ternary gradients with two stronger elution solvents in a weaker one. These gradients may be helpful to improve the resolution of samples whose separation selectivity in binary gradient elution is too low. In a ternary gradient, the concentrations of two stronger solvents i and j, φ_i and φ_j, in a ternary mobile phase are changed simultaneously, in most the simple case in a linear manner:

$$\varphi_i = A_i + B_i V \tag{14}$$
$$\varphi_j = A_j + B_j V$$

The effect of a ternary gradient on separation is illustrated by the example of reversed-phase gradient-elution of a nine-component mixture of phenols (Fig. 7) [34]. Here, neither a linear binary gradient of methanol in water nor a gradient of acetonitrile in water provided satisfactory separation. The separation selectivity for the early eluting compounds is better with a gradient of acetonitrile in water, but the separation selectivity of the last two eluting compounds is better with a gradient of methanol in water; a ternary gradient with increasing concentration of methanol and simultaneously decreasing concentration of acetonitrile improved significantly the resolution of the sample mixture.

In practice, two specific types of ternary gradients are probably most useful and the easiest to describe theoretically [31]:

1. The "elution strength" ternary gradients, where the concentration ratio of the two strong solvents, $\varphi_i/\varphi_j = r$, is constant and the sum of the concentrations, $\varphi_T = \varphi_i + \varphi_j$, changes in a linear manner during the elution:

$$\varphi_T = A_T + BV \tag{15}$$

2. The "selectivity" ternary gradients, where the sum of the concentrations of the two strong solvents, φ_T, is constant during the elution, but their concentration ratio, X, changes in a linear manner:

$$\frac{\varphi_i}{\varphi_T} = X = \frac{\varphi_{0i}}{\varphi_T} + BV = X_0 + BV \tag{16}$$

The "elution strength" ternary gradients have a similar effect on solute retention as binary gradients. Hence, Eqs. (4), (5), or (9) with φ_T instead of φ can be used to describe the retention. Both in reversed-phase [31,32,34,35] and in normal-phase [11,12,33] systems, with ternary mobile phases, the parameters a_T, b_T, k_{BT}, k_{wT}, and m_T of Eqs. (4), (5), and (9) should be determined from the experimental data for various φ_T at a constant concentration ratio r, to be used instead of a, b, k_B, k_w, and m in Eqs. (6), (7), and (10) for calculations of the elution volumes in ternary "elution strength" gradient elution using the same optimization approach as with binary solvent gradients. The optimization of a ternary "elution strength" gradient is illustrated in Fig. 8 for reversed-phase separation of a mixture of phenylurea herbicides, whose optimized separation using binary gradient elution is shown in Fig. 3. The optimized "elution strength" ternary gradient provides the separation in approximately half the time necessary for the separation with optimized binary gradient of methanol in water [35].

On the other hand, different approaches should be used for the prediction of retention and optimization of separation with "selectivity" ternary gradients in reversed-phase and in normal-phase systems. In reversed-phase systems where the following retention equations apply in binary mobile phases comprised of water and organic solvent **i** and of water and organic solvent **j**, respectively [31,32,35]:

$$\begin{aligned} \log k_i &= a_i - m_i \varphi_i \\ \log k_i &= a_j - m_j \varphi_j \end{aligned} \tag{17}$$

the elution volumes in the elution with linear "selectivity" ternary gradients can be calculated from Eq. (9) with the constants $A = -\varphi_T/(1 + A_i/A_j)$, $a = a_i - m_i\varphi_T$, $m = (a_j - a_i)/\varphi_T + m_i - m_j$. A_i and A_j are the initial concentrations of the polar solvents **i** and **j**, respectively, at the start of the gradient [31,32,35].

In normal-phase systems, the retention in ternary mobile phases was found to be controlled by the equation $1/k = a + bX + cX^2$ at a constant sum of concentrations of the two polar

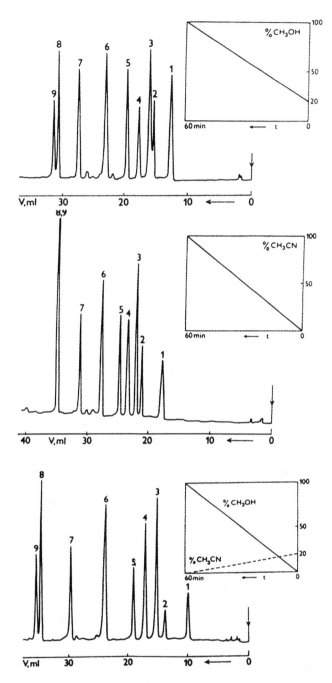

Figure 7 Reversed-phase gradient-elution separation of a mixture of phenols using binary linear gradients of methanol in water and of acetonitrile in water and a ternary gradient of methanol and acetonitrile in water optimized to attain improved separation of the pairs of compounds 2 and 3, 8 and 9. Column: LiChrosorb RP-C 18, 5 μm, 300 × 4 mm I.D., flow rate = 1 mL/min, UV detection at 254 nm. Sample compounds: 4-cyanophenol (1), 2-methoxyphenol (2), 4-fluorophenol (3), 3-fluorophenol (4), *m*-cresol (5), 4-chlorophenol (6), 4-iodophenol (7), 2-phenylphenol (8), and 3-*tert*-butylphenol (9). (Adapted from Ref. 34.)

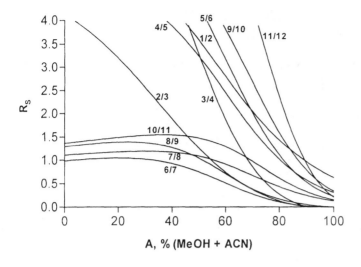

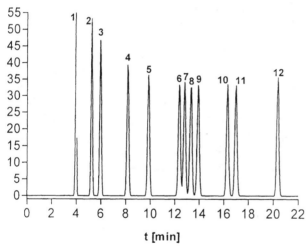

Figure 8 (Top) Resolution window diagram for the ''elution strength'' ternary gradient-elution separation for a mixture of 12 phenylurea herbicides in dependence on the initial sum of concentrations of methanol and acetonitrile in water at the start of the gradient, with the optimized concentration ratio of acetonitrile, $X = \varphi_{acetonitrile}/(\varphi_{acetonitrile} + \varphi_{methanol}) = 0.4$, and gradient volume $V_G = 31$ mL. Column and sample compounds as in Fig. 4. (Bottom) Separation of the 12 phenylurea herbicides with optimized ternary gradient, 18.6% methanol + 12.4% acetonitrile in water–60% methanol + 40% acetonitrile in water in 73 min, 1 mL/min. (Adapted from Ref. 35.)

solvents, **i** and **j**, $\varphi_T = \varphi_i + \varphi_j$. The net retention volumes, $V_R' = V_R - V_0$, in ''selectivity'' ternary gradients controlled by Eq. (16) can be calculated from Eq. (18) [11,12,33]:

$$\frac{(V_R')^3}{3} cB^2 + \frac{(V_R')^2}{2}(b + 2\,cX_0)\,B + V_R'(a + bX_0 + cX_0^2) = V_0 \qquad (18)$$

The constants a, b, c in Eq. (18) depend on the solute, on the chromatographic system, and on the preset sum of concentrations of the two polar solvents, **i** and **j**, φ_T, and can be determined in isocratic experiments with ternary mobile phases.

VII. POSSIBLE PITFALLS AND ERRORS IN APPLICATION OF THE THEORY OF GRADIENT ELUTION

Our investigation of the suitability of the theory of gradient elution for prediction and optimization of experimental conditions included a study of possible sources of errors limiting the practical applications of our approach. The errors may originate either in the instrumentation for gradient elution or in the chromatographic system.

Some gradient devices do not mix precisely enough the preset volume ratios of mobile phase components, especially when the proportion of the solvents mixed are outside the range from 1:20 to 20:1 [36]. With some instruments, these errors may result in significant deviations of the experimental gradient from the preset profile in the initial and in the final parts of the gradients. To avoid irreproducible results with such instruments, when possible, the gradients should not be started at concentrations of the solvent **B** lower than 5% and terminated at concentrations higher than 95% **B**.

Another instrumental source of errors, which may account for problems when gradient methods are transferred from one instrument to another, is connected with the ''gradient dwell volume,'' which can be quite significant, even a few mL with some instruments, and may differ from one instrument to another. At the start of the gradient, the ''dwell volume'' of the instrument is filled with the mobile phase of the composition corresponding to the initial gradient conditions, and consequently the same volume of the mobile phase flows through the column before the starting gradient profile arrives at the top of the column. Hence, the expected gradient elution is delayed, and some weakly retained sample solutes migrate a certain distance along the column before they are taken by the front of the gradient. The volume of the eluate corresponding to this unintended initial isocratic step contributes to the elution volume, but, on the other hand, the part of the column length available for the gradient elution is shorter than expected. Even if the gradient dwell volume cannot be neglected, precise prediction of the gradient elution data is possible when either the sample injection is delayed with respect to the start of the gradient to compensate for the gradient dwell volume, or if the correction for the dwell volume is introduced into the calculations. To this end, the elution volumes are calculated in the same way as in a two-step elution with initial isocratic step preceding the gradient, using appropriate modifications of Eqs. (6), (7), or (10), as described in more detail elsewhere [5,10,37].

Nonideal behavior and secondary equilibria occurring in some chromatographic systems can also affect the results of gradient elution. These phenomena are most significant in normal-phase HPLC, where more polar components are preferentially adsorbed on polar adsorbents or on polar bonded stationary phases. Even trace amounts of moisture in the mobile phase can affect significantly the distribution of sample compounds between the stationary and the mobile phases. Hence, it is very important to control the moisture in the chromatographic system. The easiest way to do it is to use dried solvents.

Further, as preferential adsorption of the polar solvent **B** on the polar adsorbent occurs during gradient elution in normal phase systems, deviations of the actual composition of the mobile phase from the preset values may occur during the gradient elution. To suppress this effect, which is most significant with gradients starting in pure nonpolar solvent, gradients should be started at a nonzero concentration of the polar solvent where possible [10–12]. Finally, it is very important to work under controlled temperature conditions with a thermostated column to get reproducible results in repeated gradient-elution experiments.

Using a sophisticated gradient elution chromatograph, dry solvents, a controlled constant temperature, and taking into account the gradient dwell volume in the calculations, good reproducibility of the retention data in normal-phase systems, even after several months of column use, small differences between the calculated and the experimental elution data ≤0.25 mL or ≤2% were found, which is comparable with the precision of prediction of gradient-elution data in reversed-phase systems [10].

VIII. FUTURE PERSPECTIVES

We surveyed the results of our approach to the description of gradient elution and to the prediction of optimum gradient conditions in two review articles [5,31] and in a book [38]. Work is still in progress on the refinement of calculation algorithms. Using numerical calculations based

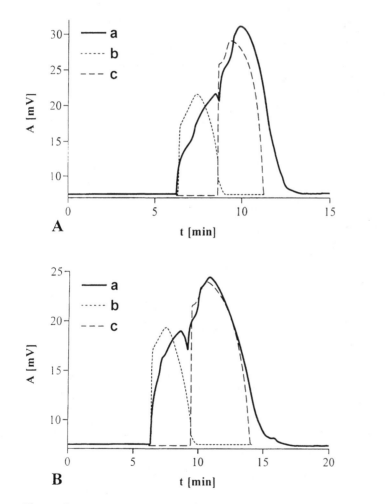

Figure 9   Calculated (dashed lines, transformed to expected detector response) and experimental (full lines, a) band profiles of phenol (first peak, b) and o-cresol (second peak, c) in the overloaded gradient-elution separation of a 5 mL sample feed containing 0.04 mol/L (0.2 mmol) of each component dissolved in water. (A) 0–100% methanol in 10 min; (B) 0–100% methanol in 20 min. Column: Separon SGX C18, 7 μm, 150 × 3.3 mm I.D., flow rate = 1 mL/min. A = response of the UV detector at 289 nm; t = time from the start of the feed injection. (From Ref. 40.)

on the finite difference method, whole band profiles can be predicted, not only for conventional gradient elution, but also for gradient elution separations on overloaded columns under conditions used in preparative HPLC. Presently we are investigating this interesting field in a joint project with George Guiochon and his group at the University of Tennessee, Knoxville [39,40]. An example of overloaded separation run in a conventional analytical column under conditions common in preparative HPLC is shown in Fig. 9. Other area in which detailed research is necessary is the description of gradient elution in chromatographic systems utilizing simultaneous change in concentrations of strong solvent **B** and of specific mobile phase additives affecting the selectivity of separation.

REFERENCES

1. P Jandera, H Engelhardt. Liquid chromatographic separation of organic acid compounds. Chromatographia 13:18–23, 1980.
2. LR Snyder. Gradient elution. In:C Horvath, ed. High-Performance Liquid Chromatography. Advances and Perspectives, Vol. 1. New York: Academic Press, 1980, pp. 208–316.
3. B Drake. Contributions to the theory of gradient elution analysis. Arkiv Kemi 8:1–21, 1955.
4. P Jandera, J Churáček. Gradient elution in liquid chromatography II. Retention characteristics in solvent-programmed chromatography—theoretical considerations. J Chromatogr 91:223–235, 1974.
5. P Jandera, J Churáček. Liquid chromatography with programmed composition of the mobile phase. Adv Chromatogr 19:125–260, 1980.
6. P Jandera, J Churáček. Gradient elution in liquid chromatography I. The influence of the composition of the mobile phase on the capacity factor in isocratic elution—theoretical considerations. J Chromatogr 91:207–221, 1974.
7. LR Synder. Principles of Adsorption Chromatography. New York: Dekker, 1968, pp. 185–239.
8. P Jandera, M Janderová, J Churáček. Gradient elution in liquid chromatography VIII. Selection of the optimal composition of the mobile phase in liquid chromatography under isocratic conditions. J Chromatogr 148:79–97, 1978.
9. P Jandera, J Churáček. Gradient elution in liquid chromatography III. Verification of the theoretical relationships for elution characteristics in isocratic and gradient-elution chromatography on silica. J Chromatogr 93:17–39, 1974.
10. P Jandera, M Kučerová. Prediction of retention in gradient-elution normal-phase HPLC with binary solvent gradients. J Chromatogr A 759:13–25, 1997.
11. P Jandera, L Petránek, M Kučerová. Characterisation and prediction of retention in isocratic and gradient-elution normal-phase HPLC on polar bonded stationary phases with binary and ternary solvent systems. J Chromatogr A 791:1–19, 1997.
12. P Jandera, M Kučerová, J Holíková. Description and prediction of retention in normal-phase HPLC with binary and ternary mobile phases. J Chromatogr A 762:15–26, 1997.
13. P Jandera. Simultaneous optimisation of gradient time, gradient shape and initial composition of the mobile phase in the HPLC of homologous and oligomeric series. J Chromatogr A 845:133–144, 1999.
14. Y Baba, M Fukuda, N Yoza. Computer-assisted retention prediction system for oligonucleotides in gradient anion-exchange chromatography. J Chromatogr 458:385–394, 1988.
15. J Hildebrand, RL Scott. Regular Solutions. Engelwood Cliffs, NJ: Prentice-Hall, 1962.
16. PJ Schoenmakers, HAH Billiet, R Tijssen, L de Galan. Gradient selection in reversed-phase liquid chromatography. J Chromatogr 149:519–537, 1978.
17. P Jandera, J Churáček, L Svoboda. Gradient elution in liquid chromatography X. Retention characteristics in reversed-phase gradient-elution chromatography. J Chromatogr 174:35–50, 1979.
18. K Karch, I Sebastian, I Halasz, H Engelhardt. Optimization of reversed-phase separations. J Chromatogr 122:171–184, 1976.

19.   LR Snyder, JW Dolan, JR Gant. Gradient elution in HPLC I. Theoretical basis for reversed-phase systems. J Chromatogr 165:3–30, 1979.
20.   PJ Schoenmakers, HAH Billiet, L de Galan. Influence of organic modifiers on the retention behavior in reversed-phase liquid chromatography and its consequences for gradient elution. J Chromatogr 185:179–195, 1979.
21.   C Horvath, W Melander, I Molnar. Solvophobic interactions in liquid chromatography with nonpolar stationary phases. J Chromatogr 125:129–156, 1976.
22.   WR Melander, C Horvath. Reversed-phase chromatography. In: C Horvath, ed. High-Performance Liquid Chromatography. Advances and Perspectives. Vol. 2. New York: Academic Press, 1980, pp. 113–319.
23.   P Jandera, H Colin, G Guiochon. Interaction indices for prediction of retention in reversed-phase liquid chromatography. Anal Chem 54:435–441, 1982.
24.   H Colin, G Guiochon, P Jandera. Interaction indices and solvent effects in reversed-phase liquid chromatography. Anal Chem 55:442–446, 1983.
25.   P Jandera. Methods for characterization of selectivity in reversed-phase liquid chromatography I. Derivation of the method and verification of the assumptions. J Chromatogr 352:91–110, 1986.
26.   P Jandera. Methods for characterization of selectivity in reversed-phase liquid chromatography II. Possibilities for the prediction of retention data. J Chromatogr 352:111–126, 1986.
27.   P Jandera. Reversed-phase liquid chromatography of homologous series. A general method for prediction of retention. J Chromatogr 314:13–36, 1984.
28.   P Jandera. Methods for characterization of selectivity in reversed-phase liquid chromatography IV. Retention behaviour of oligomeric series. J Chromatogr 449:361–389, 1988.
29.   JW Dolan, LR Snyder, MA Quarry. Computer simulation as a means of developing an optimized reversed-phase gradient-elution separation. Chromatographia 24:261–275, 1987.
30.   P Jandera, J Churáček. Gradient elution in liquid chromatography XII. Optimization of conditions for gradient elution. J Chromatogr 192:19–36, 1980.
31.   P Jandera. Predictive calculation methods for optimization of gradient elution using binary and ternary solvent gradients. J Chromatogr 485:113–141, 1989.
32.   P Jandera. Predictive optimization of gradient-elution liquid chromatography. J Liquid Chromatogr 12(1&2):117–137, 1989.
33.   P Jandera. Optimization of gradient elution in normal-phase HPLC. J Chromatogr A 797:11–22, 1998.
34.   P Jandera, J Churáček, H Colin. Gradient elution in liquid chromatography XIV. Theory of ternary gradients in reversed-phase liquid chromatography. J Chromatogr 214:35–46, 1981.
35.   P Jandera, B Prokeš. Predictive optimization of the separation methods of phenylurea pesticides using ternary mobile phase gradients in reversed-phase HPLC. J Liquid Chromatogr 14(16&17):3125–3151, 1991.
36.   P Jandera, J Churáček, L Svoboda. Gradient elution in liquid chromatography XIII. Instrumental errors in gradient-elution chromatography. J Chromatogr 192:37–51, 1980.
37.   P Jandera, J Churáček. Gradient elution in liquid chromatography IX. Selection of optimal conditions in stepwise-elution liquid chromatography. J Chromatogr 170:1–10, 1979.
38.   P Jandera, J Churáček. Gradient Elution in Liquid Column Chromatography. Amsterdam: Elsevier, 1985.
39.   P Jandera, D Komers, G Guiochon. Effects of the gradient profile on the production rate in reversed-phase gradient elution overloaded chromatography. J Chromatogr A 760:25–39, 1997.
40.   P Jandera, D Komers, G Guiochon. Optimization of the recovery yield and of the production rate in overloaded gradient elution reversed-phase chromatography. J Chromatogr A 796:115–127, 1998.

14

Solvent Selection for Optimal Separation in Liquid Chromatography

Haleem J. Issaq

SAIC Frederick, NCI-Frederick Cancer Research and Development Center, Frederick, Maryland

I. INTRODUCTION

The selection of a mobile phase that will give optimum resolution in liquid chromatography is not a simple matter. The most important considerations are the properties of the mixture being separated and the stationary phase. The mobile phase can be selected only when these two factors have been defined. A trial-and-error approach is generally used to find a mobile phase that will satisfactorily resolve all the components of the mixture. In liquid chromatography (LC) the mobile phase is generally a mixture of two or more pure solvents. This is especially true when the reversed-phase mode is employed, where the mobile phase is composed of water and an organic modifier. If the composition of the mobile phase remains the same during the entire procedure, we speak of *isocratic* elution; however, if the composition of the mobile phase is continuously changing within a predetermined time period, we speak of *gradient* elution.

This chapter will deal with the selection of the mobile phase using systematic statistical and graphical approaches. Although the discussion will concentrate on reversed-phase LC, the methods discussed are applicable to other chromatographic modes, such as partition, adsorption, ion exchange, or chiral.

This project has been funded in whole or in part with federal funds from the National Cancer Institute, National Institutes of Health, under Contract No. NO1-CO-56000.

By acceptance of this article, the publisher or recipient acknowledges the right of the U.S. Government to retain a nonexclusive, royalty-free license and to any copyright covering the article. The content of this publication does not necessarily reflect the views or policies of the Department of Health and Human Services, nor does mention of trade names, commercial products, or organizations imply endorsement by the U.S. Government.

II. ROLE OF THE MOBILE PHASE

The mobile phase in gas chromatography (GC) is an inert gas (hydrogen, nitrogen, helium, etc.), which carries the solute (in vapor form) through the stationary phase and does not greatly affect, or contribute to, the separation process. Retention in GC is, therefore, the result of direct interaction between the vaporized solute and the stationary phase. Separations are achieved by solute polarity differences, volatility differences, molecular shape differences, or a combination of these and other factors.

In contrast, the mobile phase in LC, thin-layer chromatography (TLC), and high-performance liquid chromatography (HPLC) plays an important, if not a major, role in the separation process. The mobile phase determines not only the separation of the components in a mixture, but the degree of resolution (R_s) (how far the peaks are from each other), selectivity (the order of elution), and elution times. Mobile phase strength (polarity) determines retention time (R_t), and mobile phase composition determines selectivity and resolution.

Selection of the mobile phase is based on the properties of the stationary phase, which in turn is chosen after consideration of the properties of the solute mixture to be separated. This will determine the chromatographic process, i.e., adsorption, partition, ion exchange, etc. When a single solvent is used, it is selected from the elutropic series, in which hexane is the least polar and water the most polar. The elutropic series is not used when selecting a mobile phase that consists of a mixture of pure solvents. Snyder [1] developed an equation for determining the polarity of the mixed mobile phase (see below). In liquid chromatography, the relation between the solute and the mobile and stationary phases may be represented by the peaks of a triangle. This means that there are three types of interaction: solute–mobile phase, mobile phase–stationary phase, and solute–stationary phase. Note that when a mixture of solvents is used, solvent-solvent interaction and solvent demixing (in TLC) must also be considered. Separation of a mixture is achieved when an optimum balance is reached between these three interactions. If the interaction between one pair (e.g., solute–mobile phase) is much stronger than between the others, poor separation or no separation at all results. In other words, to resolve the components of a mixture, the chromatographer must create conditions under which the solutes are each forced to favor both mobile and stationary phases differently. To achieve this, the stationary phase or the mobile phase must be altered. When a mixture of compounds does not elute off the column as a result of strong solute–stationary phase interaction or weak solute–mobile phase interaction, the mobile phase or the column material should be changed to achieve stronger solute–mobile phase interaction and weaker solute–stationary phase interaction. If, on the other hand, the mixture elutes with the solvent peak, then the mobile phase or the column material should be changed so that solute-mobile phase interaction is weaker than before. It is, of course, cheaper and easier to change the mobile phase than the column.

This chapter discusses methods for selecting an optimum mobile phase for isocratic elution and gradient. These methods are based on sound statistical approaches which eliminate operator intuition or guessing and result in savings in time and material.

III. THEORETICAL CONSIDERATIONS

Selection of a pure solvent in liquid chromatography is, as mentioned earlier, based on the eluotropic series. However, the selection of a binary solvent mixture, ternary, and so on, and prediction of the solution and elution times of the components of a mixture are more complex. Snyder [1] presented the following equation for calculating the polarity of a solvent mixture:

$$P' = \Sigma \Phi_i P_i \tag{1}$$

where Φ_i and P_i are the volume fractions in the solvent mixture that has a polarity of P'. This relation and polarity of pure solvents do not apply to normal phases, where the calculations are more complicated [1].

The relation between retention time and solvent strength is described by the following equation [1]:

$$\log(K_1/K_2) = \alpha A_s(\varepsilon_2^\circ - \varepsilon_1^\circ) \tag{2}$$

where K' is the capacity factor, ε° is the solvent strength parameter of solvents 1 and 2, A_s is the molecular area of the adsorbed sample, α is the adsorbent surface activity function, and

$$K' = (R_t - R_{t0} - R_{t0})R_{t0} \tag{3}$$

where R_t and R_{t0} are the retention times of retained and unretained solutes. It was found that the relation in Eq. (2) does not always hold [2–4]. For example, when two mobile phases (acetonitrile-water and methanol-water) that have the same solvent strength calculated according to Eq. (1) were used to elute the same solutes (naphthalene and biphenyl) on the same solid phase, the retention times were not the same [4]. A term was later added to Eq. (2) to account for solvent-solute interactions [2,3]. Since different solvents give different selectivities [5–8], changing the solvent composition may result in a different elution order, depending on the properties of the sample mixture and the solvent chosen. For a mobile phase mixture, solvent strength (polarity) in general determines elution time of the solvents (i.e., R_t), while mobile phase composition determines its selectivity. The composition of the mobile phase would determine the degree of separation α, between two adjacent peaks i and ii, where

$$\alpha = K_2'/K_1' \tag{4}$$

Based on Snyder's theory [1], Saunders [7] presented a graphical representation based on ε° for selecting a solvent for adsorption liquid chromatography. The application of these graphs is rapid and provides a reasonable first approximation to a solvent mixture appropriate for a given sample. It must be stressed that the results are approximate, and in some cases the solvent mixture will not be ideal [7].

For a given sample and adsorbent, long K' varies linearly with ε°. This is generally true for K' values between 1 and 10, which is an acceptable working range following separation of a component from a mixture, but which does not lead to dilution of the sample or long retention times. Solvent strength (polarity) gives a general indication of solute retention, but it may not predict the correct retention times [5,6,9]. Resolution between two adjacent peaks is defined by

$$R_S = (R_{t2} - R_{t1})/1/2(W_1 + W_2) \tag{5}$$

where R_t is the time of elution of peak maximum and W is the baseline width of the peak in units of time.

Also, resolution in liquid chromatography has been defined [6] by the following equation:

$$R_S = 1/4(\alpha - 1)N^{1/2}[K'/(1 + K')] \tag{6}$$

where the number of theoretical plates, N, is defined as

$$N = 16(R_t/W)^2 \tag{7}$$

Note that all the above factors are a function of R_t [Eqs. (3)–(7)].

The three terms in Eq. (6) should be optimized to achieve maximum resolution; however, if the experimental conditions (flow rate, column dimensions, particle size, and properties of the sample) are kept constant, the only parameter affecting separation will be the mobile phase.

It is important to have a solvent that will give reasonable retention times for all components of the mixture, R_t between 5 and 40 minutes, in HPLC, and an R value of 0.2 to 0.8 in thin-layer chromatography.

IV. BINARY, TERNARY, QUATERNARY SOLVENT MOBILE PHASES

Two solvent mobile phases may consist of two binary mixtures (85% methanol-water and 70% acetonitrile-water) or of a binary mixture and a pure solvent (60% methanol-water and pure acetonitrile).

Multisolvent mobile phases may be selected to give optimal separation of a mixture based on a statistical approach or graphical representation. Both approaches are sound and can give optimum results. The statistical approach requires the use of a computer, while the graphical one may not.

A ternary solvent mixture may be three pure solvents, three binary mixtures or a mixture of both as mentioned earlier. In reversed-phase chromatography where methanol-water, acetonitrile-water, and tetrahydrofuran (THF)-water are thought of as three binary mixtures, they might also be considered as a quaternary solvent mixture, methanol-acetonitrile-THF-water.

V. STATISTICAL APPROACHES TO MOBILE PHASE OPTIMIZATION

Belinky [10] used a simplex statistical design strategy to achieve an isocratic mobile phase for the separation of a mixture of polycylic aromatic hydrocarbons (PAH). Using reversed-phase columns and methanol-acetonitrile-water, Belinky assumed that each solvent would occupy the vertex of a triangle (Fig. 1). Since the PAH will not elute from the column with water, a 65% methanol-water mixture was chosen as the weakest mobile phase along the AC axis and 55% acetonitrile-water as the weakest mobile phase along the AB axis. These points are labeled C' and B', respectively. This results in four vertices, CBB'C', where C' and B' are pseudo components of methanol-water and acetonitrile-water, respectively. A simplex design for a three-component mixture requires a triangular coordinate system with three vertices.

In this case two triangular systems are obtained: C'CB and B'C'B (Fig. 1), which may be represented by two simplex designs having three components each. Belinky's objective was to generate response surfaces that would allow the prediction of resolution at any point within C'CB and B'C'B.

The response surface is described by the following equation:

$$Y = b_1X_1 + b_2X_2 + b_3X_3 + b_{12}X_1X_2 + b_{13}X_1X_3 + b_{23}X_2X_3 + b_{123}X_1X_2X_3 \qquad (8)$$

where Y is the capacity factor or resolution, which is the sum of the individual contributions of each component or pseudo component in the system X_1, X_2, and X_3. The b coefficients are determined experimentally; Y values are measured at each of the following concentrations for each of the simplex approaches. The first seven generated values will be used to find the values of the b coefficients. The last three points were used to measure the fit.

The calculation of the b coefficients is shown in the following equations:

$$b_1 = Y_A \qquad (9)$$

$$b_2 = Y_B \qquad (10)$$

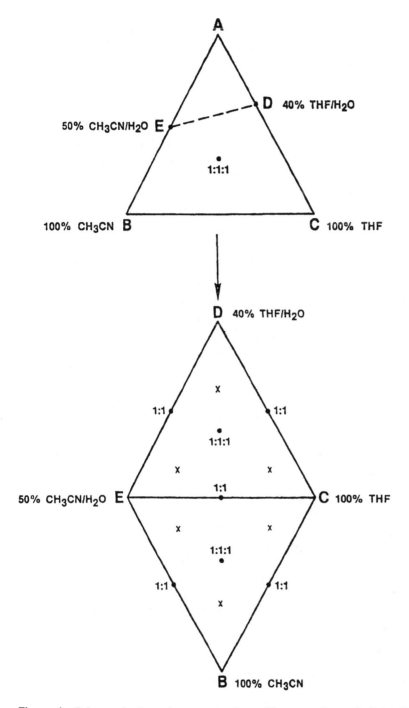

Figure 1 Solvent selection using two organic modifiers according to Belinky. See text for details.

$$b_3 = Y_C \tag{11}$$

$$b_{12} = 4(Y_{AB}) - 2(Y_A + Y_B) \tag{12}$$

$$b_{13} = 4(Y_{AC}) - 2(Y_A + Y_C) \tag{13}$$

$$b_{23} = 4(Y_{BC}) - 2(Y_B + Y_C) \tag{14}$$

$$b_{123} = 27(Y_{ABC}) - 12(Y_{AB} + Y_{AC} + Y_{BC} + 3(Y_A + Y_B + Y_C) \tag{15}$$

Substituting the b values in Eq. (8) will allow the prediction of Y for any solvent composition within the confines of the model, which is assumed to be an accurate representation of the response surface. A total of 17 different mobile phase compositions were required to generate the response surfaces for the two simplex designs. An optimum mobile phase was found which separated the PAH mixture.

One disadvantage of this approach, which is generally sound, is that 17 different chromatographic experiments are required in order to find an optimum mobile phase. This is time-consuming, since the chromatographer has to identify the elution order of the peaks in each of the 17 runs.

Two statistical approaches similar to that of Belinky [10] have been published. Glajch et al. [11] and Issaq et al. [12] discussed a statistical design based on peak pair resolution and overlapping resolution mapping (ORM) for selecting a mobile phase that will give optimum resolution of the components of a mixture. Both works are based on a statistical design by Snee [13]. The reader is referred to Snee's work [13] for the equations and the procedure used for the ORM of each pair of peaks. The procedure used by Issaq et al. [12] will be described here in detail. For the computer programs used, see Appendixes 1 and 2 in Ref. 12.

A combination of the three initial solvents is devised according to Table 1. Other combinations may also be used. The initial solvents A, B, and C may be pure, a mixture of two organic solvents (normal phase), or a mixture of water and an organic modifier for reversed phase. After selecting the solvents and proportions to be used, 10 data points, one for each solvent combination, are collected. These are used to calculate the resolutions of each pair of compounds in the mixture. For a mixture of four solutes, if no peak crossover takes place, the resolution between each pair (1–2, 2–3, 3–4) is used. If peak crossover does occur, the resolution between all the peaks is calculated (1–2, 1–3, 1–4, 2–3, 2–4, 3–4) and used in determining the mobile phase that will give optimum separation.

Two computer programs are used to predict optimum solvent composition. The first (Appendix 1 of Ref. 12) is a FORTRAN program (PEAKIN) which rearranges resolutions to correct for crossover and produces a data file suitable for use in the next program. The second program (Appendix 2 of Ref. 12) is a Statistical Analysis System (SAS, version 79.5) rout [14]. A DATA paragraph converts the three-dimensional solvent compositions to a two-dimensional triangle representation as used by Snee [13]. The data are fitted into a cubic model for a three-dimensional

Table 1 Ratios (v/v) of Solvents C', C, and B Used in Belinky's Study

Solvent	Ratios									
C'	1.00	0.00	0.00	0.50	0.50	0.00	0.33	0.67	0.16	0.16
C	0.00	1.00	0.00	0.50	0.00	0.50	0.33	0.16	0.67	0.16
B	0.00	0.00	1.00	0.00	0.50	0.50	0.33	0.16	0.16	0.67

Source: Ref. 10.

system. The parameters of the cubic equation for each set of peak resolutions are computed using the general linear model (GLM) procedure.

The PRINT procedure lists predicted resolutions of each peak pair for all solvent combinations, varying each solvent from 0 to 100% by 2% increments. Contour plots of the region where the predicted resolution above a desired level determined by the analyst are produced (see, e.g., Fig. 2) using the PLOT procedure. The union of these plots showing the region where all resolutions are above this level and plots showing the area of maximum total resolution are also produced using PLOT. A flow chart of the procedure is shown in Fig. 3.

Ideally, where a combination of three modifiers and a base solvent is used, the region of the optimum mobile phase mixture found from the ORM calculations will be in the center of the triangle (Fig. 2). If one of the solvents (e.g., C) is not ideal, the optimum mixture will be composed of the other two solvents (A and B) with only a small amount of C (see Fig. 4). Therefore, the optimum region can indicate which of the three solvents is a poor choice. The

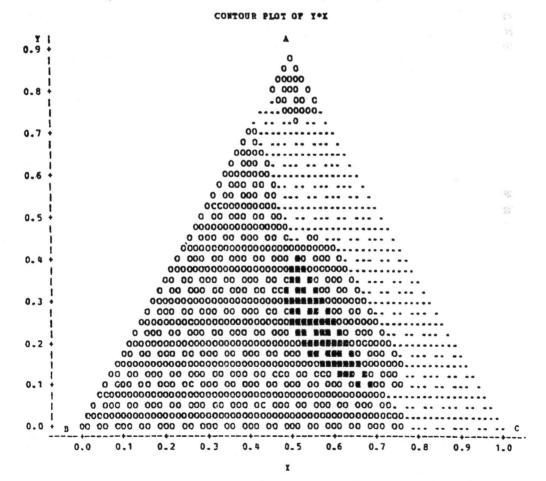

Figure 2 Contour plot showing areas of optimum solvent composition (●) where the resolution between each pair is greater than 5.0.

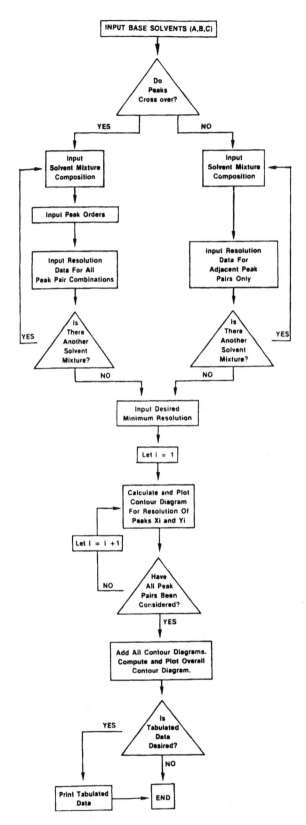

Figure 3 Flow chart of the procedure used for the ORM method of solvent optimization.

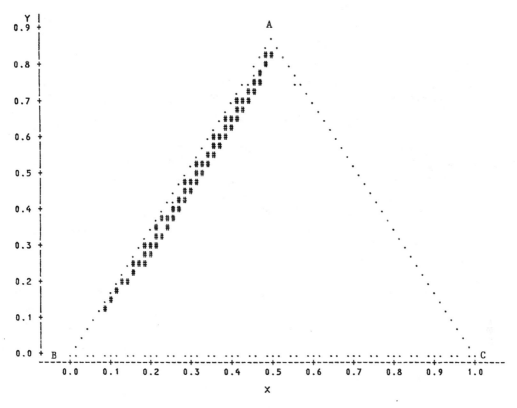

Figure 4 Contour plot of X vs. Y showing areas of optimum mobile phase composition, where the resolution between two adjacent peaks is greater than 2.5.

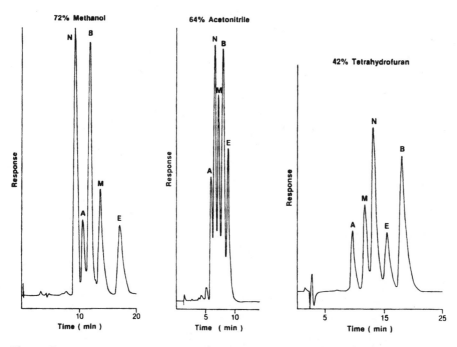

Figure 5 Chromatograms of the HPLC separation of a mixture of naphthalene (N), anthraquinone (A), biphenyl (B), methylanthraquinone (M), and ethylanthraquinone (E) using three different mobile phase compositions.

base solvents are water for reversed phase and hexane for normal phase [11]. Other solvents for normal phase may also be used.

It is also possible to select one solvent (e.g., B), which gives better resolution of the components of a mixture than the other two solvents (A and C). The contour plot will show a bias toward solvent B. In this case, other solvents should be substituted for A and C. These examples show that the initial selection of the individual mobile phases is an important step which can lead to good resolution using the three organic modifiers.

The HPLC and TLC results indicate that this solvent selection system can be successfully applied to mobile phase optimization. Peak crossover due to different solvents can easily be handled by this method for both HPLC and TLC. Figure 5 shows the separation of a naphthalene-biphenyl-anthraquinone-methylanthraquinan-ethylanthraquinone mixture. Note the peak crossover in each of the solvents used. Figure 6 shows the separation using the predicted mobile phase mixture on reversed-phase C_8 columns.

Today manufacturers are equipping their instruments with pumps and microprocessors that will generate any mobile phase mixture from up to four different solvents, without operator supervision. This makes the chromatographer's job much easier. Also, instruments that are equipped with automatic injection will inject a set of standards so that the peak elution order can be determined, that is, which peak corresponds to what component. The ORM procedure

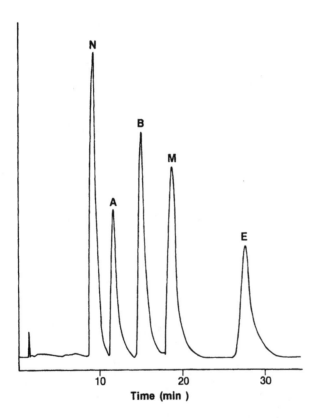

Figure 6 Chromatogram of the separation of the test mixture using a predicted mobile phase of 64% acetonitrile/water, 72% methanol/water, 42% tetrahydrofuran/water (10:67:23).

was used for the separation of 26 fentanyl homologs and analogs [15], for the separation of mixtures of selected steroids [16], and for optimum TLC separations [17].

VI. GRAPHIC PRESENTATION OF MOBILE PHASE OPTIMIZATION

The ORM approach works extremely well when three organic modifiers and a base solvent are necessary to achieve optimum resolution of all components of a complex mixture.

Another approach to solvent optimization has been published based on the linear relationship between log K' and the log mole fraction of the solvent [18]. This approach is not as sound or as generally applicable as the statistical approaches discussed earlier. Recently a more practical approach to mobile phase optimization with two organic modifiers has been published [19]. Only five chromatographic runs were required for the database, and the subsequent mathematical treatment of the data is much less involved. The method is based on the window diagram technique originally developed by Laub and Purnell [20,21] for the optimization of separations in gas-liquid chromatography. Recently a review of the window diagram application to gas chromatography was published [22]. The method is not limited to linear retention behavior or to two-component solvent systems. Also, peak crossovers are easily handled. The method was successfully applied to the optimization of separation of a five-component mixture in reversed-phase HPLC with two organic modifiers and water-based solvents [22]. The initial ratios of organic-water selected are approximately 70–75% methanol-water, 60–65% acetonitrile-water, and 40–50% tetrahydrofuran-water. The strategy for selecting two of these three is illustrated below. For simplicity, assume that a five-component mixture is to be separated. The sample is first injected where 60% acetonitrile-water is the mobile phase. Should four peaks be obtained, standards are used to identify the two coeluting peaks. Only these two are then reinjected and eluted using a different solvent, for example, 50% tetrahydrofuran-water. Should the two components be separated, the different mobile phase compositions are prepared using 60% acetonitrile as solvent A and 50% THF as solvent B. Should 50% THF-H_2O fail to separate the coeluting pair, the percentage of THF is adjusted or another organic modifier is selected. This approach is simple and time saving because the analyst has only to separate the pair not resolved; also, the identification of two components is simpler than identifying all components in a mixture.

This separation strategy was used to separate a mixture of anthraquinone, 2-methylanthraquinone, 2-ethylathraquinone, naphthalene, and biphenyl. The sample solution was chromatographed with 60% acetonitrile-H_2O. Only three peaks were observed. Anthraquinone and naphthalene coeluted, as did 2-methylanthraquinone and biphenyl. However, both solute pairs were separated with 40% THF-H_2O. This demonstrates that each of the four pairs had been resolved in at least one of the initial solvents.

After selecting the initial solvents and the proportions of each in the three-solvent combination, the retention times database is generated by recording the retention times of each solute in each of the different solvent combinations: 75% A, 50% A, and 25% A, in addition to 100% A and 100% B.

Table 2 shows the composition of solvents used and the retention times for each of the five solutes with each of the five different mobile phases. The retention data as a function of mobile phase composition were fit to a polynomial of the fourth order by least-squares analysis.

Figure 7 shows plots of the calculated retention times for each solute as a function of mobile phase composition. Note that, in contrast to gas chromatography, the plots are not linear. This is due to the complicated nature of solute–mobile phase, solute–stationary phase, and mobile phase–stationary phase interactions [8].

Table 2 Retention Times for Each Solute at Different Mobile Phase Compositions

Solvent	Mobile phase composition		Actual component percentage			Retention time (min)					
	A[a] (%)	B[b] (%)	AN	THF	H₂O	Anthraquinone	2-Methyl-anthraquinone	Naphthalene	2-Ethylanthraquinone	Biphenyl
1	100	0	60	0	40	7.23	8.98	7.23	10.94	8.98
2	75	25	45	10	45	8.26	10.30	9.16	13.27	11.77
3	50	50	30	20	50	9.57	11.98	11.30	15.82	15.13
4	25	75	15	30	55	11.65	14.75	14.67	19.98	20.38
5	0	100	0	40	60	16.44	21.20	22.41	30.19	34.03

[a] 60% acetonitrile (AN)–water.
[b] 40% tetrahydrofuran (THF)–water.

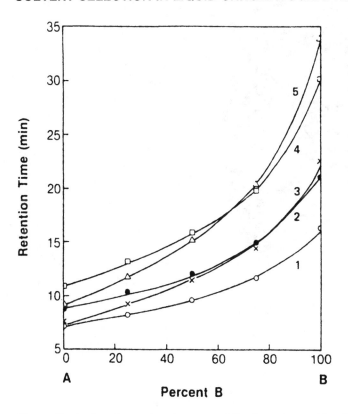

Figure 7 A plot of retention time vs. mobile phase composition for five solutes. Solvent A is 60% acetonitrile/water and solvent B is 40% tetrahydrofuran/water. Solutes are anthraquinone (1), methylanthraquinone (2), ethylanthraquinone (3), naphthalene (4), biphenyl (5).

Figure 8 is a window diagram showing plots of retention time ratios versus mobile phase composition for all 10 pairs of the five solutes. The region of retention time ratio values that are higher than the minimum found at each mobile phase composition is shaded. Note that when the relative retention is calculated to be less than unity (peak crossover), the reciprocal is taken such that the ratio is always greater than or equal to unity. The tops of the windows represent the mobile phase composition giving the best separation for the least-separated pair. Two windows are seen in Fig. 8: one at 21% B with a minimum retention ratio = 1.1, and a considerably smaller window (poorer separation) at 100% B. Thus, the optimum mobile phase composition for this particular separation is predicted to be 27% B (10.8% THF: 43.8% AN: 44.4% H_2O), which does, in practice, give baseline separation of the components of the mixture (Fig. 9).

The theoretical measure of separation of a solute pair in chromatographic techniques is the relative retention α, which in HPLC is defined as the ratio of the capacity factor of the more retained solute to the less retained solute. A plot of K' versus mobile phase composition gave exactly the same results as those obtained from Fig. 9. However, here the analyst can obtain the minimum value of α (1.15 at 27% B). Using this value, we can calculate the minimum number of plates (N_{req}) for separation according to Purnell's equation [23]. In this instance, and assuming a capacity factor of five, N_{req} is calculated to be approximately 2350 plates. Note that accurate measurement of the column "dead volume" in HPLC is a difficult problem with no

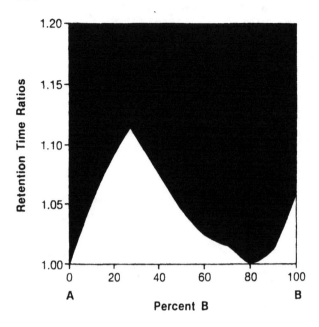

Figure 8 Window diagram for all solute pairs based on data from Fig. 7.

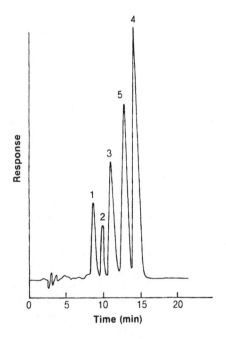

Figure 9 Chromatogram of the test mixture separated using predicted mobile phase composition of 43.8 acetonitrile, 10.8% tetrahydrofuran, and 44.4% water as predicted in Fig. 8.

easy solution [24]. Any optimization techniques dependent on k' data suffer from the unavailability of accurate methods for the determination of column dead volume. The window diagram method presented here does not require the accurate determination of k'. As demonstrated earlier, the optimum solvent composition can be determined from raw retention data. When retention time is plotted against mobile phase composition, it can be seen that a total of five runs (Fig. 8) will give the mobile phase that will separate all the components of a simple mixture. For complex mixtures the ORM approach is preferred.

Deming and Turoff [25] used the window diagram technique to select an optimum pH of a buffer for the separation of a mixture of benzoic acids on reversed-phase HPLC. The procedure consisted of measuring the retention time of the weak acid of interest at pH 3 or higher, values in an appropriate buffer, after which models were fitted to the data and the model parameters were used to construct window diagrams from which an optimum pH could be selected. Deming and coworkers used the window diagram technique to study the effect of pH on other liquid-chromatographic separations [26,27]. Later they extended the simple factor study in which values of pH and the concentration of ion interaction reagent were chosen to give optimum separation [28,29]. Otto and Wegscheider [30] used a multifactor model to optimize the separation of dibasic substances on reversed-phase chromatography. Elution strength, ionic strength, and pH were studied in order to construct a three-dimensional semiempirical model window diagram for predicting retention times of dibasic compounds. The original article [30] should be consulted for details.

The optimization of temperature and solvent strength for the separation of phenylthiohydantoin amino acids on reversed-phase columns was achieved using the window diagram technique by plotting the logarithm of retention time versus solvent strength or temperature [31].

Lindberg et al. [32] discussed the application of factorial design and response surface calculations to the optimization of a reversed-phase ion-pair chromatographic separation of morphine, codeine, noscapine, and papaverine.

Kowalski [33] has reviewed the above statistical approaches in detail and therefore they will not be discussed here. The reader can also refer to the original article of Lindberg et al. [32].

Schoenmakers et al. [34] devised a simple procedure for the optimization of reversed-phase separations with ternary mobile phase mixtures. The procedure is based on the use of isoeluotropic mixtures, that is, mixtures that are expected to yield the same capacity factor for a hypothetical average solute [35] and the "window diagram" technique. Their procedure can be summarized as follows: a gradient run from pure water to pure methanol would show if all solutes in the mixture can be eluted isocratically within a given time, a K' value between 1 and 10 [36]. The sample is then run in two binary isoeluotropic mixtures (acetonitrile/water and tetrahydrofuran/water), and the peaks are identified. Then a linear plot is constructed of ln K' versus the solvent composition, and the ternary mixture that gives the best separation is selected. The sample is run using the optimum mobile phase. If the experimental results are not satisfactory, the calculations are repeated until there is no change in the optimal composition and the best separation has been achieved. It is assumed that in K' versus mobile phase composition is linear, which may or may not be true. Our results [18] indicate that K' versus mobile phase composition is not linear. In a later study, Dronen et al. [37] utilized the chromatographic response function (CRF) to Morgan and Deming [38,39] to develop a more general approach to the selection of mixed mobile phases. The CRF procedure is described in detail in Ref. 38. Berridge used the sequential simplex procedure, CRF, to optimize reversed-phase [40] and normal-phase [41] HPLC separations. He also tested the quality of the separation by examining the second derivative of the final chromatogram. If two peaks coeluted, the second derivative will reveal it, which will then require a rerun of the optimization procedure using different criteria and/or a different selection of solvents. The CRF is one that describes the separation quality in terms of the numbr of peaks, resolution, and time of analysis. The CRF used by Berridge [41] is defined as follows:

$$CRF = \sum_{i=1}^{L} R_i + L^a - b(T_L - T_A) + c(T_1 - T_0) \tag{16}$$

where R is the resolution between adjacent peaks, L is the number of peaks detected, T_A is the specified analysis time (in minutes), T_L is the retention time of the last peak, T_1 is the retention time of the first peak, T_0 is the specified minimum retention time, and a, b, and c are operator-selectable weightings (set to 2). Using this above procedure, the separation of a five-component mixture required 23 experiments in which two of the five peaks were not baseline separated. Although the procedure is automated, we feel that 23 experiments are a waste of time and material unless the separation is an extremely difficult one.

Toon and Rowland [42] proposed a simple graphical method for the selection of a binary mobile phase that will give optimum separation. Six barbiturates were to be separated on a revised-phase column using acetonitrile-water. The solution, as stated, lay in the empirical observation that for all barbiturates studied, a linear relationship existed between $\log(1 - R_0/R_1)$, designated as R_Q, and the mobile phase, where R_1 and R_0 are retained and nonretained solutes. Optimum resolution is defined as complete separation of the solute peaks at their base in the shortest time. By plotting R_0 values, estimated for both the front and the back of solute peaks against percentage solvent composition, one can find the optimum mobile phase. Optimum resolution is defined as the point where the plots of the R_0 front of one solute intersect with the R_0 back of another solute. It is claimed that this procedure worked for multicomponent (*n*) mixtures after some trial and error, since there will be $n - 1$ points of intersection. It is much easier, in our opinion, to adopt the window diagram approach or a gradient run from pure water to pure acetonitrile to select an isocratic mobile phase. However, this procedure of Toon and Rowland [42] was criticized by Tomlinson [43], who found that R_0 is always negative and at variance with Fig. 1, 3, 5, and 8 in Ref. 42. Other points were also raised by Tomlinson [43], which will not be discussed here. Rafel [44] compared the resolution of nitroaromatics and flavones by three different methods: the simplex method [45], the extended Hooke-Jeeves direct search method [46], and the Box-Wilson steepest ascent path [47]. The results showed that the optimal resolutions obtained by any of these methods are similar; however, the Hooke-Jeeves method, following a 2k-factorial design, appeared to be a faster and easier approach.

VII. OPTIMIZATION OF GRADIENT ELUTION

Gradient elution is mainly used in two cases: (a) when an isocratic mobile phase fails to give adequate resolution of all the components in a mixture, due to the components' properties, and (b) when an isocratic mobile phase gives satisfactory resolution of all the components but those peaks eluting last are wide and far from each other. In the first case gradient elution is used to facilitate the separation, while in the second case it is used to bring the late eluting peaks closer together, which results in better sensitivity and shorter analysis time. (For detailed discussion of gradient elution, see Ref. 8.) The gradient shapes generally used are linear, convex, or concave (see Fig. 10). Although these gradient shapes may improve resolution for each adjacent pair of components, the retention times between peaks is not the same and may not be optimum, which will result in a long experiment time. Gradient separations that result in the same resolution between the adjacent pairs will be defined as optimum gradient. To achieve an optimum gradient,

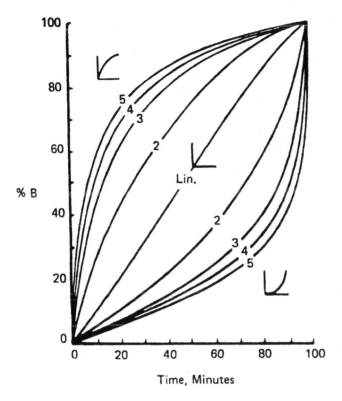

Figure 10 Gradient shapes used in HPLC.

the chromatographer should treat each adjacent pair of components as a separate experiment in order to optimize the separation between the first pair, then the second pair, and so on, so that the sum of these series of pair optimization experiments is the optimum gradient. This seems to be very complex, but it is not if a computer is used to store the results of these series of experiments, to select the mobile phase for each pair based on the empirical data, and to compile the final mobile phase. It is clear from the above that the gradient shape will be different from those shown in Fig. 10. The gradient shape predicted will consist of a series of straight lines (Fig. 11). Each straight line represents a mobile phase composition that will result in the same resolution between all the pairs in the mixture. These straight lines may or may not have the same slope. To achieve an optimum gradient, a series of experiments are needed, the number of which depends on the complexity of the sample (number of components) and difficulty of separation. An average procedure will require five runs. The runs are of two types, and it is up to the chromatographer to select the one he prefers. The first type is shown in Fig. 12, where the mobile phase composition is constant but the time period is variable. Assume that the gradient is linear and will run from 100%A (the weak solvent) to 100%B (the strong solvent) in 10 minutes, the next gradient will run from 100%A to 100%B in 20 minutes, the third one from 100%A to 100%B in 30 minutes, and so on until all the components in the mixture are resolved. Solvents A and B should be miscible. Figure 13 shows the second type of gradient mobile phase composition experiments needed to determine the optimum gradient. In this case the mobile phase remains constant. The linear gradient is run from 0%B to 20%B (100%A to 80%A) in 30 minutes, the next time it is run from 0%B to 40%B in 30 minutes, and third run from 0%B

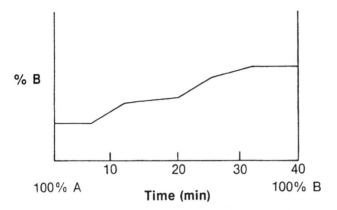

Figure 11 Step gradient.

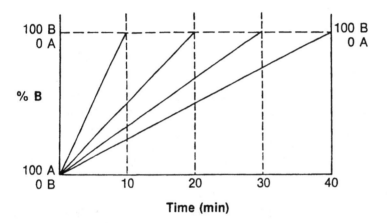

Figure 12 Gradient mobile phase using variable time but the same mobile phase composition, 100A to %A.

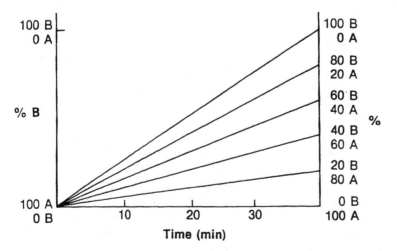

Figure 13 Gradient mobile phase using constant time (40 min) but variable mobile phase composition as shown.

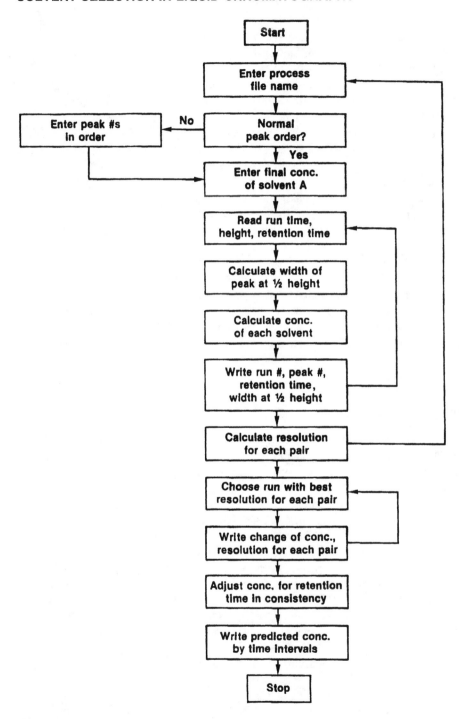

Figure 14 Flow chart of the computer program used for peak identification.

to 60%B in 30 minutes, and so on. In each case the computer will calculate the resolution between the peaks and print it in a table at the end of the experiment. The chromatographer will specify in the computer program the resolution needed between the adjacent pairs. The computer will at the end of the run search for the mobile phase composition that will give the predetermined resolution between each adjacent pair and print it in a table and in a graphical form as shown in Fig. 11. It is also possible that one of these five experiments may result in a reasonable resolution. In such a case the computer prediction will not be required.

A flow chart of the computer program is presented in Fig. 14. The program assumes that the peak areas and retention times are saved on the system's processed data file and the signal from the instrument is available on the system's raw data files.

The program operates as follows: for each run the user enters the processed data file name; the peak's elution order, in case of peak crossover; and the final concentration of solvent A, where the starting mobile phase composition is 0%A and 100%B. The processed data file gives the runtime and the peak retention times and heights. The peak width at half-height is calculated from the raw data file. The concentration for each solvent at the peak retention time is computed, and the resolution between each peak pair is calculated. Data for up to 10 runs with 10 peaks per run can be processed; this can be easily expanded to a larger number of peaks. For each peak pair, the retention times and mobile phase compositions that produced the best resolution are selected as recommended at those times. If there are retention time conflicts between adjacent peaks (i.e., the best resolved retention times between peak 1 and 2 overlap the retention times for peaks 2 and 3), then a regression is fit to the four time-concentration points to smooth the data. However, no corrections are done on the mobile phase compositions if there are no retention time conflicts, which may result in inconsistent composition recommendations.

VIII. PEAK IDENTIFICATION

One of the most difficult and time-consuming aspects of searching for a mobile phase that will give optimal separation is the identification of the eluted peaks. To establish such a mobile phase, the analyst, depending on the method chosen, must run 5–23 experiments using different solvent combinations. Since these various methods require the use of solvents of different selectivities, this can lead to different elution orders of the components, that is, peak reversal. Thus, the analyst must identify each eluted peak at the end of each chromatographic run, which is normally done by spiking the sample with a known standard or injecting each of the standards separately, a time-consuming procedure. Issaq and McNitt [48] developed a computer program whereby the eluted peaks are identified by the area ratio of each peak compared with the other eluted peaks in that experiment. The computer program takes into consideration peak reversal, peak coalescence, and peak splitting. The analyst can also use the ratio of the areas under the peaks generated using two different wavelengths, for example, 254 and 280 nm. This computer program is also suited for use with radiolabeled compounds, since the number of counts is directly proportional to the amount of radioactivity.

A more accurate method for identifying the eluted components is one based on the molecular and physical properties of the compounds under study. The use of mass spectrometry (HPLC/MS) would give accurate results, but this is relatively expensive. Another method would be to use Fourier transform infrared spectroscopy (FTIR), but this is not without problems such as background interference from the solvent and the sample. These problems may be overcome if the analyst can subtract interference with the aid of a computer.

A detector that can record the spectra in 1/100 the time it takes FTIR to record the spectra, and at half the price, is the photodiode array (PDA) detector, first introduced by Hewlett-Packard.

The model 1040 PDA detector records the ultraviolet-visible spectra of the eluting component in 10 ms without interrupting the chromatographic process, unlike stopped-flow ultraviolet detectors. This detector is based on the combination of a series of photodiodes, which monitor all wavelengths (190–600 nm) simultaneously, with a data acquisition processor network for fast analog-to-digital conversion and fast data storage. A detailed description of the detector can be found in Ref. 49.

Two other computer-assisted methods have been published; the EluEx expert system has been developed for optimizing the reversed phase and ion-pair chromatography methods [50]. The program predicts a starting mobile phase based on the chemical structure of the solutes. In this system the polarity of each compound is determined according to the logarithm of the 1-octanol/water partition coefficients. For details see Ref. 50.

The PRISMA optimization model is made up of three parts [51]. Basic parameters such as stationary phase and initial single solvents are selected in the first part. The second part determines the optimal conditions of the selected solvents. The third part concerns the transfer of the optimized mobile phase between the various column and planar chromatographic techniques and the adjustment of parameters such as flow rate, development mode, and particle size. For details see Ref. 51.

The DryLab program [52] was developed for both isocratic and gradient methods. The method is based on input data from two isocratic or gradient experiments, which models the separations for a given mixture and simulates new experiments under different mobile phase conditions. For details see Ref. 52. The DryLab programs are commercially available.

IX. CONCLUSIONS

This chapter has dealt with methods for selecting an optimum mobile phase for isocratic elution. It is clear that the window diagram technique is a very simple, uncomplicated approach to mobile phase optimization using two binary mobile phases. It is equally clear that the ORM approach is a very efficient method for selecting a ternary or quaternary mobile phase system. The ORM approach requires a computer, while the window diagram does not. In most cases a graphic presentation is all that is needed. Today the analyst can buy a computer program for isocratic and gradient mobile phase optimization. For details of statistical approaches, the reader should consult Refs. 53 and 54. These approaches are systematic and require a defined set of conditions and a minimum number of experiments. Their use should be encouraged, since they will save the chromatographer time and money.

REFERENCES

1. LR Snyde. Principles of Adsorption Chromatography. New York: Marcel Dekker, 1968.
2. M Jaroniec, JK Rozzylo, JA Jaroniec, B Oxcik-Mendyk. J Chromatogr 188:27, 1980.
3. DE Martire, RE Boehm. J Liquid Chromatogr 3:753, 1980.
4. HJ Issaq. Pittsburgh Conference on Analytical Chemistry and Applied Spectroscopy, Atlantic City, 1981.
5. LR Snyder. J Chromatogr Sci 92:223, 1974.
6. SR Baklayer, R McIlwrick, and E Roggendorf, J Chromatogr 14, 343, 1977.
7. DL Saunders, Anal Chem 46, 470, 1974.
8. LR Snyder and JJ Kirkland, Introduction to Modern Liquid Chromatography, 2nd ed. New York: Wiley, 1979.
9. D Rogers. Am Lab 12:49, 1980.

10. BR Belinky. Analytical Technology and Occupational Health Chemistry. ACS Symposium Series, Volume 220, American Chemical Society, Washington, DC, 1980, pp. 149–168.
11. JL Glajch, JJ Kirkland, KM Squire. J Chromatogr 199:57, 1980.
12. HJ Issaq, JR Klose, KL McNitt, JW Kaky, GM Muschik. J Liquid Chromatogr 4:2091, 1981.
13. RD Snee. Chem Technol 9:702, 1979.
14. SAS Institute. Raleigh, NC: SAS User's Guide, 1979.
15. IS Lurie, AC Allen, HJ Issaq. J Liquid Chromatogr 7:463, 1984.
16. PE Antle. Chromatographia 15:277, 1982.
17. J Sherma, S Charvat. Am Lab 15:138, February 1983.
18. S Hara, K Kunihiro, H Yamaguchi, E Soczewinski. J Chromatogr 239:687, 1982.
19. HJ Issaq, GM Muschik, GM Janini. J Liquid Chromatogr 6:259, 1983.
20. RJ Laub, JH Purnell, PS Williams. J Chromatogr 134:249, 1977.
21. RJ Laub, A Pelter, JH Purnell. Anal Chem 51:1878, 1979.
22. RJ Laub. Am Lab 13:47, 1981.
23. JH Purnell. J Chem Soc 1268, 1960.
24. E Grushka, H Colin, G Guiochon. J Liquid Chromatogr 5:1297, 1982.
25. SN Deming, MSH Turoff. Anal Chem 50:546, 1978.
26. WP Price, Jr, R Edens, DL Hendrix, SN Deming. Anal Biochem 93:233, 1979.
27. WP Price, Jr, SN Deming. Anal Chim Acta 108:277, 1979.
28. RC Kong, B Sachok, SN Deming. J Chromatogr 199:307, 1980.
29. B Sachok, RC Kong, SN Deming. J Chromatogr 199:317, 1980.
30. M Otto, W Wegscheider. J Chromatogr 258:11, 1983.
31. CM Noyes. J Chromatogr 266:451, 1983.
32. W Lindberg, E Johansson, K Johansson. J Chromatogr 211:201, 1981.
33. BR Kowalski. Anal Chem 52:112R, 1980.
34. PJ Schoenmakers, ACJH Dronen, HAH Billiet, L de Galan. Chromatographia 15:688, 1982.
35. PJ Schoenmakers, HAH Billiet, L de Galan. J Chromatogr 218:261, 1981.
36. PJ Schoenmakers, HAH Billiet, L de Galan. J Chromatogr 205:13, 1980.
37. ACJH Dronen, HAH Billiet, PJ Schoenmakers, L de Galan, Chromatographia 16:48, 1983.
38. SL Morgan, SN Deming. J Chromatogr 112:267, 1975.
39. SL Morgan, SN Deming. Sep Purif Methods 5:333, 1976.
40. JC Berridge. J Chromatogr 244:1, 1982.
41. JC Berridge. Chromatographia 16:174, 1973.
42. S Toon, M Rowland. J Chromatogr 298:341, 1981.
43. E Tomlinson. J Chromatogr 236:258, 1982.
44. J Rafel. J Chromatogr 282:287, 1983.
45. W Spendley, JR Hext, FR Himsworth. Technometrics 4:441, 1962.
46. R Hooke, TA Jeeves. J Assoc Comput Mach 8:221, 1961.
47. GEP Box, KB Wilson. RJ Stat Soc B 13:1, 1951.
48. HJ Issaq, KL McNitt. J Liquid Chromatogr 5:1771, 1982.
49. JM Miller, SA George, GB Willis. Science 218:241, 1982.
50. PP Csokan, F Darvas, F Csizmadia, K Valko. LC GC Int 6:361, 1993.
51. S Nyiredy, K Dallenbach-Tolke, O Sticher. J Liq Chromatogr 12:95, 1989.
52. LR Snyder, JW Dolan, DC Lommen. J Chromatogr 485:65, 1989.
53. JL Glajch, LR Snyder. Computer Assisted Method Development for HPLC. New York: Elsevier, 1990.
54. PJ Schoenmakers. Optimization of Chromatographic Selectivity, A Guide to Method Development. New York: Elsevier, 1994.

15

My Life in Separation Sciences: Study of Separation Mechanisms

Kiyokatsu Jinno

Toyohashi University of Technology, Toyohashi, Japan

I. HISTORICAL PERSPECTIVE

I received my doctorate in 1973 under Professor Daido Ishii, Department of Applied Chemistry, Nagoya University, a well respected scientist who is considered one of the best Japanese separation scientists. My project was not on separation sciences. My degree was awarded on radioanalytical chemistry; my thesis was entitled ''Nondestructive Determination of Various Elements by Neutron Activation Analysis.'' After receiving the degree, I worked at Toshiba Research and Development Center until 1978, when I moved to an academic position. At Toshiba I was working in lithography for super large-scale integrated circuits, an active and interesting field. In 1977 I met Professor Tsugio Takeuchi, a member of the Department of Applied Chemistry, Nagoya University, at that time, and he asked me about my work at Toshiba. A few months after our meeting I received a letter from the Japanese Ministry of Education and Culture stating that I was selected as one of the committee members to discuss new concepts for a new technical institution that would be built in Toyohashi. Since Professor Takeuchi's name was on the members' list, I phoned him inquiring about this offer. He said that since I had told him that university work would be nicer because we could do what we wanted, he chose me to be a faculty member of a new national technological university. After a long period of discussion between Toshiba and Professors Ishii and Takeuchi, I finally moved to the Toyohashi University of Technology (TUT) as an associate professor in April 1978, when the university has just opened. Professor Takeuchi had expected me to work in radiochemistry using an accelerator to be built on the new campus. Unfortunately, the machine never appeared in our TUT campus because Professor Takeuchi, the chief investigator of this project, died suddenly in the summer of 1979. No one continued this accelerator project, and I had to find a new project for my research at TUT. Professor Takeuchi purchased many chromatographic instruments, and my mentor, Professor Ishii, recommended that I start my research work on microcolumn liquid chromatography (micro-LC), which was his main topic at that time. I accepted his advice and started working on

253

micro-LC. However, several people had already started working in this area, and in order to do original research work I had to find the room to do so. I found two areas in separation sciences, especially in micro-LC, to pursue. One was the hyphenation between spectroscopy, such as infrared (FT-IR) and atomic emission (ICP), and micro-LC. The other was rather difficult, but basic and theoretical, dealing with retention mechanism in LC. I will focus on the latter topic in this chapter in order to show my contributions to the developments of separation sciences in the last 20 years.

II. MY CONTRIBUTIONS TO SEPARATION SCIENCES

As described above, my contributions to the development of separation sciences can be divided into at least two main categories: hyphenated techniques in liquid phase separations and the study of retention mechanisms in liquid phase separations. I will discuss the second topic in this section in detail to show how I thought, how I made progress, and how I solved the problems of our research projects in this area.

The first questions I asked when I started my chromatographic projects in 1979–1980 were why people working in chromatography did not try to understand the separation of various compounds and what is the meaning of retention in chromatography.

As you know, chromatography in any system can only give us information on retention; sometimes it is a retention time, sometimes it is a retention volume, and sometimes it is a so-called retention factor. This information is not an accurate identification tool, it just gives us quantitative information, which is defined by the peak area in the chromatogram. It also gives the analyst rough qualitative information by the confirmation using some analytes that can give the same retention as the standard solutes. Therefore, chromatography in itself is not an accurate analytical tool, rather it is a separation tool. The process of identifying the components resolved by chromatography requires the use of spectroscopic methods. In order to improve this situation, one has to develop ideas to explain the details of the retention mechanism and then use this idea to employ chromatography as a tool for accurate identification in addition to its ability to resolve the analytes in a mixture.

A. QSRR Approach

Based on the above-mentioned idea, I started the work to define the retention in reversed-phase LC separations. The approach I selected was QSRR (quantitative structure-retention relationships) [1], a method used to identify the properties of the solutes to control their retention by correlating the retention factors (k) measured at certain separation conditions in LC and solutes' properties (Pi) such as physicochemical or geometrical factors. Generally, k can be described by Pi as the following function:

$$k = f(Pi) \tag{1}$$

If one can find Pi to give the highest correlation coefficient for Eq. (1), one can assume the most important factors to control the retention and then the retention mechanism can be understood.

Many different compounds were chosen as test probes, including aromatic compounds, polycyclic aromatic hydrocarbons (PAHs), amino acids, amino acid derivatives, oligopeptides, drugs, pesticides, polymer additives, and so on. For Pi there are many factors related to the retention based on the idea of quantitative structure-activity relationships (QSAR) for drug design. I selected various parameters of Pi as follows: physicochemical descriptors (e.g., hydrophobicity), log P (partition coefficient between 1-octanol and water) and hydrophobic substitution constant, geometrical descriptors [e.g., van der Waals volume (Vw) and van der Waals surface area (Aw) and shape parameter or length-to-breadth ratio (L/B)], topological descriptors [e.g., connectivity index χ and correlation factor (F)], electronic descriptors, Hammett's constant (σ, HD, and HA), which are proton-donating and proton-accepting properties. By the multiregression analyses of these retention data with the above-mentioned descriptors, the most dominant properties are found by the regression coefficients. The results we have found show that the most important descriptor in reversed-phase LC separations is hydrophobicity. The next most dominant descriptors are the shape and size of the molecules for nonpolar compounds and electrical descriptors for polar compounds. This means that if one can generate some equations to describe the retention, like Eq. (1) where Pi is defined in terms of log P, π, L/B and σ, we can predict retention of other solutes if Pi values for those compounds are known.

B. Construction of Computer-Assisted Separation System

Based on these facts, found by the basic QSRR studies, one can produce a system using a computer that can have several functions, as stated below, since the retention can be described by several descriptors such as log P, L/B, and σ as

$$\log k = f(Pi, Xi) \tag{2}$$

where Xi indicates the separation conditions such as mobile phase composition or column temperature.

To generalize the retention descriptions for n different conditions, the following n equations should be obtained by the same procedure with multiple regression analysis. The relationships for each compound's group and LC system can be described in general form as follows:

$$
\begin{aligned}
X = X_1 & \qquad \log k_1 = a_1 P_1 + b_1 P_2 + C_1 \\
X = X_2 & \qquad \log k_2 = a_2 P_1 + b_2 P_2 + C_2 \\
& \qquad\qquad \vdots \\
X = X_n & \qquad \log k_n = a_n P_1 + b_n P_2 + C_n
\end{aligned}
\tag{3}
$$

where X is the volume fraction of organic modifier in the mobile phase and a and b are the coefficients corresponding to descriptors P_1 and P_2, respectively, c is the intercept, and n is the number of experimental conditions examined. If a, b, and c can be expressed as functions of X, i.e., X-a, X-b, and X-c are highly correlated, the following equation can be solved by the multiple regression analyses:

$$\log k = f_1(X)P_1 + f_2(X)P_2 + f_3(X) \tag{4}$$

This equation means that if X, the concentration of organic modifier in the mobile phase, and P_1 and P_2, descriptors of a compound, are given, the logarithm of the retention factor, log k, can be determined for any chromatographic conditions. This is the basic concept for predicting retention. Based on this concept, the following three functions can be predicted:

1. Retention of any solutes can be produced under a certain separation condition
2. Optimization of the separation for selected solutes based on retention prediction
3. Identification of the peaks appearing on the chromatogram measured at a certain separation condition by the properties estimated using the measured retention factor matching to those in the data base of such properties of compounds

We derived the computer-assisted *micro*computer *a*ssisted *s*eparation *sy*stem (MCASYST) [2], which can be useful in the following three analytical areas:

1. Analytical system for large PAHs in diesel engine particulate matter [3]
2. Analytical system of toxic drugs in human fluids for clinical, toxicological, and forensic analytical problems [4]
3. Estimation of amino acid sequencing of oligopeptides by LC retention information [5]

In order to show the applicability of the above systems, I will give here one practical example, namely, that of drug analysis in human fluids in a hospital emergency room. The basic structure of MCASYST for this drug analysis is shown in Fig. 1.

The actual analytical process is as follows: The human fluid specimen was obtained from a patient who was brought to the emergency room of a hospital in Tokyo. After some sample preparation procedures were performed, the sample was ready for LC injection. Inputting the desired analysis time and the desired column for the analysis into MCASYST, the system set-up, and the optimized separation conditions, we started the analysis. After the chromatogram was recorded, the peaks were searched on the display and the calculated retention factors for individual peak brought into the retention prediction function in the system in one step. In another step the spectral data of each peak were measured by the photodiode array detector, recorded, and searched through the UV spectral library. In the first process, input of the retention factors of each peak to Eq. (5) were used to calculate the log Pe, which is the estimated hydrophobicity of the analytes that appeared as peaks in the chromatogram:

$$\log k = f_1(X) \log Pe + f_2(X) \tag{5}$$

where X is the separation condition.

The system's data base of log Pe stores the log Pe numerical values of many compounds. The compounds in this data base that have log Pe values similar to those estimated by the measured retention factor are listed in their order of matching as possible candidates. Again the measured UV/VIS spectra of the peaks are compared to the spectra in the UV/VIS spectral library and the best matched candidates are listed in their order of matching. Combining the candidates from the retention search and the spectral search, the most promising candidates for

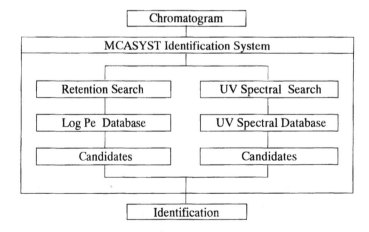

Figure 1   Flowchart of identification process for toxic drug compounds by MCASYST.

the peaks in the chromatogram are selected. Two matching processes to the data base systems can confirm the accurate identification of the compounds that eluted as peaks in the chromatogram. A typical example of this identification process is summarized in Fig. 2. In this example we tried to identify peak No. 4 in the three-dimensional chromatogram. By the retention matching process the most promising candidate for this identification was flunitrazepam, and by the UV/VIS spectral matching process again the most promising candidate was flunitrazepam. Based on these two procedures, the other five peaks in the chromatogram have been identified. The results are listed in the table in Figure 2. Comparison of the drug names prescribed leads one to conclude that the toxic condition of this patient had been induced by some accidental or suicidal act in which she consumed a large amount of medicines prescribed by her doctor. This example clearly indicates the potential of this computerized system for practical applications.

C. The Concept of Tailored Phase Based on Molecular Recognition Mechanism

Further extension of the above-mentioned research has progressed to a somewhat different but very important area of separation science—how to investigate the retention mechanism in chromatography based on the interpretation of molecular level interactions.

Chromatography is a technique that is used to separate various components from complex mixtures based on differences at the molecular level of the interactions between the solutes, solids or liquid, the stationary phase, and carrier or mobile phase. Depending on the kind of carrier, one can group chromatography into three different modes: gas chromatography (GC) when the mobile phase is gas, liquid chromatography (LC) when the mobile phase is liquid, and supercritical fluid chromatography (SFC) when the mobile phase is supercritical fluid. Chromatographic separations are generally based on the elution time difference depending on the strength of the interactions between the solute and the stationary phase in GC because the mobile phase is an inert gas. However, in LC and SFC, the interactions between the solute, the stationary phase, and the mobile phase are all dominant factors that control the elution time (retention). In order to understand the contribution of each molecular level interaction—solute-stationary phase, solute-mobile phase, stationary phase-mobile phase—various theoretical and experimental approaches have been applied. For the interpretation of the stationary phase structures, spectroscopic instruments are used. For the interpretation of the solute structures, computer calculations for molecular modeling are usually used. For the contribution of the mobile phase to the solute and the stationary phase structures, several spectroscopic and solvathochromic methods have been proposed. Compilation and analysis of the information related to chromatographic retention mechanisms will help in understanding the separation process.

In chromatography the retention mechanisms are generally not yet known. However, to design the best separation system for a particular analytical problem, the mechanism should be known and the interactions at the molecular level interpreted. To understand and interpret such mechanisms is a hard task, particularly in LC systems, because the interactions are so complex and some of the interactions are sometimes dominant and at other times not dominant in controlling the retention. In such cases one creates a model for the interaction in chromatographic separation processes by using information about solutes and stationary phase structures. The molecular recognition mechanism is such a model in separation processes. The molecule can be recognized by the stationary phase, and the separation process can be induced by different molecular level interactions between the stationary phase and the molecules that have a different shape and size from each other. This mechanism induces enhanced selectivity for the separation of such molecules, and selectivity enhancement is the most important aspect in recent separation science.

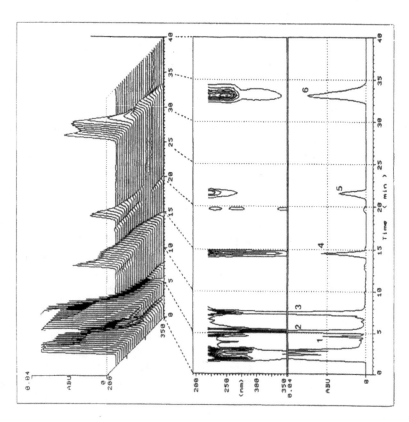

Figure 2 Typical example of toxic drug analysis of a patient's gastric contents by MCASYST.

In order to enhance selectivity, various methods have been proposed. In these methods the design of new stationary phases, which can offer higher selectivity for particular compounds groups, is a chemistry-oriented approach. Based on molecular-molecular interactions between solutes and stationary phase moiety, one achieves enhanced selectivity if one can produce such an interaction field in the separation process. This mechanism is known as molecular recognition in separation systems. For this kind of research, microcolumn LC is the most versatile technique, since it requires a very small amount of stationary phase materials and mobile phase solvents. This feature of the technique makes the evaluation procedures very effective, economical, and environmentally favorable. Microcolumn LC permits the small-scale synthesis of stationary phase materials for the evaluation of the chromatographic performance, and it requires about 100 times less material and organic solvent compared to conventional LC. For these reasons the author has used microcolumn LC in his research on all these projects.

One of the examples described here is the separation of fullerenes. The target compounds, fullerenes, have recently attracted attention from materials scientists, chemists, and other scientists because their shape is very unique and very interesting from a chemistry point of view. The molecules are bulky, steric, and have different sizes depending on their molecular weight (carbon atom numbers contained in the molecules), and the cage of the molecules can have metal atoms inside which produce a different chemistry compared to that based on elements. Separations of such molecules, which are a most interesting and difficult analytical problem, are best resolved by micro-LC, a method that is well suited to solving such a problem.

The structures of typical small fullerenes (MW < 936) are shown in Fig. 3. The key to the molecular recognition mechanism for fullerene separation is the shape and size of the molecules. Once we recognize the shape and size of these molecules, a stationary phases is then tailored for their separation. In order to pursue this purpose, evaluation of typical stationary phases in LC have been performed before synthesizing new stationary phases based on structures and functions. Typical ODS phases with different functionalities, which are polymeric or monomeric, were evalu-

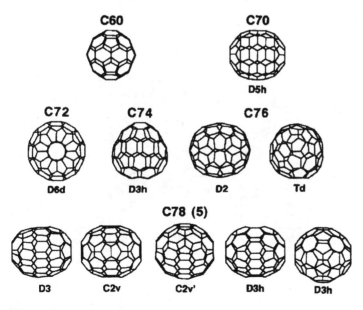

Figure 3 Structures of fullerenes smaller than molecular weight 936. (From *http://shachi.cochem2. tutkie.tut.ac.jp/Fuller/Fuller.html.*)

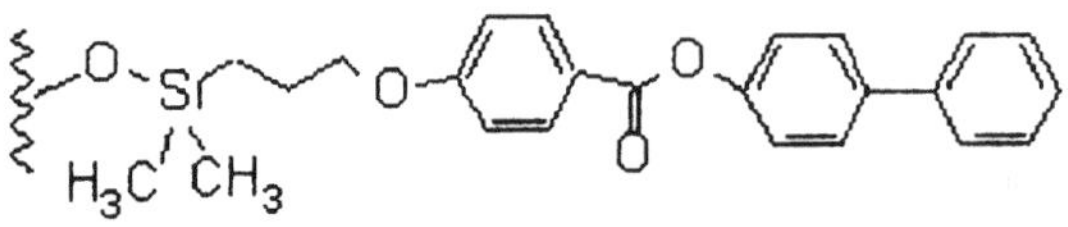

Figure 4 Chemical structure of a liquid-crystal bonded silica stationary phase.

ated. The results have indicated that the polymeric ODS phase has better recognition capability than the monomeric ODS phases for the shape and size of the fullerenes. Decreasing the temperature enhances such recognition power. We found that the most important parameter is not the functionality of the ODS phases, but the distance between the bonded phases on the silica surface and the rigidity of these bonded phases. The distance between each bonded phase should be concomitant with the diameter of the fullerene. Using this information, one can synthesize rigid phases that have an interval between the bonded moieties similar to the size of the desired solutes. The liquid-crystal bonded phase, the chemical structure of which is shown in Fig. 4, is most suitable for this purpose. Although the interval is not the correct size, the spacing between each moiety can be increased by increasing the temperature, as shown in Fig. 5. The molecular model clearly shows that this phase can retain C_{60} at higher temperature, in which the liquid-crystal moiety moves more freely than that at lower temperature, where the molecules have a more restricted movement. In the latter case, the phase can exclude fullerenes from the chromatographic interaction, since the distance between the bonded moieties is not enough to permit fullerenes to penetrate into the space. This mechanism produces the large difference observed between the retention of C_{60} and C_{70}. The selectivity for these two fullerenes with the liquid-crystal bonded silica phase in LC increases with increasing temperature, as seen in several chromatograms obtained at different temperatures (Fig. 6). This behavior is not typical in LC, where the selectivity and retention at high temperature are generally lower than at lower temperature.

Similar discussions will be made regarding other stationary phases designed and synthesized for the selectivity enhancement of fullerene separations based on molecular-molecular interac-

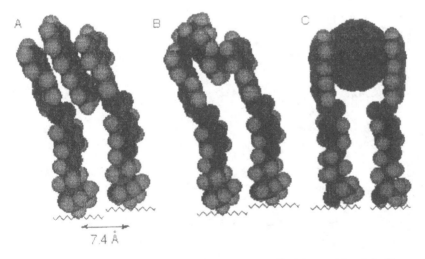

Figure 5 Molecular-molecular interactions between a liquid-crystal bonded silica stationary phase and some solutes such as (A) triphenylene (planar), (B) o-Terphenyl (nonplanar), and (C) C_{60} (bulky) molecules.

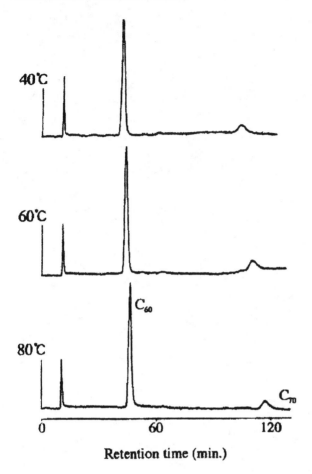

Figure 6 Separation of C_{60} and C_{70} with a liquid-crystal bonded silica stationary phase at various temperatures.

tion. One of the products of such discussions is the possibility of using C_{60} fullerene as a stationary phase in LC separation [8]. The bulky structure and curvature shape of the fullerene would produce uncommon retention characteristics for some compounds. We synthesized various C_{60} bonded silica stationary phases and evaluated their performance for the separation of polycyclic aromatic hydrocarbons (PAHs) and other compounds. As an example, one of the C_{60} bonded silica (Fig. 7) was used to enhance the selectivity of the separation of calixarenes, since the different sizes of the cavity of calixarenes will produce different degrees of interaction between the solutes and the stationary phase. For the typical separation of t-butyl calix[4]arene, -[6]arene, and -[8]arene, this C_{60} phase can give much higher selectivity than a typical ODS phase, as seen in Fig. 8. The reason for such selectivity enhancement can be explained based on the interaction model between the C_{60} bonded moiety and calixarenes, as demonstrated in Fig. 9.

One can conclude the above basic investigation by stating that the concept of tailored stationary phases is the most important aspect of designing analytical systems, since sample preparation processes can also be designed in a similar fashion, whereby selective extraction can be performed prior to selective separation based on this concept.

H
CH3
Si-CH3
O
Si

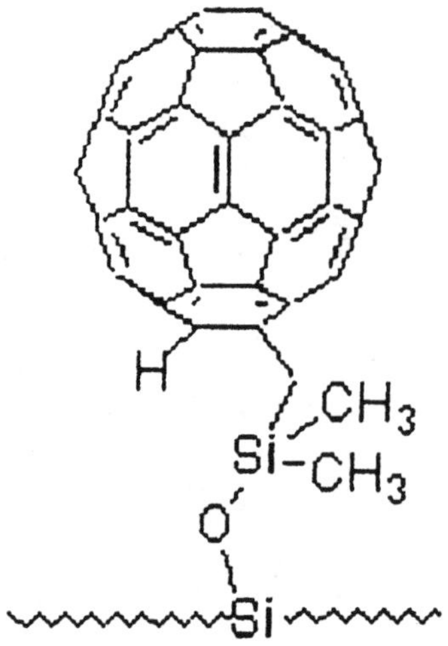

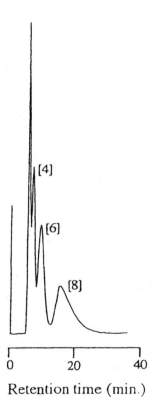

Figure 7 Structure of C_{60} bonded silica stationary phase.

[4]
[6]
[8]
0
20
40
Retention time (min.)

Figure 8 Typical chromatogram for the separation of calixarenes with the C_{60} bonded silica stationary phase. (From Ref. 8.)

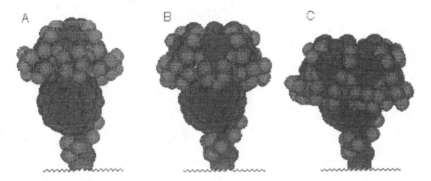

Figure 9 Molecular modeling scheme of the proposed interaction between three calixarenes and C_{60} bonded silica stationary phase by a space filling model: (A) *t*-butyl calix[4]arene; (B) *t*-butyl calix[6]arene; (C) *t*-butyl calix[8]arene with C_{60} phase. (From Ref. 8.)

ACKNOWLEDGMENTS

So many people have made contributions to my work in the separation sciences that I cannot list all of their names. However, the following names should be mentioned: Professor Leslie S. Ettre, Yale University; Professor Pat Sandra, University of Gent, Belgium; Professor Milton L. Lee, Brigham Young University; Professor Peter R. Griffiths, University of Idaho; Professor Marja-Liisa Riekkola, University of Helsinki, Finland; Professor Joseph J. Pesek, San Jose State University; Dr. John Fetzer, Chevron Research and Development Company; Professor Daido Ishii, Professor Emeritus, Nagoya University, Japan; the late Professor Shinichi Sasaki, Previous President, Toyohashi University of Technology, Japan; and the late Professor Hiroyuki Hatano, Professor Emeritus, Kyoto University, Japan. Of course, my students who graduated and are now in my group and my colleagues have contributed very much to my research life. And finally, I express my deepest and sincerest appreciation to my wife, Sadako, and my two children, Kyoko and Kiyoshi, since I could not continue this wonderful life without their heartfelt support.

REFERENCES

1. R Kaliszan. Quantitative Structure-Chromatographic Retention Relationships. New York: John Wiley & Sons, 1987.
2. K Jinno. A Computer-Assisted Chromatography System. Heidelberg: Huethig, 1990.
3. JC Fetzer, WR Biggs, K Jinno. HPLC analysis of the large polycyclic aromatic hydrocarbons in a diesel particulate. Chromatographia 21:439–442, 1986.
4. K Jinno, M Kuwajima. Microcomputer-assisted liquid chromatographic separation system application to toxic compounds identification in poisoned human fluids. J Chromatogr Sci 27:57–62, 1989.
5. K Jinno, Y Ban. Computer-assisted prediction of small peptides sequencing in reversed-phase liquid chromatography. Chromatographia 30:51–57, 1990.
6. K Jinno (ed.). Chromatographic Separations Based on Molecular Recognition. New York: Wiley-VCH, 1996.
7. K Jinno (ed.). Separation of Fullerenes by Liquid Chromatography. Cambridge, UK: The Royal Society of Chemistry, 1999.
8. K Jinno, K Tanabe, Y Saito, H Nagashima, RD Trengove. Retention Behavior of Calixarenes With Various C60 Bonded silica phases in microcolumn liquid chromatography. Anal Commun 34:175–178, 1997.

16

Travels of a Chromatographer

Fred E. Regnier

Purdue University, West Lafayette, Indiana

I. NEBRASKA

I was born and grew up in a little farming community of roughly 400 people in Nebraska. My high school had 78 students at the time I graduated and the distinction of having won 72 football games without a loss. I never played on a football team that lost a game. My French-German ancestors arrived in America in the mid-1800s; some from Paris, others from near Trier. Growing up in the country was great. By 5 years of age I recall having a horse and dog, a lot of freedom, and a long list of responsibilities, most of which I liked because I could do them with my horse and dog. Among my chores were to get wood for my mother to cook, bring the milk cows home from the pasture in the late afternoon, take water and lunch to my father in the fields, periodically check the cattle in the summer pasture to see if any needed attention, and gather eggs for my mother. But there was also time to play and dream. I loved to build and tinker with things. Along with my friends I built log "forts" in the woods and played all kinds of pretend games. My father was a good mechanic, welder, and blacksmith, all of which he taught me. I remember overhauling a single cylinder pump engine at 10 years of age. It took 2 weeks, but I managed to get the engine back together and it ran. By 16 years of age my interests had evolved to putting very large V/8 engines salvaged from wrecks into cars my friends and I built from scrap yard parts. Judicious modification of these engines allowed us to produce 300–350 horsepower in a car the size of a Volkswagen beetle. Few production cars today would beat these little rockets in a test of raw acceleration.

My college career and life in chemistry began in a funny way. The fall after I graduated from high school, several of my friends suggested that I go to a small state teachers college with them. Having nothing else to do and not relishing working outside all winter, I joined them. When we arrived I discovered that by taking a lot of courses in physics, chemistry, and math I could minimize the number of courses I had to take in humanities. By my second year in college, I became a teaching assistant. My relationship with Professor Hanford Miller was a life-altering experience. In fact, he changed many people's lives. Seventy-five percent of his chemistry majors eventually earned a Ph.D. From my generation alone, four of his students became university professors, two went to major pharmaceutical companies, and several joined the FDA. That's remarkable for a little college of 1200 students. In addition to introducing me to the beauty of chemistry, he led me to research and teaching. He allowed me to give lectures in his general chemistry classes, not because he wanted or needed help, but to mentor me. I loved organic chemistry and natural products. I worked in the library and became familiar with all the science books in the holdings of the college. This allowed me ample time to browse through all the chemistry books in the library when I had completed restacking books. It was in this way that I stumbled onto the small number of books on electrophoresis and chromatography the library held and became fascinated with separations. My senior thesis was on paper electrophoresis. (Polyacrylamide gels hadn't been invented yet.)

II. GRADUATE SCHOOL AND POSTDOCTORAL EXPERIENCES

After graduating from college in 1960, I taught high school for a year in the small community of Humboldt, Nebraska. My neighbor, the local veterinarian, was the one who convinced me to go to graduate school. The fact that I had to teach chemistry, physics, biology, algebra, and general science along with coaching junior high sports hastened the decision. Of the 10–15 universities that sent application forms, the one from Oklahoma State University (OSU) was the shortest, and it was the only school to which I applied. I was admitted into the biochemistry program in 1961, where I became involved in the study of natural product biosynthesis, first working on carotenoids with Professor George Odell and then on monoterpenes with Professors George Waller in biochemistry and Peter Eisenbraun in organic chemistry. All of these labs required and maintained state-of-the-art separation systems for their research. My "problem" was that I was more interested in separations than solving problems in isoprenoid biosynthesis. From my engine- and car-building days, I still loved building things and was fascinated with putting chromatography systems together. I recall putting pumps on packed liquid chromatography (LC) columns in 1961, before the advent of "flash chromatography" and commercial LC instruments.

In addition to working on terpenes, Professor Eisenbraun's laboratory synthesized hydrocarbon standards for the American Petroleum Institute, Conoco Petroleum, and Phillips Petroleum. Purification of these materials was frequently a problem. Although we had state-of-the-art spinning band distillation systems, we became fascinated with preparative gas chromatography. The most radical system we built, in collaboration with the folks at Conoco, had 2–6 inch diameter columns and used crushed firebrick as the support with Apiezon L as the liquid phase. Carrier gas consumption was so enormous that it had to be recycled. This was achieved by using carbon dioxide as the carrier gas and a large compressor for recycling. Although all this was great fun, it obviously wasn't a good idea, because it never became popular.

Monoterpenes are volatile and therefore good candidates for gas chromatography (GC). For this reason, we used packed column GC extensively in our studies of monoterpene biosynthesis. By 1965, Professor Waller had obtained a prototype of the LKB 9000 GC-MS from the Karolinska Institute that became "my" instrument. I promptly blocked the first molecular separator, as advised by LKB, and built a 150 m stainless steel capillary for the separation and analysis of natural products. This was my first experience with a system that could separate more than 300 components and produce mass spectra for all of them in less than 2 hours. These systems were incredible for research in natural product and flavor chemistry. The problem was in trying to identify the hundreds of components. You could get a mass spectrum of any component but generally had little idea of the structures. This led me to "reaction gas chromatography," a technique popularized by the laboratories of Beroza in the United States and Berezkin in the

U.S.S.R. The essence of reaction GC is to execute chemical reactions in either the injection port or on the column of a GC system. We used silylation, acetylation, reduction, hydrogenation, and hydrogenolysis to elucidate the structure of 100 ng quantities of a volatile substance. Mass spectra of the products of these reactions gave one an idea of both the functionality and the carbon skeleton of a compound. This era in George Waller's lab was a really exciting experience, of which I have very fond memories. I had no responsibilities except to do research and pass courses. It was great.

In fact, my whole stay at OSU was wonderful in retrospect. At the time I thought it difficult and stressful, complaining vociferously to my fellow graduate students. OSU seriously stretched me during the course of introducing me to new worlds of science. In addition to giving me world class instruments, George gave me money and allowed me to do pretty much anything I wanted. One was to be a ''course freak'' and take almost every course in the graduate chemistry curriculum. Molecular biology and regulatory biology were in their early days, protein chemistry was developing rapidly, and all kinds of new analytical strategies were being developed for biological compounds. Taking courses in organic synthesis, physical organic, stereochemistry, steroids, natural products, physical chemistry, statistical thermodynamics, microbiology, entomology, plant physiology, and animal physiology gave me an extremely rich background that I drew on for many decades. In view of the fact that I ultimately became an analytical chemist, it is ironic I never took a single analytical course as a graduate student.

Because many insect pheromones are terpenes, my interests turned to pheromones during the end of my stay in graduate school. This led to my postdoctoral work with Professor John Law at the University of Chicago and Professor Edward Wilson at Harvard. John and Ed had collaborated before John moved from Harvard to the University of Chicago. I became the shuttle between the two labs after John moved, in addition to bringing them a GC/MS capability. Almost everything about chemical communication in animals and insects was new at that time, and it was easy to make new discoveries. The years with John and Ed were an incredible experience. The vision of these extraordinary individuals was awesome to behold. It was during this time that Ed was developing his concepts of sociobiology, population dynamics, and the other things for which he became so famous and John was breaking so much new ground in insect biochemistry.

John and Ed also introduced me to the politics of science. Ed experienced a lot of nastiness in the late 1970s from people who didn't like his brand of science. Some would have liked to seen him replaced at Harvard with a molecular biologist. I was to have a similar experience in the Department of Biochemistry at Purdue. John in particular mentored me in how to get promoted. I remember him telling me to ''publish within the first year, work your way into the fabric of the department, and never offend a full professor or person of power in your field if you can avoid it. Making enemies can come back to haunt you.'' He was right, even though I didn't always heed his advice.

III. THE MOVE TO PURDUE

When I was preparing to join the Biochemistry Department at Purdue in 1968 the question was whether to become a biochemist/molecular biologist or to pursue my love of analytical biochemistry in general and separations in particular. Ed pointed out to me that solving most large biological problems generally requires some new type of measurement strategy, technique, or technology and that by dedicating at least a portion of your time to the measurement science aspects of biological problems, you will always be relevant. He also noted that people in general don't like to develop new techniques, tending instead to use established procedures in the literature. This means you can gain an edge in the competition to solve problems by developing new tools. Again Ed was a prophet.

Although I had played with liquid chromatography for many years, I got into high-performance liquid chromatography (HPLC) indirectly through Ed. He had introduced me to fire ant pheromones in 1967 during my stay at Harvard. Fire ants have a nasty sting and had become a major problem in the southern part of the United States. It was thought that it might be possible to control them with pheromones. These ants are prodigious layers of odor trails, and the trail pheromone would be an obvious candidate to use as bait. The problem was that the chemical nature of the trail pheromone was unknown. Ed and I used to go to Florida to collect large numbers of ants, from which I would extract and go about trying to identify the trail substance. By 1969 we were able to collect a hundred pounds of ants in a single trip.

When I moved to Purdue I chose to continue working on the fire ant trail substance. Very early in the work we found the trail substance to be an unsaturated sesquiterpene hydrocarbon based on thin layer chromatography and mass spectrometry. From my work with carotenoids at OSU, I knew the quickest way to isolate trail substance for identification was to extract with hexane and run the extract on a silica gel column using hexane as the mobile phase. This was an ideal problem for HPLC. HPLC was still in its infancy at that time and instruments were very expensive. Having a limited research budget, I built a gradient elution instrument using two ISCO syringe pumps, a machine shop–crafted valve, and an absorbance monitor from Nester-Faust.

IV. SURFACE COATINGS FOR CHROMATOGRAPHY AND ELECTROPHORESIS

During my early years at Purdue, a graduate student named Dale Deutch was in a lab next to mine and I was asked to teach the graduate level course in analytical biochemistry. Dale was interested in doing affinity chromatography on controlled porosity glass (CPG) and brought me all the papers relating to the attachment of organic species to the surface of glass through organosilane chemistry. At the same time I was explaining soft gel cellulosic and dextran supports for the chromatography of proteins in the analytical biochemistry class. It occurred to me that if we used silane chemistry to attach dextrans to the surface of high-porosity glass or silica, we would have a material equivalent to the soft gels in separation properties but with pressure stability to greater than 200 bar. Within 2 weeks we had applied a dextran coating to 44–64 μm CPG of 200 Å pore diameter (10μm particles were not available yet), packed a 1 m long column, and used it to separate proteins and DNA fragments in roughly 10 minutes.

Thus began in the spring of 1973 the 15-year era in my lab of developing new chromatographic media for protein separations. We quickly learned that applying semi-rigid dextran polymers to the surface of porous inorganic supports was not the best way to prepare packing materials. Dextrans did not completely cover the surface silanols, and the chromatography of basic proteins was terrible. "Diol coatings" prepared by attaching γ-glycidoxypropyltrimethoxy silane to silica particles followed by hydrolysis of the oxirane to a diol gave a far better coating. The problem with the diol coating was that it eroded from the surface within a few weeks when used at neutral pH. It just didn't have the requisite stability. This led us to thin polymer coatings prepared by depositing multifunctional oxiranes, such as diglycidyl ethylene glycol on the surface of a γ-glycidoxypropyl silane derivatized support and then polymerizing with boron trifluoride. During the course of polymerization, the glycidoxypropyl silane groups were incorporated into the polymer skin at many sites. The epoxy coatings were of vastly superior stability. They also shielded proteins from contact with the silanol-rich surface underneath. Another attractive feature of these coatings was that they allowed attachment of a wide variety of stationary phase groups either during polymerization or afterward.

The major contributions of Klaus Unger to our work must also be noted. The support materials that he developed were enormously valuable to us and did more than any other single thing I can identify to catalyze modern HPLC of biological macromolecules. They were the standard by which everything else has been measured. By 1974, Klaus had developed macroporous, controlled porosity, 10 μm silica particles that were a dream to coat and pack. CPG particles in contrast differed in density and were extremely difficult to pack. But beyond being the source of great particles, Klaus has been a good and loyal friend for decades. Evenings spent going through a few bottles of wine with Klaus, Milton Hearn, Bengt Osterland, Jan-Christer Janson, and Tim Wehr at ISPPP planning meetings are fond memories.

After playing with different coating processes for 3 years, ShungHo Chang, Karen Gooding, Rodney Noel, and I published our first papers on HPLC of proteins in 1976. One of these papers, taken from the remarkable Ph.D. thesis of ShungHo, is probably the one in my whole career of which I am proudest. This paper showed that the five isoenzymes of lactate dehydrogenase and three of creatinine phosphokinase could be separated in 5 and 3 minutes, respectively, with full retention of biological activity. This was at least a 60 times faster than possible with soft gels of the time. It was also established in this work that enzymes could withstand the pressure and turbulence encountered during HPLC. This seems ridiculous today, but it was a serious question at the time. A second major question was whether polymer coatings would shrink support pore diameter to the extent that intraparticle diffusion of proteins would be greatly restricted. Obviously, this wasn't a problem. This work was so important to us because it validated the concept that proteins could be separated in an adsorption mode in minutes with resolution superior to that experienced with soft gels. We had a reason to go ahead with the development of a whole array of packing materials for the chromatography of proteins.

Chromatography of polymers was an important issue in the 1970s, particularly in quality control and environmental analyses. Among the more difficult to analyze were the polyamines. We were never able to elute them from any of our coated silica columns under reasonable conditions. After grappling with this problem for 6 months, it began to occur to us that if they were so difficult to elute from columns, perhaps they could be used to prepare coatings. I gave this problem to Andrew Alpert in the fall of 1976 and he showed within weeks that polyethylene-imine (PEI) adsorbs so tenaciously to the surface of silica that after cross-linking with a multi-functional oxirane it provided a weak anion exchange coating of exceptional stability. In fact, the underlying silica matrix leaches away long before any of the PEI coating can be eroded from silica particles. This was by far the simplest, most reproducible coating we ever made. The simplicity of the coating process is one of the reasons multiple suppliers of commercial materials have used it. By varying the molecular weight and concentration of PEI in the coating solution, it is even possible to vary the fuzziness and binding capacity of the polyamine layer. Further reaction with methyl iodide produced a strong cation exchanger. Others and we used these two ion exchangers widely in protein separations for the next two decades. Andy and Tim Schlabach published a series of clinical applications of PEI ion exchangers in monitoring heart attacks and identifying abnormal hemoglobins.

But beyond the actual coating and the coating process, the PEI and epoxy coating work of the late 1970s established a series of principles we and most manufacturers of commercial materials have exploited for the last two decades. One was that a thin layer of polymer with an attached stationary phase could be inserted (or synthesized) in the pores of a support without causing serious mass transfer limitations. The second major principle established by the work of Andy Alpert, George Vanecek, and Bill Kopaciewicz was that the surface area and protein loading capacity of a sorbent can be extended up to an order of magnitude by attaching ion exchange groups to a fuzzy polymer layer inside the support pores. Bill particularly optimized the PEI coating process for high capacity. Almost all the very high-capacity, high-performance process-scale media used in biotechnology today exploit this principle.

It wasn't until the mid-1980s that we found the PEI coating to be of equal efficacy with porous poly(styrene-divinylbenze) (PSDVB) supports. Dyno had developed and began to supply Pharmacia Fine Chemicals with macroporous PSDVB particles in the early 1980s for production of their monobead packings, but these particles were not available on the open market. We had to wait until the mid-1980s to obtain the Dyno material and similar materials from Polymer Laboratories. Actually, the Polymer Laboratories particles turned out to be superior to the Dyno materials in our hands. Mary Ann Rounds and Bill Kopaciewicz showed that after PSDVB is sulfonated, PEI can be applied in a manner almost identical to silica. Mary Ann also showed there was no difference in the performance of PEI-coated silica and PSDVB-based packings of the same physical characteristics. The great advantage of the PSDVB-based anion exchanges over silica-based materials was that they were chemically stable from pH 1 to pH 14, gave high recovery of proteins, and were pressure stable to greater that 100 bar.

PSDVB was obviously a very attractive support material but too hydrophobic for any type of protein fractionation except reversed-phase separations. The problem with the PEI coating was that it was really only good for the production of anion exchangers. In contrast, the epoxy coating process could be used to prepare sorbents for a number of different modes of chromatography, but we found it couldn't be applied readily to PSDVB. Our problem in 1986 was to find a coating procedure for PSDVB that would allow us to prepare chromatography sorbents for separating proteins in all the known modes. After a great deal of effort by YanBo Yang, Steve Cook, Lazlo Varady, and Ning Mu, the idea of using the water-hydrocarbon interface at the surface of PSDVB particles to orient amphophilic monomers, or oligomers, finally emerged. It occurred to us that as in the case of a surfactant, the hydrophobic portion of an amphophilic monomer (or oligomer) would turn inward toward the PSDVB surface while the hydrophilic portion turned out toward the aqueous environment. In addition, the surface would be closely packed with monomers and the distance between neutral hydrophilic groups would be small. If these monomers are then polymerized, or oligomers cross-linked, a dense coating would be obtained in which the PSDVB surface is coated and the portion of the coating in contact with the solution would be hydrophilic. The preferred hydrophilic moiety was a hydroxyl group. Because the coating was adsorbed to the surface at a large number of sites, it was impossible to remove. This coating proved to be very successful, serving as the base coating for a wide variety of the commercial POROS materials used in the biotech industry for purification of human therapeutic proteins.

The 15-year period from 1973 to 1988, in which the lab was oriented so heavily toward preparing surface coatings, was one of the most enjoyable and gratifying of my career. The graduate students of that era and I went on a priceless scientific adventure that will live with me forever.

V. CAPILLARY ELECTROPHORESIS

Jim Jorgenson's discovery of capillary electrophoresis (CE) in the early 1980s was very exciting. Because polyacrylamide gel electrophoresis (PAGE) was so widely used by biochemists for the analysis of proteins, it appeared that CE would displace slab and tube gel electrophoresis and eventually dominate protein analysis. The only problem was to overcome protein adsorption at the walls of the fused silica capillaries. Jim tried the diol coating described above and found it to lack stability, as we had in the case of HPLC. This caused us to think of trying the polymeric coatings we had used in HPLC. Jim Bao and John Towns found that our old epoxy, PEI, and adsorbed surfactant coatings all worked well in CE. But an even better and simpler coating could be obtained with polyvinyl alcohol as described by Gerhard Shomburg. Unfortunately, CE turned out not to be so powerful for protein separations. Chromatography still dominates protein separations.

VI. ELECTROPHORETICALLY MEDIATED MICROANALYSIS

We had a lot of fun with this technique. Biochemists love PAGE, particularly because they can carry out assays in the gel following the separation. After the separation, reagents are added to the gel and assays performed on protein or DNA zones. When the reaction is complete the extent of reaction in the various gel spots is recorded in some type of picture. This led us to the question of how one could perform a similar operation in CE. Jim Bao and I began to examine this issue in 1989 and quickly realized that differential electrophoretic mobility could be used to merge zones in a CE system and initiate chemical reactions. This would permit us to introduce reagents into the capillary during or, after a separation, merge these reagents electrophoretically with analytes, and sweep reaction products past the detector. Because people like to refer to techniques as an acronym, there was a long debate about what this technique should be called. We finally settled on the name electrophoretically mediated microanalysis (EMMA), using a feminine name as was done in naming immunological assays and hurricanes. During the next decade a series of people worked on the problem, describing a number of ways to execute chemical reactions in capillary systems using pL–nL volumes of sample. Dan Wu detected 500 molecules of alkaline phosphatase with an absorbance detector. [Ed Yeung and Norman Dovichi ultimately showed that it was possible to detect single enzyme molecules with the technique using laser-induced fluorescence.] Dale Paterson and Brian Harmon modeled EMMA. Brian also discovered that moving boundary EMMA was 10 times more sensitive than zonal EMMA. Inkeun LeeSong found that through zero potential incubation, selectivity could be easily varied to overcome sample matrix interference. An enormously attractive feature of EMMA is that it allows both separations and chemical characterization reactions to be executed in the same capillary. This ''chemistry-in-a-capillary'' was a forerunner of ''chemistry-on-a-chip'' in that voltage was used to move and mix reagents in a capillary system, albeit it a cylindrical instead of a rectangular channel. The advantage of the chip format over fused silica capillaries for chemical reactions is that they allow faster analysis and the introduction of re-agents at any desired point before, during, or after a separation. The disadvantage of microfabri-cated systems is that reagents are generally introduced laterally and mixing is more difficult to achieve than in a fused silica capillary where reagents are introduced as bands. The problem with side entry introduction in chips is that mixing is largely by diffusion. Brian Burke overcame this problem of chips for EMMA by injecting samples and reagents as would be done with any CE separation.

VII. PROTEOMICS

As noted above, biochemists love gel electrophoresis. It is thus no surprise that when protein chemists gave birth to proteomics in the late 1990s they would choose 2-D gel electrophoresis as the preferred separation system, even through they are labor intensive and do not couple well with mass spectrometry. The fact that thousands of spots must be visualized and cut from a gel, tryptic digested individually, each sample aliquoted onto a MALDI plate and then analyzed by MALDI is a massive operation. Obviously no self-respecting chromatographer could allow this approach to stand unchallenged, and we didn't.

My lab had a fascination with multidimensional chromatography for years, going back to the work of Bill Kopaciewicz in 1981 and Lin Janis in 1986. Both of these folks worked on systems in which proteins were separated by one type of column in a first dimension and then directly transferred to another column where they were separated in a second dimension. Their work led to the BioCAD Work Station and INTEGRAL instruments at PerSeptive Biosystems (PBIO). The availability of these automated, multivalved instruments and electrospray mass spectrometers at PBIO in the summer of 1994 permitted us to play some interesting and enjoyable games. One was to see how long it would take to confirm the primary structure of a protein in a complex mixture. Using a combination of an antibody column, a mixed-bed ion exchange column, an immobilized enzyme column, a desalting column, and a reversed-phase chromatography column in series with an ESI-MS, we were able to analyze the primary structure of haemoglobin derived from a serum sample within 90 minutes. Our *Analytical Chemistry* paper reporting these results in 1996 was probably the first on automated proteomics, appearing 2 years before the term "proteomics" was coined.

Our early forays into proteomics at Purdue began in 1998, focusing on global internal standard technology for quantification of protein in concentration flux, signature peptide identification of proteins, and multidimensional chromatographic methods for peptide identification. The thing that most distinguished our work from others was our emphasis on the study of posttranslational modification and quantification of changes in regulatory sites in proteins. Bob Gaehalen pointed out to us that almost 50% of all proteins are posttranslationally modified in some way and that in many cases these modifications are involved in regulating cellular processes. Clearly identifying proteins wasn't enough. You had to be able to identify and quantify those forms involved in regulation. We chose to do this by developing affinity selection methods for specific types of modification and labeling methods to facilitate both quantification and peptide identification.

VIII. MICROFABRICATED SYSTEMS

My experiences with microfabricated systems began in 1994 at PerSeptive Biosystems (PBIO). Noubar Afeyan and I started a "microfab" program at PBIO in 1994 to build microdevices for clinical chemistry. Martin Fuchs was recruited to run that program. In slightly more than a year we had shown it was possible to build a 32-channel immunological assay system on a small wafer, fondly called the "pizza chip." This name arose from the fact that all the sample and buffer wells were around the outside of the circular wafer and the channels lead to a central detection zone, giving the chip the look of a pizza. The pizza chip was featured in the 1995 annual report of PBIO. Several things became clear early on. One was that getting samples onto a device this small was difficult. Although it was easy to create a microchannel world on the chip, it was not easy to go from the macroscopic human world to the micro-world on the chip. Second, miniaturization per se conveyed no real advantage in immunological assays. Although sample volume was smaller, getting a mL of blood for a macroscopic assay really wasn't a problem. We were never able to interest any of the major suppliers of clinical instrumentation in this technology.

In the spring of 1996, a particularly talented and ambitious student by the name of Bing He came to my lab who wanted to do a Ph.D. thesis on microfabricated systems. I had been toying with ideas for microfabricating chromatography columns for several years and suggested he might start on this problem. Bing plunged himself into this problem and came up with a solution in which he microfabricated the entire chromatographic system, including the "particles," a 15 nL column, a 100 pL mixer, filters, the sample inlet, and buffer wells. Feature sizes in the chip went down to 1.5 μm, and the channel walls were almost vertical. Support structures for stationary phase immobilization were etched into the wafer in rows of small cubes, leading to the name "collocated monolith support structures" (COMOSS). The COMOSS columns produced roughly 700,000 plates per meter in the reversed-phase capillary electrochromatography (CEC) mode. As part of his applications work with this chip, Bing was the first to show CEC of peptides from a tryptic digest. One of the problems of the COMOSS systems is that they had to be produced by deep reactive ion etching, making them expensive. Fabrication costs were in the range of thousands of dollars per column. In an effort to overcome the cost issue, we began molding COMOSS columns in 2000. The difficulty of molding channels 1.5 μm wide and 20–40 μm deep caused us to go to 5 μm wide channels. Even with 5 μm channels the molded columns still produce more than 200,000 plates per meter. Microfabrication clearly holds great potential for producing large numbers of chromatography columns on a small wafer that can be operated in parallel.

Although there is great excitement about "labs-on-a-chip" and they get good press coverage, acceptance is slow. One has to wonder why. Perhaps the fact that they are so small and difficult to interface with the outside world is the problem. Acceptance may depend on the development of supporting technology and the formation of sample aliquoting and detection systems. A more disturbing reason might be that the need for and utility of microfabricated analytical devices is much smaller than their press coverage. Time will tell.

IX. COMPANIES

A significant part of my career was spent with several companies I helped found. The first was BioSeparations Incorporated (BSI). This was a company that Barry Karger and I started around 1984, to the best of my recollection. BSI provided courses and seminars in analytical biochemistry for the biotechnology community. Biotechnology was expanding rapidly in traditional pharmaceutical companies at the time, and a number of people had not been exposed to HPLC, CE, and mass spectrometry of peptides and proteins. Recognizing the need for compact, "in-house" courses that exposed people to the latest literature and thinking on HPLC and CE of proteins and peptides, we developed a series of courses tailored to the unique needs of particular groups of people. After consulting with a company, we would design a course that most nearly met their needs. Courses were 2–3 days in length and restricted to 30–50 people. The most difficult thing about teaching these courses was having to talk for 2 days, even in 1-hour segments. I learned many things from these courses. Although I had been involved in protein and peptide separations for many years, I learned a lot from listening to Barry's lectures. I also learned a great deal about biotechnology from the attendees. As I recall, we taught this course to more than 1000 people in the United States and Europe over the course of 4 years. The success of the course encouraged us to start an annual symposium on analytical biotechnology. We brought in all the leading people in the field. The formula was simple—if you could fill the program with really great people you would really love to hear yourself, attendees would be happy. In some respects, this symposium was like the IBC conferences of today. The positive part of BSI was that it was a great learning experience. The downside was that it required a lot of travel and the two of us were the whole show. We had no one to delegate things to like we could in the lab. Although Barry and I found it easy to work together and enjoyed the teaching, all the travel and organizational issues eventually wore us out and we dissolved the company.

I met Bob Dean and Noubar Afeyan in 1987 through our mutual interest in process-scale purification of proteins. Noubar was just finishing a Ph.D. with Daniel Wang in the Bioprocess Engineering Center at MIT and a Masters Degree in business from the Sloan School. Although he had a large number of job opportunities, he was interested in starting a company. Bob had been a professor of mechanical engineering at Dartmouth and founder of at least four companies, the most recent being one that specialized in continuous fermentation. We met at Dartmouth in the late summer of 1987 and decided to found a company directed at process-scale separations. Bob chose the name Synosis, which we later changed to PerSeptive Biosystems (PBIO).

During the several hour drive from Dartmouth back to Boston, Noubar and I had formulated the general outline of a plan we were to follow for the first 5 years of the company. Major components of this plan were to develop a new family of chromatographic media and equipment to use these materials in process development and manufacturing. Our problem was how to fund the company. The first year was filled with an endless series of business plans and meetings with investors, all of whom were very skeptical about a start-up guided by a young man with a brand new Ph.D. and two gray-haired academics. The first home of PBIO was an office in the basement of an old warehouse. Within a year Bob decided the company was going to be too chemical and he left. By 1990, we had obtained funding, moved into a real lab in the MIT research park, hired some people, and began to execute our plan.

Process-scale chromatography prior to the 1990s was a slow, low-resolution technique. We had the idea at PBIO that if we could decrease separation time without losing efficiency we could get a significant share of the process scale market for protein separations. This led us to develop "perfusion chromatography" and introduce the POROS™ product line. Perfusion chromatography is based on the fact that supports with pores of 6000–8000 Å allow a small amount of mobile phase to flow through the particle, transporting proteins into and out of pore matrices by convection. The flow of mobile phase through particles has relatively little impact on performance until intraparticle transport by convection exceeds that by diffusion. This is most likely to occur (a) with large molecules that diffuse slowly and (b) at high mobile phase velocity where stagnant mobile phase mass transfer limitations are the greatest problem. Enhancing mass transfer in this way allowed preparative separations to be executed on a time scale normally associated with analytical separations. The POROS materials are now widely used for process chromatography in biotechnology.

By 1993 PBIO had become a public company and we were looking for new ways to grow. We already produced a line of chromatography equipment in addition to the POROS materials, but we realized there was a substantial need for new discovery tools in the biotechnology/pharmaceutical industry and academia. This led to discussions with Marvin Vestal in which Vestec merged with PBIO. It was through Marvin and Vestec that PBIO got into mass spectrometry. Our venture into oligonucleotide synthesis, peptide synthesis, and immunological assay equipment occurred by acquisition of the Milligen group from Millipore. Acquisition of an immunological assay product line from Advanced Magnetics filled out our immunological assay offerings for the research market. Although the products all these companies brought to PBIO were of great value, the people were much more valuable. On the science side, Marvin Vestal, Bill Carsons, George Vella, and Carl Paul come to mind as being particularly gifted. An even larger number could be named in management, sales, and marketing.

In the spring of 1995 PBIO moved to a huge new facility in Framingham, MA, and set about integrating all these people and products into a smooth-running unit. The company eventually grew to over 500 people and had established an international presence by the fall of 1997, when it was acquired by Perkin Elmer.

I thought this would be the end of companies for me until my old friend Noubar started Newcogen, a company that starts companies. We are in business once again with a little start-up called Beyond Genomics (BG). In some respects it's like raising children—you learn something about the process after you've done it several times, but it still worries and stretches you. BG looks like a strong child, but a lot of things can happen to a child today on the way to maturity.

X. MEMORABLE MEETINGS

I still remember in great detail the first separations meeting I ever attended. It was the Second International Symposium on Column Liquid Chromatography held in Wilmington, Delaware, in 1975. Jack Kirkland had invited me to make a presentation on our protein HPLC work. Csaba Horvath, J. F. K. Huber, Barry Karger, George Guiochon, Lloyd Snyder, Jack Kirkland, and Klaus Unger were there and sitting near the front when I gave the paper. I even remember the questions Csaba and J. F. K. asked. It was the first time I had met real separations people. At last I had found a group of people who spoke the same language and liked the same things I did. I found the experience fascinating, exciting, and humbling, all at once. The fascination and excitement came from meeting all the people whose work I had admired for years. The humbling part came from the realization that if I wanted to belong to this field I had to produce on their scale. It wasn't clear that was possible, and still is not today.

The second most memorable meeting in the International Symposium on Column Liquid Chromatography series is the one I ran in Baltimore in 1992. Having attended all these meetings after the Wilmington one, I had concluded that the meetings were becoming too formal. It would be fun to lighten them up a little. The meeting fell on the anniversary of Columbus's discovery of America. I thought it would be funny if we opened the meeting with some sort of skit playing on this fact. We opened the meeting with a skit in which Columbus's ship captains came strolling in, greeting people as they went to the stage where they reported to Queen Isabella what they had found in America. After a great deal of convincing, Barry Karger, Csaba Horvath, and Phyllis Brown agreed to dress up in Spanish costumes of the time and be the ship captains. I was the fourth ship captain. We claimed there was a fourth ship captained by a woman and that it had been hidden for all these years out of male bigotry. We also claimed that the only way Columbus got the grant for the trip from the Queen was because of Phyllis and the "old-girls" network with the Queen. Louise Beaver was Queen Isabella and Mercedes deFutos announced the captains to the Queen. The captains' reports had a scientific spin. I remember Barry espousing the theory that the world was constructed like a piece of capillary tubing while I claimed it perhaps had huge through-pores. But probably the funniest of all was seeing the men dressed up in big, fluffy hats, puffy little short pants, and tight fitting stockings with swords at their sides. It was like something out of a comic book. Videos of the event were absolutely hilarious.

Another memorable meeting was the first International Symposium on Proteins, Peptides, and Polynucleotides (ISPPP) in roughly 1982–83. During the International Symposium on Column Liquid Chromatography organized in Boston by Barry, Milton Hearn and I were discussing the fact that HPLC of biological macromolecules was accelerating rapidly and it was perhaps time to start a symposium focusing on that subject alone. We convinced Klaus Unger to join us to assure a European point of view. Our problem was that I had neither the requisite funds nor experience in America to organize and run a symposium. We went to Tim Wehr and Lenna Chow at Varian, who agreed to help us with both. Pharmacia later joined them. The first symposium in Washington, DC, attracted several hundred people and was held in an old bus station. [This was all that was available on very short notice.] Thus began a series that continues today.

During the course of the next 15 years, I was to develop a close bond with Milton, Klaus, Tim, Bengt Osterlund, and Jan Chister Janson. The ISPPP planning meetings were particularly fun. I remember staying up half the night drinking wine and talking about chromatography, the politics of the field, people, and sharing dreams of what we were trying to do.

XI. CONCLUSIONS

I began my trip as a little boy on the plains of Nebraska, where I had dreams of traveling to other lands, seeing great things, and meeting great people. *The story of my life is that my boyhood dreams came true.* Science allowed me to travel to most of the countries in the world, meet the finest minds in science, and hear what and how they thought. I experienced many cultures and came to realize the strength and beauty of human diversity, both in thought and life style. It was a priceless trip that I treasure.

17

The Role of High-Performance Liquid Chromatography and Capillary Electrophoresis in Life Sciences

Phyllis R. Brown

University of Rhode Island, Kingston, Rhode Island

My career in chromatography was not planned; it started accidentally when I was doing postdoctoral work in the pharmacology section of Brown University. The head of the section saw a "nucleic acid analyzer" at a meeting and thought it would be helpful in our research. This instrument was actually a high-performance liquid chromatograph (HPLC) dedicated to the separation of nucleotides. If it had been called by its real name, HPLC, I never would have become a chromatographer because our section head did not know what an HPLC was or what it could do for us.

My career has been a checkered one. I went to Simmons College for my first 2 years of college. Simmons was a small college in Boston dedicated to train women to take their place in the work world. Because there was a science school at Simmons, it never occurred to me that, as a woman, I could not become a scientist, and I received a solid foundation in chemistry, physics, math, and biology. Most importantly, we had to take an excellent English composition course, public speaking, and, yes, typing. Who would have guessed how important typing would be in those precomputer days? After two years at Simmons, I became a wartime bride and moved to Washington, DC, where my husband was stationed with the Navy Department.

I transferred to George Washington University and received my B.S. in chemistry in 1944. I worked at NIH, the Geological Laboratory of the Carnegie Institution of Washington, and finally at the Harris Research Laboratory until the end of World War II. We moved back to Providence, RI, where my husband started a plastics molding company and I applied to Brown University to work for an advanced degree in chemistry. However, Brown was not interested in accepting a woman into their chemistry program, especially since there was a large pool of returning veterans. Since there were no chemistry jobs available in Providence at the time, I settled down to being a housewife, mother, and community volunteer. By the time my youngest child was in school all day, I had been active in or president of almost all of the community organizations in our area. After considerable discussion and with my husband's urging and support, I went back to school.

In the early 1960s, the chairman of the Chemistry Department at Brown took me on as a personal challenge to see if someone could do graduate work in chemistry after being out of the field for over 18 years. Much to my surprise, 6 years later, I received my Ph.D. in chemistry from Brown and did postdoctoral work in the pharmacology section for 3 years. I then became an instructor and the following year a research assistant professor in the same section. Unfortunately, I encountered discrimination at Brown and felt I had to leave.

Thanks to Dr. Lily Horning, the wife of the president of Brown, I heard of an opening at the University of Rhode Island. It was a "one-year-only" appointment in the Department of Chemistry, but it turned out to be a great career move for me, and I have been there ever since. I have enjoyed being at URI and was able to have a very productive scientific career. Moreover, my training in pharmacology proved to be an invaluable asset in my research.

My early work, first at Brown and then at URI, involved the application of HPLC to biochemical and pharmacology problems. We optimized many assays for nucleotides [1], nucleosides, and their bases [2] and showed how powerful HPLC could be for these analyses. We then investigated mechanisms of retention in order to facilitate the determination of the best conditions for good separations.

In the early 1970s, most chromatographers were interested mainly in chromatographic theory and instrumentation, while the biologists and biochemists did not think HPLC would work. In fact, it took about a decade before HPLC was used routinely in pharmaceutical, biochemical, and medical laboratories. Therefore, our work was unique because we applied HPLC to studies involving real samples such as red blood cells, plasma, platelets, etc.

To accomplish our goals using real samples, we had to investigate three very important parameters: sample preparation, detection, and peak identification. The only detectors available in the early 1970s were the single wavelength UV detector and the refractive index detector. Both of these detectors had limitations; for UV detection, the analytes had to contain UV-absorbing chromophores or be capable of being derivatized with a chromaphore and the refractive index detector did not have the sensitivity required in our pharmacological studies. In addition, with neither detector could we obtain characterization and unequivocal identification of the peaks of the analytes, nor could we determine the purity of the peaks.

Although a great deal of research was carried out in the 1980s, it was not until the early 1990s that good interfaces were developed so that mass spectrometry (MS) could be used routinely as a detector. In the latter part of the 1990s, the possibility of interfacing nuclear magnetic resonance (NMR) to HPLC became a reality, and more recently a new interface has been developed so that FT-IR can be used for detection and characterization of analytes.

In the past few years, much of the work done in our laboratory has been to assess the usefulness of HPLC-FT-IR, and we have applied it to many diverse solutes. The major advantages of HPLC-FT-IR are that non–UV-absorbing species can be detected, a full IR spectrum of each analyte is obtained in real time, and positive identification of known compounds can be made with library matching. We demonstrated this clearly in our work on sphingoid bases. In addition unknown analytes can be characterized, impurities in peaks determined and often identified, and the sample is not destroyed so that further analyses can be carried out.

Although our interests have been broad, from analyzing the blood of the Tasmanian devil to collaborative studies of heart disease, cancer, and immunodeficiencies, our major body of work has been on the analysis of purines and pyrimidines and their nucleosides and nucleotides. We established concentrations for the nucleotides in red blood cells of a normal population and for the nucleosides and bases in serum or plasma. We were the first to use microparticle chemically bonded anion exchange packings for the HPLC of the nucleotides [1] and microparticle chemically bonded C18 packings for the nucleosides and bases [2]. Over the years, we refined these assays and applied them to other physiological matrices. We also optimized assays for analyses of derivatives of nucleosides and bases, which are used as chemotherapeutic agents for diseases such as cancer, heart disease, and AIDS.

We established erythrocyte nucleotide and serum nucleoside and base profiles for many species ranging from humans down to primitive species. Because of this work, we were asked to collaborate with the University of Tasmania on studies they were doing on the Tasmanian devil and other marsupials. We found the profiles unique for each species that was studied, and we were able to place many almost extinct marsupial species on the appropriate family tree.

In the mid-1970s, my work with Anne Fausto-Sterling of Brown University showed that HPLC could be successfully applied to studies of DNA and RNA synthesis in the fruit fly [3]. Our work was one of the developments that made possible the human genome study.

Since URI did not have a medical school, I collaborated with researchers from many medical institutions such as NIH, Sloane-Kettering Hospital, Rhode Island Hospital, Miriam Hospital, as well as other schools such as MIT, Brown University, Hebrew University, and the University of Tasmania and industrial companies such as Waters Associates, Perkin Elmer, Gillette, and Pfizer Inc.

At URI, one of our first studies was a kinetic study of the acid-catalyzed breakdown of NADH. This study was important because it showed which of two pathways proposed by Oppenheimer and Kaplan was correct. More importantly, it showed that HPLC could be useful in kinetic studies [4].

Later on in the early 1980s, we did enzyme assays using computer programs and showed that the activities of not only one but two enzymes in a metabolic pathway could be optimized and determined simultaneously.

The majority of our work in the 1970s was in the optimization of assays for nucleotides and their metabolites and analogs and in application of these assays in pharmaceutical and biomedical investigations. Much of our work has been summarized in the books we either wrote or edited [5-7].

In 1979, I spent a sabbatical semester at the Hebrew University in Jerusalem with Professor Eli Grushka. It was there that we investigated structure-retention relationships in the RPLC of purine and pyrimidine compounds and formulated general rules by which we could predict the retention order of nucleosides and bases [8]. With members of my group, we later investigated mechanisms of retention and effect of stationary phases in these separations. With one of my graduate students, we quantified these structure retention relations and investigated the use of ion-pairing reagents for nucleotides as well as nucleosides and their bases. Thus, we had a solid base for the analysis of purine and pyrimidine compounds. On this basis, we were able to use chemometrics to classify patients with acute and chronic leukemias and we could separate clearly normal populations from those with leukemia [9].

Since a total analysis is only as good as the weakest step, we investigated sample preparation methods. Unfortunately, there was no one general method of preparing a sample for HPLC analysis. We also found that peak identification was very important in a reliable HPLC analysis for solutes in complex matrices.

For many years before MS became available as a detector, a combination of on-line and off-line methods had to be used to confirm the identity of a peak. In those days, "the enzyme peak shift" method I developed for the identification of biologically active molecules, especially those in the purine and pyrimidine metabolic pathways, was one of the few ways of unequivocally identifying and determining the purity of a peak [10].

The work in my laboratory has continued and expanded in other areas. We have developed and optimized HPLC assays for polyaromatic hydrocarbons, amino acids, peptides, proteins, sphingoid bases, enantiomers, vitamins, triglycerides, biogenic amines, and many type of drugs including antidepressants and antiviral and anti-AIDS drugs. We have developed and/or optimized conditions for HPCL/ESI/MS and HPLC/ESI/MS/MS as well as HPLC/FT-IR and HPLC/ELSD separations, and we have worked on new packings that give better separations with these detectors. We have scaled down equipment for microbore and short columns, but unfortunately we have not gotten into "separations on a chip," which I see as a way of the future.

In the last decade, we have gotten into capillary electrophoresis (CE), and we were the first to see the potential of CE for the separation of cations. We also investigated the parameters that affect the CE analysis of nucleic acid constituents, amino acids, peptides, enantiomers, and fullerenes. Recently we started an exciting research project on coating capillaries with conducting polymers. These polymers can be changed from conducting polymers to insulating ones and from being hydrophobic to hydrophilic by a change in pH; thus they are tunable and can be controlled by a "pH switch."

My career has been full, rich, and challenging. I was fortunate to be active in the development of separation techniques that paved the way for breakthroughs in the human genome project, pharmaceutical, biotechnology, bioengineering, and medical research.

It is exciting that analytical laboratories have created and tested the rapid, sensitive, selective, and reliable techniques that are vital to the life sciences as well as to other disciplines and ultimately to our health. In my first book in the early 1970s [5], I predicted that "HPLC would open new horizons for the analysis of non-volatile, thermally labile molecules that are so important in biochemistry and medical research." That prediction, I am happy to say, has come true.

I could not close this account without acknowledging my wonderful group of students. Their enthusiasm, hard work, and creativity have led me into new projects, attacking analytical problems with innovation and enthusiasm! I was indeed fortunate to have been active in research when the great potential of HPLC and then CE for use in the life sciences was realized and to work in a department that allowed me the latitude for my research. I was also blessed to have many fine collaborators, colleagues, and friends to whom I could turn to for advice and help. In addition, I learned a great deal by working with Cal Giddings and Eli Grushka as an editor on the "Advances in Chromatography" series, which kept me on the cutting edge of the research in separations for the past three decades. It has been an exciting time to be in the field of chromatography, and I am grateful to all my friends, students, colleagues, and coworkers for helping make possible a most fascinating and productive career.

REFERENCES

1. RA Hartwick, PR Brown. Performance of microparticle chemically-bonded anion exchange packings in the analysis of nucleotides. J Chromatogr 112:651–661, 1975.
2. RA Hartwick, PR Brown. Evaluation of microparticle chemically-bonded reverse phase packings in the high pressure liquid chromatographic analysis of nucleosides and their bases. J Chromatogr 126: 679–691, 1976.
3. A Fausto-Sterling, LM Zhentlin, PR Brown. Rates of RNA synthesis during embryogeneses in *Drosophila melanogaster* Dev Biol 40:78–83, 1974.
4. JM Miksic, PR Brown. The reactions of reduced nicotinamide adenine dinucleotide in acid: studies by reversed phase high pressure liquid chromatography. Biochemistry 17:2234–2238, 1978.
5. PR Brown. High Pressure Liquid Chromatography: Biochemical and Biomedical Applications. New York: Academic Press, 1973.

6. AM Krstulovic, PR Brown. Reverse Phase Liquid Chromatography: Theory, Practise and Biomedical Applications. New York: Wiley Interscience, 1982.

7. A Weston, PR Brown. HPLC and CE: Principles and Practice. New York: Academic Press, 1997.

8. PR Brown, E Grushka. Structure retention relations in the reversed-phase high performance chromatography of purine and pyrimidine compounds. Anal Chem 52:1210–1215, 1980.

9. HA Scoble, M Zakaria, PR Brown, HF Martin. Liquid chromatographic profile classification of acute and chronic leukemias. Computers Biomed Res 16:300–310, 1983.

10. PR Brown. The rapid separation of nucleotides in cell extracts using high pressure liquid chromatography. J Chromatogr 52:257–272, 1970.

18

Half a Century in Separation Science

Hamish Small

HSR, Leland, Michigan

I. BEGINNINGS

I had an early introduction to separation science. In 1951, almost fresh from university, I joined the physical chemistry group at the Atomic Energy Research Establishment (AERE), at Harwell, England. This was a time when ion-exchange resins were just coming into widespread use, and one of the projects of this group was defining the fundamental properties of these new materials. My assignment was to study the electrochemical properties of a commercial ion-exchange resin, Dowex 50, so I spent the next couple of years developing techniques for measuring the electrical conductances and ion mobilities of this cation exchanger in various counterionic forms. The project was full of interest and challenge. At the experimental level there was the challenge of designing and building novel apparatus to make the conductance and mobility measurements on these particulate materials. At the theoretical level, finding a relationship between the conductance of a suspension of ion-exchange particles in an electrolyte and the separate conductances of the ion exchanger and electrolyte phases was a challenge until I discovered that Maxwell, some 150 years earlier, had developed a theory about the conductance of conducting spheres in a conducting medium. When I applied the Maxwell theory to our ion-exchange systems, the agreement between experiment and theory was remarkably good. Discovering a connection between my work and what had been done many years before by the great Maxwell added to the delight that I was beginning to feel in doing research.

My supervisor during this time was Geoffrey C. Barker, a scientist with a rare combination of superb theoretical insight, inventiveness, and great experimental skill; he would eventually become one of the leading pioneers of modern analytical electrochemistry. Daily contact with this outstanding man and observing his scientific methods had, I am sure, an influence on my future research style. For example, in the British scientific culture where most senior scientists considered lab work to be the job of underlings, Barker was unusual in being very much a hands-on investigator. He designed and built his own equipment and instruments, he made measurements himself, and he taught me the virtue of first-hand observation. It was from these beginnings and my association with Barker that I learned the power of experimentation and the importance of being ''at the bench'' as much as possible.

The last project that I was to work on while at Harwell was classified as top secret at the time. With the coming of nuclear weapons and nuclear power, and uranium becoming a material of great strategic importance, someone at a fairly high level within the British atomic energy establishment decided that Harwell should embark on a project for the large-scale extraction of uranium from seawater. If successful it could have eliminated Britain's dependence on remote sources of uranium. Uranium occurs in seawater at about two parts per billion, and the amount of uranium in a cubic mile of seawater—about 9 metric tons—is tantalizingly large. So the project was launched on the premise that if an absorber could be designed that had a sufficiently high selectivity for uranium, then large-scale processes would soon follow. Apart from some discussions about how the absorber might be contacted with the seawater in a large-scale process, I do not recall anyone even mentioning how mass transfer rates might affect the productivity of any anticipated process. At that time I was very new to ion exchange, but as my knowledge and understanding deepened, I concluded that the large-scale extraction of uranium from seawater was a seriously flawed endeavor. Calculation of the kinetics of mass transfer using some simple models and the most optimum of conditions showed that the rate of accumulation of uranium onto an absorber, regardless of its affinity, would be far too slow to concentrate uranium at anything approaching a useful rate. But this insight came many years later, long after I had left Harwell for the United States.

II. EARLY YEARS AT DOW

In the 1950s the United States was a great magnet for a young scientist with ambition, so with my wife and an infant daughter I left Britain for good in December 1955. In looking for employment in the United States I had written to a number of universities, and I still have the letters from Professors Kolthoff and Martell offering me research fellowships. I recall how impressed I was with the replies that these obviously busy men prepared for me, not just explaining the details of the fellowships but offering help in many personal and family matters as well. It was not the sort of treatment to which I was accustomed, and it helped no end in convincing me that I would find a warm welcome in the United States. However, timing and an inclination toward applied research persuaded me to decline the university offers and to accept a job with the Dow Chemical Company instead. Dow was by then one of the leaders in ion-exchange technology, and the comfort offered by such a familiar landmark no doubt influenced my decision to go to Midland, Michigan.

I began work in Dow in what was known as the Physical Research Laboratory (PRL) or more often as the ''Physics Lab.'' The Physics Lab had a great reputation throughout the company for invention and innovation; polystyrene, polyethylene, saran, polymer foams, and magnesium from seawater were just a few of the major Dow technologies that had been pioneered there. The director of the laboratory was Bill Bauman. One of the company's great scientist-managers, he had been the leader in establishing Dow as a major producer of ion-exchange resins. Shortly after I arrived at Dow, Bill moved out of the lab to a higher position in Dow, but we continued to keep in touch.

My immediate boss at Dow was Bob Wheaton. Bob is an individual of exceptional graciousness and patience, and it was fortunate indeed for me that my early career at Dow coincided with his leadership of the ion-exchange group. Bob is also a fine scientist and had already made his mark on ion-exchange technology. For example, he and Bill Bauman had invented ion exclusion, a technique for large-scale processing that would later find analytical application.

Another member of PRL was Mel Hatch, as imaginative an organic chemist as he was a creator of the most terrible puns. Mel had invented ion retardation, a process that could separate electrolytes using water as an eluent, and was also exploring its large-scale application to various Dow processes.

In many fields, but especially in ion-exchange technology, analyzing for a variety of inorganic cations and anions is a common problem and, at that time, involved much tedious and time-consuming "wet chemistry." In determining species such as the alkali metal ions, the alkaline earth ions, the ammonium ion, the halide ions, sulfate, nitrate, phosphate, carbonate, etc., a separate analytical method had to be employed for each ionic species—or, at best, a small set of ions. Therefore, many methods with widely differing chemistry and instrumentation had to be used. Furthermore, by today's standards much of inorganic analysis was relatively insensitive. So in the late 1950s a few of us in the Physics Lab foresaw the benefits of replacing these many methods with a single chromatographic technique that would be universally applicable to all ionic species.

The chromatographic separation of the inorganic analytes did not seem to be a major problem for there was an abundance of ion-exchange methods that we could draw on. But detection of the resolved analytes presented difficulties. Since many of the ions of interest lacked suitable chromophores or convenient means of generating chromophores, they were deemed to be undetectable by light-absorption methods (later we showed that commonly held idea to be false [1]). Electrical conductivity monitoring had promise; electrical conductance was a universal property of ions in solution, and it had a simple dependence on ion concentration, at least in dilute solutions. But conductivity detection following directly on ion-exchange separation had what we perceived to be a serious drawback, namely, distinguishing the small conductance peaks of the analytes from the conductance "noise" of the highly conducting ionic eluents that ion-exchange chromatography demanded.

Mel Hatch's ion retardation resins offered a possible solution; they would allow us to separate the electrolytes using water as eluent, in which case the analyte ions would be easily detectable in this low-conductivity, essentially noise-free background. But although Mel's resins worked well in large-scale applications where column loading was high, we found that the few unbalanced ionic sites in the resins were enough to irreversibly bind the small amounts of electrolyte that were typical in analytical applications. Since we were unable to devise a suitable stationary phase, the project went into a sort of limbo, although I recall that Bob Wheaton, in his annual research planning statement for upper management, always included some remarks on what we had now begun to call "inorganic chromatography."

So what would eventually become ion chromatography (IC) had a very quiet start. No task forces or crash programs were set up at Dow, nor was there any clamor at scientific meetings for chromatography to be expanded to inorganic analysis. It was just a dream that a few of us shared as far back as the late 1950s, and many years would pass before we made the breakthrough that would start IC on its way to success.

III. THE MENTAL ATTIC

Where we store experiences I refer to as our "mental attic," and I visit mine frequently in search of inspiration. Two experiences stored in my mental attic in the late 1950s would be keys in the evolution of IC some 10 years later.

Shortly after joining Dow in 1955 I became aware of a problem known as "resin clumping." Most large-scale ion-exchange deionizers use a single bed containing an intimate mixture of stoichiometrically equivalent amounts of cation and anion exchangers in, respectively, the hydronium and hydroxide forms. When the bed's capacity to capture salts is exhausted, the bed is gently backwashed and it stratifies into two layers with the less dense anion exchanger on top. Using appropriately embedded headers, two very different regeneration chemistries can then be carried out on the two layers without removing any resin from the bed—a huge advantage. However, in the early days the two resins formed "clumps" on mixing, refused to come apart in a water backwash, and the only way to disengage the resins was to flood the bed with strong brine, a cumbersome and time-consuming procedure. Oddly, at that time there was no satisfactory explanation for the clumping phenomenon, but it became clear to me that it must be caused by the electrostatic interaction between polymer chain segments on the oppositely charged cation and anion exchangers. With the help of this insight I invented a surface treatment of the particles that prevented their clumping in the first place. This experience and the success of the project had two effects. The experience impressed on me the tenacity of electrostatic bonding between ion-exchange particles of opposite charge; the success impressed my bosses enough to allow me more leeway in choosing my own research projects. With this new freedom I went to work on an idea that I called gel-liquid extraction.

In gel-liquid extraction I was trying to marry the best features of liquid-liquid extraction and chromatography. In liquid-liquid extraction systems the extractants often had very selective extraction chemistry, but they were used in devices of low efficiency. Chromatography, on the other hand, was characterized by high efficiency. To combine the positive features of the two processes, I proposed immobilizing the extractants in polymer (cross-linked polystyrene) beads and then using these extractant-swollen beads as the stationary phase in a chromatographic process. But the swollen polymer beads were so hydrophobic that they clustered together in the aqueous mobile phases and led to poorly packed, very inefficient columns and poor separations. Adding surfactants had a beneficial but fugitive effect, but what I discovered that worked very well and was permanent was to lightly sulfonate the surface of the polymer beads. As a result they became hydrophilic and free-flowing in an aqueous environment even when their core was swollen with a very hydrophobic extractant.

I devised a number of applications for gel-liquid extraction, but, successful though it was in the lab, I was unsuccessful in promoting its application to large-scale processes. I applied for patents, wrote a paper [2], and around 1960 I terminated the project and moved on. But one of the last things I did was save a quantity of the surface-sulfonated polystyrene beads.

IV. BREAKTHROUGH

In the fall of 1971 I was exploring a new idea on the inorganic chromatography problem. I had synthesized a type of ion retardation resin in which the charges were perfectly balanced and was having some success using it with water eluent and conductivity detection, when I learned that Bill Bauman, who was now in Dow's Texas division in Freeport, had expressed an interest in inorganic chromatography. Knowing that I was likely to get a thoughtful response from Bill, I summarized my thoughts and ideas in a letter to him. In the exchange of correspondence that followed, we made one of the key breakthroughs in what would eventually be called ion chromatography.

Instead of an ion retardation approach, we would use ion-exchange chromatography to separate the analyte ions and conductivity to detect them. And in order to detect the smallest conductance changes contributed by the analyte peaks, we introduced a device called the "stripper," later called the "suppressor." The suppressor was a column of ion-exchange resin placed between the separator and the detector that converted eluent to a low-conductivity form, while leaving the conductance of the analytes relatively unaffected—or in some cases enhancing it [3,4]. We would reduce the background noise simply by reducing the background signal.

The suppressor was a key concept in the evolution of IC, but it was a component of the chromatographic system, whose capacity to function was partly consumed in each run and therefore required periodic interruptions of the chromatography to restore it to its original capacity. This idea of a consumable component was radically new to chromatography, and some doubted that it would be accepted. Because I shared that skepticism, I knew that we must make the interruptions for suppressor regeneration as unobtrusive as possible if our technique was to have any chance of success. To this end the new separating phases that we developed were critical.

There were two properties that a useful ion-exchange separator needed above all others: a low capacity to match the low concentration of eluent that suppressor longevity dictated, and stability in the strongly basic (or acidic) eluents that we proposed to use. In 1971 the only commercially available low-capacity ion exchangers had been developed for the then burgeoning field of HPLC, but they were based on silica and I doubted that they would be stable in the eluents we proposed to use. It was in my mental attic that I found more rugged alternatives.

I recalled the experience with gel-liquid extraction and thought that the surface-sulfonated resin that I had used to immobilize the extractants might be a good candidate for a low-capacity cation exchanger. So I loaded a single column with a top layer of the surface-sulfonated resin that I had saved about 10 years before and a bottom layer of a strong base anion exchanger in the hydroxide form. Using hydrochloric acid as eluent and a conductivity cell/meter as detector, I injected a mixture of lithium, sodium, and potassium salts and obtained the chromatogram of Fig. 1. Because I had used a good deal more separating resin than was necessary, the chromatograph took a long time to complete, but otherwise the experiment was a great success. I recall being tremendously excited by this result; we had demonstrated a workable concept for inorganic chromatography and the way ahead looked bright indeed. The date was November 9, 1971.

For anion chromatography I proposed to separate the analytes on a low-capacity anion exchanger using sodium hydroxide as the mobile phase and to remove the sodium hydroxide on the hydronium form of a strong acid cation exchanger such as Dowex 50. But finding an anion exchanger of suitably low capacity that would be stable in sodium hydroxide was a problem. The commercial silica-based materials did not look promising. I thought of surface quaternizing polystyrene but couldn't figure out a way to keep the anion-exchanging layer sharp in the way that surface sulfonation yielded a sharp cation-exchanging layer, and I felt that a nondiffuse active layer was essential for efficient chromatography. The resin-clumping problem some 15 years earlier gave me the clue that I needed. That experience had taught me that anion-exchange particles would adhere strongly to oppositely charged cation-exchange particles, so drawing on that background I prepared the first successful anion exchanger for IC by treating surface-sulfonated polystyrene-DVB substrate particles with a colloidal dispersion of anion-exchange particles. When the colloidal particles made contact with the oppositely charged substrate, they became firmly attached and formed a thin monolayer of anion exchanger, the active stationary phase for anion chromatography. Initially we prepared these colloidal dispersions by grinding large anion-exchange particles, but later we prepared anion-exchange resins in latex form, a superior way of controlling the size, uniformity, and ion-exchange properties of the colloidal materials. A colleague, Jitka Solc, developed the emulsion polymerization route to DVB–cross-linked polyvinylbenzylchloride latex particles, which we then quaternized with various amines to give colloidal sized anion exchangers.

Some had doubts about the physical stability of these pellicular composites, but I had developed faith in the tenacity of the electrostatic linkage between colloid and substrate; in some

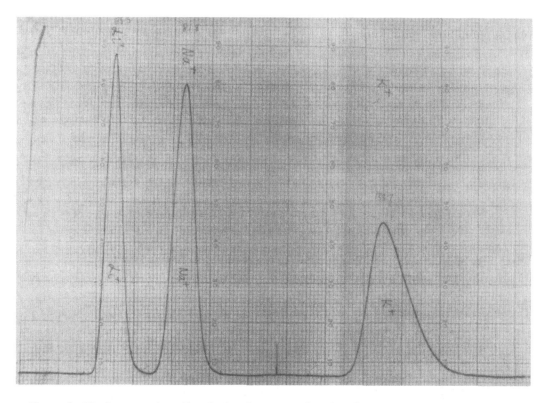

Figure 1 The first separation of ions by ion chromatography using eluent suppression and conductometric detection, carried out on November 9, 1971.

serious attempts to separate them I have never been successful. As well as providing products with excellent physical and chemical stability, this approach to anion-exchange separators has a number of advantages from a manufacturing point of view. From a liter of resin latex one can prepare an enormous number of columns, and reproducibility of the ion-exchange chemistry of the stationary phase from column to column is absolutely assured. This route to resin preparation was eventually adopted and very successfully elaborated by the Dionex Corporation and is much used to this day as the source of Dionex columns.

Developing suitable eluents was another key to IC's successful development, particularly for anion analysis. Sodium hydroxide was our eluent of first choice since it suppressed to water, the background with lowest conductivity. But hydroxide was a weak displacing ion for many of the analytes of interest, and relatively high concentrations were necessary to elute the more intractable species, thus shortening the lifetime of the suppressor. The use of phenate ion instead of hydroxide alleviated the problem immensely; phenate was a high-affinity anion and it suppressed to phenol, a very weak acid whose solutions were feebly conducting. While this led to great improvements in the speed and efficiency of IC for anions, phenate was never used in commercial IC instruments, although some similar eluents such as cyano-phenate were. Perplexed by the anomalously high displacing power of a certain "sodium hydroxide" solution, we discovered that its unusual potency was due to carbonate contamination, and it was from that accidental discovery that we developed the carbonate-bicarbonate system of eluents [5] that became the workhorse anion eluent for many years to come.

In these early days we showed how the ion-exchange, eluent-suppression technique, while greatly simplifying inorganic analysis, also enabled the analysis of many organic ions that had previously been inaccessible to chromatography because they lacked suitable chromophores. Aliphatic acids, amines, and quaternary ammonium ions were notable examples. Today suppressed conductometric detection, because of its universal ability to detect most ionic species with high sensitivity, is often the chosen method of detection regardless of whether the species are inorganic or organic.

In the pioneering work on IC we also demonstrated how ion-exchange membranes could provide continuous suppression, thus avoiding interruptions for regeneration [6]. Today, the majority of ion chromatographs employ membrane-based suppression devices that suppress continuously without interrupting the chromatography.

Having the new analytical technique accepted by the analytical community in Dow was a major hurdle. Tim Stevens, at that time a young analytical chemist with responsibility for ion analysis by the old methods, attended a seminar that I gave on the new technique and he recognized its potential. So he joined our small group in the Physics Lab, and together we improved and devised new eluents for cation analysis by IC. After several months Tim had accomplished enough to convince his management in Dow's analytical laboratories that what we had developed was a sound and useful technique, and he returned to the analytical laboratories where he had an important influence in establishing IC within Dow.

When IC began to show promise as a new universal tool for ion analysis, commercialization moved to the top of the list of priorities, and in 1975 Dow established a licensing agreement with Durrum Chemical to exploit the Dow technology, which was then described in several key patents. This was the beginning of Dionex Corporation. In September 1975 IC got its first public exposure at the fall meeting of the American Chemical Society, where Dionex showed the first commercial instrument and we presented the much-cited paper in *Analytical Chemistry* [3].

In the years following the transfer of technology to Durrum, I continued research and invention on IC. In 1979, using light-absorbing ions in the eluent of ion-exchange chromatography, I demonstrated that transparent analytes could be detected and quantified from the vacancies they produced in the UV-absorbing background. This new detection technique, indirect photometric detection, upset the prevailing dogma that photometers could only detect species containing chromophores [1]. Indirect photometric detection has had limited use in IC, but the principles are widely applied in capillary electrophoresis.

V. A NEW DIRECTION

In 1983 I left Dow, and in the fall of the same year Dionex expressed an interest in retaining me as a consultant. My association with Dionex has allowed me to continue research in separation science, particularly IC, and the collaboration has produced a number of useful inventions, some of them recent.

I had frequently returned to my earliest experience with ion exchangers in the hope of finding a use for their conducting properties, and as early as 1974 I had proposed using electrically polarized ion-exchange beds in IC. My notes from that time record an idea—"Possibilities for continuously regenerating an ion chromatography stripper bed by electrical means"—and they show a proposed device comprising an ion-exchange bed clamped between membranes and electrodes. But there were more pressing priorities at that time, and I never reduced any of the concepts to practice. Others, notably Speigler, had shown how an ion-exchange bed could be regenerated electrochemically, but it became apparent to me that if this regeneration was carried out under conditions of constant current, then such a polarized bed could be a source of base or acid, whose concentration would depend on the current passing. Thus we would have a convenient means for producing eluent of precisely controlled concentration simply by passing water through an electrically polarized bed of ion-exchange resin [7]. I also discovered a new way of operating the polarized bed so that it was not only a source of base (acid) but also a means of recycling or suppressing it [8]. Since the manner of operating the bed and the counterflowing ion streams had parallels in distillation under refluxing conditions, I called the technique ion reflux. In one embodiment of ion reflux I was able to demonstrate eluent generation, separation, and suppression, all within a single, electrically polarized bed using simply water as the pumped phase. Thus IC had in a way come full circle. In the very early days at Dow we had considered water as the ideal eluent but had never been able to come up with a suitable stationary phase and had abandoned that approach. Now with the aid of some new electrochemistry we were able to carry out IC simply by pumping water.* Dionex is now developing products from these new concepts and inventions, and I look forward to the day when some embodiment of ion reflux will become an important part of the way we practice ion chromatography.

VI. THE CHROMATOGRAPHY OF COLLOIDAL PARTICLES

The Dow Chemical Company produces large quantities of polymer latexes, a component of water-based paints, adhesives, and paper coatings. Knowing the particle size of the polymer phase is essential to understanding the behavior of latexes and to adapting them to their manifold uses. I became interested in particle size measurement in the late 1960s, when my management asked me to develop a fast method for determining the size of a polymer latex whose particles were around 1 μm in diameter. At that time electron microscopy and various light scattering techniques were the principal methods for particle size measurement, and each had limitations. Microscopy was accurate but slow, and it required an expensive instrument and a skilled operator to make the measurements.

Light scattering techniques of that time were very fast but were useless when the colloid was polydisperse, and many of the materials of interest were just that. So, with this background, I started research to see if I could extend chromatography beyond the molecular region into the supermolecular or colloidal region. That work led to several new discoveries concerning the transport of particles through porous media and to a particle sizing technique called hydrodynamic chromatography (HDC) [9]. With regard to colloid transport in packed beds of solid spherical particles, I discovered that under certain conditions larger colloidal particles were carried through packed beds faster than smaller particles while under different conditions the reverse applied. I was also able to demonstrate the separation of colloidal particles based on their chemical composition; a mixture of polystyrene and polymethylmethacrylate latexes of essentially identical size could be resolved into two well-separated peaks with the PMMA preceding the PS (Fig. 2). Some colloid theoreticians took up the challenge of finding a quantitative mechanism for these new separations, and in a number of cases their derived models are in remarkable agreement with the experimental observations [10].

The chromatographic technique known as HDC is a fast and extremely precise means of determining the particle size of a wide variety of colloidal materials. As applied to polymer latexes it has been used to follow emulsion polymerization kinetics, to detect particle aggregation, and to measure the size of particles whose volume is sensitive to such environmental factors

---

* I emphasize, however, that the mechanism of separation in ion reflux is still ion exchange. The eluent is an electrolyte, as it must be, but it is generated and suppressed in situ.

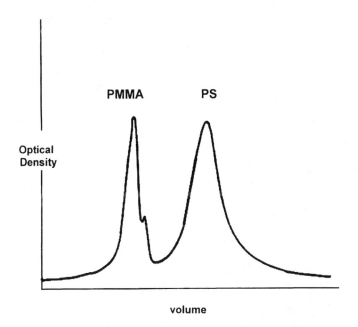

Figure 2 A separation of polystyrene particles (PS) from poly(methylmethacrylate) (PMMA) particles of essentially identical size. The PS particles were of very uniform particle size distribution and 0.234 μm average diameter. The PMMA particles were similarly uniform in size and 0.24 μm average diameter. The separation was carried out on a 1 m column filled with 20 μm polystyrene particles using 0.4 M sodium chloride as eluent.

as pH, surfactants, and organic solvents. In a number of these applications HDC displays unique capabilities. The Dow Chemical Company applies HDC widely and profitably in its polymer latex business, where it provides product definition, assists in quality control, and is an important tool in controlling the synthesis of latexes.

VII. REFLECTIONS

A long career in anything is inevitably rewarding. In looking back over my 50 or so years in research and invention, I would have to conclude that our work in ion chromatography has been the most satisfying in many different ways. The extent of its impact on modern chemical analysis, although difficult to measure, is extremely gratifying. It has brought benefits to the chemical, biological, and environmental sciences. There are few industries in which IC is not used, whether it is to control processes, to assure product quality, or to monitor waste streams. In the electrical power–generating industry, for example, IC monitoring of boiler feed waters protects massively costly installations from the effects of corrosive ions. In this application, where potentially damaging ions are successfully determined at the parts per trillion level, IC displays unprecedented sensitivity. And the diversity of its applications continues to grow with such examples as determining the ion content of electroplating baths, the acids in wine, toxic ions in infant foods, corrosive ions in packaging film, the composition of acid rain, the ions in soil, trapped ions in polar ice caps, the ions in body fluids.

There have, of course, been the awards and the medals. Receiving the American Chemical

Society Award in Chromatography and the recognition by professional colleagues that it implies is especially cherished. But my greatest rewards have come in other ways. Seeing the jobs and careers that ion chromatography has given to others, having our work validated in textbooks as part of analytical science, new colleagues discovered, new places visited, are a rich part of my career in separation science.

I still do research and invent, and I hope that I can continue to do so for some time.

REFERENCES

1. H Small, TE Miller. Indirect photometric chromatography. Anal Chem 54:462–469, 1982.
2. H Small. Gel liquid extraction. The extraction and separation of some metal salts using tri-n-butyl phosphate gels. J Inorg Nucl Chem 18:232–244, 1961.
3. H Small, TS Stevens, WC Bauman. Novel ion exchange chromatographic method using conductimetric detection. Anal Chem 47:1801–1809, 1975.
4. H Small. Ion Chromatography. New York: Plenum Press, 1989.
5. H Small, J Solc. Ion chromatography—principles and applications. In: The Theory and Practice of Ion Exchange. London: Society of Chemical Industry, 1976.
6. TS Stevens, JC Davis, H Small. Hollow fiber ion-exchange suppressor for ion chromatography. Anal Chem 53:1488, 1981.
7. H Small, Y Liu, N Avdalovic. Electrically polarized ion-exchange beds and ion chromatography: eluent generation and recycling. Anal Chem 70:3629–3635, 1998.
8. H Small, J Riviello. Electrically polarized ion-exchange beds in ion chromatography: ion reflux. Anal Chem 70:2205–2212, 1998.
9. H Small. Hydrodynamic chromatography. A technique for size analysis of colloidal particles. J Colloid Interface Sci 48:147–161, 1974.
10. H Small. Discoveries concerning the transport of colloids and new forms of chromatography. Accounts Chem Res 25:241–246, 1992.

19

Development of Countercurrent Chromatography and Other Centrifugal Separation Methods

Yoichiro Ito

National Heart, Lung, and Blood Institute, National Institutes of Health, Bethesda, Maryland

I. ORIGIN OF THE COIL PLANET CENTRIFUGE (1963–1968)

Countercurrent chromatography (CCC) [1–6] originated in our laboratory at the National Institutes of Health in the late 1960s, but it really started to bloom more recently. Development of CCC was preceded by the invention of the coil planet centrifuge (CPC), which dates back to the early 1960s [7].

In the 1950s, when I was a student at Osaka City University Medical School in Japan, I decided to conduct research on lymphocytes because at that time their function was not well described. They were morphologically classified into large, medium, and small lymphocytes. After graduating from the medical school, I served one year in a rotating internship at the Yokosuka U.S. Naval Hospital, Yokosuka, Japan, where I was fortunate to be sponsored by Dr. John Featherstone, Commander and radiologist. When I asked him to help me select the hospital where I could do such research, he kindly wrote many letters of application and delivered them to me, only asking for my signature! I chose a 4-year residency in pathology at Cuyahoga County Metropolitan General Hospital in Cleveland, Ohio, because it promised to give this research opportunity in my second year of residency. One year later I started research on a lymphocyte-stimulating factor from irradiated rats with my peer resident, Dr. Fay B. Weinstein, and we published a preliminary report on the subject in the *Journal of the National Cancer Institute* [8]. In order to continue research on the lymphocyte function, it was first necessary to isolate them. Since the conventional density gradient method with a short centrifuge tube was found to be inefficient, I conceived the idea of a centrifuge embodying a long coiled tube that underwent planetary motion, subjecting the cells to an Archimedean screw force. The motion of particles through a coiled tube in this device, called a coil planet centrifuge (CPC), was then mathematically analyzed by Fay's husband, Dr. Marvin A. Weinstein, a brilliant physicist.

Two years later I moved to Michael Reese Hospital in Chicago, where I received help from Dr. Lloyd Arnold, a professor of biology at Loyola University in Chicago. In spite of his busy

(A)

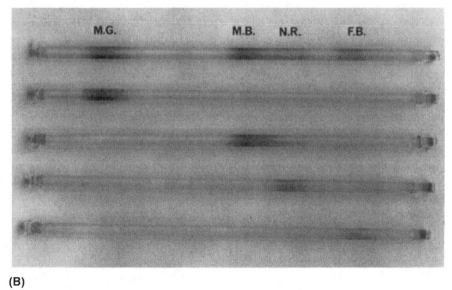

(B)

schedule, he tried to construct a CPC at his own expense. Unfortunately, the project was prematurely terminated due to the expiration of my Fulbright Scholarship travel fund.

In 1963 I returned as a lecturer in the Department of Physiology at Osaka City University Medical School, Osaka, Japan. Professor Eiichi Kimura, the head of the department, introduced me to Mr. Yonezo Kubota, the founder and president of Kubota Centrifuge Corporation in Tokyo. Mr. Kubota enthusiastically designed and constructed a prototype of the CPC with his own hands and delivered it to our research laboratory in the spring of 1964. I still remember my excitement on that day when Professor Eiichi Kimura, Dr. Ichiro Aoki (an associated professor), and I were chatting together in front of this first prototype. I said, "tomorrow I will start the experiment with the CPC and continue it for a long time that may surprise you." After Mr. Kubota retired, I handed the project to a small company called Sanki Engineering, Ltd, Kyoto, Japan. I owe many thanks to Mr. Reizo Matsumoto, who helped me to find this excellent company. Its president, Mr. Kan-ichi Nunogaki, and his younger brother, Yoshiaki Nunogaki, were very interested in the device and quickly built a prototype, which surpassed the original in both design and function (Fig. 1A). The separation of latex particles of cell size was successful, and the results were published in *Nature* in 1966 [7]. At that time Dr. Aoki had been working on a countercurrent distribution method, and we started to explore the capabilities of the CPC in this direction (Fig. 1B) [9].

In 1967 Dr. K. Nunogaki and I presented a paper on the CPC at the 7th International Conference on Medical and Biological Engineering held in Stockholm, Sweden. During the conference, I met Dr. Robert L. Bowman from the National Institutes of Health and accepted his invitation to visit his laboratory in Bethesda, MD. After I delivered a talk on the CPC, Dr. Bowman invited me to join his research group. In 1968 I became a Visiting Scientist working in his laboratory to develop a new chromatographic method, and the first paper on countercurrent chromatography was published in *Science* in 1970 [1]. Since then the development of CCC has continued steadily.

Over these three decades the research on lymphocytes advanced remarkably, revealing their important immunological functions as B cells and T cells. Unfortunately, the CPC originally built for purification of these cells missed the opportunity to be part of this research; instead it has evolved into a chromatographic technique called "high-speed CCC" and is used for separation and purification of a wide variety of natural and synthetic products. I hope that the utility of CCC will be extended in the coming century.

II. DEVELOPMENT OF COUNTERCURRENT CHROMATOGRAPHY

A. Development of Hydrostatic CCC Systems (1969–1973)

In the Laboratory of Technical Development headed by Dr. Bowman, I started the development of a simple hydrostatic CCC system [1] because introduction of the flow-through mechanism to the original CPC was very troublesome. As illustrated in Fig. 2, this basic system uses a stationary coil filled with one phase of an equilibrated two-phase solvent system. The other phase introduced into the coil percolates through the stationary phase on one side of the coil.

Figure 1 The coil planet centrifuge. (A) the original prototype of the coil planet centrifuge designed by Sanki Engineering, Kyoto, Japan. (B) Separation of dyes by the coil planet centrifuge. M.G., Methyl green; M.B., methylene blue; N.R.: neutral red; F.B.: basic fuchsin. (From Ref. 9.)

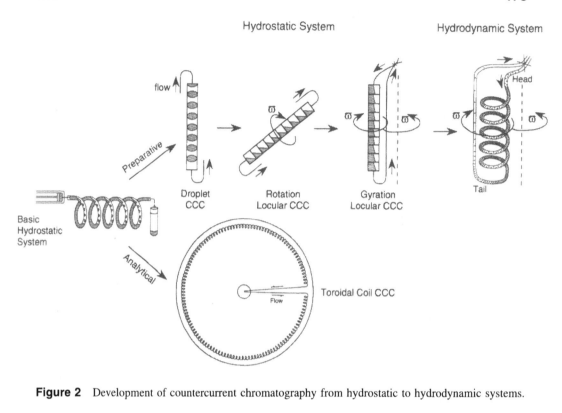

Figure 2   Development of countercurrent chromatography from hydrostatic to hydrodynamic systems.

Consequently, solutes are partitioned between the two phases in each helical turn and eluted from the coil in the order of their partition coefficients.

In this hydrostatic CCC system the efficiency is increased by simply decreasing the dimensions of the coil, which is subjected to centrifugal force in order to facilitate the countercurrent flow of two solvent phases through a narrow tubing to achieve analytical separation (toroidal coil CCC or helix CCC). Alternatively, the preparative system is created by reducing one side of the coil entirely occupied by the mobile phase while the efficient side of the coil is replaced by a straight tubular column. This droplet CCC [10] is further modified by dividing the tubular space into multiple compartment called *locules*. In rotation locular CCC [11], a tilted locular column is rotated around its axis to mix the two solvent phases in each locule. In order to improve the mixing effect, gyration locular CCC [11] was devised in which a bundle of locular columns is held vertically and gyrated around their axes, the motion being similar to that of a test tube on a vortex mixer that can be supported by fingers. Consequently, the system permits continuous elution without twisting the flow tubes. Nevertheless, this system required preparation of elaborate locular columns, which must be tightly secured onto the gyrating shaft. This problem led me to consider the alternative use of a coiled column as in the original CPC (Fig. 2, upper right). I was gratified to see that the coiled column mounted onto the gyration shaft surpassed the partition efficiency of locular CCC [12,13]. This important breakthrough opened the door to a new CCC system called the "hydrodynamic CCC system," which utilizes an Archimedean screw force to enhance the solute partitioning process in the coiled column. In order to fully utilize the capability of this system I focused my effort on the development of seal-less flow-through CPC systems.

B. Development of Seal-Less Flow-Through Centrifuge Systems (1971–1983)

Figure 3 schematically illustrates a series of seal-less flow-through centrifuge systems for performing CCC [14]. Each system consists of a cylindrical coil holder with a bundle of flow tubes, one end of which is supported at the top of the centrifuge. They are classified into three groups: synchronous, nonplanetary, and nonsynchronous according to the motion of the coil holder. The type I synchronous system (left top) produces a planetary motion analogous to the vortex mixer mentioned earlier, where a vertical coil holder rotates about its own axis and synchronously revolves around the central axis of the centrifuge in the opposite direction. This synchronous counterrotation of the holder steadily unwinds the twist of the tube bundle caused by revolution, thus eliminating the need for the rotary seal. This principle works equally well for the rest of the synchronous systems with tilted (types I-L and I-X), horizontal (types L and X), dipping (types J-L and J-X), or even inverted (type J) orientation of the holder. Here, type J is a transi-

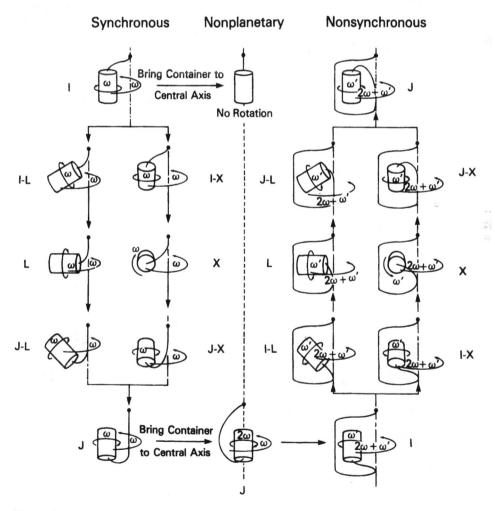

Figure 3 Series of seal-less flow-through centrifuge systems for performing countercurrent chromatography.

tional form to the nonplanetary scheme. When the position of the holder is shifted to the central axis of the centrifuge, the holder rotates at a doubled rate around the central axis of the centrifuge as in the ordinary centrifuge system, while the tube bundle is shifted toward the periphery to revolve around the holder at a rate of one half that of the holder. (This nonplanetary system has been used for apheresis centrifuges, as described later.) The system also provides a base for the nonsynchronous system on which the holder is again shifted toward the periphery to undergo synchronous planetary motion as shown in the figure. In all these nonsynchronous systems, the rate of rotation and the revolution of the coiled column becomes independently adjustable.

A decade of effort had produced a variety of CCC schemes based on these seal-less centrifuge systems as summarized in Fig. 4. Among those, high-speed CCC is selected and described below since it is the most important milestone in the development of CCC.

C. Development of High-Speed CCC (1980–1988)

During the mid-1970s, I was using the type J synchronous CPC with a set of coiled tubes eccentrically affixed onto the column holder. This horizontal flow-through CPC [15] was extensively used for separation of natural products and synthetic peptides. However, the method required relatively long separation times because application of a high flow rate would carry over stationary phase resulting in loss of peak resolution. This was a common complication in all CPC schemes that existed at that time.

A breakthrough that overcame this problem came quite unexpectedly when I wound Teflon tubing *directly* onto the coil holder hub of the horizontal flow-through centrifuge. Under this coaxial coil orientation on the rotary shaft the two phases were quickly separated along the coil

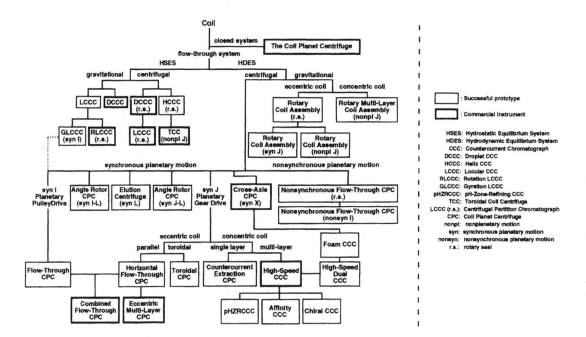

Figure 4   Summary of various CCC instruments developed since 1960.

to establish *bilateral hydrodynamic distribution*, where one phase entirely occupied the head side and the other phase the tail side [16]! (The head-tail relationship is defined according to an Archimedean screw effect: the object different in density in a rotating coil is driven toward the head of the coil.) This discovery prompted me to develop an efficient hydrodynamic system called high-speed CCC [17,18].

The principle of high-speed CCC is illustrated in Fig. 5, where each coil is drawn as a straight line to indicate the distribution of the two phases. The top diagram (Fig. 5A) shows the *bilateral hydrodynamic distribution* of the two phases in an end-closed coil with a white phase (head phase) on the head side and a black phase (tail phase) on the tail side. This hydrodynamic distribution of the two phases suggests that the white phase, if introduced from the tail end, would quickly move toward the head, and similarly the black phase introduced from the head would quickly move toward the tail. This hydrodynamic motion can be effectively utilized for performing CCC, as shown in Fig. 5B. The upper coil is first entirely filled with the black phase followed by elution with the white phase through the tail. Then, the white phase quickly moves toward the head leaving a large volume of the black phase as a stationary phase. The lower coil is filled with the white phase followed by elution with the black phase through the head. Then the black phase similarly moves through the coil toward the tail, leaving a large volume of the stationary white phase in the coil. In either case, the solutes introduced locally into the mobile phase will be efficiently partitioned between the two phases and quickly eluted from the coil. Consequently, high retention of the stationary phase under a high flow rate, which had been beyond reach of all other CCC schemes, was achieved by the present system.

This system also permits simultaneous introduction of the two phases through the respective end of the coil as shown in Fig. 5C. This dual CCC system requires an additional flow tube at

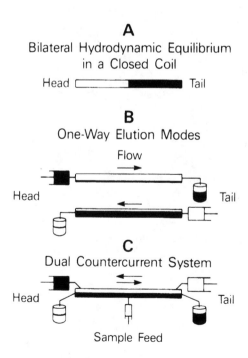

Figure 5 Principle of high-speed CCC.

each end of the coil and, if desired, a sample feed line at the middle portion of the coil. The system has been successfully used for liquid-liquid dual CCC [19] and gas-liquid dual CCC or foam CCC [20–23].

The first prototype of the high-speed CCC centrifuge held a coil holder on one side and a counterweight holder on the other side to balance the centrifuge system. This multilayer CPC became the first commercial model available through P.C. Inc., Potomac, MD, in the early 1980s. This design was later improved by eliminating the counterweight and mounting three identical coiled columns symmetrically around the rotary frame to balance the centrifuge system (Fig. 6) [24–27]. This improved model first became commercially available through Pharma-Tech Research Corporation, Baltimore, MD, in the late 1980s.

For stroboscopic observation of the hydrodynamic motion of the two solvent phases in the rotating coil, I designed a spiral column with a transparent cover and invited Dr. Walter D. Conway of the School of Pharmacy at the State University of New York at Buffalo to work on this with me. We then observed an interesting hydrodynamic motion of the two phases in the spiral column as illustrated in Fig. 7 [3,4]. The upper diagram schematically shows the distribution of the two phases in the rotating spiral column. About one fourth of the area near the center of the revolution exhibits vigorous mixing of the two phases (mixing zone), while in the rest of the area the two phases are completely separated in such a way that the lighter phase (white) occupies the inner portion and the heavier phase (black) the outer portion of the coil (settling zone). The lower diagram shows the motion of the mixing zone through the stretched spiral column at four labeled positions. It clearly indicates that the mixing zone travels through the spiral column at a rate of one turn per one revolution of the column. This finding implies the

Figure 6   An advanced model of high-speed CCC centrifuge.

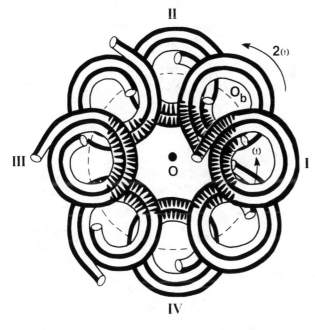

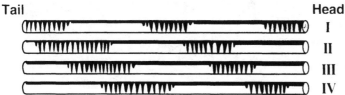

Figure 7 Motion of the two solvent phases in a rotating spiral column in high-speed CCC centrifuge.

important fact that at every portion of the spiral path the solutes are subjected to a repetitive partition cycle of mixing and settling at a very high frequency of 13 times per second at 800 rpm of column revolution. This finding explains its high partition efficiency.

Since 1980s the type-J multilayer CPCs have been used for separation of various natural and synthetic products using a conventional organic/aqueous solvent systems [4]. However, this system failed to retain satisfactory amounts of the stationary phase of aqueous-aqueous polymer phase systems useful for separation of macromolecules [28]. In order to overcome this problem, we developed a "cross-axis CPC," which was based on the X, L, and their hybrid synchronous planetary motion (see Fig. 3) [29–33]. This system allows retention of the viscous polymer phase system in the coiled column under the strong centrifugal force acting across the diameter of the coiled tube. This cross-axis CPC has been successfully used for purification of lipoproteins and recombinant proteins with PEG/potassium phosphate solvent systems [34,35].

D. pH-Zone–Refining CCC (1990–1994)

In 1990 during the purification of a thyroxin derivative (bromoacetyl T$_3$) by high-speed CCC, I observed an unusually sharp peak of the product [36–38]. It was very strange, because a nearby eluted impurity peak showed normal broadness. By investigating the cause of this sharp peak formation, I finally found that the sharp peak was associated with the elution of an organic acid

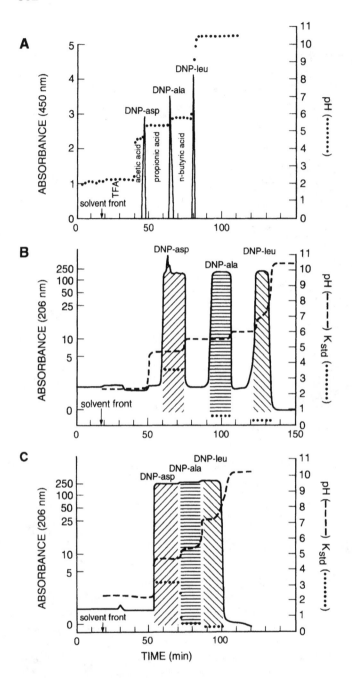

Figure 8 Development of pH-zone-refining CCC. (From Ref. 37.)

(bromoacetic acid) present in the sample solution. This phenomenon is also reproduced by adding organic acids to the stationary phase. Using a set of four organic acids in the stationary phase, a milligram quantity of three dinitrophenyl amino acids was beautifully resolved each forming a sharp peak between the spacer acids (Fig. 8A). To my surprise, when the sample size was increased, these sharp peaks were each transformed into a rectangular peak associated with a specific pH (Fig. 8B). Then, elimination of three spacer acids caused fusion of all three peaks into a large rectangular peak where the specific pH zones were still preserved (Fig. 8C) [37–39]. These results clearly suggested that the method might be useful for preparative separation of ionizable compounds. Because of the formation of pH zones, this method was named the pH-zone–refining CCC.

The preparative capability of this method was first demonstrated on the separation of D& C orange No. 5 [39]. Three major components were well resolved with minimum mixing zones where the % yield of each component *increased* with the increased sample size. During the past several years, the method has been successfully applied to a variety of compounds including derivatives of amino acids and peptides [40–43], alkaloids [44,45], lignans with an anti-HIV activity [46], etc. When the present method was applied to chiral separation by dissolving a chiral selector in the stationary phase, gram quantities of enantiomers were well resolved [47]. Because pH-zone–refining CCC yields a large amount of pure fractions at very high concentration (near saturation), I believe that the method will become an important purification method in industry.

III. CENTRIFUGAL PRECIPITATION CHROMATOGRAPHY (1998–)

The CCC technique uses countercurrent movement of mutually immiscible two solvent phases in an open column to carry out a separation. I wondered then if it might be possible to utilize two mutually miscible solvents when they are partitioned by a semipermeable membrane. Preliminary studies indicated that this new system, called centrifugal precipitation chromatography, may open a rich domain of applications including cell separation by density gradient, fractionation of biopolymers with a salt and/or pH gradient, and fractional precipitation of small molecules and their polymers by organic solvents.

This chromatographic system was first successfully applied for the separation of proteins based on ammonium sulfate precipitation [48,49]. Figure 9A shows the principle and design of the apparatus. The method uses a pair of channels divided by a dialysis membrane. A concentrated ammonium sulfate solution is eluted through one channel and water through the other channel in the opposite direction. Then, diffusion of ammonium sulfate through the membrane forms an exponential concentration gradient of ammonium sulfate along the water channel. Under a centrifugal force field, proteins introduced into the water channel are precipitated according to their solubility in ammonium sulfate solution. After all proteins are precipitated, chromatographic elution is initiated by gradually reducing the ammonium sulfate concentration in the upper channel. This causes a proportional decrease of ammonium sulfate concentration in the gradient in the water channel so that the once precipitated proteins are redissolved and again precipitated at slightly advanced locations. Consequently, proteins repeat dissolution and precipitation along the channel and finally elute from the column in the order of their solubility in ammonium sulfate solution.

The design of the separation column is illustrated in Fig. 9B, which shows a pair of plastic discs equipped with a mutually mirror imaged spiral grooves. A dialysis membrane is sandwiched between these discs to form a desired column structure shown in Fig. 9A. These discs are tightly bolted together and mounted on a seal-less continuous flow centrifuge shown in Fig. 9C.

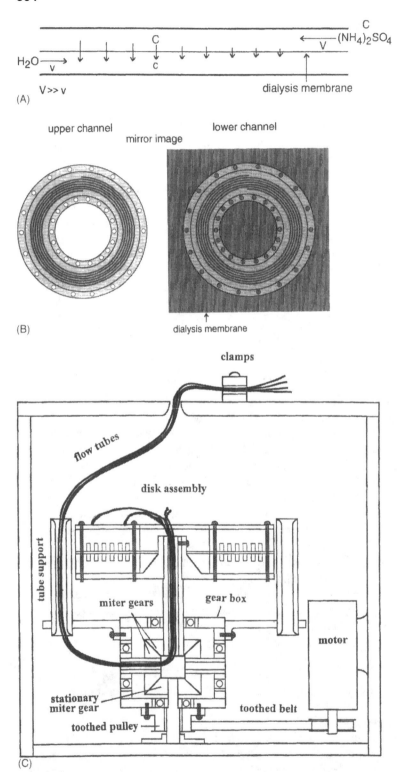

Figure 9 Principle and design of the apparatus for centrifugal precipitation chromatography. (A) Twin channel divided by the dialysis membrane. (B) Design of the separation disk. (C) Cross-sectional view of the apparatus. (From Ref. 49.)

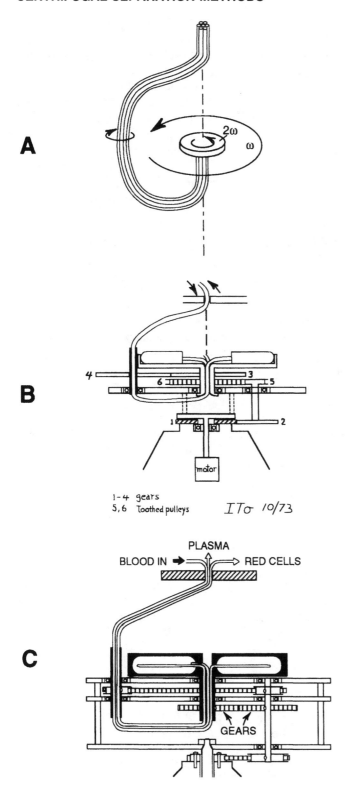

Figure 10 Principle and design sketches of the first seal-less continuous flow centrifuge used for apheresis. (From Ref. 52.)

The capability of the method was first demonstrated by separation of monoclonal antibodies against human mast cells in collaboration with Dr. Tadashi Okada of Department of Physiology, Aichi Medical University, Aichi, Japan. IgM fractions showed enhanced antibody activity over 10 times that of the original culture supernatant. The separation of monoclonal antibodies from ascitic fluid was also successful. The method is also applicable to affinity precipitation analogous to affinity chromatography. Recombinant ketosteroid isomerase was directly purified from a crude *E. coli* lysate adding an affinity ligand, estradiol-PEG(polyethylene glycol)-5000, to the sample solution.

Using ammonium sulfate in an open column, the method eliminates irreversible adsorption and denaturation of proteins caused by solid support. It also has the capability of concentrating and recovering a small amount of proteins due to the osmosis effect. Because of these advantages, the method may eventually be used in research laboratories and industrial plants.

IV. SEAL-LESS CONTINUOUS FLOW APHERESIS CENTRIFUGE (1975)

During the development of various seal-less flow-through centrifuge systems, I had an opportunity to participate in improving the apheresis centrifuge system by eliminating the conventional rotary seal which caused detrimental loss of platelet in the sheep peripheral blood [50–52]. In collaboration with Drs. Jacques Suaudeau and Theodore Kolobow of our laboratory (Laboratory of Technical Development, National Heart, Lung, and Blood Institute), I designed a seal-less continuous flow centrifuge for apheresis of sheep blood.

The principle and design of the apparatus is illustrated in Fig. 10. Figure 10A shows unique planetary motion, which prevents twisting the flow tubes. The centrifuge bowl is rotated at a double speed relative to the rotation of the rotary frame which fulfills the requirement shown in Fig. 3 of a nonplanetary system. A sketch of the first mechanical design, Fig. 10B, was improved in the design shown in Fig. 10C. A set of gears and pulleys produces the desired motion described above. This first model of seal-less continuous flow centrifuge was successfully applied to apheresis of sheep blood with minimum platelet deterioration [52]. Currently about 30,000 units of this seal-less continuous flow centrifuge system are being used for apheresis and cell separation in blood banks in the United States, Europe, and Japan.

ACKNOWLEDGMENTS

The author wishes to express deep gratitude to those who have participated in the development and application of CCC. Names and contributions of those individuals are found in the chapters of multiauthor monographs on CCC cited in Refs. 2,4–6. The author thanks Dr. Henry M. Fales of the National Heart, Lung, and Blood Institute, NIH, Bethesda, MD, for editing the manuscript.

REFERENCES

1. Y Ito, RL Bowman. Science 167:281–283, 1970.
2. NB Mandava, Y Ito, eds. Countercurrent Chromatography: Theory and Practice. New York: Marcel Dekker, 1988.
3. WD Conway. Countercurrent Chromatography: Apparatus, Theory and Applications. New York: VCH, 1990.

4. AP Foucault, ed. Centrifugal Partition Chromatography. Chromatographic Series, Vol. 68. New York: Marcel Dekker, 1994.
5. Y Ito, WD Conway, eds. High-Speed Countercurrent Chromatography. New York: Wiley Interscience, 1996.
6. JM Menet, T Thiebaut, eds. Countercurrent Chromatography. Chromatographic Science Series, Vol. 82. New York: Marcel Dekker, 1999.
7. Y Ito, MA Weinstein, I Aoki, R Harada, E Kimura, K Nunogaki. Nature 212:985–987, 1966.
8. Y Ito, FB Weinstein. J Natl Cancer Inst 29:229–237, 1962.
9. Y Ito, I Aoki, E Kimura, K Nunogaki, Y Nunogaki. Anal Chem 41:1579–1584, 1969.
10. T Tanimura, J Pisano, Y Ito, RL Bowman. Science 169:54–56, 1970.
11. Y Ito, RL Bowman. J Chromatogr Sci 8:315–323, 1970.
12. Y Ito, RL Bowman. Science 173:420–422, 1971.
13. Y Ito, RL Bowman. Anal Chem 43:69A–75A, 1971.
14. Y Ito. In: E Heftmann, ed. Chromatography, Fundamentals and Applications of Chromatography and Related Differential Migration Methods. Elsevier: Amsterdam, 1992, pp. A69–A107.
15. Y Ito, RL Bowman. Anal Biochem 82:63–68, 1977.
16. Y Ito. J Chromatogr 207:161–169, 1981.
17. Y Ito. J Chromatogr 214:122–125, 1981.
18. Y Ito, JL Sandlin, W Bowers. J Chromatogr 244:247–258, 1982.
19. YW Lee, C Cook, Y Ito. J Liq Chromatogr 11:37–53, 1988.
20. Y Ito. J Liq Chromatogr 8:2131–2152, 1985.
21. Y Ito. J Chromatogr 403:77–84, 1987.
22. H Oka, K Harada, M Suzuki, H Nakazawa, Y Ito. Anal Chem 61:1998–2000, 1989.
23. H Oka, Y Ikai, J Hayakawa, K Harada, M Iwaya, H Murata, M Suzuki, Y Ito. J Chromatogr A 791: 53–63, 1997.
24. Y Ito, H Oka, J Slemp. J Chromatogr 475:219–227, 1989.
25. Y Ito, H Oka, YW Lee. J Chromatogr 498:169–178, 1990.
26. Y Ito, H Oka, E Kitazume, M Bhatnagar, Y Lee. J Liq Chromatogr 13:2329–2349, 1990.
27. Y Ito, E Kitazume, J Slemp. J Chromatogr 538:81–85, 1991.
28. PÅ Albertsson. Partition of Cell Particles and Macromolecules, 3rd ed. New York: Wiley Interscience, 1986.
29. Y Ito. Step Sci Technol 22:1971–1988, 1987.
30. Y Ito. Sep Sci Technol 22:1989–2009, 1987.
31. Y Ito, TY Zhang. J Chromatogr 449:135–151, 1988.
32. Y Ito, TY Zhang. J Chromatogr 449:153–164, 1988.
33. Y Ito, TY Zhang. J Chromatogr 455:151–162, 1988.
34. Y Shibusawa, Y Ito, K Ikewaki, DJ Rader, H Brewer, Jr. J Chromatogr 596:118–122, 1992.
35. YW Lee, Y Shibusawa, FT Chen, J Myers, J Schooler, Y Ito. J Liq Chromatogr 15:2831–2841, 1992.
36. Y Ito, Y Shibusawa, H Fales, H Cahnmann. J Chromatogr 625:177–181, 1992.
37. Y Ito. pH-peak focusing and pH-zone-refining countercurrent chromatography. In: Y Ito, WD Conway, eds. High-Speed Countercurrent Chromatography. Chemical Analysis Series, Vol. 132. Wiley Interscience, New York, 1996, pp. 121–175.
38. Y Ito, Y Ma. J Chromatogr A 753:1–36, 1996.
39. A Weisz, AL Scher, K Shinomiya, H Fales, Y Ito, J Am Chem Soc 116:704–708, 1994.
40. Y Ito, Y Ma. J Chromatogr A 162:101–108, 1994.
41. Y Ma, Y Ito. J Chromatogr A 678:233–240, 1994.
42. Y Ma, Y Ito. J Chromatogr A 702:197–206, 1995.
43. Y Ma, Y Ito. J Chromatogr A 771:81–88, 1997.
44. Y Ma, Y Ito, EA Sokoloski, H Fales. J Chromatogr A 685:259–262, 1994.
45. Y Ma, EA Sokoloski, Y Ito. J Chromatogr A 724:248–353, 1996.
46. Y Ma, L Qi, J Gnabre, RCC Huang, FE Chou, Y Ito. J Liq Chromatogr 21:171–181, 1998.

47. Y Ma, Y Ito, AP Foucault. J Chromatogr A 704:75–81, 1995.

48. Y Ito. J Liq Chrom Rel Technol 22:2825–2836, 1999.

49. Y Ito. Anal Biochem 277(1):143–153, 2000.

50. Y Ito, J Suaudeau, RL Bowman. Science 189:999–1000, 1975.

51. J Suaudeau, Y Kolobow, R Vaillancourt, A Carvalho, Y Ito, A Erdmann III. Transfusion 18:312–319, 1978.

52. Y Ito. Sealless continuous flow centrifuge. In: B McLeod, TH Price, MJ Drew, eds. Apheresis: Principle and Practice. Bethesda, MD: AABB Press, 1997, pp. 9–13.

20

Gel Permeation Chromatography/ Size Exclusion Chromatography

Jack Cazes

Florida Atlantic University, Boca Raton, Florida

I. INTRODUCTION

Gel permeation chromatography/size exclusion chromatography (GPC/SEC) is a liquid column chromatographic technique that segregates molecules based on their effective molecular sizes in solution, i.e., their hydrodynamic volumes. The stationary phase particles (often polymeric, but sometimes based on silica) contain pores of various sizes. Solute molecules that are smaller than the available pores are able to enter the stationary phase particles and, thus, reside within the particles for a finite time, then leave the pores and enter the mobile phase where they can travel along the column, carried by the mobile phase, until they enter pores further along the column. Molecules that are larger than the available pores cannot enter the pore structure. They travel through the column exclusively in the mobile phase and elute first. Very small molecules can enter virtually every pore they encounter and, therefore, elute last. Intermediate-size molecules travel through the column at a rate that is in proportion to the fraction of the time they spend in the mobile phase. The sizes, and sometimes the shapes, of the mid-size molecules regulate the extent to which they can leave the mobile phase and enter the pores. So the largest molecules in a sample will elute first, followed by smaller molecules, sequentially, according to size, and finally by the smallest molecules. GPC/SEC is particularly well suited to separate high molecular weight substances, including biological macromolecules and synthetic polymers. The resultant data are generally used to calculate the molecular weight averages (number-average, weight-average, viscosity-average, z-average, z + 1-average) as well as the molecular weight distribution of the molecules contained in a sample.

A. Reminiscing

During the 1960s, this author was involved in developing new engineering polymers for industrial applications. As part of this work, there developed a need to visualize the molecular weight distributions of polymeric materials of interest as well as their molecular weight averages. These data are excellent predictors of the physical properties of polymers.

At that point in time, the molecular weight averages were determined by physical and chemical measurements involving the bulk polymer, e.g., osmometry, light scattering, end-group analysis, etc. Molecular weight distribution was obtained either by calculating ratios of various molecular weight averages (e.g., $\overline{M}_w/\overline{M}_n$, $\overline{M}_z/\overline{M}_w$, etc.) or by physically fractionating the polymer using a technique known as the Baker-Williams technique, which often involved a difficult, slow process with a superimposed temperature and solvent gradient. The bulk property measurements were crude estimates of molecular weight distribution, at best, and the Baker-Williams fractionation was slow and very difficult to control, often requiring a couple of days to a week for analysis of a single sample.

When a commercial GPC instrument came along, it revolutionized our approach to polymer characterization. We were able to obtain a fractionated molecular weight distribution in a couple of hours (later on, within minutes)—or so we thought! The GPC process was plagued with errors, mostly resulting from axial dispersion within the early GPC columns and difficulty in calibrating the columns in terms of molecular weight of a polymer of interest. These problems will be discussed later in this chapter.

The first attempts to calibrate the GPC column involved injection of a series of polymer samples, of the same type as the ones being analyzed, having very narrow molecular weight distributions; with a narrow distribution, it could be assumed (to a first approximation) that the known (measured) molecular weights of the standards would be located at the apex of the GPC curve for each injected standard sample. Thus, we would plot log(mol wt) versus retention volume at the apex for each standard, and obtain a useable calibration curve. This approach worked quite well (if axial dispersion was not too great), but there were no standards available for any polymer other than polystyrene. Therefore, this approach was very limited.

Later it was suggested that the GPC calibration curve for a polymer of interest could be obtained by using a calculation involving a "Q-factor." The Q-factor was a number that represented the length, in Angstroms, of the extended chain of the polymer of interest. The factor was initially estimated (guessed??) from molecular models held in their extended conformations, and later by comparison of GPC data for an unknown polymer sample with the GPC results for a polymer whose Q-factor was known. This, of course, was a very poor, artificial way to go, but it became somewhat useful for comparison of molecular weight distributions of polymer samples of the same type, but not for the determination of absolute molecular weights.

Eventually, newer, much better calibration techniques were developed which are used today for processing GPC/SEC data. They will be discussed later. However, even with its problems, the technique rapidly became the method of choice for determining molecular weights and molecular weight distributions of polymers.

B. History

Apparently, the first effective separation of polymers based on *gel filtration* chromatography appears to be the one reported by Porath and Flodin [1]. They employed insoluble, lightly cross-linked polydextran gels to separate various water-soluble macromolecules. Gel filtration generally employs aqueous solvents and hydrophilic column packings, which swell extensively in water. However, due to the fact that, at high flow rates and pressures, these lightly cross-linked soft gels have low mechanical stability and collapse easily, gel filtration is generally used with very low flow rates to minimize high backpressures, which would otherwise cause the gels to collapse. The technique is primarily used for fractionation of biomolecules in aqueous media [2].

Moore, at the Dow Chemical Company, described an improved polymer separation technique and called it gel permeation chromatography in 1964 [3]. GPC utilizes organic solvents and rigid hydrophobic stationary phases. Moore developed rigid polystyrene gels, cross-linked to varying degrees with divinylbenzene for separating synthetic polymers. These gels are able to withstand high pressures and flow rates. The more rugged GPC quickly flourished in industrial laboratories, where characterization of polymers is of primary concern [4].

Gel permeation chromatography has also become known as size exclusion chromatography, sieving chromatography, steric exclusion chromatography, etc. Many of the names reflect a particular facet of the perceived mechanisms for the separation; in this chapter, we will call the technique GPC/SEC. The reader is referred to the literature for a discussion of alternate names [5].

C. Mechanism

A very basic, simplistic mechanism was described above and in an early paper by Cazes [6]. The mechanism of the separation differs from many other types of liquid chromatography, which generally involve differential adsorption and desorption. GPC/SEC is a partitioning process in which solute molecules are partitioned between the liquid contained in the pores of the stationary phase and the mobile phase liquid. Although the liquid, in both cases, is the same one, there is a steric factor that controls entry into the pores and exit into the mobile phase. Barth et al. [7] discussed several theories of GPC/SEC separations.

The chromatogram shown in Fig. 1 represents a GPC curve that is often obtained for a polymer. The chromatogram is a broad "envelope" representing the range of molecular sizes

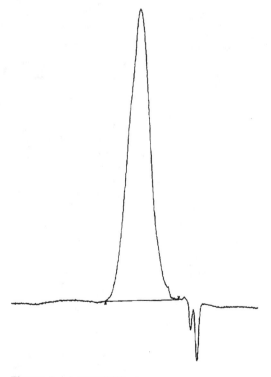

Figure 1 A GPC/SEC chromatogram.

present in the sample. An envelope is seen instead of individual peaks for the various species present in the polymer sample because of limited resolution and overlapping of the peaks belonging to individual species.

The weight-average molecular weight of a polymer is defined as [6]:

$$\overline{M}_w = \frac{\sum_{i=1}^{\infty} W_i M_i}{\sum_{i=1}^{\infty} W_i} \tag{1}$$

while the number-average molecular weight is defined as:

$$\overline{M}_n = W \Big/ \sum_{i=1}^{\infty} N_i = \sum_{i=1}^{\infty} M_i N_i \Big/ \sum_{i=1}^{\infty} N_i \tag{2}$$

where:

    W = total weight of polymer
    W_i = weight fraction of a given molecule, i
    N_i = number of moles of each species, i
    M_i = molecular weight of each species, i

The ratio M_w/M_n, known as the polydispersity, is an estimate of the molecular weight distribution for the polymer. A large polydispersity represents a broad molecular weight distribution. A polymer with a polydispersity approaching 1 primarily contains molecules having the same molecular weight.

D. Exclusion Limit

A given GPC/SEC column is limited to analyzing polymers for which the molecular sizes (proportional to molecular weights) of its component molecules fall within a limited range. A typical GPC/SEC calibration curve is shown in Fig. 2. The sharp slope of the calibration at the high

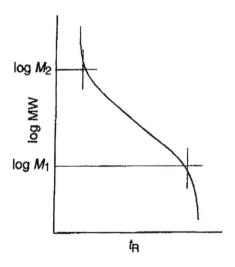

Figure 2   A typical GPC/SEC calibration curve.

molecular weight end is caused by a dramatic decrease of the magnitude of the partition coefficient with increasing polymer molecule dimensions beyond the sizes of the available pores in the column packing. Polymer chains larger than M_2 are larger than the largest available pores. Since it is physically impossible for these molecules to enter the stationary phase pores, at almost every plate in the column they are forced to remain in the mobile phase; they elute, with little or no separation, at the void volume of the column. The dependence of the partition coefficient on the sizes and shapes of the molecules leads to separation of polymer molecules of different molecular weights from each other and elution from the column at different times.

In actual practice, there is a small but finite amount of separation of molecules, which are totally excluded from the pores. This is not the result of a GPC/SEC mechanism; rather, it derives from the existence of a velocity gradient of the mobile phase and the population gradient of the polymer molecules that reside near the stationary phase particles. The linear velocity of the mobile phase is slower near the particle surface as a result of the "no-slip" boundary condition of the fluid at the surface of a solid. In fact, this mode of molecular fractionation is known as "hydrodynamic chromatography."

E. Effect of Intermolecular Interactions

It is often observed that, as the concentration of the polymer solution injected increases at a constant injection volume, the retention curve shifts toward a longer time [8]. This effect, called "overloading," is manifested universally when the mobile phase is a good solvent for the polymer being analyzed and becomes more significant as polymer molecular weight increases. Simple thermodynamics explains this effect [21]: As c_M (concentration in the mobile phase) increases, the solution starts to deviate from the ideally dilute solution, and it becomes difficult to neglect the second virial coefficient A_2. For a solution in a good solvent, a positive A_2 makes μ_M (the chemical potential in the mobile phase) larger than the chemical potential in the ideally dilute solution of the same concentration, as given by the following equation:

$$\mu_M = \mu^0 + k_B T \ln(c_M/c^0)$$

where μ^0 and c^0 are the chemical potential and the concentration, respectively, at an appropriate reference state, k_B is the Boltzmann constant, and T is the temperature.

The same change applies to μ_S. Since $c_S < c_M$, the increase in μ_M is greater than the increase in μ_S; this results in an increase in K, the partition coefficient. At higher concentrations, the polymer molecule is retained longer, thereby delaying the retention volume. The concentration at which the effect of A_2 becomes not negligible is around the overlap concentration c*, which can be defined as c*$[\eta] = 1$, where $[\eta]$ is the intrinsic viscosity. The actual concentration of c* is low, especially for high molecular weight polymers. This is why a concentration as low as 0.1 wt% sometimes exhibits a concentration-dependent chromatogram.

Intermolecular association is another phenomenon that leads to a change in the retention curve with concentration. In a solution in which A_2 is negative, for example, a solution in a poor solvent, a polymer molecule tends to associate with other molecules at increased concentrations. In effect, the pore in the GPC/SEC column "sees" suspensions of a larger dimension, causing the retention curve to shift to a shorter time.

II. BAND BROADENING

In GPC/SEC, band broadening is a complex process, more so than with classical chromatographic modes. We observe only an envelope resulting from many overlapping chromatographic

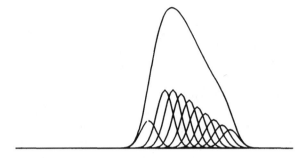

Figure 3 Chromatographic envelope observed for a group of overlapping, incompletely resolved, peaks.

peaks. This is illustrated in Fig. 3. Following the early work of Moore [8], Tung presented a general mathematical treatment of the problem of band broadening [9]. Because there was extensive band broadening with early low-efficiency GPC columns, several simple broadening correction procedures were suggested. But, later, during the 1970s, a significant improvement in column design and resolution rendered the peak broadening problem almost insignificant. As more and more detailed information was demanded from the GPC data, the discussion of broadening was renewed.

A. Experimental Estimation of Band Broadening

A simple approach to estimating the extent of band broadening is to work with a single, low molecular weight molecule which is, without doubt, eluted by a size separation (GPC) process. The number of theoretical plates, N, is determined from V_r, the retention volume at the apex of the chromatographic peak:

$$N = V_r/\sigma$$

and σ is the standard deviation that can be calculated from the width at 10% of the peak height:

$$\sigma = W_{0.1}/4.3$$

Octadecane (for organic systems) and saccharose (for aqueous systems) are the preferred solutes.

For polymers, one may use very narrow distribution standards at low concentrations where viscous effects are insignificant. Thus, for polymers <500,000 molecular weight, a concentration of <0.1 mg/mL should be used. Of course, the concentration must be decreased by an order of magnitude for polymers >500,000. With modern GPC/SEC columns, peaks are not truly Gaussian; rather, they are skewed Gaussian functions. Good results are observed for exponentially modified Gaussian functions [10]. Two parameters, σ and τ, define the shape of the peak and allow quantitative mapping of band spreading characteristics for subsequent correction.

For a polymer whose polydispersity approaches 1, and for which the effect of molecular weight distribution is negligible, the contributions of band broadening may be considered to be the result of

Extra-column effects
Eddy dispersion
Static dispersion
Mass transfer

The classical Van Deemter equation is not sufficient; it assumes that extra-column effects are negligible and peaks that are Gaussian. Observations indicate that peaks are skewed and, consequently, each contribution must be treated individually.

B. Extra-Column Effects

Often, extra-column effects are not considered in detail; rather, a chromatographic system is selected that makes these effects unimportant. If extra-column effects are to be considered at all, then the observed dispersion increases, generally, with high molecular weight solutes whose diffusion coefficients are small. Consider solutes passing through a length of instrument tubing, which is, actually, a cylindrical tube. Molecules entering near the center of the tube remain in the high-velocity area, and those that enter near the tube's wall reside in the low-velocity zone. This introduces skewing, which is most important for large, high molecular weight molecules.

C. Eddy Dispersion

This effect is the result of the tortuous path molecules must endure while traveling in the mobile phase as they pass through a packed column. The spaces (channels) through which the molecules must travel are the interstitial spaces between the particles of column packing material. These pathways correspond to a variety of shapes, lengths, and flow velocities. The resultant distribution is Gaussian and is somewhat dependent upon mobile phase velocity.

D. Static Dispersion

Dispersion of this type is caused by axial diffusion of the sample, due to Brownian motion, during the time of residence in the interstitial spaces of the column. The spreading is Gaussian, and its standard deviation, σ, is related to the diffusion coefficient, D, of the solute

$$\sigma = (2Dt_0)^{1/2}$$

In GPC/SEC of polymers, the diffusion coefficients are small and the spreading contribution becomes negligible.

E. Mass Transfer

It is necessary to account for the exchange that occurs between the pores and the interstitial space [11]. A model was suggested in which the pores are considered to be long cylinders, and residence time corresponds to a single dimension Brownian motion [12]. A mathematical treatment for this process was suggested by Karatzas and Shreve [13]. For a solute molecule, the number of visited pores corresponds to a Poisson distribution and, for a small number of visits, the elution volume distribution becomes broadened and skewed. This is in agreement with the observation that strong peak distortion is observed for molecules that elute near the exclusion limit.

III. CALIBRATION OF GPC SYSTEMS

A. Calibration Using Narrow Molecular Weight Distribution Standards

The most direct approach to calibration of GPC/SEC columns involves injection of a series of narrow molecular weight distribution (MWD) standard samples of the polymer of interest (i.e., the same polymer whose molecular weight is to be measured by GPC/SEC). The molecular weight range of the standard samples should be at least as broad as the distribution of the unknown polymer sample. The polydispersities of the standards should not exceed $M_w/M_n =$ 1.05, except for very high molecular weights ($>10^6$, where they may be 1.1 or sometimes as large as 1.2), and their peak molecular weights must be accurately known. As an alternative to experimentally determined peak molecular weights, root mean square molecular weights, i.e., the square root of ($M_w \times M_n$), of the standard samples may be used as a very close approximation of the peak molecular weight.

There is a very limited number of polymers for which narrow standards are available. They include polystyrene, poly-(methylmethacrylate), polyisoprene, polybutadiene, poly-(dimethylsiloxane), dextran, and a few others.

Polymer standard solutions must be extremely dilute to avoid concentration dependence of their retention volumes. Too high a concentration will lead to increased retention volumes. Generally, when high-efficiency GPC/SEC columns are used, the concentrations of the narrow standards should be $\leq 0.025\%$ for molecular weights greater than 10^6, $\leq 0.05\%$ between 10^6 and 10^5, and $\leq 0.1\%$ below 10^5. Multiple standard samples may be dissolved in the same solution only if their molecular weights are far enough apart to yield baseline resolution between the resultant peaks. Typically, two standards are injected for each decade of the calibration curve. All instrumental operating conditions must be the same during calibration as those used to analyze unknown polymer samples. An extremely constant flow rate is essential. Even slight uncorrected flow variations with invalidate the calibration. A small amount of a low molecular weight internal standard may be added to the polymer standard solutions to serve as a monitor of flow variations and to correct retention volumes for flow variations.

For each standard, the logarithm of the peak molecular weight (or the root mean square molecular weight) is plotted versus the retention volume (V_r), and a smooth line is drawn through the data points to produce the calibration curve (see Fig. 4). The resultant curve defines the exclusion limit, as well as the lower extremity of the resolving range of the column (often called the "total volume"), and the useful resolving range of the column. Often, the calibration curve is represented by an nth-order polynomial of the form:

$$\log MW = A + bV_r + cV_r^2 + dV_r^3 + \cdots$$

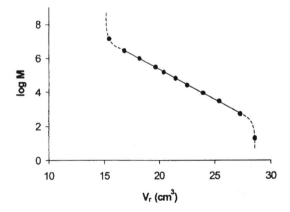

Figure 4 An example of a narrow calibration curve constructed with narrow molecular weight distribution polymer standards.

The coefficients are generally computed directly by regression on the experimental data points. It is rarely necessary to go beyond a third-order fit. Higher order equations can be very misleading since they often produce strangely oscillating functions, which make no sense from a chromatographic standpoint. Occasionally, higher order fits will lead to curves that actually double back on themselves, in a fashion similar to a "snake biting its own tail"!

Often several GPC/SEC column packing materials having different resolving ranges (i.e., ranges of available pore sizes) will be blended to yield a column with a linear calibration across a broad resolving range for the column. Using such columns guarantees equal resolving ability over several decades of molecular weight.

B. Universal Calibration of GPC/SEC Columns

Although calibration of GPC/SEC columns with narrow molecular weight distribution standards is useful and generally yields excellent results when used to determine molecular weights of polymeric substances, as we have commented previously, there are very few polymers for which such standards are available. Occasionally, a polymer sample having a broad molecular weight distribution will be fractionated using GPC/SEC to prepare such standards for a polymer of interest, but this is time consuming and often does not yield standards with a narrow enough molecular weight distribution to be useful for calibration purposes.

Early on it was recognized that the property of a polymer that is responsible for GPC/SEC separations is its effective molecular size in a dilute solution. Several size parameters have been suggested to describe the dimensions of polymer molecules, including the radius of gyration, end-to-end distance, mean external length, and others. However, we must consider that the polymer molecular size is a result of interactions of chain segments with the surrounding solvent in a dilute solution. Thus, polymer molecules may be represented as equivalent hydrodynamic spheres [14] to which the Einstein viscosity equation applies. It follows that the hydrodynamic volume of a polymer molecule is a function of its intrinsic viscosity and its molecular weight. This is expressed by the familiar Mark-Houwink expression:

$$[\eta] = KM^a$$

where K and a are Mark-Houwink constant and exponent, respectively, for the polymer at a specific experimental condition (solvent, temperature, etc.). These constants and exponents have been tabulated for a large number of substances [15]. Some of them are given in Table 1. The intrinsic viscosity is the value of the reduced viscosity (at infinite dilution, i.e., at zero concentration:

$$[\eta] = \lim(c \to 0) \frac{\eta_{sp}}{c}$$

where η_{sp} is the specific viscosity and c is the polymer concentration. The intrinsic viscosity is determined by measuring the specific viscosity at several concentrations, plotting the resultant data, and extrapolating the curve to zero concentration, i.e., where the line intersects the specific viscosity axis.

The use of $[\eta]M$ as the size parameter for a "universal GPC/SEC calibration" was first proposed by Benoit et al. [16] in 1967. They showed this approach to be a valid one for homopolymers and copolymers with various chemical compositions and geometric molecular shapes. Included in their data were linear, branched, comb-shaped, and star-shaped polymers. When their GPC/SEC retention volumes were plotted against the logarithms of their hydrodynamic

Table 1 Mark-Houwink Constants for Selected Polymers

Polymer	Temp (°C)	$K \times 10^2$ cm^3/g	a
Polystyrene	23	1.11	0.723
Poly(methyl styrene)	25	4.2	0.608
Poly(vinyl chloride)	25	1.50	0.77
Poly(vinyl acetate)	25	3.50	0.63
Poly(methyl methacrylate)	23	0.93	0.69
	25	1.08	0.702
Poly(ethyl methacrylate)	25	1.549	0.679
Poly(butyl methacrylate)	25	0.503	0.758
Poly(methyl acrylate)	25	0.388	0.82
Poly(isobutylene)	40	5.79	0.593
Polycarbonate	25	3.99	0.77
Polybutadiene	30	2.56	0.74
Polyisoprene	25	1.77	0.753
Butyl rubber	25	0.85	0.75
Poly(vinyl butyral)	25	1.4	0.80
Poly(2-vinyl pyridine)	25	2.23	0.66
Poly(dimethyl siloxane)	25	0.65	0.77
Cellulose nitrate	25	25.0	1.00
Poly(DL-lactic acid)	31.15	5.49	0.639
Poly(ethylene-co-vinyl acetate) (27–29% VA)	20	9.7	0.62
Poly(ethylene-co-propylene-co-ethylidene norbornene) (EPDM: 27% PP, 11.5%: ENB)	35	27.4	0.54

volumes (i.e., [η]M), all of the data points fell on a single calibration line, regardless of chemistry or geometry! Their universal calibration is shown in Fig. 5.

The use of the universal calibration requires a primary column calibration with narrow molecular weight distribution polymer standards in a manner as described in the previous section. Thus, for example, for GPC/SEC using tetrahydrofuran as the mobile phase, polystyrene (PS) standards are used, particularly since there are numerous suitable narrow molecular weight distribution polystyrene standards commercially available. A plot of log [η]$_{PS}$M$_{PS}$ vs. retention volumes, V$_r$, can be constructed. Average molecular weights and molecular weight distributions of any polymer sample eluted from the same column(s) via an exclusion mechanism, under the same experimental conditions, may be calculated since, at any retention volume, the relationship holds:

$$[\eta]_i M_i = [\eta]_{PS,i} M_{PS,i}$$

and, therefore,

$$M_i = [\eta]_{PS,i} M_{PS,i}/[\eta]_i.$$

If we substitute, in the denominator, the Mark-Houwink expression, [η] = KM^a, we obtain

$$M_i = ([\eta]_{PS,i} M_{PS,i}/K)^{(1/1+a)}$$

It is seen, then, that the molecular weight of each fraction may be obtained and average molecular weights may be calculated by use of the appropriate summations of the data for the fractions.

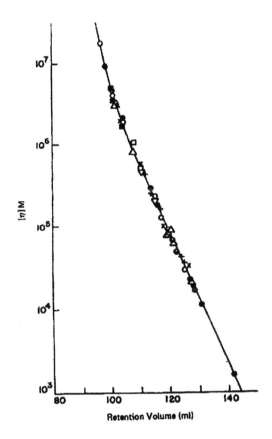

Figure 5 Universal calibration curve, based on the hydrodynamic volumes of the polymer standards. (From Ref. 16.)

Experimental details for the GPC/SEC determination of molecular weight averages and molecular weight distributions using the universal calibration concept are described in an official ASTM method [17]. Further details concerning validity of the universal calibration technique have been discussed by Dawkins [18].

IV. EFFECT OF EXPERIMENTAL CONDITIONS

A. Sample Concentration

Sample concentration is, perhaps, the most important operational variable in GPC/SEC. Retention volumes of polymers increase with increased concentration of the polymer in the sample solution. This concentration dependence is a well-known phenomenon, with the effect becoming more important with increasing molecular weight. It is not very important for polymers with molecular weights of $<10^4$ daltons. The increase in retention volume due to an increased concentration is usually attributed to a decrease in the hydrodynamic volume of the polymer in solution.

B. Injection Volume

The retention volume of a polymer increases with increased injection volume [19]. This suggests that a constant injection volume must be used when analyzing polymers by GPC/SEC, i.e., when constructing the calibration curve as well as when analyzing unknown polymer samples. The use of an injection valve that uses a loop to measure and deliver the sample solution to the column is highly recommended. Use of a loop that has a volume equal to the volume to be injected is the most foolproof way to operate a GPC/SEC system. If injection volumes are permitted to vary from injection to injection, significant errors will be introduced into the calculated molecular weight averages.

C. Flow Rate

Retention volume will increase with increasing flow rates. Equilibrium effects are responsible for this, since polymer diffusion in a GPC column between the internal pores and the mobile phase is slow enough so that equilibrium cannot be established at each plate in the column.

D. Column Temperature

The variations in retention volume with varying temperature are mainly due to (a) expansion/ contraction of the mobile phase in the column and (b) secondary effects involving the solute and the stationary phase. As mobile phase temperature changes and the liquid expands or contracts when it enters the column, the actual flow rate inside the column changes. This flow variation will be translated into errors in the estimated molecular weight averages, as discussed above. Accurate, precise molecular weight averages can only be obtained if column temperature is maintained constant throughout the entire system, during the entire analysis.

E. Additional Factors

Other factors that can affect retention volume and, consequently, molecular weight averages, include mobile phase viscosity, sizes of the available stationary phase pores, and effective hydrodynamic volumes of the polymer molecules. In practice, using modern instrumentation and columns, the effects of the first two are negligible. However, the effective size of solute molecules may change with a change in temperature. Intrinsic viscosity changes resulting from temperature changes were studied [20]. In many cases, the effects were significant.

V. ESTIMATION OF INTRINSIC VISCOSITY BY GPC/SEC

The intrinsic viscosity, $[\eta]$, of a polymer is often used to characterize and to predict processing and end-use properties [22]. It represents the viscosity a polymer would exhibit if it were possible to make the measurement at infinite dilution, i.e., approaching zero concentration of the polymer in solution, where polymer molecules do not interact with each other.

GPC data can be readily converted to intrinsic viscosity, since GPC analysis of polymers is generally performed in very dilute solution and because GPC results and viscosity are both functions of the effective molecular size in solution, i.e., the hydrodynamic volume, of the polymer. The intrinsic viscosity, $[\eta]$, of a polymer may be estimated using the familiar Mark-Houwink relationship:

$$[\eta] = KM^a$$

in which K generally assumes values between 10^{-1} and 10^{-6}, and a usually falls between 0.5 and 0.9. Values of K and a, which are dependent upon polymer type, solvent condition, and temperature, may be found in Ref. 15; a few are given in Table 1.

The viscosity-average molecular weight may be calculated from GPC data with the following equation:

Table 2 GPC/SEC vs. Viscometric Intrinsic Viscosities
of Polymers

Polymer sample	Intrinsic viscosity [η]		Ref.
	Viscometric	From GPC	
Polystyrenes			
2,000,000[a]	3.92	4.16	2
670,000[a]	1.79	1.73	2
411,000[a]	1.28	1.23	2
411,000[a]	1.24	1.16	3
258,000 (NBS-706)	0.931	1.00	2
179,000 (NBS-705)	0.740	0.728	2
160,000[a]	0.62	0.68	2
160,000[a]	0.627	0.582	3
97,200[a]	0.424	0.404	3
51,000[a]	0.28	0.24	2
51,000[a]	0.265	0.269	3
19,800[a]	0.138	0.136	2
19,800[a]	0.142	0.155	3
10,300[a]	0.086	0.085	3
Broad MWD #1	0.746	0.799	3
Broad MWD #2	0.970	0.976	3
Broad MWD #3	1.01	1.01	3
Broad MWD #4	0.806	0.810	3
Polycarbonate			
A	0.511	0.516	2
B	0.544	0.544	2
Polyethylene			
NBS-1475 (linear)	1.01	1.00	2
NBS-1476 (branched)	0.902	0.899	2

[a] Nominal molecular weight.
Source: Ref. 23.

$$M_v = \left[\frac{\sum X_i (M_i)^a}{\sum X_i} \right]^{1/a}$$

where X_i is the height or area of each incremental slice of the experimental GPC distribution. Some results are given in Table 2 [23]. Agreement with viscometrically determined intrinsic viscosities is excellent.

VI. POLYMER DEGRADATION IN GPC/SEC COLUMNS

It is often observed that ultra–high molecular weight polymers will degrade, by shearing or elongational forces, as they pass through tightly packed GPC/SEC columns. The highest molecular weight polymer that can be analyzed by GPC/SEC without degradation occurring depends upon the width of the molecular weight distribution of the polymer, upon the columns used, and upon the experimental conditions, especially the linear velocity of the mobile phase.

Whereas narrow MWD polymers of 10^7 molar mass have been successfully analyzed, if the sample has a broad MWD, the upper limit is somewhere below 10^6.

Degradation of ultra–high molecular weight polymers is generally accompanied by concentration effects, anomalous flow patterns, poor column resolution, and extremely poor reproducibility. Often, when polymer degradation occurs in the column, repetitive injections of the same sample will lead to chromatograms that actually look different from each other. Peak shapes of degraded samples are often abnormally broad, with severe tailing. In some cases, when dealing with polymers whose degradation products are relatively polar, nonsteric, adsorptive effects are evident.

It is sometimes seen that very large polymeric components of a solution of an ultra–high molecular weight sample will actually be retained on the inlet frit of the GPC/SEC column. Of course, this will be accompanied by increased backpressure—sometimes gradually, other times catastrophic.

It has been suggested that, to avoid degradation, one should use a "theta" solvent for the polymer as the mobile phase. The dimension of a macromolecule depends upon the degree to which the polymer's chains are solvated and/or extended/contracted. Ideal theta solvents are known for many polymers. For instance, cyclohexane at 34°C is a theta solvent for polystyrene. The use of ideal solvents has been generally avoided, perhaps due to the fact that use of such solvents can lead to absorption of the polymer onto the stationary phase. For GPC/SEC analysis of ultra–high molecular weight polymers, one should use specially designed GPC/SEC columns with large particle sizes and ultra–large pore sizes. One should also work at low enough flow rates, where shear forces are minimized.

VII. APPLICATIONS

A. Nonionic Surfactants

Nonionic surfactants are amphiphilic molecules composed, most often, of poly(oxyethylene) blocks as the water-soluble group, and fatty acids, alkylphenols, or various synthetic polymers as the hydrophobic segment. They are available under many different trade names, e.g., Tween, Igepal, Brij, Pluronic, Triton, and others. Good solvents for these surfactants, which are often used for GPC/SEC analysis, are tetrahydrofuran and chloroform. A differential refractive index detector is generally used. Columns are, most often, polystyrene/divinylbenzene cross-linked matrix, e.g., Styragel (Waters, USA), Phenogel (Phenomenex, USA), PSS Gel (Polymer Standards Co., Germany), and PL-Gel (Polymer Labs, UK).

The use of a universal calibration with on-line viscometric detection is highly recommended. Direct use of a calibration constructed with narrow MWD polystyrene standards has sometimes led to low apparent molecular weights. The universal calibration approach has yielded accurate molecular weight information for most linear and comb-shaped graft copolymers of poly(oxyethylene).

B. Polyesters

Thermoplastic polyesters are very useful as engineering polymers because of their outstanding chemical and thermal resistance. With glass fiber reinforcement or mineral fillers, polyesters can be used to replace metals, ceramics, and other engineering materials. Aromatic polyesters, due to their crystalline structure and polarity, require hostile solvents and/or high temperatures to dissolve the polymer [24]. Early work with polyesters employed *m*-cresol or *o*-chlorophenol as the mobile phase. Due to their high viscosities, temperatures as high as 135–145°C had to be used; polymer solutions allowed to stand too long often led to some hydrolytic degradation of the polymers [25]. More recently, hexafluoroisopropanol (HFIP) has been successfully used as a room-temperature solvent for both polyesters and polyamides [26]. The solvent is extremely expensive; therefore, it is generally repeatedly recovered from the eluate by distillation.

It should be noted that absolute molecular weights have proven to be inaccurate when using HFIP as the mobile phase, probably due to non–size exclusion interactions occurring within the GPC system. Most often, relative molecular weight comparisons are reported when HFIP is used. Absolute molecular weights can be obtained with light-scattering detection.

GPC/SEC is routinely used to evaluate new polyesters and for monitoring their weatherability in outdoor applications.

C. Polycarbonates

Polycarbonates are noncrystalline thermoplastics with outstanding toughness, impact strength, clarity, and heat deflection temperatures. Generally, they are linear aromatic polyesters of carbonic acid formed by the combination of difunctional phenols (called bisphenols) and carbonate linkages.

Direct calibration may be performed using commercially available polycarbonate standard samples (Polymer Standard Service Co., Mainz, Germany; American Polymer Standards Co., Mentor, OH).

D. Polyamides

Synthetic polyamides exhibit outstanding engineering properties, including high strength, toughness, stiffness, abrasion resistance, and retention of physical and mechanical properties across a broad range of temperatures. It is their semi-crystalline morphology and the intermolecular hydrogen bonding of their amide groups that are responsible for their outstanding engineering properties.

Aliphatic polyamides are generally known as Nylons, whereas aromatic polyamides, or aramids, are marketed as Nomex and Kevlar.

High cohesive strength makes many polyamides very solvent resistant; as with polyesters, they can only be dissolved under very aggressive conditions. Successful GPC/SEC analysis requires a solvent that is a good one for the polymer, chemically inert, and compatible with the stationary phase. Several high-temperature solvents have been used for GPC/SEC analysis, e.g., m-cresol at 135°C, o-chlorophenol at 100°C, hexamethylphosphoramide at 85°C, and benzyl alcohol at 130°C. Often, when using such mobile phases, degradation and non–size exclusion effects become significant.

Hexafluoroisopropanol and trifluoroethanol have been successfully used at room temperature for selected polyamides. Recently, a room temperature eluent comprising methanesulfonic acid +5% methanesulfonic anhydride +0.1 M sodium methanesulfonate has been recommended for analysis of Kevlar and Technora polyester fibers. Mixtures of HFIP with chloroform and dichloromethane have been successfully tested as GPC/SEC mobile phases.

E. Natural Rubber

Natural rubber is a very high molecular weight polymer with a very complex microstructure, which is gradually destroyed when the polyisoprene is dissolved in a solvent. Often a portion of the rubber remains insoluble; this insoluble material is called a "gel" or "macrogel." The soluble polyisoprene fraction often contains low levels of aggregates comprising "microgel."

In most cases, tetrahydrofuran is used as the mobile phase, although occasionally, cyclohexane has been used [27]. Since cyclohexane does not require an added stabilizer, use of this solvent permits the use of UV detection at 220 nm. Sensitivity of this detector is much greater than a differential refractometer; very low concentrations may, therefore, be used to avoid deleterious concentration effects. Light-scattering and viscometric detectors can be used to study branching rates of these very high molecular weight polymers.

Calibration in cyclohexane requires use of polyisoprene standards. Calibration with polystyrenes in tetrahydrofuran generally leads to overestimated molecular weights. Thus, in THF it is necessary to perform universal calibration to obtain valid results.

VIII. SUMMARY

GPC/SEC analysis of polymeric substances, both natural and synthetic, has become the method of choice to elucidate the molecular weights and molecular weight distributions of these high molecular weight substances. With the advent of this technique, under suitable experimental conditions, one can readily fractionate a polymer, obtain an actual "picture" of its mass distribution, and see if it is unimodal, bimodal, or multimodal. Before GPC/SEC, one would be required to depend on measurements made with a bulk, unfractionated polymer and then estimate the shape of the mass distribution function for the polymer. With GPC, with suitable massaging of the data, we actually see the distribution, and we are able to mathematically reassemble the separated GPC/SEC fractions in any manner that fulfills our need for information.

REFERENCES

1. J Porath, P Flodin. Nature 183:1657–1659, 1959.
2. AV Danilov, IV Vagenina, LG Mustaeva, SA Moshnikov, EY Gorbunova, VV Cherskii, MB Baru. J Chromatogr A 773:103–114, 1997.
3. JC Moore. J Polym Sci Part A 2:835, 1964.
4. VS Lafita, Y Tian, D Stephens, J Deng, M Meisters, L Li, B Mattern, P Reiter. Proc. Int. GPC Symp. 1998, Waters Corp., Milford, MA, 1998, pp. 474–490.
5. J Johnson, R Porter, M Cantow. J Macromol Chem Part C 1:393–434, 1966.
6. J Cazes. J Chem Educ 43:A567, 1966.
7. HG Barth, BE Boyes, C Jackson. Anal Chem 70:251R–278R, 1998.
8. JC Moore. J Polym Sci A2:835, 1964.
9. LH Tung. J Appl Polym Sci 10:375, 1966
10. MS Jeansonne, JP Foley. J Chrom Sci 29:258, 1991.
11. DH Kim, AF Johnson. Computer model for GPC of polymers. In: T Provder, ed. Size Exclusion Chromatography. ACS Symposium Series. Washington, DC: American Chemical Society, 1984.
12. JP Busnel. Band broadening in size-exclusion chromatography. In: J Cazes, ed. Encyclopedia of Chromatography. New York: Marcel Dekker, Inc., 2001, p. 79.
13. I Karatzas, S Shreve. In: Brownian Motion and Stochastic Calculus. Springer-Verlag, New York, 1991.
14. PJ Flory. Principles of Polymer Chemistry. Ithaca, NY: Cornell University Press, 1953.
15. M Kurata, Y Tsunashima, M Iwama, K Kamada. In: J Brandrup, EH Immergut, eds. Polymer Handbook. New York: John Wiley & Sons, 1975, pp. 1–60.
16. Z Grubisic, P Rempp, H Benoit. J Polym Sci B 5:753, 1967.
17. ASTM D3593-80. Standard Test Method for Molecular Weight Averages and Molecular Weight Distribution of Certain Polymers by Liquid Size-Exclusion Chromatography (Gel Permeation Chromatography-GPC) Using Universal Calibration. American Soc. for Testing & Materials, West Conshohoken, PA.

18. JV Dawkins. Calibration of separation systems. In: J. Janca, ed. Steric Exclusion Chromatography of Polymers. New York: Marcel Dekker, Inc., 1984, pp. 53–116.
19. S Mori. J Appl Polym Sci 21:1921, 1977.
20. S Mori, M Suzuki. J Liq Chromatogr 7:1841, 1984.
21. I Teraoka. Progr Polym Sci 21:89, 1996.
22. MY Hellman. Paper presented at the International Symposium on Liquid Chromatographic Analysis of Polymers & Related Materials, Houston, Texas, 1976.
23. J Cazes, RJ Dobbins. Polym Lett 8:785–788, 1970.
24. S Mori. Anal Chem 61:1321, 1989.
25. WW Yau, JJ Kirkland, DD Bly. Modern Size Exclusion Chromatography. New York: John Wiley & Sons, Inc., 1979, pp. 390–394.
26. TQ Nguyen. Paper presented at the International GPC Symposium '98, 1998, pp. 135–153.
27. H Bartels, ML Hallensleben, G Pampus, G Schulz. Angew Makromol Chem 180:73, 1990.

21

From Metal Organic Chemistry to Chromatography and Stereochemistry

Volker Schurig

Institute of Organic Chemistry, University of Tübingen, Tübingen, Germany

I. THE STONY WAY TO THE FIRST ENANTIOMERIC SEPARATION BY COMPLEXATION GAS CHROMATOGRAPHY— RAMIFICATIONS HIDDEN IN THE ANNALS

My involvement in chromatography started in 1969 by mere coincidence. I had been educated in chemistry at Tübingen University in Germany. I joined the group of Ernst Bayer in the Department of Organic Chemistry in 1966, yet my thesis work was purely inorganic, i.e., nitrogen fixation and reduction to ammonia with titanium catalysts [1,2]. This challenging topic, i.e., to conduct the Haber-Bosch process under physiological conditions as done by nature, is still elusive at the beginning of the new millennium. My supervisor Ernst Bayer embarked on a very broad spectrum of scientific endeavors [3]. His experiments on the extraction of gold from seawater, revisiting the unsuccessful experiments of Nobel laureate Fritz Haber after World War One in an effort to ease Germany's war debts, earned him a Robert A. Welch professorship at the University of Houston in Texas. At about the same time Emanuel Gil-Av (Zimkin) [4,5] worked together with Binyamin Feibush and Rosita Charles-Sigler at the Weizmann Institute of Science, Rehovot, Israel, on another highly challenging topic, namely on the separation of optical isomers (enantiomers) employing chiral amino acid derivatives as chiral stationary phases (CSPs) by gas chromatography. According to Emanuel Gil-Av [6], this topic was in 1966 in "a state of frustration: nobody believed it could be done. In fact, people were convinced that there could not possibly be a large enough difference in the interaction between the D- and L-solute with an asymmetric solvent. This was the feeling people had, even those known as unorthodox thinkers." Nowadays the situation is almost reversed. There are only very few classes of chiral compounds that are not yet amenable to separation into enantiomers by chromatography. However, the way to this significant achievement of separation science has been cumbersome, at least in the experience of the present author.

When Emanuel Gil-Av et al. published the first example of a fully reproducible separation of derivatized amino acid enantiomers by GC in 1966 [7], the Manned Spacecraft Center at

NASA invited him to Texas to determine organogenic elements and compounds from lunar materials retrieved by the crew of Apollo 11 from the surface of the Sea of Tranquility [8] (with J. Oró et al.) in an effort to find chiral amino acids in extraterrestrial space using the exciting novel tool of enantiomer analysis (the result turned out to be negative at the nanogram detectability range). In the laboratory of Albert Zlatkis at the University of Houston, Emanuel Gil-Av and Ernst Bayer got to know each other. Since Bayer was also an established peptide chemist, he supplied some di- and tripeptides to Gil-Av for use as chiral stationary phases. Coworkers of E. Bayer from Tübingen, W. Parr and W. A. König, conducted experiments in Houston, the name of the latter being invariably connected with important work in the area of chiral separations ever since. I entered the stage when Emanuel Gil-Av (Fig. 1) looked for a young postdoctoral fellow for investigations on enantiomeric separation at the Weizmann Institute in Israel. I finished the Ph.D. in 1968, and when reading the lucid invitation letter of Emanuel Gil-Av, I was immediately struck by this intriguing topic, although many contemporary scientists were still sceptical or even ignorant about the importance and relevance of such work, culminating in remarks such as: "what for separating identical compounds such as optical isomers"! It was clearly overlooked until the 1970s that enantiomers possess different chemical and physical properties in a chiral biogenic environment, and, after the thalidomide tragedy, drug stereochemistry became an eminently important topic in contemporary research. Thus, according to new estimates, the chiral technology today amounts to $150 billion for drugs, $8 billion for agrochemicals, and $150 millions for intermediates per annum with a 15% annual growing rate (J. Winkler, personal communication, 1999, Business Communications Co., Inc., Norwalk, CT).

Figure 1 Ernst Bayer, Emanuel Gil-Av, and Volker Schurig (in the back, Vadim Davankov) at the opening ceremony of the 3rd International Symposium on Chiral Discrimination, Tübingen, 1992.

But back to 1969: E. Gil-Av asked me to synthesize the tripeptide (L-Val)$_3$ and its enantiomer (D-Val)$_3$ as mirror image CSPs. My Tübingen colleague Günther Jung taught me peptide synthesis and protecting group (BOC) chemistry. Since D-valine was very expensive I rehearsed the synthesis of the tripeptide many times with natural L-valine. Yet for an inconceivable reason I failed in the mirror-image chemistry when preparing the tripeptide using unnatural D-valine, and, very frustrated, I decided to drop peptide chemistry and return to my original metal organic roots. E. Gil-Av and J. Shabtai had previously extensively investigated the separation of olefins and aromatics in oil shale deposits of the Dead Sea area as a possible source of liquid fuel for the State of Israel. They used GC columns coated with silver nitrate in ethylene glycol for selective separations of isomeric olefins including geometric *cis/trans*-alkenes ("argentation gas chromatography"). Thus, I decided to try to separate for the first time a chiral olefin on an optically active metal coordination compound, a method later coined "enantioselective complexation gas chromatography" [9]. For this purpose palladium(II) and platinum(II) complexes containing chiral phosphite ligands were prepared. These efforts coincided with the development of chiral rhodium(I) phosphane complexes for asymmetric hydrogenation of prochiral olefins. However, the enantioselective approach totally failed in chromatography. The compounds prepared did not cause any selective π-complexation with olefins, and one apparent resolution into two peaks was just caused by olefin isomerization.

A breakthrough came when B. Feibush suggested the use of optically active acylated terpeneketonates, such as 3-trifluoroacetyl-1R-camphor (TFA-Cam) as chiral ligand. I found a simple method for the preparation of metal chelates of TFA-Cam via a barium intermediate ("the barium method") [10], which was later extended to the synthesis of chiral paramagnetic NMR shift reagents [11], and I prepared dicarbonylrhodium(I)–, nickel(II)–, and europium(III)–TFA-Cam chelates, which were used as selective stationary phases in GC after dissolution in squalane.

Ln = Eu, Dy, Yb   M = Mn, Co, Ni   M = Mn, Co, Ni

Unfortunately, these chiral stationary phases did not separate any racemic mixtures that I probed during my time at the Weizmann Institute in 1969–71. In retrospect, it was a period of missed opportunities, delaying the field for almost 6 years. As it was found much later, the selection of the metal complexes was fortunate but the random choice of racemic test compounds was ill-fated and the GC instrument, a Barber-Colman running with a tritium β-ionization cell fed by argon as detector, was archaic. Nevertheless, by working with selective rhodium(I)-containing GC stationary phases, very useful achiral applications [12] emerged, and the lanthanide tris-chelates of TFA-Cam became very versatile paramagnetic shift reagents in NMR spectroscopy [11] including the first use for a chiral metal complex, i.e., tetracarbonyliron fumarate [13]. Incidentally, this advance was extended to the concept of heterotopism (homo-, enantio-, and diastereotopic positions) in metal organic compounds [14], which found its way into stereochemistry textbooks.

In 1971 Albert Zlatkis invited me to the University of Houston to continue the investigations on enantiomeric separation by complexation gas chromatography (see Sec. V). I thoroughly checked the chemistry store for racemic chiral compounds, but the winner target was not among them. After my return to Tübingen, I changed the research topic. Ernst Bayer, previously engaged with the Merryfield synthesis of peptides, invented the liquid phase method. The idea consisted of using a soluble carrier for the growing peptide chain. Because of my knowledge of transition metal chemistry, we combined efforts and set out to introduce the concept of polymeric homogeneous metal catalysis. This approach consisted of utilizing a linear phosphine-modified polystyrene as ligand for rhodium(I) and separating the product from the polymeric catalysts by membrane filtration in solution [15]. Interestingly, this concept has been reestablished very recently [16]. In further activities, my first research student, C. Mark, developed chiral "binap"-ligands for asymmetric hydrogenation of prochiral amino acids precursors [17]. The gas-chromatographic methodology based on the Gil-Av approach and developed further by Bayer was at our disposal, while other groups still used polarimetry for the determination of optical purity at this time.

II. THE FIRST ENANTIOMERIC SEPARATION OF AN OLEFIN BY COMPLEXATION GAS CHROMATOGRAPHY ON A CHIRAL RHODIUM(I) COORDINATION COMPOUND—AN ALMOST UNEXPECTED RESULT AFTER AN 8-YEAR DEADLOCK

In my Habilitation thesis I presented in 1975 a comprehensive study of stability constants of 40 acyclic and cyclic C2–C7 olefins and the rhodium(I) and rhodium(II) coordination compounds as well as the classical silver nitrate/ethylene glycol system based on the concept of the retention increase R' [18].

R = phenyl

Rh(I)   Rh(II)   Ag(I)

A striking difference in the π-complexation selectivity due to steric, strain, and electronic effects was observed between the three metal ions Rh(I), Rh(II), and Ag(I) [18]. In the course of these investigations a remarkable correlation between the rate of rhodium(I)-catalyzed hydroformylation of monoolefins and the measured stability constants on Rh(I) was found [19]. The extraordinary selectivity of the rhodium(I) complex dissolved in squalane toward olefins is

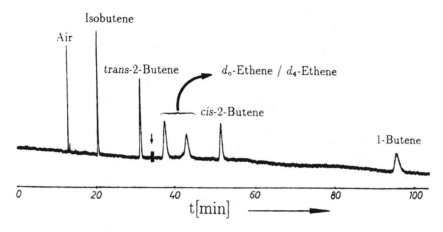

Figure 2 Pronounced chemoselectivity toward olefins in complexation gas chromatography. Column: 100 m × 0.25 mm (i.d.) nickel capillary coated with 0.06 m dicarbonyl-rhodium(I)-3-TFA-camphorate in squalane at 25°C; carrier gas: 40 psig nitrogen. (From Ref. 20.)

shown in Fig. 2. Noteworthy is the extreme separation factor between isobutene and 1-butene (the isomers with the same boiling point cannot be separated on pure squalane). Also, ethene is eluted after n-pentane [20]! Due to an inverse isotopic effect, the rhodium(I) complex could quantitatively separate all deuterated ethylenes C_2H_4, C_2H_3D, $C_2H_2D_2$, C_2HD_3, and C_2D_4 (see Fig. 3, top) [21].

Since terminal olefins showed the strongest interaction with rhodium(I), I selected the smallest chiral aliphatic olefin, i.e., 3-methyl-1-pentene, as a test substrate for enantiomeric separation. Nowadays, however, it is well recognized that enantioselectivity is independent of the strength of chemical interaction between the selectand and the selector. Frequently enantiomers are separated despite the fact that the association with the chiral selector is marginal, e.g., due to steric hindrance, which is important for chiral recognition. This earlier misconception delayed the finding of the first case of enantioselective complexation gas chromatography! The goal was virtually abandoned after 6 years of unsuccessful experiments. Yet a suspicious peak-splitting was unexpectedly observed for the chiral cyclic olefin 3-methylcyclopentene during the determination of stability constants on the rhodium(I) complex [19]. This coincidental finding created an exciting activity in the lab, and within 48 hours I was able to confirm the first enantiomeric separation by complexation gas chromatography [22]. To exclude an artefact, another specimen of the target olefin was ordered (by express service), which indeed exhibited the same resolution. To further prove separation of optical isomers, two important control experiments were devised: the (R)-enantiomer of 3-methylcyclopentene was prepared together with the selector having the opposite chirality obtained from unnatural (−)-camphor using the fast "barium method" [10] within a day. Indeed, peak inversion (reversal of the elution order) was observed when the opposite configurated selector was used. Then it was reasoned that on the racemic CSP comprising equimolar amounts of the (+)- and (−)-selector peak coalescence should commence. In the course of this decisive control experiment (which turned out to be positive as only one peak was observed), a very unexpected discovery relevant to solid-state stereochemistry was made by mere serendipity. When the racemic dicarbonyl-rhodium(I)-TFA-($1R,1S$)-camphorate was prepared by the same 'barium method' [10], a deep red-green dichroic solid was obtained while the enantiomers of the same complex prepared just

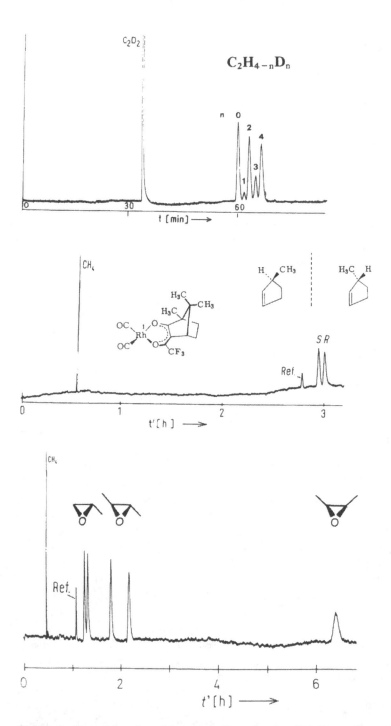

Figure 3 (Top) Separation of deuterated ethenes by complexation gas chromatography. Column: 200 m × 0.5 mm (i.d.) stainless steel capillary coated 0.15 m dicarbonyl-rhodium(I)-3-TFA-camphorate in squalane at 22°C; carrier gas: 1.2 mL nitrogen/min. (From Ref. 21.) (Middle) First enantiomer separation of a chiral olefin (3-methylcyclopentene) by complexation gas chromatography on 0.15 m dicarbonyl-rhodium(I)-3-TFA-(1R)-camphorate in squalane (conditions as stated above). (From Ref. 22.) (Bottom) Separation of enantiomers and *cis/trans*-diastereomers of aliphatic oxiranes by complexation gas chromatography. Column: 200 m × 0.5 mm (i.d.) nickel capillary coated with 0.15 m nickel(II) bis[3-TFA-(1R)-camphorate] in squalane at 50°C; carrier gas 2.1 mL/min nitrogen. (From Ref. 28.) The long retention times invoked in these historical chromatograms were later reduced to minutes.

beforehand consisted of yellow crystals. Yet all spectroscopic data were alike. Moreover, when the yellow crystals of the *R* and *S* enantiomers were mixed, the red color of the *R,S* racemate was gradually formed in a solid-state reaction. I was not aware of any previous example in the literature in which pure enantiomers possessed a different color than the racemate. Indeed, this observation, called chirodichroism, aroused the interest of Kurt Mislow, and it was the starting point of another research project devoted to chiral phenomena in the solid state [23]: a columnar structure with close metal-metal contacts and leading to dichroism is built up in the racemate, while only a zig-zag arrangement rather than a metal chain is formed in the single-handed enantiomers.

The gas-chromatographic enantiomer separation of 3-methylcyclopentene immediately led to interesting applications, i.e., the first investigation of the kinetic resolution of an olefin in the presence of a chiral rhodium catalyst [24]. Together with Emanuel Gil-Av it was shown that the specific rotation of 3-methylcyclopentene was more than double that reported in the literature, and based on the new value, chirooptical calculations by Brewster as well as the enantiomeric excess obtained via asymmetric borane chemistry of H. C. Brown had to be revised for this chiral olefin [25]. Incidentally, the first observed enantiomer separation of an olefin on a rhodium(I) coordination was mainly due to the high efficiency of the capillary column used (see Fig. 3, middle). Based on an observed separation factor of $\alpha = 1.02$ at room temperature, the enantioselectivity (as expressed by $-\Delta_{R,S}(\Delta G)$) amounted to just 0.02 kcal/mole! Although the scope of enantiomeric separation of olefins on rhodium(I) coordination compounds was limited, the finding commanded academic interest in view of the increasing activities in rhodium(I)-catalyzed asymmetric transformations of prochiral olefins at this time [26,27]. Moreover, this kind of research attracted my first and very gifted research students C. Mark, B. Koppenhöfer, R. Weber, K. Hintzer, and D. Wistuba, who contributed greatly to the further development of enantioselective complexation gas chromatography, described in the following sections.

III. EXTENSION OF THE SCOPE OF ENANTIOSELECTIVE COMPLEXATION GAS CHROMATOGRAPHY TO OXYGEN-, NITROGEN-, AND SULFUR-CONTAINING RACEMIC COMPOUNDS

The scope of separation of enantiomers by complexation gas chromatography was extended to chiral oxygen-, nitrogen-, and sulfur-containing compounds using various chiral 1,3-diketonate *bis* chelates of manganese(II), cobalt(II), and nickel(II) derived from perfluoroacylated terpene-ketones such as camphor, 3- and 4-pinanone, thujone, nopinone, menthone, isomenthone, carvone, and pulegone [28–30]. The metal chelates were able to separate one of the smallest chiral class of compounds, namely aliphatic aziridines, oxiranes, and thiiranes. The enantiomeric separation of methyloxirane and *trans*-2,3-dimethyloxirane together with the diastereomeric separation of *cis/trans*-2,3-dimethyloxirane is shown in Fig. 3 (bottom) [28].

The possibility of determining the enantiomeric composition of chiral oxiranes in our laboratory gave us a lead in the study of metal-catalyzed asymmetric oxidations. Thus the first enantioselective epoxidation of aliphatic olefins by a chiral molybdenum-peroxo complex with enantiomeric excesses up to 40% was discovered by C. Mark in cooperation with Kagan and Mimoun [31], followed by investigations on the kinetic resolution of the oxiranes by the same molybdenum complex doped with a chiral diol by F. Betschinger [32]. These innovative advances were milestones for the well-known systems titanium/tartrate (Katsuki-Sharpless epoxidation) and manganese(II)-salen (Jacobson kinetic resolution), used worldwide in the synthesis of enantiomers. The analytical tool available to us also proved invaluable in a comprehensive investigation of the cytochrome-P450–mediated epoxidation of xenobiotic simple olefins and kinetic resolution of the oxiranes formed by epoxide-hydrolase and glutathione-S-transferase [33,34], as well as in the synthesis of highly enantiomerically enriched aliphatic oxiranes by the "chiral pool" approach [35].

The enantioselective system chiral oxirane/chiral metal coordination compound turned out to be of interest in its own right. The main principles of enantioselective chromatography could be studied in this simple system. Thus, four different enantioselective processes were discovered in complexation gas chromatography, including enantiomerization phenomena (see Sec. V), kinetic resolution accompanying chromatographic resolution, and enthalpy/entropy compensation (see Section VI) [36,37], and as many as six peak coalescence phenomena have been distinguished and verified in enantioselective chromatography [36,38].

Before the advent of modified cyclodextrins, enantioselective complexation gas chromatography represented a very useful tool for chiral analysis of volatile compounds such as flavors [39–41] and insect pheromones [42]. The enantiomeric purity of natural pheromones and of specimens prepared in our group by K. Hintzer and in other laboratories (W. Francke, K. Mori) were readily determined by complexation gas chromatography. It turned out that the synthetic products often had a higher enantiomeric purity compared to the natural pheromones, which rarely exceeded the margin of a 99% enantiomeric excess! A typical application related to the spiroketal olean, the principal pheromone of the olive fly, a serious pest in the Mediterranean, will be recalled here. In biological trials performed in Greece, G. Haniotakis found that males responded only to the (R)-enantiomer, serving as a sex attractant, whereas females responded just to the (S)-enantiomer, functioning as a short-range arrestant throughout the day and as an aphrodisiac in the process of mating. In order to investigate the stereochemistry of the natural pheromone, W. Francke, who collaborated on the project, brought a dry-ice–cooled n-pentane solution of the charcoal-absorbed volatiles from a stripping experiment involving 4000 living females within the shortest possible time (since the spiroketal may be prone to racemization) from Greece via Bulgaria to the Stuttgart airport. To our surprise, the sample contained only *racemic* olean. Obviously, nature employs stereochemistry in a very economic fashion by using the racemic mixture whereby each single enantiomer elicits different responses in the two sexes [43].

In the mid-1970s, two timely approaches to enantioselective gas chromatography were pursued independently at Tübingen University, i.e., complexation gas chromatography in my group and hydrogen-bonding gas chromatography in the group of Ernst Bayer. In a pioneering advance, Frank, Nicholson, and Bayer linked the valine diamide selector of Gil-Av and Feibush to a polysiloxane-copolymer [44]. The resulting *chiral* poly (*si*loxane) containing *val*ine (Chirasil-Val), commercialized by Chrompack International, Middelburg, The Netherlands, combined the enantioselectivity of the amino acid selector with the universal gas-chromatographic properties of polysiloxanes. Other groups were also active in this novel approach [45,46]. We adopted the strategy to link the chiral metal-containing selector to a polymeric backbone in complexation gas chromatography. As mentioned before, I gained experience in this approach when preparing soluble polymeric catalysts [15]. Indeed, a limiting factor of coordination-type CSPs was the low temperature range of operation (25–120°C). The thermostability was therefore increased by the preparation of immobilized polymeric CSPs whereby the metal chelate was linked via the 10-methyl group of the camphor moiety to dimethylpolysiloxane (Chirasil-Nickel) [47]. Chirasil-Nickel proved useful in supercritical fluid chromatography (SFC) for less volatile racemates employing carbon dioxide as a mobile phase [48].

Chirasil-Europium

Chirasil-Europium could be employed in three different approaches: (a) in enantioselective complexation gas chromatography, (b) as an enantioselective polymeric NMR shift reagent, and (c) as an enantioselective catalyst for a hetero-Diels-Alder reaction [49]. In the latter work, we reinforced our previous strategy to employ a homogeneous and polymeric catalyst, which can be separated from products by mere precipitation and recycled for further catalytic runs [15].

The demand of complexation gas chromatography for enantiomer analysis warranted the establishment of the service CCC & CCC (commercial capillary columns for chiral complexation chromatography) for potential users in industry and academia. As an example, popular targets were alkyl 2-chloropropanoates obtained biotechnologically via lactic acid. From the orders that CCC & CCC received one could easily deduce the interconnections between the various leading companies engaged in manufacturing a chiral herbicide. Moreover, it was striking to witness how potential customers became aware of the service despite the absence of any advertizing activities. With the advent of the modified cyclodextrins for enantioselective gas chromatography, which are applicable to virtually all classes of compounds, the service was discontinued and our know-how on inclusion gas chromatography (see Sec. VII) was subsequently shared with Chrompack International, Middelburg, The Netherlands.

IV. METHODOLOGIES, GUIDELINES, AND RECOMMENDATIONS FOR THE ANALYSIS OF ENANTIOMERS

Along with the rapid development of chiral technology following the thalidomide tragedy, spurred by the allegation that only one enantiomer of the sedative drug was teratogenic, a growing demand for efficient techniques devoted to the precise analysis of enantiomers up to high enantiomeric purities arose. While the first instances of the separation of enantiomers by chromatography were devised for entirely academic interest, soon commercial applications for determining enantiomeric compositions were desperately sought in industry and academia to replace the imprecise measurement of optical rotations. Consequently, multivolume treatices devoted to asymmetric synthesis were preceded by articles concerned with the analysis of enantiomers. Based on my experience in the field, I was able to contribute to these endeavours at a very early stage and ever since [24,50,51]. These efforts were complemented recently by the comparison between the various definitions in enantiomeric analysis, i.e., enantiomeric excess *ee*, enantiomeric ratio *er*, and enantiomeric composition *ec* along with their diastereomeric counterparts [52].

V. ENANTIOMERIZATION—"DYNAMIC" ENANTIOSELECTIVE CHROMATOGRAPHY

In the treatice *75 Years of Chromatography—A Historical Dialogue* (which can be considered as a precedent to the present book), A. Zlatkis remarked [53]:

> In August of 1955 I was invited to join the chemistry department of the University of Houston as an assistant professor and I readily accepted. Since I was interested in stereochemistry and had an experimental background in gas chromatography, I decided to attempt to solve a classical problem, i.e., to separate the enantiomers of a compound which contained an asymmetric nitrogen atom. This seemed feasible for an aziridine molecule from an energetic viewpoint, but low temperatures would be necessary. Gas chromatography should provide the answer using an optically active stationary phase. . . . Three months were spent in attempting the resolution of amines without success.* (. . .* In collaboration with Dr. V. Schurig of the University of Tübingen, we have just recently solved this problem. The resolution of 1-chloro-2,2-dimethylaziridine was achieved using a 100 m × 0.5 mm nickel capillary column coated with an optically active nickel complex of 3-trifluoroacetyl camphorate. The separation factor of 1.43 was indeed large.)

The footnote (*) did not mention that this result was delayed for many years, comprising the second missed opportunity during my scientific career: A. Zlatkis invited me to Houston after my stay with Emanuel Gil-Av because he was obviously still interested in the above topic. Yet for inconceivable reasons he never mentioned the aziridine problem, nor did he provide me with the sample. As mentioned in Section IV, 7 years later, following my return to Tübingen, we found the first separation of enantiomers by the nickel chelate [28] and A. Zlatkis immediately sent 10 g of the highly toxic and explosive aziridine (by regular airmail), which finally provided the result mentioned above and was published together with Colin F. Poole [54]. The separation factor was indeed large enough for enantiomeric separation with the kind of packed columns I used in Houston for accumulating thermodynamic data of molecular association [55]. Indeed we later used 1-chloro-2,2-dimethylaziridine to demonstrate the feasibility of preparative enantiomeric separation by complexation gas chromatography by a packed column. On the isolated enantiomers we determined chiroptical data and the nitrogen inversion barrier, and we corrected the literature assignment of the absolute configuration at the stereogenic nitrogen atom [56]. The aziridine, as suggested by A. Zlatkis, fully compensated me with the delay of its resolution because it turned out to be an intriguing molecule for dynamic gas chromatography. At the end of the manuscript I stated: "Furthermore, complexation chromatography may develop into a simple method for the investigation of slow inversion processes. If inversion takes place during the resolution process the resulting phenomena of peak coalescence should lend themselves to kinetic analysis" [53]. I asserted that complexation gas chromatography should not only represent a useful quantitative method to measure thermodynamic data of enantioselectivity, but should be a useful tool for the kinetic study of slow enantiomerization processes [57]. The term enantiomerization [9,57] implied a process in which the individual stereoisomers of a racemic (or enriched) mixture of enantiomers undergo inversion of configuration (or conformation) during the chromatographic separation of the enantiomers. The dynamic behavior of interconverting enantiomers should give rise to intrinsic distortions of the elution curves amenable to the acquisition of kinetic data by peak form analysis. The equilibria of enantiomerization, i.e., $R \rightleftharpoons S$, which independently occur in the mobile and stationary phase, are appealing since they are

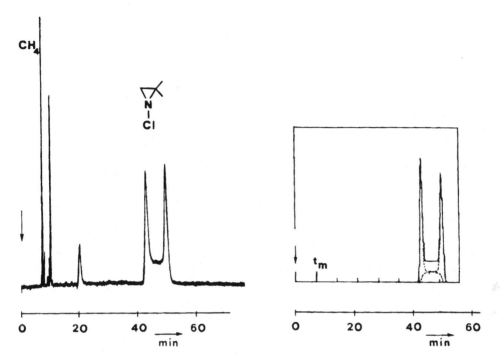

Figure 4 Plateau-formation due to enantiomerization (nitrogen inversion) of 1-chloro-2,2-dimethylaziridine by gas chromatography on nickel(II) bis[3-TFA-(1R)-camphorate] [57]. (Left) Experimental chromatogram. (Right) Simulated chromatogram.

governed by entropic changes only and represent a simple case of a reversible unimolecular reaction. Indeed, I detected a typical interconversion profile characterized by the appearance of an overlapping zone, called plateau and caused by inverted molecules between the terminal peaks of noninverted molecules for 1-chloro-2,2-dimethylaziridine by complexation gas chromatography on the chiral nickel(II) chelate [57] (see Fig. 4, left). Clearly, if enantiomerization is rapid at the chromatographic time scale, only one peak will be observed (peak coalescence).

The observed interconversion profiles represented not only an important diagnostic tool for the recognition of inversion of chirality, but I felt that such interconversion profiles could be simulated, and by comparison of simulated and experimental chromatograms it should be possible to obtain kinetic data of activation, i.e., to quantify the inversion barrier. H. Karfunkel, a physicochemist, wrote a computer program that simulated the peak form of interconverting enantiomers (see Fig. 4, right) and allowed the determination of rate constants of interconversion of enantiomers in a chromatographic column [57]. This topic received high priority in our group in the following years, and the program was steadily improved by W. Bürkle, M. Jung (who developed the SIMUL program), M. Fluck, and very recently by O. Trapp (who developed the program CHROMWIN, which operates on a personal computer). The comprehensive program CHROMWIN is currently used also for reaction chromatography. A prerequisite for calculations involving enantiomerization is the cyclic process obeying the principle of microscopic reversibility in the mobile and the stationary phase [57].

$$
\begin{array}{ccc}
\mathbf{A}_m & \xrightleftharpoons[k^m]{k^m} & \mathbf{B}_m \\
K_A \big\updownarrow & & \big\updownarrow K_B \qquad \text{mobile phase} \\
\hdashline
& & \qquad\qquad \text{stationary phase} \\
\mathbf{A}_s & \xrightleftharpoons[k^s_{-1}]{k^s_1} & \mathbf{B}_s
\end{array}
$$

The approach of dynamic chromatography for investigating interconversion processes during chromatography has been adopted and extended to other chromatograpic techniques by Mannschreck, Veciana/Crespo, Gasparrini, Allenmark, König, Pirkle, and W. Lindner, to mention a few. In our group we studied stereochemically attractive molecules like diaziridines [58], Trögers base [59], homofuran [60] (see Fig. 5), drugs like thalidomide and oxazepam, and the atropisomers of polychlorinated biphenyls. For the latter class of compounds a new technique was introduced, published in the *Analytical Separation 2000* proceedings edited by Colin F. Poole, called stopped-flow multidimensional gas chromatography [61]. By this method, em-

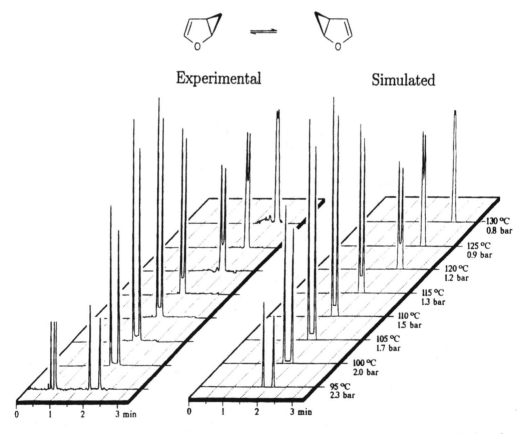

Figure 5 Comparison of experimental and calculated chromatograms displaying enantiomerization of homofuran by dynamic complexation gas chromatography on Chirasil-Nickel. (From Ref. 60.)

ploying two enantioselective columns (1 and 3) and one reactor column (2) in series (1-2-3), interconversion barriers up to 200 kJ/mole can be determined at high temperature. The investigations on enantiomerization barriers were recently stimulated by legislative demands. Thus, the U.S. Food and Drug Administration (FDA) requires that the stability protocol for enantiomeric drug substances and drug products include a method or methods capable of assessing the stereochemical integrity of the drug substance and drug product. Dynamic chromatography and stopped-flow techniques require only minute amounts of the racemic sample rather than isolated enantiomers for the acquisition of inversion barriers.

VI. ENTHALPY-ENTROPY COMPENSATION— THE ISOENANTIOSELECTIVE TEMPERATURE

The partition equilibrium in chromatography is governed by thermodynamics, while efficiency is controlled by kinetics of mass transfer, assumed to be rapid. Hence the separation process is controlled by the Gibbs free energy, which depends on enthalpy, entropy, and temperature according to the Gibbs-Helmholtz equation $-\Delta G = -\Delta H + T\Delta S$. Feibush and Beitler appeared to be the first to consider the role of enthalpy and entropy in enantioselective gas chromatography [62]. In his thesis at Tübingen, Koppenhöfer predicted the existence of a temperature T_s [63], later coined the isoenantioselective temperature, T_{iso} [64], at which no enantiomer separation is possible because of enthalpy-entropy compensation. Below T_{iso}, enantiomer separation is governed by the enthalpy contribution and enantioselectivity increases with decreasing temperature, whereas above T_{iso}, enantiomer separation is governed by the entropy contribution and enantioselectivity increases with increasing temperature. When traversing T_{iso} the elution order is reversed (peak inversion). When my coworker R. Link investigated a new nickel(II) chelate containing an unsaturated camphor ligand for the preparation of Chirasil-Nickel, he noted an unusual increase of enantioselectivity when raising the temperature from 70 to 90°C for the Z-enantiomeric pair of the insect pheromone chalcogran (2-ethyl-1,6-dioxaspiro[4,4]nonane) and for iso-

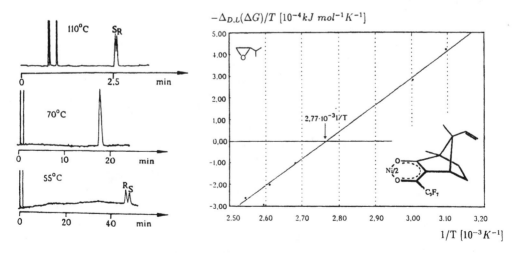

Figure 6 (Left) Temperature-dependent reversal of enantioselectivity for the enantiomers of isopropyl-oxirane by complexation gas chromatography on nickel(II)-*bis*-[3-(heptafluoro-butanoyl)-(1R)-8-(methylene)-camphorate]. (Right) Linear van't Hoff plot and determination of the isoenantioselective temperature T_{iso} (70°C). (From Ref. 37.)

propyloxirane, respectively, and he remarked: "an explanation is not yet available." It took us some time to realize that we observed probably the first entropy-driven enantiomer separation in enantioselective chromatography, again by mere serendipity! In a painstaking effort I investigated the system by myself, accumulating a host of data applying the approach to the determination of concise thermodynamic data through the concept of the retention-increase R' [18]. In both examples the van't Hoff plot $-\Delta_{R,S}(\Delta G)/T$ against $1/T$ was strictly linear (see Fig. 6), and the measured isoenantioselective temperature could be verified by thermodynamic measurements based on the following equations [37,64]:

$$-\Delta_{R,S}(\Delta G) = RT \ln \frac{K_R}{K_S} = -\Delta_{R,S}(\Delta H) + T \cdot \Delta_{R,S}(\Delta S)$$

$$T_{\text{iso}} = \frac{\Delta_{R,S}(\Delta H)}{\Delta_{R,S}(\Delta S)} \qquad (\text{for } \Delta_{R,S}(\Delta G) = 0)$$

At the same time Gil-Av et al. [64] reported on another case of the change in the elution order of enantiomers on hydrogen-bonding CSPs, and many examples are now known in enantioselective liquid chromatography.

VII. SWITCHING FROM COMPLEXATION TO INCLUSION— A PARADIGMATIC CHANGE WITH AN UNEXPECTED STRONG IMPACT

Molecules such as halocarbons and saturated hydrocarbons are devoid of any function prone to hydrogen bonding and metal coordination. They therefore resisted the separation into enantiomers by Chirasil-Val- and Chirasil-Metal-type CPSs, respectively. E. Gil-Av suggested trying cyclodextrins as inclusion-type CSP in gas chromatography, a topic well developed by Eva Smolková-Keulemansová in the achiral regime. Thus, the postdoctoral fellow Young Hwan Kim of South Korea came via Rehovot to Tübingen to investigate the use of permethylated β-cyclodextrin, dissolved in the polysiloxane OV 101 (20%) and coated onto a 26 m × 0.25 mm stainless steel capillary, for the enantiomer separation of the chiral aliphatic hydrocarbon 2,2,3-trimethylheptane (methyl-*n*-butyl-*tert*-butyl-methane). The attempt failed and Kim continued to work on the use of dirhodium tetraacetate, a dimeric Rh(II)-complex, for HPLC separations of olefins. This apparent failure was the third missed opportunity in my research endeavors in view of the eminent importance that peralkylated cyclodextrins gained subsequently in enantioselective gas chromatography as was later shown by us and by others. Today, modified CDs are capable of separating almost any class of chiral compounds, thus largely displacing hydrogen bonding and complexation CSPs. Our approach to concentrate only to unfunctionalized hydrocarbons was an academic challenge (which has been solved in the meantime [65]) but it was rather narrow-minded from a practical point of view: we just failed to inject any of those compounds routinely investigated by Chirasil-Val already in 1981! Thus, this important advance was delayed for another 6 years. Meanwhile Kościelski, Sybilska, and Jurczak demonstrated in 1983 the separation of enantiomers of the apolar racemic hydrocarbons α- and β-pinene on packed columns coated with a mixture of native α-cyclodextrin and formamide [66]. Despite large separation factors α, the columns had a limited lifetime and efficiency was poor. In 1986 my coworker H.-P. Nowotny extended the investigations of microsomal epoxidations to aromatic olefins such as styrene, and he needed a new protocol for the efficient enantiomer separation of styrene oxide (phenyloxirane), which tended to be sluggish with the existing metal chelates. Thus he suggested to me using permethylated cyclodextrins dissolved in polysiloxanes for this purpose, which earned my opposition in view of our previous results with H. Y. Kim. Yet occasionally coworkers may be well advised not to listen to the supervisor! It was a favorable coincidence that he found in a drawer left by my predecessor Holm Pauschmann—who until today is an ardent collector of chromatographic items and materials—various samples of polysiloxanes like OV 17. The breakthrough came when he acquired the new polar polysiloxane OV-1701, a phenyl-cyanopropylsilicone, which exerted excellent properties as a solvent for alkylated cyclodextrins. In his thesis, H.-P. Nowotny investigated higher alkylated cyclodextrins including chiral *sec*-pentyl groups [67]. I presented the first results in September 1987 at the Symposium of Advances in Chromatography organized by A. Zlatkis at a special session celebrating Ernst Bayer's sixtieth anniversary in West Berlin. The publication of the Berlin Proceedings was delayed and appeared in 1988 [68]. Also in 1987, Juvancz, Alexander, and Szejtli used undiluted permethylated β-cyclodextrin for the separation of enantiomers at high temperatures [69], whereas König et al. and subsequently Armstrong et al. employed undiluted *n*-pentylated/acetylated cyclodextrins, which are liquid at ambient temperature [70]. The strategy to dissolve the cyclodextrins in polysiloxanes, adopted from complexation gas chromatography, is regarded as the method of choice in terms of column efficiency combined with enantioselectivity [71], and today the approach is commercially available from all leading column manufacturers. According to the GC-Chirbase data bank compiled by B. Koppenhoefer, Tübingen, cyclodextrins accounted for as many as 15,000 entries in the literature as by the end of 1998!

In our first account on the enantioselective approach to inclusion gas chromatography I added [68]: "Systematic studies on the influence of the size of the CD cavity and the chemical structure of the racemates on enantioselectivity in combination with correlation of the order of elution and absolute configuration, may allow insights into pertinent mechanisms of chiral recognition. The anchoring of CD to the polysiloxane backbone represents another challenge for the future." Chemically linking permethylated CD to a polysiloxane matrix was accomplished independently by Fischer et al. [72] and by D. Schmalzing in our group [47]. The Chirasil approach was later extended by J. Pfeiffer to calixarenes [73].

Chirasil-Dex, commercialized by Chrompack-Varian, International, as GC columns, can be thermally immobilized on the fused-silica surface or on silicagel, respectively, and the CSP has been used by us not only in the GC mode but in the SFC mode, in the packed and open-tubular LC mode, as well as in the packed and open-tubular CEC mode, including the monolithic approach advanced by D. Wistuba in CEC [74]. In a unified approach it has been possible to utilize a single open-tubular column for the enantiomeric separation of hexobarbital by all existing methods: GC, SFC, LC, and CEC (see Fig. 7) [75].

The use of open-tubular CEC for enantiomeric separation on Chirasil-Dex was discovered by my coworker Sabine Mayer despite discouragement by some experts in the field on theoretical grounds and against my own scepticism. With the Chirasil-Dex coating available in open-tubular CEC, Mayer realized a dual chiral recognition system by adding another chiral selector into the mobile phase: the mobile phase additive consisted of β-cyclodextrin-sulfopropyl-ether, the first negatively charged CD additive used in CE [76]. It is odd to realize that the use of an open-tubular column for enantiomer separation in CEC, and particularly in LC, was realized only 30 years after Gil-Av et al. resorted to open-tubular columns in the GC mode!

Using our cyclodextrin approach, we pioneered the investigation of chiral atropisomeric polychlorinated biphenyls (PCBs) in terms of their enantiomeric separation and measurement of rotational barriers as well as the enantiomeric separation of inhalation anesthetics such as enflurane, isoflurane, and desflurane [77]. We found the fastest enantiomeric separation for enflurane with a miniaturized column in just 7 seconds [77]. While the separation factors in CD-mediated gas chromatography of enantiomers are usually in the range from 1.02 to 1.2, we recently detected unexpectedly high values of α = 2–8 for chlorofluoroethers on Königs Lipodex E CD phase bonded to a polysiloxane [77].

As a matter of principle (or due to the long induction period?), my scientific œuvre is essentially restricted to chirality, even though some of our advances in method development may be useful in the realm of achiral separations as well.

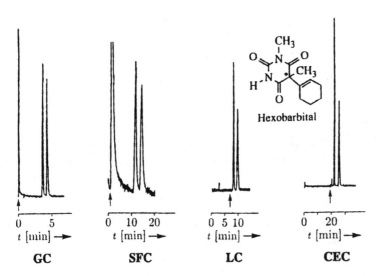

Figure 7 Unified enantioselective chromatography—enantiomeric separation of hexobarbital by GC: 145°C, 1 bar hydrogen, SFC: 60°C, ρ 0.25 g/mL carbon dioxide, LC (0.15 bar) and CEC (30 kV): (20°C, borate/phosphate buffer (pH 7)/acetonitrile 90/10, v/v). Unified column: 80 cm × 0.05 mm (i.d.) fused silica capillary coated with Chirasil-Dex (0.15 μm). The arrows denote the dead times. (From Ref. 75.)

VIII. CONCLUSION AND ACKNOWLEDGMENTS

I am indebted to three distinguished scientists—Ernst Bayer, Emanuel Gil-Av, and Albert Zlatkis—who guided and encouraged me at my early steps through chemistry and chromatography. I am especially thankful to the late Emanuel Gil-Av, who introduced me to the fascinating topic of chiral discrimination.

Emanuel Gil-Av delivered his last lecture at the 7th International Symposium on Chiral Discrimination, Jerusalem, in 1995. At the same time I held a fellowship at The Institute for Advanced Studies, The Hebrew University, Jerusalem, headed by Israel Agranat and focusing on "Chirality of Drugs and Chiral Recognition." My last publication with E. Gil-Av was related to a theoretical approach aimed at understanding the separation of excess enantiomer from the racemic mixture on an achiral stationary phase [78]. Our last scientific encounter was concerned with my predicition of an unusual elution order of R-S-R (or S-R-S) occurring during enantiomeric separation on an achiral stationary phase doped with a small amount of a chiral stationary phase [79]. He actually did not quite agree with my assertions, and the system will now be simulated by my coworker O. Trapp using the Chrom Win computer program.

Emauel Gil-Av was fascinated by an experiment of nuclear physicists intended to determine absolute configuration by the direct imaging of a small chiral molecule via Coulomb explosion ("molecular striptease") [80]. Together with the physicist Zeev Vager and my coworker O. Trapp, we hope to solve this problem in the memory of Emanuel Gil-Av [4] by carrying out the crucial experiment at the Kofler Accelerator Tower at the Weizmann Institute of Science.

I always had the good fortune to share my views and ideas with dedicated and capable coworkers, the names being mentioned in the text and in the reference list. Many discoveries are due to them, while I merely provided a creative and relaxed scientific atmosphere. Finally, I had the pleasure to be engaged in many cooperative efforts, which led to innovative approaches in many different fields. Thus, my coworker H. Hofstetter neé Diebold investigated the first use of enantioselective combinatorial selectors (cyclopeptide libraries) together with the group of G. Jung, Tübingen [81], while O. Hofstetter used antibodies raised against D- and L-amino acids in an enantioselective biosensor system [82] in a cooperation with M. Wilchek and B. Green, Rehovot, Israel. B. Gross separated an inherently chiral [60]fullerene derivative on Gil-Av's TAPA stationary phase by enantioselective HPLC together with A. Hirsch, Erlangen, Germany [83]. M. Schleimer linked Pirkle selectors to polysiloxanes and used the CSPs for enantiomeric separation by HPLC in the laboratory of W. H. Pirkle, Urbana, IL [84], and M. Juza investigated the preparative enantiomeric separation of the inhalation anesthetic enflurane by simulated moving bed gas chromatography (GC-SMB) in cooperation with M. Morbidelli, ETH, Zürich, Switzerland [85]. Finally, in a joint venture with the late W. Göpel, Tübingen, we used successfully quartz microbalances coated with chiral stationary phases for the discrimination of enantiomers [86], although the static system, contrary to chromatography, possesses just one theoretical plate!

I particularly enjoyed sharing my expertise in stereochemistry and chromatography with the scientific community through my engagements as the founding co-editor of *Enantiomer— a Journal of Stereochemistry* and as a co-editor of *Journal of Chromatography A*. In *Enantiomer* is published in the Postscriptum Section articles related to the history of stereochemistry and chirality in art and poetry.

Besides my interest in chromatography, in my leisure time I practice music spurred by my musical education in the choir Dresdener Kreuzchor. At the EUCHEM-Conference on Mechanisms of Chiral Recognition in Chromatography and the Design of Chiral Phase Systems [87] at Kungälv, Sweden, organized in 1995 by S. Allenmark, I tried to establish a (somewhat artificial) link between chromatography and chromatic tunes in music including chiral discrimination on the piano.

REFERENCES

1. V Schurig. Stickstoff-Fixierung und Reduktion zu Ammoniak mit metallorganischen Katalysatoren. PhD thesis, University of Tübingen, 1968.
2. E Bayer, V Schurig. Stickstoff-Fixierung und Reduktion zu Ammoniak mit metallorganischen Katalysatoren. Chem Ber 102:3378–3390, 1969.
3. V Schurig. Ernst Bayer's 70th birthday. Chromatographia 43:225–226, 1997.
4. V Schurig. In memoriam—Emanuel Gil-Av. J High Resolut Chromatogr 19:462–463, 1996.
5. S Sarel, S Weinstein. Emanuel Gil-Av (Zimkin). Enantiomer 1:iii–vii, 1996.
6. E Gil-Av. Present status of enantiomeric analysis by gas chromatography. J Mol Evol 6:131–144, 1975.
7. E Gil-Av, B Feibush, R Charles-Sigler. Separation of enantiomers by gas-liquid chromatography with an optically active stationary phase. Tetrahedr Lett 1009–1015, 1966.
8. J Oró, WS Updegrove, J Gibert, J McReynolds, E Gil-Av, J Ibanez, A Zlatkis, DA Flory, RL Levy, CJ Wolf. Organogenic elements and compounds in type C and D lunar samples from Apollo 11. Proc Apollo 11 Lunar Sci Conf 2:1901–1920, 1970 (cf. section 8 therein).
9. V Schurig, W Bürkle. Extending the scope of enantiomer resolution by complexation gas chromatography. J Am Chem Soc 104:7573–7580, 1982.
10. V Schurig. Chiral d-metal ion coordination compounds. The preparation of d-3-trifluoroacetyl-camphorato complexes of rhodium, palladium and nickel. Inorg Chem 11:736–738, 1972.
11. V Schurig. Chiral shift reagents for NMR spectroscopy. A simple and improved access to lanthanide-*tris*-chelates of d-3-TFA-camphor. Tetrahedr Lett 32:3297–3300, 1972.
12. E Gil-Av, V Schurig. Gas chromatography of monoolefins with stationary phases containing rhodium coordination compounds. Anal Chem 43:2030–2033, 1971.
13. V Schurig. NMR-spektroskopischer Nachweis der Chiralität π-komplexierter prochiraler Olefine. Tetrahedr Lett 16:1269–1272, 1976.
14. V Schurig, Stereoheterotopicity and stereoisomerism in heterochiral *cis* vs. *trans* olefin metal π complexes. Tetrahedr Lett 25:2739–2743, 1984.
15. V Schurig, E Bayer. A new class of catalysts. CHEMTECH 6:212–214, 1976.
16. C Bolm, A Gerlach. Polymer-supported catalytic asymmetric Sharpless dihydroxylations of olefins. Eur J Org Chem 1:21–27, 1998.
17. C Mark. Zugang zu 2,2'-Bis(diphenylphosphinomethyl)-1,1'-Binaphthyl als chiralem Komplexliganden für homogen katalysierte asymmetrische Synthesen. Diploma thesis, University of Tübingen, 1977.
18. V Schurig. Thermodynamik and Analytik der molekularen Assoziierung von Donormolekülen mit Übergangsmetallchelaten durch Komplexierungschromatographie. Habilitation thesis, Tübingen, 1975, Chapter I.
19. V Schurig. Relative stability constants of olefin-rhodium(I) vs. olefin-rhodium(II) coordination as determined by complexation gas chromatography. Inorg Chem 25:945–949, 1986.
20. V Schurig, R C Chang, A Zlatkis, E Gil-Av, F. Mikeš. Application of dicarbonyl-rhodium-trifluoro-acetyl-d-camphorate to special problems of olefin analysis by gas-liquid chromatography. Chromatographia 6:223–225, 1973.
21. V Schurig. Separation of deuterated ethylenes $C_2H_{4-n}D_n$ by complexation chromatography on a rhodium(I) complex. Angew Chem Int Ed 15:304, 1976.
22. V Schurig. Resolution of a chiral olefin by complexation chromatography on an optically active rhodium(I) complex. Angew Chem Int Ed 16:110, 1977.
23. V Schurig. Chirodichroism of different enantiomeric compositions of a planar d^8-metal complex. Angew Chem Int Ed 20:807–808, 1981.
24. V Schurig. Gas chromatographic methods for determining enantiomeric ratios. In: JD Morrison, ed. Asymmetric Synthesis. Vol. I: Analytical Methods. New York: Academic Press, Chapter 5, 1983, pp. 59–86 (translated into Russian).
25. V Schurig, E Gil-Av. Chromatographic resolution of chiral olefins. Specific rotation of 3-methyl-cyclopentene and related compounds. Israel J Chem 15: 96–98, 1976/77.

26. HB Kagan, TP Dang. Asymmetric catalytic reduction with transition metal complexes. I. A catalytic system of rhodium(I) with (−)-2,3-O-isopropylidene-2,3-dihydroxy-1,4-bis(diphenylphosphino)butane, a new chiral diphosphine. J Am Chem Soc 94:6429–6933, 1972.

27. WS Knowles, MJ Sabacky, BD Vineyard, DJ Weinkauff. Asymmetric hydrogenation with a complex of rhodium and a chiral biphosphine. J Amer Chem Soc 97:2567–2568, 1975.

28. V Schurig, W Bürkle. Quantitative resolution of enantiomers of trans-2,3-epoxybutane by complexation chromatography on an optically active nickel(II)-complex. Angew Chem Int Ed 17:132–133, 1978.

29. V Schurig, W Bürkle, K Hintzer, R Weber. Evaluation of nickel(II) bis[α-(heptafluorobutanoyl)-terpeneketonates] as chiral stationary phases for the enantiomer separation of alkyl-substituted cyclic ethers by complexation gas chromatography. J Chromatogr 475:23–44, 1989.

30. V Schurig, W Bürkle. Extending the scope of enantiomer resolution by complexation gas chromatography. J Am Chem Soc 104:7573–7580, 1982.

31. HB Kagan, H Mimoun, C Mark, V Schurig. Asymmetric epoxidation of simple olefins with an optically active molybdenum(VI)-peroxo complex. Angew Chem Int Ed 18:485–486, 1979.

32. V Schurig, F Betschinger. Kinetic resolution of aliphatic oxiranes mediated by in situ formed molybdenum(VI)(oxo-diperoxo) hydroxy acid amide/chiral diol complexes. Bull Soc Chim Fr 131:555–560, 1994.

33. V Schurig, D Wistuba. Asymmetric microsomal epoxidation of simple prochiral olefins. Angew Chem Int Ed 23:796–797, 1984.

34. D Wistuba, V Schurig. Kinetic resolution of simple aliphatic oxiranes. Complete regio- and enantioselective hydrolysis of cis-3-ethyl-2-methyl-oxirane. Angew Chem Int Ed 25:1032–1034, 1986.

35. B Koppenhoefer, V Schurig. (R)-Alkyloxiranes of high enantiomeric purity from (S)-2-chloro-alkanoic acids via (S)-2-chloro-1-alkanols: (R)-methyloxirane. Organic Syntheses 66:160–172, 1987.

36. V Schurig. Molecular association in complexation gas chromatography. In: K Jinno, ed. Chromatographic Separations Based on Molecular Recognition. New York: Wiley-VCH, Chapter 7, 1996, pp. 371–418.

37. V Schurig, F Betschinger. Metal-mediated enantioselective access to unfunctionalized aliphatic oxiranes: prochiral and chiral recognition. Chem Rev 92:873–888, 1992.

38. V Schurig. Peak coalescence phenomena in enantioselective chromatography. Chirality 10:140–146, 1998.

39. V Schurig. Enantiomer separation by complexation gas chromatography—applications in chiral analysis of pheromones and flavours. In: P. Schreier, ed. Bioflavour '87, Berlin: de Gruyter, 1988, pp. 35–54.

40. V Schurig, H Laderer, D Wistuba, A Mosandl, V Schubert, U Hagenauer-Hener. Enantiomer separation of n-alkenyl-3-acetates and of alkyl substituted 1,3-dioxolanes by complexation gas chromatography. J Ess Oil Res 1:209–221, 1989.

41. J Jauch, D Schmalzing, V Schurig, R Emberger, R Hopp, M Köpsel, W Silberzahn, P Werkhoff. Isolation, synthesis and absolute configuration of filbertone, the principal flavor component of the hazelnut. Angew Chem Int Ed 28:1022–1023, 1989.

42. R Weber, V Schurig. Complexation gas chromatography—a valuable tool for the stereochemical analysis of pheromones. Naturwiss. 71:408–413, 1984.

43. G Haniotakis, W Francke, K Mori, H Redlich, V Schurig. Sex specific activity of (R)-(−)- and (S)-(+)-1,7-dioxaspiro-[5,5]undecane, the major pheromone of dacus oleae. J Chem Ecol 12:1559–1568, 1986.

44. H Frank, GJ Nicholson, E Bayer. Rapid gas chromatographic separation of amino acid enantiomers with a novel chiral stationary phase. J Chromatogr Sci 15:174–176, 1977.

45. T Saeed, P Sandra, M Verzele. Synthesis and properties of a novel chiral stationary phase for the resolution of amino acid enantiomers. J Chromatogr 186:611–618, 1980.

46. WA König. The Practice of Enantiomer Separation by Capillary Gas Chromatography. Heidelberg: Hüthig, 1987.

47. V Schurig, D Schmalzing, M Schleimer. Enantiomer separation on immobilized Chirasil-Metal and

Chirasil-Dex by gas chromatography and supercritical fluid chromatography. Angew Chem Int Ed 30:987–989, 1991.

48. M Schleimer, M Fluck, V Schurig. Enantiomer separation by capillary SFC and GC on Chirasil-Nickel: observation of unusual peak broadening phenomena. Anal Chem 66:2893–2897, 1994.

49. F Keller, H Weinmann, V Schurig. Chiral polysiloxane-fixed metal-1,3-diketonates (Chirasil-metals) as catalytic Lewis acids of a hetero Diels-Alder reaction. Inversion of enantioselectivity upon catalyst-polymer-binding. Chem Ber/Recueil 130:879–885, 1997.

50. V Schurig. Current methods for the determination of enantiomeric compositions. Part I: Definitions and polarimetry. Kontakte (Darmstadt), 1:54–60, 1985. Part II: NMR spectroscopy with chiral lanthanide shift reagents. Kontakte (Darmstadt) 2:22–36, 1985. Part III: Gas chromatography on chiral stationary phases. Kontakte (Darmstadt) 1:3–22, 1986.

51. V Schurig. Determination of enantiomeric purity by direct methods—overview, by chemical correlation, by polarimetry, by NMR-spectroscopy and by gas chromatography. In: Houben-Weyl. Methods of Organic Chemistry, Stereoselective Synthesis. Vol. E21a. New York: Thieme, 1995, pp. 147–192.

52. V Schurig. Terms for the quantitation of a mixture of stereoisomers. Enantiomer 1:139–143, 1996.

53. Albert Zlatkis. In: LS Ettre. A Zlatkis eds. 75 years of chromatography—a historical dialogue. J Chromatogr Libr 17:473–482, 1979.

54. V Schurig. W Bürkle, A Zlatkis, CF Poole. Quantitative resolution of pyramidal nitrogen invertomers by complexation chromatography. Naturwissenschaft 66:423, 1979.

55. V Schurig, RC Chang, A Zlatkis, B Feibush. Thermodynamics of molecular association by gas-liquid chromatography. σ-Donor molecules and dimeric 3-trifluoroacetylcamphorates of Mn(II), Co(II) and Ni(II). J Chromatogr 99:147–171, 1974.

56. V Schurig, U Leyrer. Semi-preparative enantiomer separation of 1-chloro-2,2-dimethylaziridine by complexation gas chromatography—absolute configuration and barrier of inversion. Tetrahedr Asymmetry 1:865–868, 1990.

57. W Bürkle, H Karfunkel, V Schurig. Dynamic phenomena during enantiomer resolution by complexation gas chromatography. A kinetic study of enantiomerization. J Chromatogr 288:1–14, 1984.

58. M Jung, V Schurig. Determination of enantiomerization barriers by computer simulation of interconversion profiles: enantiomerization of diaziridines during chiral inclusion gas chromatography. J Am Chem Soc 114:529–534, 1992.

59. O Trapp, V Schurig. On the stereointegrity of Tröger's base—gas-chromatographic determination of the enantiomerization barrier. J Am Chem Soc 122:1424–1430, 2000.

60. V Schurig, M Jung, M Schleimer, F-G Klärner, V Schurig. Investigation of the enantiomerization barrier of homofuran by computer simulation of interconversion profiles obtained by complexation gas chromatography. Chem Ber 125:1301–1303, 1992.

61. S Reich, O Trapp, V Schurig. Enantioselective stopped-flow multidimensional gas chromatography—determination of the inversion barrier of 1-chloro-2,2-dimethylaziridine. J Chromatogr A, 2000.

62. U Beitler, B Feibush. Interaction between asymmetric solutes and solvents—diamides derived from L-valine as stationary phases in gas-liquid partition chromatography. J Chromatogr 123:149–166, 1976.

63. B Koppenhöfer. Chirales Erkennen mit Enzymen und Modellsystemen: Chromatographische Enantiomerentrennung, chemische und mikrobiologische Synthese von Glykolen und Oxiranen hoher Enantiomerenreinheit. Thesis, University of Tübingen, 1980, pp. 65–94.

64. V Schurig, J Ossig, R Link. Temperature dependent reversal of enantioselectivity in complexation gas chromatography on chiral phases. Angew Chem Int Ed 28:194–196, 1989.

65. V Schurig, H-P Nowotny, D Schmalzing, Gas-chromatographic enantiomer separation of unfunctionalized cycloalkanes on permethylated β-cyclodextrin. Angew Chem Int Ed 28:736–737, 1989.

66. T Kościelski, D Sybilska, J Jurczak, Separation of α- and β-pinene into enantiomers in gas-liquid chromatography systems via α-cyclodextrin inclusion complexes. J Chromatogr 280:131–134, 1983.

67. H.-P. Nowotny, Enantiomerentrennung durch Inklusions-Gaschromatographie an per-alkylierten Cyclodextrinen. Thesis, University of Tübingen, 1986–1989.
68. V Schurig, H-P Nowotny, Separation of enantiomers on diluted permethylated β-cyclodextrin by high-resolution gas chromatography. J Chromatogr 441:155–163, 1988.
69. Z Juvancz, G Alexander, J Szejtli, Permethylated β-cyclodextrin as stationary phase in capillary gas chromatography. J High Resol Chromatogr/Chromatogr Commun 10:105–107, 1987.
70. WA König. Enantioselective Gas Chromatography with Modified Cyclodextrins. Heidelberg: Hüthig, 1992 (and references cited therein).
71. V Schurig, H.-P Nowotny, Gas chromatographic separation of enantiomers on cyclodextrin derivatives. Angew Chem Int Ed Engl 29:939–957, 1990.
72. P Fischer, R Aichholz, U Bölz, M Juza, S Krimmer, Polysiloxan-gebundenes Permethyl-β-cyclodextrin—eine chirale stationäre Phase mit großer Anwendungsbreite in der gaschromatographischen Enantiomerentrennung. Angew Chem 102:439–441, 1990.
73. J Pfeiffer, V Schurig, Enantiomer separation of amino acid derivatives on a new polymeric chiral resorc[4]arene stationary phase by capillary gas chromatography. J Chromatogr A 840:145–150, 1999.
74. V Schurig, D Wistuba, Recent innovations in enantiomer separation by electrochromatography utilizing modified cyclodextrins as stationary phases. Electrophoresis 20:2313–2328, 1999.
75. V Schurig, M Jung, S Mayer, S Negura, H Jakubetz, Toward unified enantioselective chromatography with a single capillary column coated with Chirasil-Dex. Angew Chem Int Ed Engl 33:2222–2223, 1994.
76. S Mayer, M Schleimer, V Schurig. Dual chiral recognition system involving cyclodextrin derivatives in capillary electrophoresis. J Microcol Sep 6:43–48, 1994.
77. V Schurig, Review—separation of enantiomers by gas chromatography. J Chromatogr A, 2000.
78. E Gil-Av, V Schurig. Resolution of non-racemic mixtures in achiral chromatographic systems: a model for the enantioselective effects observed. J Chromatogr A 666:519–525, 1994.
79. V Schurig, On the intermediacy between an achiral vs. chiral stationary phase in the enantiomer discrimination of non-racemic mixtures. Enantiomer 1:429–430, 1996.
80. E Heilbronner, JD Dunitz. Reflection on Symmetry in Chemistry and Elsewhere. Basel: Verlag Helvetica Chimica Acta; Weinheim: VCH, 1993, p. 67.
81. G Jung, H Hofstetter, S Feiertag, D Stoll, O Hofstetter, K-H Wiesmüller, V Schurig. Cyclopeptide libraries as new selectors for capillary electrophoresis. Angew Chem Int Ed. 35:2148–2150, 1996.
82. O Hofstetter, H Hofstetter, M Wilchek, V Schurig, B Green. Chiral discrimination using an immunosensor. Nature Biotechnol 17:371–374, 1999.
83. B Gross, V Schurig, I Lamparth, A Hirsch. Enantiomer separation of [60]fullerene derivatives by micro-column-HPLC using (R)-(−)-TAPA as chiral stationary phase. J Chromatogr A 791:65–69, 1997.
84. M Schleimer, WH Pirkle, V Schurig. Enantiomer separation by high-performance liquid chromatography on polysiloxane-based chiral stationary phases. J Chromatogr A 679:23–34, 1994.
85. M Juza, O Di Giovanni, G Biressi, V Schurig, M Mazzotti, M Morbidelli. Continuous enantiomer separation of the volatile inhalation anesthetic enflurane with a gas chromatographic simulated moving bed (GC-SMB) unit. J Chromatogr A 813:333–347, 1998.
86. A Hierlemann, K Bodenhöfer, M Juza, B Gross, V Schurig, W Göpel. Enantioselective monitoring of chiral inhalation anesthetics by simple gas sensors. Sensors Mater 11:209–218, 1999.
87. S Allenmark, V Schurig. Chromatography on chiral carriers. J Mater Chem 7:1955–1963, 1997.

22

Developments in Surface Chemistry for the Improvement of Chromatographic Methods

Joseph J. Pesek and Maria T. Matyska

San José State University, San José, California

I. INTRODUCTION

Surface modification has played a crucial role in the advancement of separation science for solving a multitude of challenging analytical problems. Most materials used in a variety of separation processes, particularly chromatographic methods, have undergone some type of surface modification in order to improve their performance or to be more selective for specific analytes. The modification process itself can involve either a chemical reaction or the adsorption of one or more compounds to alter the surface properties of the separation material. The methods developed to accomplish this goal are too numerous to cover completely here and often can be very specific for a certain analytical problem or technique. Some of the approaches used in the modification of separation media have been adapted from chemical reaction or adsorption protocols developed for altering catalysts, surfactants, fillers, and other such materials where the surface plays a crucial role in the intended use. Some of these approaches have been more universal than others and have found use in a variety of chromatographic (gas as well as several liquid chromatographic formats) and electrophoretic (slab as well as several types of capillary formats) methods. This review will focus on two advances in surface chemistry that originated in the authors' laboratory and have resulted in improved separation materials for a variety of applications. These two developments, which will be discussed separately, are the silanization/hydrosilation reaction scheme for the modification of surfaces and the fabrication of etched open tubular columns for use in capillary electrochromatography (CEC). The silanization/hydrosilation process was first introduced in 1989 [1], while a description of etched open tubular capillaries as a viable approach for CEC was first published in 1996 [2].

II. SILANIZATION/HYDROSILATION SURFACE MODIFICATION PROCESS

The surface modification method developed in our labs is the two-step silanization/hydrosilation process that has proved to be versatile, rugged and applicable to a variety of separation techniques [1,3–5]. In the first step, the surface (porous silica, other oxides such as titania, alumina, and zirconia, or the inner walls of conventional and etched capillaries) is reacted with triethoxysilane (TES) as shown below:

SILANIZATION

$$-\underset{\overset{\displaystyle |}{O}}{\overset{\overset{\displaystyle |}{O}}{Si}}\text{-OH} + (OEt)_3Si\text{-H} \xrightarrow{H^+} \underset{\overset{\displaystyle |}{O}\;\;\overset{\displaystyle |}{OY}}{\overset{\overset{\displaystyle |}{O}\;\;\overset{\displaystyle |}{OY}}{Si\text{-O-Si-H}}} + EtOH$$

In this step most of the silanols (or OH groups on other oxide surfaces) are replaced by hydrides on the silica surface. Since these hydroxyl groups are generally responsible for strong or even irreversible adsorption of certain solutes, particularly bases, on the chromatographic support or capillary wall, this reaction results in the formation of a new surface that has more favorable properties. A high degree of cross-linking in this reaction generally results in very few silanols (hydroxyl groups) that are accessible to solutes in either chromatographic or electrophoretic separation modes. The second step, hydrosilation, attaches the organic moiety to the hydride intermediate in order to create the product that has the desired surface properties.

HYDROSILATION

$$\equiv Si\text{-H} + R\text{-CH}=\text{CH}_2 \xrightarrow{cat.} \equiv Si\text{-CH}_2-\text{CH}_2-R$$

cat = catalyst, typically hexachloroplatinic acid

As shown above, the bonded organic moiety is attached to the surface by a stable Si-C bond. This feature leads to the high stability observed in both chromatographic and electrophoretic experiments [3,5,6]. While the most common approach to date for attaching an organic species to the silica hydride utilizes a terminal olefin, it is also possible to bond molecules with the olefin in a nonterminal position [7], alkynes [8], and other functional groups such as cyano [9]. Catalysts besides hexachloroplatinic acid such as other metal ion complexes and free radical initiators [10] can also be used in the hydrosilation reaction. Some unique aspects of the silanization/hydrosilation approach that are distinctly different from conventional organosilanization are discussed in more detail below.

A. Hydrosilation of Alkynes

The availability of this synthetic pathway via the silanization/hydrosilation process can lead to unique phases that are not possible through organosilanization. In our initial evaluation of hydrosilation of alkynes [8], several bonded phase configurations were identified as possible surface structures. Four types of linkages between the organic moiety and the hydride intermediate surface are shown below:

Carbon 13 cross-polarization, magic-angle spinning (CP-MAS) NMR spectra of several bonded materials made by this approach were used to rationalize the possible existence of the above structures. The spectra contain a low-intensity broad peak near 12 ppm that is consistent with the presence of a direct Si—C bond at the surface that occurs when olefins are bonded to silica hydride via hydrosilation [3,5]. Only structures **I** and **IV** have the appropriate surface configuration to account for both the chemical shift observed as well as the broad nature of the peak. Olefin peaks in some, but not all of the spectra, indicate that either **II** or **III** or both structures may be present as well in certain bonded materials. Structure **II** might be considered as the most likely bonded moiety, since this product would be predicted from following an identical mechanism to the reaction of an olefin with silica hydride. However, the fact that some products from the hydrosilation of alkynes show little or no evidence of olefin peaks indicates that the mechanism is not the same in both cases. From a chromatographic point of view, **I**, **III**, and **IV** are the most interesting since they represent materials with two points of attachment to the surface and hence would have potentially higher stability than stationary phases with only a single bond to the silica support. The surface coverage of several different types of alkynes bonded to silica hydride was quite good, between 3.8 and 5.4 μmol/m^2. These values are near the upper limit of what might be expected for monomeric bonding to most silica surfaces [11]. The chromatographic behavior of the materials synthesized to date from alkynes is similar to that obtained from hydrosilation of olefins as well as monomeric stationary phases of equivalent structure produced by organosilanization. This result is reasonable since the main differences are near the surface rather than further along the organic chain where most interactions with solutes occur.

B. Bonded Liquid Crystals

Liquid crystals as a separation material in high-performance liquid chromatography (HPLC) are rapidly gaining interest. Our first effort involved a comparison of two terphenyl liquid crystal stationary phases to a standard ODS material for the retention of 16 polycyclic aromatic hydrocarbons (PAHs) as well as the separation of C_{60} and C_{70} fullerenes [12]. It was demonstrated that one liquid crystal phase behaved in a similar manner to the standard octadecylsilica (ODS) phase for separation of the fullerenes and had many of the same retention characteristics for the PAHs. However, the other liquid crystal phase had both higher retention and twice the resolving power (α) for the two fullerene compounds. Variable temperature studies also revealed a fundamental difference in the retention mechanism between the fullerenes and the PAHs on this liquid crystal phase since the enthalpy (ΔH) of transfer of these solutes is positive for both C_{60} and C_{70} while it is negative for the aromatic compounds. It was also shown that this particular liquid crystal phase had very good molecular shape recognition capabilities. In a follow-up study using the cholesterol stationary phase [13] and evaluating a similar group of PAHs, it was determined that this material also possessed very high selectivity for the more planar molecules. Both the terphenyl phase from the previous work and the cholesterol phase have α values that are twice as large as the ODS phase when comparing the retention of the same pair of planar and nonplanar molecules of the same size. Finally, when a series of benzodiazepines was studied on the cholesterol phase, it was found that the elution order was quite different from that of an ODS phase [14]. This difference in elution order was attributed to the molecular shape discrimination ability of the cholesterol phase. Several studies have described work on cholesterol phases [15–17] following our reports [13,14,18,19] on the synthesis, characterization and applications of this bonded material. The uniqueness of the bonded liquid crystal has been demonstrated in solid-state magnetic resonance studies [19]. This study involved a comparison of the octadecyl (C-18) stationary phase to two types of bonded liquid crystals. It was shown that the relaxation time (T_2) for the main peak in the spectrum representing most of the methylene groups in C-18 decreases regularly with a decrease in temperature, indicating an overall restriction in the molecular motion. However, monitoring appropriate locations on the liquid crystal molecules reveals a nonuniform change, in some cases an increase, in T_2 with decreasing temperature. The difference is probably due to the greater degree of association or ordering of the liquid crystal materials in comparison to the C-18 moiety. This behavior may be similar to phase transitions of the pure material but are not a true change of state since the molecule is attached at one end.

In a recent study [20] two commercially available liquid crystals, 4-cyano-4'-n-pentyl-1, 1'-biphenyl (A below) and 4-cyano-4'-n-pentoxy-1,1'-biphenyl (B below), were bonded to a silica hydride surface via hydrosilation in the presence of a free radical iniator, t-butyl peroxide.

A)

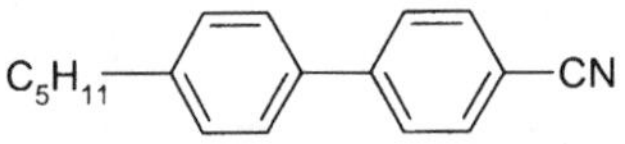

B)

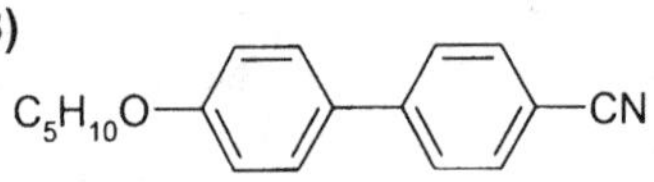

Many commercially available liquid crystal materials contain a cyano group at one end of the molecule, so utilizing these compounds is advantageous from the point of view of developing a broad range of new stationary phases. The cyano group is amenable to hydrosilation in the absence of an olefin and with a free radical initiator as the catalyst. The possible products (C and D) of hydrosilation of silica hydride with aromatic cyano compounds such as those used in this study are shown below.

C)

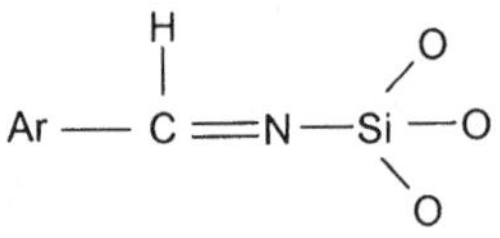

D)

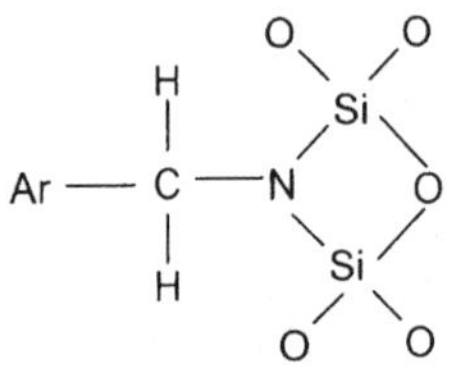

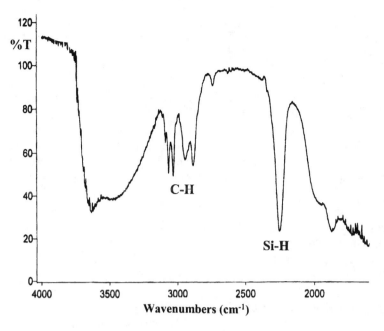

Figure 1 DRIFT spectrum of the liquid crystal 4-cyano-4'-*n*-pentyl-1,1'-biphenyl bonded to a silica hydride surface.

Elemental analysis, diffuse reflectance infrared Fourier transform (DRIFT) spectroscopy, and ^{13}C and ^{29}Si CP-MAS NMR spectroscopy are used to confirm the success of the bonding reaction. An example of the DRIFT spectrum of one of the bonded liquid crystals is shown in Fig. 1. The ^{13}C CP-MAS spectra suggest a difference in the bonded phase morphology of the two materials. Static hydrolytic stability tests indicate that these materials do not degrade significantly in both acidic and basic solutions. Chromatographic tests confirm that these two bonded phases behave differently with respect to their retention of PAHs, alkyl-substituted benzenes, and benzodiazepines.

C. Silanization/Hydrosilation of Oxides Other than Silica

On alumina the silanization reaction using TES is described as follows [21,22]:

$$\equiv Al-OH \ + \ H-Si(OEt)_3 \ \xrightarrow[\text{water, dioxane}]{HCl} \ \equiv Al-O-\overset{\overset{\textstyle O}{|}}{\underset{\underset{\textstyle O}{|}}{Si}}-H \ + \ 3EtOH$$

Under ideal conditions, the TES will cross-link with neighboring groups to produce a hydride monolayer on the alumina surface. This intermediate can be further modified via hydrosilation utilizing a terminal olefin in the presence of a suitable catalyst such as hexachloroplatinic acid (Speier's catalyst), as shown in the following reaction:

$$\equiv\!Al\!-\!O\!-\!\underset{\underset{|}{O}}{\overset{\overset{|}{O}}{Si}}\!-\!H \; + \; H_2C\!=\!CH\!-\!R \xrightarrow{\text{catalyst}} \equiv\!Al\!-\!O\!-\!\underset{\underset{|}{O}}{\overset{\overset{|}{O}}{Si}}\!-\!CH_2\!-\!CH_2\!-\!R$$

The possibility of using other unsaturated functional groups (alkynes, cyano, and nonterminal olefins) within the bonding moiety and a variety of catalysts such as other transition metal complexes and free radical initiators can also be exploited as on silica.

It was possible to characterize both the hydride intermediate and the products (octyl and octadecyl modified alumina) by DRIFT spectra for bare alumina, the hydride intermediate, and the products. In the case of bare alumina, the spectrum contains a broad peak between 3800 and 2700 cm^{-1} as a result of adsorbed water on the surface. Also evident are several sharp peaks superimposed on the broad band between 3500 and 3700 cm^{-1}. These peaks are due to the various hydroxyl groups on the surface of alumina identified in previous studies [23–25]. Upon silanization, many of the alumina hydroxyl peaks disappear indicating the reaction was successful. This result is confirmed by the appearance of a strong Si-H stretching peak near 2250 cm^{-1} and a small peak near 3750 cm^{-1} representing the silanol (Si-OH) groups formed when not all of the adjacent TES moieties cross-link during silanization. The spectrum of the final product of the hydrosilation reaction indicates success of this reaction by a decrease in the Si-H band and the appearance of strong carbon-hydrogen stretching bands between 3000 and 2800 cm^{-1}. Additional characterization was also done by ^{29}Si cross-polarization CP-MAS nuclear magnetic resonance (NMR) spectroscopy. Here the spectrum of the hydride intermediate contains a strong peak for the Si-H group at -82 ppm and small peaks near -100 ppm for the silanols (also detected by DRIFT) and for a silicon surrounded by four oxygens (-110 ppm). The latter is the result of a small amount of polymerization of the TES beyond a monolayer coverage on the surface. The products were also characterized by ^{13}C CP-MAS NMR. The spectra obtained were identical to those of the octyl [3] and octadecyl [3] silica–based materials. Finally, ^{27}Al CP-MAS NMR spectra were also obtained on the bare alumina, hydride intermediate, and octadecyl product. From the integrated intensities of the two peaks, the bare alumina had approximately 20% of the aluminum atoms in the T_d geometry and 80% in the O_h geometry. A new peak appeared near the O_h chemical shift upon formation of the hydride intermediate that increased in intensity upon hydrosilation. No change in the T_d peak was seen in either the hydride intermediate or product spectra. These results confirm the success of the two reactions as well as establish that by this process all of the chemical modification is confined to aluminum atoms possessing octahedral geometry.

These materials have undergone a limited amount of chromatographic testing [21]. A reversed-phase test mixture was run on the octadecyl material, and the results are consistent with reversed-phase retention. The peak symmetries and efficiencies obtained in this test were comparable to or better than those obtained on a commercially available octadecyl alumina column. A sample of anilines (aniline, N-methylaniline, and N, N-dimethylaniline) was also tested on this column. All solutes were easily separated with symmetric peaks. Good pH stability ($\Delta k' < 10\%$ after more than 2000 column volumes) was obtained under both acidic (pH = 2) and basic (pH = 10) conditions. These results suggest that further investigations of materials made by this method should be conducted.

It has also been demonstrated that oxides such as titania [22,26,27], zirconia[22], and thoria[22] can also be modified successfully by the silanization/hydrosilation process. These materials were also characterized by DRIFT as well as ^{13}C and ^{29}Si CP-MAS NMR, with results being similar to those described above for the modified alumina. However, only chromatographic testing of the titania bonded phases has been reported [26,27]. Again, the behavior of these alkyl-modified materials is consistent with reversed-phase behavior, with the peaks displaying good efficiency and high symmetry for a variety of solutes.

III. ETCHED CHEMICALLY MODIFIED CAPILLARIES FOR CEC

A. Background

Capillary electrochromatography is one of the newest micro-separation formats available. In CEC there are two general approaches to this hybrid technique that combines features of both HPLC and capillary electrophoresis (CE): the packed column configuration that utilizes stationary phases similar to HPLC and the open tubular format where the stationary phase is immobilized on the capillary wall. A summary of the continuum of techniques that exist between μ-HPLC and CE that can be used in the micro-column format are shown in Fig. 2. The figure also denotes on a relative scale the importance of the major separation mechanisms, solute/bonded phase interactions and electrophoretic mobility, that exist in the continuum, with k' being dominant at the μ-HPLC end and μ_{ep} being the only factor in CE. The open tubular approach (OTCEC) serves as a bridge between CE and packed column CEC. The first effective OTCEC separations indicating chromatographic interactions were demonstrated a number of years ago by Tsuda and coworkers [28] using an octadecyl modified 30 μm i.d. capillary. A more definitive way of proving that chromatographic effects are possible in the open tubular format can be provided through the separation of optical isomers. A number of chiral selectors such as cyclodextrin and several cellulose derivatives that were bonded to the inner wall of fused silica capillaries resulted in the separation of enantiomers [29–34]. Since the two optical isomers have identical electrophoretic mobilities, separation can only be achieved through differences in solute/bonded phase interactions.

Initially, further exploitation of the OTCEC technique was hampered by two fundamental problems that seriously limited its potential as a viable separation method: the low capacity of the column due to the small area available for bonding a stationary phase and the long distance that molecules would have to migrate to interact with the bonded moiety. The latter problem could be addressed by reducing the column i.d., but this is often an unsatisfactory solution because such a decrease further limits the sample size as well as the detection path length in optical systems and hence the detection limit of the analyte. A new approach has been developed to overcome these two major problems though chemical etching of the inner wall of the capillary. This process produces structural changes within the capillary that are more favorable for

voltage assisted   pressure assisted   packed capillary

μ-HPLC ⇔ μ-HPLC ⇔ CEC ⇔ **CEC** ⇔ OTCEC ⇔ **CE**

LAMINAR FLOW ⇒ ⇒ ⇒ ⇒ ⇒ ⇒ *ELECTRICALLY DRIVEN FLOW*

← ← ← ← ← **Increasing influence of solute/bonded phase interactions (k')**

Increasing influence of electrophoretic mobility (μ_{ep}) → → → → →

Figure 2   Schematic diagram representing a summary of the continuum of microscale separation techniques.

OTCEC. First, the etching process increases the overall surface are of the capillary by as much as 1000-fold [35], which allows more stationary phase to be attached to the wall, thus increasing the capacity of the column. Second, the dissolution and redeposition of silica material during etching creates radial extensions from the wall that decrease the distance a solute must travel in order to interact with the stationary phase.

B. Chemical Etching Process

A section (~2 m in length) of bare capillary with a 375 μm o.d. and a 50 μm i.d. is filled with concentrated (15 M) HCl, sealed and heated overnight at 80°C. This step removes impurities on the wall that might subsequently be deposited on the etched surface [35]. Upon opening, the tube is flushed successively with deionized water, acetone, and diethyl ether. The tube is then dried at ambient temperature for 1 hour under nitrogen flow. After drying the capillary is filled with a 5% (w/v) solution of ammonium hydrogen difluoride in methanol and allowed to stand for 1 hour. The methanol is removed by nitrogen flow for 0.5 hour. After this process the capillary is sealed at both ends and heated in a modified GC oven at temperatures between 300 and 400°C for a period of 3–4 hours. The exact combination of time and temperature determines the surface morphology that is formed on the inner wall[2].

C. Chemical Modification Process

The modification process used is the same as described above for porous oxide materials, silanization/hydrosilation. The two-step process in capillaries is started by first treating the capillary with a pH 10 ammonia solution (6 mM) for 20 hours at a flow rate of 0.1–0.2 mL/h in order to rehydroxylate the surface. The capillary is rinsed with deionized water followed by a wash with 0.1 M HCl and then a second rinsing with water follows. The tube is dried with nitrogen, filled with dioxane, and flushed with a 1.0 M TES solution in dioxane for 90 minutes at 90°C. After the TES treatment, the capillary is washed with THF for 2 hours and then with 1:1 THF-water for 2 hours. After washing, the capillary is dried with a flow of dry nitrogen gas for 0.5 hour. The resulting hydride capillary is flushed with dry toluene followed by a constant flow of an olefin as a pure liquid with Speier's catalyst (10 mM hexachloroplatinic acid in 2-propanol) or an olefin dissolved in toluene with the catalyst for a period of 45 hours at 100°C. Before the olefin/catalyst solution is started, the mixture is heated to 60–70°C for a period of 1 hour. After completion of the hydrosilation reaction, the capillary is flushed successively with toluene and THF for 1 hour. After clean-up, the capillary is dried overnight at 100°C under nitrogen flow. A summary of the complete process for producing OTCEC columns by the two major steps (etching and chemical modification) described above is shown in Fig. 3.

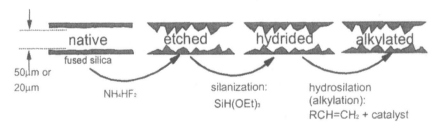

Figure 3 Summary of the complete process for producing etched, chemically modified capillaries for open tubular capillary electrochromatography.

D. Characterization of Etched, Chemically Modified Capillary Surface

A number of different approaches have been developed for characterizing the etched capillaries. In contrast to the porous particles typically used in HPLC, where the chemically modified surface is readily accessible to various spectroscopic techniques [36,37], probing the inner wall of a fused silica capillary is not a straightforward process. In most cases, it is necessary to open the capillary so that the inner surface is exposed. The procedure usually involves carefully breaking the capillary along the longitudinal axis and then examining the fragments under an optical microscope in order to select the most suitable samples for a particular characterization technique. Useful fundamental information about the properties of the etched chemically modified surface has been obtained from the five methods described briefly below.

Scanning electron microscopy (SEM) provides a picture of the surface morphology that exists on the inner wall of the capillary. This image is particularly useful for comparing the original smooth wall to the greatly roughened surface after etching as well as for comparing the effects of various etching conditions with respect to the type of new three-dimensional structures formed. A variety of 20 and 50 μm etched capillaries have been examined by SEM with respect to trying to correlate the surface morphology obtained as a function of the etching conditions [2,38]. The structures that result from the ammonium hydrogen difluoride process indicate that for short etching periods, at least 2 hours and at a temperature of 300°C, the surface appears to have the greatest amount of roughness and longer radial extensions from the surface. Using longer etching times and/or higher temperatures generally results in a diminishing of these features as the original extensions from the inner wall "melt" away to a less tortuous morphology. At best, SEM provides a good visual image of the surface but only a semi-quantitative description of the overall roughness that results from the etching process.

Atomic force microscopy (AFM) offers an alternative to SEM in providing a topological description of the etched, chemically modified surface. The two methods in combination provide complementary and confirming characterization of the inner wall that can be used to help understand the separation process in OTCEC using this format. The advantages of AFM over SEM include the ease of sample preparation because a conducting surface is not required, more precise determination of surface roughness from higher resolution in the z-direction (access to height information that is not available in SEM), and the ability to measure localized surface forces in order to identify possible solute/bonded phases interactions. The disadvantages of AFM for the characterization of OTCEC capillaries center on the difficulties in physically accessing regions of convoluted surfaces and uncertainties in rendering nonplanar surfaces. However, earlier AFM studies [39,40] have shown that some of the problems associated with nonplanar and nonuniform surfaces of capillaries coated with polyacrylamide on the inner wall can be overcome. A comparison of SEM and AFM images of the same surface for the inner bore of native, etched, and etched chemically modified capillaries revealed a good visual correspondence between the electron and force microscopy data [41]. However, there are resolution differences between the two techniques. Subtle features revealed in AFM images require high magnification by SEM that limits the scan size to areas too small to observe the larger surface components. Both AFM and SEM can easily detect the differences in surface morphology present in the four types of capillaries that are relevant to this approach to OTCEC: native, etched, etched-hydride, and etched-chemically modified. Besides images of the surface, AFM provides additional information that makes it unique as a surface characterization method for the etched, chemically modified capillaries. These measurement features include root mean square (RMS) roughness, surface area, and surface-tip forces of attraction.

In the DRIFT method, infrared radiation is reflected directly from a surface and contains the spectral information about the bulk material as well as any molecules chemically bonded or physically adsorbed. Sensitivity is enhanced through signal averaging so that many spectra can be rapidly co-added with the S/N ratio improving as the square root of the number of scans. For surface analysis, S/N can also be improved by proper choice of detector. For maximum sensitivity, the mercury-cadmium-telluride (MCT) detector is essential for the relatively low surface area materials such as the OTCEC capillaries. Analysis of the DRIFT spectrum for a cyclodextrin OTCEC capillary [42] reveals two features that are important for the characterization of the modified surface. First is the peak near 2270 cm^{-1} that can be assigned to the Si-H stretching frequency. This band indicates successful silanization of the surface in the first step of the chemical modification process of the etched inner wall. This peak does not disappear during the second step, hydrosilation, because not all Si-H groups react to produce a site with an organic moiety attached, i.e., the reaction is not 100% complete. This is essentially the same result that occurs on porous silica when a hydride surface is modified through hydrosilation [3–5]. The second important feature is the bands in the 2800–3000 cm^{-1} range that can be assigned to carbon-hydrogen stretching frequencies. These infrared absorptions indicate that the second step, hydrosilation, was also successful since these bands can only be present when the organic moiety has been attached to the surface. Neither of these features is observed on an etched capillary that has not been modified by the silanization/hydrosilation process.

Photoelectron spectroscopy, or as it is sometimes called electron spectroscopy for chemical analysis (ESCA), provides another means of characterizing the inner wall of the fused silica capillary if it is properly opened so that there is easy access to the inner bore. In the case of ESCA, a beam of x-rays must strike the surface to be analyzed so that the electrons subsequently emitted can be analyzed according to the kinetic energy they possess. The kinetic energy(ies) of the emitted electrons is characteristic of the element(s) present on the surface. Therefore, the ESCA spectrum could be used for verifying the presence of the bonded moiety on the surface as well as the components of the fused silica matrix underneath the organic layer. Indeed, ESCA spectra of etched chemically modified capillaries reveal a C (2s) peak in all cases that is indicative of the success of the hydrosilation reaction. However, more interesting results are obtained from the ESCA spectra of capillaries that have been etched but not modified. As expected, the spectra contain peaks for Si and O that are part of the fused silica matrix [9]. In addition to these two peaks, there are also present peaks for the elements F and N. The presence of these elements indicates that some part of the etching reagent (ammonium hydrogen fluoride) has been trapped in the surface matrix during the etching process. An additional ESCA experiment where the etched surface is bombarded with high-energy ions to successively remove the outermost layer of material down to a depth of ~700 Å reveals no substantial change in the elemental composition over this range. Such a result is not surprising since SEM photos reveal that surface features are as large as 3–5 μm in length as a consequence of the etching process [2,43]. The ESCA data provide unique information about the OTCEC materials and etching process that cannot be obtained from any other method.

The electroosmostic flow (EOF) of the OTCEC capillary provides an indirect means for characterizing the surface of the materials with respect to the sign and magnitude of charge on the inner wall. For example, successful modification of the surface with an organic moiety greatly reduces the number of silanol groups and hence the magnitude of the EOF. In some cases, chemical modification can introduce groups containing a positive charge, such as sulfonic acid, that reverses the direction of the EOF. First, the overall EOF of the OTCEC capillaries is quite low. The general shape of a plot of EOF as a function of pH for the OTCEC capillaries is the same for hydrophobic [38], hydrophilic [44], chiral [42], and liquid crystal [9] bonded phases. At low pH the EOF is anodic instead of cathodic. Then in the pH range of 3–4.5 the EOF changes from anodic to cathodic and remains in this direction to the highest pH measured (~8.2). The variation in pH where the EOF changes from anodic to cathodic may be a function of the etching conditions as well as the nature of the organic group bonded to the surface. The most interesting aspect of this determination of EOF as a function of pH is the anodic EOF that occurs at low pH. From the ESCA studies described above that show nitrogen as part of the surface matrix, it is probable that the ammonium moiety from the etching reagent is responsible for the anodic EOF at low pH. Under these conditions, the nitrogen would be protonated, giving a positive charge to the surface and hence the anodic EOF. As the pH is raised, the nitrogens and silanols still present lose protons so that the surface changes from a net positive to a net negative charge, resulting in the change to a cathodic EOF. The fluorine detected in the ESCA measurements could also be responsible for some negative charge in the surface matrix.

E. Applications of Etched, Chemically Modified Capillaries for CEC

While the primary advantages of CEC (high selectivity with high efficiency) were obvious early in the development of the technique, some significant limitations with respect to the analysis of basic compounds were also identified [45–50]. For the analysis of small molecules, basic compounds such as drugs and metabolites are of considerable interest. Therefore, many OTCEC investigations have focused on these compounds as a possible alternative to packed capillary CEC. Tetracyclines have been studied using OTCEC with etched chemically modified capillaries, and it was possible to obtain symmetrical peaks ($A_s < 1.5$) with reasonable efficiency (50,000–100,000 plates/m) for a variety of mixtures [2,51]. In addition, it was possible to separate by OTCEC with an etched octadecyl column mixtures of some tetracylines and their degradation products typical of practical samples that were not resolved by HPLC. A physiologically important pair of compounds also analyzed by HPLC [52], tryptamine and serotonin, were rapidly determined by OTCEC using both C_{18} and diol modified etched capillaries [53,54]. The effect of column diameter was also documented using this analyte pair. It was found that under identical conditions and using the same stationary phase (C_{18}), resolution was significantly better on a column with a 20 μm i.d. than one with a 50 μm i.d. [38]. This result also illustrates the chromatographic effect that is present in the etched chemically modified columns, since no significant change in resolution would be expected upon a decrease in capillary diameter if the compounds were separated only by electrophoretic differences. The use of an etched diol modified capillary was shown to be applicable to the analysis of aspartame, an artificial sweetener found in many food products and soft drinks [55]. The stationary phase effect has also been illustrated for another physiologically important pair of analytes, caffeine and theophylline, using two different liquid crystals bonded to the etched wall of the capillary [9]. Both longer retention and better resolution were obtained on a column utilizing a cyanopentoxy (nematic type) liquid crystal bonded material than on a column with a cholesterol moiety (cholesteric type) attached to the etched surface. A pyrimidine/purine base mixture was effectively separated by both the nematic and cholesteric (Fig. 4) type stationary phases with high efficiency [9]. As a further illustration of the chromatographic effect, the elution order changed from thymine < cytosine < adenine < guanine < adenosine on the cholesterol column to thymine < guanine < cytosine < adenine < adenosine on the cyanopentoxy column. For the separation of a mixture of the metabolic components of serotonin, the cyanopentoxy column [9] performs better than the cholesterol bonded phase. The nematic liquid crystal column displays excellent peak symmetry ($A_s \approx 1$ for all peaks) and good efficiency (N = 75,000–100,000 plates/m). In contrast, it

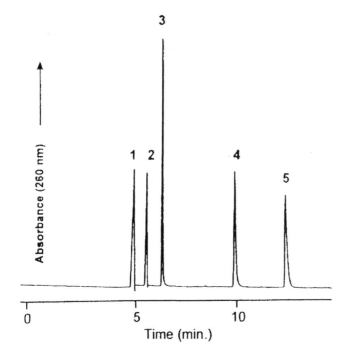

Figure 4 Separation of a pyrimidine/purine base mixture on an etched cholesterol modified capillary. Solutes: 1 = thymine; 2 = cytosine; 3 = adenine; 4 = guanine; 5 = adenosine.

has been shown that the cholesterol bonded phase is more suited to the separation of various benzodiazepine mixtures [9,18,56]. Better separation, although not necessarily better efficiency was demonstrated for a benzodiazepine mixture on an etched cholesterol modified capillary in comparison to either a bare or unetched cholesterol modified capillary [18]. The presence of solute/bonded phase interactions was also demonstrated with a benzodiazepine mixture. Retention factors (k) increased significantly on an etched cholesterol modified column with an i.d. of 50 μm in comparison to one having an i.d. of 75 μm. It was also demonstrated that resolution changed as a function of the etching conditions. Since the only aspect of the capillary that changes as a function of the etching conditions (time and temperature) is the surface area of the inner wall, this variable will control the amount of stationary phase in the column. If separation was only dependent on electrophoretic effects, changing the amount of the cholesterol moiety bonded to the wall per unit length of column should have no effect on either retention or resolution.

Similar to small basic molecules, proteins and peptides present a challenge for obtaining good analytical data in the packed capillary CEC format. The first report on the application of etched octadecyl modified capillaries for CEC describes the separation of a five-component mixture of basic peptides and proteins with all peaks having good efficiency and symmetry [2]. In addition, a detailed peak shape and migration time comparison for the peptide bradykinin was made on bare, etched, etched-hydride, and etched-chemically modified capillaries in order to verify the presence of solute bonded phase interactions. A follow-up investigation described some additional protein separations as well as a comparison of column performance for an angiotensin mixture on bare, etched diol, and etched octadecyl capillaries [57]. It was shown

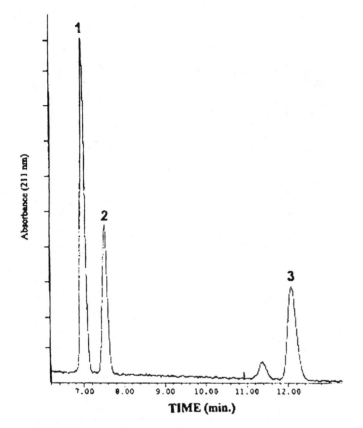

Figure 5 Separation of a mixture of angiotensins on etched diol modified capillary. Solutes: 1 = angiotensin 3; 2 = angiotensin 1; 2 = angiotensin 2.

that the diol (Fig. 5) and octadecyl capillaries had distinctly different separation capabilities under identical experimental conditions, indicating again the presence of solute/bonded phase interactions. The improved resolution that can be obtained when reducing the capillary diameter was demonstrated for a mixture of cytochrome cs (Fig. 6) on a 20 μm i.d. etched octadecyl modified capillary [38]. The main problem with using smaller-diameter capillaries is detectability for samples with limited mass or volume of material, especially with typical UV/visible absorbance detectors. Higher sensitivity detectors such as laser-induced fluorescence and mass spectroscopy would be more conducive to using the smaller-diameter capillaries. Another possible means for utilizing the etched capillary involves coating a stationary phase on the surface instead of chemically bonding an organic moiety as described above. Such an approach was investigated using Polybrene as the coating agent and testing the properties of the new surface for the separation of a protein mixture [58]. The etched liquid crystal modified capillaries were also tested for the separation of protein mixtures [9]. For the two liquid crystal moieties investigated, similar separations of the protein mixtures were obtained. This result is not surprising since previous studies in HPLC have shown that liquid crystals are effective in discriminating between small molecules based on molecular shape [13,14]. The close association of the bonded liquid crystal moieties probably helps in the resolution of small molecules but does not assist much in the separation of large molecules such as proteins.

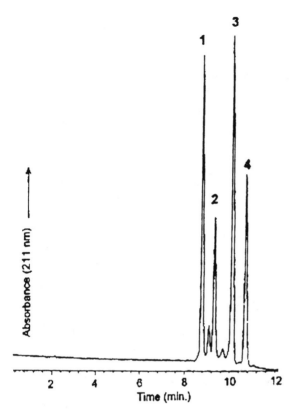

Figure 6 Separation of a mixture of cytochrome c's on a 20 μm i.d. etched octadecyl modified capillary. Solutes: 1 = horse; 2 = bovine; 3 = chicken; 4 = tuna.

Among the chiral compounds best resolved with an etched cyclodextrin capillary were benzodiazepines [42]. The four compounds that were successfully resolved included oxazepam ($\alpha = 1.05$), temazepam ($\alpha = 1.01$), diazepam ($\alpha = 1.11$), and chlorodiazepam ($\alpha = 1.04$). The enantiomeric resolution was verified by running the same sample on an achiral (octadecyl) column where only one peak was observed. As a further test of the etched open tubular approach for the analysis of optical isomers, another column was fabricated based on the selector naphthylethylamine. This column is best suited for the resolution of the optical isomers of dinitrobenzoyl methyl esters of amino acids. The best separation ($\alpha = 1.14$) was obtained for the alanine derivative. In addition, the peak symmetry and efficiency for the naphthylethylamine column was significantly better than that obtained on the cyclodextrin column.

Several types of etched capillaries have been tested with respect to within-column reproducibility. For example, the reproducibility of the migration times for 151 consecutive injections of the proteins lysozyme and ribonuclease A was tested on an etched C_{18} modified column at pH = 3.0 [53]. It was found that both solutes gave no discernible increase or decrease in migration time, and the overall reproducibility in t_M was ±1.5%. This test was conducted after more than 200 injections of other samples as well as after being installed and removed from the CE apparatus several times, which included a methanol wash. A similar test was run on a 20 μm i.d. etched C_{18} capillary using tryptamine as the test solute [38]. After the column had already been subject to more than 300 injections, the migration time of tryptamine for 30 consecutive runs was ± 1.7%. The two liquid crystal columns were also tested after considerable use for reproducibility of migration times for small basic solutes [9]. For the cholesterol capillary after more than 275 runs and for the cyanopentoxy capillary after more than 150 runs, 30 consecutive injections gave an RSD < 2% for the test solute migration time.

IV. CONCLUSIONS

Both of the developments described in this chapter, the silanization/hydrosilation process and etched chemically modified capillaries for electrochromatography, have evolved over the years since their inception and are moving toward more mature technologies. A wide variety of surfaces from porous silica and several other oxides as well as the inner walls of fused silica capillaries have been modified by the silanization/hydrosilation method. The result is a new surface generally free of the hydroxyl groups often responsible for a variety of undesirable chromatographic and electrophoretic properties. The modification method possesses several aspects that allow for the possibility of producing unique separation materials. The new open tubular approach to CEC provides a method for separating a wide variety of small basic molecules as well as peptide and protein mixtures. OTCEC will most often be complementary to packed column CEC, with each method having advantages for particular types of separation problems. Both of these technologies are still under active investigation, and further improvements in each of them are likely to further expand their potential uses.

REFERENCES

1. JE Sandoval, JJ Pesek. Anal Chem 61:2067, 1989.
2. JJ Pesek, MT Matyska. J Chromatogr 736:255, 1996.
3. JE Sandoval, JJ Pesek. Anal Chem 63:2634, 1991.
4. MC Montes, C-H Chu, E Jonsson, M Auvinen, JJ Pesek, JE Sandoval. Anal Chem 65:808, 1993.
5. JJ Pesek, MT Matyska, JE Sandoval, E Williamsen. J Liq Chromatogr Rel Technol 19:2843, 1996.
6. M Chiari, M Nesi, JE Sandoval, JJ Pesek. J Chromatogr 717:1, 1995.
7. SO Akapo, J-MD Dimandja, JJ Pesek, MT. Matyska. Chromatographia 42:141, 1996.
8. JJ Pesek, MT Matyska, M Oliva, M Evanchic. J Chromatogr A 818:145, 1998.
9. MT Matyska, JJ Pesek, A Katrekar. Anal Chem 71:5508, 1999.
10. JJ Pesek, MT Matyska, EJ Williamsen, M Evanchic, V Hazari, K Konjuh, S Takhar, R Tranchina. J Chromatogr A 786:219, 1997.
11. EF Vansant, P Van Der Voort, KC Vrancken. Characterization and Chemical Modification of Silica. Amsterdam: Elsevier, 1995.
12. Y Saito, H Ohta, H Nagashima, K Itoh, K Jinno, JJ Pesek. J Microcol Sep 7:41, 1995.
13. A Catabay, Y Saito, C Okumura, K Jinno, JJ Pesek, E Williamsen. J Microcol Sep 9:81, 1997.
14. A Catabay, M Taniguichi, K Jinno, JJ Pesek, E Williamsen. J Chromatogr Sci 36:111, 1998.
15. C Delaurent, V Tomao, AM Siouffi. Chromatographia 45:363, 1997.
16. MA Al-Haj, P Haber, R Kaliszan, B Buszewski, M Jezierska, Z Chilmonzyk. J Pharm Biomed Anal 18:721, 1998.
17. B Buszewski, M Jerierska, M Welniak, R Kaliszan. J Chromatogr A 845:433, 1999.
18. AP Catabay, H Sawada, K Jinno, JJ Pesek, MT Matyska. J Cap Electrophoresis 5:89, 1998.
19. JJ Pesek, MT Matyska, EJ Williamsen, R Tam. Chromatographia 41:301, 1995.
20. JJ Pesek, MT Matyska, S Muley. Chromatographia 52:439, 2000.
21. JJ Pesek, JE Sandoval, M Su. J Chromatogr 630:95, 1993.
22. JJ Pesek, VH Tang. Chromatographia 39:649, 1994.
23. LD Frederickson. Anal Chem 26:1883, 1954.
24. DJC Yates, PJ Luccheis. J Phys Chem 35:1346, 1961.

25. JB Peri. J Phys Chem 69:211, 1965.
26. JJ Pesek, MT Matyska, J Ramakrishnan. Chromatographia 44:538, 1997.
27. A Ellwanger, MT Matyska, K Albert, JJ Pesek. Chromatographia 49:424, 1999.
28. T Tsuda, K Nomura, G Nakagawa. J Chromatogr 248:241, 1982.
29. S Mayer, V Schurig. J High Res Chromatogr 15:129, 1992.
30. D Armstrong, Y Tang, T Ward, M Nichols. Anal Chem 65:1114, 1993.
31. S Mayer, V Schurig. J Liq Chromatogr 16:915, 1993.
32. S Mayer, V Schurig. Electrophoresis 15:835, 1994.
33. S Mayer, M Schleimer, V Schurig. J Microcol Sep 6:43, 1994.
34. E Francotte, M Jung. Chromatographia 42:521, 1996.
35. F Onuska, ME Comba, T Bistridcki, RJ Wilkinson. J Chromatogr 142:117, 1977.
36. JJ Pesek, MT Matyska. J Interface Sci 5:103, 1997.
37. JJ Pesek, MT Matyska. In: A Dabrowski, ed. Adsorption and Its Application in Industry and Environmental Protection. Amsterdam: Elsevier, 1999, Vol. I, pp. 117–142.
38. JJ Pesek, MT Matyska, S-J Cho. J Chromatogr A 845:237, 1999.
39. S Kaupp, HJ Watzig. J Chromatogr A 781:55, 1997.
40. A Cifuentes, J Diez-Mesa, J Fritz, D Anselmetti, AE Bruno. Anal Chem 70:3458, 1998.
41. PE Pullen, JJ Pesek, MT Matyska, J Frommer. Anal Chem 72:2751, 2000.
42. JJ Pesek, MT Matyska, S Menezes. J Chromatogr A 853:151, 1999.
43. JJ Pesek, MT Matyska. J Capillary Electrophoresis 4:213, 1997.
44. MT Matyska, JJ Pesek, JE Sandoval, U Parkar, X Liu. J Liq Chromatogr Rel Technol 23:97, 2000.
45. MM Dittmann, K Wienand, F Bek, GP Rozing. LC/GC 13:800, 1995.
46. MM Dittmann, GP Rozing. J Microcol Sep 9:399, 1997.
47. JJ Pesek, MT Matyska. Electrophoresis 18:2228, 1997.
48. LA Colon, Y Guo, A Fermier. Anal Chem 69:461A, 1997.
49. MM Robson, MG Cikalo, P Myers, MR Euerby, KD Bartle. J Microcol Sep 9:357, 1997.
50. MG Cikalo, KD Bartle, MM Robson, P Myers, MR Euerby. The Analyst 123:87R, 1998.
51. JJ Pesek, MT Matyska. J Chromatogr 736:313, 1996.
52. H Wakabayashi, K Shimada. J Chromatogr 381:21, 1986.
53. MT Matyska. Chem Anal (Warsaw) 43:637, 1998.
54. JJ Pesek, MT Matyska. J Liq Chromatogr Rel Technol 21:2923, 1998.
55. JJ Pesek, MT Matyska. J Chromatogr A 781:423, 1997.
56. K Jinno, H Sawada, AP Catabay, H Watanabe, NBH Sabli, JJ Pesek, MT Matyska. J Chromatogr A 887:479, 2000.
57. JJ Pesek, MT Matyska, L Mauskan. J Chromatogr A 763:307, 1997.
58. JJ Pesek, MT Matyska, S Swedberg, S Udivar. Electrophoresis 20:2343, 1999.

23

Multimodal Chromatography and Gel Electrophoresis

Haleem J. Issaq

SAIC Frederick, NCI-Frederick Cancer Research and Development Center, Frederick, Maryland

I. INTRODUCTION

The past century witnessed many advances in separation science in general and chromatography in particular. The use of visual detection was replaced by sensitive instrumental, spectroscopic, electrochemical, and other detectors. Automation, employing computers and autosamplers, allowed continuous analysis by most separation techniques. New support materials for liquid chromatography for the separation of all kinds of materials were developed. Capillary gas chromatography (GC) columns replaced packed columns. Capillary electrophoresis took over some of the applications of slab gel electrophoresis. Solid phase extraction replaced liquid/liquid extraction in many applications where sample cleaning is required prior to analysis.

Chromatographic separation is a process that involves an interaction between the stationary phase, the mobile phase, and the mixture to be resolved. Optimum separation is achieved when the interaction among these three is at an optimum. The sample has to be compatible with the other two, which means that it should be soluble to a certain extent; this is especially true in thin layer chromatography (TLC) and liquid chromatography (LC), in the mobile phase, and should not be strongly adsorbed to the stationary phase or not have any interaction with it. The mobile phase should be compatible with the stationary phase and the sample mixture.

Although many stationary phases (packing materials) and methods have been developed for TLC, GC, and high-performance liquid chromatography (HPLC), until recently there has

365

not been one stationary phase that could be used in more than one chromatographic separation mode: adsorption, partitioning, ion exchange, chiral, etc. Such phases would give the analytical chemist the capability to resolve complex mixtures of different origin and containing a wide variety of analytes as well as interfering compounds with the same column. In this chapter a discussion of multimodal separations with selected illustrative examples will be presented. The term "multimodal separations" applies to a TLC plate or prepacked HPLC column with a single or more than one support material that can be used in different chromatographic modes, whereby the only experimental parameter that changes is the mobile phase properties. By changing the mobile phase from polar to apolar the mechanism of separation changes: adsorption, partition, ion exchange, hydrophobic interaction, inclusion, etc.

Multimodal or multidimensional—which is the correct term? The author's opinion is that both terms apply to different aspects of the separation process. In certain cases both terms apply to the same system. Two examples can be cited here. A TLC plate can be developed in the first dimension in a normal phase mode, taken out, dried, turned 90°, and then developed in the second dimension in a reversed-phase mode. In two-dimensional gel electrophoresis, separation of cell proteins is carried out in the first dimension by isoelectric focusing and in the second dimension by size (see details below). Therefore, here we have two systems which are multidimensional and multimodal. "Multidimensional" will apply to the on-line coupling of two chromatographic systems or a chromatographic and a spectroscopic system. "Multimodal" is when one TLC plate or HPLC column is used in more than one separation mode, as will be discussed in the following sections.

II. THIN LAYER CHROMATOGRAPHY

A. Two Side-by-Side Stationary Phases

In the late 1970s our group developed and used a two-phase, normal phase (silica) and reversed-phase (derivatized silica), side-by-side TLC plate [1]. This gave a plate with two different selectivity phases, adsorption and partition. This plate was used for the separation of a mixture of oxidation products of cholesterol using two-dimensional development. Diethylether was used as the mobile phase for the first development on the silica gel part, while heptane-hexamethyldisilazane (48:2) was used in the second dimension, the silanized section of the plate. Later, a biphasic TLC plate, Multi-K, coated side by side with a 3 cm strip of C18 and a 17 cm strip of silica gel, was introduced by Whatman. The Multi-K plate was used for the separation of a mixture of 13 sulfonamides [2] and a mixture of conjugated bile acids [3]. The 3 cm C18 strip was used for clean-up and the 17 cm silica gel strip for the separation. Since a C18 phase requires a water-organic modifier mobile phase, some complications and limitations were encountered using this plate. Later, Whatman introduced another 20 cm precoated biphasic plate whereby the 17 cm strip is coated with reversed-phase C18 while the narrow 3 cm strip is coated with silica gel. The silica gel side is used for clean-up and/or preliminary separation, while the C18 side is used for resolving the mixture. This was a more successful and useful plate, since clean-up is done on the silica gel strip with an organic solvent that is easier to dry at room temperature than an aqueous/organic solvent. Such a plate was used in our laboratory for the resolution of the antineoplastic agent bryostatin 1 and bryostatin 2 from a *Bugula neritina* extract [4]. Clean-up was carried out on the silica gel strip, while separation of bryostatin 1 from bryostatin 2 and at least 30 other compounds was achieved on the reversed-phase side of the plate.

Another plate coated side by side with an amino modified layer with silica gel was used for the separation of a mixture of acidic and neutral glycolipids [5].

B. Single Phase Multimodal TLC Plates

Since the introduction of biphasic TLC plates for multimodal separations, single phase precoated TLC plates were produced that could be used, by changing the mobile phase, as multimodal phases. Unlike alkyl-derivatized silica gel plates, which are of hydrophobic character and are mainly used with aqueous-polar organic mobile phases in a single mode, the amino- and cyano-modified silica gel plates are of hydrophilic character and can be used as a multimodal medium for the separation of hydrophilic or charged compounds. The biphenyl-derivatized silica gel plates also exhibit multimodal properties.

1. Amino-Modified Silica Gel Plates

The chromatographic properties of this support are determined by the alkyl amino groups that are bonded to the silica surface. Because of the basic nature of the amino groups, the plate can be used as a weak ion exchanger employing aqueous buffers. The versatility of this plate is demonstrated by the separation of a mixture of adenosine and adenosine phosphates, which can be resolved in pure water or in (30:70) water-methanol, and for the separation of a mixture of naphthalene sulfonic acids in (40:60) methanol-water [6]. In addition to its ion-exchange properties, the amino-bonded precoated TLC plate was used for the separation of a mixture of phenols using a mobile phase of (50:50) acetone-chloroform [6].

2. Cyano-Modified Silica Gel Plates

The derivatization of silica gel with cyanopropyl group produces a hydrophylic stationary phase, which can be used in both reversed-phase format with aqueous-polar organic mobile phase and normal phase format with purely organic mobile phase. Jost et al. [7] illustrated the dual mechanism of separation of the cyano-modified plate by resolving a mixture of steroids, namely cholesterol, androsterone, cortexolone, corticosterone, and cortisone in (90:10) ether-acetone in the first experiment and in (60:40) acetone-water in the second experiment. In both experiments complete resolution of the five steroids was achieved.

In our laboratory we were able to resolve the anticancer agent Taxol from other closely related compounds in a complex mixture employing a cyano-modified TLC plate [8]. The methylene chloride extract of *Taxus brevifolia* bark was spotted at the corner of the cyano-precoated plate and developed in the first dimension in (45:50:5) methylene chloride–hexane–acetic acid, taken out, dried at room temperature under nitrogen, turned 90°, and developed in the second dimension in (40:25:35:0.5) water-acetonitrile-methanol-tetrahydrofuran. The cyano-modified plates were used in a multimodal format to resolve two different mixtures: nine hormone derivatives and cholesterol, and seven of its bile acid metabolites [9].

Another advantage of the cyano-modified silica gel plate is that it can be used with ion-pairing agents. For example, the separation of codeine, barbituric acid, luminal, thiogenal, and morphine hydrochloride was facilitated by developing the plate in (40:60) methanol-water and 0.1 mol/L tetraethylammonium bromide as the ion-pairing agent [7].

3. Phenyl-Modified Silica Gel Plates

As mentioned earlier, the phenyl-modified silica gel plates can be used in both normal phase and reversed-phase formats. The *Texus brevifolia* bark extract was used to demonstrate the point. After spotting, the plate was developed in (15:2:3) hexane-isopropanol-acetone in the first dimension, taken out, dried, turned 90°, and developed in the second dimension in (70:30) methanol-water [8]. Taxol was completely resolved from at least 30 other natural products in the extract.

III. HIGH-PERFORMANCE LIQUID CHROMATOGRAPHY

HPLC columns packed with amino-, cyano-, and biphenyl-modified silica particles are commercially available from several manufacturers. These columns can be run in a normal phase format using organic solvents or in a reversed-phase format using water-polar organic solvents as the mobile phases. In our efforts to develop a column support that can be used for the separation of widely different compounds, we fabricated columns made from mixed derivatized silica particles which have different separation mechanisms, i.e., multimodal properties. Mixed supports can be divided into two main groups: (a) physically mixed supports and (b) chemically mixed bonded ligands of differing properties. For example, C_8 derivatized silica and cyclodextrin derivatized silica are mixed in a predetermined ratio and packed into the same column. Conversely, a mixed ligand column is one where both ligands are bonded to the same silica particle at a predetermined ratio. In our laboratory we carried a series of studies to evaluate the utility of mixed packings and mixed ligands [10,11]. El Rassi and Horvath [12] packed a column with a mixture of anion and cation support material to achieve the separation of both anions and cations. In our laboratory a mixture of β-cyclodextrin derivatized silica and C_{18} packed in a column was used for the separation of dipeptide mixture [13]. Regnier et al. [14] observed that in protein as well as biomolecule separations, a weak cation exchange chromatography column can be made to behave in two different modes—ion exchange or hydrophobic interaction chromatography—depending on whether the salt concentration increased or decreased in the gradient mobile phase. Floyd et al. [15] simultaneously bonded ionic and hydrophobic silanes to form a mixed ligand phase, which was used for the separation of oligodeoribonucleotides by both ion exchange chromatography or hydrophobic interaction chromatography. Colmsjo and Ericsson [16] synthesized packings with two functional groups, C_{18}/cyanodecyl and C_{18}/cyanopropyl, bonded to the same silica, for the selective separation of some polycyclic aromatic hydrocarbons. Floyd and Hartwick wrote an excellent review of mixed-interaction stationary phases [17].

A. The Cyclodextrin-Bonded Silica Gel Prepacked Columns

The α-, β-, and γ-cyclodextrin prepacked HPLC columns have been shown to exhibit multimodal properties. Depending on the mobile phase used, the column can be made to behave as a reversed, normal, chiral, or ion exchange one [18]. Presented here are selected applications that show the multimodal properties of these cyclodextrin (CyD) columns. These CyD stationary phases are good candidates for use in a mixture or in series with other HPLC columns, to improve selectivity and resolution, because widely different mobile phases can be used. When mixed phases or columns of different selectivities are coupled in series, the mixed stationary phase should be compatible with the mobile phase.

1. Reversed-Phase Separations

The separations are achieved using a mobile phase of methanol/water or acetonitrile/water. The mechanism of separation is inclusion complex formation. The separation of widely different compounds, including polycyclic aromatic hydrocarbons [19], mycotoxins [19], quinones [19], cyclic [20] and acyclic nitrosoamines [21], barbiturates [22], dipeptides [23,24], and a mixture of short-chain peptides [13] was carried out. Also, the separation of aspartame (Nutrasweet) from diet soft drinks [25] and vitamins [26] were reported. In certain cases, addition of 0.1–0.01 triethylammonium acetate improved the peak shape and resolution.

2. Normal Phase Separations

These separations, as mentioned earlier, are carried out by using organic mobile phases. A mixture of substituted anilines (positional isomers) were resolved on a γ-CyD column using a mobile phase of (7.5/92.5) 2-propanol/heptane [27]. The separation of pyrazole from 4-hydroxy-pyrazole was accomplished using a β-CyD column and a mobile phase of 50% isopropanol in hexane [28]. Since then, many applications have appeared in the scientific literature employing organic phases with CyD columns.

3. Chiral and Geometrical Isomers

α-, β-, and γ-CyD columns have been used for the separation of optical, geometrical, and positional isomers as well as enantiomers, epimers, and conformers [21,22,24,27,29–34]. The CyD cavity and the inclusion complex formation are well suited for the separation of such compounds. Armstrong et al. [29] reported the separation of a series of drug stereoisomers. Others reported the separation of d- and l-amino acids [22,30,31]. The separation of the structural isomers o-, m-, and p-xylene, o-, m-, and p-cresol, and o-, m-, and p-nitroaniline, the geometrical isomers cis- and trans-clomiphene, and steroid epimers were reported [32]. Issaq et al. resolved a mixture of E and Z isomers of acyclic nitrosoamines [21] and nitrosoamino acids [33]. Ridlon and Issaq reported the separation of dipeptide stereoisomers [24].

The effects of different experimental parameters such as temperature, buffer type, concentration, and pH and organic modifier concentrations on resolution using a β-CyD columns were studied [18,30,31].

4. Separation of Ions

Armstrong et al. [27] reported the separation of chloride ion from bromide ion ($\alpha = 4.2$) and bromide ion from iodide ion ($\alpha = 1.4$) using a mobile phase of methanol/water. Nitrate was resolved from nitrite and formate from sulfate using both cyclodextrin columns and a mobile phase made of 60:40 methanol:water and 0.01% TEAA [28]. No differences in retention were observed in using the α- or the β-CyD columns, which may suggest that the separation mechanism of these ions is not through inclusion complex formation but by other forces.

5. Separation of Acids and Bases

The separation of amino acids [27], nitrosoamino acids [33], and carboxylic acids [35] were achieved using CyD columns. Abidi [36] resolved a mixture of n-alkyldimethylammonium chlorides into their homologous components using a β-CyD column and a mobile phase of methanol/water or acetonitrile/water. It was found [36] that resolution and selectivity were susceptible to changes in the percentage of the organic modifier. For example, a four-component mixture was resolved with 15% acetonitrile or 50% methanol but not in 55% acetonitrile or 70% methanol/water.

6. Separation of Organometallic Compounds

Armstrong et al. [37] reported the separation of enantiomeric pairs of ferrocenes, ruthenocenes, and osmocenes using a β-Cyd column and a mobile phase of methanol/water. Although the mechanism of separation is through complex formation, it was found that those enantiomers that formed weaker hydrogen bonds with the CyD were better resolved. Chang et al. [38] found that the mechanism of separation of several arenetricarbonylchromium (0) compounds together with the arene ligands was a mixed one where both inclusion and solvophobic interactions were involved.

7. Mixed CyD/Other Packing Materials

CyD supports are well suited, as mentioned above, due to their chemical properties, to be used with other supports to improve selectivity and resolution. Issaq et al. [10] studied the selectivity of mixed β-CyD/C18 materials in the same column. Also, a β-CyD column was used in series with a C18 reversed-phase column [10] and SCX column [39]. In both cases separations were achieved that were not possible with either column alone.

B. Alkyl Modified Silica Gel Packing Materials

Nahum and Horvath [40] and Kennedy et al. [41] observed that in reversed-phase chromatography, two different modes may be employed, depending on whether the mobile phase has a high or low content of water. This effect is due to the presence of residual silanol groups.

C. Column Switching

Complex mixtures can be resolved by HPLC employing the column-switching technique, whereby one column is used for preliminary separation and a second column for further separation. Both columns should resolve the mixture by different mechanisms to qualify as a multimodal technique. Also, column switching is used for sample cleaning on the first column, and separation is achieved with the second column. We used a multimodal column switching procedure to resolve the caffeine (1,3,7-trimethylxanthine) metabolite AFMU (5-acetylamino-6-formylamino-3-methluracil) from other metabolites in human urine [42]. The first column was a 5 μm Hypersil ODS column from Hewlett-Packard, while the second column was a Bio-Gel SEC 20-XL coumn from BioRad. For details, see original publication [42].

IV. GAS CHROMATOGRAPHY

To achieve multimodal separations in gas chromatography, two columns with different separation mechanisms are employed. A peak of interest, eluted from the first column, is transferred to the head of the second column for further separation. As in HPLC, the two columns can be used to achieve complete separation of a mixture, or the first column is used for clean-up while the second column is used for resolving the mixture. Duinker et al. [43] used two capillary columns of different polarities for the separation of polychlorinated biphenyl (PCB) congeners. Sonchick [44] used column switching for the separation of PCBs, whereby the first column (SPB-1, low polarity) resolves the PCBs from the solvent and the second column (DB-1701, medium polarity) is used for the separation. These two studies did not take advantage of different modes of GC separation, but only of differences in column polarity.

Issaq et al. [45] developed a multimodal dual column GC procedure for the separation of PCB congeners. The first column separates the solute mixture on the basis of vapor pressure, while the second column separates them based on the geometrical shape of the molecule. The heart-cut from the DB-1 column is transferred to the head of a smectic liquid crystalline open tubular capillary column. The liquid crystalline stationary phase provides enhanced separation of isomeric mixtures with nearly identical vapor pressure on the basis of solute geometry [46,47]. In other words, solutes that coelute on a column that is based on vapor pressure or polarity are resolved on a liquid crystalline column based on differences in their geometry.

V. GEL ELECTROPHORESIS

Gel electrophoresis can be used in multimodal format, as mentioned above, to achieve the separation of complex mixtures. The separation of cellular proteins by two-dimensional electrophoresis is a good example. Proteins are separated by two sequential steps: first by their charge and then by their mass—two different modes of separation. In the first step, a cell extract is spotted on a polyacrylamide gel and resolved in a pH gradient based on their isoelectric point (pI), the pH at which the net charge of the protein is zero. This technique, called isoelectric focusing (IEF), can resolve proteins that differ by one charge. Proteins that have been separated on an IEF gel can then be separated in the second dimension, based on their molecular weights, on a polyacrylamide slab gel saturated with sodium dodecyl sulfate (SDS). As many as 1000 proteins can be resolved by this multidimensional-multimodal technique. For examples, the reader is referred to a special issue of *Electrophoresis* devoted to the proceedings of the Third Siena 2-D Electrophoresis Meeting in Italy in 1988 [48].

ACKNOWLEDGMENT

This project has been funded in whole or in part with federal funds from the National Cancer Institute, National Institutes of Health, under Contract No. NO1-CO-56000.

REFERENCES

1. HJ Issaq, N Risser. Pittsburgh Conf. on Analytical Chemistry and applied Spectroscopy, March 5–9, 1979, Cleveland, OH, paper No. 305. Also J Liq Chromatogr 3:841, 1980.
2. J Sherma. Practice and Application of Thin Layer Chromatography on Whatman KC18 Reversed Phase Plates. Clifton, NJ: Whatman TLC Technical Series, Vol. 1, 1983.
3. RE Levitt, JC Touchstone. J High Resolut Chromatogr Chromatogr Commun 2:587, 1979.
4. HJ Issaq, K Seburn, P Andrews, DE Scaufelberger. J Liq Chromatogr 12:3129, 1989.
5. JG Alvarez, JC Touchstone. J Chromatogr 436:515, 1988.
6. W Jost, HE Hauck. J Chromatogr 261:235, 1983.
7. W Jost, HE Hauck, W Fischer. Chromatographia 21:375, 1986.
8. MW Stasko, KM Whiterup, T Ghiorzi, TG McCloud, SA Look, GM Muschik, HJ Issaq. J Liq Chromatogr 12:2133, 1989.
9. W Jost, HE Hauck. Personal communication, E. Merck, Darmstadt, Germany, 1988.
10. HJ Issaq, DW Mellini, TE Beasley. J Liq Chromatogr 11:333, 1988.
11. HJ Issaq, J Gutierrez. J Liq Chromatogr 11:2851, 1988.
12. Z El Rassi, C Horvath. J Chromatogr 359:255 1986.
13. HJ Issaq. Unpublished results.
14. LA Kennedy, W Kopaciewicz, FE Regnier. J Chromatogr 359:73, 1986.
15. TR Floyd, LW Lu, RA Hartwick. Chromatographia 21:402, 1986.
16. AL Colmsjo, MW Ericsson. Chromatographia 21:392, 1986.
17. TR Floyd, RA Hartwick. Mixed-interaction stationary phases for HPLC. In: Cs Horvath, ed. High Performance Liquid Chromatography, Vol. 4. Orlando, FL: Academic Press, Inc., 1986, pp. 45–90.
18. HJ Issaq. I Liq Chromatogr 11:2131, 1988.
19. DW Armstrong, A Alak, W DeMond, WL Hinze, TE Riehl. J Liq Chromatogr 8:261, 1985.
20. HJ Issaq, JH McDonnel, DE Weiss, DG Williams, JE Saavedra. J Liq Chromatogr 9:1783, 1986.
21. HJ Issaq, M Glennon, DE Weiss, GN Chmurny, JE Saavedra. J Liq Chromatogr 9:2763, 1986.
22. DW Armstrong, W DeMond. J Chromatogr Sci 22:411, 1984.
23. HJ Issaq. J Liq Chromatogr 9:229, 1986.
24. CD Ridlon, HJ Issaq. J Liq Chromatogr 9:3377, 1986.

25. HJ Issaq, DE Weiss, CD Ridlon, SD Fox, GM Muschik. J Liq Chromatogr 9:1791, 1986.
26. DW Armstrong, A Alak, K Bui, W DeMond, T Eard, TE Riehl, WL Hinze. Incl Phenomena 2:533, 1984.
27. CA Chang, Q Wu. J Liq Chromatogr 10:1359, 1987.
28. HJ Issaq. J Liq Chromatogr 11:2131, 1988.
29. DW Armstrong, TJ Ward, RD Armstrong, TE Beesley. Science 232:1132, 1986.
30. WL Hinze, TE Riehl, DW Armstrong, W DeMond, A Alak, T Ward. Anal Chem 57:237, 1985.
31. T Takenchi, H Asai, D Ishii. J Chromatogr 357:409, 1986.
32. DW Armstrong, W DeMond, A Alak, WL Hinze, TE Riehl, KH Bui. Anal Chem 57:234, 1985.
33. HJ Issaq, DW Williams, N Schultz, JE Saavedra. J Chromatogr 452:511, 1988.
34. HJ Issaq, ML Glennon, DE Weiss, SD Fox. In: WL Hinze, DW Armstrong, eds. Ordered Media in Chemical Separations. Washington, DC: ACS Symposium Series 342, 1986, pp. 260–271.
35. KG Feitsina, J Bosman, BFH Drenth, RA DeZeeuw. J Chromatogr 333:59, 1985.
36. SL Abidi. J Chromatogr 362:33, 1986.
37. DW Armstrong, W DeMond, BP Czech. Anal Chem 57:481, 1985.
38. CA Chang, H Abdel-Aziz, N Melchor, Q Wu, KH Pannell, DW Armstrong. J Chromatogr 347:51, 1985.
39. HJ Issaq, GM Muschik. Am Lab May:23–31, 1988.
40. A Nahum, C Horvath. J Chromatogr 203:53, 1981.
41. LA Kennedy, W Kapaciewicz, FE Regnier. J Chromatogr 359:73, 1986.
42. DW Mellini, NE Caporaso, HJ Issaq. J Liq Chromatogr 16:1419, 1993.
43. JC Duinker, DE Scultz, G Patrick. Anal Chem 60:478, 1988.
44. SM Sonchick. J Chromatogr Sci 24:32, 1986.
45. HJ Issaq, SD Fox, GM Muschik. J Chromatogr Sci 27:172, 1989.
46. GM Janini. Adv Chromatogr 17:231, 1979.
47. GM Janini, GM Muschik, HJ Issaq, RJ Laub. Anal Chem 60:1119, 1988.
48. Electrophoresis 20:639–1122, 1999.

24

Field Flow Fractionation

Haleem J. Issaq

*SAIC Frederick, NCI-Frederick Cancer Research
and Development Center, Frederick, Maryland*

Field-flow fractionation (FFF) is a family of chromatographic-like elution techniques in which an external field or gradient, rather than partitioning between phases, causes differential retention. FFF techniques were conceived and developed by Professor J. Calvin Giddings starting in 1965, and the theory was first published in 1968 [1]. The method has now become a powerful means for the separation of polymers, colloids, and particles from a wide variety of fields, ranging from medicine to fabrication to environmental studies. For example, FFF has been successfully used for the separation of a variety of biological materials spanning a broad molecular weight and diameter range including proteins [2], protein aggregates [3–6], protein polymer conjugates [7], lipoproteins [8], DNA [4,5,9], viruses [5,10,11], bacteria [3,12], and cells [13]. Also, FFF has been shown to be suitable for the separation of polymers and for the determination of the size distribution of polymer systems [14–16]. For other applications the reader is referred to a special issue on FFF of the *Journal of Liquid Chromatography and Related Technology* [17] honoring Professor Giddings.

FFF can also be used for measuring colloidal and macromolecular primary and secondary properties (Table 1), which are often difficult or impossible to measure using other techniques, obtained using different fields [18]. Using a variety of fields and simple force equations (see below), physicochemical data can be acquired, including particle and polymer mass and molecular weight, density, hydrodynamic diameter, equivalent spherical diameter, charge, diffusion coefficient, and thermal diffusion coefficient [18].

By acceptance of this article, the publisher or recipient acknowledges the right of the U.S. Government to retain a nonexclusive, royalty-free license and to any copyright covering the article. The content of this publication does not necessarily reflect the views or policies of the Department of Health and Human Services, nor does mention of trade names, commercial products, or organizations imply endorsement by the U.S. Government.

Table 1 Primary and Secondary Properties of Colloids and Macromolecules Measurable by FFF

Primary property	Secondary property
Mass, molecular weight (m)	Colloid aggregation
Effective spherical diameter (d)	Adsorbed films (mass, thickness)
Hydrodynamic diameter (d_h)	Particle shape
Density ($P\rho$)	Shell structure (mass, thickness)
Diffusion coefficient (D)	Particle composition
Thermal diffusion factor (α)	Particle porosity
Electrical charge (q)	Polymer molecular weight
Hydrodynamic lift force (F_L)	Polymer composition
	Surface composition
	Particle shape, deformation

Arrows indicate routes (not comprehensive) from primary to secondary properties.
Source: Ref. 17.

In addition, FFF is readily coupled to other instruments for the measurement of difficult parameters of large polymers, microgels, simple and complex colloids, various association complexes, and diverse particles ranging in size from 1nm to 100 μm [3].

The FFF separation is carried out in a thin ribbon-like channel (75–250 μm) by interaction with an external field, applied perpendicular to the channel, rather than an internal stationary phase. Separation occurs by differential retention of solute within a laminar flow stream bounded by thin parallel plates [3]. Because an open flow channel and simple physical forces are used in this technique, FFF is a widely applicable method for macromolecules with great flexibility in sample type, carrier liquid or solvent, pH, ionic strength, etc. The absence of channel packing material means that there is minimal possibility that biological materials will be altered or denatured by interaction with the surface as in liquid chromatography. The theory of FFF has been extensively discussed [3,18–20].

The principles of FFF are shown in Fig. 1. A stream of carrier liquid is introduced at one end of the channel and a small volume sample is injected. The injected sample spreads out across the channel breadth and proceeds down the channel undergoing separation. The separation process originates in the flow profile across the narrow dimension of the channel, which is parabolic in form. For parabolic flow, the flow velocity approaches zero at the walls (Fig. 1b). An external driving force is applied on the contents of the channel in a direction perpendicular to the flow axis. The injected components are driven by the applied force toward one of the walls, the accumulation wall, which causes the components of the mixture to have different velocities and thus elute at different times.

There are four main different FFF techniques determined by the different applied driving forces: (a) sedimentation FFF, (b) flow FFF, (c) thermal FFF, and (d) electrical FFF. Giddings derived a force equation for each of the four techniques [Eqs. (1)–(4)]. From these equations, he was able to determine different physicochemical properties [18].

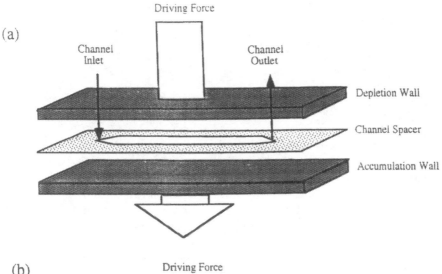

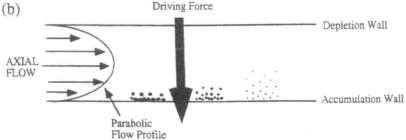

Figure 1 Schematic diagram of FFF channel (a) and separation process (b). (From Ref. 8.)

$$F \text{ (sedimentation)} = m'G = mg(\Delta\rho/P\rho) = V_p\, \Delta\rho G = (\pi/6)d3\, \Delta\rho G \tag{1}$$

$$F \text{ (flow)} = fU = 3\pi\eta dHU = (kT/D)U \tag{2}$$

$$F \text{ (thermal)} = D_t f(dT/dx) = \alpha k(dT/dx) \tag{3}$$

$$F \text{ (electrical)} = qE = \mu fE \tag{4}$$

where

    m' = effective mass m = particle mass
    V_p = particle volume
    $\Delta\rho$ = density difference
    $P\rho$ = particle density
    d = effective spherical diameter
    d_h = hydrodynamic diameter
    f = friction
    D = ordinary diffusion coefficient
    D_t = thermal diffusion coefficient
    α = thermal diffusion factor
    q = effective charge

$$\mu = \text{electrophoretic mobility}$$
$$dT/dx = \text{temperature gradient}$$
$$E = \text{electrical field strength}$$
$$G = \text{acceleration}$$
$$k = \text{Boltzman's constant}$$
$$T = \text{absolute temperature}$$
$$U = \text{cross flow velocity}$$
$$\eta = \text{viscosity.}$$

Flow FFF is a universal separation technique that can separate and measure virtually all classes of macromolecules and particles in aqueous and nonaqueous environments. In flow FFF, a cross-flow stream is applied as an external field [3,21]. The level of retention is determined by the flow rate of the cross-flow stream and the sizes of diffusion coefficients of the separated components. Because smaller components have larger diffusion coefficients, they form thicker steady-state layers and thus have a higher velocity in the parabolic flow stream. Consequently, smaller components are eluted first followed by the larger components.

Sedimentation FFF is applied for the separation and characterization of colloidal particles. Particle mass governs sedimentation force and is thus measurable by FFF. Thermal FFF has been used primarily for fractionation of polymers, especially those of high molecular weight. It is also applicable to particles in both aqueous and nonaqueous media. Electrical FFF is capable of measuring the charge residing on macromolecular and colloidal species. For recent advances and applications of electrical FFF, consult Refs. 22 and 23.

Although FFF is an excellent separation technique for macromolecules and colloids and for the measurement of many of their primary and secondary properties, its use has not become universal.

ACKNOWLEDGMENT

This project has been funded in whole or in part with federal funds from the National Cancer Institute, National Institutes of Health, under Contract No. NO1-CO-56000.

REFERENCES

1. JC Giddings. J Chem Phys 49:81, 1968.
2. M-K Liu, P Li, JC Giddings. Protein Sci 2:1520, 1993.
3. JC Giddings. Science 260:1456, 1993.
4. KG Wahlund, A Litzen. J Chromatogr 461:73, 1989.
5. A Litzen, KG Wahlund. J Chromatogr 476:413, 1989.
6. A Litzen, KG Wahlund. Anal Biochem 212:469, 1993.
7. JC Giddings, MN Benincasa, M-K Liu, P Li. J Liq Chromatogr 15:1729, 1992.
8. P Li, M Hansen, JC Giddings. J Liq Chromatogr Rel Technol 20:2777, 1997.
9. M-K Liu, JC Giddings. Macromolecules 26:3576, 1993.
10. JC Giddings, F Yang, MN Myers. J Virol 21:131, 1977.
11. A Litzen, J Wahlund. Anal Chem 63:1001, 1991.
12. JJ Kirkland, CH Dilks, Jr., SW Rementer, WW Yau. J Chromatogr 593:339, 1992.
13. BN Barman, ER Ashwood, JC Giddings. Anal Biochem 12:35, 1993.
14. JC Giddings, MN Myers, J Janca. J Chromatogr 186:261, 1979.
15. SL Brimhall, MN Myers, KD Caldwell, JC Giddings. Sep Sci Technol 16:671, 1981.
16. JC Giddings, M Martin, MN Myers. J Polym Sci Chem Ed 19:815, 1981.

17. J Liq Chromatogr Rel Technol 20:2509–2929, 1997.
18. JC Giddings. Anal Chem 67:592A–598A, 1995.
19. MA Benincasa, JC Giddings. Anal Chem 64:790, 1992.
20. JC Giddings. Analyst 118:1487, 1993.
21. JC Giddings, ed. Unified Separation Science. New York: John Wiley, 1991.
22. KD Caldwell, Y-S Gao. Anal Chem 65:1764, 1993.
23. N Tri, K Caldwell, R Beckett. Anal Chem 72:1823, 2000.

25

The Impact of Analytical Supercritical Fluid Extraction and Supercritical Fluid Chromatography on Separation Science

Jerry W. King

National Center for Agricultural Utilization Research,
Agricultural Research Service/USDA, Peoria, Illinois

I. INTRODUCTION AND HISTORICAL PERSPECTIVE

The role of supercritical fluid media, or ''critical'' fluid technology (which also includes the near critical state) in analytical separation science developed from several diverse sources, originating as a hybrid technique of gas chromatography in the mid-1960s. The most often cited origins of supercritical phenomena are from phase equilibria studies or geochemical processes conducted at high pressures and temperatures [1]. Interest in this particular ''intermediate state'' of matter also found early application in petroleum recovery [2], which was followed by pioneering research in the late 1960s on applying supercritical fluids for the extraction of solutes (e.g., caffeine) and as a versatile reaction medium (polymerizations, hydrogenations). In parallel with this latter trend was the observation that chromatographic retention parameters were effected by column pressure when utilizing nonideal fluids such as carbon dioxide as eluents. This gave rise to a different form of gas chromatography called ''dense gas chromatography'' [3], which permitted intractable solutes of limited volatility to be solubilized in the mobile phases and separated as a function of eluent pressure or density.

Without doubt the most recognizable advocate for this new separation technique was J. Calvin Giddings of the University of Utah [4], from whom the author received his initial training and introduction to the field. These early academic-based studies in dense gas chromatography (GC) were frustrating due to the lack of commercial equipment as well as problems in sample

---

Names are necessary to report factually on available data; however the USDA neither guarantees nor warrants the standard of the product, and the use of the name by USDA implies no approval of the products to the exclusion of others that may also be suitable.

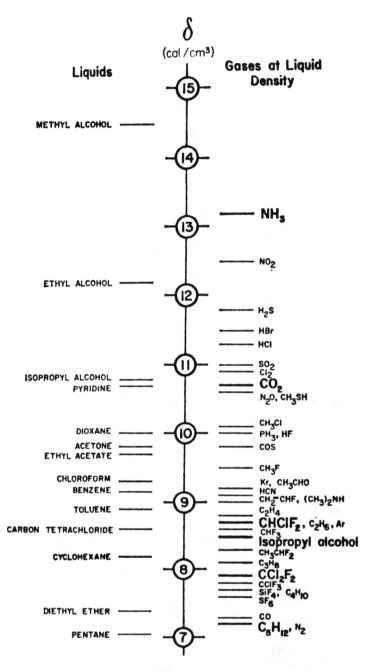

Figure 1 Solubility parameters of compressed gases at liquid-like densities compared to solubility parameters of liquids in their condensed state. (Reproduced with permission of American Association for the Advancement of Science.)

introduction and solute detection under such extreme conditions. It was not until the mid-1970s that high-performance liquid chromatography (HPLC) injection techniques were merged with bonded phase column technology [5] to produce a viable analytical chromatograph, which was eventually commercialized in a open tubular column format in the early 1980s.

One of the seminal concepts to come out of the early dense GC studies was the correlation of gas (supercritical fluid) solvent power with that exhibited by various liquids as characterized by the solubility parameter concept [6]. As illustrated in Fig. 1, dense gases can attain solubility parameters at a high level of compression that are equivalent to those of liquid solvents; however, the ability to mechanically adjust the solvent power and hence the selectivity of these fluids by regulating the applied pressure is a key feature that makes supercritical fluid chromatography (SFC) and extraction (SFE) unique. Figure 2 illustrates this principle nicely, where the solubility parameter of the fluid is shown as a function of pressure at several different temperatures. Over the given temperature range, it is apparent that CO_2 attains a higher solubility parameter than either nitrogen or helium. This is due to the convient critical temperature for CO_2, 31°C, and the nonideal fluid properties exhibited by CO_2 over this particular temperature range. Figure 2 also shows that no matter the temperature, CO_2 experiences a considerable increase in solvent power close to its critical pressure of 72 atm (1060 psi).

Although solubility parameter theory was only intended to provide an approximate estimate of a compressed fluid's solubility properties, it has been refined by this author [7,8] and others [9,10] to provide excellent quantitative prediction and correlations for many solute types in SFE and SFC. This basic concept obtained from initial SFC studies has been profusely cited in the literature up to the beginning of the new millennium and has served as a guide in both analytical as well as engineering applications of supercritical fluids.

Retention patterns of solutes in supercritical fluids are pressure or fluid density dependent, as indicated in Fig. 3, for even a simple solute such as benzene when using supercritical carbon dioxide (SC-CO_2) as an eluent. Chromatographic elution parameters such as capacity factors or retention volumes can vary over several magnitudes as a supercritical fluid eluent's pressure is

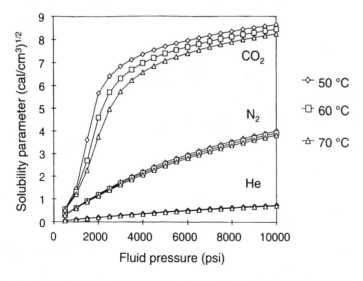

Figure 2 Solubility parameters for helium, nitrogen, and carbon dioxide versus fluid pressure. (From Ref. 35.)

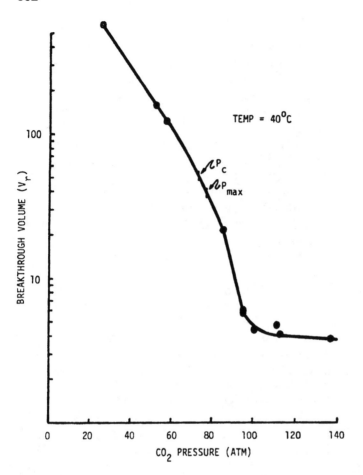

Figure 3 Retention volume for benzene versus carbon dioxide pressure on a styrene/divinylbenzene column. (From Ref. 11.)

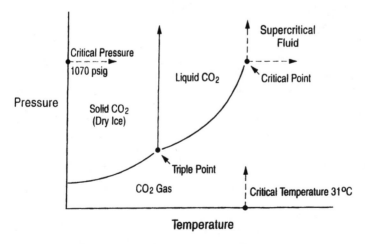

Figure 4 Phase diagram for carbon dioxide. (From Ref. 48.)

changed; a reflection of solute's solubility enhancement in the supercritical fluid or the fluid's interaction with the stationary phase in the chromatographic column [11,12]. This serves as the basis for separating solutes when chromatography is conducted in either the isobaric, isconfertic, or pressure programmed modes. Pressure programming has become by far the most popular mode when performing SFC, although modern instrumentation also allows temperature and flow programming to be enacted during the analysis. The addition of organic solvents to the mobile phase became a requisite with the introduction of packed column SFC [13], for enhancing solute solubility in the mobile phase, and for adjustment of an analyte's retention durng SFC analysis. These addition components to the mobile phase are called cosolvents or modifiers and can only be dissolved in a finite amount in the supercritical fluid at a given pressure and temperature to avoid a two-phase eluent system.

Differences in opinion have arisen over the years as to the need to preserve the fidelity to a true one-phase eluent system in SFC. As show in Fig. 4, the conventional supercritical state

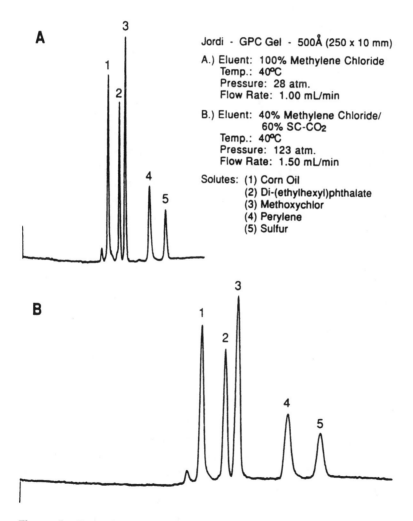

Figure 5 Comparison of conventional liquid-based SEC separation vs. SEC separation using liquid eluent and SC-CO$_2$.

for fluids such as CO_2 was often defined as that region above the fluid's critical temperature and pressure, respectively. However successful chromatographic separations have been attained in the liquid CO_2 region and by using pressures and temperatures below the critical point of eluent. This arbitrary boundary (indicated by the dashed lines in Fig. 4) for the supercritical state has been noted by Chester [14] and has perhaps retarded the use of experimental conditions outside the conventional definition of the supercritical state. Recently, solvating gas chromatography [15] has been reported under conditions outside the arbitrary boundary of the supercritical fluid state, although such conditions were also utilized by Sie and coworkers [16] over 30 years ago. King and coworkers [17] have demonstrated that two-phase eluent systems, in which one of the mobile phase components is a fluid under supercritical conditions and one a liquid, can yield useful analytical separations. This is demonstrated in Fig. 5, where the addition of the weaker eluent component, SC-CO_2, to methylene chloride does not detract but actually enhances the separation of a five-component mixture. Similar beneficial effects have been attained using a supercritical component in conjunction with a conventional liquid solvent by Olesik [18], called enhanced fluidity chromatography.

II. THE MATURATION OF SFC AND ITS IMPACT ON ANALYTICAL SEPARATION SCIENCE

SFC matured as a viable analytical technique throughout the 1980s and 1990s as documented in well-known texts by Lee and Markides [19], Berger [20], and Anton and Berger [21]. Curiously, it was the technique of packed column SFC that first became available commercially through the efforts of the Hewlett Packard Company [22]; although this development was followed by the introduction of capillary instrumentation in the United States by the Lee Scientific Company. This latter mode of SFC became quite popular and served as the basis for several well-known symposia that exhibited the potential of the technique throughout the 1980s.

The use of capillary SFC was not without its pitfalls, being limited to a practical extent by very narrow-bore columns having limited solute capacity and the use of the flame ionization detector. Nonetheless, when performed properly, capillary SFC utilizing SC-CO_2 could be the method of choice when analyzing many complex mixtures. As a trace analysis technique (ppm and ppb), capillary SFC as well as packed column SFC was limited by the analyst's ability to couple many of the hetero-element detectors (ECD, NPD, etc.) to achieve stable, reproducible analysis. Some success has been achieved using either ultraviolet (UV) or evaporative light scattering detection (ELSD), particularly when coupled with packed column SFC. The efforts of Taylor and colleagues [23] in attempting to utilize a variety of detectors in SFC are worth noting.

The true worth of a separation technque is found in its routine application to everyday analysis problems. As noted in Table 1, capillary SFC excels in several areas of analysis, most notably in the environmental analysis of petroleum derivatives, the characterization of oligomer mixtures of natural and synthetic polymers, and class separations of species having significant molecular weight differences. This has resulted in several standard methods, including the extension of simulated distillation curves over that which can be achieved using gas chromatography. The method also permits class separations of lipid species [24] and has found routine application in preference to GC and HPLC in the author's laboratory. As illustrated in Fig. 6, relative short analysis times (45 min) can be achieved using capillary SFC to resolve the components found in a deodorizer distillate sample. The elution order in this case, as with many SFC separations, is based on molecular weight differences in the solutes, thereby allowing large separation factors to be achieved between such lipid species as free fatty acids, mono-, di-, and triglycerides, and,

Table 1 Successful applications of SFC

Hydrocarbon and PAH mixtures
Petroleum compound class separations
Oligomeric mixture fractionation
Lipid class separations
Essential oils
Simulated distillation chromatography
Separation of enantiomers
Surfactants
Explosives
Pesticides

in some cases, phospholipids [25]. Packed column SFC lagged behind capillary SFC in application for many years, but in the late 1990s it has undergone a renaissance due to the development of more versatile instrumentation and niche applications in both pharmaceutical and petrochemical analysis. The high-resolution factors attained by using packed column SFC for the separation of enantiomers [26] has seen SFC establish itself as a competitive technique with HPLC, and class separations of petrochemical mixtures has become an official method. The ability to use SFC in conjunction with flame ionization detection has also provided the pharmaceutical analyst

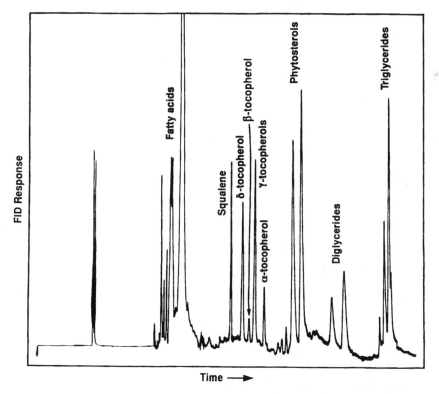

Figure 6 Separation of lipid components in deodorizer distillate by capillary SFC. (From Ref. 25.)

Table 2   Applications of SFC in Industrial Analysis

Elimination of sample preparation
Deformulation of commercial products
Raw material specifications
Monitoring of reaction products and kinetics
Support of supercritical fluid research
Analysis of minor components
Characterization of small samples (with SFE)
Analysis of thermally labile analytes
Determination of physicochemical properties

with a complementary method to HPLC analysis, particularly for the detection of nonchro-maphoric contaminents in drug formulations and raw materials.

SFC has proven itself directly applicable in industrial analysis, as noted in Table 2. The ability to elute both the components of interest as well as interfering components using pressure or density programming can reduce or eliminate sample preparation prior to analysis. Programming of the SFC mobile phase can also be utilized for deformulating specific types of products into their individual constituents, as demonstrated in Fig. 7 for the capillary SFC resolution of a lipstick formulation [27]. SFC has also proven facile in the author's laboratory in support of supercritical fluid-based research projects, i.e., the monitoring of enzyme-initiated reactions in SC-CO$_2$ and the characterization of extracts obtained via SFE with SC-CO$_2$.

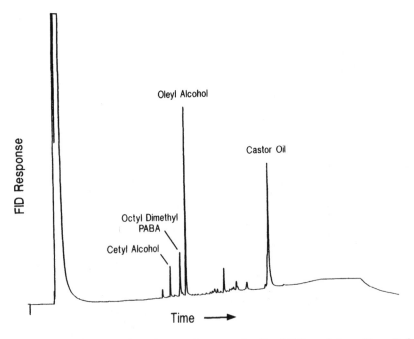

Figure 7   Separation of components in a quencher lipstick formulation. (From Ref. 27.)

Table 3 Determination of
Physicochemical Properties by SFC

Diffusion coefficients
Sorption isotherms
Phase distribution constants
Solubility measurements
Critical loci
Partial molar volumes
Virial coefficients

The measurement of physicochemical properties by SFC is an often overlooked application of the technique. Table 3 lists some of these properties that have been determined by SFC. The diffusion coefficients of dissolved solutes in dense fluids were first measured as early as 1968 by the chromatographic band broadening method, but more diffusion coefficient data need to be collected on solutes that are feasible industrially to extract with CO_2. Quasi-equilibrium properties, such as phase distribution constants or solubility measurements, can be rapidly achieved using SFC [28], particularly on a relative basis when the value of a reference compund is known beforehand. Retention parameter shifts under certain conditions will yield solute partial molar volumes or second interaction virial coefficients for solute-fluid interactions. So-called threshold pressures, first approximated by the author in enhanced migration studies with Giddings and Myers [4], can be measured by SFC, as indicated in Fig. 8 for the pesticide malathion in SC-CO_2. This critical locus was established by measuring the first appearance of a peak for malathion on a nitrogen-phosphorus detector downstream from the SFC column, which is

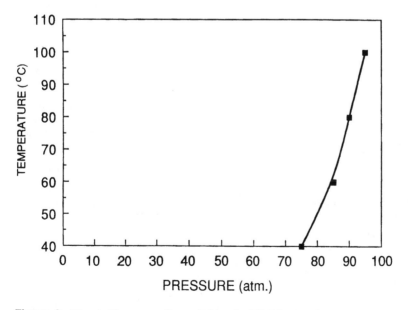

Figure 8 Threshold pressure for malathion in SC-CO_2 as a function of temperature and pressure as determined by SFC with a nitrogen/phosphorus detector.

Table 4 Detection Limits of Techniques
Used to Assess Threshold Pressures in
Supercritical Fluids

Detection principle	Sensitivity (g)
Gravimetric	10^{-3}
Infrared spectroscopy	10^{-7}
Ultraviolet spectroscopy	10^{-9}
Mass spectrometry	10^{-9}
Flame ionization	10^{-10}
TLC visualization	10^{-12}

commensurate with the solvation of the pesticide by SC-CO$_2$. However, as noted by the author [29], threshold pressures can depend on the measurement technique as shown in Table 4, with chromatographic-based methods being particularly sensitive to the dissolved solute's concentration in the fluid phase due to the detection methods employed.

Analytical-scale SFC, even at low resolution, can provide useful data for optimizing critical fluid based processes. The ease or difficulty in extracting solutes from a particular sample matrix can be assessed by using elution pulse chromatography [30], and minature sorbent-filled columns can likewise be used in the SFC mode to select sorbents for the preparative/production scale SFC of numerous compounds. The author [31] and others [32] have used this method to optimize SFC conditions for the enrichment of tocopherols or phospholipids from agriculturally derived products or to predict breakthrough of volatiles and nonvolatile compounds from sorbents. An outstanding example of this experimental philosophy was the development of appropriate conditions for the enrichment and fractionation of ethyl esters of fish oils by Lembke [33] using an 200 × 4 mm, i.d. analytical column. This process was eventually scaled up to production plant size in Tarragona, Spain.

III. THE DEVELOPMENT OF ANALYTICAL SFE

Analytical SFE was developed after SFC, although the basis of the technique was available in the engineering literature in the late 1960s. Pioneering studies by Stahl and coworkers [34] actually demonstrated the potential of SFE for both processing and analytical purposes utilizing thin layer chromatography. Analytical chromatographers using SFC initially foresaw that SFE would be a complementary on-line technique to SFC, applicable to relatively small samples that would yield extracts commensurate with the operation of either capillary or packed column SFC. However, in the early 1990s, concern about the use and disposal of organic solvents in the analytical laboratory became a focal point for government agencies such as the Environmental Protection Agency (EPA) and the National Institute for Occupational Safety and Health (NIOSH), and related agencies such as the U.S. Department of Agriculture (USDA) and Food and Drug Administration (FDA) responded in kind to demonstrate their compliance with these environmental and worker safety concerns.

This format provided our introduction into analytical SFE rather than the practice of SFC. Our laboratory had already been involved in developing processing concepts using supercritical fluids for close to a decade, hence we could incorporate concepts used in process SFE into the design of an optimal analytical SFE system for toxicant and nutritional analysis of foods and agricultural products. This was the beginning of a synergestic approach in our critical fluid research, which allowed developments in processing, or vice versa in analytical, to be transferred between these related fields [35]. Research for a related USDA agency, the Food Safety and Inspection Service (FSIS), set the guidelines for the development of our off-line SFE methodology. This required that analytical SFE be capable of replacing several liters of organic solvent used per sample in FSIS's traditional methodology, and that sample sizes of the order of 25–50 g be extracted.

Initial studies focused on the quantitative extraction of lipid (fat) from meat samples since the analytes of interest (pesticides) were contained in the peritoneal fat phase of the meat matrix. The moisture content of these sample matrices proved to be inhibitory in allowing contact between the SC-CO$_2$ and the lipid phase, and this was eventually solved by dehydrating the meat matrix initially and later simplified by the use of "extraction enhancer" called Hydromatrix (Varian, Harbor City, CA). This was the beginning of a number of original contributions of our research group in applying SFE to food and natural product analysis. Other research teams in parallel were beginning to apply SFE to the analysis of environmental samples [36], polymers [37], and drugs [38], each area characterized by unique problems that had to be solved. For example, SFE of soil samples was difficult due to sample matrix-analyte interactions, but despite these difficulties at least three EPA-approved methods were generated in a relatively short time during the early 1990s.

On-line analytical SFE was to become the dominant technique as opposed to on-line SFE due to simplicity of operation and compatibility with existing methodology. On-line methods, while inherently more sensitive, required considerable skill on the analyst's part and could not be adjusted easily for diverse applications. Figure 9 is a schematic of our generic laboratory SFE units at NCAUR, which have served as templates for commercial instrumentation (e.g., Spe-ed Unit, Applied Separations, Inc., Allentown, PA). Either liquid-cooled piston pumps or booster pumps/compressors are commonly used to deliver the CO$_2$ into an extraction cell, which is placed in a heated oven or zone to assure supercritical extraction conditions. Pressure and flow are regulated by a narrow orifice device placed downstream from the cell. This can be a tubular restrictor, micrometering valve, backpressure regulator, or an automatically controlled valve to compensate for changes in fluid flow and hence backpressure. Extracts are collected in one of several devices after the extracting fluid stream is decompressed, and these may consist of an empty or solvent-filled vial, a sorbent-filled tube or cartridge, or a subambient cooled device. The importance of adequate extract collection cannot be overemphasized, since excellent analyte recoveries via SFE can be reduced due to poor collection efficiency [39].

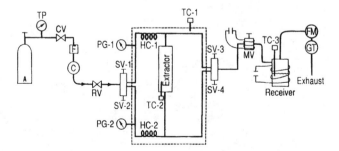

Figure 9 General laboratory SFE unit. A = CO$_2$ cylinder; TP = cylinder pressure gauge; CV = check valve; F = filter; C = air-driven gas booster compressor; RV = relief valve; SV = on/off switching valve; PG = pressure gauge; HC = equilibration coil; TC = thermocouple; MV = micrometering valve; FM = flow meter; GT = gas totalizer.

The need to process multiple samples was recognized early on by this investigator, and it seemed prudent to design an extractor that mimicked existing equipment utilized by analytical chemists in their laboratories (e.g., Soxhlet extractor). This gave rise to the development of a parallel sample extractor, which was co-developed with Marvin Hopper of the FDA laboratory in Lexena, Kansas [40]. This gave rise to several commercial offerings embracing this principle, perhaps the most succesful being the FA-100 Fat Analyzer produced by Leco, Inc. (St. Joseph, MI). The original NCAUR prototype consisted of six tubular extraction vessels, approximately 2 feet in length, $1/2$ in. internal diameter, that provided a 100 mL cell volume for large samples. Sample size in the early days of analytical SFE was quite controversial, but our design was guided by the amounts specified in traditional protocols and the need to have a representative sample when extracting agricultural commodities with localized toxicant contamination (e.g., aflatoxins). The analytical SFE industry eventually offered an assortment of sample sizes, but a 10 mL capacity was considered adequate for most needs.

Analytical SFE was considered a complex technique to apply by some due to the large number of experimental parameters that had to be considered in optimizing an analysis. Table 5 shows most of experimental parameters that impact on conducting successful SFE. Many of these parameters have been commented on previously, but those unique to SFE and not other sample preparation procedures are probably pressure and flow rate. All of the other parameters have their analogs in competing methodology, although the vernacular may be somewhat different. Analytical SFE faced stiff competition from other alternative sample preparation procedures that minimized solvent use, such as solid phase microfiber extraction (SPME), accelerated solvent extraction (ASE), and microwave-assisted extraction (MAE). However, the extraction specificity of SFE through the minipulation of pressure and temperature is perhaps what intrigued analytical chemists most about the technique.

Indeed, SFE can perform crude fractionation relative to chromatography by changing the fluid density, but it is rare to obtain a "clean" extract unless the sample matrix is insoluble in the supercritical fluid and the compounds to be isolated from each other differ substantially in their physicochemical properties. For example, the separation of fat from a foodstuff or contaminants in a soil sample can be handled quite adequately by SFE. On the other hand, the isolation of pesticides from a food sample that contains appreciable quantities of fat or water may be more problematic. This is the type of problem that faced our research team and hence was the

Table 5 Experimental Variables in Analytical SFE

Pressure
Temperature
Flow rate
Extraction time
Collection technique
Sample size
Homogeneity of sample
Choice of supercritical fluid
Choice of modifier
Amount of modifier
System leaks
Sample matrix effects
System contamination

Table 6 Options for Integrating Sample Clean-up with SFE

Fluid density–based fractonation
Supercritical fluid adsorption chromatography
Integration of selective adsorbents
Alternative fluids to carbon dioxide
On-line SFE/chromatography methods
Inverse SFE
SF-modified size exclusion chromatography (SEC)
Use of binary gas mixtures

basis for initiating research to integrate sample clean-up techniques into the basic SFE scheme. Table 6 lists some of the possibilities for sample clean-up integration with SFE and the systematic approach we took in solving the problem. In some cases, a judicious choice of extraction fluid density may provide an extract that is perfectly acceptable for analysis without the need for further clean-up. More common in SFE practice is to use a sorbent, either in the cell or after decompression, to further fractionate the extract. These sorbents tend to be normal phase chromatography adsorbents; therefore, SFC may be useful as a screening tool to choose the optimal sorbent for clean-up of the extract under SFE conditions. Another sorbent-based method invented by the author is "inverse" SFE, where a sorbent is incorporated into the extraction cell to isolate the target analyte of interest under SFE conditions, while interfering compounds are removed by the extraction fluid. This concept has been exploited in pesticide residue analysis as well as fractionation of pharmaceutical-based preparations [41].

Although SC-CO_2 reigns supreme as the prime extraction fluid in analytical SFE, other fluids can sometimes be put to advantage. For example, fluoroform, HCF_3, can be used rather than SC-CO_2, resulting in an extract with 100 times less fat than that obtained with SC-CO_2 under identical extraction conditions. Likewise, one can blend binary mixtures of fluids in their supercritical state to achieve a homogeneous extraction fluid that is more specific for analytes such as pesticide residues [42]. Recalling Fig. 2, it is apparent that using a supercritical fluid with a lower critical temperature than CO_2, such as nitrogen, the solubility parameter of the extracting fluid will be less than when using neat SC-CO_2. This is one of the reasons that a 70 mol% CO_2/30 mol% N_2 mixture will give extracts containing less than 5 mg of fat while assuring complete recovery of pesticides at the ppm level [42].

It is not possible to cover all of the aspects of analytical SFE and its optimization in this short review and memoir. The reader is referred to several primer texts [43,44] in the field and more advanced treatises or reviews [45,46] for further information. However, it is instructive to look at several of the applications of analytical SFE that demonstrate how the technique is utilized and its areas of applicability.

IV. APPLICATIONS OF ANALYTICAL SFE

Table 7 lists areas of application in which analytical SFE has been applied successfully. Within each generic class of compounds, there are certain compounds or subclasses that have not been extracted successfully using SC-CO_2, such as the β-lactam drugs. The results obtained with SFE are also somewhat matrix dependent, therefore, certain pesticides that are extracted successfully from foods may be more problematic, or require a change in conditions, for removal from soils. This is also true when using other sample preparation methods. Overall, pesticides as a com-

Table 7 Application Areas for
Analytical SFE

Pesticides
Petroleum products
Environmental samples
Fat and lipid analysis
Drugs and antibiotics
Polymer oligomers/additives
Metal analysis
Volatiles and flavors

pound class extract well using $SC-CO_2$ or $SC-CO_2$/modifier mixtures, and the reader should consult the review by Lehotay [47] on this subject.

One of major successes of analytical SFE is in the extraction of fat and similar lipid substances from foods. In this regard, the author's research group has been a major contributor to this field, with over 15 publications concerned with various aspects of fat determination. This interest did not occur by accident, since our success was based on research conducted for the extraction of oils from seed matrices [48]. Studies in our laboratory clearly showed that extraction of fat or oil triglycerides could best be accomplished at pressures approaching 10,000 psi and temperatures in the range of 70–80°C, where their solubility was maximized (see Fig. 10). This set the upper operational limit with respect to pressure for most commercial instrumentation. Our group also was successful in integrating new official methodology for total and speciated fat analysis, as mandated by the Nutritional Labeling & Education Act (NLEA), with SFE. This included the development of enzyme-catalyzed reaction under supercritical fluid conditions to form fatty acid methyl esters (FAME) required in the NLEA analysis [49]. It should be noted that lower pressures and temperatures are frequently used for selectively extracting other lipid

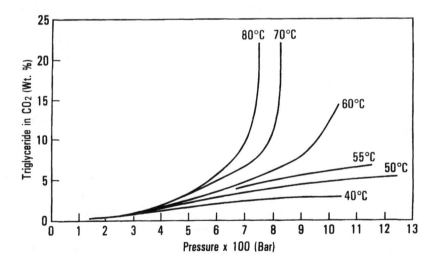

Figure 10 Effect of changing pressure and/or temperature on the solubility of triglycerides in $SC-CO_2$. (From Ref. 48.)

species such as cholesterol and fat-soluble vitamins, while phospholipids require the addition of a cosolvent (ethanol) for successful SFE. Such solubility findings were eventually incorporated into the development of a standard method for determining the oil content of vegetable seeds [50].

Analytical SFE has also experienced success when applied to the analysis of drugs in foods, biological matrices, and pharmaceutical preparations. In this field of application it is not unusual to employ a small quantity of cosolvent dissolved in SC-CO$_2$ to accelerate the extraction of the drug from the sample matrix. As noted previous, early success using SFE was recorded in the environmental analysis field, particularly in the extraction of organochlorine pesticides and dioxins, polynuclear aromatic hydrocarbons, and total petroleum hydrocarbons (TPH), which are all registered as official EPA methodology. The TPH method utilized SC-CO$_2$ as a replacement for a banned fluorocarbon previously used in running the method. Additional uses for analytical SFE are currently being developed, particularly in the areas of natural product analysis, extraction of metals, and as a more benign method for characterizing the volatile content of food and flavor components [51].

It is worth noting several examples from the author's work to illustrate some of the benefits of analytical SFE. For example, numerous "coupled" techniques have been generated using SFE

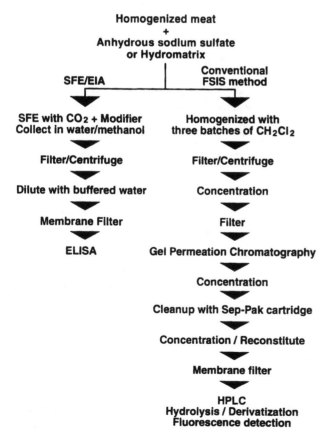

Figure 11 Analytical methodologies for the determination of carbamate pesticides in meats. (From Ref. 35.)

in combination with various forms of chromatography or spectroscopy. Most of these coupled or tandem techniques make use of on-line SFE rather than the off-line mode of extraction. An exception to this is the coupling of enzyme immunoassay (EIA) with SFE for both the qualitative and quantitative determination of toxicants in environmental and food samples. Figure 11 shows in a stepwise fashion the developed SFE-EIA method. This method employed only benign agents such as CO_2 and water and two simple filtration steps before applying the EIA for the quantitative determination of carbamate pesticides in meat products. This method is much simpler than the official method used by a regulatory agency, which is very solvent extensive, requires multiple clean-up steps, and a complex HPLC postcolumn derivatization method for determining the presence of carbamate pesticides in meats [52].

Another example from our method development studies that nicely illustrates the benefits of using either SFE or SFC is in the analysis of total fatty acid content in an industrial soapstock sample. In this case, a SFE method was developed that coupled both the extraction of the fatty

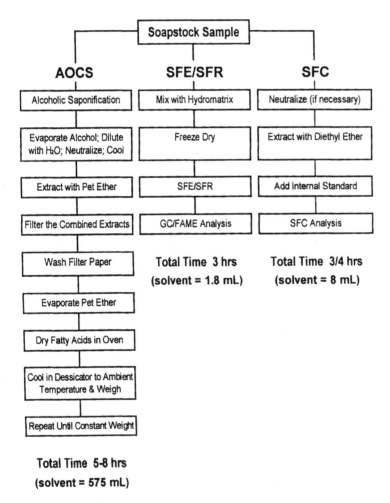

Figure 12 Comparison of AOCS Official Method G3-53 with the SFE coupled with an enzymatic-catalyzed reaction (SFE/SFR) and SFC methods. (From Ref. 53.)

acid moieties and triglycerides with a lipase-based methylation of the fatty acids to allow on-line analysis via gas chromatography [53]. This rapid method consisted of mixing the sample with the previously mentioned Hydromatrix, quickly freeze-drying this mixture, and then extracting and derivatizing the extract simultaneously using the coupled-automated SFE method. The benefits of such a technique are illustrated in Fig. 12, where the SFE/SFR (supercritical fluid reaction) method is contrasted with the conventional solvent and labor-intensive method, the American Oil Chemical Society (AOCS) Official Method G3-53, in a flow chart of the analysis. Note that the AOCS method consists of many manual steps, takes 5–8 hours depending on the analyst, and requires over 0.5 L of organic solvent. However, the SFE-based method takes only 3 hours and utilizes less than 2 mL of solvent. An alternative method, which is quite rapid but gives slightly lower results than either the AOCS or the SFE/SFR method, incorporates capillary SFC to analyze the soapstock sample. As indicated in Fig. 12, this method takes only 45 minutes and uses only 8 mL of solvent. Considering that the developed assays were applicable to monitoring tanker delivery trucks upon which time-based demurrage was being charged, the SFC method, even with its inherent inaccuracy, may be the preferred method, particularly if it can be used in a diagnostic fashion for detecting problematic shipment lots of this particular industrial by-product.

V. THE CONVERGENCE OF ANALYTICAL WITH PROCESSING SUPERCRITICAL FLUID METHODOLOGY

With the beginning of the new millennium it is important to assess the role of analytical SFE and SFC in terms of their future. Both techniques will continue to find niche analytical uses, probably SFE more than SFC, according to Smith [46]. From this author's perspective, SFC has entered an age in which standardization of methods and routine use of the technique are critical. Considering the limited number of commercial vendors for SFC, it is hoped that its use in chiral separation technology and perhaps SFC-MS (mass spectrometry) couplings will assure its continual use. Analytical SFC can be used as a template for the design and optimization of process SFC–based separations. SFC also competes favorably with normal phase liquid chromatography (LC) methods in terms of solvent reduction and column efficiency. Recently its use has been expanded in the simulated moving bed approach [54], while Taylor and King [55] have shown that coupling SFE with SFC in the preparative mode can provide greater fractionation and enrichment of targeted solutes than that achieved with supercritical fluid–based fractionation columns.

Automation in analytical SFE to date has been achieved using sequential analysis of multiple samples as typified by operation of the Isco Model 3560 extractor (Isco, Inc., Lincoln, NE). Such a unit can extract and collect 24 samples consecutively and is microprocessor controlled. The latter feature permits not only routine analysis, but the optimization of method development with respect to variables such as pressure, temperature, extraction time (fluid volume), cosolvent addition, and different sample types. This mode of operation can be viewed as an extension of combinatorial technology in which a wide variety of conditions, sample types, and phenomena can be rapidly studied and assessed. Currently this approach is being used in the author's laboratory for optimizing various processing applications of supercritical fluid technology as indicated in Table 8 [56]. The savings in time, labor, and reagent expense can be considerable using the above approach prior to scaling up a process. It is interesting to note in Table 8 that many of the processes investigated involve surveying reaction chemistry possibilities in critical fluid media. Reaction chemistry has been gradually integrated into analytical SFE, as the survey by

Table 8   Examples of Analytical Instrumentation Utilized in
Nonanalytical Applications Involving Critical Fluids

Sterol ester fractionation using sorbents
SFE and methylation of phospholipids and steryl esters
Evaluation of enzyme catalytic activity in SC-CO$_2$
Optimization of SFE of cedarwood oil in SC-CO$_2$ and LCO$_2$
Sorbent selection for preparative SFC of phospholipids
SFE/SFC for enrichment of steryl esters from corn bran
SC-CO$_2$ extraction of pheromone components from fir needles
Selective extraction of components from RBO deodorizer distillate
Optimization of enzymatic hydrolysis of fat-soluble vitamins
Feasibility of enzyme-initiated acetylation of cedrol

Field [57] indicates, and this will be a continuing trend in the author's opinion, since reaction rates are pressure dependent and kinetic processes accelerated in supercritical fluids relative to those conducted in the condensed liquid state.

In summary, it has been the author's privilege to be at the inception of analytical SFC and SFE and to participate in their development over the past 30 years. In addition, his experiences and research in this field have involved not just the analytical use of supercritical fluids in the separation sciences, but also their application in the fields of chemical engineering, food technology, and natural products processing. There are differences in approach when using the separation sciences in engineering versus analytical chemistry, but the seminal principles behind techniques such as SFE and SFC are the same regardless of the area of application. Consequently, it is his feeling and philosophy that there is much to be gained by following developments on a multidisplinary front, and even more so in applying such knowledge in a synergestic fashion across several disciplines as the need arises.

REFERENCES

1. HS Booth, RM Bidwell. Solubility measurement in the critical region. Chem Rev 44:477–513, 1949.
2. BH Sage, WN Lacey. Volumetric and Phase Behavior of Hydrocarbons. Palo Alto, CA: Stanford University Press, 1939.
3. JC Giddings, MN Myers, JW King. Dense gas chromatography at pressures to 2000 atmospheres. J Chromatogr Sci 7:276–283, 1969.
4. MN Myers, JC Giddings. Ultra-high-pressure gas chromatography in micro columns to 2000 atmospheres. Sep Sci 1:761–776, 1966.
5. PA Peaden, JC Fjeldsted, ML Lee, SR Springston, M Novotny. Instrumental aspects of capillary supercritical fluid chromatography. Anal Chem 54:1090–1093, 1982.
6. JC Giddings, MN Myers, L McLaren, RA Keller. High pressure gas chromatography of non-volatile species. Science 162:67–73, 1968.
7. JW King. Fundamentals and applications of supercritical fluid extraction in chromatographic science. J Chromatogr Sci 27:355–364, 1989.
8. JW King, JP Friedrich. Quantitative correlations between solute molecular structure and solubility in supercritical fluids. J Chromatogr 517:449–458, 1990.
9. SR Allada. Solubility parameters of supercritical fluids. Ind Eng Chem Proc Des Dev 23:344–348, 1984.

10. Y Ikushima, T Goto, M Arai. Modified solubility parameter as an index to correlate the solubility in supercritical fluids. Bull Chem Soc Jpn 60:4145–4147, 1987.

11. JW King. Supercritical fluid adsorption at the gas-solid interface. In: TG Squires, ME Paulaitis eds. Supercritical Fluids—Chemical and Engineering Principles and Applications, ACS Symposium Series #329. Washington, DC: American Chemical Society, 1987, pp. 150–171.

12. E. Klesper. Chromatographie mit überkritischen Fluidenphasen. Angew Chem 90:785–793, 1978.

13. CF Poole, JW Oudsema, TA Dean, SK Poole. Stationary phases for packed column supercritical fluid chromatography. In: B Wenclawiak, ed. Analysis with Supercritical Fluids: Extraction and Chromatography. Berlin: Springer-Verlag, 1992, pp. 116–133.

14. TL Chester. Chromatography from the mobile phase perspective. Anal Chem 69:165A–169A, 1997.

15. Y Shen, ML Lee. Solvating gas chromatography using packed columns. In: PR Brown, E Grushka, eds. Advances in Chromatography, Vol. 38. New York: Marcel Dekker, 1998, pp. 75–113.

16. ST Sie, GWA Rijnders. Chromatography with supercritical fluids. Anal Chim Acta 38:31–44, 1967.

17. JW King, SE Abel, SL Taylor. SEC for sample cleanup using supercritical fluids. Proceedings of the 5th International Symposium on Supercritical Fluid Chromatography and Extraction, Baltimore MD, 1994, p. D-24.

18. ST Lee, SV Olesik, SM Fields. Applications of reversed-phase high performance liquid chromatography using enhanced-fluidity liquid mobile phases. J Microcol (Sep. 7):478–483, 1995.

19. ML Lee, K Markides, eds. Analytical Supercritical Fluid Chromatography and Extraction. Provo, UT: Chromatography Conferences, Inc., 1990.

20. TA Berger. Packed Column SFC. Cambridge, U.K.: Royal Society of Chemistry, 1995.

21. K Anton, C Berger, eds. Supercritical Fluid Chromatography with Packed Columns. New York: Marcel Dekker, 1998.

22. DR Gere. Supercritical fluid chromatography. Science 222:253–259, 1983.

23. LT Taylor. Trends in supercritical fluid chromatography: 1997. J Chromatogr Sci 35:364–382, 1997.

24. C Borch-Jensen, J Mollerup. Applications of supercritical fluid chromatography to food and natural products. Sem Food Anal 1:101–116, 1996.

25. JW King, JM Snyder. Supercritical fluid chromatography—a shortcut in lipid analysis. In: R McDonald, M Mossoba, eds. New Techniques and Applications in Lipid Analysis. Champaign, IL: American Oil Chemical Society Press, 1997, pp. 139–162.

26. KW Phinney. SFC of drug enantiomers. Anal Chem 72:204A–211A, 2000.

27. JW King. Applications of capillary supercritical fluid chromatography-supercritical fluid extraction to natural products. J Chromatogr Sci 28:9–14, 1990.

28. KD Bartle, AA Clifford, SA Jafar, JP Kithinji, GF Shilstone. Use of chromatographic retention measurements to obtain solubilities in a liquid or supercritical fluid mobile phase. J Chromatogr 517:459–476, 1990.

29. MA McHugh, VJ Krukonis. Supercritical Fluid Extraction, 2nd ed. Boston: Butterworth-Heinemann, 1994, p. 368.

30. MEP McNally, JR Wheeler. Increasing extraction efficiency in supercritical fluid extraction from complex matrices. J Chromatogr 447:53–63, 1988.

31. SL Taylor, JW King, L Montanari, P Fantozzi, MA Blanco. Enrichment and fractionation of phospholipid concentrates by supercritical fluid extraction and chromatography. Ital J Food Sci 12:65–76, 2000.

32. M Saito, Y Yamauci. Isolation of tocopherols from wheat germ oil by recycle semi-preparative supercritical fluid chromatography. J Chromatogr 505:257–271, 1990.

33. P Lembke. Production of high purity n-3 fatty acid-ethyl esters by process scale supercritical fluid chromatography. In: K Anton, C Berger, eds. Supercritical Fluid Chromatography with Packed Columns. New York: Marcel Dekker, 1998, pp. 429–443.

34. E Stahl, KW Quirin, D Gerard. Dense Gases for Extraction and Refining. New York: Springer-Verlag, 1988.

35. JW King. Analytical supercritical fluid techniques and methodology: conceptualization and reduction to practice. J Assoc Off Chem Int 81:9–17, 1998.

36. SB Hawthorne, JW King. Principles and practice of analytical SFE. In: M Caude, D Thiebaut, eds. Practical Supercritical Fluid Chromatography and Extraction. Chur, Swizerland: Harwood Academic Publishers, 1999, pp. 219–282.

37. HJ Vandenberg, AA Clifford. Polymers and polymer additives. In: AJ Handley, ed. Extraction Methods in Organic Analysis. Sheffield, England: Sheffield Academic Press, 1999, pp. 221–242.

38. AAM Stolker, MA Sipoli Marques, PW Zoontjes, LA Van Ginkel, RJ Maxwell. Supercritical fluid extraction of residues of veterinary drugs and growth-promoting agents from food and biological matrices. Sem Food Anal 1:117–132, 1996.

39. L. McDaniel, GL Long, LT Taylor. Statistical analysis of liquid trapping efficiencies of fat soluble vitamins following SFE. J High Resolut Chromatogr 21:245–251, 1998.

40. ML Hopper, JW King, JH Johnson, AA Serino, RJ Butler. Supercritical fluid extraction (SFE): multivessel extraction of food items in the FDA total diet study (TDS). J Assoc Off Anal Chem Int 78:1072–1079, 1995.

41. WN Moore, LT Taylor. Analytical inverse SFE of polar pharmaceutical compounds from cream and ointment matrices. J Pharm Biomed Anal 12:1227–1232, 1994.

42. JW King, Z Zhang. Selective extraction of pesticides from lipid-containing matrices using supercritical binary gas mixtures. Anal Chem 70:1431–1436, 1998.

43. LT Taylor. Supercritical Fluid Extraction. New York: John Wiley, 1996.

44. MD Luque de Castro, M. Valcarel, MT Tena. Analytical Supercritical Fluid Extraction. Berlin: Springer-Verlag, 1994.

45. JR Dean, ed. Application of Supercritical Fluids in Industrial Analysis. London: Blackie Academic, 1993.

46. RM Smith. Supercritical fluids in separation science—the dream, the reality and the future. J Chromatogr A 856:83–115, 1999.

47. SJ Lehotay. Supercritical fluid extraction of pesticides in foods. J Chromatogr A 785:289–312, 1997.

48. JW King. Critical fluids for oil extraction. In: PJ Wan, PJ Wakelyn, eds. Technology and Solvents for Extracting Oilseeds and Nonpetroleum Oils. Champaign, IL: AOCS Press, 1997, pp. 283–310.

49. JM Snyder, JW King, MA Jackson. Fat content for nutritional labeling by supercritical fluid extraction and an on-line lipase catalyzed reaction. J Chromatogr 750:201–207, 1996.

50. Oil in oilseeds: supercritical fluid extraction method—AOCS Official Method Am 3-96. Official Methods of American Oil Chemical Society. Champiagn, IL: AOCS Press, 1997.

51. JM Snyder, JW King, Z Zhang. Comparison of volatile analysis of lipid-containing and meat matrices by solid phase micro- and supercritical fluid-extraction. In: CJ Mussinan, MJ Morello, eds. Flavor Analysis: Developments in Isolation and Characterization. ACS Symposium Series #705. Washington, DC: American Chemical Society, 1998, pp. 107–115.

52. K. Nam, JW King. Supercritical fluid extraction and enzyme immunoassay for pesticide detection in meat products. J Agric Food Chem 42:1469–1474, 1994.

53. JW King, SL Taylor, JM Snyder, RL Holliday. Total fatty acid analysis of vegetable oil soapstocks by supercritical fluid extraction/reaction (SFE/SFR). J Am Oil Chem Soc 75:1291–1295, 1998.

54. T Giese, M Johanssen, G Brunner. Separation of stereoisomers in a SMB-SFC plant: determination of isotherms and simulation. Proceeding of the International Meeting of the GVC-Fachausschuss Hochdruckverfahrenstechnik, Karlsruhe, Germany, 1999, pp. 275–278.

55. SL Taylor, JW King. Optimization of the extraction and fractionation of corn bran oil using analytical supercritical fluid instrumentation. J Chromatogr Sci 38:91–94, 2000.

56. JW King, FJ Eller, SL Taylor, AL Neese. Utilization of analytical critical fluid instrumentation in non-analytical applications. Proceedings of the 5th International Symposium on Supercritical Fluids, Atlanta, GA, 2000, pp. 1–20.

57. J Field. Coupling chemical derivatization reactions with supercritical fluid extraction. J Chromatogr A 785:239–249, 1997.

26

Solid Phase Microextraction

Janusz Pawliszyn

University of Waterloo, Waterloo, Ontario, Canada

I. INTRODUCTION

Solid phase microextraction (SPME) was developed to address the need to facilitate rapid sample preparation both in the laboratory and on-site where the investigated system is located [1]. In the technique a small amount of extracting phase dispersed on a solid support is exposed to the sample for a well-defined period of time. In one approach a partitioning equilibrium between the sample matrix and extraction phase is reached. In this case convection conditions do not affect the amount extracted. In a second approach utilizing short-time preequilibrium extraction, if convection/agitation are constant, then the amount of analyte extracted is related to time. Quantification can then be performed based on timed accumulation of analytes in the coating. Figure 1 illustrates several implementations of SPME that have been considered. They include mainly open bed extraction concepts such as coated fibers, vessels, and agitation mechanism disks, but in-tube approaches are also considered. Some best address issues associated with agitation, and others ease of implementing sample introduction to the analytical instrument. It should be noted that solid phase microextraction was originally named after the first experiment using an SPME device, which involved extraction on solid fused silica fibers, and later, as such, as a reference to the appearance of the extracting phase, relative to a liquid or gaseous donor phase, even though it is recognized that the extraction phase is not always technically a solid.

The configurations and operation of the SPME devices are very simple. For example, for the coated fiber implementation of the technology, one who knows how to use a syringe is able to operate the SPME device. In the case of automated in-tube extraction for HPLC, fitting a piece of the GC capillary into the system and then turning on the autosampler is all that is required to start its operation. The technology is designed to greatly simplify sample preparation. This feature, however, creates a false impression that the extraction is a simple, almost trivial process. This misunderstanding frequently results in disappointment. It should be emphasized that the fundamental processes involved in solid phase microextraction are similar to more traditional techniques, and therefore challenges to develop successful methods are analogous. The nature of target analytes and complexity of sample matrix determine the level of difficulties in

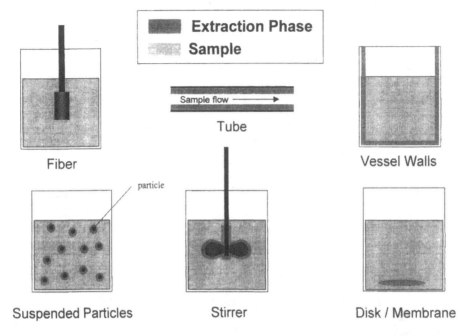

Figure 1 SPME configurations.

accomplishing a successful extraction. The simplicity, speed, and convenience of the extraction devices primarily impacts the costs of practical implementation and automation of the developed methods [2]. The objective of this chapter is to emphasize the fundamental principles of the technique that define the advantages and limitations of SPME technology.

II. PRINCIPLES OF SOLID PHASE MICROEXTRACTION

In SPME a small amount of extracting phase associated with a solid support is placed in contact with the sample matrix for a predetermined amount of time. If the time is long enough, a concentration equilibrium is established between the sample matrix and the extraction phase. When equilibrium conditions are reached, exposing the fiber for a longer time does not result in the accumulation of more analytes. Two different implementations of the SPME technique have been explored extensively to date. One is associated with a tube design and the other with fiber design. The tube design can use very similar arrangements as SPE, but the primary difference, in addition to volume of the extracting phase, is that the objective of SPME is never an exhaustive extraction. This substantially simplifies the design of systems. For example, in-tube SPME for analysis of liquids uses 0.25 mm i.d. tubes and about 0.1 µL of extraction phase, because concern about breakthrough is not relevant since exhaustive extraction is not an objective. In fact, the objective of the experiment is to produce full breakthrough as soon as possible, since this indicates that equilibrium extraction has been reached.

A more traditional approach to SPME involves coated fibers. The transport of analytes from the matrix into the coating begins as soon as the coated fiber has been placed in contact with the sample. There is a substantial difference in performance between the liquid and solid coatings (Fig. 2). A comparison of adsorptive versus absorptive equilibrium extraction is useful. In both

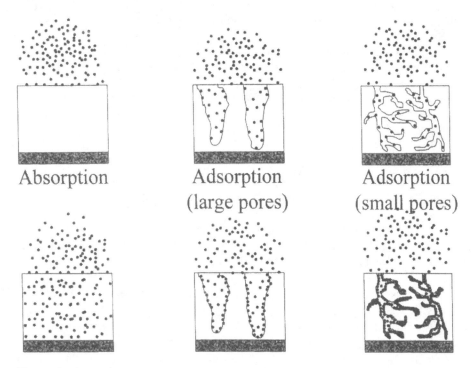

Absorption   Adsorption   Adsorption
                    (large pores)   (small pores)

Figure 2 Schematic representation of absorptive versus adsorptive extraction and adsorption in small versus large pores.

cases, the extraction process begins by adsorption of analytes at the extraction phase/matrix interface. Then diffusion of analytes into the bulk of the extraction phase follows. If the diffusion coefficients of the analytes in the extraction phase are high, then the analytes partition fully between the two phases, and absorptive extraction is accomplished. This process is aided by thin extraction phase coatings or the convection of the sample matrix (if flowing liquid). On the other hand, if the diffusion coefficient is low the analyte remains at the interface and adsorption results. The principal advantage of absorption extraction (partitioning) is a linear isotherm over a wide range of analyte and interference concentrations, since the property of the extraction phase does not change substantially until the extracted amount is about 1 wt% of the extraction phase. On the other hand, in adsorption extraction the isotherm is highly nonlinear for higher concentrations when the surface coverage is substantial. This causes a particular problem in equilibrium methods since the response of the fiber for the analyte at high sample concentrations will depend on the concentrations of both analytes and interferences. The advantages of the solid sorbents include higher selectivity and capacity for polar and volatile analytes.

III. MULTIPHASE EQUILIBRIA

Solid phase microextraction is a multiphase equilibration process. Frequently the extraction system is complex, as in a sample consisting of an aqueous phase with suspended solid particles having various adsorption interactions with analytes, plus a gaseous headspace. In some cases specific factors have to be considered, such as analyte losses by biodegradation or adsorption on the walls of the sampling vessel or stirring mechanism. In the discussion below we will consider only two phases: the fiber coating and a homogeneous matrix such as pure water or air.

Typically, SPME extraction is considered to be complete when the analyte concentration has reached distribution equilibrium between the sample matrix and the fiber coating. In practice, this means that once equilibrium is reached, the extracted amount is constant within the limits of experimental error and it is independent of further increase of extraction time. The equilibrium conditions can be described as [3]:

$$n = \frac{K_{fs} V_f V_s C_0}{K_{fs} V_f + V_s} \tag{1}$$

where n is the mass of analyte extracted by the coating, K_{fs} is a fiber coating/sample matrix distribution constant, V_f is the fiber coating volume, V_s is the sample volume, and C_0 is the initial concentration of a given analyte in the sample.

Strictly speaking, this discussion is limited to partitioning equilibrium involving liquid polymeric phases such as poly(dimethylsiloxane). The method of analysis for solid sorbent coatings is analogous for low analyte concentration, since the total surface area available for adsorption is proportional to the coating volume if we assume constant porosity of the sorbent. For high analyte concentrations, saturation of the surface can occur, resulting in nonlinear isotherms, as discussed later. Similarly high concentration of a competitive interference compound can displace the target analyte from the surface of the sorbent.

Equation (1), which assumes that the sample matrix can be represented as a single homogeneous phase and that no headspace is present in the system, can be modified to account for the existence of other components in the matrix by considering the volumes of the individual phases and the appropriate distribution constants. The extraction can be interrupted and the fiber analyzed prior to equilibrium. To obtain reproducible data, however, constant convection conditions and careful timing of the extraction are necessary.

Simplicity and convenience of operation make SPME a superior alternative to more established techniques in a number of applications. In some cases, the technique facilitates unique investigations. Equation (1) indicates that, after equilibrium has been reached, there is a direct proportional relationship between sample concentration and the amount of analyte extracted. This is the basis for analyte quantification. The most visible advantages of SPME exist at the extremes of sample volumes. Because the setup is small and convenient, coated fibers can be used to extract analytes from very small samples. For example, SPME devices are used to probe for substances emitted by a single flower during its life span; the use of submicrometer diameter fibers permits the investigation of single cells. Since SPME does not extract target analytes exhaustively, its presence in a living system should not result in significant disturbance. In addition, the technique facilitates speciation in natural systems, since the presence of a minute fiber, which removes small amounts of analyte, is not likely to disturb chemical equilibria in the system. It should be noted however, that the fraction of analyte extracted increases as the ratio of coating to sample volume increases. Complete extraction can be achieved for thick coatings and small sample volumes when distribution constants are reasonably high. This observation can be used to advantage if exhaustive extraction is required. It is very difficult to work with small sample volumes using conventional sample preparation techniques. Also, SPME allows rapid extraction and transfer to an analytical instrument. These features result in an additional advantage when investigating intermediates in the system. For example, SPME was used to study biodegradation pathways of industrial contaminants [4]. The other advantage is that this technique can be used for studies of the distribution of analytes in a complex multiphase system [5] and to speciate different forms of analytes in a sample [6].

In addition, when sample volume is very large, Eq. (1) can be simplified to:

$$n = K_{fs}V_fC_0 \qquad (2)$$

which points to the usefulness of the technique for field applications. In this equation, the amount of extracted analyte is independent of the volume of the sample. In practice there is no need to collect a defined sample prior to analysis, because the fiber can be exposed directly to the ambient air, water, production stream, etc. The amount of extracted analyte will correspond directly to its concentration in the matrix without being dependent on the sample volume. When the sampling step is eliminated, the whole analytical process can be accelerated, and errors associated with analyte losses through decomposition or adsorption on the sampling container walls will be prevented. This advantage of SPME could be enhanced practically by developing portable field devices on a commercial scale.

IV. RAPID SAMPLING

In the case of solid sorbents the coating has a well-defined crystaline glass structure, which, if dense, substantially reduces the diffusion coefficients within the structure. Therefore, within the experimental time the extraction occurs only on the surface of the coating. This can be demonstrated by considering extraction of proteins illustrated in Fig. 3. The original mixture contains three compounds: myoglobin, cytochrome, and lysozyme. During fiber extraction with polyacrylic acid, compounds with weaker affinity are only observed at short extraction times. When extraction time is longer, the displacement of analytes with lower affinities occurs. In this case, lysozyme with a stronger affinity for the coating replaces the other two compounds during extraction. This effect is associated with the fact that there is only limited surface area available for adsorption. If this area is substantially occupied, then the displacement effects occur [7,8] and the equilibrium amount extracted can vary with concentrations of both the target and other analytes. In extraction of analytes with liquid coatings, on the other hand, partitioning between

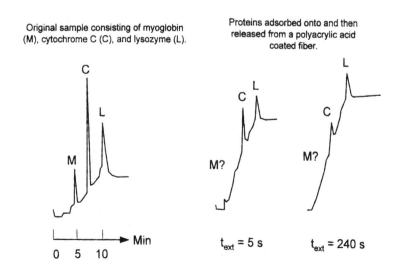

Figure 3 Protein extraction using poly(acrylic) acid–coated fiber. Demonstration of myoglobin and cytochrome c displacement by lysozyme with time.

the sample matrix and extraction phase occurs. In this case, equilibrium extraction amounts vary only if the coating property is modified by the extracted components, which only occurs when the amount extracted is a substantial portion (a few percent) of the extraction phase. This is very rarely observed since SPME is typically used to determine trace contamination samples.

The only way to overcome this fundamental limitation of the porous coatings is, as the above figure suggests, to use an extraction time much less than the equilibrium time so that the total amount of analytes accumulated onto the fiber is substantially below the saturation value. When performing such experiments, not only is it critical to precisely control extraction times, but convection conditions must also be monitored to ensure that they are constant or can be compensated for. One way of eliminating the need for compensation of convection is to normalize (use the same) aggitiation conditions. For example, the use of stirring means at well-defined rotation rates in the laboratory or fans for field air monitoring will ensure consistent convection. The short-time exposure SPME measurement described has an advantage associated with the fact that the rate of extraction is defined by diffusivity of analytes through the boundary layer (Fig. 4) of the sample matrix, and their corresponding diffusion cofficients, rather than distribution constants. The mass of extracted analyte can be estimated from the following equation [9]:

$$n(t) = \frac{2\pi D_g L}{\ln\left(\dfrac{b + \delta}{b}\right)} C_g t \tag{3}$$

where n is the mass of extracted analyte over sampling time t, D_g is the gas-phase molecular diffusion coefficient (cm^2/s), b is the outside radius of the fiber coating (cm), L is the length of the coated rod (cm); δ is the thickness of the boundary layer surrounding the fiber coating

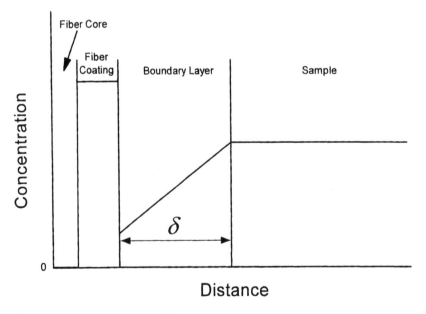

Figure 4 Boundary layer model.

(cm), and C_g is analyte concentration in the bulk air (ng/mL). The linear uptake of analytes with time for short exposures is demonstrated in Fig. 5.

A precise understanding of the definition and thickness of the boundary layer in this sense, is useful. The thickness of the boundary layer (δ) is determined by both rate of convection (agitation) in the sample and analyte diffusion [10]. Thus, in a single sample, the boundary layer thickness will be different for different analytes. Strictly speaking, the boundary layer is a region where analyte flux is progressively more dependent on analyte diffusion and less on convection, as the extraction phase is approached. For convenience, however, analyte flux in the bulk of the sample (outside of the boundary layer) is assumed to be controlled by convection, whereas analyte flux within the boundary layer is assumed to be controlled by diffusion. δ is defined as the position where this transition occurs, or the point at which convection in is equal to diffusion away. At this point, analyte flux from δ towards the extraction phase (diffusion controlled) is equal to the analyte flux from the bulk of the sample towards δ, controlled by convection. The differences in diffusion coefficients between compounds are small compared to the differences in distribution contants. This makes it easier to calibrate the system. Because of the large differences in distribution constants between analytes, the resulting chromatograms are characterized by small peak areas for compounds with small distribution constants and large areas for those with large constants. With uptake dependent on diffusion coefficients, all compounds in a chromatogram with similar molecular masses will have similar peak areas, given similar detector responses. Also, it is relatively simple to calculate the diffusion coefficient for a given analyte and therefore correct for the small differences in it. It must be understood that this system is only suitable for trace analysis. When sample concentrations become too high, saturation of the active sites occurs, and uptake rates are no longer linear. Shorter exposure times, where smaller amounts are extracted, can solve this problem. Also, at these higher concentrations, samples are easily extracted and analyzed with PDMS fiber using conventional SPME extraction methods. Results of extraction by the diffusion type of approach are shown in Fig. 6. Accumulation of volatile components on the solid coating in 10 s is much larger compared to the 10-minute

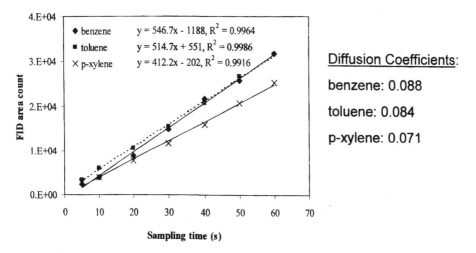

Figure 5 Correlation of uptake rate with diffusion coefficient for short sampling time (nonequilibrium) extraction of VOCs by PDMS-DVB fiber.

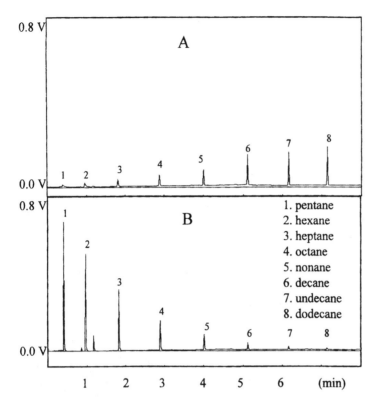

Figure 6 Comparison of VOC mass loading between PDMS and PDMS/DVB. A: PDMS, 10 min; B: PDMS/DVB, 10 s.

equilibrium extraction on PDMS. This approach to extraction is not limited to devices using the fiber geometry, but is generally applicable.

V. EXTRACTION MODES WITH COATED FIBER SPME

SPME sampling can be performed in three basic modes: direct extraction, headspace extraction, and extraction with membrane protection. Figure 7 illustrates the differences between these modes. In direct extraction mode (Fig. 7a), the coated fiber is inserted into the sample and the analytes are transported directly from the sample matrix to the extracting phase. To facilitate rapid extraction, some level of agitation is required to transport the analytes from the bulk of the sample to the vicinity of the fiber. For gaseous samples, natural flow of air (e.g., convection) is frequently sufficient to facilitate rapid equilibration for volatile analytes, but for aqueous matrices more efficient agitation techniques, such as fast sample flow, rapid fiber, or vial movement, stirring, or sonication, are required to reduce the effect of the ''depletion zone'' produced close to the fiber as a result of slow diffusional analyte transport through the stationary layer of liquid surrounding the fiber.

In the headspace mode (Fig. 7b), the analytes are extracted from the gas phase equilibrated with the sample. The primary reason for this modification is to protect the fiber from adverse effects caused by nonvolatile, high molecular weight substances present in the sample matrix

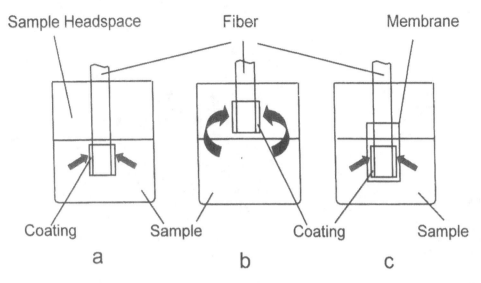

Sample Headspace   Fiber   Membrane

Coating   Sample   Coating   Sample

a   b   c

Figure 7 Modes of SPME operation: (a) direct extraction, (b) headspace SPME, (c) membrane-protected SPME.

(e.g., humic acids or proteins). The headspace mode also allows matrix modifications, including pH adjustment, without affecting the fiber. In a system consisting of a liquid sample and its headspace, the amount of an analyte extracted by the fiber coating does not depend on the location of the fiber, in the liquid phase or in the gas phase, therefore the sensitivity of headspace sampling is the same as the sensitivity of direct sampling as long as the volumes of the two phases are the same in both sampling modes. Even when no headspace is used in direct extraction, a significant sensitivity difference between direct and headspace sampling can occur only for very volatile analytes. However, the choice of sampling mode has a very significant impact on the extraction kinetics. When the fiber is in the headspace, the analytes are removed from the headspace first, followed by indirect extraction from the matrix. If the Henry's constant of a given compound is high, then the concentration of analytes in the headspace is high, resulting in very rapid extraction since the extracted analytes originate primarily from the gaseous headspace (see Fig. 8a). On the other hand, if the Henry's constants are low, then the extraction is long since the analytes need to diffuse from the condensed phase before they reach the fiber (Fig. 8b). Therefore, in the case of extraction of aqueous samples, volatile and nonpolar analytes are extracted much faster than semivolatiles or polar volatiles. Temperature has a significant effect on the kinetics of the process, since it determines the vapor pressure of analytes above the condensed phase. In general, the equilibration times for volatile compounds are shorter for headspace SPME extraction than for direct extraction under similar agitation conditions, because of the following three reasons: a substantial portion of the analytes is present in the headspace prior to the beginning of the extraction process, there is typically large interface between sample matrix and headspace, and the diffusion coefficients in the gas phase are typically higher by four orders of magnitude than in liquids. The concentration of semivolatile compounds in the gaseous phase at room temperature is small, and headspace extraction rates for those compounds are substantially lower. They can be improved by using very efficient agitation or by increasing the extraction temperature. Figure 9 illustrates equilibration time profiles obtained for extraction of methamphetamine from urine sample at various temperatures. At 22°C and 40°C the equili-

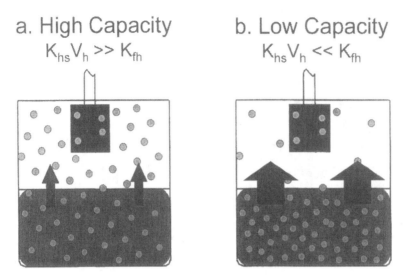

Figure 8 Headspace SPME of compounds characterized by high (a) and low (b) Henry constant.

bration is very long, exceeding 100 minutes, as indicated in this graph. It drops to about 20 minutes when the extraction temperature is 60°C and to only a few minutes when the temperature is 73°C. The dramatic change with the equilibration time is associated with the fact that an increase in temperate results in an increase of the analyte's Henry's constant, an increase in diffusion coefficient as well as a decrease of amount extracted at equilibrium. This decrease is associated with the fact that the distribution constant decreases with temperature increase. Therefore, it is important to carefully optimize the extraction temperature for shortest equilibration

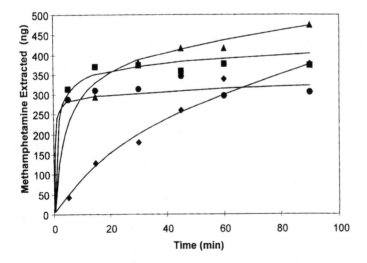

Figure 9 Temperature dependance of the absorption time profiles obtained for methamphetamine.

times and acceptable sensitivities. In most SPME applications, equilibrium extraction is performed. In many cases, however, when the equilibration times are long, preequilibrium quantification can be considered. It is important in such experiments to ensure constant agitation conditions and good timing of extraction times to obtain good precision.

In the third mode (SPME extraction with membrane protection; Fig. 7c), the fiber is separated from the sample with a selective membrane, which lets the analytes through while blocking the interferences. The main purpose for the use of the membrane barrier is to protect the fiber against adverse effects caused by high molecular weight compounds when very dirty samples are analyzed. While extraction from headspace serves the same purpose, membrane protection makes possible the analysis of less volatile compounds. The extraction process is substantially slower than direct extraction because the analytes need to diffuse through the membrane before they can reach the coating. Use of thin membranes and an increase in extraction temperature result in shorter extraction times.

VI. EXTRACTION MODES WITH IN-TUBE SPME

There are two fundamental approaches to in-tube SPME: (a) active or dynamic, when the analytes are passed through the tube, and (b) passive or static, when the analytes are transferred into the sorbent by diffusion. In either of these approaches, the coating may be supported on a fused silica rod or coated on the inside of a tube or capillary. Below is a discussion of the theoretical aspects of the extraction processes that use these geometric arrangements.

A. Dynamic In-Tube SPME

In this system we assume the use of a piece of fused silica capillary, internally coated with a thin film of extracting phase (a piece of open tubular capillary GC column) or that the capillary is packed with extracting phase dispersed on an inert supporting material (a piece of micro-LC capillary column). During introduction of the sample the front of analyte migrates through the capillary with a speed proportional to the linear velocity of the sample and inversely related to the partition ratio [11,12]. For short capillaries with a small dispersion, the extraction time can be assumed to be similar to the time required for the center of the band to reach the end of the capillary. The extraction time is proportional to the length of the capillary and inversely proportional to the linear flow rate of the fluid. Extraction time also increases with an increase in the coating/sample distribution constant and with the thickness of the extracting phase, but decreases with an increase in the void volume of the capillary. An increase in the coating/sample distribution constant produces an increase in absolute amount extracted. It has been observed that increases in amounts extracted can be achieved, in many cases, by preconditioning the capillary with methanol or some other appropriate solvent, prior to extraction. Enhancement has even been observed when a plug of methanol is aspirated into the capillary before the sample is drawn in and follows the sample plug in the capillary during the extraction aspirate/dispense steps. This is analogous to the solvent preconditioning used in SPE to enhance extraction.

In practice, in-tube SPME is implemented by replacing a section of the tubing in a commercially available autosampler and then programming the autosampler to pass sample in and out of the extraction capillary until equilibrium or a suitable extraction level has been reached.

It should be emphasized that the above discussion is valid only for direct extraction when the sample matrix passes through the capillary. This approach is limited to particulate-free gas and clean water samples. The headspace SPME approach can broaden the application of in-tube SPME. In that case, careful consideration to the mass transfer between sample and headspace should be given in order to describe the process properly. Also, if the flow rate is very rapid producing turbulent behaviour and the coating/sample distribution constant is not very high, then perfect agitation conditions are met and equation 4 can be used to estimate equilibration times. In this case equilibration time, t_e, is assumed to be achieved when 95% of the equilibrium amount of analyte is extracted from the sample:

$$t_e = t_{95\%} = \frac{(b - a)^2}{2D_f} \tag{4}$$

where $b - a$ refers to the thickness of the sorbent material and D_f refers to analyte diffusion in the sorbent.

Removal of analytes from a tube is an elution problem analogous to frontal chromatography and has been discussed in detail [12]. In general, if the desorption temperature of a GC is high and thin coatings are used, then all the analytes are in the gas phase as soon as the coating is placed in the injector, and the desorbtion time corresponds to the elution of two void volumes of the capillary. For liquid desorption, the desorbtion volume can be even smaller since the analytes can be focused at the front of the desorption solvent [13].

B. Static In-Tube SPME Time-Weighted Average Sampling

In addition to the analyte concentration measurement at a well-defined place in space and time, obtained by using the approaches discussed above, an integrated sampling is possible with a simple SPME system. This is particularly important in field measurements when changes of analyte concentration over time and place variations must often be taken into account.

When the extracting phase is not exposed directly to the sample, but is contained in a protective tubing (needle) without any flow of the sample through it (Fig. 10a), the extraction occurs through the static gas phase present in the needle. The integrated system can consist of extraction phase coating the interior of the tubing, or it can be an externally coated fiber withdrawn into the needle. These geometric arangements represent a very powerful method able to generate a response proportional to the integral of the analyte concentration over time and space (when the needle is moved through the space) [14]. In these cases, the only mechanism of analyte transport to the extracting phase is diffusion through the gaseous phase contained in the tubing. During this process, a linear concentration profile is established in the tubing between the small needle opening, characterized by surface area A and the distance Z between the needle opening, and the position of the extracting phase. The amount of analyte extracted, dn, during time interval dt can be calculated by considering Fick's first law of diffusion [15]:

$$dn = AD_g \frac{dc}{dz} dt = AD_g \frac{\Delta C(t)}{Z} dt \tag{5}$$

where $\Delta C(t)/Z$ is a value of the gradient established in the needle between needle opening and the position of the extracting phase, Z, $\Delta C(t) = C(t) - C_z$, where $C(t)$ is a time-dependent concentration of analyte in the sample in the vicinity of the needle opening, and C_z is the concentration of the analyte in the gas phase in the vicinity of the coating. C_z is close to zero for a high coating/gas distribution constant capacity, i.e., $\Delta C(t) = C(t)$. The concentration of analyte at the coating position in the needle, C_z, will increase with integration time, but it will be kept low compared to the sample concentration because of the presence of the sorbing coating. Therefore, the accumulated amount over time can be calculated as:

$$n = D_g \frac{A}{Z} \int C(t)d \tag{6}$$

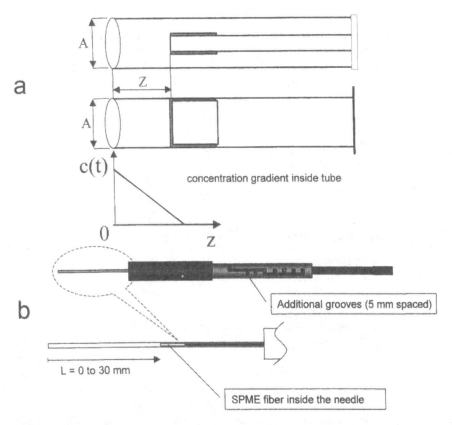

Figure 10 Use of SPME for in-tube time-weighted average sampling: (a) schematic, (b) adaptation of commercial SPME manual extraction holder.

As expected, the extracted amount of analyte is proportional to the integral of the sample concentration over time, the diffusion coefficient of analytes in gaseous phase, D_g, in the area of the needle opening, A, and inversely proportional to the distance of the coating position in respect of the needle opening, Z. It should be emphasized that Eq. (6) is valid only in a situation in which the amount of analyte extracted onto the sorbent is a small fraction (below RSD of the measurement, typically 5%) of the equilibrium amount in respect to the lowest concentration in the sample. To extend integration times, the coating can be placed further into the needle (larger Z), the opening of the needle can be reduced by placing an additional orifice (smaller A), or a higher capacity sorbent can be used. The first two solutions will result in a low measurement sensitivity. An increase of sorbent capacity presents a more attractive opportunity. It can be achieved by increasing either the volume of the coating or its affinity towards the analyte. An increase in the coating volume will require an increase of the device size. The optimum approach to increased integration time is to use sorbents characterized by large coating/gas distribution constants.

The exploitation of restricted access to the absorbing medium allows the implementation of SPME for time-weighted average (TWA) sampling. Where diffusion to the sorbent surface is limited, the sorbent can act as a sort of ''zero sink'' such that extraction is very far from equilibrium under normal sampling conditions. In practice then, any analytes reaching the sorbent surface are absorbed, essentially exhaustively. The rate of diffusion, however, is still dependent on the sample concentration, so the total amount absorbed by the coating is proportional to the average of analyte concentrations over time, hence time-weighted average sampling is achieved. This has been implemented to date with the conventional fiber assembly by retracting the fiber a known distance inside the needle (Fig. 10b). The small size of the needle orifice limits diffusion to the sorbent surface, and the ultimate diffusion rate is also a function of the distance between the fiber tip and the end of the needle. Depending on the volatility and concentration of the analyte of interest, the fiber may be positioned either closer to or further from the end of the needle to achieve the desired degree of nonequilibrium extraction and sensitivity. It would also be possible to implement this type of sampling with the sorbent coated on the interior wall of a capillary. To date, however, the retractable needle implementation has gained the most attention, due to its ease of use and adjustability for the analyte and sample at hand.

VII. PREDICTION OF DISTRIBUTION CONSTANTS

In many cases the distribution constants present in Eqs. (1) and (2), which determine the sensitivity of SPME extraction, can be estimated from physicochemical data and chromatographic parameters. This approach eliminates the need for calibration. For example, distribution constants between a fiber coating and gaseous matrix (e.g., air) can be estimated using isothermal GC retention times on a column with stationary phase identical to the fiber coating material [16]. This is possible because the partitioning process in gas chromatography is analogous to the partitioning process in solid phase microextraction, and there is a well-defined relationship between the distribution constant and the retention time. The nature of the gaseous phase does not affect the distribution constant, unless the components of the gas, such as moisture, swell the polymer, thus changing its properties. A most useful method for determining coating-to-gas distribution constants uses the linear temperature programmed retention index (LTPRI) system, which indexes compounds' retention times relative to the retention times of n-alkanes. This system is applicable to retention times for temperature-programmed gas-liquid chromatography. The logarithm of the coating-to-air distribution constants of n-alkanes can be expressed as a linear function of their LTPRI values. For PDMS, this relationship is $\log K_{fg} = 0.00415 \times$ LTPRI-0.188 [17]. Thus, the LTPRI system permits interpolation of the K_{fg} values from the plot of $\log K_{fg}$ versus retention index. The LTPRI values for many compounds are available in the literature; hence this method allows estimation of K_{fg} values without experimentation. If the LTPRI value for a compound is not available from published sources, it can be determined from a GC run. Note that the GC column used to determine LTPRI should be coated with the same material as the fiber coating.

Estimation of the coating/water distribution constant can be performed using Eq. (3). The appropriate coating/gas distribution constant can be found by applying the techniques discussed above, and the gas/water distribution constant (Henry's constant) can be obtained from physicochemical tables or can be estimated by the structural unit contribution method [18].

Some correlations can be used to anticipate trends in SPME coating/water distribution constants for analytes. For example, a number of investigators have reported the correlation between octanol/water distribution constant K_{ow} and K_{fw}. This is expected, since K_{ow} is a very general measure of the affinity of compounds to the organic phase. It should be remembered, however, that the trends are valid only for compounds within homologous series, such as aliphatic hydrocarbons, aromatic hydrocarbons, or phenols; they should not be used to make comparisons between different classes of compounds because of different analyte activity coefficients in the polymer.

VIII. EFFECT OF EXTRACTION PARAMETERS

Thermodynamic theory predicts the effects of modifying certain extraction conditions on partitioning and indicates parameters to control for reproducibility. The theory can be used to optimize the extraction conditions with a minimum number of experiments and to correct for variations in extraction conditions, without the need to repeat calibration tests under the new conditions. For example, SPME analysis of outdoor air may be done at ambient temperatures that can vary significantly. The relationship that predicts the effect of temperature on the amount of analyte extracted allows calibration without the need for extensive experimentation [19]. Extraction conditions that affect K_{fs} include temperature, salting, pH, and organic solvent content in water. An extraction temperature increase causes an increase in the extraction rate but simultaneously a decrease in the distribution constant. In general, if the extraction rate is of major concern, the highest temperature that still provides satisfactory sensitivity should be used.

Adjustment of the pH of the sample can improve the sensitivity of the method for basic and acidic analytes. This is related to the fact that unless ion exchange coatings are used, SPME can extract only neutral (nonionic) species from water. By properly adjusting the pH, weak acids and bases can be converted to their neutral forms, in which they can be extracted by the SPME fiber. To make sure that at least 99% of the acidic compound is in the neutral form, the pH should be at least two units lower than the pK_a of the analyte. For the basic analytes, the pH must be larger than pK_a by two units.

The volume of the sample should be selected based on the estimated distribution constant K_{fs} [20]. The distribution constant can be estimated by using literature values for the target analyte or a related compound, with the coating selected. K_{fs} can also be calculated or determined experimentally by equilibrating the sample with the fiber and determining the amount of analyte extracted by the coating. Care must be taken to avoid analyte losses via adsorption, evaporation, microbial degradation, etc., when very long extraction times are required to reach the equilibrium.

The sensitivity of the SPME method is proportional to the number of moles of the analyte n extracted from the sample and, for direct extraction, is given by Eq. (1). As the sample volume V_s increases, so does the amount of analyte extracted until the volume of the sample becomes significantly larger than the product of the distribution constant and volume of the coating (fiber capacity $K_{fs} \ll V_s$).

IX. OPTIMIZATION OF SPME METHODS: PRACTICAL ASPECTS OF SPME CALIBRATION AND OPTIMIZATION

A properly optimized method ensures good accuracy and precision together with low detection limits. Below I briefly summarize the most important steps that should be considered when developing SPME methods.

A. Selection of Fiber Coating

The chemical nature of the target analyte determines the type of coating used. A simple general rule, "like dissolves like," applies very well to the liquid coatings. Selection of the coating is based primarily on polarity and volatility characteristics of the analyte. Poly(dimethylsiloxane) (PDMS) is the most useful coating and should be considered first. It is very rugged and able to withstand high injector temperatures, up to about 300°C. PDMS is a nonpolar liquid phase, thus it extracts nonpolar analytes very well with wide linear dynamic range. However, it can also be applied successfully to more polar compounds, particularly after optimizing extraction conditions. An additional advantage of this phase is the possibility of estimating the distribution constants for organic compounds from retention parameters on PDMS-coated GC columns.

Both the coating thickness and the distribution constant determine the sensitivity of the method and the extraction time. Thick coatings offer increased sensitivity but require much longer equilibration times. As a general rule, to speed up the sampling process the thinnest coating offering the sensitivity required should be used.

B. Selection of Derivatizing Reagent

Derivatization performed before and/or during extraction can enhance sensitivity and selectivity of both extraction and detection as well as enable SPME determination of analytes normally not amenable to analysis by this method. Postextraction methods can only improve chromatographic behavior and detection. Incorporation of the derivatization step complicates the SPME procedure, and therefore should only be considered when necessary. Selective reactions producing specific analogs result in less interference during quantitation. This approach can be used for analyte determination in complex matrices. Also, sensitivity enhancement can be achieved when the derivatizing reagent contains moieties that enhance detection.

C. Selection of Extraction Mode

Extraction mode selection is based on the sample matrix composition, analyte volatility, and its affinity to the matrix. For very dirty samples the headspace or fiber protection mode should be selected. For clean matrices, both direct and headspace sampling can be used. The latter is applicable for analytes of medium to high volatility. Headspace extraction is always preferential for volatile analytes because the equilibration times are shorter in this mode compared to direct extraction. Extraction conditions for many compounds, including polar and ionic ones, can be improved by matrix modifications. Application of headspace SPME can be extended to semivolatile compounds and analytes strongly bound to the matrix by increasing the extraction temperature. Fiber protection should be used only for very dirty samples in cases where neither of the first two modes can be applied.

D. Selection of Agitation Technique

Equilibration times in gaseous samples are short and frequently limited only by the rate of diffusion of the analytes in the coating. A similar situation occurs when analytes characterized by large air/water distribution constants are determined in water by the headspace technique. When the aqueous and gaseous phases are at equilibrium prior to the beginning of the sampling process, most of the analytes are in the headspace. As a result, the extraction times are short even when no agitation is used. However, for aqueous samples, agitation is required in most cases to facilitate mass transport between the bulk of the aqueous sample and the fiber.

Magnetic stirring is most commonly used in manual SPME experiments. Care must be taken when using this technique to ensure that the rotational speed of the stirring bar is constant and that the base plate does not change its temperature during stirring. This usually implies the use of high-quality digital stirrers. Alternatively, with cheaper stirrers the base plate should be thermally insulated from the vial containing the sample to eliminate variations in sample temperature during extraction. Magnetic stirring is efficient when fast rotational speeds are applied.

The needle vibration technique uses an external motor and a cam to generate a vibrating motion of the fiber and the vial. This technique has been implemented by Varian in their SPME autosampler. This technique provides good agitation, resulting in equilibration times similar to those obtained for magnetic stirring, but good performance is limited to small vials and direct extraction mode. This technique can be conveniently applied to process a large number of samples since the sample vials do not require any manipulations, such as the introduction of stirring bars.

E. Selection of Separation and/or Detection Technique

So far, most SPME applications have been developed for gas chromatography, but other separation techniques, including HPLC, capillary electrophoresis (CE), and supercritical fluid chromatography (SFC), can also be used in conjunction with this technique. The complexity of the extraction mixture determines the proper quantitation device. Regular chromatographic and CE detectors can normally be used for all but the most complex samples, for which mass spectrometry should be applied. As selective coatings become available, the direct coupling to MS/MS and ICP/MS becomes practical.

F. Optimization of Desorption Conditions

Standard gas chromatographic injectors, such as the popular split-splitless types, are equipped with large-volume inserts to accommodate the vapors of the solvent introduced during liquid injections. As a result, the linear flow rates of the carrier gas in those injectors are very low in splitless mode, and the transfer of the volatilized analytes onto the front of the GC column is also very slow. No solvent is introduced during SPME injection; therefore, the large insert volume is unnecessary. Opening the split line during SPME injection is not practical, since it results in reduced sensitivity. Efficient desorption and rapid transfer of the analytes from the injector to the column require high linear flow rates of the carrier gas around the coating. This can be accomplished by reducing the internal diameter of the injector insert, matching it as closely as possible to the outside diameter of the coated fiber. Narrow bore inserts for SPME are commercially available from Supelco for a range of GC instruments.

Parameters that control the desorption process in the HPLC interface are analogous to those in GC applications. In addition to temperature and flow rate, the composition of the mobile phase also affects the process. In many cases, it is possible to use the mobile phase for desorption without any modifications. In some instances, addition of an appropriate solvent to the interface will assist desorption. The linear flow rate of the mobile phase should be maximized by choosing a small i.d. tubing for the desorption chamber. Temperature also plays an important role in accelerating the desorption.

G. Optimization of Sample Volume

The volume of the sample should be selected based on the estimated distribution constant K_{fs}. The distribution constant can be estimated by using literature values for the target analyte or a related compound, with the coating selected. K_{fs} can also be calculated or determined experimentally by equilibrating the sample with the fiber and determining the amount of analyte extracted by the coating. Care must be taken to avoid analyte losses via adsorption, evaporation, microbial degradation, etc., when very long extraction times are required to reach the equilibrium.

The sensitivity of the SPME method is proportional to the number of moles of the analyte n extracted from the sample and, for direct extraction, is given by Eq. (1). As the sample volume V_s increases, so does the amount of analyte extracted until the volume of the sample becomes significantly larger than the product of the distribution constant and volume of the coating (fiber capacity $\times K_{fs} \ll V_s$).

H. Determination of Extraction Time

An optimal approach to SPME analysis is to allow the analyte to reach equilibrium between the sample and the fiber coating. The equilibration time is defined as the time after which the amount of analyte extracted remains constant and corresponds within the limits of experimental error to the amount extracted after infinite time. Care should be taken when determining the equilibration times, since in some cases a substantial reduction of the slope of the curve might be wrongly taken as the point at which equilibrium is reached. Such a phenomenon often occurs in headspace SPME determinations of aqueous samples, where a rapid rise of the equilibration curve corresponding to extraction from the gaseous phase only is followed by a very slow increase related to analyte transfer from water through the headspace to the fiber. Determination of the amount extracted at equilibrium allows the calculation of the distribution constants.

When equilibration times are excessively long, shorter extraction times can be used. However, in such cases the extraction time and mass transfer conditions have to be strictly controlled to assure good precision. At equilibrium, variations in the extraction time do not affect the amount of the analyte extracted by the fiber. On the other hand, at the steep part of the curve, even small variations in the extraction time may result in significant variations of the amount extracted. The shorter the extraction time is, the larger the relative error is. Autosamplers can measure the time very precisely and the precision of analyte determination can be very good, even when equilibrium is not reached in the system. However, this requires that the mass transfer conditions and the temperature remain constant during all experiments.

I. Calculation of the Distribution Constant

The target analyte's distribution constant defines the sensitivity of the method. It is not necessary to calculate the fiber coating/sample matrix distribution constant, K_{fs}, when the calibration is based on isotopically labeled standards or standard addition, or when identical matrix and headspace volumes are used for the standard and for the sample with external calibration. However, it is always advisable to determine K_{fs} since this value gives more information about the experiment and aids optimization. K_{fs} can be used for calculation of the headspace volume, sample volume, and coating thickness required to reach the desired sensitivity.

The distribution constant for direct extraction mode can be calculated from the following dependence obtained from Eq. (6):

$$K_{fs} = \frac{nV_s}{V_f(C_0 V_s - n)} \tag{7}$$

J. Optimization of Extraction Conditions

An extraction temperature increase causes an increase in the extraction rate but, simultaneously, a decrease in the distribution constant. In general, if the extraction rate is of major concern, the highest temperature that still provides satisfactory sensitivity should be used.

Adjustment of the pH of the sample can improve the sensitivity of the method for basic and acidic analytes. This is related to the fact that unless ion exchange coatings are used, SPME can extract only neutral (nonionic) species from water. By properly adjusting the pH, weak acids and bases can be converted to their neutral forms, in which they can be extracted by the SPME fiber. To make sure that at least 99% of the acidic compound is in the neutral form, the pH should be at least two units lower than the pK_a of the analyte. For the basic analytes, the pH must be larger than pK_a by two units.

K. Determination of Linear Dynamic Range of the Method

Modification of the extraction conditions affects both the sensitivity and the equilibration time. It is advisable, therefore, to check the previously determined extraction time before proceeding to the determination of the linear dynamic range. This step is required if substantial changes of the sensitivity occur during the optimization process.

SPME coatings include polymeric liquids such as PDMS, which by definition have a very broad linear range. For solid sorbents, such as Carbowax/DVB or PDMS/DVB, the linear range is narrower because of the limited number of sorption sites on the surface, but it still can span several orders of magnitude for typical analytes in pure matrices. In some rare cases when the analyte has extremely high affinity towards the surface, saturation can occur at low analyte concentrations. In such cases, the linear range can be expanded by shortening the extraction time.

L. Selection of Calibration Method

Standard calibration procedures such as external calibration can be used with SPME. The fiber blank should be first checked to ensure that neither the fiber nor the instrument causes interferences with the determination. The fiber should be conditioned prior to the first use by desorption in a GC injector or a specially designed conditioning device. This process ensures that the fiber coating itself does not introduce interferences. Fiber conditioning may have to be repeated after analysis of samples containing significant amounts of high molecular weight compounds, since such compounds may require longer desorption times than the analytes of interest.

When the matrix is simple (e.g., air or groundwater), the distribution constants are very similar to those for pure matrix. It has been shown, for example, that typical moisture levels in ambient air, as well as the presence of salt and/or alcohol in water in concentrations lower than 1%, usually do not change the K values beyond the 5% RSD typical for SPME determinations. In many such instances, calibration might not be necessary since the appropriate distribution constants that define the external calibration curve are available in the literature or can be calculated from chromatographic retention parameters. External calibration can also be used successfully when the matrix is more complex but well defined (e.g., process streams of relatively constant composition). Of course, calibration standards have to be prepared in such cases in the matrix of the same composition rather than in the pure medium (e.g., water).

A special calibration procedure, such as isotopic dilution or standard addition, should be used for more complex samples. In these methods, it is assumed that the target analytes behave similarly to spikes during the extraction. This is usually a valid assumption when analyzing homogeneous samples. However, it might not be true when heterogeneous samples are analyzed, unless the native analytes are completely released from the matrix under the conditions applied. Moreover, whenever any of these methods is used, an inherent assumption is made that the response is linear in the concentration range between the original analyte concentration and the spiked concentration. While this is usually true for fibers extracting the analytes by absorption (PDMS, PA) and detectors with wide linear range (FID), problems may arise when porous polymer fibers (PDMS/DVB, Carbowax/DVB) are used or when the detector applied has a narrow linear dynamic range. It is important, therefore, to check the linearity of the response using standard solutions before applying standard addition or isotopic dilution for calibration. To improve the accuracy and precision, multipoint standard addition should be used whenever practical.

M. Precision of the Method

The most important factors affecting precision in SPME are as follows:

Agitation conditions

Sampling time (if nonequilibrium conditions are used)

Temperature

Sample volume (small sample sizes)

Headspace volume (small sample sizes)

Vial shape

Condition of the fiber coating (cracks, adsorption of high molecular weight species)

Geometry of the fiber (thickness and length of the coating)

Sample matrix components (salt, organic material, humidity, etc.)

Time between extraction and analysis

Analyte losses (adsorption on the walls, permeation through Teflon, absorption by septa)

Geometry of the injector

Fiber positioning during injection

Condition of the injector (pieces of septa)

Stability of the detector response

Moisture in the needle

To ensure good reproducibility for SPME measurement, the experimental parameters listed above should be kept constant.

N. Automation of the Method

SPME is a very powerful investigative tool, but it can also be a technique of choice in many applications for processing a large number of samples. To accomplish this task would require automation of the methods developed. As automated SPME devices with more advanced features and capabilities become available, automation of the methods developed becomes easier. The currently available SPME autosampler from Varian enables direct sampling with agitation of the sample by fiber vibration and static headspace sampling. In some cases custom modifications to the commercially available systems can facilitate operation of the method closer to optimum conditions.

X. CONCLUSIONS

One can draw a number of paralells between developments and applications of SPME with electrochemical methods. The coulombetric technique corresponds to the total extraction method. Although the most precise, this technique is not used frequently because of the time required to complete it. SPME is capable of producing exhaustive extraction when the volume of the extraction phase is large enough combined together with high distribution constants. In fiber geometry, the larger volume translates into thicker coatings, which results in long extraction times. The alternative approach is to disperse the whole volume of the extraction phase onto a larger surface area, resulting in a thinner coating and faster equilibration times. For example, solid support material may include particulate matter, a stirring mechanism, or the vessels walls (see Fig. 1). In this case, however, there would be more handling required to conveniently introduce the extraction phase into the sample introduction system (GC or HPLC). It might necessitate the use of organic solvent to desorb the analytes from the extraction phase. Equilibrium potentionmetric techniques are more frequently used (pH electrode), particularly in cases where the sample is a simple mixture and/or selectivity of the membrane is sufficient to quantify target analyte in complex matrices. The equilibrium SPME method has some advantages in this regard, since the technique is typically coupled with separation and/or mass spectrometry detection methods, which allows identification and quantification of many components simultaneously. The advantage of electrochemical methods is response time because of low capacities of electrodes.

Some electrochemical methods, like amparometry, are based on mass transport through the boundary layers as in preequilibrium SPME. Analoguously, in SPME calibration based on diffusion coefficients can be accomplished as long as the agitation conditions are constant and the extraction times are short and the coating has a high affinity towards the analytes. Figure 6 illustrates the results related to the 10s extraction times using solid coating. In some implementations of the technology, the rate of mass transfer to the extraction phase can be purposely restricted by placing the phase in the needle and therefore achieving the time-weighted average measurements of concentration in a specific time period.

The potential savings in analysis time, reduced solvent use, and apparent simplicity of SPME techniques will continue to attract the interest of analytical chemists searching for improved analysis methods. As long as analysts have a sound understanding of the theory and principles behind this technique, good accuracy and precision will follow.

REFERENCES

1. J Pawliszyn. Solid Phase Microextraction. Theory and Practice. New York: Wiley, 1997.
2. J Pawliszyn, ed. Applications of Solid Phase Microextraction. Cambridge, U.K.: RSC, 1999.
3. D Louch, S Motlagh, J Pawliszyn. Anal Chem 64:1187, 1992.
4. J Al-Hawari. In: J Pawliszyn, ed. Applications of Solid Phase Microextraction. Cambridge, U.K.: RSC, 1999, pp. 609–621.
5. J Poerschmann, F-D Kopinke, J Pawliszyn. Environ Sci Technol 31:3629–3636, 1997.
6. Z Mester, J Pawliszyn. Rapid Comun Mass Spectrom 13:1999–2003, 1999.
7. J-L Liao, C-M Zeng, S Hjerten, J Pawliszyn. J Microcolumn (Sept. 8):1–4, 1996.
8. T Gorecki, X Yu, J Pawliszyn. Analyst 124:643–649, 1999.
9. J Koziel, M Jia, J Pawliszyn. Anal Chem 72:5178–5186, 2000.
10. CV King. J Am Chem Soc 57:828–831, 1935.
11. J Crank. Mathematics of Diffusion. Oxford: Clarendon Press, 1989, p. 14.
12. J Pawliszyn. J Chromatogr Sci 31:31, 1993.
13. R Eisert, J Pawliszyn. Anal Chem 69:3140–3147, 1997.
14. M Chai, J Pawliszyn. Environ Sci Technol 29:693, 1995.
15. J Crank. Mathematics of Diffusion. Oxford: Clarendon Press, 1989, p. 2.
16. Z Zhang, J Pawliszyn. J Phys Chem 100:17648, 1996.
17. P Martos, A Saraullo, J Pawliszyn. Anal Chem 69:1992, 1997.
18. R Schwarzenbach, P Gschwend, D Imboden. Environmental Organic Chemistry. New York: John Wiley & Sons, 1993, pp. 109–123.
19. P Martos, J Pawliszyn. Anal Chem 69:206, 1997.
20. T Gorecki, J Pawliszyn. Analyst 124:643–649, 1999.

27

Uppsala School in Separation Science: My Contributions and Some Personal Reflections and Comments*

Stellan Hjertén

Uppsala University, Uppsala, Sweden

I. HISTORICAL PERSPECTIVE

In Nature there are no disciplines, only phenomena.

The Svedberg

The best can be accomplished only if it is pleasant and involves a creator's joy.

Arne Tiselius, 1969, in a letter when the author was appointed a professor in biochemistry

In his invitation letter, the editor of this volume, Dr. Haleem J. Issaq, asked me "to write a chapter about my experiences in and contributions to the advances in separation science." However, in several sections, including this one, I will—according to my and his wish—also refer to some episodes in my research career and, at the same time, make a few personal reflections and comments. They should be considered as an attempt to shed light on incidents and attitudes which are of *general* interest to scientists and perhaps also to scholars in the history of science and ideas.

A negative attitude toward new approaches is rather common in our (scientific) community. The examples I will give may stimulate (young) researchers not to give up their ideas as soon as they encounter resistance. The objective of my personal reflections and comments is also to stimulate discussion and to make young researchers aware of scientific pitfalls. They are absolutely not intended to express general negativism or pessimism. I can assure that almost always I have gone and still go to my laboratories every morning with joy, clouded by no hostile feelings

---

* A summary of my scientific production is found in Ref. 1.

421

whatever toward anybody or anything, to devote myself to my hobby: research. My father, who was an extremely skillful cabinetmaker and woodcutter, assured me that if he could relive his life he would have chosen the same profession. To me, separation science has the same attraction, although my wish would be to orient myself more toward physical aspects.

The birth of the Uppsala School in separation science coincides with Theodor (The) Svedberg's design and development of the ultracentrifuge for sedimentation studies of proteins [2–7]. For the first time it became possible to analyze proteins and determine accurately their molecular weights. Svedberg drew the incorrect conclusion that the molecular weights of all proteins were a multiple of 35,000. I mention this curiosity to underline immediately what Francis Crick replied when a researcher said that some of Linus Pauling's ideas have not turned out to be right: "A man who is always right almost never says anything significant." The Svedberg, remembered as "the father of the ultracentrifuge," was awarded the Nobel Prize in 1926 "for his work on disperse systems." In the presentation of his scientific contributions the sedimentation studies were mentioned but were perhaps not appreciated as much as they deserved. In 1930, the year Tiselius defended his thesis (see below), Kai O. Pedersen joined Svedberg's research team and was to devote all his scientific life to sedimentation studies. With Svedberg he wrote the classical volume *The Ultracentrifuge* [8].

At the time when Svedberg introduced the ultracentrifuge, many researchers were of the opinion that proteins were some kind of colloids with varying particle size. However, this hypothesis was given up (although slowly by some scientists) when it became clear that a protein sedimented as a sharp boundary blurred only by diffusion, which means that all molecules of each particular protein species are the same size.

Upon The Svedberg's suggestion, his pupil, Arne Tiselius, decided to study the behavior of proteins in an electric field [4,6,7,9–13]. The birth of the moving boundary electrophoresis method in its modern quantitative design made it possible to show that all molecules of a particular protein migrated as a sharp boundary also in this technique, i.e., that all molecules of a given protein had the same net charge density. This observation, along with the analogous one made by The Svedberg regarding the size of proteins, convinced even the most skeptical researchers that all molecules of a given protein were practically identical. To honor and commemorate these two great Swedish scientists, sedimentation constants are given in Svedberg units (10^{-13} s) and electrophoretic mobilities in Tiselius units (10^{-5} cm^{2}s^{-1} V^{-1}) [14]. Tiselius was awarded the Nobel Prize in chemistry in 1948 for his research on electrophoresis and chromatography and for the establishment of the heterogeneity of serum proteins. His thesis, "The Moving Boundary Method of Studying the Electrophoresis of Proteins," was published in 1930. It should be mentioned that a few simple moving boundary electrophoresis experiments had been published during the nineteenth century.

Some contemporaries thought that the Prize should have been shared with Tiselius's pupil, Harry Svensson (who later changed his family name to Rilbe), because of his important contributions to the theory of moving boundary electrophoresis and the optical detection of the boundaries. However, in my opinion, a researcher who has developed a method so successfully that it can be considered as established should enjoy a greater reputation and be more honored than those who later contribute with improvements, however important. The situation is more complicated when the improvements are of such decisive importance that the original method would be of dubious value without them. I will return to these viewpoints from a different angle in connection with a discussion of the definition of the term "pioneer" (Sec. III.A.7.b). I would strongly emphasize that Harry Rilbe's later contributions to the development of isoelectric focusing at the California Institute of Technology and Karolinska Institute in Stockholm are outstanding and remarkable [15] (see Sec. III.A.7.b). Interestingly, Rilbe and—owing to his justified authority—all other researchers investigating isoelectric focusing had the idea that the pH gradients are stable infinitely until I showed theoretically that this is not the case [16].

Although this chapter is primarily concerned with research in separation science performed at Uppsala University, I will underline the importance of the unique synthesis of the carrier ampholytes developed by Olof Vesterberg at Karolinska Institute, Stockholm [17]. The success isoelectric focusing enjoys today depends to a great extent on his synthesis method. My friend and colleague, Jerker Porath, devoted a great part of his thesis to large-scale electrophoresis and made very important contributions to this field [18]. He is, however, more known for his studies of gel filtration [19] (although without my discovery of the molecular-sieving properties of cross-linked dextran, the Sephadex™ gels would probably never have been developed; see Sec. III.A.6.a), bioaffinity chromatography [20], including immobilized metal ion affinity chromatography (IMAC) [21]—methods used in biochemical laboratories all over the world.

Rilbe, Porath, and I were not the only pupils Tiselius had in the field of separation science. Per Flodin (Sec. III.A.6.a) and Per-Åke Albertsson made very important contributions. Albertsson developed in an ingenious way nondenaturing two-phase systems for the separation of cell particles and biopolymers, a technique used worldwide [22]. After his dissertation in 1960 and a sabbatical year in the United States, he left Uppsala in 1965 for a professorship at Umeå University. Karin Caldwell, a pupil of Jerker Porath, moved a few years after her thesis defense to the United States, where she made very important contributions to our understanding of surface chemistry. Therefore, it was quite natural that she in 1998 became the first holder of the professorship in Surface Biotechnology at Uppsala Biomedical Center. When Professor Porath retired in 1987, the head of the Department of Biochemistry (in 1986 the Institute of Biochemistry changed its name to the Department of Biochemistry) tried to retain the character of Porath's chair, i.e., the development of new separation methods for biopolymers, but failed. The utterance from the Faculty of Science at Stockholm University was typical: research in this area is successfully performed at companies and should, therefore, not be done at universities. As far as I know, no separation methods—or extremely few—have their origin in laboratories outside the university world.

My chair was announced in biochemistry, especially separation science, upon my retirement in 1995. The Faculty of Science at Uppsala University was as unenthusiastic as Stockholm University about the emphasis on separation science. Under the cloak of democracy its negative view was imparted to two international boards, the members of which had no (or little) experience in separation science. The Board of Professors at the California Institute of Technology and that at the University of Cambridge gave the same answer—almost literally—as our Faculty: "... with all due respect to The Svedberg, Arne Tiselius, Jerker Porath and Stellan Hjertén, I must assert that the area of separation science, as practiced by these great chemists, is over" and "... the time has come to break with traditions," respectively [23]. The latter board also stated that Hjertén's group "has declined in productivity in recent years" [23], which is completely wrong and only reflects the negative attitude of the Faculty of Science at Uppsala University (including the Chemistry Section) toward separation science. Proper evaluation of all research activities is an absolute requirement, as is the impartiality and competence of the reviewers. Did the above two review teams fulfill these demands? Has a university a sound research policy when it overnight abolishes a discipline which for decades had been—and still is—one of the flagships of the Faculty? It is the duty of any university professor and, therefore, also my duty to raise questions of this type at the least suspicion about improprieties, particularly because there still are pleaders for the same evaluation approach who want to do evaluations in a similar spirit, even on a level higher than the Faculty. I leave it to the reader to decide whether the conclusion of the Board, i.e., that of the Faculty, that separation science in the way that I have practiced (Sec. III) and am continuing to practice it (Sec. IV) should be discontinued, was based on objectivity. This issue is of little importance, considering the relatively small area this branch of science represents in the (life) sciences as a whole. However, one should pay more attention to it in case it reveals a general, exaggerated belief in the potential and desire of industry to solve problems within basic research (the discussions in Secs. II and III.A.6.a and b show that this sentence should not be interpreted to mean that researchers of the Uppsala School are not interested in cooperation with industry).

My impression is that there is an obvious trend to neglect the importance of basic research not only at Swedish, but also at several foreign universities. An indication that this trend may be broken—at least in Sweden—is a recent official declaration from the present Swedish Minister of Education: "The 1990s were in many respects the decade of applied and industrial research. The government has now decided that the balance should be shifted back toward basic research in recognition of the fundamental role of free, curiosity-driven research in the development of society. The free search after knowledge is important just because no one can foresee which new discoveries will become useful in the long term and because knowledge itself has its own value. Resources for basic research and research training should therefore be increased." No representative of any Swedish university has for many years expressed so clearly the importance of fundamental research. Have our universities lost their original identity?

For more information about the history of the development of electrophoresis in the Uppsala School, see Ref. 24.

II. SOME CHARACTERISTIC FEATURES OF BASIC RESEARCH IN SEPARATION SCIENCE IN THE UPPSALA SCHOOL

Teamwork:

Coming together is a beginning, keeping together is progress, working together is success.

There is no limit to what can be accomplished if it doesn't matter who gets the credit.

Separation scientists belonging to the Uppsala School have all developed new methods based on sedimentation, partition, electrophoresis, or chromatography to separate and characterize biomolecules and particles, particularly macromolecules such as proteins. This is in sharp contrast to most analytical chemists who use known methods, for instance, commercially available HPLC columns, and optimize the separation conditions for a particular type of sample, usually containing low molecular weight compounds such as drugs, seldom biopolymers. Of course, we need both types of approaches.

I have often been asked the question: Why don't you also apply your methods to a purely biological or medical problem? My answer is simple: The methodological issues are so intricate that there is no time for anything else, although my working day is still 10–12 hours (I realized this problem in the 1970s when I studied both separation methods and membrane proteins). I have given a similar answer to the common question of why I don't have a firm of my own to get some financial gain from all of the widely used methods I have introduced. My opinion is that one has to choose between research and business, and I made an early decision that I do not regret.

Looking back at their contributions to science, some researchers say in interviews that they remember with pleasure a certain day because of a discovery. Such statements add zest to a story, but may be an after-the-fact rationalization. No doubt the final success is, in most cases, a consequence of a series of consecutive advances and improvements and, therefore, not concentrated in one or a few days, which does not mean that the feeling of satisfaction experienced isn't very strong when a scientific problem approaches a solution. Arne Tiselius called it happiness in a speech at the 1947 Nobel banquet: "When a new thought is born, or when one of the deep secrets of Nature yields to the searching scientist—in this very act of creation—there is a pure and primitive happiness deeper than anything of this kind which can ever be granted a human being to experience."

Many of my former and present colleagues of the Uppsala School certainly share with me the opinion that a new separation technique will become widely used only if it is easy to handle and commercially available (observe, however, that the scientific quality of a paper describing a new method is independent of commercialization of the method). In recent years an additional feature, automation, is increasingly important. Collaboration with researchers and engineers in both academia and industry is necessary to satisfy all these diversified requirements. Cooperation—a word of honor in research applications and at banquet speeches—is seldom utilized optimally in practice, which would be possible only (a) if it does not matter who gets the credit for a success, (b) if companies were more interested in long-term projects, and (c) their researchers had more influence over the selection of projects. It is tempting to play with the thought of what could be accomplished if the top researchers in a particular field in industry and academia worked together to solve a common scientific problem. This is, at present, a utopian dream. However, I hope, but am not convinced, that in the future many of the problems associated with such teamwork can be solved. A prerequisite is, however, that they are openly and frankly discussed. A good example of an almost ideal situation is the internationally recognized and fruitful cooperation between Pharmacia Fine Chemicals and the Institute of Biochemistry during the 1960s and 1970s. To revive this spirit, more power should be given to managers with good scientific background who have the capability to look into their crystal ball for important profitable long-term projects.

I have also had stimulating contacts with Bio-Rad Laboratories (Hercules and Richmond, CA) whose President, Dr. David Schwartz, has a remarkable feeling for which scientific ideas can be transferred to profitable commercial products. Unfortunately, his company, like Pharmacia Biotech and probably most other companies, has lost the previously predominant American pioneer spirit, where the readiness to take risks was considerably greater than at today's companies, which are forced to give priority to safe, short-term projects to satisfy their shareholders' demand for rapid returns. One cannot request a company to be interested in unprofitable apparatus and chemicals, even if they have great potential to lead to important scientific advances. Therefore, international organizations with financial support from foundations, governments, etc., should be established with the goal of manufacturing such products.

Tiselius often pointed out the importance of developing *gentle* methods for the separation of proteins in order to minimize the risk of denaturation. Although correct in many cases, a somewhat less negative attitude among the researchers in the Uppsala School toward the use of organic solvents, surfactants, extreme pH and temperatures, etc., had certainly broadened our repertoire of new separation techniques.

For more information on achievements in the Uppsala School, see Refs. 24 and 25.

III. THE MOST IMPORTANT SEPARATION METHODS DEVELOPED IN MY LABORATORIES AND SOME COMMENTS

The secret of science is to ask the right question and it is the choice of problem more than anything else that marks the man of genius in the scientific world.

Sir Henry Tizard

We cannot adjust the wind but we can adjust the sails.

Breakthroughs in biochemistry, biology, medicine, and related disciplines often run parallel to the introduction of new analytical and preparative separation methods, provided that they have been commercialized, as stated in the previous section. The latter requirement is increasingly important, since in most of today's laboratories even a simple chemical synthesis is seldom done or simple equipment constructed. The reasons for this, and the obvious consequence—a delay in the progress of life sciences—should be discussed seriously. Such a discussion is, however, outside the scope of this chapter. The establishment of international organizations of the kind suggested in Sec. II may be a partial solution to these problems.

In this summary of my research I will complement the description of the scientific activities with comments on some events and incidents, hopefully, of some general interest, especially since Uppsala at that time was called the Mecca of Biochemistry—in appreciation of the many separation methods that originated continuously from the Institute of Biochemistry. My scientific career has been centered around the development of new separation methods and improvements of existing ones (although in the 1970s, as mentioned above, membrane proteins were also studied; see Ref. 25 and Ref. T5 in Sec. VII). The projects dealt with are, accordingly, many. I am very grateful to all my coworkers for their important contributions to the success of these projects. Their names appear in Sec. VII and the references. Some of them are also mentioned in the text in connection with a particular event. Owing to space limitations I have chosen to concentrate on methods I have developed and which are used in many life science laboratories. Several of them have been commercialized.

A. Commercialized Separation Methods in Worldwide Use

1. *Chromatography on Hydroxyapatite (First Trade Name Bio-Gel HT, Later Several Others)*

In 1954, shortly after my studies for a bachelor of science in mathematics, chemistry, and physics, Professor Arne Tiselius approached me in a corridor in his newly created building for research in the biochemical field and asked whether I would like to join his group. I understood that Tiselius was interested in physicochemical aspects of biochemistry, which appealed to me— and still does (in fact, I have found that the most successful Ph.D. students in biochemistry are often those who also have a good background in physical chemistry or physics). Later on I heard that a faculty colleague of Tiselius, a professor in physics, had recommended me (I was a laboratory teacher at the Institute of Physics at that time). Tiselius suggested that I should study the chromatographic behavior of proteins on brushite ($CaHPO_4 \cdot 2H_2O$), an adsorbent that he had recently introduced. One of the projects was to find out how temperature affected the adsorption. In an attempt to equilibrate a brushite column at 40°C with a phosphate buffer of pH 6.9, I found that the pH of the effluent was around 5 and remained there for days. I also observed that a very thin layer formed at the top of the column had a somewhat different appearance than the rest of the column. X-ray analysis showed that this layer consisted of hydroxyapatite. From then on we abandoned the apparently unstable brushite for hydroxyapatite, which I prepared from brushite by boiling at high pH for rapid removal of the phosphoric acid released (which had caused the low pH of the effluent in the above temperature studies). Tiselius wrote the paper, with Östen Levin and me as coauthors. Östen was a friend of mine who I suggested join Tiselius's group (in fact, we were his only students at that time and the last before his retirement in 1968) [26]). The paper became a "This Week's Citation Classic" [27]. The reason for the popularity of hydroxyapatite as an adsorbent is its unique adsorption mechanism: (a) low molecular weight compounds are seldom adsorbed; (b) the number of available carboxylic and phosphate groups in the sample constituents (often proteins and DNA) usually determines the degree of adsorption [28]; (c) usually only phosphate ions can be used for desorption. As a consequence of point (a), small molecules, for instance in serum, can often easily be separated from proteins that may disturb an analysis. The presence of salts (e.g., 10% sodium chloride) usually has little or no effect on the interaction. The adsorption mechanism of hydroxyapatite thus differs markedly from that of ion exchangers. Therefore, sequential use of these two kinds of adsorbents often gives a very high degree of purification of proteins. Hydroxyapatite and DEAE-cellulose [29] are the classical and still widely used chromatographic beds for the purification of biopolymers.

2. Free Zone (Capillary) Electrophoresis

In the introduction of my thesis on automated narrow-bore tube electrophoresis, I stated that there are two methods to perform carrier-free electrophoresis in capillaries: one being based on rotation of the horizontal capillary around its long axis, the other on the use of a stationary capillary with a very narrow bore [30]. The latter method is nowadays called high-performance capillary electrophoresis (HPCE). My work was focused on the first method, mainly because at that time (1960s) no UV monitors sensitive enough to detect zones in a tube with a diameter much below 1 mm were available, nor were fused silica capillaries. In fact, in close collaboration with Lars Mattson, the head of our mechanical workshop, and Göte Eriksson, a very able instrument maker, a method was developed to grind and polish both the outside and the inside of commercial quartz tubes in order to get less noise at the baseline in the electropherograms (observe that the tube was *scanned* at predetermined intervals of time). To reduce the noise further, a detector was designed which permitted analyses at much lower concentrations than did the commercially available ones. It was based on a patented, unique method to measure the ratio of the absorbancies at two different wavelengths using only one photomultiplier. In the thesis, I (a) showed, starting from an equation I derived for the electroendosmotic mobility, which is more general than the classical Helmholtz equation, how to proceed in practice in order to eliminate electroendosmosis by static or dynamic coating of the capillary wall (when I started the free zone electrophoresis experiments, it was stated that "electroosmotic streaming cannot be eliminated or counteracted by any known method . . ." [31]; by coating I eliminated it and by hydrodynamic compensation I counteracted it); (b) derived an equation for thermal zone deformation; (c) showed that constant current, but not constant voltage, gives a migration velocity which is independent of temperature variations in the buffer and is therefore preferable to constant voltage in most electrophoresis experiments in order to get reproducible results; (d) traced the light when it transverses a capillary tube to derive the relationship between the transmission of the sample in a capillary and the concentration of the sample; (e) introduced indirect detection; (f) discussed the influence of the sample concentration and the mobility of the buffer ion on peak asymmetry; (g) separated inorganic and organic ions, proteins, nucleic acids, virus, and whole cells (bacteria) by capillary zone electrophoresis (CZE), both on an analytical and micropreparative scale; (h) described isoelectric focusing of proteins (the first focusing experiments performed in the absence of an anticonvection medium). These eight points are, of course, relevant to all electrophoresis experiments in capillaries, independently of their diameters, i.e., even in the range 0.025–0.1 mm, which includes the dimensions of the stationary electrophoresis tubes commonly used in HPCE. As a curiosity it may be mentioned that NASA's electrophoresis experiments in space did not become successful until the electrophoresis tube was coated according to point (a). Many of the problems in modern HPCE—and in electrophoresis in chips— were, accordingly, solved as early as 1967. Therefore, I am glad that colleagues call me the father of capillary electrophoresis, just as my tutor, Arne Tiselius, is remembered as the father of moving boundary electrophoresis and his tutor, The Svedberg, as the father of the ultracentrifuge. Of course, there are several distinguished separation scientists in the Uppsala School, although they have not been favored with an honorific title. Among these I would like to mention Per-Åke Albertsson, Per Flodin, Jerker Porath, and Harry Rilbe.

Sometimes the success of a new method comes several years after the initial publication. An example is the above-described free zone electrophoresis in narrow-bore tubes. The first paper was published in 1958 [32], and about 20 papers on this and other separation methods were published before I defended my thesis in 1967 [30]. As mentioned above, I pointed out in my thesis that one can do free zone electrophoresis experiments in stationary, very narrow separation channels. I intended to publish a paper dealing with such experiments, partly because I observed that dyestuffs following their separation did not sediment when the rotation was stopped, but never got around to it. In fact, only a few years after the publication of my thesis, Virtanen [33] and Mikkers et al. [34] used the latter approach to run free zone electrophoresis experiments in 0.2–0.4 mm plastic or glass capillaries. However, nobody—or very few people—applied the technique to real samples. A possible explanation could be that given above— namely, that commercial fused silica capillaries, which are superior to plastic and glass capillaries for capillary electrophoresis (and chromatography), were not available at that time (introduced by Bente et al.) [35], nor were high-sensitivity UV and fluorescence detectors. We may thus use the common, vacuous cliché: the time was not ripe for this approach until 1981, when Jorgenson and Lukacs came up with the ingenious idea to employ the commercial fused silica capillaries [36], originally introduced for gas chromatography [35]. This paper made many researchers aware of the great potential of HPCE and resulted in a large number of publications.

Most analytical chemists had little or no experience with electrophoresis and became very enthusiastic when they suddenly realized that they had a new unexplored tool at their disposal, whereas biochemists were not equally impressed because they had routinely used electrophoresis (and isoelectric focusing) for decades, particularly polyacrylamide gel electrophoresis, for the separation of proteins and remain quite satisfied with that technique, because of its high resolution, especially in the 2-D mode, and the possibility to analyze several samples in a single run with simple, reliable equipment. The shorter analysis time is more important in the pharmaceutical industry for quality control of drugs than in basic research in biochemistry. However, this great enthusiasm—which, of course, I personally enjoyed very much—among some less experienced chemists had a dark side: several reinvented-wheel papers on capillary electrophoresis were accepted even in such renowned journals as *Analytical Chemistry*.

I was the first to introduce in HPCE a polymer coating to eliminate adsorption and electroendosmosis [37] (the coating is based on the equation for electroendosmotic velocity I derived in my thesis and the practical approach described [30], which, in fact, has been used for most coatings published later), capillary gel electrophoresis of low molecular weight compounds and proteins [38–41], capillary isoelectric focusing along with methods to mobilize the focused protein zones [42], micropreparative capillary electrophoresis based on continuous elution [40,41,43] (a similar technique is now widely used to combine HPCE with mass spectrometry), multipoint detection using *one* UV monitor [44], and low-conductivity buffers permitting field strengths up to 2000 V/cm or more if one can avoid electric shock [45] (these buffers can be used in all modes of analytical and preparative electrophoresis to decrease the run times and thus suppress the diffusional zone broadening). We were also the first to point out that an electrophoresis tube bent into loops gives greater broadening than does a straight tube (this broadening was discussed at two HPCE meetings before it was published [44]). Many analytical problems cannot be solved by capillary electrophoresis because of the low concentration of the sample. The difficulty can be overcome by off- and on-tube concentration. We have developed several such methods, for instance, those based on the use of isoelectric focusing; zone electrophoresis toward a small-pore gel or toward a gradient in effective cross-sectional area, viscosity, or electrical conductivity; a combination of displacement electrophoresis and a counterflow [46,47] (compared to other similar approaches, ours has the great practical advantage of not requiring a careful adjustment of the counterflow); Donnan effects, utilized in gels [48] or hollow fibers [49]; or continuous evaporation of water from a hollow fiber and simultaneous transport of sample into the fiber [49]. We were able to achieve a more than 1000-fold enrichment of proteins.

As early as the 1950s I solved the problem of detecting low concentrations of proteins in paper electrophoresis by developing an on-line enrichment technique. It was based on an abrupt increase of the cross-sectional area of the electrophoresis medium achieved by placing the paper on top of a dialysis bag [50].

It was not difficult for me to foresee the broad application range and immense success free zone electrophoresis enjoys today because the earlier widely used Tiselius free moving boundary method permits reliable analyses only of macromolecules, particularly proteins, affords only partial separation, and requires as much as 50–100 mg of a protein for an analysis, whereas the former method demands only ng quantities or less and is applicable to all kinds of samples, such as inorganic ions, low molecular weight organic substances (e.g., nucleotides and drugs), proteins, DNA, viruses, cells, and cell particles [30,51]. Capillary electrophoresis made it possible to finish the Human Genome Project earlier than anticipated—an indication of the strength of the method.

3. Polyacrylamide Gels for Molecular-Sieve Electrophoresis

My first electrophoresis experiments in polyacrylamide gels were done in glass tubes with phycocyanin, a blue protein from the alga *Ceramium rubrum* as sample (this protein and phycoerythrin, a red, strongly fluorescent protein from the same alga, were routinely used in Tiselius's laboratories for methodological investigations, particularly in the initial steps). The phycocyanin was resolved into several bands, so sharp that I had never seen anything like it before. I showed this experiment to Tiselius, my tutor, but surprisingly he had no comments. He was more interested in the birds in the birdfeeder I had mounted outside the window of our laboratory [both of us were fascinated bird-watchers (and amateur photographers)]. However, later on when I gave him material for his report to the European Research Office of the U.S. Department of the Army, he was very enthusiastic [24]. Or was Tiselius still "above all interested in the quantitative aspects" and "showed a lack of foresight"—as he expressed it in his autobiography [9] in an attempt to explain why he did not publish the experiments he performed in 1927 on the separation of "red phycoerythrin from blue phycocyanin by electrophoresis in a slab of gelatin, obtaining beautiful narrow migrating zones"?

Rolf Mosbach and I also employed polyacrylamide gels for molecular-sieve chromatography [52]. At that time I was already aware of the analogy between electrophoresis and chromatography and suggested, therefore, the terms molecular-sieve chromatography and molecular-sieve electrophoresis to emphasize the analogy [53]. Much later I described this analogy in the form of an equation valid for all separation methods based on differential transport of the solutes, i.e., for both electrophoresis, chromatography, and centrifugation [54,55]. In the latter paper I derived an H-function that is more accurate and general than the Kohlrausch ω-function. I strongly emphasize that both Raymond and Ornstein also are pioneers in the field of polyacrylamide gel electrophoresis (see Ref. 24). Probably they—like myself—searched for a substitute for Smithies' starch gels [56], which had some undesirable properties. The starch and polyacrylamide gels gave a resolution of proteins not achieved before. No doubt, Smithies is the true pioneer in the field of electrophoretic molecular sieving.

When I introduced cross-linked polyacrylamide for molecular-sieve electrophoresis of proteins, I could not predict the great significance it was to have for the analysis of DNA (the enormously important Human Genome Project is based on this technique). I am convinced that Raymond and Ornstein also neglected the importance of DNA analyses. This is an example of the difficulties in foreseeing the consequences of fundamental research—and its indirect potential impact on our society (see Sec. I). Nor could I predict that also the agarose gels that I introduced for electrophoresis of proteins (see below) were to be so important for analysis of DNA.

We also designed a preparative polyacrylamide gel electrophoresis apparatus with continuous elution which gave the *same* high resolution as analytical polyacrylamide gel electrophoresis, even when as much as 1 g of protein was applied [57].

For micropreparative molecular-sieve electrophoresis, a dextran gel and a copolymer gel of dextran and polyacrylamide were synthesized (I still have in my laboratory the allyl dextran made by Pharmacia for the preparation of the copolymer). Following a run the gel was cut and the protein of interest was released by dextranase [58]. Pharmacia was not interested in this approach. I also suggested that this company manufacture beads of the copolymer gel for molecular-sieve chromatography because it is more rigid than are polyacrylamide and dextran gels and could, therefore, be expected to afford a higher flow rate. However, Pharmacia was at that time of the opinion that the price would be too high, although it must have changed its mind when it later commercialized gel beads of this copolymer under the trademark Sephacryl.

4. Agarose for Nonsieving Electrophoresis

The original reason for my studies of agarose gels was quite the opposite of that mentioned for polyacrylamide gels. I was interested in finding a gel that gave electropherograms of proteins similar to those obtained in my free zone electrophoresis experiments, which I had started a few years earlier, i.e., a gel without molecular-sieving properties. Agar gels had been used for electrophoresis by Wieme [59] and others for many years, but the presence of charged groups made them far from ideal (my attempts at making them noncharged by esterification were not successful). Studying many articles on the chemical structure of agar, I learned that a Japanese researcher, Araki, had found as early as 1937 that agar is composed of two polysaccharides: a charged one, which he called agaropectin, and a noncharged one, agarose [60]. I contacted him and he wrote a letter to me in Japanese in which he explained how he had arrived experimentally at this composition. (Upon my request he later gave me an English translation and I convinced myself by repeating his experiments that his statement was correct [61]. The advantages of agarose over agar for immunochemical experiments were emphasized in a paper [62] published in the same issue as the article on zone electrophoresis in agarose [61]. Agarose in the form of a gel suspension was used for preparative electrophoresis [63]. The suspension competed favorably with cellulose powder as anticonvection medium in experiments aimed at separation of proteins, due to the low concentration of agarose (0.11–0.15%, w/v) and its biocompatibility (water content $> 99.8\%$) [64]. The method also permits purification of particles, such as viruses and ribosomes.

Araki's method for the fractionation of agar was intended only for the elucidation of its composition. Therefore I developed some simpler and gentler techniques for larger-scale preparation of agarose [65,66].

In a review on the structure of agar [67] published 16 years after Araki's observation that agar consists of agarose and agaropectin, one might have expected that reports on structural determinations of agar should include separation of its two major components, which should have facilitated the interpretation of experimental data, but no such approach is mentioned. The term "agarose" does not appear in the subject index of this journal (*Advantages in Carbohydrate Chemistry*) until 1969. I have many times asked myself the question: Why did not structural researchers take advantage of the known fact that agar is composed of two polysaccharides? The reason cannot be ascribed to a language barrier, since several Japanese chemists have had their interest focused for many years on the structure of agar. Perhaps the hypothesis about the influence of different types of conservative thinking put forward in Secs. III.A.5 and III.A.7.d can also be applied to this issue, since the original conception of the structure of agar seems to be that it is a polysaccharide, each polymer chain of which contains sulfate (and carboxylic?) groups. In 1956 Araki showed that agarose is built up of alternate residues of 3,6-anhydro-L-galactose and D-galactose [68].

5. Derivatized Agarose Gels for Molecular-Sieve Electrophoresis

Take advantage of a disadvantage! Or, as Einstein formulated it: In every difficulty there is an opportunity. This optimistic attitude is often very fruitful and is certainly practiced unconsciously by many researchers. The introduction of derivatized agarose is an illustrative example. Derivatization of agarose above the gel point, i.e., of solubilized agarose, gives upon lowering the temperature a gel with lower gel strength, which is a drawback from a *chromatographic* point of view. The cause is that the groups attached to the agarose chains prevent the chains from coming close to each other and forming hydrogen bonds. In other words, the macroporous structure characteristic of agarose gels is lost when the chains adopt a random coil conformation, i.e., the structure becomes similar to that of cross-linked polyacrylamide or dextran. When we tested the derivatized (methoxylated) agarose gels for use in molecular-sieve electrophoresis, it appeared that they are superior to polyacrylamide gels: the resolution is higher, the UV transmission is higher, and they are replaceable, which is a prerequisite for automated analyses by capillary electrophoresis [69]. In addition, they are not toxic. In other words, what was a disappointment in chromatography was turned to an advantage in electrophoresis. Despite their excellent electrophoretic properties, these agarose gels are seldom used for analysis of proteins. Cross-linked polyacrylamide is after 40 years still the dominant medium. The explanation may be conservatism in the form of disinclination to abandon a well-known method or/and to ascribe a modified product properties other than it originally had. Both types of conservatism are fanned by the impact of words on our thinking.

The reason I mention these examples of "the law of inertia" is that they illustrate the difficulties nowadays of getting new methods to be widely accepted. Another example with more far-reaching consequences is the negative attitude a few reviewers (and editors) of some journals, including methodological ones like the *Journal of Chromatography*, have toward novel techniques (I will not discuss the reasons). For instance, my experience is that a descriptive paper is sometimes treated more positively than an original, odd article, particularly when disposed in a way that does not follow the formal rules proposed by the journal (although new techniques often require another outline). This influence of formalists on research seems to be ever-increasing. For instance, it has become more and more common to use a set of standard parameters to judge the competence of a researcher—the number of articles published, the journal used for publication, etc. To me an examiner is not competent if he or she employs these bureaucratic criteria.

It is difficult to know whether politicians, bureaucrats, or researchers are responsible for the trend to back a particular project more than an able scientist—a philosophy also cherished by many money-granting authorities. I believe this policy may have fatal future consequences. For instance, many researchers are not interested in general (standard) projects formulated by others, since they have their own ideas about which research fields are important and, by virtue of their skill or for other reasons, already have the economic support they need, whereas researchers who lack money have a chance here and turn to an area where they have limited experience and in which they perhaps have little interest. The achievements may, therefore, be meager. A colleague told me that this happened in the United States when research on cancer was a preferred area.

6. Gel Filtration (Synonyms: Molecular-Sieve Chromatography, Size-Exclusion Chromatography, and Others)

a. Discovery of the Molecular-Sieving Properties of Dextran Gels (Sephadex™)   During the 1950s a great part of the research activities at Tiselius's Institute of Biochemistry was focused on the development of preparative electrophoresis with glass beads and later cellulose as stabilizing media. Per Flodin and Jerker Porath devoted much effort to minimizing the adsorption of solutes onto cellulose by esterification of its carboxylic groups. Although they succeeded very well, they searched for a still better medium. Flodin was interested in testing dextran gels synthesized by cross-linking with epichlorohydrin by a method developed earlier by Björn Ingelman. These gels were distributed to different researchers at the Institute for *electrophoretic* test experiments. I asked for a gel suspension as *transparent* as possible because I was designing an equipment for UV monitoring of the effluent from hydroxyapatite columns using mercury and zinc lamps as light sources and photo tubes as detectors [70] and wanted to employ the same apparatus for *on-tube* detection of the zones obtained upon zone electrophoresis in a column filled with gel grains of dextran (now we know that very transparent dextran gels have large pores and are, therefore, appropriate for the separation of large molecules, such as proteins). Upon chromatographic elution of the colored sample components (following the electrophoretic separation), I observed visually that there was a separation according to the size of the analytes. I showed my experiment to Jerker Porath, who was so interested that he visited my laboratory several times that day and also showed me that Lathe and Ruthven [71] had made similar observations on columns packed with gel particles of starch. A few days later Flodin visited the Institute, and when I asked him whether Porath had told him about my experiments his answer was "no." From this time onward dextran gels were investigated solely by Flodin and Porath, and only their *chromatographic* properties. The distribution among other researchers for electrophoretic studies ceased. Flodin and Porath asked me later whether I was of the opinion that I should be a co-inventor of a patent on the dextran gels. As a young student I had no idea about the importance of patents and answered "no." I now know that the one who makes a discovery that leads to a patent should own the patent. On the same occasion they admitted that they had not done experiments before mine to show that dextran gels separate substances according to size. Their explanation was that it was unnecessary to do experiments since it was obvious that the gels had such properties. If so, Flodin and Porath were not aware of the great potential of the gels for chromatography before I showed them. In connection with the celebration in 1977 of the 500th anniversary of Uppsala University, Porath and I (and other researchers at the Institute) described previous and present research activities at our Institute. I wrote about my discovery of the molecular-sieving properties of dextran gels as outlined above, and sitting together with Porath to prepare the final manuscript he raised no objections, but in the printed version important aspects had been omitted [25]. His explanation was that not he, but his secretary, along with a Ph.D. student, had removed parts of my description.

I asked Tiselius for a discussion with him, Flodin, and Porath about my contribution to the development of the dextran gels (trade name Sephadex™), but he refused—also about a year later—with the motivation that he thought that I had forgotten the whole issue. An explanation of Tiselius's disinclination to shed light on the Sephadex story could be that he either did not understand in the beginning the importance of the gels for molecular-sieving (maybe because the expertiments described in Ref. 19 are less impressive than those in Ref. 71 and, therefore, paid less attention to them) or was very eager to keep a calm and tranquil atmosphere at the Institute because he often pointed out that a relaxed surrounding is a prerequisite for a researcher to utilize his or her optimum capacity. The title of an interview with him in 1968 in a local newspaper when he retired is significant: "Interplay Research-Industry is Necessary: An Atmosphere of Anti-Stress Important in Science."

This title also reveals that Tiselius was very anxious to keep good contacts with industry (and Pharmacia, the manufacturer of Sephadex™, particularly). An expression of this may be the following incident. The day(s) before Flodin was to defend his thesis [72], Helge Laurell, at that time working on his doctoral thesis, pointed out to the opponent, Kai Pedersen (Tiselius's friend, as previously mentioned famous for his studies of sedimentation of proteins), that Deuel and Neukom in 1954 had cross-linked locust bean gum with epichlorohydrin, granulated the gel, dried the gel particles, rehydrated them, and packed them in a column and separated small and large molecules by chromatography [73]. Their method for the preparation of the gels is, thus, strikingly similar to that of Sephadex. Their hypothesis about the separation mechanism, based on the assumption that the smaller the molecules, the easier they can penetrate the beads, is still valid. Interestingly, the opponent did not mention Deuel and Neukom's paper. One can only speculate about the reason for this, but what is nearest to hand is that a reference to their article had probably prevented Pharmacia from getting a patent. In the introduction of one of my papers on molecular-sieve chromatography, I referred to Deuel and Neukom's contributions to the development of this method, but Tiselius deleted this part of the introduction (he examined all papers from the Institute before they were submitted for publication).

The above description of some episodes at the Institute of Biochemistry should not be interpreted to mean that I had bad relations with Tiselius (I was probably his spoiled child, and as his personal research assistant we published several papers together) and Porath (we have always been good friends and colleagues). Tiselius insisted on being a co-author with Frederick Sanger (who was awarded two Nobel Prizes) of an article in *Nature*, which shocked Sanger, but they did not bear any grudge [74]. Personally, I cannot understand Sanger's criticism, since his insulin studies were based on a frontal analysis method developed by Tiselius. Some conflicts are natural when two individuals meet, particularly among scientists with their strong personal engagement in their research. However, characteristic of successful persons (like Tiselius, Porath, and, hopefully, myself), and one reason why they are successful, is that they pay more attention to positive properties and incidents than negative ones and, therefore, seldom retain grudges against other people.

b. Soft Gels of Cross-Linked Polyacrylamide (Trade Name Bio-Gel P)   Referring to the discussion in the above section, it was natural for me to turn to studies of gels other than those of dextran. Having been aware of the analogy between electrophoresis and chromatography and having used polyacrylamide gels for electrophoretic molecular-sieving (Sec. III.A.3), it was also natural that I chose these gels for chromatographic molecular-sieving. Polyacrylamide gels (and agarose gels) might never have been employed for molecular sieving (which certainly had hampered the development of life sciences) if I also had been engaged in the development of the dextran gels (Sephadex™). This is only one example of a "rule" which I have empirically found to be valid in research, as well as in life in general: An experiment or an incident which *in nascendi* seems to be negative may later, in the long run—when you view it in the proper perspective—in combination with other experiments and incidents turn out to give positive consequences. Rolf Mosbach and I did experiments on chromatographic molecular-sieving on cross-linked polyacrylamide in the early 1960s. Tiselius referred to them in 1961 [75], and in

1962 we published all the details [52]. These experiments are historical in the sense that they showed that very large proteins could be separated by gel filtration and before Sephadex™ was introduced for the same purpose, as pointed out by Tiselius [75] and at a time when many researchers, including Tiselius, had doubts about whether proteins could be separated by molecular-sieving due to their high molecular weights and attendant slow diffusion into and out of a gel particle, which could lead to serious tailing. Jerker Porath was also impressed by our data and suggested to me to contact Pharmacia with the motivation that they might otherwise ''put a stick in the wheel.''

Later I described a suspension-polymerization procedure for the preparation of gel particles of polyacrylamide in beaded form [76]. About 500 (!) experiments were required to obtain beads of uniform size. This technique formed the basis for the commercial production of the beads by Bio-Rad. My colleague, Rolf Mosbach, took a patent on a similar procedure which Tiselius disliked, maybe in his eagerness to protect the Pharmacia product Sephadex™. However, he was very receptive to the suggestion from David Schwartz, President of Bio-Rad Laboratories, established a few years earlier, that I assist Bio-Rad in the preparation of beads on a process scale. From then on I have been fascinated by the rapid, efficient American way to commercialize a laboratory product—in sharp contrast to the European policy (nowadays the difference is negligible). It may be of some interest to know that a representative of Pharmacia Fine Chemicals upon an animated discussion suddenly handed me his telephone and asked me to phone David Schwartz and suggest that the cooperation with Bio-Rad should be interrupted.

The competition between Pharmacia Fine Chemicals (manufacturing the dextran gels) and Bio-Rad (manufacturing the polyacrylamide gels, Bio-Gel P) certainly contributed to a continuous improvement of these media.

Several equations have been derived for the relationship between the molecular weight of a protein and its retention volume, assuming that a gel bead has channels in the form of cones or cylinders. To avoid the use of similar relatively naive physical models, the correctness of which may be questioned, I made a thorough thermodynamic study and arrived at equations that could be experimentally verified [77]. The advantage of thermodynamic treatments is that they are not based on any physical model. Only the initial and final states are of interest.

c. Soft Agarose Gels (Trade Names Sepharose, Bio-Gel A, etc.)   Compared to gels of dextran and polyacrylamide, those of agarose have three distinct advantages: (a) they are mechanically harder and thus afford higher flow rates; (b) they have larger pores, i.e., they can be used for the separation not only of proteins, but also of particles, such as viruses and ribosomes; (c) agarose-based ion exchangers do not swell or shrink upon a decrease or increase of the ionic strength of the buffer.

The commercial agarose gels are prepared in beaded form by a method similar in principle to that I described in 1964 [78]. Shortly after their commercial introduction in 1967, the agarose gels became the most profitable chromatographic medium of Pharmacia—and still are. Their popularity on the laboratory, as well as process scale, is due to the difficulties in finding a substitute with the same or better chromatographic properties, such as biocompatibility, large pore size, ease of derivatization, and pH stability (following cross-linking). Later, Pharmacia Fine Chemicals introduced agarose columns for ion-exchange chromatography, hydrophobic-interaction chromatography (in cooperation with my group), and affinity chromatography (in cooperation with Jerker Porath's group), particularly because the chromatographic properties of agarose gels are superior to those of dextran gels (and polyacrylamide gels) for these chromatographic modes (see above). I introduced in 1973 the term hydrophobic-interaction chromatography, which now is generally accepted [79]. The original idea for this method came from our studies of membrane proteins. The solubilization of these proteins with surfactants takes place via interactions between the hydrophobic patches on the protein and the hydrophobic part of the surfactant molecule. Knowing this it was quite natural for us to try to develop a separation method based on this type of interaction. First we adsorbed sodium dodecylsulfate (SDS) electrostatically to DEAE-Sephadex, and then we coupled alkyl amines covalently to Sepharose by the cyanogen bromide technique, developed in Porath's laboratories. However, ligands that are both charged and nonpolar give mixed-mode chromatography, which has the disadvantage that some proteins cannot be desorbed because a decrease in the ionic strength of the eluting buffer decreases the hydrophobic, but increases the electrostatic interaction. Therefore, we turned to synthesize agarose beads with *noncharged*, nonpolar ligands; for a review, see Ref. 80.

Gels of agar [81] and cross-linked polyvinylethyl carbitol and polyvinyl pyrrolidone [82] have also been suggested for molecular-sieve chromatography, but for different reasons they have not been widely used. The preparation of gels of polyacrylamide is also described in Ref. 82.

d. Rigid Agarose Gels for HPLC (Trade Names of Products Similar to Those We Prepared: Superose, Superdex) In 1981 we made the agarose beads small (2–10 μm) and very rigid, following thorough cross-linking studies, to be used for HPLC in different modes [83–94]. These extremely hard gels are sometimes incorrectly classified as soft gels, like the original, truly soft, agarose gels—perhaps as a consequence of the tendency to stick to an old term for a product, even if its properties have been improved. Proteins almost always require gradient elution for high resolution. Refs. 91 and 92 are also of interest from the viewpoint that they show both theoretically and experimentally that proteins can be eluted isocratically provided that the ligand density is low.

Surprisingly, Pharmacia Biotech was not interested in cooperation on our HPLC project but wanted rather to develop the columns themselves (for molecular-sieving later called Superose™) with the official motivation that the project leader was not interested in cooperation. The top leaders had no (or did not want to have) influence on the development of this product. The Swedish model of democracy could celebrate another victory: much, including the quality of a product, can be sacrificed for leveling the right of determination (this model also prevailed at Swedish universities during the 1970s and 1980s and is still fostered, although to a lesser extent). Also many of my colleagues have—even officially—regretted the half-hearted interest Pharmacia, including Biotech, has shown in cooperation with Uppsala University.

Composite agarose-dextran gels have a matrix of agarose with the pores filled with dextran covalently linked to the matrix. This approach permits the design of gels of tailored pore size. They have the high rigidity of cross-linked agarose [94]. Superdex is the trade name of a similar product, manufactured by Pharmacia Biotech.

7. Unique HPLC Media

a. Compressed Rigid Beds of Nonporous Agarose and Coated Silica Beads for Different Modes of HPLC of Proteins None of these beds have been commercialized, which is surprising, since they have many good chromatographic properties, e.g., a resolution that is constant or increases upon an increase in flow rate. These columns are forerunners of those discussed in Sec. 7.b.

During the last two decades, one has seen an ever-increasing demand for more rapid runs and, at the same time, higher resolution. According to classical chromatographic theory, such experiments cannot be designed. This "dogma" in chromatography may have delayed its progress. The theory is based on the assumption that the flow is laminar and that the beads are nondeformable. The situation becomes quite different, however, if we abandon these two assumptions and introduce a third one, namely, that the beads are covalently linked to each other and, for narrow-bore tubes, also to the wall (the third assumption will be treated below in Secs. b and c). Compression of a bed means deformation of the beads, resulting in shorter distances between them and thus shorter times for a solute to move from one bead to another, thereby affording higher resolution. It is, of course, not difficult to compress "soft" gels (e.g., gels of dextran and polyacrylamide). Such experiments leading to an increase in resolution have been described in the literature [95], but to gain in resolution the flow rate had to be sacrificed. In practice, to retain a high flow rate, a novel method for the preparation (including cross-linking) of beads had to be designed such that they were neither excessively hard nor excessively soft. We succeeded in developing such a method. The beads were made nonporous to eliminate the slow, resolution-suppressing diffusion of solutes into and out of a pore, and their surface was rendered rough to increase the protein capacity and, hopefully, to produce nonlaminar flow. It was a long process to understand theoretically which parameters were of particular importance and a still longer one to transfer the theoretical requirements to a practical synthesis method. The first positive experiments were presented in 1987 at a symposium in Oberammergau, Germany. There are two events in connection with the development of these gels that I remember with pleasure. A Chinese Ph.D. student, Yao Kunquan, showed me late one evening an experiment on an agarose-based ion exchanger that made me very happy: upon increasing the flow rate, the column became shorter and at the same time the resolution *increased* considerably (Fig. 2 in Ref. 96). We thus had beads that could be compressed and yet allowed a high flow rate. Later I asked another Chinese Ph.D. student, Jia-Li Liao, to do some experiments the night before I planned to go to the above symposium in Oberammergau. One day earlier he had done hydrophobic-interaction chromatography experiments on agarose beads, which indicated—but not conclusively—that the resolution increased with an increase in flow rate, even when the bed height was the same at all flow rates, which was not the case in the above ion-exchange experiments. At six o'clock in the morning, he phoned to tell me that his experiments were very successful and I could bring another chromatogram to the symposium (Fig. 3 in Ref. 96). We presented our results at several meetings, but some distinguished researchers (analytical chemists) were very skeptical about our results, both regarding the experimental fact that the resolution increased with the flow rate and our explanation that the rough surface of the beads probably generated a nonlaminar flow—deliberately a vague term. Could this frigid reception originate from a feeling that a cat (a biochemist) sneaked in among the pigeons (the established analytical chemists), especially as the holy cow (the silica column) might find a competitor in a cat's offspring (the new agarose-based HPLC column)? (Maybe those who do not believe in this interpretation mean—and I have no strong objections—that it rather reflects my interest in psychological explanations.) This joke refers to *a few* analytical chemists. The great majority of them are probably more positive toward new techniques than are other life scientists. An example is their great interest in HPCE (see Sec. III.A.2).

No doubt, it was encouraging to me to learn about the observations Professor Hermann Bauer, at the University of Tübingen, presented in a poster at the 7th International Symposium on Chromatography in Vienna in 1988. In open tubes (without ligands), the plate height became lower when the inner surface was etched, i.e., rough, compared to a nonetched, smooth surface. When I referred to Professor Bauer's experiments as support for our hypothesis that the increased resolution on the compressed agarose beds upon an increase in flow rate *may* be caused by nonlaminar flow (observe, not necessarily turbulent flow), I got the answer that we cannot trust his experiments. We experienced the same skepticism from a few colleagues when we introduced rigid agarose gels as HPLC media for chromatographic molecular-sieving (see Sec. III.A.6.d) and the continuous beds (see Sec. III.A.7.b).

The main reason why I mention these episodes is that they represent a negative attitude toward novel ideas, which is not common, but yet too common to be neglected (see also Sec. III.A.5). We should welcome new ideas with open arms, even if they later turn out to be wrong (which was not the case with the above columns), since people who often are skeptical about nonconventional approaches can do a great deal of damage in all creative activities as well as the "yes-but" people, as pointed out by one of the former presidents of IBM, who used this term to characterize coworkers who first politely say "yes," but then employ all their energy to find reasons to kill any new idea. See also Section I as to F. Crick's defense of researchers who put forward bold hypotheses.

   b.   *Compressed Continuous Polymer Beds (Monoliths) (Bio-Rad's Trade Name UNO)*   In the late 1980s I decided to start a project with the objective to synthesize a bed in situ in the chromatographic tube, thus eliminating the cumbersome and expensive preparation of beads and subsequent packing. I had some experience in preparing beds in situ, since I synthesized such beds for molecular-sieve chromatography in the 1960s. They were prepared from acrylamide and bis-acrylamide. The latter monomer (cross-linker) was used at high concentrations, since from our studies of electrophoresis in polyacrylamide gels we knew that under such conditions the pore size becomes large [57]. I discontinued these studies because the flow rate was low. Owing to important contributions from Jia-Li Liao, a Ph.D. student, the synthesis of these polyacrylamide gels could be modified successfully (the concentration of the cross-linker is still high). The gel bed can be compressed to afford a high resolution while retaining a low backpressure, and the resolution increases upon an increase in flow rate, i.e., the bed could be designed to have the same attractive, unique property as had the above compressible agarose bed. A very important part of the synthesis protocol is the inclusion in the reaction mixture of ammonium sulfate or any other agent (such as a polymer) that causes aggregation (precipitation) of the polymer chains formed. These so-called continuous beds, which consist of 0.2–0.5 μm *nonporous* particles that are covalently linked to each other, are described in a review article [97]; see this article for references to most of our publications on continuous beds. They are used not only in standard columns but also in capillary columns with inside diameters down to 15 μm without loss in resolution compared to columns with larger bores [97,98] (see Sec. c below). The eddy diffusion in continuous beds is low due to the small dimensions of the particles. (In an attempt to reduce the eddy diffusion to zero, and thus obtain still higher resolution, we have recently introduced entirely homogeneous gel beds operating in the electrochromatographic mode [99] (see Sec. IV.A). Three years following our first paper on in situ preparation of continuous beds [100], Svec and coworkers published a method for the preparation of a continuous bed from less polar monomers using organic solvents and replacing ammonium sulfate with porogens, agents often employed to create pores in beads [101]. The beds were called continuous rods of porous polymer. Continuous beds have been given many other names: continuous column supports, continuous porous silica rods, monoliths, continuous polymer rods [102–105], etc. The nomenclature is, therefore, very confusing. The term monolith is relatively often used, although it is logical to reserve it for the homogeneous gels discussed in Sec. IV.A, since—according to several encyclopedias—a monolith consists of a single block that is massive and impermeable (all other beds consist of particles and thus do not consist of a single piece); for discussion of nomenclature, see Ref. 105a. My personal opinion is that one should not introduce new notations for a method but try to retain the original one (provided that there are no particular reasons to change it). Otherwise, there is a risk that the uninitiated will erroneously believe that the methods are basically different. One should also honor the pioneers by retaining their terminology (e.g., J. A. P. Martin, awarded the Nobel Prize in 1952 for his chromatographic achievements, told me that he was sorry to see the term "displacement electrophoresis," introduced by him, being replaced by "isotachophoresis").

Certainly, many chromatographers have reflected on the possibility to synthesize a bed directly in the chromatographic tube. Some of them probably also tried but failed and did not report the negative results. This is probably the reason why only a few in situ preparation methods have been published and none have become widely used. To understand the problem encountered in an in situ synthesis, I will refer to something I recently wrote [97]: The difficulties are "illustrated very well by the following quotation from a letter I received shortly after our first paper on the continuous beds was published: 'These papers are of great interest to me: For years I have been considering "continuous" beds, but Prof. XX and Dr. YY warned me that the high resistance and slow flow-rate would be difficult to overcome.' My answer, based on the experiments described below, is certainly also illuminating. 'As your colleagues have pointed out, it is not easy to synthesize continuous beds with low flow resistance. I believe that many researchers have tried to prepare such beds, but the difficulties to achieve both high flow rates *and* high resolution are not so easy to overcome.' " Bearing these difficulties in mind, I am happy that our in situ synthesized columns are easy and rapid to prepare and compete favorably as regards both low flow resistance and high resolution with the best commercial columns packed with presynthesized beads, properties that have generated a great interest in in situ syntheses of many different types of continuous beds. The main reason we succeeded was certainly that we combined experiments with theoretical considerations. The same approach has been used by many researchers in the Uppsala School, for instance, by Harry Rilbe and Arne Tiselius, when they were developing isoelectric focusing and moving-boundary electrophoresis, respectively.

A colleague of mine brought up some time ago the question of how to define the term "pioneer" in separation science. Should Tiselius, Rilbe, and myself be considered pioneers in the fields of moving boundary electrophoresis, isoelectric focusing, and in situ preparation of continuous beds, respectively? However, none of us was the first to study these methods, but the first to make such important contributions that the methods became widely used. The definition of the word "pioneer" in the books I have consulted seems to be so vague that it does not distinguish between these two categories of inventors. Perhaps the reader could suggest two appropriate terms.

The compressed beds of the cross-linked, *nonporous* beads are useful for anion-exchange, cation-exchange, reversed-phase, hydrophobic-interaction, boronate, immuno-affinity and dye-ligand affinity and chiral-recognition chromatography and chromatofocusing [97]. Naked silica columns are known to give a strong nonspecific interaction and to be degraded at pH above 8. For decades, much effort has been laid down to eliminate these disadvantages. The first *simple* and efficient method is probably that developed in my group. It is based on an easy coating with glycidol or acrylamide derivatives [106]. Probably some companies use this method or a similar one, since several pH-stable silica columns are now commercially available (the chemical composition of the coating is, however, seldom revealed). It should be emphasized that the coating procedure made the 30–45 μm silica beads nonporous and *compressible*, which resulted in a resolution comparable to that obtained on rigid 1.5 μm beads. These silica columns share with the compressed continuous beds and the compressed agarose columns the attractive relation between resolution and flow rate (see above).

c. Noncompressed Continuous Polymer Beds for Capillary Chromatography and Capillary Electrochromatography It is difficult to compress a continuous bed uniformly when the diameter is 0.5 mm or less. Upon increasing the monomer concentration above a critical value, the bed becomes very rigid, i.e., noncompressible, and more uniform. We have used this technique for capillary chromatography and electrochromatography (CEC) in tubes with inside diameters in the range 0.015–0.4 mm.

Upon electroendosmosis in *straight* capillaries, all (neutral) molecules move at the same speed (except in a thin layer at the capillary wall), whereas in a hydrodynamic flow the velocity distribution is parabolic. Although very often stated, this does not necessarily mean that electrochromatography gives much narrower zones than does chromatography (we assume that the two experiments are performed on the same column using the same sample and mobile phase). The reason is that the flow channels in a bed are *not straight*, but are curved and interconnected, which gives rise to variations in field strength in a channel and, thereby, different electroendosmotic velocities with attendant deformation of the zones. This can be verified experimentally by visual inspection of the migration of colored, neutral substances in nonstraight coupled tubes of different diameters filled with a charged transparent homogeneous gel. Besides these variations in field strength, there are other factors—all discussed in Ref. 107—which make the electroendosmotic flow nonuniform. Therefore, it is not surprising that the plate heights for pressure- and electroendosmosis-driven chromatography on continuous beds differ only slightly; the fact that these beds are composed of nonporous extremely small beads (0.2–0.5 μm) contributes to the small difference. Theoretically, one can expect the plate heights in electrochromatography, compared to those in chromatography, to be smaller the larger the pores in a bead (up to a certain pore size), because the combined electroendosmotic and diffusional transport of a solute in a pore is faster than the diffusional transport alone [97]. This has been verified experimentally by Stol et al. [108], who also showed that nonporous beds (like continuous beds) give the lowest plate height in capillary electrochromatography. Despite the above facts, most researchers still state that electroendosmosis-driven chromatography always gives much lower plate heights than does pressure-driven chromatography. How to explain this false message? The only answer I have is that if something wrong is repeated often enough it will sooner or later be accepted as true. Evidently, we researchers, who by our profession should be expected to be more objective than people in general, can be manipulated in this way, i.e., by the same approach as that used in commercial announcements (advertisements).

References are given in Ref. 97 to methods for the preparation of the continuous beds for pressure- and electroendosmosis-driven chromatography along with applications. Gradient elution of proteins in the CEC mode requires special precautions [109]. For the use of continuous beds in chips, see Sec. IV.A.

d. Purification of Hydrophobic Membrane Proteins and Solubilization of Membrane Proteins   During the development of hydrophobic-interaction chromatography on agarose gels using *neutral*, nonpolar ligands, I found that serum proteins could be desorbed by ethylene glycol in the presence of high salt concentrations, which was surprising, because hydrophobic interaction increases with an increase in ionic strength [79] (see Sec. III.A.6.c, which explains why we did not use charged ligands as many of my colleagues did). If one describes this observation differently, saying that the well-known property of ethylene glycol to break hydrophobic interactions is also retained at high salt concentrations, one immediately draws the important conclusion that we now have at our disposal an aqueous solution with the remarkable property to suppress at the *same* time both hydrophobic (by virtue of ethylene glycol) and electrostatic interactions (by virtue of high salt concentration). For instance, since the bonds between antigens and antibodies and between hydrophobic membrane proteins often are predominantly of electrostatic and hydrophobic nature, it is obvious that these bonds can be weakened or disrupted in an aqueous buffer solution containing ethylene glycol and sodium chloride at appropriate concentrations. This approach was employed to purify the human growth hormone (HGH) by coupling its antibodies to an agarose column for immuno-affinity chromatography [110] and to solubilize hydrophobic membrane proteins [111]. To facilitate crystallization of (membrane) proteins, we have developed a capillary electrophoretic method for rapid analysis of the protein solution used for crystallization, the surrounding mother liquor and the protein crystals (following their solubilization) prior to and after exposure to the x-ray beam [112]. The method gives information about whether low resolution in the diffraction pattern is caused by denaturation of the proteins during crystallization. I wish to recall that hydrophobic membrane proteins in the double layer are surrounded by lipids that have a nonpolar "tail" and a hydrophilic "head." In other words, the natural milieu of hydrophobic proteins is an aqueous solution containing a surfactant. One might therefore expect at least some of the most hydrophobic proteins to be biologically active (e.g., not denatured) in the presence of SDS (sodium dodecyl sulfate), a surfactant feared by protein chemists for its strongly denaturing properties. To test this hypothesis I went to our library and found that the hypothesis was correct. Encouraged by this finding we tested several membrane enzymes for activity in the presence of SDS and found that SDS even increased the activity of a neuraminidase from one strain of influenza virus [113]. Other enzymes lost some or all activity in a buffer containing SDS, but could be reactivated by addition of a mild, *neutral* surfactant at concentrations considerably higher than that of SDS. The mixed micelles formed thus contained low concentrations of SDS, which decreases its denaturing properties. Some membrane proteins can be solubilized by no other surfactant than SDS. If so, they are extremely hydrophobic and are, therefore, likely to be active in SDS or can be renatured as outlined. Such proteins can, with advantage, be purified in the presence of SDS, and I believe that one could crystallize at least part of them by slow removal of SDS. Very probably a protein that regains its biological activity when SDS is removed (e.g., an enzyme) also regains its structure. The risk that the structure following renaturation is not the native one is perhaps not greater than the risk that the structure of a water-soluble protein in a crystal differs from that in free solution.

Many membrane proteins are soluble in a buffer supplemented with ethylene glycol and salt (see above), i.e., in the absence of surfactants (in fact, ethylene glycol has a stabilizing effect on several enzymes). It is obvious that this buffer has the potential to permit a gentle crystallization of some membrane proteins. I would with pleasure give more information to those who are interested in trying the two crystallization approaches suggested.

Part of my research on membrane proteins is described in Ref. 25. See also Ref. T5 in Sec. VII.

8. *Thermal Micropump*

When we introduced the continuous beds, the difficulties of filling a chromatographic tube of any diameter, as well as chips, uniformly with a stationary phase disappeared. There was no longer any hindrance to rapid progress in the field of microchromatography [97,114], provided that a pump could be designed to deliver a *pulse-free flow* in the nanoliter to microliter range, without splitting. Our approach was to construct such a pump based on the expansion of a liquid upon heating. A great advantage is that this design does not involve any moving parts, which guarantees a pulse-less flow. Columns for hydrophobic-interaction chromatography of proteins gave the same resolution for 15 and 320 μm wide columns, using this pump at flow rates in the (nl–μl)/min range [115]. It can also be used as a liquid delivery system in chips (Sec. IV.A).

IV. PRESENT RESEARCH ACTIVITIES

Our spirit lies in daring. Throughout the centuries there were men who took first steps down new roads armed with nothing but their vision.

Ayn Rand

What kind of man could live where there is no daring? I don't believe in taking foolish chances, but nothing can be accomplished without taking any chance at all.

Charles Augustus Lindbergh

A. Electrophoresis, Chromatography, and Electrochromatography in Capillaries, Chips, and Thin-Layer Rectangular Chambers Filled with Noncompressed Continuous Beds or Homogeneous Gels—and Any Combination of These Methods for Flexible 2-D Experiments

1. We are developing a new method to coat fused silica capillaries. Based on a novel technique, the coating permits repeated washings at extreme pH values.

2. We were the first to separate proteins in chips by chromatography and electrochromatography employing gradient elution and UV detection [114]. These techniques are being improved. The chips are made of quartz or diamond (diamond has even higher UV transmission than has quartz and considerably higher heat conductivity, i.e., higher field strengths can be used without risk of thermal zone broadening). The stationary phase is a continuous bed, which is easy to prepare in situ (the difficulties to pack the channels in chips uniformly with beads are obvious). As expected, chip and capillary chromatography on continuous beds give the same resolution [114]. We are planning a series of experiments in chips.

3. Although continuous beds have a very low eddy diffusion (the first term in the van Deemter equation), it is still lower (zero) in the homogeneous gels we are investigating. The longitudinal diffusion (the second term) is smaller than diffusion in free solution, and the resistance to mass transfer (the third term) is lower than that in a packed and continuous bed [99]. These gels, accordingly, approach the ideal electrochromatographic medium. We are studying them not only for analysis in capillaries and chips but also in thin gel slabs for a project on multidimensional separations using a new simple approach to combine all the micromethods we have developed and are developing.

4. We are studying a novel technique to determine the diffusion coefficient and the charge of a solute (and the electronic charge) by HPCE using frontal analysis.

5. A homogeneous polyacrylamide gel crosslinked with allyl-β-cyclodextrin is being studied for capillary electrophoresis of DNA for two important reasons: (a) it gives extremely high plate numbers ($450,000-1,600,000$ m^{-1}); (b) the capillary can be used repeatedly without destruction of the gel in sharp contrast to conventional polyacrylamide gels crosslinked with N, N'-methylene-bisacrylamide (due to their instability they are usually replaced by polymer solutions, which, unfortunately, give a lower resolution).

B. Gels Mimicking Antibodies (Imprinting?)

Most presentations at symposia on molecular imprinting deal with low molecular weight compounds. The most likely reason that imprinting of proteins is seldom treated is that those who have tried (and they are probably many) have not succeeded very well. Therefore, some years ago I took it as a challenge to develop a method for selective recognition of proteins.

One can assume that many researchers have tried to use the conventional preparation method for imprinting of small molecule also for proteins. The obvious question was, therefore: Why is this method not applicable to proteins? The answer was equally obvious: most functional monomers used are charged, which causes the beds obtained by polymerization to behave as ion exchangers and the desired selectivity for macromolecules is thus lost. The strong electrostatic interactions must, accordingly, be exchanged for weak interactions, such as hydrogen bonds. However, the number of weak bonds must be large in order to create a strong overall binding and a high degree of specificity. Our approach is to synthesize a gel of acrylamide and bisacrylamide in the presence of the protein to be adsorbed selectively (these monomers form, for instance, hydrogen bonds with proteins or dipole-dipole interactions). Following granulation, packing the gel grains in a glass tube and washing away the protein, the gel column should have the desired selectivity. The method for the preparation is, accordingly, similar to (a) that used for immobilization of proteins by entrapment and (b) that used for imprinting of small molecules, with the important difference that the functional monomers are neutral. By this technique we have adsorbed selectively eight proteins (among them myoglobin from horse, the specificity being so high that myoglobin from whale was not adsorbed) [116]. These gels thus behave like "artificial antibodies." Their advantages compared to native antibodies are, for instance, that no experimental animals are required and that they are very stable (a dehydrated gel regains its selectivity upon rehydration). We are investigating these gels further, for instance, to (a) decide rapidly what monomers are potential candidates for the synthesis of a selective gel; (b) establish the nature and the strength of the bonds between the gel and the proteins; (c) develop an analogous electrophoretic and electrochromatographic technique; and (d) to extend the application range to peptides, viruses, and cells (bacteria).

V. APPLIED RESEARCH OUTSIDE THE FIELD OF SEPARATION SCIENCE

The knowledge and the experiences we gained in methodological electrophoretic and chromatographic studies have been employed in many different areas, for instance, for the development of (a) wound dressings, (b) agents against diarrhea, (c) catheters to suppress bacterial adhesion, (d) toothpastes and mouthwash to prevent bacteria from adsorbing to the teeth, and (e) a simple method for the determination of bacteria in urine.

VI. TEACHING YOUNG STUDENTS

To be a winner . . . all you need to give is all you have.

If it is to be it is up to me.

A. Course in Modern Separation Methods

As a university professor, one has the duty to teach students. Professor Paul Roos, my very best friend at the Institute of Biochemistry, and I fulfilled this obligation by giving for about 25 years an annual, 10-week course on modern biochemical separation methods for six foreign and six Swedish participants (see Fig. 1). [Some of my lectures are found in S. Hjertén. Analytical electrophoresis. In: G Milazzo, (ed). Topics in Bioelectrochemistry and Bioenergetics, Vol. 2. New York: John Wiley and Sons, 1978, pp. 89–128.] Roos is famous for the development of an effective, gentle method for the isolation of human growth hormone (HGH) from pituitaries. His preparation was employed for the treatment of pituitary dwarfism and formed the basis for KABI's purification of HGH on a process scale. HGH was for many years the most profitable among all the products of this Swedish pharmaceutical company. Roos also prepared human pituitary gonadotropins, which were used by Professor Carl Gemzell, Uppsala University Hospital, to treat female sterility.

During the 1970s and 1980s we maintained high requirements in the written examination

UPPSALA SEPARATION SCHOOL

January 22 – March 23, 1979

Institute of Biochemistry
University of Uppsala
Sweden

ORGANIZERS:
Professors Stellan Hjertén
and Paul Roos

BIOCHEMICAL SEPARATION METHODS

The course is centered around modern analytical and preparative methods for the separation of cells, virus, proteins and nucleic acids, and their characterization. It consists of lectures and laboratory work dealing with the following methods: moving boundary electrophoresis, free zone electrophoresis, zone electrophoresis in both sieving and non-sieving anticonvection media, isoelectric focusing, displacement electrophoresis (isotachophoresis), molecular sieve chromatography, hydroxyapatite chromatography, hydrophobic interaction chromatography, covalent chromatography, bioaffinity chromatography, counter-current distribution (liquid phase partition), analytical and preparative centrifugation methods (centrifugation in different kinds of density gradients, determination of sedimentation coefficients and of molecular weights), immunodiffusion, immunoelectrophoresis, determination of diffusion coefficients, light scattering and bioluminescence.

Knowledge of biochemistry and mathematics, corresponding to a basic university degree, is required. Good knowledge of English is necessary. The number of participants is limited to 12, 6 from Sweden and 6 from abroad.

Course fee: US $300.

Living expenses to cover food and accommodation in student rooms will be a minimum of US $1,000. No fellowships are available through the organizers.

The closing date for applications is November 15, 1978

**Application forms can be obtained from
Secretary
Institute of Biochemistry
University of Uppsala
Box 576
S-751 23 UPPSALA
Sweden**

Figure 1

in this international course, but later we were gradually forced to lower them considerably to allow most of the students to pass the examination. During a few years preceding my retirement in 1995, the contents of my lectures were reduced by about 60%. I was obliged to do so because the students' basic knowledge in physics, physical chemistry, and mathematics was far from satisfactory (observe that I refer only to *basic* knowledge). This finding, and the fact that many students in molecular biology have expressed a desire not to learn any chemical formulas in our courses in biochemistry, may have the same explanation: courses and projects that require some logical considerations are less popular among many life science students than those of a more descriptive nature. If this statement is correct, it has a negative impact not only on disciplines like separation science, which require the above basic knowledge, but indirectly also on many other disciplines since training in logical thinking has a positive influence in all realms of activities, including research, not the least in the writing of scientific papers. Students (researchers) with this background knowledge have the capacity to achieve a creative interplay between theory (if necessary a mathematical treatment) and practical experiments—often a very fruitful approach.

B. Special Lectures: Personal Views on Issues Crucial for Success in Research

It is certainly important for all of us, both as scientists and human beings, to raise questions like this: What would happen if you weren't doing what you now are doing? My answer has surprisingly often been: ''Nothing!'' With a good conscience I have then done something else. Another question of importance: What successful things have you done today (this month, this year), and what mistakes have you made? These and many other questions, including my personal viewpoints on what research is all about, served as starting points for lectures for students about issues of importance in research (admittedly, a topic where opinions are divergent). From these lectures, delivered both in Sweden and abroad, I have borrowed the quotations found in the beginning of each main section. Those lacking the name of the author are taken from magazine announcements and advertisements in areas outside the realm of science.

VII. SOME ADDITIONAL RESEARCH ACTIVITIES, PRESENTED IN THE FORM OF TITLES OF THE CORRESPONDING PAPERS (AND SOME SHORT COMMENTS WHEN APPROPRIATE)

Basketball: You'll always miss 100% of the shots you don't take.

Do not follow where the path may lead . . . go instead where there is no path and leave a track.

T1 S Hjertén, S Jerstedt, A Tiselius (1965) Electrophoretic ''Particle Sieving'' in Polyacrylamide Gels as Applied to Ribosomes, Anal Biochem 11, 211–218. Interestingly, large particles, such as virus, can also be size-sieved in polyacrylamide gel electrophoresis, provided that the gel concentration is very low.

T2 S Hjertén, S Jerstedt, A Tiselius (1965) Some Aspects of the Use of ''Continuous'' and ''Discontinuous'' Buffer Systems in Polyacrylamide Gel Electrophoresis, Anal Biochem 11, 219–223. In this paper we show both theoretically and experimentally that conventional buffers can give the same high resolution in polyacrylamide gel electrophoresis of proteins as do the discontinuous buffer systems introduced by Ornstein.

T3 S Hjertén (1971) A New Method for the Concentration of High- and Low-Molecular Weight Substances and for Their Recovery Following Gel Electrophoresis, Partition and Precipitation Experiments, Biochim Biophys Acta 237, 395–403.

T4 BG Johansson, S Hjertén (1974) Electrophoresis, Crossed Immunoelectrophoresis and Isoelectric Focusing in Agarose Gels with Reduced Electroendosmotic Flow, Anal. Biochem. 59, 200–213. The dynamic coating, i.e., the addition of a neutral polymer to the buffer to suppress electroendosmosis (and adsorption) that I introduced in 1958 (Arkiv för Kemi 13 (1958), 151–152) for free zone electrophoresis in narrow-bore tubes and described more extensively in my thesis (J. Chromatogr. Rev. 9 (1967), 122–219) is useful also in gel electrophoresis. Dynamic coating is now widely used in HPCE.

T5 The following papers describe part of our studies on membrane proteins. K.-E. Johansson, S. Hjertén (1974) Localization of the Tween 20-Soluble Membrane Proteins of *Acholeplasma laidlawii* by Crossed Immunoelecophoresis, J. Mol. Biol. 86, 341–348; O. J. Bjerrum, P. Lundahl, C.-H. Brogren, S. Hjertén (1975) Immunoabsorption of Membrane-Specific Antibodies for Determination of Exposed and Hidden Proteins in Human Erythrocyte Membranes, Biochim. Biophys. Acta 394, 173–181; K.-E. Johansson, I. Blomqvist, S. Hjertén (1975) Purification of Membrane Proteins from *Acholeplasma laidlawii* by Agarose Suspension Electrophoresis in Tween 20 and Polyacrylamide and Dextran Gel Electrophoresis in Detergent-Free Media, J. Biol. Chem. 250, 2463–2469; L. Liljas, P. Lundahl, S. Hjertén (1976) The Major Sialoglycoprotein of the Human Erythrocyte Membrane. Release with a Non-Ionic Detergent and Purification, Biochim. Biophys. Acta 426, 526–534; K.-E. Johansson, H. Pertoft, S. Hjertén (1979) Characterization of the Tween 20-Soluble Membrane Proteins of *Acholeplasma laidlawii*, Int. J. Biol. Macromol. 1, 111–118; G. Fröman, F. Acevedo, S. Hjertén (1980) A Molecular Sieving Method for Preparing Erythrocyte Membranes, Prep. Biochem. 10, 59–67; G. Fröman, F. Acevedo, P. Lundahl, S. Hjertén (1980) The Glucose Transport Activity of Human Erythrocyte Membranes. Reconstruction in Phospholipid Liposomes and Fractionation by Molecular Sieve and Ion Exchange Chromatography, Biochim. Biophys. Acta 600, 489–501; S. Hjertén, H. Pan, K. Yao (1982) Some General Aspects of the Difficulties to Purify Membrane Proteins, in H. Peeters (Editor), Protides of the Biological Fluids, Proceedings of the 29th Colloquium, Brussels 1981, Pergamon Press, New York, pp. 15–27.

T6 S. Hjertén (1977) Fractionation of Proteins by Hydrophobic Interaction Chromatography, with Reference to Serum Proteins, Proceedings of the International Workshop on Technology for Protein Separation and Improvement of Blood Plasma Fractionation, Reston, Virginia, pp. 410–421. The experiments described in this paper and those in S. Hjertén, U. Hellman (1980) Chromatographic Desalting, Deproteinization and Concentration of Nucleic Acids on Columns of Polytetrafluoroethylene, J. Chromatogr. 202, 391–395, are probably the first where polytetrafluoroethylene (PTFE) was employed as stationary phase. A similar PTFE column has been commercialized by Du Pont under the trade name NENSORB 20. Their POLY F™ is used for reversed-phase chromatography.

T7 C. J. Smyth, P. Jonsson, E. Olsson, O. Söderlind, J. Rosengren, S. Hjertén, T. Wadström (1978) Differences in Hydrophobic Surface Characteristics of Porcine Enteropathogenic *Escherichia coli* with or without K88 Antigen as Revealed by Hydrophobic Interaction Chromatography, Infect. Immun. 22, 462–472. Using agarose columns with the *neutral*, hydrophobic ligands we had earlier introduced, we made the very important observation that only pathogenic bacteria were adsorbed. This finding prompted Torkel Wadström, Sten Björnberg, and me to develop a wound dressing consisting of a hydrophilic layer, which sucks liquid through a hydrophobic layer where the pathogenic bacteria become adsorbed (but not the hydrophilic substances needed for the healing process). This wound dressing is used worldwide.

T8 S. Hjertén, L.-G. Öfverstedt, G. Johansson (1980) Free Displacement Electrophoresis (Isotachophoresis): An Absolute Determination of the Kohlrausch Functions and Their Use in Interaction Studies, J. Chromatogr. 194, 1–10. We determined experimentally the value of the Kohlrausch functions and verified that they had the same value for *moving* boundaries. For these experiments, the first ever done, we employed the capillary electrophoresis apparatus described in my thesis (an instrument with the high accuracy required for mobility determinations had not been designed earlier). The importance of this verification is better understood if we recall that many equations describing electrophoretic phenomena are based on the Kohlrausch function.

T9   S.K. Tylewska, T. Wadström, S. Hjertén (1980) The Effect of Subinhibitory Concentrations of Penicillin and Rifampicin on Bacterial Cell Surface Hydrophobicity and on Binding to Pharyngeal Epithelial Cells, J. Antimicrob. Chemother. 6, 292–294. Using hydrophobic-interaction chromatography and free zone electrophoresis, we found that treatment of bacteria with antibiotics diminished the surface hydrophobicity of the bacteria and made their surface charge more negative. Accordingly, besides the well-known effect of antibiotics to disturb the protein synthesis of bacteria, they also change their surface properties so that the adhesion to mucosa, for instance, decreases—a new important finding (the electrostatic repulsion increases since probably all biological surfaces are negatively charged and the hydrophobic interaction decreases).

T10   S. Hjertén, T. Wadström (1990) What Types of Bonds Are Responsible for the Adhesion of Bacteria and Viruses to Native and Artificial Surfaces? In: Pathogenesis of Wound and Biomaterial-Associated Infections (T. Wadström, I. Eliasson, I. Holder, Å Ljungh, eds.), Springer-Verlag, London, pp. 245–253. The article gives background information of importance for the preceding paper.

T11   T. Wadström, A. Faris, M. Lindahl, K. Lövgren, B. Ågerup, S. Hjertén (1980) Prevention of Enterotoxigenic E. coli (ETEC) Diarrhea by Hydrophobic Gels: A Preliminary Study. In: Proceedings of the 3rd International Symposium on Neonatal Diarrhea, VIDO, October 6–8, 1980, Saskatoon (S.D. Acres, A. J. Forman, H. Fast, eds.), pp. 237–249. We know from the experiments discussed above (see Ref. T7) that pathogenic bacteria often have a hydrophobic surface structure. This finding prompted us to test whether agarose beads to which hydrophobic groups were attached could suppress or prevent diarrhea caused by enterotoxigenic E. coli. Experiments performed in rabbits clearly indicated that the hypothesis was correct.

T12   L.-G. Öfverstedt, G. Johansson, G. Fröman, S. Hjertén (1981) Protein Concentration and Recovery from Gel Slabs by Displacement Electrophoresis (Isotachophoresis) and the Effects of Electroosmosis and Counter Flow, Electrophoresis 2, 168–173. This concentration technique has later been used widely in HPCE in both the off-and on-tube version. We also showed that different electroendosmotic flows in a tube causes rotation of the buffer. A similar rotation must occur in electrochromatography upon gradient elution since the gradient creates different electroendosmotic flow rates in different sections of the capillary. Will the rotation give rise to an increase or a decrease in resolution? The former alternative cannot be excluded if the transport of an analyte from one adsorption site to another takes place by a fast electroendosmotic flow rather than by slow diffusion.

T13   M. Lindahl, A. Faris, T. Wadström, S. Hjertén (1981) A New Test Based on ''Salting Out'' to Measure Relative Surface Hydrophobicity of Bacterial Cells, Biochim. Biophys. Acta 677, 471–476. We used hydrophobic-interaction chromatography in our first approach to determine the surface hydrophobicity of bacteria (see Ref. T7). The disadvantage of this technique is that bacteria can become mechanically entrapped in the column bed. I suggested, therefore, another method based on the plausible assumption that the more hydrophobic the bacteria, the lower the concentration of ammonium sulfate required to precipitate them. The method worked well, although some improvements were later introduced.

T14   S. Hjertén, K. Yao, M. Ogunlesi (1981) Immobilization of Enzymes on Columns of Brushite, J. Chromatogr. 215, 25–30. The advantages of this affinity chromatography technique are: (a) the columns are easy and inexpensive to prepare; (b) the immobilization takes place by adsorption of the proteins, not by covalent binding; (c) if required for the enzyme activity, high concentrations of any buffer (except phosphate buffer) can be used without desorption of the protein, which is in sharp contrast to immobilization on ion-exchangers; (d) the enzyme can be desorbed for renaturation and adsorbed again. The lifetime of the column can, if the renaturation is successful (or a fresh enzyme sample is immobilized), be longer than that of columns with covalently attached proteins.

T15   S. Hjertén, Z.-Q. Liu, S. L. Zhao (1983) Polyacrylamide Gel Electrophoresis: Recovery of Non-Stained and Stained Proteins from Gel Slices, J. Biochem. Biophys. Methods 7, 101–113. Compared to other recovery techniques this one has the advantages that the protein upon its migration out of the gel (a) will not become diluted, but rather concentrated; and (b) is subjected to an additional purification step.

T16   S. Hjertén, H. Pan (1983) Purification and Characterization of Two Forms of a Low-Affinity Ca^{2+}-ATPase from Erythrocyte Membranes, Biochim. Biophys. Acta 728, 281–288. The purpose of this investigation was two-fold: to purify this protein and to show that the stationary phase in electrophoresis and chromatography may give rise to *artificial* protein complexes. Our hypothesis was that such complexes may form at the surface of gel beads because the polymer chains of the gel beads are more mobile at the surface than inside the bead, and the proteins may therefore aggregate there (it is well known that proteins aggregate in the presence of free polymers or even precipitate when the polymer concentration is high enough). Using such experimental conditions that this problem did not arise, we could show that the enzyme was not part of a huge complex, consisting of several different proteins, as was stated in the literature. It is likely that some other reported protein complexes also are artifacts.

T17   L.-G. Öfverstedt, K. Hammarström, N. Balgobin, S. Hjertén, U. Pettersson, J. Chattopadhyaya (1984) Rapid and Quantitative Recovery of DNA Fragments from Gels by Displacement Electrophoresis (Isotachophoresis), Biochim. Biophys. Acta 782, 120–126. This technique has been widely used, as pointed out by C. Hammann and M. Tabler (1999) Bio Techniques, 26, 422–424.

T18   F. Kilár, S. Hjertén (1989) Fast and High Resolution Analysis of Human Serum Transferrin by High Performance Isoelectric Focusing in Capillaries, Electrophoresis 10, 23–29. This technique can be employed to diagnose abuse of alcohol.

T19   S. Hjertén, L. Valtcheva, K. Elenbring, D. Eaker (1989) High-Performance Electrophoresis of Acidic and Basic Low-Molecular-Weight Compounds and of Proteins in the Presence of Polymers and Neutral Surfactants, J. Liq. Chromatogr. 12(13), 2471–2499. The paper shows that aromatic compounds interact with polymers (no ligands are required). We were the first to use consciously this "aromatic adsorption" for HPCE analyses.

T20   S. Hjertén, M. Kiessling-Johansson (1991) High-Performance Displacement Electrophoresis in 0.025- to 0.050-mm Capillaries Coated with a Polymer to Suppress Adsorption and Electroendosmosis, J. Chromatogr. 550, 811–822. It has been pointed out that the resolution would be higher if very narrow capillaries were used, provided that the electroendosmosis could be eliminated (F. M. Everaerts, J. L. Beckers, T. P. E. M. Verheggen. In: Journal of Chromatography Library, Vol. 6. Isotachophoresis. Theory, Instrumentation and Applications. Amsterdam: Elsevier, 1976, p 396). Our paper was the answer.

T21   K.-O. Eriksson, A. Palm, S. Hjertén (1992) Preparative Capillary Electrophoresis Based on Adsorption of the Solutes (Proteins) onto a Moving Blotting Membrane as They Migrate Out of the Capillary, Anal Biochem. 201, 211–215. The method is very appropriate for mass spectrometry of the adsorbed fractions, since salts can easily be washed away. A similar apparatus is now commercially available from LC Packings, The Netherlands.

T22   J.-L. Liao, C.-M. Zeng, A. Palm, S. Hjertén (1996) Continuous Beds for Microchromatography: Detection of Proteins by a Blotting Membrane Technique. Anal. Biochem. 241, 195–198. See comments on the above paper.

T23   J. Mohammad, Y.-M. Li, M. El-Ahmad, K. Nakazato, G. Pettersson, S. Hjertén (1993) Chiral-Recognition Chromatography of β-Blockers on Continuous Polymer Beds with Immobilized Cellulase as Enantioselective Protein, Chirality 5, 464–470. Very rapid separations can be obtained (within 30–60 s). The plate height of the fast enantiomer *decreased* when the flow rate was increased, as was the case in the electrochromatography experiments in homogeneous gels described in Á. Végvári, A. Földesi, C. Hetényi, M. G. Schmid, V. Kudirkaite, A. Maruška, S. Hjertén (2000) A New Easy-to-Prepare Homogeneous Continuous Electrochromatographic Bed for Chiral Recognition, Electrophoresis, 21, 3116–3125.

T24   S. Hjertén, D. Eaker, K. Elenbring, C. Ericson, K. Kubo, J.-L. Liao, P.-A. Lidström, C. Lindh, A. Palm. T. Srichaiyo, L. Valtcheva, R. Zhang (1995) New Approaches in the Design of Capillary Electrophoresis Experiments. Jpn. J. Electrophoresis 39, 105–118. Several novel approaches in microseparation techniques are presented, for instance, electrochromatography in continuous beds, and scanning of agarose gels confined in capillaries.

T25   Y. Zhang, C.-M. Zeng, Y.-M. Li, S. Hjertén, P. Lundahl (1996) Immobilized Liposome Chromatography of Drugs on Capillary Continuous Beds for Model Analysis of Drug-Membrane Interactions. J. Chromatogr. A, 749, 13–18.

REFERENCES

You must either find a way or make one.

Hannibal

Do not fear the winds of adversity. Remember: a kite rises against the wind rather than with it.

1. S Hjertén. In: CW Gehrke, RL Wixom, E Bayer eds. Chromatography—A Century of Discovery. The Bridge to the Sciences and Technology. New York: Elsevier, 2001.
2. T Svedberg, H Rinde. The ultracentrifuge, a new instrument, the determination of size and distribution of size of particles in amicroscopic colloids. J Am Chem Soc 46:2677–2693, 1924.
3. B Elzen. The failure of a successful artifact. The Svedberg ultracentrifuge. In: S Lindqvist, ed. Center on the Periphery, Historical Aspects of 20th-Century Swedish Physics. Science History Publications/USA, 1993, pp. 347–377.
4. KO Pedersen. The Svedberg och ultracentrifugen. Arne Tiselius och proteinforskningen. Ronden 4:33–52, 1975.
5. A Lundgren. The ideological use of instrumentation. The Svedberg, atoms and the structure of matter. In: S Lindqvist, ed. Center on the Periphery, Historical Aspects of the 20th-Century Swedish Physics. Science History Publications/USA, 1993, pp. 327–346.
6. KO Pedersen. The Svedberg and Arne Tiselius, the early development of modern protein chemistry at Uppsala. In: G Semenza, ed. Selected Topics in the History of Biochemistry: Personal Recollections. Elsevier, 1983, pp. 233–281.
7. A Lundgren. Naturvetenskaplig institutionalisering: The Svedberg, Arne Tiselius och biokemin. In: S. Widmalm, ed. Vetenskapsbärarna. Naturvetenskapen i det svenska samhället, 1880–1950. Gidlunds förlag, 1999, pp. 117–143.
8. T Svedberg, KO Pedersen. The Ultracentrifuge (UC). Oxford: Clarendon, 1940. Reprinted by the Johnson Reprint Corporation, New York, 1959.
9. A Tiselius. Reflections from both sides of the counter. Ann Rev Biochem 37:1–24, 1968.
10. S. Hjertén. A Tiselius (1902–1971). J Chromatogr 65:345–348, 1972.
11. S Hjertén. Dedication to Professor Arne Tiselius. Ann NY Acad Sci 209:5–7, 1973.
12. S Hjertén, A Tiselius—pionjär i biokemisk metodologi. Kemisk Tidskrift (1):36–37, 1997.
13. S Hjertén. Arne Tiselius' Impact on (Separation) Science, The Physico-Chemical Biology, Centennial Memorial Issue of the Birth of the Late Dr. Keizo Kodama. 35(4):239–241, 1991.
14. N Catsimpoolas, S Hjertén, A Kolin, J Porath. The Tiselius unit for electrophoretic mobility. Nature 259:264, 1976.
15. H Rilbe. A scientific life with chemistry, optics and mathematics. Electrophoresis 5:1–17, 1984.
16. S Hjertén, J-L Liao, K Yao. Theoretical and experimental study of high-performance electrophoretic mobilization of isoelectrically focused protein zones. J Chromatogr 378:127–138, 1987.
17. O Vesterberg. Synthesis and isoelectric fractionation of carrier ampholytes. Acta Chem Scand 16: 456–466, 1962.
18. HI Gedin, J Porath. Studies of zone electrophoresis in vertical columns. II Zone electrophoresis of serum proteins. Biochim Biophys Acta 26:159–169, 1957.
19. J Porath, P Flodin. Gel filtration: a method for desalting and group separation. Nature 183:1657–1659, 1959.
20. R Axén, J Porath, S Ernback. Chemical coupling of peptides and proteins to polysaccharides by means of cyanogen halides. Nature (London) 214:1302–1304, 1967.
21. J Porath. Immobilized metal ion affinity chromatography. Review. Protein Expression Purif 3:263–281, 1992.
22. P-Å Albertsson. Partition of Cell Particles and Macromolecules, 2nd ed. Uppsala:Almquist & Wiksell, 1971.
23. Evaluation of the Faculty of Science and Technology at Uppsala University, BOT, November 1995, pp. 44 and 63.

24. S Hjertén. The history of the development of electrophoresis in Uppsala. Electrophoresis 9:3–15, 1988.

25. S Hjertén, J Porath. Chemistry. In: L-O Sundelöf, ed. Acta Universitatis Upsaliensis, Uppsala University 500 Years, 9, Faculty of Science at Uppsala University. Uppsala: Almqvist and Wiksell, 1976, pp. 27–66.

26. A Tiselius, S Hjertén, Ö Levin. Protein chromatography on calcium phosphate columns. Arch Biochem Biophys 65:132–155, 1965.

27. A Tiselius, S Hjertén, Ö Levin. Protein Chromatography on Calcium Phosphate Columns. Current Contents, Citation Classics, 28(46), 1985, p. 21.

28. G Bernardi. Chromatography of nucleic acids on hydroxyapatite. Nature 206:779–783, 1965.

29. HA Sober, EA Peterson. Chromatography of proteins on cellulose ion-exchangers. J Am Chem Soc 76:1711–1712, 1954.

30. S Hjertén. Free zone electrophoresis. Chromatogr Rev 9:122–219, 1967.

31. CB Brinton, MA Lauffer. In: M Bier, ed. Electrophoresis. Theory, Methods, and Applications. New York: Academic Press, 1959, p. 411.

32. S Hjertén. Free zone electrophoresis. Preliminary note. Arkiv Kemi 13:151–152, 1958.

33. R Virtanen. Acta Polytech Scand Chem 123:1–67, 1974.

34. FEP Mikkers, FM Everaerts, TPEM Verheggen. High-performance zone electrophoresis. J Chromatogr 169:11–20, 1979.

35. PF Bente, EH Zerenner, RD Dandenau, U.S. patent 21, 293, 415 (Oct. 6, 1981).

36. JW Jorgenson, KD Lukacs. High-performance separations based on electrophoresis and electroosmosis. J Chromatogr 218:209–216, 1981.

37. S Hjertén. High-performance electrophoresis, elimination of electroendosmosis and solute adsorption. J Chromatogr 347:191–198, 1985.

38. S Hjertén. High-performance electrophoresis: the electrophoretic counterpart of high-performance liquid chromatography. J Chromatogr 270:1–6, 1983.

39. S Hjertén. High performance electrophoresis (HPE). In: H Hirai, ed. Electrophoresis' 83. Berlin: Walter Gruyter & Co., 1983, pp. 71–79.

40. M-D Zhu, S Hjertén. High-performance electrophoresis. An alternative detection technique. In: V Neuhoff, ed. Electrophoresis '84. Weinheim: Verlag Chemie, 1984, pp. 110–113.

41. S Hjertén, M-D Zhu. High-performance electrophoresis—an alternative and complement to high-performance liquid chromatography. In: B Rånby, ed. Pure and Applied Chemistry, Physical Chemistry of Colloids and Macromolecules. Oxford: Blackwell Scientific Publications, 1987, pp. 133–136.

42. S Hjertén, J-L Liao, K Yao. Theoretical and experimental study of high-performance electrophoretic mobilization of isoelectrically focused protein zones. J Chromatogr 387:127–138, 1987.

43. S Hjertén, M-D Zhu. Micropreparative version of high-performance electrophoresis—the electrophoretic counterpart of narrow-bore HPLC. J Chromatogr 327:157–164, 1985.

44. T Srichaiyo, S Hjertén. Simple multi-point detection method for high-performance capillary electrophoresis. J Chromatogr 604:85–89, 1992.

45. S Hjertén, L Valtcheva, K Elenbring, J-L Liao. Fast, high-resolution (capillary) electrophoresis in buffers designed for high field strengths. Electrophoresis 16:584–594, 1995.

46. S Hjertén, J-L Liao, R Zhang. New approaches to concentration on a microliter scale of dilute samples, particularly biopolymers with special reference to analysis of peptides and proteins by capillary electrophoresis. I. Theory. J Chromatogr A 676:409–420, 1994.

47. J-L Liao, R Zhang, S Hjertén. New approaches to concentration on a microliter scale of dilute samples, particularly biopolymers with special reference to analysis of peptides and proteins by capillary electrophoresis. II. Applications. J Chromatogr A 676:421–430, 1994.

48. S Hjertén, L Valtcheva, Y-M Li. A simple method for desalting and concentration of micro-liter volumes of protein solutions with special reference to capillary electrophoresis. J Cap Electrophoresis 1:83–89, 1994.

49. R Zhang, S Hjertén. A simple micro method for concentration and desalting utilizing a hollow fiber, with special reference to capillary electrophoresis. Anal Chem 69:1585–1592, 1997.

50. S Hjertén. Zone sharpening in paper electrophoresis, a method allowing application of dilute protein solutions. Biochim Biophys Acta 32:531–534, 1959.

51. S Hjertén, K Elenbring, F Kilár, J-L Liao, A Chen, C Siebert, M-D Zhu. Carrier-free zone electrophoresis, displacement electrophoresis and isoelectric focusing in a high-performance electrophoresis apparatus. J Chromatogr 403:47–61, 1987.

52. S Hjertén, R Mosbach. "Molecular-sieve" chromatography of proteins on columns of cross-linked polyacrylamide. Anal Biochem 3:109–118, 1962.

53. S Hjertén. Molecular and particle sieving (chromatographic and electrophoretic) on agarose and polyacrylamide gels. In: H. Peeters, ed. Protides of the Biological Fluids, Proceedings of the 14th Colloquium, Bruges 1966. Amsterdam: Elsevier Publ. Co., 1967, pp. 553–561.

54. S Hjertén. Zone broadening in electrophoresis with special reference to high-performance electrophoresis in capillaries: an interplay between theory and practice. Electrophoresis 11:665–690, 1990.

55. S Hjertén, C Ericson, Y-M Li, R Zhang. Capillary electrophoresis and capillary chromatography. Theoretical analogies and practical differences. Biomed Chromatogr 12:1–2, 1998.

56. O Smithies. Molecular size and starch gel electrophoresis. Arch Biochem Biophys Suppl 1:125–131, 1962.

57. S Hjertén, S Jerstedt, A Tiselius. Apparatus for large scale preparative gel electrophoresis. Anal Biochem 27:108–129, 1969.

58. S Hjertén. Preparative gel electrophoresis: recovery of products (with a note on free zone electrophoresis). In: E Reid, ed. Methodological Developments in Biochemistry. Vol 2: Preparative Techniques. London: Longman, 1973, pp. 39–48.

59. RJ Wieme. Studies on Agar Gel Electrophoresis, Techniques—Applications. Brussels: Arscia Uitgaven N.V., 1959.

60. C Araki. J Chem Soc Jpn 58:1338, 1937.

61. S Hjertén. Agarose as an anticonvection agent in zone electrophoresis. Biochim Biophys Acta 53:514–517, 1961.

62. S Brishammar, S Hjertén, B von Hofsten. Immunological precipitates in agarose gels. Biochim Biophys Acta 53:518–521, 1961.

63. S Hjertén. Zone electrophoresis in columns of agarose suspensions. J Chromatogr 12:510–526, 1963.

64. J Porath, S Hjertén. Some recent developments in column electrophoresis in granular media. In: D Glick, ed. Methods of Biochemical Analysis, Vol. 9. New York: Interscience Publ. Inc., 1962, pp. 193–210.

65. S Hjertén. A new method for preparation of agarose for gel electrophoresis. Biochim Biophys Acta 62:445–449, 1962.

66. S Hjertén. Some new methods for the preparation of agarose. J Chromatogr 61:73–80, 1971.

67. T Mori. Seaweed polysaccharides. In: Advances in Carbohydrate Chemistry, Vol. 8. New York: Academic Press, 1953, pp. 315–350.

68. C Araki. Bull Chem Soc. Jpn 29:543, 1956.

69. S Hjertén, T Srichaiyo, A Palm. UV-transparent, replaceable agarose gels for molecular-sieve (capillary) electrophoresis of proteins and nucleic acids. Biomed Chromatogr 8:73–76, 1994.

70. S Hjertén. Calcium phosphate chromatography of normal human serum and of electrophoretically isolated serum proteins. Biochim Biophys Acta 31:216–235, 1959.

71. GH Lathe, CRJ Ruthven. The separation of substances and estimation of their relative molecular sizes by the use of columns of starch in water. Biochem J 62:665–674, 1956.

72. P Flodin. Dextran gels and their application in gel filtration. Thesis, University of Uppsala. Halmstad: Meijels Bokindustri, Sweden, 1962.

73. H Deuel, H Neukom. Some properties of locust bean gum. Paper presented at the 122nd meeting American Chemical Society, September 1952; published in Nat Plant Hydrocolloids 11:51–61, 1954.

74. F Sanger. Sequences, sequences and sequences. Ann Rev Biochem 57:1–28, 1988.

75. A Tiselius. Einige neue Trennungsmethoden und ihre Anwendung auf biochemische und organischchemische Probleme. Experientia 17:1–11, 1961.

76. S Hjertén. Chromatography on polyacrylamide spheres. In:MW Chase, CA Williams, eds. Methods in Immunology and Immunochemistry, Vol. II. New York: Academic Press, 1968, pp. 142–150.

77. S Hjertén. Thermodynamic treatment of partition experiments with special reference to molecular-sieve chromatography. J Chromatogr 50:189–208, 1970.

78. S Hjertén. The preparation of agarose spheres for chromatography of molecules and particles. Biochim Biophys Acta 79:393–398, 1964.

79. S Hjertén. Some general aspects of hydrophobic interaction chromatography. J Chromatogr 87: 325–331, 1973.

80. S Hjertén. Hydrophobic-interaction chromatography of proteins, nucleic acids, viruses, and cells on non-charged amphiphilic gels. In: D Glick, ed. Methods of Biochemical Analysis, Vol. 27. New York: John Wiley & Sons, Inc. 1981, pp. 89–108.

81. A Polson. Fractionation of protein mixtures on columns of granulated agar. Biochim Biophys Acta 50:565–567, 1961.

82. DJ Lea, AH Sehon. Preparation of synthetic gels for chromatography of macromolecules. Can J Chem 40:159–160, 1962.

83. S Hjertén, K Yao. High-performance liquid chromatography of macromolecules on agarose and its derivatives. J Chromatogr 215:317–322, 1981.

84. S Hjertén. HPLC of biopolymers on columns of agarose. Acta Chem Scand B 36:203–209, 1982.

85. S Hjertén. Agarose gels in HPLC separation of biopolymers. TrAC 3:87–90, 1984.

86. S Hjertén, K-O Eriksson. High-performance molecular sieve chromatography of proteins on agarose columns: the relation between concentration and porosity of the gel. Anal Biochem 137:313–317, 1984.

87. S Hjertén, Z-Q Liu, D Yang. Some Studies on the resolving power of agarose-based high-performance liquid chromatographic media for the separation of macromolecules. J Chromatogr 296: 115–120, 1984.

88. S Hjertén, D Yang. High-performance liquid chromatographic separations on dihydroxyboryl-agarose. J Chromatogr 316:301–309, 1984.

89. K-O Eriksson, S Hjertén. Estimation of peptide/protein molecular weights by high-performance molecular-sieve chromatography on agarose columns in 6 M guanidine hydrochloride. J Pharm Biomed Anal 4:63–68, 1986.

90. S Hjertén, K Yao, Z-Q Liu, D Yang, B-L Wu. Simple method to prepare noncharged, amphiphilic agarose derivatives, for instance for hydrophobic interaction chromatography. J Chromatogr 354: 203–210, 1986.

91. S Hjertén, K Yao, K-O Eriksson, B Johansson. Gradient and isocratic high-performance hydrophobic-interaction chromatography of proteins on agarose columns. J Chromatogr 359:99–109, 1986.

92. K Yao, S Hjertén. Gradient and isocratic high-performance liquid chromatography of proteins on a new agarose-based anion exchanger. J Chromatogr 385:87–98, 1987.

93. T Sindhuphak, I Svensson, U Hellman, V Patel, S Hjertén. Hydrophobic interaction chromatography of incompletely methylated transfer RNA from Escherichia coli on octyl-sepharose. J Chromatogr 368:113–124, 1986.

94. S Hjertén, B-L Wu, J-L Liao. An high-performance liquid chromatographic matrix based on agarose cross-linked with divinyl sulphone. J Chromatogr 396:101–113, 1987.

95. ML Fishman, RA Barford. Increased resolution of polymers through longitudinal compression of agarose gel columns. J Chromatogr 52:494, 1970.

96. S Hjertén, Y Kunquan, J-L Liao. The design of agarose beds for high-performance hydrophobic-interaction chromatography and ion-exchange chromatography which shows increasing resolution with increasing flow rate. Macromol Chem Macromol Symp 17:349–357, 1988.

97. S Hjertén. Standard and capillary chromatography, including electrochromatography, on continuous polymer beds (monoliths), based on water-soluble monomers. Ind Eng Chem Res 38:1205–1214, 1999.

98. C Ericson, S Hjertén. Pump based on thermal expansion of a liquid for delivery of a pulse-free

flow particularly for capillary chromatography and other microvolume applications. Anal Chem 70:366–372, 1998.

99. S Hjertén, Á Végvári, T Srichaiyo, H-X Zhang, C Ericson, D Eaker. An approach to ideal separation media for (electro) chromatography. J Cap Elec 5:13–26, 1998.

100. S Hjertén, J-L Liao, R Zhang. High-performance liquid chromatography on continuous polymer beds (note) J Chromatogr 473:273–275, 1989.

101. F Svec, JMJ Fréchet. Continuous rods of macroporous polymer as high-performance liquid-chromatography separation media. Anal Chem 64:820–822, 1992.

102. SM Fields. Silica xerogel as a continuous column support for high-performance liquid chromatography. Anal Chem 68:2709–2712, 1996.

103. H Minakuchi, K Nakanashi, N Soga, N Ishizuka, N Tanaka. Octadecylsilylated porous silica rods as separation media for reversed-phase liquid chromatography. Anal Chem 68:3498–3501, 1997.

104. I Gusev, X Huang, Cs Horváth. Capillary columns with in situ formed porous monolithic packing for micro high-performance liquid chromatography and capillary electrochromatography. J Chromatogr A 855:273–290, 1999.

105. J Matsui, T Kato, T Takeuchi, M Suzuki, K Yokoyama, E Tamiya, I Karube. Molecular recognition in continuous polymer rods prepared by a molecular imprinting technique. Anal Chem 65:2223–2224, 1993.

105a. Á. Végvári, A. Földesi, Cs. Hetényi, O. Kochegarova, M. Schmid, V. Kudirkaite, S. Hjertén. A new easy-to-prepare homogeneous continuous electrochromatographic bed for enantiomer recognition. Electrophoresis 21:3116–3125, 2000.

106. S Hjertén, J Mohammad, KO Eriksson, JL Liao. General methods to render macroporous stationary phases nonporous and deformable, exemplified with agarose and silica beads and their use in high-performance ion-exchange and hydrophobic-interaction chromatography of proteins. Chromatographia 31:85–94, 1991.

107. C Ericson, J-L Liao, K Nakazato, S Hjertén. Preparation of continuous beds for electrochromatography and reversed-phase liquid chromatography of low-molecular-mass compounds. J Chromatogr A 767:33–41, 1997.

108. R Stol, WT Kok, H Poppe. Capillary electrochromatography with macroporous particles. J Chromatogr A 855:45–54, 1999.

109. C Ericson, S Hjertén. Reversed-phase electrochromatography of proteins on modified continuous beds using normal-flow and counter-flow gradients. Theoretical and pratical considerations. Anal Chem 71:1621–1627, 1999.

110. J-P Li, S Hjertén. High-performance liquid chromatography of proteins on deformed nonporous agarose beads: immuno-affinity chromatography, exemplified with human growth hormone as ligand and a combination of ethylene glycol and salt for desorption of the antibodies. J Biochem Biophys Methods 22:311–320, 1991.

111. S Hjertén, H Pan, K Yao. Some general aspects of the difficulties to purify membrane proteins. In: H. Peeters, ed. Protides of the Biological Fluids, Proceedings of the 29th Colloquium, Brussels 1981. New York: Pergamon Press, 1982, pp. 15–27.

112. J Sedzik, R Zhang, S Hjertén. Ups and downs of protein crystallization:studies of protein crystals by high-performance capillary electrophoresis. Biochim Biophys Acta 1426:401–408, 1999.

113. S Hjertén, M Sparrman, J-L Liao. Purification of membrane proteins in SDS and subsequent renaturation. Biochim Biophys Acta 939:476–484, 1988.

114. C Ericson, J Holm, T Ericson, S Hjertén. Electroendosmosis- and pressure-driven chromatography in microchips, using continuous beds. Anal Chem 72:81–87, 2000.

115. C Ericson, S Hjertén. Pump based on thermal expansion of a liquid for delivery of a pulse-free flow, particularly for capillary chromatography and other microvolume applications. Anal Chem 70:366–372, 1998.

116. S Hjertén, J-L Liao, K Nakazato, Y Wang, G Zamaratskaia, H-X Zhang. Gels mimicking antibodies in their selective recognition of proteins. Chromatographia 44:227–234, 1997.

28

A Decade of Capillary Electrophoresis

Haleem J. Issaq

SAIC Frederick, NCI-Frederick Cancer Research and Development Center,
Frederick, Maryland

I. INTRODUCTION

Electrophoresis as an analytical technique was first introduced by Tiselius in 1930 [1]. In his thesis he described the separation of blood plasma proteins, namely albumin from α, β and γ-globulin, using electrophoresis. For this pioneering work, Tiselius was awarded the Nobel Prize in 1948. Since then, many advancements in electrophoresis, such as paper and gel electrophoresis, have been introduced. Filter paper has been widely used as a supporting medium since the late 1940s for the separation of ionized compounds such as amino acids, lipids, nucleotides, and charged sugars [2]. Two-dimensional paper electrophoresis was carried out to enhance the resolution by changing the buffer's pH, i.e., first-dimension electrophoresis at a particular pH and second-dimension at a different pH.

Gelatin and agar gels as supports in electrophoresis have been used for at least a hundred years [2]. A major advance in gel electrophoresis took place when it was realized that polyacrylamide (PAA) gels under denaturing conditions using urea or sodium dodecyl sulfate (SDS) can be employed for the resolution of proteins [3]. SDS-polyacrylamide gel was not only used to separate proteins but also to estimate their molecular weight. Two-dimensional gel electrophoresis is used to separate cellular proteins having relatively large differences in molecular weight and charge density in two sequential steps: first by their charge and then by their mass [4]. Other historical advances of gel electrophoresis are discussed in [2].

In 1967, Hjerten showed that it was possible to carry out electrophoretic separations in a 300 μm glass tube and to detect the separated compounds by ultraviolet absorption (UV), and called it free zone electrophoresis [5]. Although other researchers used electrophoresis in tubes, glass, and Teflon [6–8], electrophoresis in a tube did not become popular until 1981, when Jorgenson and Lukacs published their work in which they demonstrated the high resolving power

Reprinted with permission from Electrophoresis, 21:1921–1939, 2000.

of capillary zone electrophoresis (CZE) [9]. The simple and efficient instrumental setup, which is the basis of most commercial instruments on the market today, used a narrow internal diameter fused silica capillary, less than 100 μm, high voltage, 30 kV, and on-column UV detection for the separation of ionic species. In 1984, Terabe introduced micellar electrokinetic chromatography (MECC) for the separation of neutral compounds by adding a micelle, SDS, to the buffer solution [10]. In 1988, the first commercial instrument was marketed. Since then, many advances and applications have taken place [11–119]. A review of the application of capillary electrophoresis (CE) in biological sciences was published by Karger et al. in 1989 [11]. For a more recent historical review of advancement in CE instrumentation, the reader is referred to the excellent write-up by Camilleri [2].

This chapter will present examples of the application of CE in different fields. During the last 10 years our laboratory has been involved in the development and application of CE for the separation of small as well as large biomolecules; therefore, where applicable examples, of our work will be presented. Otherwise, examples from the literature will be used to illustrate the utility and application of CE. This chapter is not meant to be a comprehensive one; rather it is presented here to show the speed of advancement and applicability of this microanalytical technique in many different chemical, biological, and analytical laboratories. (For recent reviews see Refs. 12 and 13.) The chapter is divided into the following sections: Why CE?; inorganic ions and organic acids; amino acids, peptides, and proteins; natural products; nucleic acids; clinical applications; illicit drugs; chiral separations; miscellaneous separations; physicochemical studies; capillary electrochromatography; and future directions.

II. WHY CAPILLARY ELECTROPHORESIS?

The first question we have been asked on numerous occasions is: Why use capillary electrophoresis when other separation techniques are available? What is so special about CE, and what are the advantages of using it considering that it is only a microanalytical technique?

The advantage of CE over gas chromatography (GC), high-performance liquid chromatography (HPLC), thin layer chromatography (TLC), and slab gel electrophoresis (SGE) is its applicability for the separation of widely different compounds, inorganic ions, organic molecules, and large biomolecules, using the same instrument and in most cases the same column while changing only the composition of the running buffer. This cannot be said about any of the other separation techniques. In addition, CE possesses the highest resolving power of any liquid separation technique, due to its plug flow and minimal diffusion. The CE resolving power is illustrated in Fig. 1, which compares the separation of a needle extract of the Western yew, *Taxus brevifolia* (taxacaea), by HPLC and CE, and Fig. 2, which compares the separation of a 66-residue peptide by both CE and HPLC. Compared to SGE, CE is faster, easier, simpler, can be automated, and gives quantitative data. Unlike SGE and TLC, where two-dimensional separations are achieved easily, in CE, like HPLC and GC, such separations cannot be performed. However, CE can be used on-line with other instrumental techniques such as HPLC, mass spectrometry (MS), and nuclear magnetic resonance (NMR). The amount of material needed for a CE experiment is very small—nanoliters of sample and microliters of buffer—compared to the other liquid separation techniques, which require microliters of sample and milliliters of solvent. The resulting CE waste, mostly an aqueous buffer, but sometimes with a small percentage of an organic modifier, is environmentally safe and can be discarded without any danger to the environment. This is not true when TLC or HPLC is used, where large amounts of organic solvent waste are produced, or with GC, where volatile compounds, which are sometimes toxic, escape into the environment.

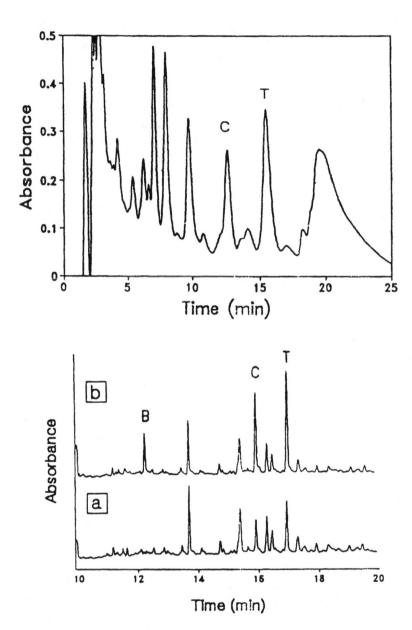

Figure 1 Comparative separation of a Western yew needle extract by HPLC (upper) and CE (lower). (a) Unspiked and (b) spiked needle extract. T = Taxol; C = cephalomannine; B = baccatin III. (From Ref. 63.)

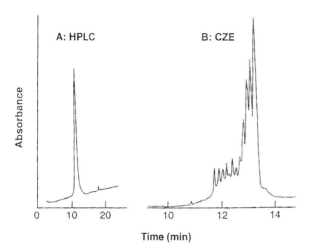

Figure 2   Comparative separation of a 66-residue peptide by (A) HPLC and (B) CEZ. (From Ref. 63, with permission.)

III. INORGANIC IONS AND ORGANIC ACIDS

Capillary electrophoresis is well suited for the separation of inorganic ions using indirect UV detection. Hjerten [5] in his initial publication presented data showing the separation of Bi from Cu ions. Since then many studies have been published dealing with the separation of inorganic ions, acids, and bases. A notable one is the separation of 36 anions in 3 minutes [14]. In another study Jones et al. [15] resolved 37 anions in 12 minutes (Fig. 3). The separation of metal ions was first reported by Hjerten [2]. Weston et al. [16] resolved 11 metal ions from each other in under 8 minutes. In our laboratory we developed a method for the separation of nitrate from nitrite at pH 3 where they are widely separated and detected at 214 nm where they absorb strongly, while most other negative ions have negligible absorbance. The method was applied for the analysis of nitrate and nitrite in urine and water (Fig. 4) [17,18].

IV. AMINO ACIDS, PEPTIDES, AND PROTEINS

The separation of amino acids, peptides, and proteins by CE has been a popular one. Many studies have been published in the last decade dealing with this subject.

A. Amino Acids

The 20 amino acids possess different chemical properties—basic, acidic, polar, hydrophilic, and hydrophobic—which make their separation by CZE according to only their charge density using a simple buffer difficult. Therefore, micellar electrokinetic chromatography is the method of choice for the separation of a mixture of amino acids where SDS is added to the buffer to effect the separation by introducing partitioning as an extra mechanism of separation in addition to charge density.

The sensitive detection of amino acids is achieved by laser-induced fluorescence (LIF). In our laboratory, amino acids were resolved by MECC and detected with pulsed UV laser–induced fluorescence after derivatization with 9-fluorenylmethyl chloroformate [19]. A recent review by Issaq and Chan discussed the separation and detection of amino acids and their enantiomers by

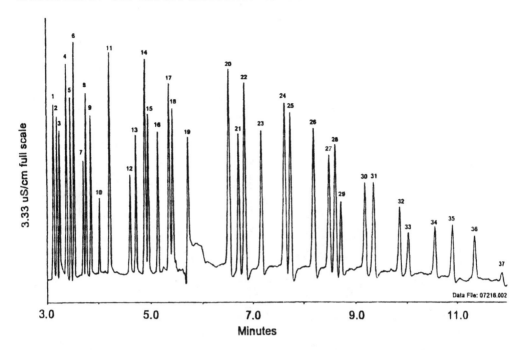

Figure 3 An electropherogram of the CZE separation of 37 anions in a 60 cm × 50 μm using a 50 mM CHES, 20 mM lithium hydroxide monohydrate, 0.03% Triton X-100, pH 9.2 at an applied potential of 25 kV (negative polarity). Pressure injection at 25 mbar for 12 seconds. Analysis was carried out at room temperature. For peak identification and concentration, see Ref. 15. (From Ref. 15.)

CE [20]. The separation of the D- and L-amino acid enantiomers can be accomplished in three ways: (a) addition of a chiral compound, such as cyclodextrin (CD) to the running buffer (Fig. 5) (b) by reacting the amino acids with a chiral reagent, such as 1-(9-fluorenyl)ethyl chloroformate (FLEC), to form two stereoisomers which can be easily resolved (Fig. 6) [21], or (c) by using a chiral column. The three aromatic amino acids, tryptophan, phenylalanine, and tyrosine, can be detected with UV-LIF without derivatization with a fluorescent tag using a KrF laser detection system, which was built in our laboratory (Fig. 7) [20,22]. The separation of tryptophan and its metabolites [23] and hydroxy proline and other secondary amino acids in biological samples [24] was reported.

B. Peptides

Many studies have dealt with the separation of peptides. In our laboratory, CE was used to separate a group of bioactive peptides (Fig. 8) and is used routinely to check the purity of synthetic peptides (see, for example, Fig. 2). Peptides are detected after separation by UV absorption or by LIF after derivatization with a florescent reagent [24]. Pulsed UV-LIF was used for the detection of peptides without derivatization if the peptides contained one of the three aromatic amino acids (Fig. 9) [22]. We also used CE to predict the mobility of peptides [26,27] and the relation between mobility of dipeptides and charge density (Fig. 10) [28].

Capillary electrophoresis was used for peptide mapping of proteins after enzymatic digestion [29]. For a comprehensive review of CE of peptides, see Ref. 30.

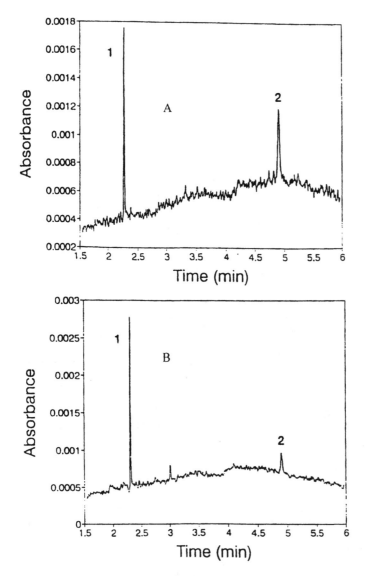

Figure 4 Electropherogram of the separation of nitrate and nitrite at pH 3 (A) water and (B) urine. Buffer: 25 mM phosphate containing 0.5% DMMAPS and 1.0% Brij-35; applied voltage: −15 kV; column: 10%, polyacrylamide-coated fused-silica [T = (g acrylamide + g N,N'-methylene ois-acrylamide)/100 mL solution]; column dimensions: L_{total} = 57 cm, $L_{detection}$ = 50 cm; i.d. = 75 μm; instrument: Beckman Model P/ACE System 2000. Detection: 214 nm. Solutes: 1 = nitrate (2.5 μg/mL); 2 = nitrite (2.5 μg/mL). (From Ref. 18.)

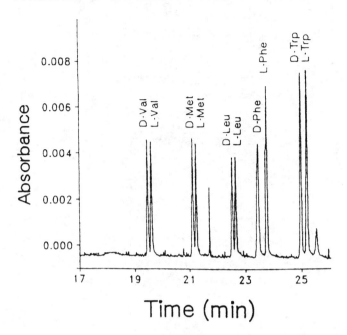

Figure 5 MECC separation of a racemic amino acid mixture derivatized with FMOC. Buffers: 5 mM sodium borate (pH 9.2), 150 mM SDS, and 40 mM γ-CD. Capillary: 50 μm × 67 cm; voltage: 10 kV. UV detection at 200 nm. (From Ref. 19.)

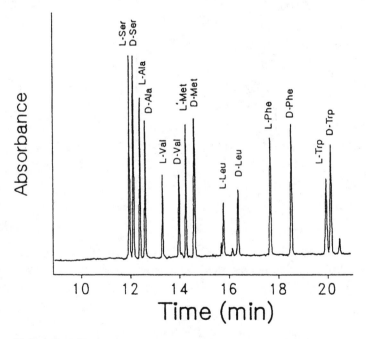

Figure 6 MECC separation of a racemic amino acid mixture derivatized with (+)-FLEC. Buffers: 10 mM sodium phosphate (pH 6.8), 25 mM SDS, and 15% ACN. (From Ref. 21.)

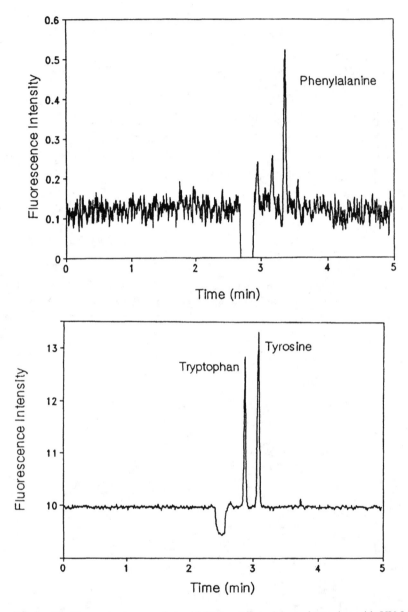

Figure 7 Electropherogram of phenylalanine, tryptophan and tyrosine with UV-LIF detection. Buffer: 5 mM sodium borate (pH 9.1). Capillary: 75 μm i.d. × 70 cm total length, 60 cm effective length. Applied voltage: 25 kV. Samples: 1×10^{-4} phenylalanine, 5×10^{-8} tryptophan and 5×10^{-6} M tryosine. Injection: 15 s, gravity. (From Ref. 22.)

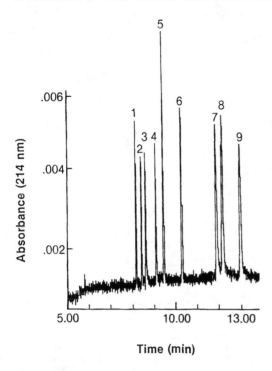

Figure 8 Electropherogram of bioactive peptides. (1) Bradykinin, (2) bradykinin fragment 1–5, (3) substance P, (4) [Arg^8]-vasopressin, (5) luteinizing hormone–releasing hormone, (6) bombesin, (7) leucine enkephalin, (8) methionine enkephalin, and (9) oxytocin. (From Ref. 8.)

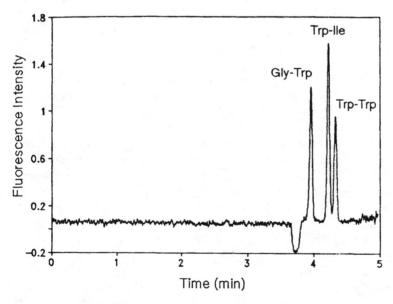

Figure 9 Electropherogram of dipeptides with LIF detection. Buffer: 10 mM sodium phosphate (pH 7.0). Capillary: 75 μm i.d. × 65 cm total length, 55 cm effective length. Applied voltage: 23 kV. Samples: 2.1×10^{-7} M Gly-Trp, 8.1×10^{-8} M Trp-Ile, and 6.9×10^{-8} M Trp-Trp. Injection: 15 s gravity. (From Ref. 22.)

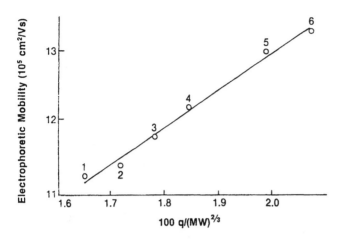

Figure 10 Electrophoretic mobility for dipeptides versus charge-to-size parameter. Solutes: 1 = FF; 2 = FD; 3 = FL; 4 = FV; 5 = FA; 6 = FG. (From Ref. 28.)

C. Proteins

The electrophoretic separation of proteins in untreated fused silica capillaries has not been very successful due to the interaction of the protein with the silanol groups on the inner capillary wall. Therefore, many studies used buffer modifiers or neutrally coated capillaries to eliminate or suppress the protein/wall interaction. A solution to the wall interaction (adsorption) of proteins was suggested by Lauer and McManigill [31], who stated that CZE of proteins in untreated fused silica capillaries is possible if coulombic repulsion between protein and the capillary can overcome adsorption tendencies. This coulombic repulsion can be achieved either by raising the pH of the buffer solution above the isoelectric point of the protein or by dynamically modifying the interfacial double layer between the wall and the bulk solution with selected ions. Adsorption of proteins to the capillary wall is eliminated by physically coating the capillary wall with methyl cellulose [6], polyacrylamide [6,32,33], polyethyleneimine [34], polyethyleneglycol [35], nonionic surfactant coatings [36], or via silane derivatization [6,37,38]. Another procedure for minimizing the adsorption of proteins is through the use of additives to the buffer. For example, one could add a high concentration of alkali salts [39], 1,3-diaminopropane [40], and cationic surfactant FC 135 [41] to the running buffer. The separation of a mixture of glycoproteins was resolved using a borate buffer containing 1 mM putrescine [42]. Capillary isoelectric focusing is used in our laboratory for two purposes: to check the purity of the protein and to determine its isoelectric point. Also, CIEF is a powerful technique whereby the analyst can determine the protein isoforms. We recently determined the isoforms of antibodies by CIEF (work in progress). Proteins are detected by their UV absorption or by LIF with or without derivatization with a fluorescent tag. In our laboratory we used UV-LIF, employing a KrF pulsed UV laser for the detection of native proteins that contained aromatic amino acid residues. Figure 11 compares the detection of two proteins in their native form by UV absorption and UV-LIF [22]. In Sec. XI we will discuss the application of CE for the determination of protein/drug and protein/DNA interaction. For a comprehensive review of protein separations, the reader is referred to Chapter 9 in Ref. 30.

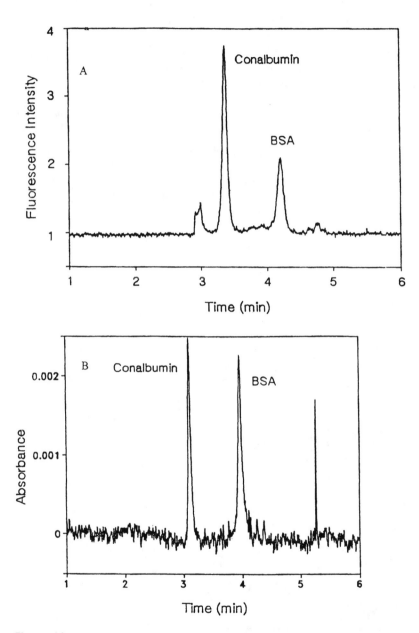

Figure 11   Electropherogram of proteins with (A) KrF-LIF detection and (B) UV absorption detection at 214 nm. Capillary: 75 μm i.d. × 57 cm total length. Applied voltage: 22 kV. Samples: 1.3×10^{-6} M conalbumin, 1.5×10^{-6} M BSA. Injection: 3 s pressure for UV detection and 2.5×10^{-8} M conalbumin and 3×10^{-8} M BSA for LIF detection. (From Ref. 22.)

V. NATURAL PRODUCTS

Natural products are the organic and inorganic compounds found in nature: in plants (leaves, needles, bark, roots, flowers, and seeds), in marine organisms (plants, animals, and microbes), in the microbial fauna found in highly diverse and sometimes extreme environments, and in the soil. The pharmaceutical industry in its drug discovery effort has relied heavily on natural products, if not as a source of drugs, then as sources of novel bioactive chemotypes that can be developed into new drugs; an excellent example is penicillin.

CZE and MECC were used for the separation of widely different compounds from natural sources, including antibiotics, flavonoids, isoflavonoids, steroids, coumarins, alkaloids, illicit drugs, nicotine, caffeine and related compounds, toxins such as aflatoxins [43], mycotoxins, heptapeptide toxins, Chinese herbal preparations, mineral elements, humic substances, and many other compounds. Two reviews of the subject have been published [44,45].

The advantage of CZE and MECC for the analysis of natural products is their high resolving power and applicability to a versatile group of compounds. In our laboratory, MECC was used

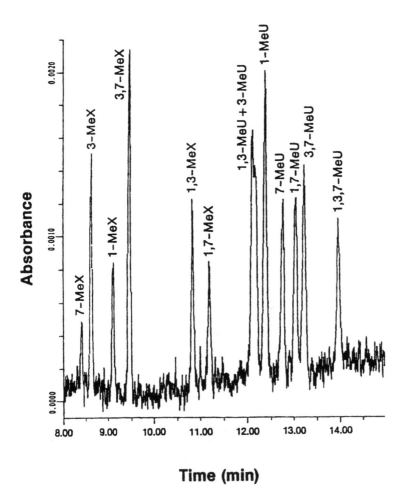

Figure 12 Electropherogram of a mixture of seven methyl-substituted xanthines and six methyl-substituted uric acids. (From Ref. 48.)

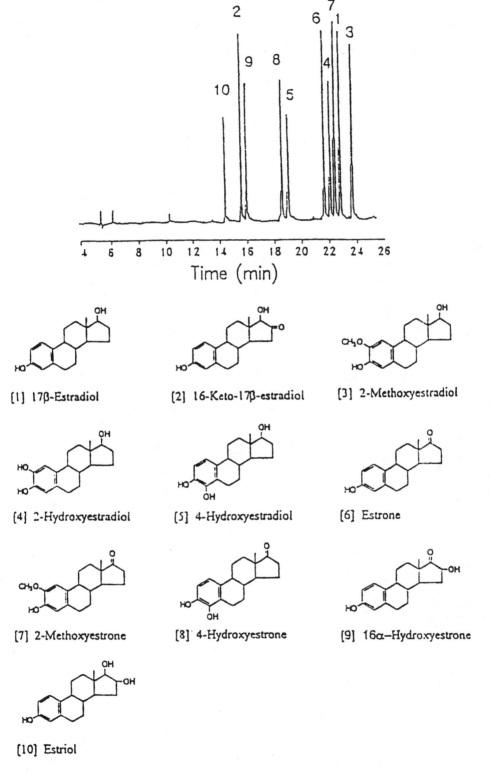

Figure 13 MECC of 10 estrogens. Buffers: 10 mM sodium phosphate (pH 7.0) containing 50 mM SDS and 20% methanol; capillary: 50 μm × 57 cm; voltage: 20 kV; pressure injection: 2S; detection: absorption at 200 nm. (From Ref. 49.)

for the separation of a bark and needle extract for the assay of taxol, an anticancer agent, from needle and bark extracts of the Western yew [46]. Also, CE with LIF detection was used for the quantification of Michellamine, which has shown activity against HIV in vitro, from plant material [47]. The separation of methyl substituted xanthines and uric acids (caffeine metabolites) (Fig. 12) [48] and estrogen and nine related compounds (Fig. 13) was reported by our group [49]. Other examples are given in Refs. 44 and 45.

VI. NUCLEIC ACIDS

Capillary gel electrophoresis (CGE) is a powerful analytical technique for the separation of double-stranded DNA fragments, single-stranded DNA, and polymerase chain reaction products (PCR) (Fig. 14). The success of CE as an analytical technique for the separation of nucleic acids is based on gel-filled capillaries. The capillary is filled with a replaceable liquid gel, also known as entangled polymers, which separates the fragments by a sieving mechanism and can be replaced after each analysis. Hydroxyethylcellulose, hydroxymethylcellulose, methylcellulose, polyvinylalcohol, liquid polyacrylamide, and others have been used as liquid polymers. For a comprehensive review of the subject, history, theory, and applications of capillary gel electrophoresis, the reader my consult Ref. 50. Issaq et al. [51] examined the effect of different parameters on migration time, resolution, and speed of analysis of DNA fragments and PCR products. These parameters include column length, applied voltage, gel type and concentration, and buffer ionic strength. The results show that a 1 cm capillary at an applied voltage of 185 V/cm filled with commercial gel was adequate for the separation of small DNA fragments and PCR products. Resolution of large fragments is better at lower field strength at constant column length and is directly proportional to column length at the same field strength. In a previous work from this laboratory we were able to resolve the wild-type (400 bp) and mutant (375 bp) PCR products in under 60 seconds using 1–2 cm effective length capillaries (Fig. 15) [52,53]. In a recent CE study [54] we compared the effect of experimental temperature on the separation of DNA digest. The results showed that better resolution of the DNA fragments was obtained at 25°C than at 40 or 50°C. Array capillary electrophoresis with gel-filled capillaries has been used for DNA sequencing [55], and recently this technique has gained acceptance as the method of choice for DNA sequencing of the human genome. Also, microchip array technology is being used for DNA sequencing [56].

VII. CLINICAL APPLICATIONS

Applications of CE in the clinical laboratory have been reported [57]. The determination of different compounds in biological fluids may not be at times a simple matter, due to the fact that biological fluids often have a complex matrix. For example, while a compound of interest may be present in blood or serum at the low mg/L, sodium and protein are present at 10,000 mg/L and 100,000 mg/L, respectively [58]. Also, the matrix may affect reproducibility, peak shape, and quantitation. However, selective detection may overcome some of the matrix effects. In our laboratory CE has been used for the separation of homovanillic and vanillylmandelic acids in baby urine [59] (Fig. 16), hydroxyproline in human urine and serum [60], and polysaccharides from human urine (Fig. 17) [61]. Also, we have reported the separation of retinoic acid isomers [62], caffeine metabolites [48], separation of nicotine and three of its metabolites in human urine (Fig. 18) [63], and separation of 10 estrogens and detection by UV (Fig. 13) [63], and the determination of estrogens by CE with electrochemical detection and microdialysis sampling [64]. Those interested in the CE analysis of drugs should consult Ref. 65.

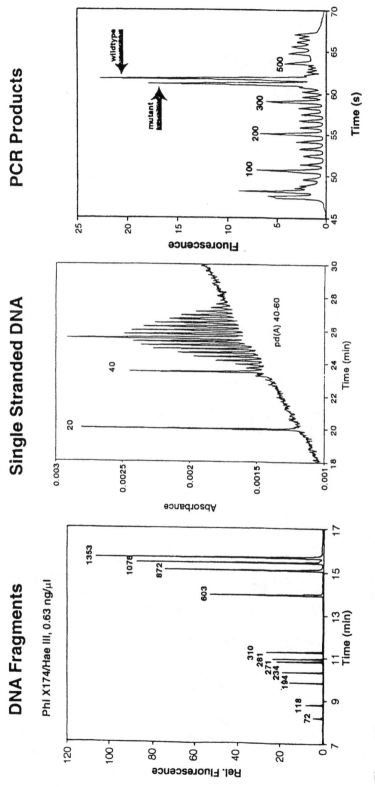

Figure 14 Electropherograms of the separation of DNA fragments (left), single-stranded DNA (middle), and PCR products (right). (From Ref. 51.)

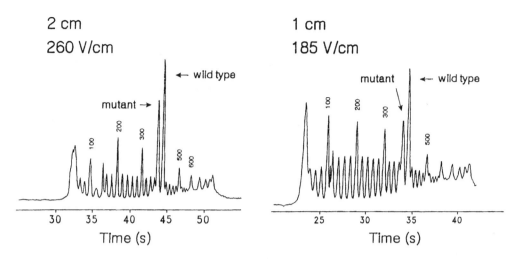

Figure 15 High-speed separation of PCR products using 1 cm and 2 cm effective length capillaries. (From Ref. 53.)

VIII. ILLICIT DRUGS

The earliest work on the use of CE for the separation of illicit drugs was reported by Weinberger and Lurie for the separation of heroin impurities employing a micellar buffer and UV detection [66]. Later, Lurie et al. used MECC with UV/LIF detection for the separation and detection of impurities in heroin and cocaine (Fig. 19) [67,68]. An improvement of 1000-fold in sensitivity over UV detection was achieved (Fig. 20). In the meantime, few studies have been published dealing with the separation of illicit drugs [69].

IX. CHIRAL SEPARATIONS

The first to report on chiral separations by CE were Gassmann et al. [70]. They resolved a mixture of dansylated D- and L-amino acids by adding Cu(II)-L-histidine to the running buffer. Since then many studies have been published dealing with chiral separations, employing different complexing agents and materials. The most widely used compounds for chiral separations are the cyclodextrins and their modified derivatives. The first use of cyclodextrins in CE was reported by Terabe [71,72] for the separation of positional isomers. Later, Guttman et al. [73] obtained chiral separations by incorporating cyclodextrin within a polyacrylamide gel-filled capillary, and Fanali [74] reported on chiral CE separations by adding cyclodextrins to the running buffer. Issaq [75] compared the separation of chiral compounds by CE with other methods. For a detailed discussion of chiral separations by CE and MECC the reader is advised to consult Ref. 76.

In our laboratory few procedures were used for the separation of D- and L-amino acids. FMOC-derivatized amino acids were resolved by MECC in an SDS-CD phosphate buffer (Fig. 5) [19,20]. The second approach involved reacting the chiral amino acid mixture with the chiral fluorescent agent FLEC to produce two diastereoisomers, which can be resolved without the addition of a chiral reagent to the running buffer in a bare fused silica capillary (Fig. 6) [20,21]. In both cases the resolved compounds were detected by UV/LIF. In another study, a polyacrylamide coated capillary was used in a reversed-flow MECC mode (RF-MECC) to resolve a

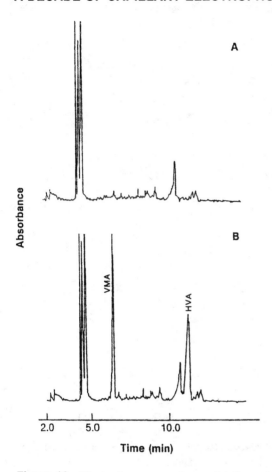

Figure 16 Electropherograms of normal infant urine: (A) untreated normal infant urine, (B) normal infant urine spiked with vanillylmandelic acid (VMA) and homovanillic acid (HVA). (From Ref. 59.)

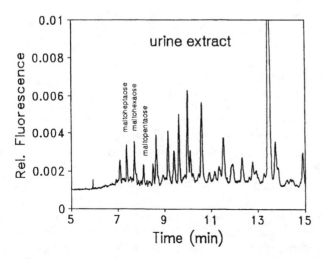

Figure 17 Electropherogram of the separation of three polysaccharides in urine. (From Ref. 61.)

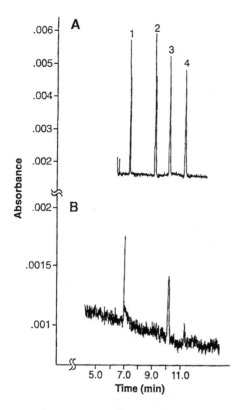

Figure 18 Separation of nicotine and three of its main metabolities. (A) Standard mix and (B) smoker's urine sample. Buffer: 50 mM sodium acetate; pH: 5.7; voltage: 20 kV; pressure injection: 3S at 0.5 psi; detection: absorption of 260 nm; instrument: Beckman Model P/ACE System 5510; column: 10% T poly-acrylamide-coated fused-silica; column dimensions: L_{total} = 47 cm, $L_{detection}$ = 40 cm; i.d. = 75 µm; solute concentration: 2–5 µg/mL water; solutes: 1-nicotine, 2-demethylcotinine, 3-cotinine, and 4-*trans*-3-hydroxycotinine. (From Ref. 63.)

racemic mixture of dansylated amino acids and dipeptides using 1% beta CD-sulfobutyl ether in 10 mM phosphate buffer at pH 3.1 [43].

X. MISCELLANEOUS SEPARATIONS

Capillary electrophoresis in all its formats, CZE, CGE, CIEF, MECC, and RF-MECC, have been applied to the separation of hundreds of widely different compounds. Because so many books and journal articles dealing with CE separations have been published, it is hard to cover such a wide topic in a short review. Therefore, this review is limited to topics on which we have conducted research and published.

RF-MECC is a procedure that was developed in our laboratory for the separation of hydrophobic compounds. The principle is very simple: electrokinetic chromatography with negatively charged CD or SDS added to the buffer was conducted in polyacrylamide-coated fused silica capillaries under suppression of electroosmotic flow with reversed polarity; i.e., the flow to the detector is from the negative electrode to the positive electrode. In this procedure the

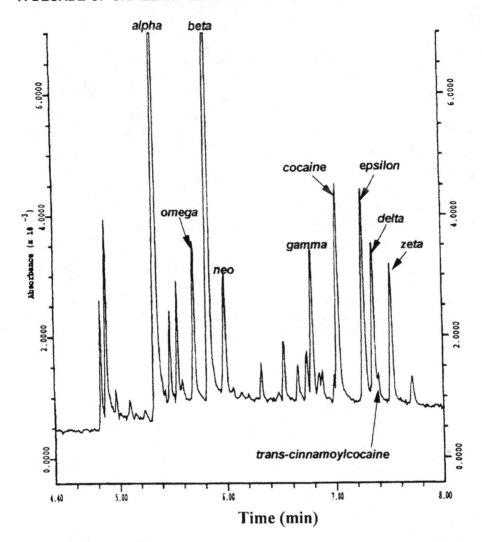

Figure 19 Electropherogram of an illicit cocaine HCI HPLC size exclusion extract. The run buffer contained 10% methanol and 90% (10 mM β-CD-SBE-(IV), 7 mM phosphate(dibasic), pH 8.6). (From Ref. 68.)

most hydrophobic solutes are eluted first, unlike MECC, where the most hydrophobic solutes elute last. This procedure was used for the separation of a mixture of polycyclic aromatic hydrocarbons (Fig. 21) [79], and a mixture of steroids (Fig. 22) [79], in addition to other hydrophobic compounds [77–80], and a racemic mixture of amino acids, dipeptides, aflatoxins, and chlorophenols [43].

Our group demonstrated the separation of pyridinecarboxylic acid isomers and related compounds [81], the separation of a mixture of catecholamine metabolites (Fig. 23) [59], the advantages of subambient temperature, 5°C, for the separation of heterocyclic nitrosoamino acid conformers (Fig. 24) [82], the separation of pyridinecarboxylic acid isomers and related compounds [83], and the separation of a retinoic acid mixture (Fig. 25).

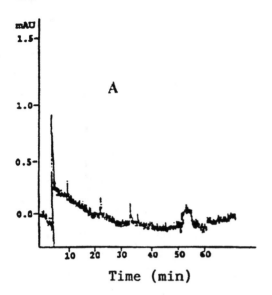
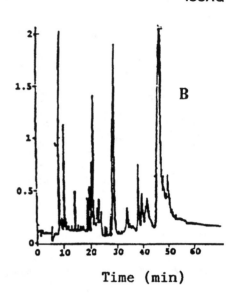

Figure 20 Comparison of (A) UV detection at 260 nm and (B) LIF detection with krypton-fluoride laser for the CE analysis of a refined Southwest Asian heroin hydrochloride exhibit. (From Ref. 67.)

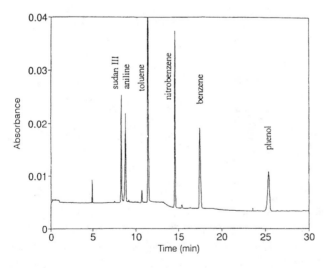

Figure 21 Order of migration versus solute hydrophobicity in RF-MECC mode. Column, 10% linear polyacrylamide-coated fused-silica. Column dimensions: L_{total} = 57 cm, $L_{detection}$ = 50; i.d. = 75 μm. Buffer: 10 mM acetate and 50 mM SDS, pH 4.2; applied voltage: −20 kV; current: 30 μA; detection: 214 nm. (From Ref. 79.)

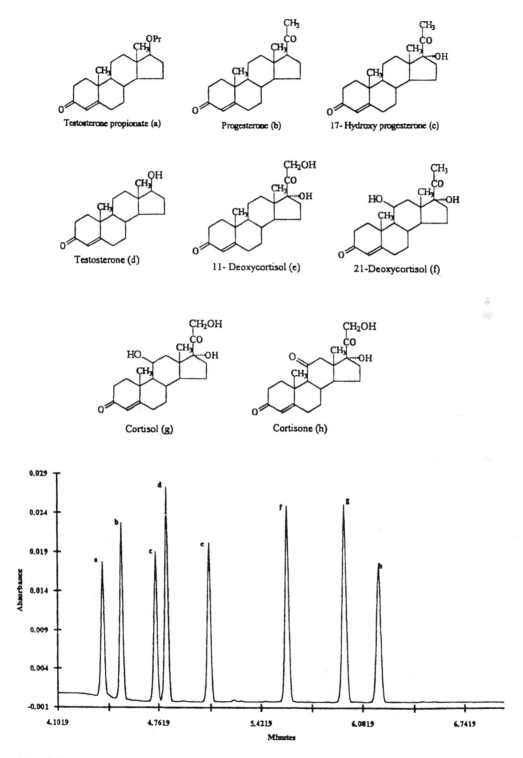

Figure 22 Separation of steroids on neutral eCAP capillary. Buffer: 100 mmol/L SDS, 20% (v/v) aceto-nitrile and 20 mmol/L ME, pH 6; voltage applied: 15 kV; temperature: 16°C. (From Ref. 79.)

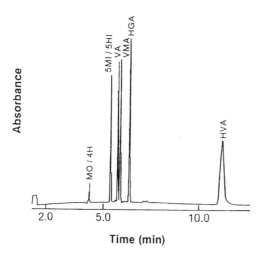

Figure 23 Electropherogram of a standard mixture of catecholamine metabolites. Buffer: 200 mM acetate; pH: 4.10; applied voltage: 25 kV; current: 115 µA. Solutes: MO = mesityl oxide; 4H = 4-hydroxy-3-methoxyphenyl-glycol; 5 Mi = 5-methoxyindole-3-acetic acid; 5 Hi = 5-hydroxyindole-3-acetic acid; VA = 4-hydroxy-3-methoxy benzoic acid; VMA = vanillylmandelic acid; HGA = homogentisic acid; and HVA = homovanillic acid. (From Ref. 59.)

XI. PHYSICOCHEMICAL STUDIES

In this section we try to answer the following question: Is CE a separation technique only, or can it be used to determine physicochemical parameters? To date, a search of the literature reveals that CE can indeed be used to determine physicochemical parameters. CE has been used for the determination of [84]: (a) pK values of weak electrolytes [85] and amphoteric compounds; (b) pI values of proteins by CIEF; (c) the mobility and its relation to charge/mass ratio in peptides and proteins; (d) mobility of nucleic acids; (e) viscosity; (f) binding and dissociation constants [86,87]; (g) the approximate molecular weight of proteins; (h) the approximate number of base pairs of DNA fragments and PCR products; (i) detection of DNA mutants [93–95]; (j) insulin content and excretion from single islets of Laugerhans [88]; (k) enzyme activity and quantity in single human erythrocyte [89]; and (l) enzyme assay and activity [90,92]. In our laboratory, CE was used to predict the mobility of peptides [26–28] and to determine protein/drug interaction and protein/DNA interaction (Fig. 26) [32]; RF-MECC was used to determine the partition coefficients and hydrophobicity of organic compounds [78]; CGE to estimate the size and number of base pairs in PCR products and DNA fragments [96]; CIEF to determine the purity and isoelectric point (pI) of proteins [40].

XII. CAPILLARY ELECTROCHROMATOGRAPHY

The column type (bare or coated), along with what it is filled with, determines the mode of CE separation. A bare fused silica capillary and a simple buffer are used for CZE. The addition of a micelle to the buffer to resolve mainly neutral compounds is known as MECC, while reversed-flow MECC uses a neutrally coated fused silica capillary and a micelle buffer for the separation of hydrophobic compounds. The addition of a chiral reagent to the buffer, simple or micelle, is used for the separation of racemic mixtures. The filling of a neutrally coated capillary with

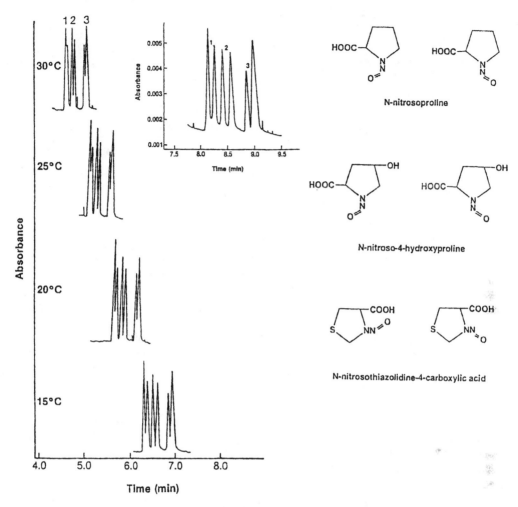

Figure 24 Electropherograms showing the separation of syn and anti conformers of selected nitroso-amino acids as a function of temperature. Solutes: 1 = *N*-nitrosothiazolidine-4-carboxylic acid, 2 = *N*-nitrosoproline, and 3 = *N*-nitroso-4-hydroxyproline; instrument: Beckman Model P/ACE System 5510; detection: 235 nm; cp;i,m 1−% T polyacrylamide-coated fused-silica; column dimensions: L_{total} = 57 cm, $L_{detection}$ = 50 cm, i.d. = 75 μm; buffer:10 mM phosphate containing 2 mM DMMAPS, and 0.1% Tween 20; pH = 7.2; applied voltage: −25 kV; solute concentration: 2–5 μg/mL. (From Ref. 82.)

an ampholyte is used for the determination of the pI of proteins by IEF. Filling the neutrally coated capillary with a liquid gel, entangled polymers, is used in CGE for the separation of DNA fragments and PCR products. In 1974, Pretorius and coworkers [97] showed that electroosmotic flow, not a pump, can be used to drive the mobile phase through a 1 mm glass LC column packed with 75–175 μm particles. Jorgenson and Lukacs [98] demonstrated the feasibility of using electroosmotic flow and 170 μm i.d. Pyrex glass tube, 68 cm long, packed with 10 μm ODS particles, for the CEC separation of a mixture made of 9-methylanthracene and perylene. When CEC is compared to HPLC, better overall resolution, higher efficiency, and zero backpressure, which will allow the use of up to 1 μm silica particles, are obtained with CEC. CEC have been used for the separation of different groups of compounds. Lurie et al. [99] applied CEC

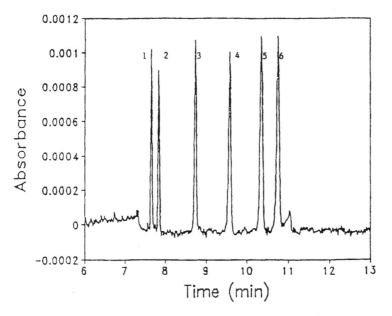

Figure 25 Separation of retinoic acid mixture from spiked rat plasma extract. Capillary: 75 μm × 47 cm; buffer: 20 mM tris-borate (pH 8.5); 25 mM SDS, 20% acetonitrile; applied voltage: 13 kV; injection: 4S, pressure; peaks: 1 = 13-*cis*-4-oxo-RA; 2 = 4-oxo-RA; 3 = 13-*cis*-acitretin; 4 = 13-*cis*-RA; 5 = 9-cis-RA; 6 = all-trans-RA. (From Ref. 62.)

for the separation of drugs of forensic interest, while Yang and El Rassi [100] applied CEC for the separation of a mixture of urea herbicides (Fig. 27). The CEC separation of nucleosides and bases [101], small and large nucleic acids [102], carbohydrates [103], and polycyclic aromatic hydrocarbons were reported [104]. The application of CEC has been reviewed very recently by Dermaux and Sandra [105]. A comparison of CEC with CE reveals that CE is simpler, easier to perform, gives higher efficiency, and has wider application. To date, although CEC has been used for the separation of different groups of compounds, it still has not been fully utilized due to the problems of generating reproducible column inlet and outlet frits and the packing of the capillary column. The introduction of open tubular columns by Pesek et al. [106] may solve some of the above problems, but the column capacity will be lower than that using a packed column. Another alternative is to mimic the packed bed by etching an array of support particles into a quartz substrate to produce an array of collocate monolith support structures. Such a microfabricated system was used for the CEC separation of peptides [107].

XIII. FUTURE DIRECTIONS

The future of capillary electrophoresis looks very promising. In the last 10 years CE was evaluated and many theoretical studies have been carried out. Our group studied the effect of buffer and buffer modifiers on resolution, mobility, and Joules heating [28,108–119] and, as mentioned above, applied CE not only as a separation technique but to answer physicochemical questions. Today CE is an established and useful microanalytical technique with widespread applications. As a matter of fact, CE has grown phenomenally: at least 15 books and hundreds of articles have been published; two specific conferences, HPCE and the Frederick CE Conference, have

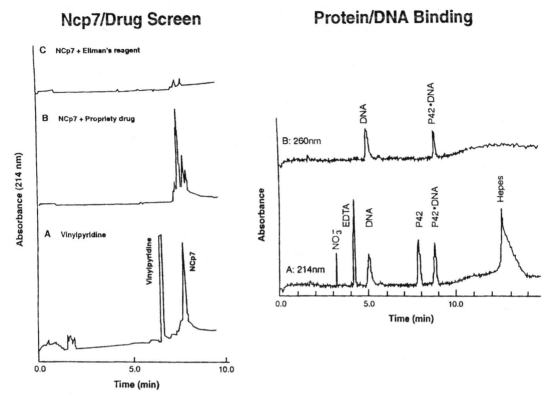

Figure 26 Electropherograms of Ncp7/drug (left) and protein/DNA (right) interaction. (From Ref. 32.)

been held in the United States; and many sessions in major meetings have been devoted to CE theory, instrumentation, and applications. There is no doubt that CE is an acceptable analytical technique. CE has found a strong footing in molecular biology; determination of DNA fragments, PCR products, and the sequencing of the human genome, in addition to studies of single cells, protein separations, and peptide mapping all use CE. The future of CE is very bright, as the number of studies, development, and application continue to multiply. In the next decade, CE will move into ultrafast separations—in seconds, not minutes—using microchip technology; and high-efficiency, high-throughput analyses employing narrower and shorter capillaries (5–10 μm i.d. and 1–5 cm long) will be used in array format to achieve high throughput analyses by CE. Instruments with 96 and 384 capillaries will be produced, not only to sequence DNA and the human genome, but to be used for efficient and simultaneous multisample analysis. The era of analyzing one sample at a time will be out of fashion. We will witness fundamental changes for laboratory instrumentation. For example, market studies already project high growth for "lab-on-a-chip" technologies. A key component of such technology involves forcing fluids to move through a chip's channels; the principles involved are borrowed from CE. As an analytical technique, CE will become an established method in the pharmaceutical industry, the clinical laboratory, and the forensic laboratory, where high speed and accurate results are required. In the forensic laboratory, CE will be used for DNA typing, which at the present is being done by slab gel electrophoresis, and for gun powder analysis. CE is well suited for the forensic laboratory because of its versatility, small sample requirements, and speed of analysis. Apart

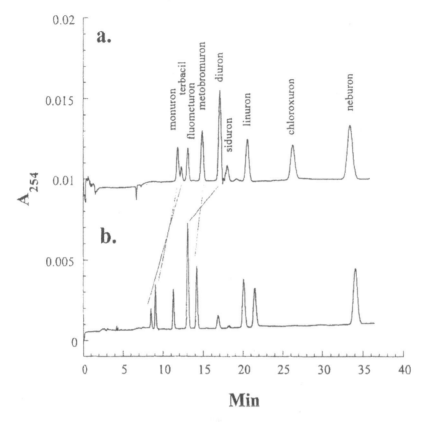

Figure 27 Electrochromatograms obtained with (a) methanol- and (b) acetonitrile-containing mobile phases. Mobile phases: (a) 30% v/v 5 mM NaH_2PO_4, pH 6.0, mixed with 70% v/v methanol; (b) 50% v/v 5 mM NaH_2PO_4, pH 6.0, mixed with 50% v/v acetonitrile; column, ODS-Zorbax; length, 27 cm; detection window at 20 cm; running voltage, 20 kV; pressure injection, 20x at 20 psi; temperature, 15°C; UV detection, 254 nm. Elution order in (a): 1, monurone; 2, terbacil; 3, fluometuron; 4, metobromuron; 5, diuron; 6, siduron; 7, linuron; 8, chloroxuron; 9, neburon; in (b): 1, terbacil; 2, monuron; 3, fluometuron; 4, diuron; 5, metobromuron; 6, siduron; 7, linuron; 8, chloroxuron; 9, neburon. (From Ref. 100.)

from its use as an analytical tool, we envision a continued and expanded role for CE in the determination of physicochemical parameters. Multidimensional separations employing separation-separation and separation-spectroscopic techniques will be routinely used for the separation and identification of complex mixtures, biological and otherwise.

XIV. CONCLUSION

It is abundantly clear that CE has gained a respectable place in different fields in which analysis is carried out. In this review, mainly of our CE research and application in the last decade, we also, where appropriate, presented work that was done in other laboratories. Because we could not possibly include in a short review all the applications that have been done using CE, the reader should refer to the more than 15 books and score of reviews, especially the biannual reviews in *Analytical Chemistry*, published on the applications of CE.

ACKNOWLEDGMENT

The author would like to thank Drs. George Janini and King Chan, SAIC Frederick, and Professor Ziad El Rassi, Oklahoma State University, for constructive and helpful comments. This project has been funded in whole or in part with federal funds from the National Cancer Institute, National Institutes of Health, under Contract No. NO1-CO-56000.

By acceptance of this article, the publisher or recipient acknowledges the right of the U.S. Government to retain a nonexclusive, royalty-free license and to any copyright covering the article. The content of this publication does not necessarily reflect the views or policies of the Department of Health and Human Services, nor does mention of trade names, commercial products, or organizations imply endorsement by the U.S. Government.

REFERENCES

1. A Tiselius. The moving boundary method of studying the electrophoresis of proteins. Thesis, Nova Acta Regiae Sociates Scientiarum Upsaliensis 1930, Ser. IV, Vol. 7, Number 4.
2. P Camilleri. In: P Camilleri, ed. Capillary Electrophoresis, Theory and Practice, 2nd ed. Boca Raton, FL: CRC Press 1997, pp. 1–22.
3. RT Swank, KD Munkers. Anal Biochem 39:462–466, 1971.
4. PH O'Farrell, Cell 1978, 14, 545–557.
5. S Hjerten. Chromatogr Rev 9:122–219, 1967.
6. V Neuhoff, WB Schill, H Sternbach. Biochem J 117:623–627, 1970.
7. R Virtanen. Acta Polytech Scand (Helsinki) 123:1, 1974.
8. FE Mikkers, FM Everaerts, TPEM Verhegen. J Chromatogr 169:11–20, 169.
9. J Jorgenson, KD Lukacs. Anal Chem 53:1298–1302, 1981.
10. S Terabe, K Otsuka, K Ichikawa, A Tsuchuya, T Ando. Anal Chem 56:111–113, 1984.
11. BL Karger, AS Cohen, A Gutman. J Chromatogr 492:585–614, 1989.
12. Z El Rassi, ed. Electrophoresis, 20:2987–3330, 1999.
13. Z El Rassi, ed. Electrophoresis, 18:2121–2504, 1997.
14. WR Jones, P Jandik. J Chromatogr 608:385–393, 1992.
15. WR Jones, J Soglia, M McGlynn, C Haber, J Reineck, C Krstanovic. Am Lab 28:25–29, 1996.
16. A Weston, P Brown, P Jandik, W Jones. J Chromatogr 608:395–402, 1992.
17. GM Janini, GM Muschik, HJ Issaq. J Capillary Electrophoresis 1:116–120, 1994.
18. GM Janini, KC Chan, GM Muschik, HJ Issaq. J Chromatogr B 657:419–423, 1994.
19. KC Chan, GM Janini, GM Muschik, HJ Issaq. J Chromatogr 653:93–97, 1993.
20. HJ Issaq, KC Chan. Electrophoresis 16:467–480, 1995.
21. KC Chan, GM Muschik, HJ Issaq. Electrophoresis 16:504–509, 1995.
22. KC Chan, GM Janini, GM Muschik, HJ Issaq, J Liq. Chromatogr 16:1877–1890, 1993.
23. KC Chan, GM Muschik, HJ Issaq. J Chromatogr A 718:203–210, 1995.
24. KC Chan, GM Janini, GM Muschik, HJ Issaq. J Chromatogr 622:269–273, 1993.
25. GM Janini, J Lukso, HJ Issaq. J High Resolut Chromatogr 17:102–103, 1994.
26. CJ Metral, GM Janini, GM Muschik, HJ Issaq. J High Resolut Chromatogr 22:373–378, 1999.
27. GM Janini, CJ Metral, HJ Issaq, GM Muschik. J Chromatogr A 848:417–433, 1999.
28. HJ Issaq, GM Janini, IZ Atamna, GM Muschik, J Lukszo. J Liq Chromatogr 15:1129–1142, 1992.
29. HJ Issaq, KC Chan, GM Janini, GM Muschik. Electrophoresis 20:1533–1537, 1999.
30. TV deGoor, A Apffel, T Chakel, W Hancock. In: JP Landers, ed. Capillary Electrophoresis of Peptides in Handbook of Capillary Electrophoresis, 2nd ed. Boca Raton, FL: CRC Press, 1998, pp. 213–258.
31. HH Lauer, D McManigill. Anal Chem 58:166–170, 1986.
32. GM Janini, RJ Fisher, LE Henderson, JH Issaq. J Liq Chromatogr 18:3617–3628, 1995.
33. S Hjerten. J Chromatogr 347:191–198, 1985.

34. JK Towns, FE Regnier. J Chromatogr 516:69–78, 1990.
35. GJM Bruin, JP Chang, RH Kuhlman, K Zegers, JC Kraak, H Poppe. J Chromatogr, 471:429–436, 1989.
36. JK Towns, FE Regnier. Anal Chem 63:1126–1132, 1991.
37. JW Jorgenson, KD Lukacs. Science (Washington, DC) 222:266–272, 1983.
38. M Gilges, H Usmann, H Kleemiss, SR Motsch, G Schomburg. J High Resolut Chromatogr 15: 452–457, 1992.
39. JS Green, JW Jorgenson. J Chromatogr 478:63–70, 1989.
40. JA Bullock, L-C Yuan. J Microcol Sep 3:241–248, 1991.
41. WGH Muijselaar, CHM de Bruijn, FM Everaerts. J Chromatogr 605:115–123, 1992.
42. JP Landers, RP Oda, BJ Madden, TC Spelsberg. Anal Biochem 205:115–124, 1992.
43. GM Janini, GM Muschik, HJ Issaq. Electrophoresis 17:1575–1583, 1996.
44. HJ Issaq. Electrophoresis 18:2438–2452, 1997.
45. HJ Issaq. Electrophoresis 20:3190–3202, 1999.
46. KC Chan, AB Alvarado, MT McGuire, GM Muschik, HJ Issaq, KM Snader. J Chromatogr B 657: 301–306, 1994.
47. KC Chan, F Majadly, TG McCloud, GM Muschik, HJ Issaq, KM Snader. Electrophoresis 15:1310–1315, 1994.
48. IZ Atamna, GM Janini, GM Muschik, HJ Issaq. J Liq Chromatogr 14:427–436, 1991.
49. KC Chan, GM Muschik, HJ Issaq, PK Siiteri. J Chromatogr A 690:149–154, 1995.
50. PG Righetti, C Gelfi. In: PG Righetti, ed. Capillary Electrophoresis of DNA in Capillary Electrophoresis in Analytical Biotechnology. Boca Raton, FL: CRC Press, 1996, pp. 431–476.
51. HJ Issaq, KC Chan, GM Muschik. Electrophoresis 18:1153–1158, 1997.
52. KC Chan, GM Muschik, HJ Issaq, KJ Garvey, PL Generlette. Anal Biochem 43:133–139, 1996.
53. KC Chan, GM Muschik, HJ Issaq. J Chromatogr B 695:113–115, 1997.
54. HJ Issaq, H Xu, KC Chan, MC Dean. J Chromatogr B 738:243–248, 2000.
55. NJ Dovichi. In: JP Landers, ed. Capillary Gel Electrophoresis for Large Scale DNA Sequencing: Separation and Detection in Handbook of Capillary Electrophoresis, 2nd ed. Boca Raton, FL: CRC Press, 1996, pp. 545–565.
56. AT Woolley, GF Sensabaugh, KA Mathies. Anal Chem 69:2182–2186, 1997.
57. RP Oda, VJ Bush, JP Lander. In: JP Landers, ed. Clinical Applications of Capillary Electrophoresis in Handbook of Capillary Electrophoresis, 2nd ed. Boca Raton, FL: CRC Press, 1996, pp. 639–673.
58. ZK Shihabi. In: JP Landers, ed. Handbook of Capillary Electrophoresis, 2nd ed. Boca Raton, FL: CRC Press, 1996, pp. 457–477.
59. HJ Issaq, K Kelviks, GM Janini, GM Muschik. J Liq Chromatogr 15:3193–3201, 1992.
60. KC Chan, GM Janini, GM Muschik, HJ Issaq. J Chromatogr 653:93–97, 1993.
61. KC Chan, HJ Issaq. Unpublished results.
62. KC Chan, KC Lewis, JM Phang, HJ Issaq. J High Resolut Chromatogr 16:558–562, 1993.
63. HJ Issaq, KC Chan, GM Muscik, GM Janini. J Liq Chromatogr 18:1273–1288, 1995.
64. M Perkins, M Zhong, HJ Issaq, M Davies, S Lunt. Am Assoc Pharm Scientists Meeting, Boston, Nov. 6, 1997.
65. AS Cohen, S Terabe, Z Deyl, Capillary Electrophoresis of Drugs. Elsevier, Amsterdam: Elsevier, 1996.
66. R Weinberger, IS Lurie. Anal Chem 63:823–827, 1991.
67. IS Lurie, K Chan, TK Spartley, JF Casale, HJ Issaq. J Chromatogr B 669:3–13, 1995.
68. IS Lurie, PA Hays, JF Casale, JM Moore, DM Castel, KC Chan, HJ Issaq. Electrophoresis 19:51–56, 1998.
69. FV Heeren, W Thorman. Electrophoresis 18:2415–2426, 1997.
70. E Gassmann, JE Kuo, RN Zare. Science 230:813–815, 1985.
71. S Terabe. Trends Anal Chem 8:129–134, 1989.

72. S Terabe, H Ozuaki, K Otsuka, T Ando. J Chromatogr 332:211–217, 1985.
73. A Gutman, A Paulus, AS Cohen, N Grinberg, BL Karger. J Chromatogr 448:41–53, 1988.
74. S Fanali. J Chromatogr 474:441–446, 1989.
75. HJ Issaq. Instrument Sci Technol 22:119–149, 1994.
76. K Otsuka, S Terabe. In: NA Guzman, ed. Chiral Separations by Capillary Electrophoresis and Electrokinetic Chromatography in Capillary Electrophoresis Technology. New York: Marcel Dekker, 1993, pp. 617– 619.
77. GM Janini, GM Muschik, HJ Issaq. J Chromatogr B 683:29–35, 1996.
78. GM Janini, GM Muschik, HJ Issaq. J High Resolut Chromatogr 18:171–174, 1995.
79. GM Janini, HJ Issaq, GM Muschik. J Chromatogr A 792:125–141, 1997.
80. HJ Issaq, GM Janini, GM Muschik. HPCE 96, Orlando, FL, January 21–25, 1996.
81. GM Janini, KC Chan, JA Barnes, GM Muschik, HJ Issaq. J Chromatogr 653:321–327, 1993.
82. GM Janini, GM Muschik, HJ Issaq. J High Resolut Chromatogr 17:753–755, 1994.
83. GM Janini, KC Chan, JA Barnes, GM Muschik, HJ Issaq. J Chromatogr 653:321–327, 1993.
84. HJ Issaq. Eastern Analytical Symposium, Somerset, NJ, November 15–20, 1998.
85. J Cai, TJ Smith, Z El Rassi. Electrophoresis 15:30–32, 1992.
86. NHH Heegard. Electrophoresis 19:367–464, 1998.
87. FA Robey. In: JP Landers, ed. Handbook of Capillary Electrophoresis, 2nd ed. Boca Raton, FL: CRC Press, 1996, pp. 591–609.
88. NM Schultz, L Huang, RT Kennedy. Anal Chem 67:924–929, 1995.
89. W Tan, ES Yeung. Anal Biochem 226:74–79, 1995.
90. D Wu, FE Regnier. Anal Chem 65:2029–2035, 1993.
91. Q Xu, ES Yeung. Nature 373:681–683, 1995.
92. NM Schultz, L Tao, DJ Rose, RT Kennedy. In: JP Landers, ed. Handbook of Capillary Electrophoresis, 2nd ed. Boca Raton, FL: CRC Press, 1996, pp. 611–637.
93. K Khrapko, JS Hanekamp, WG Thilly, A Belenkii, F Foret, BL Karger. Nucl Acid Res 22:364–369, 1994.
94. C Gelfi, PG Righetti, M Trav, S Father. Electrophoresis 18:724–731, 1997.
95. X-C Li-Sucholeiki, K Khrapko, PC Andre, LA Marcelino, BL Karger, WG Thilly. Electrophoresis 20:1224–1232, 1999.
96. HJ Issaq, GM Janini, KC Chan, GM Muschik. Eastern Analytical Symposium, Somerset, NJ, November 16–21, 1997.
97. V Pretorius, BJ Hopkins, JD Shieke. J Chromatogr 99:23–29, 1974.
98. JE Jorgenson, KD Lukacs. J Chromatogr 218:209–216, 1981.
99. IS Lurie, TS Conver, RP Meyers, CG Bailey. 10th Annual Frederick Conf. on CE, Frederick, MD, October 18–20, 1999.
100. C Yang, Z El Rassi. Electrophoresis 20:2337–2342, 1999.
101. M Zhang, Z El Rassi. Electrophoresis 20:31–36, 1999.
102. M Zhang, C Yang, Z El Rassi. Anal Chem 71:3277–3282, 1999.
103. C Yang, Z El Rassi. Electrophoresis 19:2061–2067, 1998.
104. C Yan, R Dadoo, H Shao, RN Zare. Anal Chem 67:2026–2029, 1995.
105. A Dermaux, P Sandra. Electrophoresis 20:3027–3065, 1999.
106. JJ Pesek, M Matyska, S Menezes. J Chromatogr 853:151–158, 1999.
107. H Bing, J Junyan, FE Regnie. J Chromatogr A 853:257–262, 1999.
108. HJ Issaq, IZ Atamna, CJ Metral, GM Muschik. J Liq Chromatogr 13:1247–1259, 1990.
109. IZ Atamna, CJ Metral, GM Muschik, HJ Issaq. J Liq Chromatogr 13:2517–2527, 1990.
110. IZ Atamna, CJ Metral, GM Muschik, HJ Issaq. J Liq Chromatogr 13:3201–3210, 1990.
111. HJ Issaq, IZ Atamna, GM Muschik, GM Janini. Chromatographia 32:155–161, 1991.
112. IZ Atamna, HJ Issaq, GM Muschik, GM Janini. J Chromatogr 588:315–320, 1991.
113. GM Janini, HJ Issaq. J Liq Chromatogr 15:927–960, 1992.
114. HJ Issaq, GM Janini, IZ Atamna, GM Muschik. J Liq Chromatogr 15:1129–1136, 1992.
115. GM Janini, HJ Issaq. In: NA Guzman, ed. The Buffer in Capillary Zone Electrophoresis in Capillary

Electrophoresis: Theory, Methodology and Applications. New York: Marcel Dekker, Inc., 1993, pp. 119–160.

116. GM Janini, KC Chan, JA Barnes, GM Muschik, HJ Issaq. Chromatographia 35:497–502, 1993.
117. GM Janini, KC Chan, GM Muschik, HJ Issaq. J Liq Chromatogr 16:3591–3607, 1993.
118. HJ Issaq, GM Janini, KC Chan, ZE Rassi. In: PR Brown, E Grushka, eds. Advances in Chromatography, Vol. 35. New York: Marcel Dekker, Inc., 1995, pp. 101–170.
119. HJ Issaq, PL Horng, GM Janini, GM Muschik. J Liq Chromatogr Rel Technol 20:167–182, 1997.

29

Electrophoretic Separation–Based Sensors

Julie A. Lapos and Andrew G. Ewing

The Pennsylvania State University, University Park, Pennsylvania

I. INTRODUCTION

As micro- and nano-technology continue to develop for both capillaries and chips, an important new application in separation science is in the development of sensors. Separation-based sensors, in which sample is separated and analyzed in a continuous fashion, can be used to rapidly obtain dynamic information. Offering an advantage over traditional sensors, separation-based sensors reduce interference by separating analytes from their complex sample matrix. Furthermore, identification of unexpected analytes can be determined since these sensors are not analyte specific. Electrophoresis is an attractive method for separation-based sensors because of its high efficiency, fast separation time, small sample requirements, and widespread application to biological analysis [1]. This chapter presents an overview of the application of capillary electrophoresis (CE) and chip electrophoresis to single cell analysis aimed at developing separation-based sensors. Herein, work from our laboratory will be the primary focus, in particular, the development of electrochemical detection for CE, sampling and analyzing whole cells and release from cells, and the development of continuous analysis techniques for future application as separation-based sensors.

A. Capillary Electrophoresis

The concept of applying a potential across a tube and separating analytes, the precursor to modern high-performance CE, was first realized by Hjerten as early as 1967 [2]. He showed the feasibility of performing electrophoresis in a 3 mm coated quartz tube to separate and accurately quantitate a wide range of analytes including organics, inorganics, proteins, nucleic acids, viruses, subcellular particles, and erythrocytes [2]. However, because of the large inner diameter (i.d.), the tube had to be rotated to reduce convection problems. In 1974, Virtanen introduced smaller bore Pyrex tubes (0.2–0.5 mm), which eliminated previous convection problems and simplified the design [3]. Four years later, Mikkers et al. employed narrow-bore capillaries and demonstrated CE as an analytical tool with low plate heights and picomole-size injections [4]. Early on, it was apparent that smaller i.d. capillaries aided in heat dissipation and improved separations; however, CE was hindered by the lack of technology in manufacturing small, reproducible narrow bore capillaries. As technology progressed and smaller-sized capillaries were produced, CE faced new problems such as introducing small injection plugs onto these columns and sensitive detection in spite of the narrow path length. Jorgenson and Lukacs used laser-induced fluorescence (LIF) detection to demonstrate highly efficient separations of fluorescently derivatized amines in 1981 [5]. This pivotal work, which achieved plate heights approaching 1 μm, introduced CE as a high-performance technique comparable to high-performance liquid chromatography (HPLC) and gas chromatography (GC).

Since this time, the use of CE has exploded in analytically based separations [6,7]. This increase is due to the simplicity and nature of the separation method, which offers advantages over traditional separation techniques. CE is accomplished by suspending a capillary between two buffer reservoirs and applying an electric field. The simplicity of this system permits a variety of analytes ranging from small molecules to large proteins and DNA to be separated merely by changing the buffer composition and potential field. As the driving force for analyte migration, the applied electric field influences both electrophoretic mobility and electroosmotic flow. The electrophoretic mobility, which is governed by the analytes' charge-to-size ratio, is the basis behind the separation of ions while in the applied electric field. Because the separation mechanism is based on charge, CE offers an alternative method of separating compounds orthogonal to other separation techniques. Additionally, the electric field also generates a double layer at the capillary wall. The migration of the solvated positive charges near the capillary wall cause a net movement of solution toward the cathode. This phenomenon, referred to as electroosmotic flow, presents some inherent benefits to CE. Because of electroosmotic flow, CE has a flat flow profile that produces little longitudinal spreading of the analyte zones as they traverse the capillary. This leads to the high efficiencies commonly obtained with CE, which typically are 10^5–10^7 theoretical plates. Since there is little dispersion of the analyte bands, small sample sizes are necessary to prevent column overload. Sample volumes in the nanoliter (nL) to femtoliter (fL) range are standard in CE. These small sample volumes are beneficial for studying biological systems, such as single cells, where sample size is limited. The applied potential field also has an influential role in the performance of the separation. Unlike other separation methods, CE's separation efficiency is not length dependent, but is instead voltage dependent [5]. Thus, increasing the potential field in CE causes a twofold gain. First, the efficiency of the separation will increase barring heating effects. Second, the analysis time will decrease because compounds migrate more rapidly. The length independence allows shorter capillaries to be employed with high potential fields and reduced separation time. For example, separations in less than 1 mm have been accomplished in under 140 ms using a field strength of 2.5 kV/cm [8]. Thus, the ability to separate a variety of molecules, achieve high separation efficiencies, employ small sample sizes, and perform extremely fast separations make CE readily applicable for investigating complex, dynamic microenvironments such as single cells.

B. Electrophoresis in Microfabricated Chips

Separations in microfabricated devices were first performed using gas chromatography in 1979 [9]. However, it was not until the early 1990s that lithographically made microfabricated chips were applied to electrophoretic separations [10–12]. Electrophoresis is ideal for fluid manipulation in chips because electroosmotic flow can be employed to drive buffer through the small channel diameters. Although ceramics [13] and plastics [14–17] are receiving greater attention for chip fabrication, glass is the most common substrate utilized for electrophoresis chips due to its electrical and thermal properties. To fabricate glass separation chips, lithographic techniques are applied to transfer a channel image onto a glass plate. Etching is used to create the channel by removing micrometers of exposed glass. The etched plate is then covered by a top plate with drilled holes to allow buffer and electrode introduction. The optical properties of glass make LIF detection the method of choice; however, electrochemical detection [18,19], Raman spectroscopy [20], and mass spectrometry [21–24] have been coupled to microfabricated chips as well.

There are many attractive reasons for this shift to microfabricated separation devices, including rapid time scales and the ability to multiplex separations on a single chip. Short separation distances, although desirable for rapid separations, are disadvantageous in capillaries because sample injection and capillary manipulation is difficult. However, in microfabricated chips, these problems are negligible because the fabrication process allows very short channels to be constructed on plates that can be easily managed. An illustration of this is the 200 μm channel length fabricated on a centimeter-size chip by the Ramsey group [25]. This chip with the small separation distance has performed separations in less than 1 ms. The fabrication process also allows some inherent advantages to the chip separation including flexibility in design. For example, high sample throughput has been achieved by fabricating 96 separation channels on a single 4 inch wafer [26,27]. Additionally, designs such as cross channels, which are seemingly ubiquitous in chip-based separations as a means of sample introduction, have been more difficult to construct using traditional capillaries [28]. This ease of fabrication has been exploited with the integration of pre- [29] and postcolumn reactors [30,31] and PCR systems [32] directly on a chip.

Because of the versatility of chip fabrication, a variety of designs and applications have been demonstrated, as reviewed recently [33,34]. The ultimate goal of chip-based technology is to develop a ''lab on a chip'' or a ''micro-total analysis system'' that encompasses sample preparation, separation, and detection on a single device [35,36]. By integration of these steps onto a single chip, the error in sample handling will be reduced and these devices can be employed as separation-based sensors in the field or at the bedside. Since the mid-1980s, our laboratory has been interested in developing electrophoretic separation–based sensors. With the advances in electrophoretic separations, especially in microfabrication technology and the miniaturization of CE systems, the development of rugged, small volume, biologically compatible analytical devices for application as sensors for bedside or field diagnostics should be attainable.

II. DEVELOPMENT AND APPLICATION OF SEPARATION-BASED SENSORS

The importance of understanding exocytosis, the process by which cells communicate with each other and their environment, is a longstanding goal in the neuroscience field and is of particular interest to our group. We have approached this fundamental question using many different methods; however, of importance to the subject of this chapter are the advancements using separations. Electrophoretic separations offer advantages in quantitative and qualitative study. By developing electrophoretic separation–based sensors, information can be gained that is currently unavailable using methods such as microscopy, mass spectrometry (MS), and electrochemistry. Our laboratory has been instrumental in pushing the limits to understand the fundamentals behind exocytosis, to study both the temporal and chemical nature of this process. The following section focuses on the application of single-cell analysis techniques and the development of novel electrophoretic separation–based sensors for use in monitoring cellular dynamics.

A. CE with Electrochemical Detection

To study cellular and subcellular domains to gain further insight into cellular processes such as cell differentiation, intracellular communication, neurotransmission, and the physiological effects of external stimuli on single cells, numerous analytical methods have been developed (for review, see Refs. 1, 37–39). In order to meet the demanding goals of single-cell analysis, i.e., fL sample sizes, low attomole (amol) concentrations, and the complex matrix of the cell, it is advantageous to combine CE, which separates analytes from potential matrix interferences, with sensitive detectors. MS, fluorescence, and electrochemical detection, among other techniques, have been coupled to CE for single-cell analysis [39]. MS has successfully been used to identify components from a single cell and offers the advantage of identifying separated components based on their molecular weight and structural fragmentation [40], although the sensitivity of MS is often limited unless expensive Fourier transform mass spectrometry methods are employed. Fluorescence detection methods, including indirect [41,42] and native fluorescence detection [43], have also been employed for cell analysis using CE. To date, however, the most sensitive detectors for CE analysis are LIF with sample derivatization and electrochemical detection, which have achieved low zeptomole and subattomole detection, respectively [44]. Because our research group studies electroactive species involved in neurochemical processes, we have actively pursued electrochemical detection. Electrochemical detection is advantageous because of its low mass detection limits. Additionally, detection only of electroactive analytes causes less matrix interference, and the lack of need for prelabeling reduces dilution effects of the sample. In this subsection, the progress made in electrochemical detection coupled to CE is discussed. For additional information on other detection techniques for cell analysis see Refs. 37, 39 and 45.

Because many neurotransmitters, such as norepinephrine, epinephrine, and dopamine, are electroactive, amperometric detection via carbon fibers has been routinely utilized for single-cell analysis. Amperometric detection has been performed with capillaries having as small as a 2 μm i.d. for analysis of volume-limited samples [46]. To perform electrochemical detection with CE, it is advantageous to isolate the separation capillary from the working electrode. This isolation prevents the separation voltage from affecting the electrode and causing extraneous noise. Presently, there are three different methods of isolating the electrode from the separation potential: off-column, end-column, and optimized end-column detection.

1. Off-Column Detection

Isolation of the electrode from the separation potential was first accomplished with off-column electrochemical detection using a porous glass decoupler as shown in Fig. 1A [47]. To construct the off-column detector, the capillary is cracked near the detector end and a porous glass decoupler is used to join and stabilize the broken capillary. An electrode is placed in the detection segment of the capillary in close proximity to the crack. The glass decoupler and cracked capillary are immersed in buffer and held at ground potential. This adequately eliminates the potential field in the electrode portion of the capillary, isolating the working electrode. Electroosmotic flow generated in the separation portion of the capillary is sufficient to push the fluid and analytes past the decoupler for detection at the electrode with little loss in separation efficiency. Off-column detection using the porous glass decoupler has been used for cell analysis providing detection limits of 0.5 amol for a standard separation of neurotransmitters in a 5 μm i.d. capillary with a 2.0–2.5 μm carbon fiber electrode [46]. However, the fragility and high maintenance of the porous glass decoupler prompted other decoupler materials and designs to be developed [48–51]. Following this concept, Park et al. constructed a Nafion joint for off-column detection [50]. This joint is made by casting a Nafion membrane around two capillaries spaced 1 mm

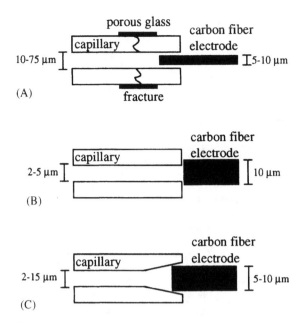

Figure 1 Electrochemical detection schemes for CE. (A) Off-column with porous glass decoupler, (B) end-column, and (C) optimized end-column electrochemical detection. (Adapted from Ref. 37.)

apart. A wire is inserted through both capillaries to aid in alignment and is removed after the Nafion joint is cast. In addition to its easy construction, the large gap between the capillaries permits a better isolation of the electric field. Detection limits of 5×10^{-10} M for hydroquinone have been achieved with this Nafion decoupler [50].

2. End-Column Detection

The use of decouplers was eliminated with a more simplified electrochemical detection technique for CE, end-column detection. In end-column detection (Fig. 1B), an electrode is positioned against the bore of the capillary [52,53]. Due to the high resistance of the small i.d. capillary, virtually all of the potential is dropped across the capillary and therefore no further isolation of the electrode is needed. Detection limits as low as 56 amol for catechol in a 5 μm-i.d. capillary with a 10 μm carbon fiber electrode have been achieved [52]. However, because detection is extremely sensitive to electrode and capillary bore alignment, this method suffers from imprecision.

3. Optimized End-Column Detection

The alignment of the electrode and capillary was improved by using an optimized end-column detection design. In this scheme (Fig. 1C) the detection end of the capillary is etched with hydrofluoric acid to achieve a conical entrance at the capillary end [54]. A carbon fiber electrode is then manipulated into the larger capillary bore. This method maintains the alignment of the electrode and capillary and permits larger electrodes to be employed. Utilizing a 2 μm i.d. capillary with a 10 μm carbon fiber electrode, a detection limit of 11 amol for catechol has been achieved [54]. This technique improves reproducibility and separation efficiency and has lower limits of detection than the previous end-column detection method. The optimized detection method combines the ease of construction of the end-column technique with improved electrode and capillary alignment, yielding detection limits similar to off-column electrochemical detection. More importantly, this method enables small capillaries to be routinely employed due to relative ease of electrode placement, which is an important advantage when studying microenvironments such as single cells.

B. Sampling Techniques for Single Cells

The function of individual cells is dependent on the identity and location of compounds within each cell. Therefore, the ability to sample from whole cells and from portions of cells is essential in understanding cellular processes. CE is ideal for single cell and subcellular analysis due to its high sensitivity and low sample volume requirements. The small i.d. capillaries are advantageous in CE to improve heat dissipation and resolution. In addition, the small size is beneficial for performing low-volume sampling because the capillary can be employed as a microinjector to sample minute quantities from whole cell and subcellular domains.

Much information can be attained when sampling whole cells, particularly when the cell line is heterogeneous, which can lead to a better understanding of cellular function as a whole. The first cell sample injection for CE was performed by sampling a cell homogenate [55]. This single-cell analysis is performed by homogenizing an invertebrate neuron and centrifuging. The supernatant is transferred to another vial, derivatizing agent is added, and the solution is then injected onto the separation capillary. This method of sampling allows a straightforward means for derivatization and sample clean-up since less cellular debris is introduced onto the capillary. Furthermore, since small sample amounts (1–100 nL) are injected, multiple injections can be drawn from a single homogenate. Soon after this initial work with cell homogenates, an intact cell was introduced directly onto a capillary to increase the mass detection limits for the analysis [56]. In order to adequately maneuver entire cells into the capillary tip and retain the cell integrity, the injection end of the capillary is etched with hydrofluoric acid to a size comparable to that of the cell as shown in Fig. 2A. The capillary is placed against the cell, and electroosmotic flow is used to inject the cell into the capillary. The cell is immobilized on the capillary wall, and thus, there is little diffusion of the cell within the capillary. Cell lysis occurs in the capillary when nonphysiological buffer is introduced, osmotically shocking the cell. Sampling the entire cell enables improved detection limits for low amounts of neurotransmitters since the entire cell is sampled. Similar work with whole mammalian cells has been accomplished by the Yeung group with erythrocytes [41]. However, instead of electrokinetically injecting the whole erythrocyte onto a capillary, a vacuum injection technique is employed. A vacuum is applied to the opposite end of the capillary to permit control of the amount of media introduced onto the capillary with the cell, and the amount of injection is easily manipulated by adjusting the vacuum. Following this work, the Ewing group introduced other mammalian cells including human lymphocytes [57], rat pheochromocytoma cells [58], as well as red blood cells [40] directly into a capillary electrokinetically for CE analysis. More recently, Dovichi's group applied chemical cytometry to whole cell sampling [59]. In this technique, a light-transparent capillary holder is mounted on a microscope allowing the capillary to be positioned vertically over the cell. Tapering of the capillary injection end is not necessary due to the close proximity of the capillary bore to the cell. This method has the advantages of visualizing the cell and capillary position during the injection process and precise controlling of the volume of media introduced with the cell.

Another sampling method for single-cell analysis is subcellular sampling. As noted by studies sampling populations of intact cells, there is a vast heterogeneity between individual cells of a specific line [43,60]. This information can lead to insights into how this heterogeneity affects the function of the cellular system within the organ as a whole. However, due to the

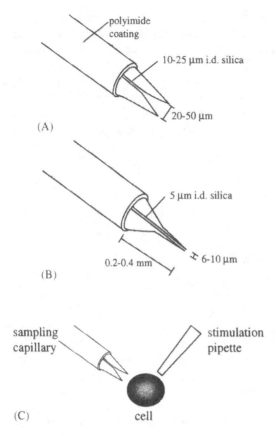

Figure 2 Capillary tip designs used for single cell analysis. (A) Tip employed for injecting whole cells into the capillary. (B) Tip employed for introducing cytoplasm onto the capillary for subsequent analysis. (C) Schematic representation of the capillary and cell configuration for extracellular sampling where the sampling capillary and stimulation pipette are in close proximity to the cell.

heterogeneity, information on sampling between control and treated cells of a given population may be misleading on a single-cell level. Thus, repeated subcellular sampling from an individual cell is more conducive to pharmacological studies in which an untreated cell could act as its own control. In addition, subcellular sampling permits specific components in a cell to be studied by sampling only a portion of the cell. By introducing an etched capillary through the cell membrane into the cell, cytoplasmic sampling of single cells has been performed [61]. To accomplish capillary insertion into a cell, the outside of the capillary tip is etched in hydrofluoric acid maintaining the i.d.; however, the outer diameter (o.d.) of the capillary is etched to as small as 6 μm with a 5 μm i.d. capillary (Fig. 2B) [46,61]. When employing a 5 μm i.d. capillary, only 0.9% of the cytoplasm has been removed from the *Planorbis corneus* cell [46]. However, to achieve cytoplasmic sampling of smaller cells such as mammalian cells, even smaller capillary i.d.s are necessary to maintain low disturbance of the cell. Currently, the application of nano-meter-diameter capillaries with amperometric detection to neurochemical analysis is underway for even smaller volume sampling [62].

    In addition to investigating the internal composition of the cell, much beneficial information about the environment surrounding a cell can be provided by sampling the extracellular fluid.

The chemical composition of the extracellular fluid surrounding a cell can influence its functions, including processes such as exocytosis or apoptosis. Exocytosis has recently been studied using small electrochemical probes due to the rapid sampling that can occur when the electrode is placed directly on a cell. Separation methods applied to exocytosis offer a unique advantage in that multiple components of release can be separated and detected. A schematic of extracellular sampling for CE analysis is shown in Fig. 2C. Both a capillary and a stimulating pipette are placed next to the cell. The pipette is used to deliver secretagogue near the cell to elicit exocytosis. The capillary then electrokinetically samples the extracellular fluid and separation follows. This technique was employed by Chen et al. to sample release 4 seconds after stimulation from a single invertebrate cell [63]. A modification of extracellular sampling has been performed by Lillard et al. where a mammalian cell is immobilized on the capillary wall [64]. Stimulant is introduced onto the capillary and allowed to react with the cell. The fluid is then separated, permitting quantification of exocytosis from mast cells. As demonstrated by the aforementioned sampling techniques, there are many different approaches to perform cell sampling. By sampling whole cells or subcellular components, such as cytoplasm or extracellular fluid, unique information about the cell environment can be obtained.

C. Single-Cell Separations and Analysis

Early applications of single-cell analysis using CE involved invertebrate neurons from *P. corneus* [61,65] and *Helix aspersa* [55] due to the large size of the cells and ease of handling. This initial work in the late 1980s prompted increased attention in applying CE to single cells to gain understanding in cellular biochemistry. The sensitivity, high resolution, and small sampling size of CE make it an idea method for single-cell analysis. Single-cell electrophoretic separations have spanned from large cell systems, such as invertebrate cells, to small mammalian cells and subcellular components. In the following sections, various examples of initial single-cell work showing the applicability of CE to analyze both whole and subcellular components of single cells are outlined.

1. Whole Invertebrate Cell

Analysis of a single neuron from *P. corneus* with CE coupled to off-column electrochemical detection has been used to provide evidence supporting a two-compartment model of neurotransmitters in a nerve cell [66,67]. Using a 25 μm i.d. etched capillary, electroosmotic flow was used to inject the dopamine neuron from *P. corneus* into the capillary followed by lysis with nonphysiological buffer [66]. Depending on the lysing time, a second cationic peak appeared in the electropherogram (Fig. 3) that did not correspond to any known electroactive species present in the cell. By increasing the lysis time of the cell in the capillary, the second peak disappeared while the first peak increased in size. The results suggested two compartments of dopamine storage in the cell. One compartment that was easily lysed contained dopamine available for immediate release. This corresponded to the first dopamine peak. A second more internalized compartment for long-term storage did not readily lyse and thus explained the delayed second dopamine peak. Upon treatment of these cells with reserpine, a pharmacological agent known to deplete vesicular stores of dopamine, only the first peak appeared, further supporting the two-compartment model. Additional verification of both the first and second peak as dopamine was confirmed using scanning electrochemical detection, which has the ability to identify analytes based on oxidation potential [67]. Furthermore, the use of scanning electrochemical detection permitted quantitation of the dopamine yielding similar values for dopamine concentration of the two peaks combined for a 1-minute lyse and for the lone peak when a 10-minute

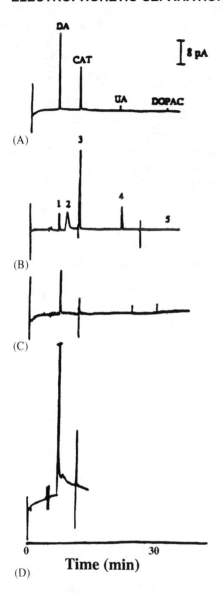

Figure 3 Electropherograms of a whole cell analysis from the dopamine cell of *P. corneus* illustrating evidence for two compartments of dopamine storage. (A) Separation of a standard solution of dopamine (DA), catechol (CAT), uric acid (UA), and dihydroxyphenylacetic acid (DOPAC). (B) Separation of components from the dopamine cell following lysis for 1 minute. Peaks 1 and 2 correspond to dopamine. (C) Separation of components from the dopamine cell following lysis for 5 minutes where only the first dopamine peak is observed. (D) Separation of components from the dopamine cell incubated with reserpine prior to injection followed by a 1-minute lysis. (From Ref. 66.)

lyse was performed [67]. Thus, this technique has proven its ability to obtain subcellular information through the sampling of whole cells.

2.   Cytoplasmic Sampling of an Invertebrate Cell

Subcellular information about the chemical composition of a cell can also be obtained by direct sampling of the cytoplasm. Sampling of cytoplasm was first demonstrated in 1988 by Wallingford et al. [65]. Although no peaks were identified, the ability to insert an etched capillary through the cell membrane and inject sample was an important first step in cytoplasmic sampling. Later work in the Ewing group on serotonin and dopamine neurons in *P. corneus* identified and quantitated the neurotransmitter amounts in each cell [46,61]. Cytoplasmic sampling was accomplished by inserting a 5 μm i.d. capillary etched down to a 6 μm o.d. into the cell. Potential was applied to the capillary to inject approximately 50 pL of cytoplasm using electroosmotic flow. After separating the cytoplasmic sample, electrochemical detection was employed due to its ease in coupling to the small i.d. capillary and for detection of low levels of analytes such as those found in cytoplasm. Figure 4 compares the separations of cytoplasmic contents with standards, identifying the presence of serotonin, catechol, and a neutral in the cytoplasm of the serotonin cell. A concentration of 3.1 μM for serotonin in the cytosol of the serotonin cell in *P. corneus* has been determined [46]. Additionally, cytoplasmic sampling of the identified dopamine cell from *P. corneus* has been analyzed, and dopamine was found to have an average

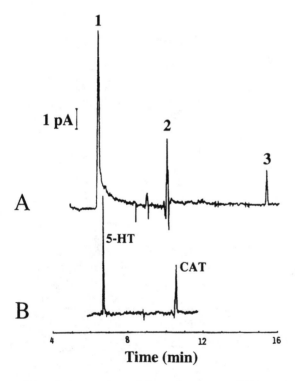

Figure 4   Comparison of electropherograms of (A) a cytoplasmic sample from the serotonin neuron in *P. corneus* to (B) a standard solution of serotonin (5-HT) and catechol. Peaks 1 and 3 in the top electropherogram correspond to the calculated mobilities of serotonin and an unidentified anion, respectively. (Adapted from Ref. 46.)

concentration of 2.2 µM in the cytoplasm [61]. The small capillary i.d. permits very small sample volumes to be introduced onto the capillary, which is beneficial when only minute sample amounts are available or when there is a need to refrain from disturbing the microenvironment too drastically.

3. Whole Mammalian Cell

Although invertebrate analysis provides insight into basic cellular neurochemistry, the analysis of mammalian cells provides a more direct model of human neuronal function. However, sampling from mammalian cells poses a slightly harder problem than invertebrate cells because of the smaller volume of the cell (~80 fL). A plethora of mammalian cells have been investigated with CE, including blood cells, neurons, adrenal medullary cells, and mast cells. Examples of whole mammalian cell analysis using CE are briefly mentioned here.

Single-cell CE separations of mammalian cells were first accomplished with the analysis of erythrocytes by the Yeung group [41]. Single erythrocytes were pulled onto a 20 µm i.d. capillary using a vacuum. Using indirect and direct LIF detection, glutathione and small ions were detected in individual cells [41]. Later, LIF was improved to increase the sensitivity of the detection method and proteins within single erythrocytes were observed and quantitated [43]. These results from a population of cells were shown to have a wide variation in cellular contents, which may be indicative of the vast heterogeneity among the cells or, more likely, due to age differences of the erythrocytes.

Individual lymphocytes have also been analyzed with CE, leading to the discovery of endogenous catecholamines [57]. Berquist et al. introduced a lymphocyte (white blood cell) onto a 10 µm i.d. capillary, lysed the cell, and then used optimized end-column electrochemical detection to discover the presence of catecholamines. Previously, catecholamines were not known to be present in lymphocytes. This information has led to the hypothesis that catecholamines may provide a means of communication between the nervous and immune systems and may explain the immunological privilege of the brain [57,68].

Pheochromocytoma cells, an immortalized cell line commonly used as a neuronal model, have been analyzed with CE coupled to LIF detection. To reduce problems associated with derivatization prior to cell sampling, such as incubation of the cell with a fluorophore or homogenization followed by derivatization of the cell in a nanovial, our group developed on-column fluorescence derivatization [58]. On-column derivatization was accomplished by first injecting the cell into the capillary, then introducing lysing and derivatizing solutions. The capillary was utilized as a subnanosize reaction chamber for the derivatization process, and dilution of the cell contents was minimized to subnanoliter volumes. From this, quantitative analysis of whole rat pheochromocytoma cells was obtained. Internal standards were added to the cell culture media to correct for analytes in the media and variability in the volume of cell suspension fluid injected. Using on-column derivatization with LIF detection, efficient separations and sensitive detection were achieved for analysis of a single pheochromocytoma cell. The results demonstrated the technique's applicability to the sensitive analysis of analytes from a single cell.

The identification of analytes based solely on migration times in fluorescence and amperometric detection can lead to problems in accurate peak determination. To provide additional assistance in identifying components from a cell separation, CE has been coupled to electrospray ionization Fourier transform ion cyclotron resonance MS for the analysis of human erythrocytes [40,69]. By employing MS, a better fingerprint of the analyte can be obtained based on the fragmentation patterns and molecular sizes. Briefly, a whole erythrocyte was introduced onto a 20 µm i.d. capillary using electroosmotic flow. Osmotic shock lysed the cell membrane at the capillary end and the components were separated. Eluent was electrosprayed into a Fourier transform ion cyclotron resonance cavity for analyte detection. With this technique the α and β subunits of hemoglobin were observed and approximately 450 amol of hemoglobin was detected [40]. This initial work demonstrated that MS could be coupled to CE for single-cell analysis to provide sensitive detection and enhance molecular identification.

4. Extracellular Sampling at an Invertebrate Cell

In addition to sampling the cytoplasm for subcellular information, sampling of fluid around the cell has provided information regarding exocytosis. Extracellular samples of a single neuron in *P. corneus* were analyzed using CE and electrochemical detection following stimulation [63]. A micropipette containing the stimulating solution was placed next to the dopamine neuron along with a capillary. Following a 6-second stimulation, fluid surrounding the cell was electrokinetically injected into the separation capillary for 30 seconds. Figure 5 shows the electropherograms of the extracellular fluid sample and standards. Only two major peaks were present in the electropherogram. These peaks were identified as dopamine and a neutral based on electrophoretic mobility. Although this technique was able to separate and identify one of the secreted

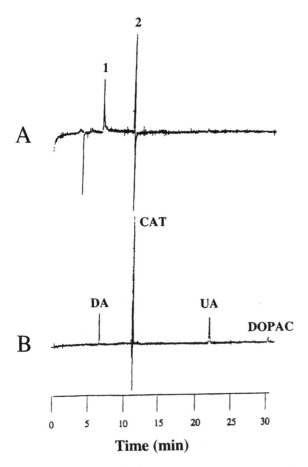

Figure 5 Electropherograms of (A) an extracellular sample taken from the dopamine cell in *P. corneus* following stimulation and (B) a standard solution of dopamine, catechol, uric acid, and DOPAC. Peaks 1 and 2 correspond to dopamine and a neutral compound, respectively. (Adapted from Ref. 63.)

compounds as dopamine, it was unable to provide any dynamic information because of its incremental sampling and separation nature.

The disadvantage of using traditional CE for studying dynamic processes such as exocytosis is that it is a serial technique. This serial nature poses a hindrance when monitoring rapid chemical changes over time since information is gathered in discrete sampling intervals. This obstacle has prompted the development of electrophoretic separation techniques that can be used for continuously monitoring rapid changes. These dynamic changes can be monitored with sensors, which provide low-volume, continuous sampling through rapid parallel or continuous separations and a detection scheme that can maintain the spatial and temporal integrity of the separation. Developments toward these separation-based sensors for cell analysis are discussed in the following sections.

D. Open Microchannels

In an effort to attain separation-based sensors, our laboratory has developed continuous separations in open microchannels [70,71]. The dynamic system consists of a sampling capillary interfaced with a rectangular channel (Fig. 6). The capillary is used to transfer sample from the place of interest and introduce the sample continuously onto the channel by sweeping across the channel entrance. The channels are constructed using two glass plates separated by spacers ranging from 109 to 0.6 μm in height. Potential is applied across both the capillary for electrokinetic sample introduction and the rectangular channel for separation. Both fluorescent [72,73] and electrochemical detection [18] have been employed to preserve the spatial integrity of the separation. Limits of detection for these detection schemes coupled to the open microchannels are in the fmol range.

Due to the continuous sampling and separation nature, open microchannel separations are amenable to monitoring dynamic processes. The ability of this system to monitor such events was first simulated with a flow injection system shown in Fig. 7A [74]. Three amino acids were alternately added to the flowing stream of buffer. The stream was continuously sampled by the capillary electrokinetically. The capillary was scanned across the channel entrance, and sample was deposited in the open channel. The results of this dynamic simulation are shown in Fig. 7B, where the three labeled amino acids are clearly resolved [74]. The capillary/channel system has also been applied by Sweedler's group to study reaction kinetics [73]. Using this method, derivatization of amino acids with a fluorophore was conducted and product formation was observed. From this data, the reaction rates for the derivatization process were calculated and compared to known values [73]. Both of these experiments, the flow injection system and the study of reaction kinetics, have demonstrated the applicability of continuous capillary sampling with open microchannels to monitor multicomponent, dynamic changes from small environments.

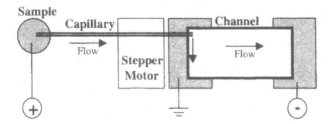

Figure 6 Microchannel electrophoretic separation system with capillary sample introduction.

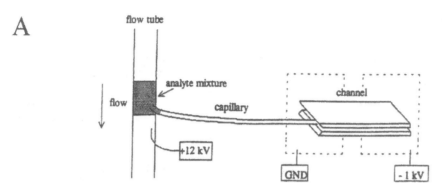

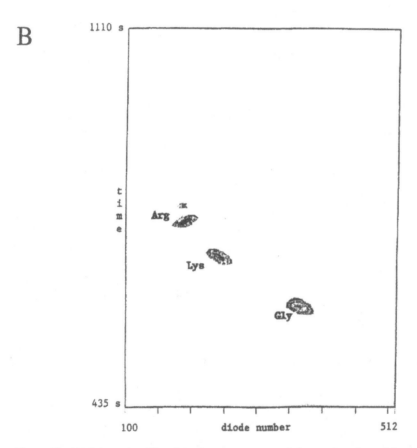

Figure 7   (A) Schematic of flow injection apparatus coupled to a microchannel for dynamic monitoring. (B) Electropherogram obtained by the staggered addition of dansylated arginine, lycine, and glycine through the flow tube, which were sampled with the capillary, separated on the channel, and analyzed with LIF detection. (From Ref. 74.)

After realizing the high temporal resolution and excellent chemical sensitivity of this technique to monitor real samples, the dynamic analysis of cellular systems has been explored. Experiments with open microchannels for cell analysis were first conducted on adrenal medullary cells [75]. The cells were stimulated and the extracellular fluid was continuously sampled with the capillary following exocytosis. The sample was deposited continuously on the channel, separated, and analyzed using end-column electrochemical detection. The results showed an increase in signal intensity ~630 s after stimulation lasting for a duration of ~40 s. Although a single peak was observed, the identity was unknown. More recently, a separation of single cell release was achieved by Liu et al. from an *Aplysia california* neuron employing a modification of this technique [76]. The cell was isolated at the outlet of the sampling capillary and held in the capillary tip behind a porous frit. Stimulation and inhibition solutions were passed through the capillary washing over the cell. The fritted capillary outlet was scanned across the end of the channel introducing sample continuously. Dynamic monitoring following perturbation revealed differences in temporal events. In addition, the presence of a multitude of peaks over a 180 s time span was observed for stimulated cell release, although there was difficulty in assigning peak identities. These examples of dynamic analysis show the potential of open microchannels to continuously monitor changing cellular events such as exocytosis, but improvements in the performance of the system are necessary.

Improvements in the microchannel system are currently being conducted to make this technique sensitive and more conducive for single-cell analysis. First, postchannel derivatization, which avoids double labeling and improves detection limits, has been developed for open microchannels for future application to cell analysis [77]. Second, ultra-small submicrometer channel heights (0.6 μm) have been employed to facilitate small sample volumes and improve heat dissipation and resolution capabilities [18]. Third, a microfabricated chip with an array of channels has been used in place of the open microchannel to improve lateral resolution [78]. Lateral dispersion has been a concern in the open microchannels and investigations as to the origin of the dispersion including field effects and capillary-channel size mismatch have produced inconclusive results [79]. The improvements mentioned—postcolumn derivatization, smaller channel heights, and microfabricated channel arrays—are expected to improve the ability of open microchannels with capillary sample introduction as a separations-based sensor for dynamic, single-cell monitoring.

E. Optically Gated Electrophoresis on a Chip

As an approach to separation-based sensors, our group has also been developing optically gated electrophoresis to continuously, rapidly sample and separate analytes on microfabricated chips. Optically gated electrophoresis was introduced in 1991 by Monnig and Jorgenson as a method to perform reproducible narrow injections for fast CZE analysis in capillaries [80]. Because optical gating in capillaries has been used to monitor rapid processes [81–85], our laboratory has applied this technique to chips [86]. This allows the advantages of chips such as cell compatibility and multiplexing ability to be combined with the advantages of optical gating, including narrow injection plugs and rapid sampling.

Chip-based optically gated electrophoresis is comprised of a high-powered laser, an electrophoresis system, and a microfabricated chip (Fig. 8). The glass chip, in which the separation occurs, possesses a single etched lane 10 to 50 μm in depth. Fluorescently labeled sample is continuously run through the separation channel via an electrophoretic potential. A laser beam is split into a gating and a detection beam. The gating beam, located near the entrance to the channel is used to continuously photobleach the sample. Injection is performed by briefly

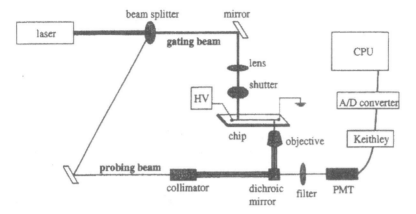

Figure 8   Diagram of system used for optically gated electrophoresis on a chip. (From Ref. 86.)

blocking the gating beam, allowing a small plug of unbleached sample through. The unbleached sample is then separated along the channel and the probe beam is utilized for LIF detection.

Initial experiments with optically gated electrophoresis on a chip have shown the ability of this system to perform reproducible, efficient separations [86]. In addition, high throughput is achieved due to the rapid sampling and fast separations performed serially. Serial injections of fluorescently labeled amino acids are shown in Fig. 9, where six separations have been easily accomplished in 30 s. Faster analysis time is necessary for optical gating on a chip to be used in studying cellular systems. This should be attainable considering optical gating in capillaries has shown separations as rapid as 140 ms, which is within the realm of some cellular dynamics, such as exocytosis [8]. By performing rapid consecutive separations, optically gated electrophoresis on a chip can be applied as a tool for dynamic cell study. As a separations-based sensor this technique can be utilized in monitoring cellular release following stimulation to gain both temporal and chemical information.

III. SUMMARY

CE has evolved throughout the last half of the twentieth century from its infancy, involving large tubes and complex instrumentation. With the help of technology, the field has progressed, implementing micrometer and sub micrometer diameter capillaries and microfabricated chips to improve the performance and simplicity of CE. The advantages of this technique have offered high-performance CE a niche in the analytical field for monitoring small environments. The high efficiency and small sample requirements of CE have facilitated the use of this technique to cellular sampling, leading to insights in cell biochemistry and new hypotheses about cellular functions. The use of electrophoresis with dynamic separation methods, such as open channel electrophoresis and optical gating on a chip, will enable further insight into the dynamics of cellular processes. As sensors, these techniques will permit dynamic monitoring as sensitive, rugged devices requiring low sample amounts.

The concept of separation-based sensors is best accomplished by techniques that provide a rapid temporal response with high sensitivity and low sample volumes. Electrophoresis in capillaries and chips provides the best approach to this goal. As temporal resolution is pushed to the millisecond domain and sensitivity to the zmol level, the separation-based sensor approach

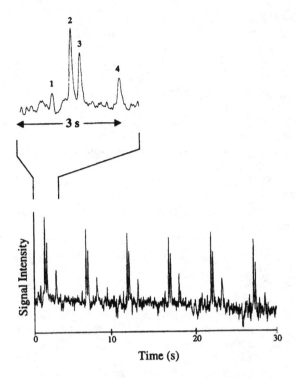

Figure 9 Six serial separations of 4-chloro-7-nitrobenzofurazan labeled amino acids using optically gated fluorescence injection on a chip. The inset shows a 3-second separation of amino acids. The amino acids were (1) arginine, (2) phenylalanine, (3) glycine, and (4) glutamic acid. (From Ref. 86.)

will become a valuable tool in our attempts to understand single-cell neurochemistry. Furthermore, this area of separations promises to play a significant role in diagnostics and nanoscience in the twenty-first century.

ACKNOWLEDGMENTS

The work cited in this chapter was funded, in part, by the National Science Foundation and the National Institutes of Health. We gratefully acknowledge our coworkers, past and present, for the work cited in this chapter.

REFERENCES

1. JA Jankowski, S Tracht, JV Sweedler. Assaying single cells with capillary electrophoresis. Trends Anal Chem 14:170–176, 1995.
2. S Hjerten. Free zone electrophoresis. Chromatogr Rev 9:122–219, 1967.
3. R Virtanen. Zone electrophoresis in a narrow-bore tube employing potentiometric detection. Acta Polytechnica Scand 123:1–67, 1974.
4. FEP Mikkers, FM Everaerts, TPEM Verheggen. High performance zone electrophoresis. J Chromatogr 169:11–20, 1979.
5. J Jorgenson, KD Lukacs. Zone electrophoresis in open-tubular glass capillaries. Anal Chem 53:1298–1302, 1981.

6. SC Beale. Capillary electrophoresis. Anal Chem 70:279R–300R, 1998.

7. RT Kennedy, I German, JE Thompson, SR Witowski. Fast analytical-scale separations by CE and LC. Chem Rev 99:3081–3131, 1999.

8. AW Moore, Jr., JW Jorgenson. Study of zone broadening in optically gated high-speed capillary electrophoresis. Anal Chem 65:3550–3560, 1993.

9. DC Terry, JH Herman, JB Angell. A gas chromatographic air analyzer fabricated on a silicon wafer. IEEE Trans Elec Dev 26:1880–1886, 1979.

10. DJ Harrison, A Manz, Z Fan, H Ludi, HM Widmer. Capillary electrophoresis and sample injection systems integrated on a planar glass chip. Anal Chem 64:1926–1932, 1992.

11. K Seiler, JD Harrison, M Manz. Planar glass chips for capillary electrophoresis: Repetitive sample injection, quantitation, and separation efficiency. Anal Chem 65:1481–1488, 1993.

12. SC Jacobson, R Hergenroder, LB Koutny, JM Ramsey. Open channel electrochromatography on a microchip. Anal Chem 33:2369–2373, 1994.

13. CS Henry, M Zhong, SM Lunte, M Kim, H Bau, JJ Santiago. Ceramic microchips for capillary electrophoresis-electrochemistry. Anal Commun 36:305–307, 1999.

14. CS Effenhauser, GJM Bruin, A Paulus, M Ehrat. Integrated capillary electrophoresis on flexible silicone microdevices: analysis of DNA restriction fragments and detection of single DNA molecules on microchips. Anal Chem 69:3451–3457, 1997.

15. L Martynova, LE Locascio, M Gaitan, GW Kramer, RG Christensen, WA MacCrehan. Fabrication of plastic microfluid channels by imprinting methods. Anal Chem 69:4783–4789, 1997.

16. RM McCormick, RJ Nelson, MG Alonso-Amigo, DJ Benvegnu, HH Hopper. Microchannel electrophoretic separations of DNA in injection-molded plastic substrates. Anal Chem 69:2626–2630, 1997.

17. MA Roberts, JS Rossier, H Girault. UV laser machined polymer substrates for the development of microdiagnostic systems. Anal Chem 69:2035–2042, 1997.

18. PF Gavin, AG Ewing. Characterization of electrochemical array detection for continuous channel electrophoretic separations in micrometer and submicrometer channels. Anal Chem 69:3838–3845, 1996.

19. AT Woolley, K Lao, AN Glazer, RA Mathies. Capillary electrophoresis chips with integrated electrochemical detection. Anal Chem 70:684–688, 1998.

20. IPA Waker, MD Morris, MA Burns, BN Johnson. Isotachophoretic separations on a microchip. Normal Raman spectroscopy. Anal Chem 70:3766–3769, 1998.

21. Q Xue, F Foret, Y Dunayevskiy, PM Zavracky, NE McGruer, BL Karger. Multichannel microchip electrospray mass spectrometry. Anal Chem 69:426–430, 1997.

22. RS Ramsey, JM Ramsey. Generating electrospray from microchip devices using electroosmotic pumping. Anal Chem 69:1174–1178, 1997.

23. D Figeys, Y Ning, R Aebersold. A microfabricated device for rapid protein identification by microelectrospray ion trap mass spectrometry. Anal Chem 69:3153–3160, 1997.

24. N Xu, Y Lin, SA Hofstadler, D Matson, CJ Call, RD Smith. A microfabricated dialysis device for sample cleanup in electrospray ionization mass spectrometry. Anal Chem 70:3553–3556, 1998.

25. SC Jacobson, CT Culbertson, JE Daler, JM Ramsey. Microchip structures for submillisecond electrophoresis. Anal Chem 70:3476–3480, 1998.

26. PC Simpson, D Roach, AT Woolley, T Thorsen, R Johnston, GF Sensabaugh, RA Mathies. High-throughput genetic analysis using microfabricated 96-sample capillary array electrophoresis microplates. Proc Natl Acad Sci 95:2256–2261, 1998.

27. Y Shi, PC Simpson, JR Scherer, D Wexler, C Skibola, MT Smith, RA Mathies. Radial capillary array electrophoresis microplate and scanner for high-performance nucleic acid analysis. Anal Chem 71:5354–5361, 1999.

28. SL Pentoney, X Huang, DS Burgi, RN Zare. On-line connector for microcolumns: Application to the on-column o-phthaldialdehyde derivatization of amino acids separated by capillary zone electrophoresis. Anal Chem 60:2625–2629, 1988.

29. SC Jacobson, R Hergenroder, AW Moore, Jr., JM Ramsey. Precolumn reactions with electrophoretic analysis integrated on a microchip. Anal Chem 66:4127–4132, 1994.

30. SC Jacobson, LB Koutny, R Hergenroder, AW Moore, Jr., JM Ramsey. Microchip capillary electrophoresis with an integrated postcolumn reactor. Anal Chem 66:3472–3476, 1994.

31. K Fluri, G Fitzpatrick, N Chiem, DJ Harrison. Integrated capillary electrophoresis devices with an efficient postcolumn reactor in planar quartz and glass chips. Anal Chem 68:4285–4290, 1996.

32. AT Woolley, D Hadley, P Landre, AJ deMello, RA Mathies, MA Northrup. Functional integration of PCR amplification and capillary electrophoresis in a microfabricated DNA analysis device. Anal Chem 68:4081–4086, 1996.

33. V Dolnik, S Liu, S Jovanovich. Capillary electrophoresis on microchip. Electrophoresis 21:41–54, 2000.

34. CS Effenhauser, GJM Bruin, A Paulus. Integrated chip-based capillary electrophoresis. Electrophoresis 18:2203–2213, 1997.

35. A Manz, DJ Harrison, EMJ Verpoorte, JC Fettinger, A Paulus, H Ludi, HM Widmer. Planar chips technology for miniaturization and integration of separation techniques into monitoring systems. J Chromatogr 593:253–258, 1992.

36. A Manz, E Verpoorte, CS Effenhauser, N Burggraf, D Raymond, DJ Harrison, HM Widmer. Miniaturization of separation techniques using planar chip technology. J High Resol Chromatogr 16:433–436, 1993.

37. DM Cannon Jr., N Winograd, AG Ewing. Quantitative chemical analysis of single cells. Annu Rev Biophys Biomol Struct 29:239–263, 2000.

38. G Chen, AG Ewing. Chemical analysis of single cells and exocytosis. Crit Rev Neurobiol 11:59–90, 1997.

39. FD Swanek, SS Ferris, AG Ewing. Capillary electrophoresis for the analysis of single cells: Electrochemical, mass spectrometric and radiochemical detection. In: JP Landers, ed. Handbook of Capillary Electrophoresis, 2nd ed. New York: CRC Press, 1997, pp. 495–521.

40. SA Hofstadler, JC Severs, RD Smith, FD Swanek, AG Ewing. Analysis of single cells with capillary electrophoresis electrospray ionization Fourier transform ion cyclotron resonance mass spectrometry. Rapid Commun Mass Spectrom 10:919–922, 1996.

41. BL Hogan, ES Yeung. Determination of intracellular species at the level of a single erythrocyte via capillary electrophoresis with direct and indirect fluorescence detection. Anal Chem 64:2841–2845, 1992.

42. Q Xue, ES Yeung. Indirect fluorescence determination of lactate and pyruvate in single erythrocytes by capillary electrophoresis. J Chromatogr A 661:287–295, 1994.

43. TT Lee, ES Yeung. Quantitative determination of native proteins in individual human erythrocytes by capillary zone electrophoresis with laser-induced fluorescence detection. Anal Chem 64:3045–3051, 1992.

44. AG Ewing, RA Wallingford, TM Olefirowicz. Capillary electrophoresis. Anal Chem 61:292A–300A, 1989.

45. SJ Lillard, ES Yeung. Capillary electrophoresis for the analysis of single cells: Laser-induced fluorescence detection. In: JP Landers, ed. Handbook of Capillary Electrophoresis, 2nd ed. New York: CRC Press, 1997, pp. 523–544.

46. TM Olefirowicz, AG Ewing Capillary electrophoresis in 2 and 5 µm diameter capillaries: application to cytoplasmic analysis. Anal Chem 62:1872–1876, 1990.

47. RA Wallingford, AG Ewing. Capillary zone electrophoresis with electrochemical detection. Anal Chem 59:1762–1766, 1987.

48. YF Yik, HK Lee, SYO Li, SB Khoo. Micellar electrokinetic capillary chromatography of vitamin B_6 with electrochemical detection. J Chromatogr 585:139–144, 1991.

49. TG O'Shea, RD Greenhagen, SM Lunte, CE Lunte, MR Smyth, DM Radzik, NJ Watanable. Capillary electrophoresis with electrochemical detection employing an on-line Nafion joint. J Chromatogr 593:305–312, 1992.

50. S Park, SM Lunte, CE Lunte. A perfluorosulfonated ionomer joint for capillary electrophoresis with on-column electrochemical detection. Anal Chem 67:911–918, 1995.

51. W Lu, RM Cassidy. Background noise in capillary electrophoretic amperometric detection. Anal Chem 66:200–204, 1994.

52. X Huang, RN Zare, S Sloss, AG Ewing. End-column detection for capillary zone electrophoresis. Anal Chem 63:189–192, 1991.

53. C Haber, I Silvestri, S Roosli, W Simon. Potentiometric detector for capillary zone electrophoresis. Chimia 45:117–121, 1991.

54. S Sloss, AG Ewing. Improved method for end-column amperometric detection for capillary electrophoresis. Anal Chem 65:577–581, 1993.

55. RT Kennedy, MD Oates, BR Cooper, B Nickerson, JW Jorgenson. Microcolumn separations and the analysis of single cells. Science 246:57–63, 1989.

56. TM Olefirowicz, AG Ewing. Capillary electrophoresis for sampling single nerve cells. Chimia 45:106–108, 1991.

57. J Bergquist, A Tarkowski, R Ekman, A Ewing. Discovery of endogenous catecholamines in lymphocytes and evidence for catecholamine regulation of lymphocyte function via an autocrine loop. Proc Natl Acad Sci USA 91:12912–12916, 1994.

58. SD Gilman, AG Ewing. Analysis of single cells by capillary electrophoresis with on-column derivatization and laser-induced fluorescence detection. Anal Chem 67:58–64, 1995.

59. SN Krylov, DA Starke, EA Arriaga, Z Zhang, NW Chan, MM Palcic, NJ Dovichi. Instrumentation for chemical cytometry. Anal Chem 72:872–877, 2000.

60. BR Cooper, JA Jankowski, DJ Leszczyszyn, RM Wightman, JW Jorgenson. Quantitative determination of catecholamines in individual bovine adrenomedullary cells by reverse-phase microcolumn liquid chromatography with electrochemical detection. Anal Chem 64:691–694, 1992.

61. TM Olefirowicz, AG Ewing. Dopamine concentration in the cytoplasmic compartment of single neurons determined by capillary electrophoresis. J Neurosci Meth 34:11–15, 1990.

62. LA Woods, AG Ewing. Capillary electrophoresis in sub-micron capillaries. Anal Chem (in press).

63. G Chen, PF Gavin, G Luo, AG Ewing. Observation and quantitation of exocytosis from the cell body of a fully developed neuron in *Planorbis corneus*. J Neurosci 15:7747–7755, 1995.

64. SJ Lillard, MA McCloskey, ES Yeung. Monitoring exocytosis and release from individual mast cells by capillary electrophoresis with laser-induced fluorescence detection. Anal Chem 68:2897–2904, 1996.

65. RA Wallingford, AG Ewing. Capillary zone electrophoresis with electrochemical detection in 12.7 μm diameter columns. Anal Chem 60:1972–1975, 1988.

66. HK Kristensen, YY Lau, AG Ewing. Capillary electrophoresis of single cells: observation of two compartments of neurotransmitter vesicles. J Neurosci Meth 51:183–188, 1994.

67. FD Swanek, G Chen, AG Ewing. Identification of multiple compartments of dopamine in a single cell by CE with scanning electrochemical detection. Anal Chem 68:3912–3916, 1996.

68. J Bergquist, A Tarkowski, A Ewing, R Ekman. Catecholaminergic suppression of immunocompetent cells. Immunol Today 19:562–567, 1998.

69. SA Hofstadler, FD Swanek, DC Gale, AG Ewing, RD Smith. Capillary electrophoresis-electrospray ionization Fourier transform ion cyclotron resonance mass spectrometry for direct analysis of cellular proteins. Anal Chem 67:1477–1480, 1995.

70. KM Bullard, PF Gavin, AG Ewing. Electrophoretic separation schemes in ultrathin channels with capillary sample introduction. Trends Anal Chem 17:401–410, 1998.

71. PF Gavin, AG Ewing. Continuous separations by electrophoresis in rectangular channels. In: JP Landers, ed. Handbook of Capillary Electrophoresis, 2nd ed. New York: CRC Press, 1997, pp. 741–764.

72. JM Mesaros, G Luo, J Roeraade, AG Ewing. Continuous electrophoretic separations in narrow channels coupled to small-bore capillaries. Anal Chem 65:3313–3319, 1993.

73. YM Liu, JV Sweedler. Nanoliter volume kinetic assays. J Am Chem Soc 117:8871–8872, 1995.

74. JM Mesaros, PF Gavin, AG Ewing. Flow injection analysis using continuous channel electrophoresis. Anal Chem 68:3441–3449, 1996.

75. PF Gavin. Development of channel electrophoresis with electrochemical array detection. PhD dissertation, The Pennsylvania State University, University Park, PA, 1997.

76. YM Liu, T Moroz, JV Sweedler. Monitoring cellular release with dynamic channel electrophoresis. Anal Chem 71:28–33, 1999.

77. CE MacTaylor, AG Ewing. Continuous electrophoretic separations in narrow channels with post-channel derivatization and laser-induced fluorescence detection. J Microcol Sep 12:279–284, 2000.

78. EM Smith, H Xu, AG Ewing. DNA separations in microfabricated devices with automated capillary sample introduction. Electrophoresis 22:363–370, 2001.

79. PB Hietpas, KM Bullard, AG Ewing. Characterization of electrophoretic sample transfer from a capillary to an ultrathin slab gel. J Microcol Sep 10:519–527, 1998.

80. CA Monnig, JW Jorgenson. On-column sample gating for high-speed capillary zone electrophoresis. Anal Chem 63:802–807, 1991.

81. AW Moore, Jr., JW Jorgenson. Rapid comprehensive two-dimensional separations of peptides via RPLC-optically gated capillary zone electrophoresis. Anal Chem 67:3448–3455, 1995.

82. AW Moore, Jr., JW Jorgenson. Comprehensive three-dimensional separation of peptides using size exclusion chromatography/reversed phase liquid chromatography/optically gated capillary zone electrophoresis. Anal Chem 67:3456–3463, 1995.

83. AW Moore, Jr., JW Jorgenson. Resolution of cis and trans isomers of peptides containing proline using capillary zone electrophoresis. Anal Chem 67:3464–3475, 1995.

84. JE Thompson, TW Vickroy, RT Kennedy. Rapid determination of aspartate enantiomers in tissue samples of microdialysis coupled on-line with capillary electrophoresis. Anal Chem 71:2379–2384, 1999.

85. L Tao, JE Thompson, RT Kennedy. Optically gated capillary electrophoresis of o-phthalaldehyde/β-mercaptoethanol derivatives of amino acids for chemical monitoring. Anal Chem 70:4015–4022, 1998.

86. JA Lapos, AG Ewing. Injection of fluorescently labeled analytes in microfabricated chips using optically gated electrophoresis. Anal Chem 72:4598–4602, 2000.

30

High Temperature and Temperature Programming in Pressure-Driven and Electrically Driven Chromatography

Nebojsa M. Djordjevic

Novartis Pharma AG., Basel, Switzerland

I. INTRODUCTION

My first introduction to chromatography and the role of temperature in chromatographic separations was in the early 1980s while working for my Ph.D. degree in analytical chemistry with Professor Richard Laub at San Diego State University. My experimental research was on a very popular and elegant method of chromatography known as inverse gas-solid chromatography. In inverse gas chromatography, the species of interest is the stationary phase (in contrast to conventional analytical gas chromatography, where the stationary phase is of interest only as far as its ability to separate the injected compounds is concerned).

Data from molecular probe experiments, carried out at different column temperatures, are usually presented in the form of a retention diagram showing the relationship between ln k vs. 1/T. From these plots, valuable information can be obtained regarding the solute-solvent interactions of the chromatographic process.

During my postdoctoral years with Professor Milton L. Lee at Brigham Young University, I had an opportunity to further explore the role of temperature, this time in capillary gas and supercritical fluid chromatography. In 1988 I moved to Sandoz Pharma (now Novartis Pharma), in Basel, Switzerland, where I started as a laboratory head in their Analytical Research and Development Department.

In the late 1980s and early 1990s capillary zone electrophoresis (CZE) emerged as a new, highly efficient separation technique. Many separation scientists were predicting the demise of liquid chromatography (LC), which was deemed inferior to CZE with regard to speed and efficiency. We at Sandoz believed that not all the parameters that could effect speed and efficiency in liquid chromatography had been sufficiently explored. Temperature was one of the parameters I knew could significantly influence retentive properties (knowing its importance in gas and supercritical fluid chromatography) and one that to date had not been fully utilized in liquid chromatographic separations. We explored high temperatures (up to 200°C) in order to perform fast and very efficient separations in liquid chromatography. We achieved a separation where one million plates were generated in less than 55 minutes. This separation represented a record for high-efficiency separations in liquid chromatography in 1991. This achievement remains one of the most memorable moments of my scientific career. We successfully proved that temperature should be regarded as a parameter that is as important as mobile phase composition, particle size, and column dimensions in accomplishing fast and efficient separations in liquid chromatography. Today we routinely use temperature (and temperature programming) as an optimization tool in both pressure-driven and electrically driven chromatography.

II. PRESSURE-DRIVEN CHROMATOGRAPHY

The use of miniaturized separation techniques, such as capillary electrophoresis, capillary electrochromatography, and capillary liquid chromatography, in analytical chemistry continues to attract much interest [1,2]. The major limitations in capillary liquid chromatography are the lack of reliable instrumentation for the delivery of micro flow rates and the excessive column pressure drop linked to the use of small particles in combination with small column dimensions. The applicability of high flow rates is limited by the backpressure that different parts of the chromatographic system (pump, injector, and column) can withstand. The pressure drop across a packed column can be approximated by Eq. (1) [3]:

$$\Delta P = \phi L \eta u / d_p^2 \tag{1}$$

where ΔP is the pressure drop, ϕ is the flow resistance factor, L is the column length, η is the mobile phase viscosity, and d_p is the particle diameter. The maximum pressure of the pump in routine work is between 150 and 500 bar, although the use of higher pressures (up to 1400 bar) has been described [4]. The equipment capable of sustaining such high pressures is sophisticated and expensive.

III. HIGH TEMPERATURE

In LC, elevated column temperature can be used as a tool to overcome the flow rate problem associated with high backpressure, allowing the use of flow rates that otherwise could not be applied. The pressure reduction is due to a decrease in eluent viscosity with increasing temperature. It is often a good approximation to assume the relationship between viscosity and temperature as shown in Eq. (2) [5]:

$$\ln \eta = a + (b/T) \tag{2}$$

where a and b are empirically determined constants.

High-temperature separation has been shown to improve analyte resolution by decreasing mobile phase viscosity and by increasing the diffusion rate of sample species [6,7], thus increasing mass transfer of the analyte to the stationary phase. The temperature dependence of analyte diffusivity can be represented by

$$D_{SB}/\eta_B T = \text{const.} \tag{3}$$

where D_{SB} is the diffusion coefficient of solute S at very low concentrations in solvent B at temperature T. The height equivalent to a theoretical plate (H) of an open-tubular column is given by the Golay equation:

$$H = (2D_m/u) + [(1 + 6k + 11k^2)/96(1 + k)^2](d_c^2 u/D_m)$$
$$+ [2k\, d_f^2 u/3(1 + k^2)D_s] \tag{4}$$

where D_m is the diffusivity of the solute in the mobile phase, D_s is the diffusivity of the solute in the stationary phase, d_c is the column inner diameter, d_f is the film thickness, u is the velocity of the mobile phase, and k is the retention factor of the solute. As can be seen from Fig. 1, where a fused silica column (85 cm × 75 μm i.d.) was used to show the influence of the column temperature on column efficiency, the H vs. u curves become much flatter in the high-velocity region when the column temperature increased from 20 to 150°C. The optimum flow rate increased from 0.03 cm/s at 20°C to 0.19 cm/s at 150°C. In conclusion, an increase in column temperature induces rise of solute diffusion in the mobile phase and decrease of the mobile phase viscosity, which allows higher linear velocities, longer columns with smaller i.d., and packing materials with smaller particle size to be applied.

One million plates have been realized in the separation of chlorobenzenes (Fig. 2, trace b) on a 19.6 m × 50 μm i.d. SB-methyl-100 column operated at 200°C. Trace a in Fig. 2 shows a chromatogram obtained under the same chromatographic conditions as trace b except for

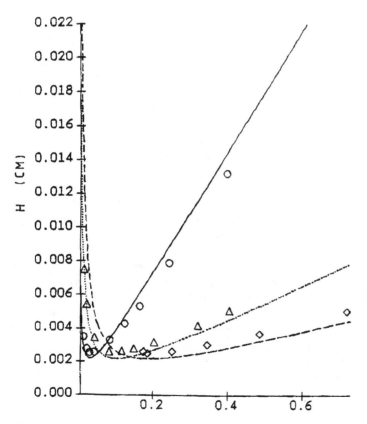

Figure 1 Influence of temperature on column efficiency. Column 0.85 m × 75 μm i.d. fused silica; mobile phase, methanol; sample, acetophenone in methanol. Solid line, theoretical curve at 22°C; ○ = experimental data at 22°C. Dotted line, theoretical curve at 100°C; △ = experimental data at 100°C; Dashed line, theoretical curve at 150°C; ◇ = experimental data at 150°C. (From Ref. 6.)

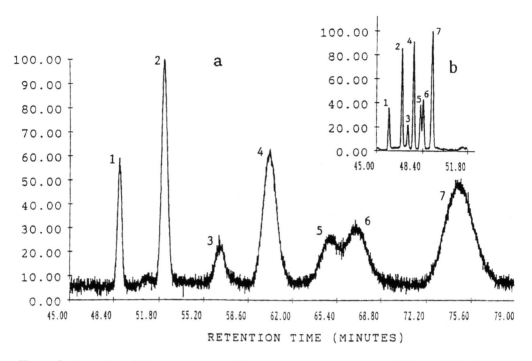

Figure 2 Separation of chlorobenzenes at different column temperatures: (a) 22°C; (b) 200°C. Column SB-methyl-100, 20 m × 51 μm; i.d. d_f = 0.25 μm mobile phase: acetonitrile/water (1 : 1); flow rate 0.75 μL/min. Peaks: 1, tropolone; 2, benzene; 3, 1,4-dichlorobenzene; 4, 1,2,4-trichlorobenzene; 5, 1,3,5-tri-chlorobenzene; 6, 1,2,4,5 tetrachlorobenzene; 7, pentachlorobenzene. (From Ref. 6.)

the column temperature (22°C). Lower column temperature had significantly deteriorated the efficiency of the separation system.

Besides this high efficiency achieved using high temperature, the possibility of using temperature as a parameter to tune selectivity is foreseen. Although nearly all of the physical parameters that play a role in liquid chromatographic separation are a function of temperature, temperature has not yet been adequately explored as a parameter to tune separation and shorten analysis times in liquid chromatography [8–21]. The lack of proper column heating systems and poor temperature stability of the silica-based packing materials (the most frequently used stationary phase in reversed-phase LC) are the most commonly listed reasons for the lack of interest in using temperature as a tool to optimize separation in LC [6,22]. The rate of advance of a solute band through a column depends on the distribution ratio K (the ratio of the concentration of the solute in the stationary phase to its concentration in the mobile phase). A solute is mobile (moves through the column) only when it is dissolved in the mobile phase. The partition ratio is proportional to the solubility of the solute in the liquid phase and solute sorption in the stationary phase. It decreases with increasing temperature according to the van't Hoff equation:

$$\ln K = -\Delta H°/RT + \Delta S°/R \qquad (5)$$

where $\Delta H°$ is the enthalpy change associated with the transfer of the solute from mobile phase to the stationary phase, $\Delta S°$ is the corresponding entropy change, R is the gas constant, and T is absolute temperature. The change in solute distribution with temperature can be put in a form, which represents the distribution factor ratio K_1/K_2 at two different temperatures T_1 and T_2:

$$K_1/K_2 = \exp[H^\circ(T_2 - T_1)/RT_1T_2] \tag{6}$$

The equation indicates that changing the column temperature can easily modulate the magnitude of retention. The effect of temperature variation on the partitioning process will strongly depend on the solute enthalpy. For the same temperature change, a solute with a large H° will be strongly affected, whereas a solute with a small H° temperature change will not produce as significant a variation in solute distribution between the stationary and mobile phases. In chromatography, the distribution coefficient is related to the solute retention factor k:

$$k = K\beta \tag{7}$$

where $\beta = V_{st}/V_{mo}$ is the phase ratio, and V_{st} and V_{mo} are volumes of the stationary and the mobile phase, respectively. The solute retention factor is a measure of the relative time spent by the solute in the stationary and mobile phases and is calculated according to:

$$k = (t_r - t_o)/t_o \tag{8}$$

where t_r is the solute elution time. If the solute is not sorbed by the stationary phase, its retention time is solely the dead-space time, t_o, for the particular column. Improved solubility in a mobile phase at elevated temperatures allows elution of very hydrophobic solutes with less organic modifiers at reasonable time. In Fig. 3 is depicted the separation of a mixture of chlorobenzenes in high-temperature open tubular liquid chromatography. The separation of the hydrophobic

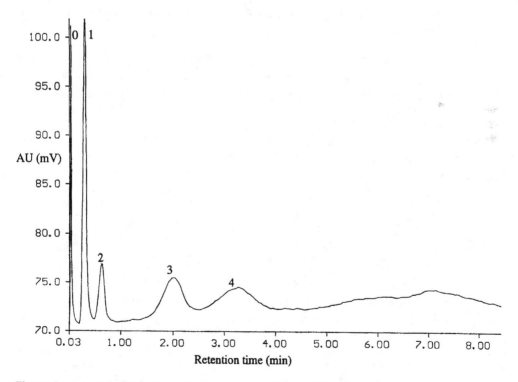

Figure 3 Separation of a mixture of chlorobenzenes. Column SB-methyl-100, 27 cm × 50 μm i.d., d_f = 0.25 μm; mobile phase 1.5 mM phosphate buffer pH = 7, flow rate = 2.3 μL/min; UV detection at 210 nm; column temperature 140°C. Peaks: 0, thiourea; 1, chlorobenzene; 2, 1,4-dichlorobenzene; 3, 1,2,4-trichlorobenzene; 4, 1,2,4,5 tetrachlorobenzene. (From Ref. 23.)

compounds was achieved on a reversed-phase stationary phase utilizing a 100% aqueous mobile phase. Increased solubility power of water at elevated temperature (140°C) and favorable phase ration (in open tubular LC the volume of the mobile phase is much larger than the volume of the stationary phase) has enabled elution of compounds with high retention factors ($k' = 30$) in a very short time [23].

IV. TEMPERATURE PROGRAMMING

Real-life samples often confront us with the problem that the same conditions are not best for all the components of the mixture. The more complex the system to be analyzed, the greater are the chances that its chromatography in a constant environment may lead to unsatisfactory results. The idea of programmed analysis is to vary the operating conditions during the analysis, so that all components of the sample may be eluted under optimum conditions. At present, there are four possible techniques for programming gradient formation in liquid chromatography: solvent programming, stationary phase programming (coupled columns), temperature programming, and flow programming. The main concerns found when selecting one programmed technique over another are the relative resolution obtained per unit time provided by each programmed technique and the ease of variation of the programmed quantity during a separation run. Furthermore, experimental considerations such as simplicity and convenience, additional equipment requirements for a given technique, and the compatibility with a detection system should be taken into consideration when selecting a programming mode. The nature of the separation problem and sample complexity should be also considered when deciding on a programming approach.

Solvent strength programming is a typical programming method in LC to control the separation where fast and highly efficient analysis is desirable. Temperature programming is not often used in LC, mainly because it is easy to manipulate the mobile phase composition (solvent strength programming). Scott and Lawrence [24] used temperature programming with conventional size LC columns in a normal phase separation. Biggs and Fetzer [25] demonstrated the use of temperature gradients to elute various structural isomers of anthraquinone disulfonic acid. McNair et al. [26], Greibrokk et al. [27], and Djordjevic et al. [28] evaluated the potential of temperature programming in conjunction with microbore and capillary LC columns. Renn and Synovec [29] investigated the effect of temperature on separation efficiency in size exclusion chromatography, while Djordjevic et al. [30] demonstrated separations in reverse and normal phase modes in open tubular capillary LC with linear and step temperature gradients. Djordjevic et al. [31] evaluated the effect of temperature and flow programming on overall chromatographic performance in reversed-phase LC by varying the temperature and the mobile phase flow rate simultaneously.

Two types of devices for achieving a temperature program exist. The first type varies the temperature as a function of time, t, and the temperature of the whole column is the same at any point along the column length, 1, ($dT/dl = 0$; $dT/dt \neq 0$). This temperature-programming technique is the type most commonly applied in chromatography (gas and liquid chromatography). In the second type, the temperature along the column varies ($dT/dl \neq 0$), while the temperature at a certain point in the column remains constant with time ($dT/dt = 0$). This approach has been used to introduce temperature gradients along the column length by flow programming of a preheated mobile phase [8].

In linear temperature programming, assuming $dT/dl = 0$, the column temperature at time t is given by

$$T = T_0 + rt \tag{9}$$

where T_0 is the initial temperature and r defines the theoretical program. The initial and final temperatures and the programming rate define the theoretical program. The use of temperature programming requires fast heat transfer from the heating medium into a chromatographic system. In practice, the actual column temperature will always lag behind the set temperature, mainly because of heat transfer delay. The difference between the theoretical and the actual program is given as the time lag (t_L) of the column temperature behind the theoretical temperature. A dashed line in Fig. 4 represents the ideal linear temperature programming profile. On this line a point was identified that represents temperature halfway through the gradient (80°C).

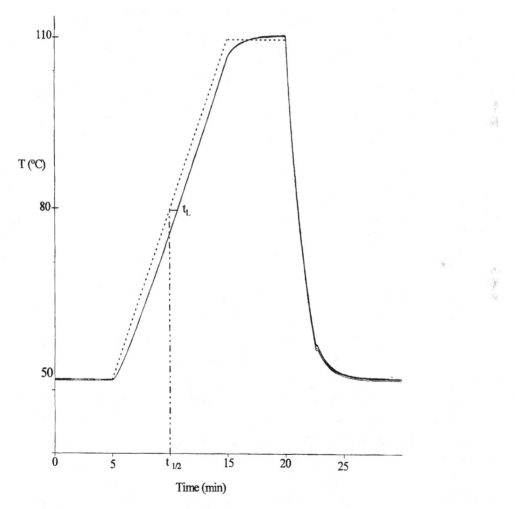

Figure 4 Reproducibility of the temperature gradient profiles obtained utilizing GC oven. Column: 250 mm × 0.1 mm Grom-sil ODS1, 3 μm particle size, mobile phase ACN/H_2O (50:50 v/v), flow rate = 1.2 μL/min. Temperature program: isothermally at 50°C for 5 min then at a 6°C/min to 110°C. The solid lines represent the actual temperature gradient profiles. The program was run five times. Dashed line represents the ideal temperature profile. Temperature lag, t_L = 0.7 min. (From Ref. 28.)

From this point on, the gradient trace, a perpendicular line, was drawn to the time axis to obtain $t_{1/2}$. The temperature lag is defined as

$$t_L = t_{1/2} - (t_G/2) \tag{10}$$

where t_G (gradient time) = final gradient time (t_f) − starting time (t_i).

Care must be taken to avoid a large temperature difference between the column and the entering solvent. It is essential to preheat the mobile phase in order to minimize t_L. The magnitude of t_L depends on the configuration used for the instrument and the heating system applied. It also depends on program rates, column dimensions (column radius and the mass of the column), and mobile phase flow rates used in the experiments [28].

The heat capacity of a column heater and its oven is important from the standpoint of applicable heating and cooling rates, effects of thermal losses, and the time required for attainment of temperature equilibration. When the equipment associated with the column heater has a low heat capacity, the apparatus permits high-speed heating and cooling. Designs with low

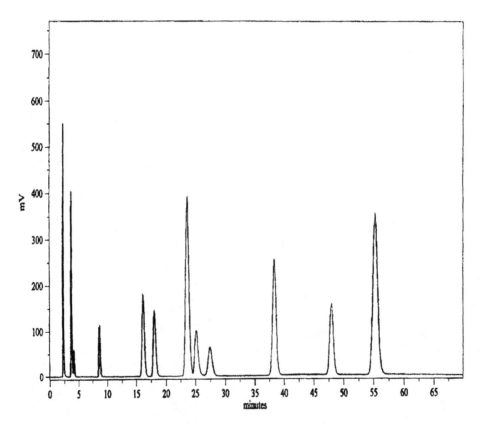

Figure 5 Separation of a test mixture. Microbore column, 150 mm × 1.0 mm, Spherisorb ODS2, particle size 3 μm; ACN/H₂O 40 : 60 (v/v), flow rate = 35 μL/min; UV detection at 210 nm. Temperature program: 20 min isothermal at 50°C, the temperature was increased to 110°C with a programming rate of 6°C/min. The final temperature was maintained constant until the end of the run. Peaks: 1, thiourea; 2, benzyl alcohol; 3, methyl benzoate; 4, toluene; 5, benzophenone; 6, naphthalene; 7, 1,4-dichlorobenzene; 8, phenothiazine; 9, biphenyl; 10, 1,3,5-trichlorobenzene; 11, 1,2,4,5-tetrachlorobenzene. (From Ref. 28.)

heat capacity permit high heating rates to be obtained readily; they also permit short column-cooling times, a feature of importance in routine analysis. A satisfactory column heater should faithfully reproduce the desired program. Figure 4 depicts a temperature profile completed from the temperature measurements on the outer column wall of a capillary LC column when a gas chromatographic oven was used to heat the column. The superimposed traces of five consecutive temperature programs with a heating rate of 6°C/min (t_i = 50°C; t_f = 110°C) overlapped to form a single trace, indicating good system reproducibility. The figure also shows a very small time delay (t_L = 0.7 min) of the column wall temperature behind the ideal temperature profile (dashed line) for the capillary LC column (250 mm × 100 μm), with an applied flow rate of 1.2 μL/min. In addition, the system exhibited very fast cooling and short equilibration time (<10 min). The equilibration time in the temperature gradient mode is compatible with the time necessary to equilibrate the column in the mobile phase gradient mode. Figure 5 depicts an overlay of two chromatograms performed under the same separation conditions. Relative standard deviations for the retention factors (n = 5) was <1%. This degree of system reproducibility

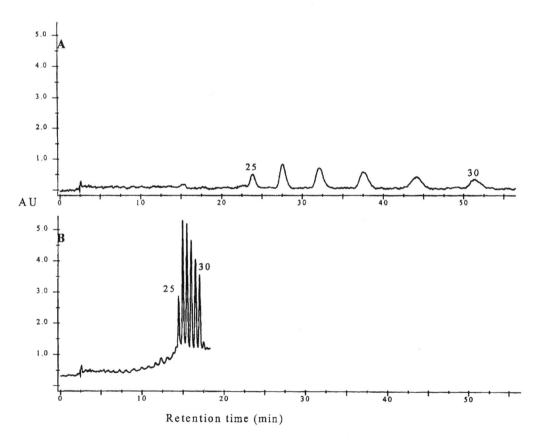

Figure 6 Isothermal and temperature programmed separation of a mixture of pd(A)$_{25-30}$ oligonucleotides. Column ODS-AQ, pore size 200 Å, particle size 3 μm. Mobile phase composition (A:B) 85.5/14.5% v: v; (A = 10 mM TEAA in water; B = ACN/10 mM TEAA in water 1:1) flow rate 0.3 mL/min; UV detection at 260 nm. (A) Column temperature 50°C; (B) temperature program: 50°C isothermally for 6.5 min, temperature gradient then 6°C/min to 100°C. (From Ref. 32.)

and precision was achievable because of the rapid temperature equilibration of the low mobile phase volumes used, small mass of the micro-LC column, and a good heating power provided by the preheating system mounted in the oven.

The separation procedure in a temperature-programmed run has to be optimized in a such manner that by applying the appropriate temperature changes, elution of strongly retained compounds will be accelerated while a complete baseline separation of earlier eluting compounds

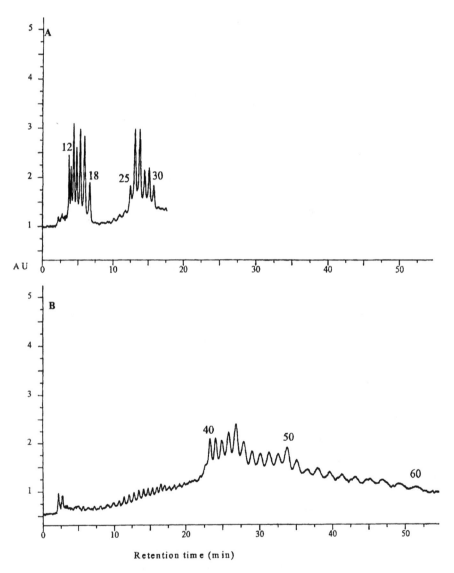

Figure 7   Temperature programmed separation of oligonucleotides. Column ODS-AQ, pore size 200 Å, particle size 3 μm. Mobile phase composition (A:B) 86/14% v:v; (A = 10 mM TEAA in water; B = ACN/10 mM TEAA in water 1:1); flow rate 0.3 mL/min; UV detection at 260 nm. Temperature program: 60°C isothermally for 5 min, temperature gradient then 6°C/min to 80°C. (A) A mixture of pd(A)$_{12-18}$ and pd(A)$_{25-30}$ oligonucleotides; (B) pd(A)$_{40-60}$ oligonucleotides. (From Ref. 32.)

is secured. Temperature-programmed separation should normally start at a low temperature where less retained compounds will elute; the column temperature should then be increased to elute more strongly retained compounds. When the elution process is started at a relatively low temperature, the hydrophobic solutes will spend most of their time in the stationary phase (due to their low solubility in mobile phase). Meanwhile, components with higher solubility in the mobile phase will move along the column. In effect, for each compound in the mixture there is an optimum elution temperature (for a given mobile and stationary phase). In programmed-temperature operation, the temperature of the column is increased during analysis.

Djordjevic et al. [32] applied temperature programming to achieve baseline separation of oligonucleotides. Figure 6A shows the separation of a mixture of oligonucleotides $pd(A)_{25-30}$ in isocratic (14.5%B) and isothermal (50°C) mode. Although the baseline separation was achieved, an excessive run time (50 min) makes this method unacceptable for routine applications. A temperature gradient was applied to reduce the analysis time and still preserve baseline separation between the pairs. Figure 6B shows the separation of the $pd(A)_{25-30}$ sample in a temperature gradient mode. A 17-minute total analysis time represents a reduction of 75% over the length of the isothermal run. In addition to shorter retention times all peaks exhibited higher peak sensitivities (the ratio of peak height to peak width). Temperature programming can be used to extend the applicability of the isocratic mode for the separation of small and larger oligonucleotides. A complete separation was achieved for a mixture of $pd(A)_{12-18}$ and $pd(A)_{25-30}$ oligonucleotides as depicted in Fig. 7A. Isothermal separation was conducted for 5 minutes followed by the temperature gradient (heating rate of 6°/min) to 80°C. The same gradient was utilized for the separation of $pd(A)_{40-60}$ sample (Fig. 7B). A complete separation of this complex mixture was done in 50 minutes.

V. HIGH TEMPERATURE AND TEMPERATURE PROGRAMMING IN ELECTRICALLY DRIVEN CHROMATOGRAPHY

The possibility of making electroosmotic flow (EOF) work as a pumped system prompted interest in the so-called electroseparation techniques. The separation of uncharged analytes in capillary electrochromatography (CEC) is based on partitioning, but the separation of charged analytes is also possible based on partitioning and electrophoretic mobility. The power of electrochromatography as separation technique lies in the combination of high efficiency and resolution (arising from electroosmotic flow profile) with the high selectivity, versatility, and universality of LC. In CEC as in LC, there are several ways to optimize selectivity, improve resolution, and shorten analysis time. When selecting the mobile phase and the stationary phase in CEC, care should be taken that both high selectivity and high electroosmotic flow (EOF) can be achieved. Mobile phase parameters commonly used to adjust selectivity and generate sufficient EOF include the type of aqueous buffer components and organic solvent, proportion of organic modifier, pH, and ionic strength [33].

The electroosmotic flow, u_{EOF}, generated in a packed capillary column when the electric field is applied can be represented by the following equation:

$$u_{cof} = \varepsilon_0 \varepsilon_r \zeta E / \eta \tag{11}$$

where ε_0 is the permittivity in a vacuum, ε_r is the relative permittivity or dielectric constant of the mobile phase, η is the viscosity of the fluid, E is the electric field strength, and ζ is the zeta potential. From this equation it follows that the u_{EOF} is proportional to the electric field strength and inversely proportional to the mobile phase viscosity. Figure 8 depicts the increase in elec-

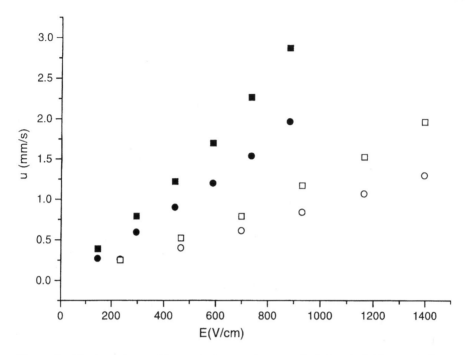

Figure 8 The dependence of liner velocity (mm/s) on applied electric field strength (V/cm). Column: Grom-Sil 100 ODS1, particle size 3 μm, total column length = 33.5 cm, packed bed length = 25 cm, i.d. = 100 μm; mobile phase: 1.5 mM phosphate buffer pH 7/acetonitrile (7/3 v : v); column set temperature 60°C (■) and 20°C (●). Column: SB-methyl-100 column, total column length = 30.0 cm, effective column length = 21.5 cm, i.d. = 50 μm, d_f = 0.25 μm; mobile phase: 1.5 mM phosphate buffer pH 7/acetonitrile (6/4 v : v); column set temperature 60°C (□) and 30°C (○). (From Ref. 34.)

troosmotic flow when the applied electric field strength is increased in a packed capillary (at 20 and 60°C) and in a coated tube (at 30 and 60°C). If fast separation is required, one should work not only at high field strengths, but also at elevated temperature. Temperature increase will influence solute partitioning and will also help to increase the electroosmotic flow [34]. In Eq. (11) the ratio ε_r/η is temperature dependent. When the column temperature was increased (at the same applied electric fields strengths, 0.9 kV/cm) from 20 to 60°C, the u_{EOF} increased approximately 50%. Most of the observed increase in EOF can probably be ascribed to the decrease in viscosity of the buffered mobile phase, although contribution of temperature induced pH changes should not be ignored.

In electrochromatography and capillary electrophoresis, temperature is often regarded as a parameter that has an adverse effect on the separation efficiency and promotes the formation of gas bubbles in the separation column. Until recently, the primary aim of controlling capillary temperature was to remove Joule heat; nevertheless, temperature adjustment can also be used as a means of optimizing selectivity and run times in electrochromatographic separations. In some instances temperature increases will yield faster separation times but a co-elution of early eluting peaks with low k values will occur. Figure 9a depicts a separation of the sample mixture at 60°C. The last peak in the chromatogram eluted after 18.5 min. Reduction of analysis time of more than 50% was achieved when compared with a chromatographic run obtained at 20°C (Fig. 9b). Although separation was faster at 60°C, not all constituents in the mixture were sepa-

DAD1 A, Sig=210,16 Ref=off (FIONA\FFA10126.D)

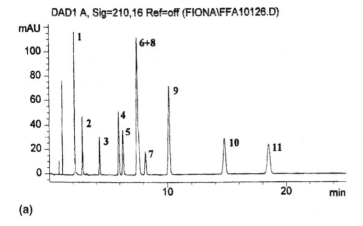

(a)

DAD1 A, Sig=210,16 Ref=off (FIONA\FFA10120.D)

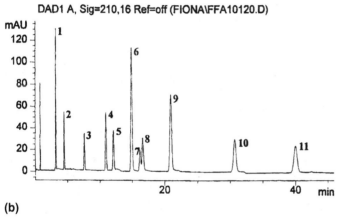

(b)

Figure 9 (a) Column: Hypersil C18 (packed bed length 25 cm, total length 33.5 cm). The running buffer was acetonitrile/5 mM phosphate buffer at pH 7 (50:50 v/v); applied voltage = 25 kV; detection wavelength = 210 nm. Peaks: 1, thiourea; 2, benzyl alcohol; 3, methyl benzoate; 4, toluene; 5, benzophenone; 6, naphthalene; 7, 1,4-dichlorobenzene; 8, phenothiazine; 9, biphenyl; 10, 1,3,5-trichlorobenzene; 11, 1,2,4,5-tetrachlorobenzene. Column temperature: 60°C. (b) Column temperature: 20°C. Other conditions and peak identification as above. (From Ref. 37.)

rated; the bands labeled 6 and 8 co-eluted. By examining solute retention factors at four different temperatures, a correlation between temperature and solute retention was established. It was evident from the plot of ln k vs. 1/T (Fig. 10) that temperature did not have the same influence on all the constituents in the mixture. Phenothiazine has the highest slope, which means that its retention is the more prone to temperature effects than the retention of other solutes in the mixture. In the graph, the line representing the retention of phenothiazine intersected with the line of 1,4-dichlorobenzene and naphthalene at 25 and at 60°C, respectively, which indicates that these peaks will co-elute under these conditions.

To optimize the separation of this complex mixture (with a broad range of retention factors 0.41 < k < 11.65 at 20°C), a gradient (programmed separation parameter) approach is needed. Varying the ratio of aqueous to organic component in the mobile phase is the most common approach for altering selectivity in LC and CEC. Recently, controlled temperature variations

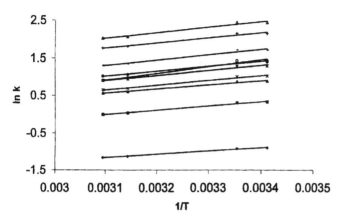

Figure 10 Plot of ln k vs. 1/T (K^{-1}) for the solutes: ◆, benzyl alcohol; ■, methyl benzoate; ▲, toluene; ✕, benzophenone; *, naphthalene; ●, 1,4-dichlorobenzene; ○, phenothiazine; −, biphenyl; +, 1,3,5-tri-chlorobenzene; △, 1,2,4,5-tetrachlorobenzene. (From Ref. 37.)

were utilized to optimize separation in capillary gel electrophoresis [35], MEKC [36], and CEC [37].

Because selectivity as well as the analysis time of the components in the mixture were very responsive to temperature change, a temperature gradient mode was selected to reduce the run time and achieve a complete separation of all solutes in the CEC mode.

As shown in Fig. 11, the overall analysis time was reduced to approximately 22 minutes, or 45% less than the isothermal separation at 20°C. The initial and the final column temperatures were 25 and 60°C, respectively, with a gradient rate of 3°C/min. In spite of this significant reduction of analysis time, all compounds were baseline separated.

Flow rate variations introduced by temperature programming also helped to speed up the analysis time. Flow rate programming mode is seldom used in chromatography. In electrically driven chromatography, it can be generated either by voltage or temperature programming. Volt-age programming will not affect solute partitioning, but this programming mode would require

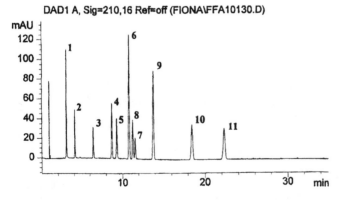

Figure 11 Separation of a test mixture. Linear temperature gradient from 25°C to 60°C with a rate of 3°C/min. Other conditions and peak identification as in Fig. 9a. (From Ref. 37.)

large changes in the electric field strengths to achieve a meaningful reduction in the analysis time of strongly retained compounds. Temperature programming during the separation runs induced changes in the mobile phase viscosity and as a consequence electroosmotic flow was increased. The changes of u_{eof} for the stationary phase and mobile phase used have been related to temperature by the empirical relationship [37]:

$$u_{eof} = a + bT \tag{12}$$

For the system employed in this work, parameters a and b were determined (0.0155 and 1.0041, respectively) by a linear fit (correlation coefficient 0.99998) of measured u_{eof} in a temperature range from 20 to 60°C.

Flow rate variation in combination with temperature programming makes it possible to considerably reduce the retention of very strongly retained compounds without having a detrimental affect on the column performance. The separations obtained under combined temperature and flow gradients show high reproducibility comparable to the reproducibility normally seen with a mobile phase gradient.

VI. CONCLUSION

Controlled column temperature variations in chromatographic separations have much potential for adjusting the retention factors, improving the resolution of difficult-to-separate compounds, and accelerating elution of more strongly retained compounds. Increase of solvent solubility power at elevated temperatures can be exploited as a way to reduce the use of organic solvents in separation. It has been shown that in pressure-driven and electrically driven chromatography programming of column temperature decreases analysis time without sacrificing high resolution and run-to-run reproducibility. Temperature programming is especially promising as an alternative to solvent gradient elution in capillary chromatography because it bypasses the technical difficulties associated with solvent gradient elution in capillary columns.

REFERENCES

1. H Poppe. J Chromatogr A 778:21, 1997.
2. MV Novotny. J Chromatogr B 778:70, 1997.
3. L Snyder, JJ Kirkland. Introduction to Modern Liquid Chromatography. New York: John Wiley & Sons, 1979.
4. JE MacNair, KC Lewis, JW Jorgenson. Anal Chem 69:983, 1997.
5. EN da C Andrade. Nature 125:309, 1930.
6. G Liu, NM Djordjevic, F. Erni. J Chromatogr 592:239, 1992.
7. H Chen, C Horváth. Anal Meth Instr 1:213, 1993.
8. LK Moore, RE Synovec. Anal Chem 65:2663, 1993.
9. F Kamiyama, K Yamazaki, K Kawamura, M Kohara. Biomed Chromatogr 10:105, 1996.
10. A Nahum, C Horáth. J Chromatogr 203:53, 1981.
11. BL Karger, JR Gant, A Hartkopf, PH Weiner. J Chromatogr 128:65, 1976.
12. Y Baba, N Yoza, S Ohashi. J Chromatogr 350:119, 1985.
13. A Tchapla, S Heron, H Colin, G Guiochon. Anal Chem 60:1443, 1988.
14. L Han, W Linert, V Gutamann. J Chromatogr Sci 30:142, 1992.
15. H Zou, Y Zhang, M Hong, P Lu. J Liq Chromatogr 15:2289, 1992.
16. FM Yamamoto, S Rokushika, H Hatano. J Chromatogr Sci 27:704, 1989.
17. MR Smith, JR Burgess. Anal Commun 33:327, 1996.
18. HJ Issaq, SD Fox, K Lindsey, JH McConnel, DE Weis. J Liq Chromatogr 10:49, 1987.

19. HJ Issaq, ML Glennon, DE Weis, SD Fox. In: W Hinze, DA Armstrong, eds. Ordered Media in Chemical Separations. Washington, DC: ACS publications, 1987, pp. 260–271.

20. HJ Issaq, M Jaroniec. J Liq Chromatogr 12:2067, 1989.

21. B Yan, J Zhao, JS Brown, J Blackwell, PW Carr. Anal Chem 72:1253, 2000.

22. G Sheng, Y Shen, ML Lee. J Micro Sep 9:63, 1997.

23. NM Djordjevic, F Houdiere, P Fowler. Biomed Chromatogr 12:153, 1998.

24. RPW Scott, JG Lawrence. J Chromatogr Sci 8:619, 1970.

25. WR Biggs, JC Fetzer. J Chromatogr 351:313, 1986.

26. HM McNair, J Bowermaster. J High Resolut Chromatogr Chromatogr Commun 10:27, 1987.

27. R Trones, A Iveland, T Greibrokk. J Micro Sep 7:505, 1995.

28. NM Djordjevic, PWJ Fowler, F Houdiere. J Micro Sep 11:403, 1999.

29. CN Renn, RE Synovec. Anal Chem 64:479, 1992.

30. K Ryan, NM Djordjevic, F Erni. J Liq Chrom & Rel Technol 19:2089, 1996.

31. F Houdiere, PWJ Fowler, NM Djordjevic. Anal Chem 69:2589, 1997.

32. NM Djordjevic, F Houdiere, PWJ Fowler. Anal Chem 70:1921, 1998.

33. HJ Issaq, GM Janini, IZ Atamna, GM Muschik, J Lukszo. J Liq Chromatogr 15:1129, 1992.

34. NM Djordjevic, PWJ Fowler, F Houdiere, G Lerch. J Liq Chrom Rel Technol 21:2219, 1998.

35. EN Fung, H-M Pang, ES Yeung. J Chromatogr A 806:157, 1998.

36. NM Djordjevic, F Fitzpatrick, F Houdiere, G Lerch. J High Resol Chromatogr 22:443, 1999.

37. NM Djordjevic, F Fitzpatrick, F Houdiere, G Lerch, G Rozing. J Chromatogr A 887:245, 2000.

31

Gradient Elution in Capillary Electrophoresis, Micellar Electrokinetic Chromatography, Capillary Electrochromatography, and Microfluidics

Haleem J. Issaq

SAIC Frederick, NCI-Frederick Cancer Research and Development Center, Frederick, Maryland

Gradient elution is routinely used in high-performance liquid chromatography (HPLC) to achieve the resolution of a mixture that could not be resolved using isocratic elution. Unlike isocratic elution, where the mobile phase composition remains constant throughout the experiment, in gradient elution the mobile phase composition changes with time. The change could occur continuously or in steps, i.e., "step gradient." In continuous gradient the analyst can pick one of three general shapes: linear, concave, or convex. Gradient elution in HPLC is achieved using two pumps, two solvents, and a solvent mixer. Step gradient is accomplished by periodic change of the mobile phase composition at predetermined times to achieve the resolution of a mixture. In capillary electrophoresis (CE) electroosmotic flow is used in place of a mechanical pump and controls the flow of the mobile phase, in most cases an aqueous buffer. Electroosmotic flow is controlled by many parameters, including buffer type, concentration, ionic strength, pH, and buffer modifiers such as organic solvents, surfactant and polymers, capillary surface chemistry, and applied voltage. Another approach to control electroosmotic flow is by applying an external electric field that can change the direction and rate of electroosmotic flow by external voltage.

A manual step gradient was used by Balachunas and Sepaniak [1] to separate a mixture of amines by micellar electrokinetic chromatography (MECC). Stepwise gradients were produced by pipetting aliquots of a gradient solvent to the inlet reservoir, which was filled with 2.5 mL of running buffer. A small magnetic stirring bar was used to ensure thorough mixing of the added gradient solvent with the starting mobile phase. The gradient elution solvent was manually added in four 0.5 mL increments spaced 5 minutes apart, 5 minutes after start of the experiment.

Bocek and his group [2] developed a method for controlling the composition of the operational electrolyte directly in the separation capillary in isotachophoresis (ITP) and capillary zone electrophoresis (CZE). The method is based on feeding the capillary with two different ion species from two separate electrode chambers by simultaneous electromigration. The composition and pH of the electrolyte in the separation capillary is thus controlled by setting the ratio of two electric currents. This procedure can be used, in addition to generating mobile phase gradients, for generating pH gradients [3]. Sustacek et al. [4] dynamically modified the pH of the running buffer by steady addition of a modifying buffer. Sepaniak et al. [5–7] produced continuous gradients of different shapes—linear, concave, or convex—by using a negative polarity configuration in which the inlet reservoir is at ground potential and the outlet reservoir at negative high potential. This configuration allows two syringe pumps to pump solutions into and out of the inlet reservoir. Tsuda [8] used a solvent program delivery system similar to that used in HPLC to generate pH gradients in CZE. A pH gradient derived from temperature changes has also been reported [9]. Chang and Yeung [10] used two different techniques—dynamic pH gradient and electroosmotic flow gradient—to control selectivity in CZE. Dynamic pH gradient from pH 3.0 to 5.2 was generated by HPLC gradient pump (Fig. 1). Electroosmotic flow gradient was produced by changing the reservoirs containing different concentrations of cetylammonium bromide for injection and running.

Capillary electrochromatography (CEC) is a separation technique that combines the advantages of micro HPLC and CE. In CEC the HPLC pump is replaced by electroosmotic flow. Behnke and Bayer [11] developed a microbore system for gradient elution using 50 and 100 µm fused silica capillaries packed with 5 µm octadecyl reversed-phase silica gel and voltage gradients, up to 30,000 volts across the length of the capillary. A modular CE system was combined with a gradient HPLC system to generate gradient CEC, as shown in Fig. 2. Enhanced column efficiency and resolution were realized. Zare and coworkers [12] used two high-voltage power supplies and a packed fused silica capillary (Fig. 3) to generate an electroosmotically driven gradient flow in an automated manner. A mixture of polycyclic aromatic hydrocarbons was resolved in the gradient mode, which were not separated when the isocratic mode was employed (Fig. 4). Others [13–16] used gradient elution in combination with CEC to resolve

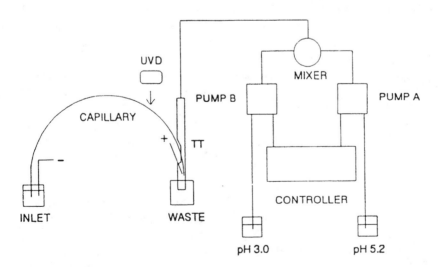

Figure 1 Schematic of the electrophoresis equipment for generating dynamic pH gradient. UVD = Ultraviolet detector; TT = PTFE tube, in which a small groove was cut in the middle to insert the electrode and the capillary. (From Ref. 10.)

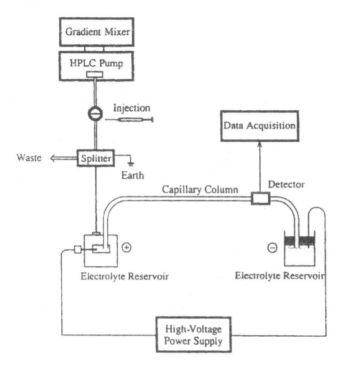

Figure 2 Schematic representation of an electrochromatographic system. (From Ref. 11.)

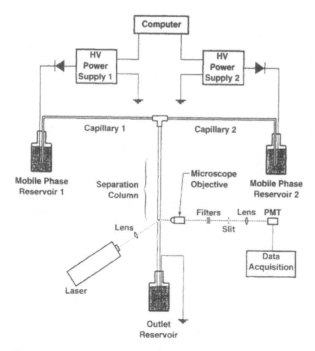

Figure 3 Schematic of the solvent gradient elution CEC apparatus. (From Ref. 12.)

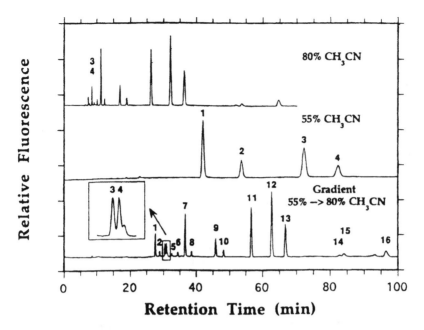

Figure 4 Electrochromatograms showing the comparison of isocratic and gradient elution for 16 polycyclic aromatic hydrocarbons. (From Ref. 12.)

different mixtures. Taylor et al. [14] and Taylor and Teale [15] designed a sampling interface for CEC using a gradient HPLC system to deliver samples and mobile phase to the column. The interface, which consists of a steel tee connector and restriction capillary, was connected to the 30 kV power supply that was continuously purged with mobile phase from the HPLC system and a constant voltage of 25 kV applied throughout the analysis. This instrumental setup was used for the CEC separation of 10 corticosteroids [15].

Multiple, intersecting narrow channels can be formed on a glass chip to form a manifold of flow channels in which CE can be used to resolve a mixture of solutes in seconds. Harrison and coworkers [17] showed that judicious application of voltages to multiple channels within a manifold can be used to control the mixing of solutions and to direct the flow at the intersection of channels. The authors concluded that such a system, in which the applied voltages can be used to control the flow, can be used for sample dilution, pH adjustment, derivatization, complexation, or masking of interferences. Ramsey and coworkers [18] used a microchip device with electrokinetically controlled solvent mixing for isocratic and gradient elution in MECC. Isocratic and gradient conditions are controlled by proper setting of voltages applied to the buffer reservoirs of the microchip. The precision of such control was successfully tested for gradients of various shapes—linear, concave, or convex—by mixing pure buffer and buffer doped with a fluorescent dye. By making use of the electroosmotic flow and employing computer control, very precise manipulation of the solvent was possible and allowed fast and efficient optimization of separation problems.

ACKNOWLEDGMENTS

This project has been funded in whole or in part with federal funds from the National Cancer Institute, National Institutes of Health, under Contract No. NO1-CO-56000. By acceptance of this article, the publisher or recipient acknowledges the right of the U.S. Government to retain nonexclusive, royalty-free license to any copyright covering the article. The content of this publication does not necessarily reflect the views of the Department of Health and Human Services, nor does the mention of trade names, commercial products, or organizations imply endorsement by the U.S. Government.

REFERENCES

1. AT Balachunas, MJ Sepaniak. Anal Chem 60:617, 1988.
2. J Popsichal, M Deml, P Gebauer, P Bocek. J Chromatogr 470:43, 1989.
3. P Bocek, M Deml, J Popsichal, J Sudor. J Chromatogr 470:309, 1989.
4. V Sustacek, F Foret, P Bocek. J Chromatogr 480:271, 1989.
5. MJ Sepaniak, DF Swaile, AC Powell. J Chromatogr 480:185, 1989.
6. AC Powell, MJ Sepaniak. J Microcol Sep 2:278, 1990.
7. AC Powell, MJ Sepaniak. Anal Instrument 21:25, 1993.
8. T Tsuda. Anal Chem 64:386, 1992.
9. CW Wang, ES Yeung. Anal Chem 64:502, 1992.
10. H-T Chang, ES Yeung. J Chromatogr 608:65, 1992.
11. B Behnke, E Bayer. J Chromatogr 680:93, 1994.
12. C Yan, R Dadoo, RN Zare, DJ Rakestraw, DS Anex. Anal Chem 68:2726, 1996.
13. K Schmeer, B Behnke, E Bayer. Anal Chem 67:3656, 1995.
14. MR Taylor, P Teale, SA Westwood, D Perrett. Anal Chem 69:2554, 1997.
15. MR Taylor, P Teale. J Chromatogr A 768:89, 1997.
16. P Gfrorer, J Schewitz, K Psecker, L-H Tseng, K Albert, E Bayer. Electrophoresis 20:3, 1999.
17. K Seller, ZH Fan, K Fluri, J Harrison. Anal Chem 66:3485, 1994.
18. JP Kutter, SJ Jacobson, JM Ramsey. Anal Chem 69:5165, 1997.

32

A History of the Use and Measurement of Affinity Interactions in Electrophoresis

Niels H. H. Heegaard

Statens Serum Institut, Copenhagen, Denmark

I. INTRODUCTION

Separations in electrical fields have been successfully applied to analytical problems in the biological sciences throughout the twentieth century, and the principle of moving charged particles by electricity has been known for almost two centuries. As early as in 1807 Reuss communicated his observations on the effects of electrical fields on colloid clay particles suspended in a solution (published in 1809 [1]). Faraday and Du Bois-Reymond characterized the phenomenon in more detail later in the same century [2], and Michaelis coined the term electrophoresis in 1909 [3]. The foundations of electrophoretic separations in tubes were subsequently laid in Scandinavia by Arne Tiselius and later by Stellan Hjertén and Rauno Virtanen [4–6]. Tiselius in his early work on moving boundary electrophoresis discussed the following features: (a) on-line detection by ultraviolet light of analytes moving in transparent quartz capillaries; (b) the increase in separation efficiency with increase in field strengths; and (c) the realization of this in the efficient heat-dissipating format of capillaries [7,8]. Even though Tiselius and his colleagues and their successors in the moving boundary electrophoresis field for technical reasons used schlieren optics of refractive index gradients for detection as wells as large U-tubes made of glass and relatively low field strengths, the features outlined by Tiselius are in fact the basis of contemporary capillary electrophoresis (CE).

Before separation science entered this stage, however, zonal separations in supporting media replaced moving boundary separations during the 1940–1950s. The zonal separations were carried out in media such as filter paper [9] cellulose acetate [10], granular starch [11] or sieving starch gels [12], agar [13,14], and the now ubiquitous gels of polyacrylamide [15] and agarose [16].

With advances in the theory and practice of open tube electrophoresis and the emergence of new technology for manufacturing fused silica capillaries, sample injectors, on-line detectors, and finally, in 1988, commercial instruments, the pendulum during the last 10–15 years has come all the way back to electrophoresis in free solution but now in a happy marriage with the zonal format, i.e., as capillary zone electrophoresis (CZE) or just capillary electrophoresis (CE) [5,6,17–25]. Basically the tubes have gotten smaller (Fig. 1). CE is free solution zone electrophoresis made possible by the high field strengths that can be used without convection disturbances because the small capillary dimensions allow rapid heat dissipation [26]. CE combines the advantages of zone methods with those of free solution moving boundary electrophoresis. It avoids the low resolution, low field strengths, high sample consumption, and lack of full separation of individual components of mixtures inherent in the classical moving boundary methods. It also avoids the denaturing conditions, difficulty of quantitation, analyte interference with the supporting media, and restricted choice of buffer conditions characteristic of high-resolution gel electrophoresis methods (Table 1).

The notion that complex formation during separation may change molecular properties and thereby migration characteristics—in electrophoresis as well as in sedimentation and partition methods—was realized almost from the beginning of the use of electrophoresis as a separation method [27]. The classical affinity electrophoretic approach is analogous to affinity chromatography because the analyte (the receptor molecule) is electrophoresed through a gel or a buffer in which the ligand is present. However, affinity electrophoresis techniques also include analysis of preequilibrated samples where molecules are still engaged in complex formation during elec-

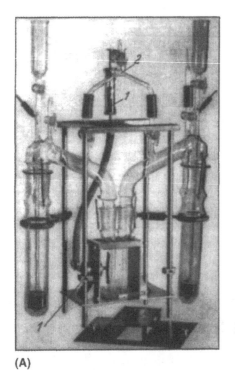

(A)   **(B)**

Figure 1 Electrophoresis in tubes, before and now. (A) Tiselius electrophoresis apparatus [237]. (B) The core of a modern capillary electrophoresis instrument (Waters Quanta), the capillary tubing is barely visible as a coil held together by a cross.

Table 1 Comparison Between Gel Electrophoresis and Capillary Zone Electrophoresis (CE)

Feature	Gel electrophoresis	CE
Analysis of small molecules	No	Yes
Analysis of macromolecules	Yes	Yes[a]
Resolution	Low	High[b]
Identification of analyte components	Many options	Fewer options[c]
Analyte quantitation	Indirect	Direct
Sample volume consumption	μL	nL
Detection limits	Moderate	Moderate-high[d]
Data processing	Manual	Computerized
Automation	No	Yes
Detection	Off-line, postelectrophoretic	On-line (-column), real time
Precision and reproducibility	Moderate	High
Parallel sample processing	Yes	No
Separation	Electrophoresis/Electrophoresis and friction	Electrophoresis and electro-osomosis
Set-up	Yes	Ready-for-use
Instrument costs	Low	High
Separation time	Slow	Fast[e]

[a] Often hampered by adsorption to the capillary wall.

[b] 10^5–10^6 theoretical plates are routinely achieved.

[c] Fraction collection is tedious and CE is essentially a nonpreparative method.

[d] With UV detection CE gives μM detection limits at best while gel electrophoresis is 100–1000 times better.

[e] Typical CE analysis times are 10–20 minutes and much faster in chip formats.

trophoresis. Obviously, all methods require a differential electrophoretic mobility of free and complexed molecules.

Affinity interactions can be used to increase the selectivity of electrophoretic separations, but the focus in the early work in moving boundary electrophoresis was to use electrophoresis as a tool to identify and characterize such interactions. Affinity zone electrophoresis in supporting media was subsequently demonstrated and native electrophoresis techniques in agarose gels developed into subfields of immunoelectrophoretic and affinity immunoelectrophoretic techniques. Modern high-resolution denaturing one- and two-dimensional gel methods are indispensable separation tools but cannot be used for affinity electrophoresis of native molecules. However, the development of the high-resolution native CE methods has resulted in new possibilities for affinity studies [28–39]. Finally, if mass spectrometry is included as a variant in vacuo of electrophoresis (the transport of charged particles in the presence of an electric field [40]), it is one of the most recent examples of electrokinetically based methods for measurement of complex formation [41]. A summary of important milestones in the use of electrophorertic methods for measurement of molecular interactions is given in Table 2.

Below, I will attempt to provide a history of affinity electrophoresis from the biased point of view of my own work in this field. We have mainly used affinity immunoelectrophoresis in agarose gels and affinity capillary electrophoresis in free solution [42]. However, I will try to ensure a logical context by including early and pertinent developments in electrophoresis and affinity electrophoresis where it seems appropriate.

Table 2 Milestones in the Development of Analytical Affinity Electrophoresis Methods

Event	Year	Ref.
Electrophoresis described	1807	1, 3
Apparatus for moving boundary electrophoresis developed	1930	7, 8
Theory of adsorption in chromatography	1940	88–90
Effect of protein interactions on electrophoretic patterns in moving boundary electrophoresis	1942	27, 49
Zone electrophoresis introduced	1946	230–232
Zone electrophoresis in gels	1950	12, 15, 16, 233, 234
Affinity interactions in zone electrophoresis	1954	67–69, 235
Development of quantitative agarose gel immunoelectrophoresis	1959	85, 121–124, 128, 129
Equations for determination of binding constants by zone electrophoresis	1970	73, 74, 96, 101
The term "affinity electrophoresis" introduced	1973	80
Modern capillary zone electrophoresis introduced	1974	6, 18–20, 24, 25
Gel shift assays for nucleic acid–binding proteins	1981	106, 107
Binding constant determination and interaction kinetics evaluation in capillary electrophoresis	1992	60, 153, 157, 158, 162, 163
Capillary electrophoresis immunoassays	1994	189, 197, 236
Complex formation demonstrated by mass spectrometry	1994	41

II. MOVING BOUNDARY ELECTROPHORESIS

Tiselius electrophoresis or free electrophoresis is a frontal separation method where complete separation never occurs because the sample fills almost the entire separation path [7,8,43]. It relies on the analysis of moving boundaries that represent partially separated sample compartments at the sample/buffer interfaces. The method is well suited for mobility measurements and was soon used for the measurement of mobility changes in analyses of low-affinity interacting components such as ovalbumin and nucleic acid [27], fatty acids and human serum albumin [44], thiocyanate anions and insulin [45], and other small ions [46,47]. It was realized that mobility and concentration data made it possible to calculate binding constants for such interactions when specific requirements were fulfilled [46,48,49]. Thus, more than 25 examples of interactions of proteins, including antibody-antigen interactions [50–55], could be listed in 1959 in a review on the applications of moving boundary electrophoresis to protein systems [56]. The majority of these interactions were examined qualitatively, i.e., by descriptions of differences in the shapes of the schlieren patterns. A notable example is the soluble antigen-antibody complexes of bovine serum albumin (BSA) and anti-BSA that were readily demonstrated in shifting proportions with different antigen-antibody ratios much in the same way (and with the same shapes) as when we almost 40 years later demonstrated antigen-antibody complex stoichiometries in agarose gel affinity immunoelectrophoresis [57] (Fig. 2).

A Tiselius electrophoresis apparatus (together with the stamina and business acumen of Niels Harboe, the founder of the Protein Laboratory) was indirectly responsible for the start of my own career in separation science because it was used to analyze patient sera for Danish hospitals. Thereby, it contributed the necessary economy for the Protein Laboratory of the University of Copenhagen to survive in the early years. Accordingly, the laboratory was therefore

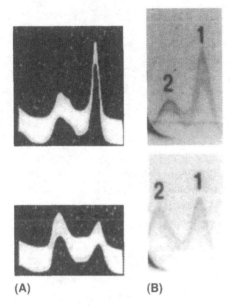

(A)   (B)

Figure 2 Antigen-antibody interactions analyzed by electrophoresis, before and now. (A) Moving boundary electrophoresis analysis of antigen-antibody complexes of varying stoichiometery [50–52]. (Modified from Ref. 50.) (B) Crossed affinity immunoelectrophoresis of α-fetoprotein–anti-α-fetoprotein complexes of different stoichiometry in agarose gels. The front complex (1) represents bivalent antigen-antibody complexes while the trailing complex (2) represents univalent complexes. (Modified from Ref. 57.)

very much in existence when I set out to do experimental work as a medical student in 1981 in Ole J. Bjerrum's group.

Today, moving boundary electrophoresis in the Tiselius format is practically never used. However, its principles have survived in the CE format in frontal analysis approaches where analyte fronts in large sample plugs containing equilibrated receptor-ligand molecules are analyzed, typically to estimate low-affinity drug binding to serum proteins [58–62]. Also, the vacancy peak method and other methods that originate in size-exclusion chromatography [63] have been employed for stoichiometry and binding constant evaluation of fast interactions [64] and are examples of the nonzonal mode of CE that may be useful in special cases [60].

III. ZONE ELECTROPHORESIS

A. Zone Electrophoresis in Solid Supports

The early years of zone electrophoresis concentrated on the application and optimization of a wide variety of solid supports such as powdered cellulose, granular starch, polyvinyl chloride resins, paper, and agar [65,66]. Molecular interactions do not appear to have been much studied by these methods except that electrophoresis was used in 1954 by Hagihawa to demonstrate antigen-antibody precipitation lines on paper strips [67] and by Nakamura and coworkers in 1958 [68], who demonstrated interactions between antigens and antibodies and between trypsin and trypsin inhibitors [69] using so-called crossing diagrams. It is interesting to note that the very low electrophoretic mobility of antibodies (γ-globulins) as compared with the bulk of other serum proteins, which is the basis of immunoelectrophoretic methods, was demonstrated by moving boundary electrophoresis already in 1937 by Tiselius [70]. The applications of affinity gel electrophoresis including the immunoelectrophoresis methods, however, did not really grow before polyacrylamide [15] and agarose [16] gels became the media of choice for sieving and nonsieving electrophoresis, respectively.

Nondenaturing polyacrylamide gel electrophoresis superseded paper [69,71] and starch [72] for the quantitative applications of affinity zone electrophoresis that were promoted by pioneering work in Japan [73–76] and elsewhere [77–79]. Despite the fact that agarose zone electrophoresis has a lower resolution than polyacrylamide gel electrophoresis, it proved very suitable for multidimensional affinity electrophoresis, especially in the affinity immunoelectrophoretic [80–82] and immunoelectrophoretic [83–86] formats that were primarily developed in Scandinavia at the time. The term "affinity electrophoresis" was first introduced by Bøg-Hansen in 1973 [80] and shortly thereafter by Takeo and coworkers [75].

1. Affinity Electrophoresis in Polyacrylamide Gels

Nondenaturing sieving polyacrylamide gel affinity electrophoresis is performed without sodium dodecyl sulfate and the separation thus depends on the size and charge-to-size ratio of the analyte. This ratio is prone to change when an analyte engages in interactions with other molecules. With few exceptions [87] the first affinity electrophoresis experiments in polyacrylamide gels were carried out so that complexed molecules had no mobility either because the ligand was covalently immobilized in the gel or because of the high molecular weight of complexes [73–75,78,79].

Quantitative aspects were launched already in the work using crossing diagrams in paper electrophoresis where estimates of relative concentrations of trypsin inhibitors in serum were presented [69]. The ideas were further developed in the ensuing publications [73,74]. It was realized that the methods were comparable to methods for interaction analysis in chromatography and the equations from the theory of adsorption in chromatography [88–90] were subsequently transformed into equations applicable to the electrophoretic interaction analysis. The analysis is based on the measurement of migration distances (d) (relative to an internal marker) as a function of the concentration of ligand in the gel. The theory is completely analogous in capillary electrophoresis, except that reciprocal peak appearance times, $1/t$, are used instead of d because of the different experimental readout (see below). In gel electrophoresis the electrophoretic mobility μ is directly proportional to the relative migration distance at a fixed time, i.e., the time at which the electrophoretic run is terminated. The assumptions behind the use of the migration shift equations are: establishment of instantaneous equilibrium, no diffusion, a homogeneously distributed ligand with a single binding site, a binding stoichiometry of 1:1, and a much higher ligand concentration than analyte concentration. An implicit assumption is that binding constants are not influenced by electric fields. Also, the analyses of gradual migration shifts require that the interaction is of low affinity, i.e., that the equilibrium is fast compared to the migration rates of the analyte and the complex, respectively. If this is true, then the two species of the analyte (the free analyte and the analyte-ligand complex) migrate as one band in the gel. The mobility (measured as d) of the detected band is then the result of the weighted average of the mobilities of the two analyte species, and the weights are determined by the equilibrium constant for the interaction and the ligand concentration [37,91–93.] The original linear equation [74] linked the dissociation constant (K_d) to the experimentally determined relative migration distances (d) as follows:

$$1/d = 1/D[1 + c/K_d)] \tag{1}$$

where D is the relative distance moved by the analyte in the absence of ligand and c is the concentration of ligand. Thus, when analyte-ligand complexes have a zero mobility, the inverse relative migration distances (1/d) plotted as a function of c yield K_d as the negative x-axis intercept of the straight line [74,75]. The first applications of electrophoresis to estimate binding constants using Eq. (1) were studies of glycan phosphorylases analyzed in glycogen-containing polyacrylamide gels [73,74]. Horejsí and coworkers were the first to report on lectin affinity electrophoresis in polyacrylamide gels in 1974 [77] and followed up with a large body of work on the practice and theory of both immobilized and free ligand affinity electrophoresis [93–96]. The methods were subsequently developed with other analyte-ligand systems such as antigen–monoclonal antibody/myeloma interactions [97,98], phosphorylases and α-amylases [75], and many others, as reviewed in Table 2 in the review by Takeo [76].

Based on their original work, Takeo and coworkers also devised a host of other approaches for affinity studies in molecular sieving gels during the 1970s [76]. Among these was the inhibition affinity approach where complexes are dissociated by competing ligands [99]. A further refinement of the technique was the two-dimensional affinity electrophoresis methods that combine isoelectric focusing and nondenaturing polyacrylamide gel electrophoresis in gels containing immobilized ligand and were employed for the quantitative investigation of the fine specificities of antibodies [100].

It seems very fitting that the founder of quantitative affinity electrophoresis in polyacrylamide gels and the inventor of affinity electrophoresis in agarose gels should join forces as they did in the work on lectin affinity agarose gel electrophoresis that resulted in the so-called general affinity equation in 1980 [101] shortly after Horejsí had described it using other terms [96]. This equation is generally applicable because it applies also to instances where the analyte-ligand complex does not have a zero mobility. This is typically the case in agarose gels where there is no sieving effect up to molecular sizes of $>10^7$ daltons [102]. The equation has the form:

$$1/(D\text{-}d) = 1/(D\text{-}d_c)[1 + K_d/c] \tag{2}$$

where d_c represents the mobility of the complex. Plots of 1/(D-d) as a function of 1/c yield $-1/K_d$. As it was pointed out later in connection with affinity CE experiments [103,104], it is actually advisable not to linearize experimental data but simply to plot them with (D-d) as a function of c according to the general rectangular hyperbolic form of the 1:1 binding isotherm [37,105] (here with relative migration distances substituted for electrophoretic mobilities):

$$(D\text{-}d) = (D\text{-}d_c)K_a c/(1 + K_a c) \tag{3}$$

From this the affinity constant (K_a) is extracted by curve-fitting programs. Equation (3) may be applied to any system involving affinity-induced mobility shifts provided the above requirements for measuring binding are met [37].

Affinity electrophoresis has had a life of its own in molecular biology where so-called gel retardation or electrophoretic mobility shift assays (EMSA) are used to reveal strong interactions between nucleic acid–binding proteins and DNA. These assays that were introduced in 1981 [106,107] are now part of the standard inventory of molecular biology cookbooks and are probably the affinity electrophoretic methods that are most used in contemporary science. The assay is usually done in polyacrylamide gels to get a clear separation of bound from free nucleic acids. Samples of preincubated radiolabeled nucleic acid mixed with protein (at different ratios) are analyzed on the gels. Protein binding alters the electrophoretic mobility of the nucleic acid and stable protein-DNA complexes are positioned as shifted bands with the exact position depending on the monospecificity of binding and on the stoichiometry of the DNA-protein complexes.

Amounts are estimated based on densitometry or radioactive counting of the bands. When DNA-protein complexes are less stable, the traditional affinity electrophoresis format with ligand in the gel may be useful [108].

2. *Affinity Electrophoresis in Agarose Gels*

Agarose is the almost neutral fraction of agar [109], and thus separations in agarose gels [16] are much less disturbed by electroosmosis and ionic interactions between the support and the analytes than separations in crude agar. Agarose gel electrophoresis in one dimension was used for affinity electrophoresis of heparin binding proteins [110]. This approach solved the problem of dissociation of intermediate affinity complexes (K_d up to 0.5 μM in this case) in traditional binding assays using preincubated samples. Two-dimensional lectin affinity electrophoresis in agarose gels followed by immunoblotting was devised by Taketa to separate α-fetoprotein glycoforms using different lectins in the first and second dimension electrophoresis [111].

Agarose gel electrophoresis was also used for the first affinophore experiments of Kasai and Shimura [112–114]. These methods introduced the principle of constructing ligands with artificially changed mobility by covalent attachment to charged carrier molecules (e.g., succinyl-poly-L-lysine). This gives a good handle in one- or two-dimensional electrophoresis to detect and separate receptor molecules even in complex sample mixtures [115].

In the rest of this section, I will concentrate on developments in the important subfield of immunoelectrophoretic and affinity immunoelectrophoretic techniques that were much practiced at the Protein Laboratory and that we found to be of considerable value for the detection, quantitation, and characterization of specific molecules in complex fluids.

The specificity and precipitating ability of polyclonal antibodies have been used with great effect in the many agarose gel electrophoresis methods that were developed at the Protein Laboratory, especially from 1970 to 1985. Numerous publications and many books attest to the fact that immunoelectrophoretic methods were the hallmark of this laboratory [83,116–118]. This is perhaps not surprising as Niels Harboe had spent the years 1946–1949 in Tiselius's laboratory perfecting the art of performing moving boundary electrophoresis. He also founded Dako Corporation, a company specializing in the production of rabbit antibodies. This company had its origin in the in-house production of antibodies at the Protein Laboratory. The combination of electrophoretic expertise, knowledge of protein chemistry, easy assess to large amounts of antibodies, and relatively meager budgets probably accounts for the intense use and development of immunoelectrophoretic techniques at this laboratory. I was subsequently heavily imprinted with these methods as a student in Bjerrum's section at the laboratory. Bjerrum himself had previously been a student of Niels Harboe at the Protein Laboratory, when he began his laboratory career as a medical student. We actually used the methods to the extent that the number of immunoelectrophoretic analyses far exceeded that of other electrophoretic separation techniques such as SDS-PAGE, and we always approached problems in protein analysis with the use of electroimmunoprecipitation methods in mind.

The methods all use electroimmunoprecipitation and are prime examples of harnessing affinity interactions, here the highly specific affinity of antibodies, to increase the selectivity of separations and to enable quantitation and identification of individual components in complex mixtures. Even though electroimmunoprecipitation is not always regarded as an affinity method, the quantitative immunoelectrophoretic techniques actually may be regarded as special examples of this class of electrophoresis methods because they involve antigen-antibody affinity interactions. The soluble antigen-antibody complexes develop into stationary precipitates in the gel, probably around the agarose helix bundles [102], but the process is not fully understood [119].

The immunoelectrophoretic analyses developed by Grabar [13] and by Macheboeuf et al. in

1953 [120] were popular during the 1960s and even later, but they are not immunoelectrophoretic methods as we define them today [84] because antigen-antibody mixing and precipitation take place by diffusion after the primary (first dimension) separation of the antigens by electrophoresis. This approach had poor quantitative capabilities but introduced the attractive concept of working in more than one dimension, e.g., for uncovering proteins previously hidden in overlapping sample zones. The first to actually demonstrate antigen-antibody precipitation in an electric field were Hagihawa, who did it on a paper strip in 1954 [67], and Nakamura and Ueta in 1958 with the crossing diagrams [68] mentioned above. These early advances gave birth to the whole field of electroimmunoprecipitation methods that were developed during the 1960s.

One-dimensional methods (rocket immunoelectrophoresis or electroimmunoassays) were introduced as a means of quantitating proteins in agar in 1959–1960 [121,122], and after the demonstration of the excellent properties of agarose for precipitation of antigen-antibody complexes [123] rocket immunoelectrophoresis [124,125] was widely used in both basic sciences and in clinical chemistry [102,126] to quantitate specific proteins. The electrophoresis step in this method is simply used to mix the differently charged reactants (antigens and antibodies). In combination with radioactively labeled antigen in a cathodic gel, the method was shown to get down to detection limits of 20 ng/mL (0.3 nM) for α-fetoprotein [126]. A useful variant of rocket immunoelectrophoresis called fused rocket immunoelectrophoresis was introduced in 1970 [127] to monitor the fate of specific proteins during a fractionation procedure.

One-dimensional affinity electrophoresis experiments have also been performed using non-precipitating antibodies, as shown in Fig. 3. Here, monoclonal antiinsulin antibodies are used

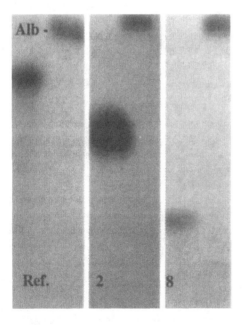

Figure 3 Autoradiograph of one-dimensional agarose gel zone electrophoresis of a mixture of human insulin and ^{125}I-albumin (Alb) in the presence of the stated amounts (μg/mL gel) of a monoclonal antibody (mAb) against bovine insulin (OX1-001, Novo Nordisk A/S). The mAb bound to the insulin band was visualized by incubating the gels with ^{125}I-labeled rabbit anti-mouse immunoglobulin antibodies. Association constants of 10^8–10^9 M^{-1} were estimated from a number of migration shifts such as those shown here. (Courtesy of Ole J. Bjerrum, Novo Nordisk A/S.)

in agarose zone electrophoresis to measure interactions with insulin in an analyte mixture of insulin and radiolabeled albumin (alb). The resolution, however, is poor and gel matrices with higher resolution (polyacrylamide [98]) or a combination with multidimensional techniques (see below) is usually required for quantitative analyses of such antigen-antibody interactions.

Two-dimensional immunoelectrophoretic methods have been very successfully used to increase the output of low-resolution one-dimensional agarose gel electrophoretic separations in the format of crossed immunoelectrophoresis as schematically shown in Fig. 4A. The principle of two-dimensional immunoelectrophoresis where electrophoresis moves the analytes in both dimensions was conceived and demonstrated by Ressler in 1960 [128]. He used a starch gel in the first dimension and an antibody-containing agar-solution in the second dimension. The application of electrophoresis instead of diffusion for the second dimension considerably shortened the time of the experiment and thereby minimized diffusion and increased resolution. In 1965 Laurell showed that the outcome was much improved by the use of agarose gel in both dimensions [85], and Clarke and Freeman in 1966 verified the quantitative aspects of the approach [129]. Laurell called the method crossed electrophoresis and Clarke and Freeman called

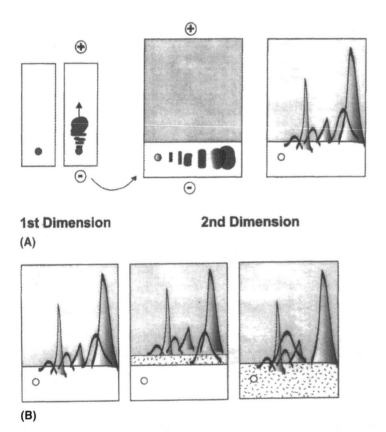

1st Dimension   2nd Dimension

(A)

(B)

Figure 4 (A) The principle of crossed immunoelectrophoresis. A mixture of antigens is separated by electrophoresis in the first dimension, transferred to the end of an antibody-containing gel, and subjected to electroimmunoprecipitation in the second dimension. (B) Principles of crossed affinity immunoelectrophoresis where reactive components are placed in an intermediate (middle panel) or in a first dimension (right panel) gel. One precipitate in the complex immunoprecipitation pattern shifts position in each case.

it immunoelectrophoresis, and it soon came to be known as crossed immunoelectrophoresis (CIE). By combining the separation by free electrophoretic mobility in the first dimension with the specificity and quantitative nature of immunoprecipitation in the second dimension, each component of complex protein mixtures is characterized by charge, quantity, and immunoreactivity. Many combinations of the basic principles (intermediate gels, tandem immunoelectrophoresis, radio-immunoelectrophoresis, crossed-line immunoelectrophoresis, and others) were developed to solve specific analytical problems as illustrated, e.g., by the work of Løwenstein on the characterization of allergens with quantitative immunoelectrophoretic methods [130].

A large body of work by Bjerrum and coworkers at the Protein Laboratory on detergent immunoelectrophoretic analysis of membrane antigens initiated in the beginning of the 1970s demonstrated that various integral membrane proteins could be solubilized and analyzed in detergent-containing agarose gels if mild nonionic detergents such as, e.g., Triton X-100, Tween 20, or Nonidet P-40 were used. The solubilized proteins were still reactive and precipitated with antibodies in the presence of these mild detergents [118,131–138]. Experiments where the immunoprecipitate of purified acetylcholinesterase from erythrocyte membranes was localized directly by esterase staining in the gel after completion of electrophoresis [139] further verified that detergent immunoelectrophoresis is a native separation method.

Three or more dimensions may be said to exist in experiments where two-dimensional immunoelectrophoresis is combined with specific interactions in an intermediate gel or in the first dimension gel (Fig. 4B). Typically, these approaches are used for low-affinity interacting molecules that influence the migration of specific analytes in the first dimension electrophoresis, which are then visualized by the second dimension immunoelectrophoresis. In the course of developing these techniques the term affinity electrophoresis was introduced [80], and the method subsequently was called crossed affinity immunoelectrophoresis. Early examples are the use of lectins to study glycoproteins in complex mixtures [80,82,101]. Lectins are proteins, often extracted from plants, that have specific but low-affinity carbohydrate-binding sites [140,141]. With lectins in solution or immobilized and added to an intermediate [80] (Fig. 4B, middle) or first dimension [82] (Fig. 4B, right) gel, it proved possible to characterize glycoproteins with respect to noncharged microheterogeneity in glycan structures and to get a prediction of the most likely behavior of the glycoproteins in lectin affinity chromatography. No qualitative differences were found between the results of analytical affinity electrophoresis and affinity chromatography [101]. Also, visualization and quantitation of glycoprotein microheterogeneity in health and disease could be achieved using these techniques, and these applications prompted the derivation of the general affinity equation to calculate binding constants in systems with free lectin [101]. Figure 5 shows the range of glycoforms of the human serum glycoprotein transferrin obtained by concanavalin A crossed affinity immunoelectrophoresis spanning from nonreactive (fraction 0) to most reactive species (fraction 3). This protein—as the only protein in the serum sample—is visualized because a monospecific antitransferrin antiserum is used in the second dimension. The microheterogeneity is very pronounced in this case because the sample is from an alcoholic patient [142,143]. The result is dependent on the differential mobility of concanavalin A compared with transferrin at the pH 8.6 of the electrophoresis buffer. It proved straightforward to ensure specificity of the observed interactions simply by adding specific displacing (competitive) monosaccharides [144], and this is also used in the second dimension to ensure liberation of all affinity precipitated material when the technique is used for quantitative estimates of the amounts of glycoforms [142].

Another example of the use of first dimension selectivity enhancement is charge shift electrophoresis where negatively or positively charged mild detergents alter analyte mobility through participating in mixed micelles with hydrophobically bound nonionic detergent. This was origi-

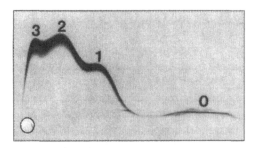

Figure 5 Crossed affinity immunoelectrophoresis with the lectin concanavalin A incorporated into the first dimension gel in an analysis of serum from an alcoholic patient. The microheterogeneity pattern of transferrin is due to the differential reactivity of differently glycosylated transferrin molelcules with the almost immobile concanavalin A and is revealed in the second dimension electroimmunoprecipitation pattern with a monospecific antibody against transferrin. Without concanavalin A in the first dimension the precipitate had a symmetrical bell-shaped curve and was positioned as fraction 0. (Based on results reported in Ref. 142.)

nally shown in one-dimensional agarose gel electrophoresis to characterize protein amphiphilicity [145]. It was introduced in combination with crossed immunoelectrophoresis [146] to enable the simultaneous characterization of charge, amphiphilicity (detergent binding), immunoreactivity, and quantity of proteins in sample mixtures.

Also, the affinity and fine specificity of monoclonal antibodies could be evaluated by crossed affinity immunoelectrophoresis by incorporating the monoclonal antibodies in solution into the first dimension electrophoresis [147]. This is an extension of the type of experiment shown in Fig. 3 because a second immunoprecipitation gel is added at a right angle to the migration path of the first gel (see also Fig. 4). Since monoclonal antibodies usually are not precipitating but of low mobility, their interactions are revealed as cathodic shifts in the positions of the precipitates [147,148]. The shifts were used to estimate binding constants for monoclonal antibodies against albumin and α-fetoprotein, and, importantly, the values were verified by independent methods [57].

Another illustration of the use of more than one set of antibodies is from our own work on the characterization of the fraction of circulating immunoglobulins that is able to bind with low affinity to a group of red blood cell membrane proteins [149–151]. These studies illustrated that some interactions will be artificially amplified by the nonphysiological low ionic strength conditions that are usually employed in gel electrohoresis [149]. Also, in that respect it was a definite improvement when capillary electrophoresis (CE) became available because CE is compatible with buffers of higher ionic strength. One example of this are the separations we showed in a study of the binding of C-reactive protein to Ca^{2+}-ions [152].

Despite the many virtues of the immunoelectrophoretic methods that seemed obvious to us and which were discussed above, they were never much used outside of specialized centers. This is probably because the techniques are dependent on the availability of good precipitating antibodies relevant for the analytical problem and because they are perceived as laborious and difficult, not very robust, and difficult to interpret.

B. Free Solution Zone Electrophoresis in Capillaries

The advent of modern capillary electrophoresis (CE) [5,6,18–20,24,25] and the introduction of the first commercial instrument in 1988 meant new possibilities for affinity electrophoresis.

First, the versatility of CE with respect to the types of molecules that can be analyzed is much greater than any other known electrophoretic technique. This is, for example, true for small molecules that are difficult if not impossible to analyze by gel electrophoresis because of lack of resolution and difficulties of detection. When I joined the group of Frank Robey at the National Institutes of Health in 1991 as a visiting fellow, CE was still a young technique and had to my knowledge not been used for affinity studies. As it happened we were interested in heparin-binding peptides and proteins [153,154], and we soon were one of the first groups to embark on measuring binding by CE. Results were published in 1992 [153] almost simultaneously with an avalanche of other papers on the uses of CE for binding studies [60,155–164]. Small molecules such as peptides were among the first classes of compounds investigated in binding studies by Whitesides and colleagues and in my own work [153,157,158,165–168]. The high and reproducible resolution of CE gives an increased versatility with respect to the types of complexes that can be separated from free molecules and a more easy use of mobility correction factors. The precision of peak appearance times also means that binding constants can be determined with much higher precision than in affinity gel electrophoresis. Thus, the significance of changes in binding constants with slight changes in buffer pH or other parameters are easier to evaluate as we did in a study on the binding of a monoclonal antibody to DNA [169]. As in the gel approaches, the affinity CE methods are employed to determine binding constants and to discover binding partners. However, the selectivity enhancement caused by even very weak affinity interactions [170] has had profound implications for separations in CE because of the high separation efficiency and broad applicability of this technique. A unique and important subfield of CE thus has emerged as the stereoselective separation of small enantiomeric molecules that is possible when weak affinity ligands (so-called additives) are included in the electrophoresis buffer [171–173].

Secondly, on-line quantitation by computer-assisted peak integration has made direct binding measurements straightforward in comparison with the indirect and imprecise methods in quantitative gel electrophoresis. Some detection strategies—e.g., laser-induced fluorescence detection that has been used from the onset of modern CE [19]—offer detection limits that are lower than those obtainable by other electrophoretic techniques and therefore enable high-sensitivity quantitation of analytes.

A unique feature of CE is its minimal sample consumption as illustrated in the extreme by the first experiments by Yeung and coworkers in which they analyzed the contents of single cells [174]. Subsequently, the detection of DNA-protein binding in large cells such as sea urchin oocytes by CE has been demonstrated [175]. Finally, there are some newer approaches using CE or hybrid CE-chromatography separation schemes with immobilized ligands or imprinted polymer matrices that probably must be considered unique for the capillary format and have been promoted especially by Staffan Nilsson and his team members [176–178].

There is no fundamental difference between the principles and uses of affinity interactions in gel electrophoresis and in CE. Consequently, most approaches for affinity electrophoresis in gels were adapted to the capillary format such as, e.g., methods with ligand addition to the electrophoresis buffer in free solution [153,155,158,163] as well as immobilized or entangled ligands (in capillary gel electrophoresis) [159,161], and the affinophoresis and affinity probe approaches [28,29,179]. Preequilibration approaches have been typical for most applications of immunoassays in the CE format [180–182]. Some points in favor of traditional gel electrophoretic techniques should not be forgotten (see also Table 1): lower detection limits of gel electrophoresis than CE with UV-based detection, relatively low cost, simultaneous analysis of multiple samples, and more easily achieved postelectrophoretic identification of components by immunochemical or other means, e.g., in the crossed immunoelectrophoresis and immunoblotting techniques. A second dimension analysis is not easily achieved in CE, and the identity of peaks is often a source of trouble. The coupling of CE with electrospray-ionization mass spectrometry as pioneered by the groups of Smith and Karger [183–187] therefore seems to be an important way ahead.

The areas of affinity applications where CE has been especially valuable, i.e., for analyte quantitation, ligand discovery, quantitative characterization of binding, and selectivity enhancement, are reviewed below.

1. Analyte Quantitation Based on Affinity Interactions in CE

Strong affinity reactions and laser-induced fluoresence detection are the basis of CE-based immunoassays for quantitation of analytes. Mostly antibodies [188] or antibody Fab fragments [28] have been used. The electrophoresis step is simply a way to separate bound from free analyte, much as is the case in the use of strong interactions in affinity gel electrophoresis. The first examples of immunochemical quantitation by CE emerged in the 1990s [188–191]. The most straightfoward approach is that exemplified by Kennedy and coworkers [192]. They used a competitive immunoassay format with a preformed complex between antibody and fluoresceinylated antigen. The signal delivered from free fluoresceinylated antigen when it is competed away by antigen in a sample is used to quantitate that antigen in the sample. This method has a high sensitivity with detection limits in the high pM range [193], low sample volume requirements, and in chip formats [190,194–196] the possibility for separations so fast that real-time monitoring of the secretion of specific components (here, insulin) from tissues or cells becomes a reality as shown by the work during the 1990s by Kennedy and his group [192,193,197,198], who also demonstrated the convenience of CE to evaluate the binding characteristics of the antibodies before using them for quantitative assays [198]. Using capillary isoelectric focusing and the affinity probe principle, a detection limit for recombinant human growth hormone of 5 pM was reported by Shimura and Karger in work using antibody fragments [28]. Picomolar detection limits were also reported by Le and colleagues in their work using a CE-based immunoassay to measure bromodeoxyuridine (as a marker of DNA damage) in DNA [199]. Specific and strongly binding compounds other than antibodies may also be used for analyte quantitation by CE such as in the innovative use of so-called aptamers (selectively binding oligonucleotides identified in combinatorial libraries) [200] and other specific binders including enzymes with strong affinity for drugs, e.g., fluorescently labeled carbonic anhydrase, that enabled the detection of dorzolamide with a 62 pM detection level in serum [201].

2. Affinity CE for the Discovery of Binding in Complex Mixtures

The high resolution of CE was used in the discovery and quantitation of affinity ligands in complex mixtures from the beginning of the 1990s [157,165]. The main difficulty lies in the final identification of the binding molecules due to the very limited sample amounts that give limited possibilities for collecting fractions or use of orthogonal identification methods. As mentioned above, this is one reason why the work on coupling affinity CE to mass spectrometry [186,187] appears very promising and is much needed. Otherwise, the principles do not differ from those developed for gel affinity electrophoretic systems. As CE is very well suited for the analysis of small molecules that are difficult to separate in other native systems, the approach seems ideal, e.g., for screening of tryptic or other proteolytic digests of proteins for active fragments as we showed with the heparin-binding protein serum amyloid P component [154,167,168]. We identified two active peptides by their shifts in appearance time when the highly anionic glycosaminoglycan heparin was added to the electrophoresis buffer. Figure 6 shows a portion of the CE analyses from which it appears that a peak (marked with a star) is

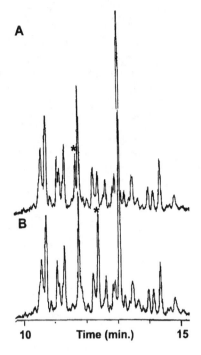

Figure 6 Affinity CE–based identification of a heparin binding peptide in a tryptic digest of serum amyloid P component. The separation in (A) is compared with the one in (B) where heparin is present in the buffer. One peptide (marked with a star) is clearly displaced to a later peak appearance time due to binding interactions of the peptide with the strongly negatively charged heparin during electrophoresis. The bulk of the remaining peptides is unaffected by the presence of heparin. (Modified from Ref. 154.)

displaced anodically (i.e., has an increased peak appearance time) when heparin is present in the electrophoresis buffer, while the many other peptides are unaffected by the negatively charged ligand. The complex peptide mixture is hereby both efficiently separated and characterized functionally, i.e., the approach is two-dimensional. The identification of this peptide relied on co-injections with purified peptides of known structure and on heparin affinity CE experiments with the HPLC-purified peptide and subfragments thereof [202,203].

3. Affinity CE for Estimation of Binding Constants

The quantitative uses of affinity CE were developed during the 1990s and have been the subject of several recent reviews [31,32,34,35,37–39,170,204,205]. Apart from the speed and precision and the much better analyte and buffer versatiliy, the considerations within this field are exactly the same as in quantitative gel affinity electrophoresis [37]. The speed of the CE approach will, however, as shown in recent years—and in contrast to gel methods—in some cases make it possible to analyze weak interactions in preincubated samples by using very fast separations as, for example, on chips [190]. CE is also well suited for the on-line study of weakly interacting systems as was shown in numerous cases for small analytes in the beginning of the 1990s [31,32,157,166]. As in affinity gel electrophoresis measurement and characterization of complexation rely solely on measurements of the analyte migration, here peak appearance time (t), in the presence of a known concentration of ligand. It was important to realize that in CE where the migration distance (d) is a constant, t is inversely proportional to the electrophoretic mobility (μ). In gel electrophoresis where the electrophoresis time is a constant, d is directly proportional to μ. As a consequence the correct substitution for d in the affinity equations is 1/t and not t, as pointed out [37,179] after we had made the error in one of our early publications [152].

Also, as in affinity gel electrophoresis the analyte does not necessarily have to be pure because of the electrophoretic separation from other components in the sample. The dissociation and association rates of the interaction must be fast enough compared with the time required for the analyte peak to appear to ensure that no peak tailing, broadening, splitting, or disappearance will occur. In practical terms the use of migration shift affinity CE has been shown to be limited to interactions that are sufficiently quick to ensure that complex dissociation half-times ($\ln 2/k_{off}$) are equal to or less than 1% of the time required to separate free from bound molecules [92,93,108]. Thus, longer separation times may enable stronger interactions to be studied by on-line affinity CE methods. Slower interactions may be quantitatively estimated by analyses of changes in the analyte zone width, as was demonstrated by Whitesides' group in 1993 [206].

Since the normalized reciprocal of the peak appearance time (1/t) is directly proportional to electrophoretic mobility, Eq. (3) was translated [179] into:

$$\Delta(1/t) = (\Delta(1/t)_{max}K_ac)/(1 + K_ac) \tag{4}$$

and linearization may be performed according to:

$$\Delta(1/t) = \Delta(1/t)_{max} - K_d(\Delta(1/t)/c) \tag{5}$$

where $\Delta(1/t)$ is the difference between the inverse normalized peak appearance time in the control experiment without ligand added and in the experiment with the ligand added to the electrophoresis buffer and K_a is the association constant. The normalization of the 1/t data is achieved by using an internal noninteracting standard. This—in much the same way as the various correction factors introduced by others [37,164,207–209]—ensures a correction of the peak appearance time data for viscosity and other changes in buffer characteristics that is a function of ligand addition but nonspecific for binding. Internal standards are equally important in quantitative affinity gel electrophoresis as pointed out by the developers of these techniques [76,101]. As noted above, it is worthwhile to realize that all approaches to linearization introduce errors so it is better to estimate binding constants by nonlinear regression methods. We and others have verified the obtained binding constants by independent methods or by comparison with litterature values [158,169].

4. Affinity in CE Used as a Selectivity Tool

A final area in which the joined forces of CE and affinity interactions have been developed to achieve unique results in comparison with other methods is in the use of very weak interactions to enhance separation selectivities. As early as 1988 Gutman and Karger reported on the use of cyclodextrins as selective complexing agents in capillary gel electrophoresis separations of optical isomers [172]. In 1991 Gutman and Cooke further reported on the selectivity increase in capillary gel electrophoretic separation of DNA fragments after adding ethidium bromide to the polyacrylamide [161]. As is clear from the very first affinity experiments in moving boundary electrophoresis, it is possible to influence the individual electrophoretic mobilities of analytes in CE by introducing specifically interacting molecules into the electrophoresis buffer. When the main purpose is to get better separations, i.e., to enhance the selectivity in CE very low affinity ligands—in this case also called additives—are commonly used at rather high concentrations. The high concentration of additives often alters the physical properties of the electrophoresis buffer such as viscosity and dielectric constant, and this must be corrected for before embarking on quantitative analyses of the interactions as mentioned above in connection with the use of internal mobility standards [207–210].

Many small organic molecules including pharamaceuticals have been separated by capillary micellar electrokinetic chromatography (MEKC) to the extent that it has become a subfield of its own [171]. The principles were first demonstrated and used by Terabe and coworkers, who increased the selectivity in the analysis of noncharged compounds by adding detergent, often SDS, or a mixture of detergents to the electrophoresis buffer [211,212]. Simultaneous separation of both ionic and uncharged and structurally similar analytes hereby became possible [213]. The separation depends on electrophoresis in combination with partition between the aqueous and pseudostationary micellar phase based on the degree of hydrophobicity of the analyte [173]. This is reminiscent of the interactions exploited in charge-shift gel electrophoresis. Many different detergents including zwitterionic types [214,215] have been demonstrated for the purpose [173], and in 1987 it was shown that also nucleotides and their building blocks could be separated using MEKC with metal additives [216].

With the addition of compounds that generate chiral micelles, the approach may also enable the separation of enantiomeric molecules [217]. Enantiomeric or chiral separations by CE and MEKC approaches have been succesful in a field where there are few alternative methods. Optical resolution in CE separations was shown for the first time by this approach already in 1985 in a separation of D and L amino acids upon addition of a Cu^{2+}–L-histidine complex to the electrophoresis buffer as a complexing agent [218]. In this field the use of different types of uncharged and charged cyclodextrins as optical selectors has also been popular after a number of publications from Terabe's group and collaborators [219–221]. These compounds are used together with micellar compounds or alone if charged cyclodextrins are used [173]. Other additives with selective—i.e., low-affinity—binding of enantiomeric compounds such as serum proteins have also been shown to be effective for chiral separations in free solution [222,223] and when covalently linked to stationary phases [176]. A practical problem that is caused by the high UV absorbance of many of these additives can be circumvented by the partial filling approach where buffer with additive does not fill the entire capillary and leaves the detector window free [224,225]. This has the added advantage of minimizing the consumption of the costly chiral selectors, especially cyclodextrins.

IV. CONCLUSIONS AND PREDICTIONS

The history of analytical affinity electrophoresis has shown that old ideas can be impressively revitalized by the combination of new technology and innovative instrument manufacturers. The principle of incorporating additives or ligands that interact with analyte(s) in an electrophoretic separation has been used almost as long as electrophoresis techniques have been in use and has led to substantial increases in separation selectivities, to possibilities for specific quantitation of analytes in complex mixtures, and to quantitative characterization of reversible binding interactions. Affinity electrophoresis separations are two-dimensional because a field (electrical) and molecular equilibria simultaneously affect analyte migration. Multidimensionality results in a very powerful separation as is illustrated by the different immunoelectrophoretic techniques and by affinity electrophoreses in capillaries. The maxim that the more things change the more they stay the same remains true for affinity electrophoresis, since open tube separations have succesfully returned and are beginning to yield new discoveries. This is true despite the presentation of a staggering amount of model systems to demonstrate that affinity CE methods actually work and despite the fact that it is very much used as a complementary tool to other methods. The affinity electrophoresis concept has attracted a very productive interplay between groups of physical chemists, chemists, and engineers and groups of biomedical researchers—the first group looking for applications and the second group looking for tools to address specific problems in biology. The challenge of lucidly outlining the potential applications of the developments in curiosity-driven basic separation science for these users has not always been met, but this now appears to be beginning to improve.

Future developments will most likely originate in existing separation technologies that will be combined in new ways to exploit the synergism of multidimensional techniques such as shown by the examples of CE–reversed-phase HPLC [226], CE-MS [183–187], and CE-NMR [227–229]. With the aid of affinity reactions, these separations will be even more powerful. It is also evident that if convenient combinations with immunodetection techniques can be developed, there will be a vast potential for increased use of the technique, as was the case with the quantitative electroimmunoprecipitation methods. Finally, lower instrument costs, new detector designs with lower detection limits, and nanotechnology developments within the fields of microfluidics and microarrays to develop miniaturized analytical devices will expand the uses of affinity CE even more.

ACKNOWLEDGMENTS

Professor Ole J. Bjerrum and Dr. Frank A. Robey created the opportunities for me to develop agarose gel and capillary electrophoresis techniques, respectively. Drs. Anne-Marie Heegaard and Peter M. H. Heegaard are thanked for their inspiring support and lasting friendship throughout the years.

REFERENCES

1. FF Reuss. Sur un nouvel effet de l'electricité galvanique. Mém Soc Impér Natural Moscou 2:327, 1809.
2. GW Gray. Electrophoresis. Sci Am 185:45–49, 1951.
3. L Michaelis. Electric transport of ferments. (I) Invertin Biochem Z 16:81, 1909.
4. S Hjertén. Free zone electrophoresis. Protides Biol Fluids Proc Colloq 7:28–30, 1959.
5. S Hjertén. Free zone electrophoresis. Chromatogr Rev 9:122–219, 1967.
6. R Virtanen. Zone electrophoresis in a narrow-bore tube employing potentiometric detection. A theoretical and experimental study. Acta Polytech Scand Chem Incl Metall 123:1–67, 1974.
7. A Tiselius. A new apparatus for electrophoretic analysis of colloidal mixtures. Trans Faraday Soc 33:524–531, 1937.
8. A Tiselius. The moving boundary method of studying the electrophoresis of proteins. Nova acta regiae societatis scientiarum upsaliensis. Ser. IV, 7, No.4, 1–107, 1930.
9. HG Kunkel, A Tiselius. Electrophoresis of proteins on filter paper. J Gen Physiol 35:89–118, 1952.
10. J Kohn. A cellulose acetate supporting medium for electrophoresis. Clin Chim Acta 2:297–303, 1957.
11. HG Kunkel, RJ Slater. Zone electrophoresis in a starch supporting medium. Proc Soc Exp Biol Med 80:42–44, 1952.
12. O Smithies. Zone electrophoresis in starch gels: group variations in the serum proteins of normal human adults. Biochem J 61:629–641, 1955.
13. P Grabar, CA Williams. Méthode permettant l'étude conjugée des propriétés électrophorétiques et immunochimiques d'un mélange de protéines. Application au sérum sanguin. Biochim Biophys Acta 10:193–194, 1953.
14. P Grabar, CA Williams. Méthode immuno-électrophorétique d'analyse de mélanges de substances antigéniques. Biochim Biophys Acta 17:67, 1955.
15. S Raymond, L Weintraub. Acrylamide gel as a supporting medium for zone electrophoresis. Science 130:711, 1959.

16. S Hjertén. Agarose as an anticonvection agent in zone electrophoresis. Biochim Biophys Acta 53: 514–517, 1961.

17. V Pretorius, BJ Hopkins, JD Schieke. Electro-osmosis. A new concept for high-speed liquid chromatography. J Chromatogr 99:23–30, 1974.

18. FEP Mikkers, FM Everaerts, TPEM Verheggen. High performance zone electrophoresis. J Chromatogr 169:11–20, 1979.

19. JW Jorgenson, KD Lukacs. Zone electrophoresis in open-tubular glass capillaries. Anal Chem 53: 1298–1302, 1981.

20. JW Jorgenson, KD Lukacs. Capillary zone electrophoresis. Science 222:266–272, 1983.

21. BL Karger. High-performance capillary electrophoresis. Nature 339:641–642, 1989.

22. MJ Gordon, X Huang, SL Pentoney, Jr., RN Zare. Capillary electrophoresis. Science 242:224–228, 1988.

23. AS Cohen, A Paulus, BL Karger. High-performance capillary electrophoresis using open tubes and gels. Chromatographia 24:15–24, 1987.

24. JW Jorgenson, KD Lukacs. Free-zone electrophoresis in glass capillaries. Clin Chem 27:1551–1553, 1981.

25. JW Jorgenson, KD Lukacs. High-resolution separations based on electrophoresis and electroosmosis. J Chromatogr 218:209–216, 1981.

26. PD Grossman. Factors affecting the performance of capillary electrophoresis separations: Joule heating, electroosmosis, and zone dispersion. In: PD Grossman, JC Colburn, eds. Capillary Electrophoresis. San Diego, CA: Academic Press, Inc., 1992, pp. 3–43.

27. LG Longsworth, DA MacInnes. An electrophoretic study of mixtures of ovalbumin and yeast nucleic acid. J Gen Physiol 25:507–516, 1942.

28. K Shimura, BL Karger. Affinity probe capillary electrophoresis: analysis of recombinant human growth hormone with a fluorescent labeled antibody fragment. Anal Chem 66:9–15, 1994.

29. K Shimura, K Kasai. Capillary affinophoresis as a versatile tool for the study of biomolecular interactions: a mini-review. J Mol Recogn 11:134–140, 1998.

30. NHH Heegaard, K Shimura. Determination of affinity constants by capillary electrophoresis. In: P Lundahl, A Lundqvist, E Greijer, eds. Quantitative Analysis of Biospecific Interactions. Amsterdam: Harwood Academic Publishers, 1998, pp. 15–34.

31. Y-H Chu, LZ Avila, J Gao, GM Whitesides. Affinity capillary electrophoresis. Acc Chem Res 28: 461–468, 1995.

32. IJ Colton, JD Carbeck, J Rao, GM Whitesides. Affinity capillary electrophoresis: a physical-organic tool for studying interactions in biomolecular recognition. Electrophoresis 19:367–382, 1998.

33. NHH Heegaard, FA Robey. The emerging role of capillary electrophoresis as a tool for the study of biomolecular noncovalent interactions. Am Lab 26:T28–X28, 1994.

34. NHH Heegaard. Biospecific interactions measured by capillary electrophoresis. In: K Standing, ed. New Methods for the Study of Molecular Aggregates. Dordrecht: Kluwer Academic Publishers, 1997, pp. 305–318.

35. NHH Heegaard. Capillary electrophoresis for the study of affinity interactions. J Mol Recogn 11: 141–148, 1998.

36. K Shimura, K-I Kasai. Affinity capillary electrophoresis: a sensitive tool for the study of molecular interactions and its use in microscale analyses. Anal Biochem 251:1–16, 1997.

37. KL Rundlett, DW Armstrong. Methods for the estimation of binding constants by capillary electrophoresis. Electrophoresis 18:2194–2202, 1997.

38. NHH Heegaard. Capillary electrophoresis. In: S Harding, PZ Chowdhry, eds. Protein-Ligand Interactions: A Practical Approach, Vol. 1. Oxford, UK: Oxford University Press, 2001, pp. 171–195.

39. NHH Heegaard, RT Kennedy. Identification, quantitation, and characterization of biomolecules by capillary electrophoretic analysis of binding interactions. Electrophoresis 20:3122–3133, 1999.

40. CR Cantor, PR Schimmel, Biophysical Chemistry. Part II: Techniques for the study of biological structure and function. New York: W. H. Freeman and Company, 1980.

41. KJ Light-Wahl, BL Schwartz, RD Smith. Observation of the noncovalent quaternary associations of proteins by electrospray ionization mass spectrometry. J Am Chem Soc 116:5271–5278, 1994.

42. NHH Heegaard. Characterization of biomolecules by electrophoretic analysis of reversible interactions. Appl Theor Electrophoresis 4:43–63, 1994.
43. RA Mosher, W Thormann. Moving boundary electrophoresis. In: RA Mosher, DA Saville, W Thormann, eds. The Dynamics of Electrophoresis. Weinheim, Germany: VCH Verlag, 1992, pp. 89–101.
44. GA Ballou, PD Boyer, JM Luck. J Biol Chem 159:111, 1945.
45. E Volkin. J Biol Chem 175:675, 1948.
46. RA Alberty, HH Marvin Jr. J Am Chem Soc 73:3220, 1951.
47. RA Alberty, EL King. Moving boundary systems formed by weak electrolytes. Study of cadmium iodide complexes. J Am Chem Soc 73:517–523, 1951.
48. RA Alberty, HH Marvin Jr. Study of protein-ion interaction by the moving boundary method. J Phys Colloid Chem 54:47–55, 1950.
49. LG Longsworth. Moving boundary electrophoresis—theory. In: M Bier, ed. Electrophoresis. Theory, Methods and Applicationa. New York: Academic Press, Inc., 1959, pp. 91–136.
50. SJ Singer, DH Campbell. Physical chemical studies of soluble antigen-antibody complexes. I. The valence of precipitating rabbit antibody. J Am Chem Soc 74:1794–1802, 1952.
51. SJ Singer, DH Campbell. Physical chemical studies of soluble antigen-antibody complexes. II. Equilibrium properties. J Am Chem Soc 75:5577–5578, 1953.
52. SJ Singer, DH Campbell. Physical chemical studies of soluble antigen-antibody complexes. III. Thermodynamics of the reaction between bovine serum albumin and its rabbit antibodies. J Am Chem Soc 77:3499–3504, 1955.
53. SJ Singer, DH Campbell. Physical chemical studies of soluble antigen-antibody complexes. IV. The effect of pH on the reaction between bovine serum albumin and its rabbit antibodies. J Am Chem Soc 77:3504–3510, 1955.
54. SJ Singer, DH Campbell. Physical chemical studies of soluble antigen-antibody complexes. V. Thermodynamics of the reaction between ovalbumin and its rabbit antibodies. J Am Chem Soc 77: 4851–4855, 1955.
55. SJ Singer, L Eggman, DH Campbell. Physical chemical studies of soluble antigen-antibody complexes. VI. The effect of pH on the reaction between ovalbumin and its rabbit antibodies. J Am Chem Soc 77:4855–4867, 1955.
56. RA Brown, SN Timasheff. Applications of moving boundary electrophoresis to protein systems. In: M Bier, ed. Electrophoresis. Theory, Methods and Applications. New York: Academic Press, Inc., 1959, pp. 317–367.
57. NHH Heegaard, OJ Bjerrum. Affinity electrophoresis used for determination of binding constants for antibody-antigen reactions. Anal Biochem 195:319–326, 1991.
58. T Ohara, A Shibukawa, T Nakagawa. Capillary electrophoresis/frontal analysis for microanalysis of enantioselective protein binding of a basic drug. Anal Chem 67:3520–3525, 1995.
59. A Shibukawa, Y Yoshimoto, T Ohara, T Nakagawa. High-performance capillary electrophoresis/frontal analysis for the study of protein binding of a basic drug. Anal Chem 83:616–619, 1994.
60. JC Kraak, S Busch, H Poppe. Study of protein-drug binding using capillary zone electrophoresis. J Chromatogr 608:257–264, 1992.
61. PA McDonnell, GW Caldwell, JA Masucci. Using capillary electrophoresis/frontal analysis to screen drugs interacting with human serum proteins. Electrophoresis 19:448–454, 1998.
62. NA Mohamed, Y Kuroda, A Shibukawa, T Nakagawa, S El Gizawy, HF Askal, ME El Kommos. Binding analysis of nilvadipine to plasma lipoproteins by capillary electrophoresis-frontal analysis. J Pharm Biomed Anal 21:1037–1043, 1999.
63. JP Hummel, WJ Dreyer. Measurement of protein-binding phenomena by gel filtration. Biochim Biophys Acta 63:530–532, 1962.
64. Y-H Chu, WJ Lees, A Stassinopoulus, CT Walsh. Using affinity capillary electrophoresis to determine binding stoichiometries of protein-ligand interactions. Biochemistry 33:10616–10621, 1994.
65. C Wunderly. Paper electrophoresis. In: M Bier, ed. Electrophoresis. Theory, Methods, and Applications. New York: Academic Press, Inc., 1959, pp. 179–223.
66. HG Kunkel, R Trautman. Zone electrophoresis in various types of supporting media. In: M Bier,

ed. Electrophoresis. Theory, Methods, and Applications. New York: Academic Press, Inc., 1959, pp. 225–262.

67. F Hagihara. Paper micro-electrophoresis. IV. Antigen-antibody reaction on filter paper (in Japanese). J Pharm Soc Jpn 74:999–1001, 1954.

68. S Nakamura, T Ueta. 'Crossing' paper electrophoresis for the detection of immune reactions. Nature 182:875, 1958.

69. S Nakamura, T Wakeyama. An attempt to demonstrate the distribution of trypsin inhibitors in the sera of various animals. J Biochem Tokyo 49:733–741, 1961.

70. A Tiselius. CLXXXII. Electrophoresis of serum globulin. II. Electrophoretic analysis of normal and immune sera. Biochem J 31:1464–1477, 1937.

71. S Nakamura, R Suzuno. Crystallization of concanavalins A and B and canavalin from japanese jack beans. Arch Biochem Biophys 111:499–505, 1965.

72. G Entlicher, M Tichá, JV Kostír, J Kocourek. Experientia 25:17, 1969.

73. K Takeo. Detection of phosphorylase on disc electrophoresis using polyacrylamide as supporting medium. Annu Rep Soc Protein Chem Yamaguchi Univ School Med 4:41–48, 1970.

74. K Takeo, S Nakamura. Dissociation constants of glucan phosphorylases of rabbit tissues studied by polyacrylamide gel disc electrophoresis. Arch Biochem Biophys 153:1–7, 1972.

75. K Takeo, A Kuwahara, H Nakayama, S Nakamura. Affinity electrophoresis of phosphorylases and α-amylases. Protides Biol Fluids Proc Colloq 23, 645, 1975.

76. K Takeo. Affinity electrophoresis. In: A Chrambach, MJ Dunn, BJ Radola, eds. Advances in Electrophoresis. Vol. I. Weinheim: VCH Verlag, 1987, pp. 229–279.

77. V Horejsí, J Kocourek. Biochim Biophys Acta 336:338, 1974.

78. M Szylit. Electrophorèse avec adsorption specifique: Interactions amylase-amidon en gel mixte di polyacrylamide-amidon. Ann Biol Clin 29:215–227, 1971.

79. SJ Gerbrandy, A Doorgeest. Potato phosphorylase isoenzymes. Phytochemistry 11:2403–2407, 1972.

80. TC Bøg-Hansen. Crossed immuno-affinoelectrophoresis. An analytical method to predict the result of affinity chromatography. Anal Biochem 56:480–488, 1973.

81. TC Bøg-Hansen, C-H Brogren. Identification of glycoproteins with one and with two or more binding sites to Con A by crossed immuno-affinoelectrophoresis. Scand J Immunol 4 (suppl.2):135–139, 1975.

82. TC Bøg-Hansen, OJ Bjerrum, J Ramlau. Detection of biospecific interaction during the first dimension electrophoresis in crossed immunoelectrophoresis. Scand J Immunol 4 (suppl.2):141–147, 1975.

83. NH Axelsen, J Kroll, B Weeke. A manual of quantitative immunoelectrophoresis. Methods and applications. Scand J Immunol 2 (suppl.1), 1973.

84. TC Bøg-Hansen. Immunoelectrophoresis. In: BD Hames, D Rickwood, eds. Gel Electrophoresis of Proteins. A Practical Approach. Oxford: Oxford University Press, 1990, pp. 273–300.

85. C-B Laurell. Antigen-antibody crossed electrophoresis. Anal Biochem 10:358–361, 1965.

86. PMH Heegaard, TC Bøg-Hansen. Immunoelectrophoretic methods. In: MJ Dunn, ed. Gel Electrophoresis of Proteins. Bristol: Wright Publishers, 1986, pp. 262–311.

87. B Lerch, H Stegemann. Gel electrophoresis of proteins in borate buffers: Influence of some compounds complexing with boric acid. Anal Biochem 29:76–83, 1969.

88. JN Wilson. A theory of chromatography. J Am Chem Soc 62:1583–1591, 1940.

89. J Weiss. On the theory of chromatography. J Chem Soc 81:297–303, 1943.

90. D DeVault. The theory of chromatography. J Am Chem Soc 65:532–540, 1943.

91. X Peng, MT Bowser, P Britz-McKibbin, GM Bebault, JR Morris, DD Chen. Quantitative description of analyte migration behavior based on dynamic complexation in capillary electrophoresis with one or more additives. Electrophoresis 18:706–716, 1997.

92. M Mammen, FA Gomez, GM Whitesides. Determination of the binding of ligands containing the N-2,4-Dinitrophenyl group to bivalent monoclonal rat anti-DNP antibody using affinity capillary electrophoresis. Anal Chem 67:3526–3535, 1995.

93. V Matousek, V Horejsí. Affinity electrophoresis: a theoretical study of the effects of the kinetics of protein-ligand complex formation and dissociation reactions. J Chromatogr 245:271–290, 1982.

94. V Horejsí, M Tichá, P Tichy, A Holy. Affinity electrophoresis: new simple and general methods of preparation of affinity gels. Anal Biochem 125:358–369, 1982.

95. V Horejsí, M Tichá. Qualitative and quantitative applications of affinity electrophoresis for the study of protein-ligand interactions: a review. J Chromatogr 376:49–67, 1986.

96. V Horejsí. Some theoretical aspects of affinity electrophoresis. J Chromatogr 178:1–13, 1979.

97. M Caron, A Faure, RG Keros, P Camillot. Application of affinity electrophoresis to the study of antigen-immunoadsorbent association equilibrium. Biochim Biophys Acta 491:558–565, 1977.

98. K Takeo, EA Kabat. Binding constants of dextran and isomaltose oligosaccharides to dextran-specific myeloma proteins determined by affinity electrophoresis. J Immunol 121:2305–2310, 1978.

99. K Takeo. Affinity electrophoresis: principles and applications. Electrophoresis 4:187–195, 1984.

100. K Takeo, R Suzuno, T Tanaka, M Fujimoto, A Kuwahara, K Nakamura. Two-dimensional affinity electrophoresis: its application to separation of individual immunoglobulins from heterogeneous antibodies. Protides Biol Fluids Proc Colloq 32:969–972, 1985.

101. TC Bøg-Hansen, K Takeo. Determination of dissociation constants by affinity electrophoresis: complexes between human serum proteins and concanavalin A. Electrophoresis 1:67–71, 1980.

102. C-B Laurell, EJ McKay. Electroimmunoassay. Methods Enzymol 73:339–369, 1981.

103. MT Bowser, DDY Chen. Monte Carlo simulation of error propagation in the determination of binding constants from rectangular hyperbolae. 2. Effect of the maximum-response range. J Phys Chem A 103:197–202, 1999.

104. MT Bowser, DDY Chen. Monte Carlo simulation of error propagation in the determination of binding constants from rectangular hyperbolae. 1. Ligand concentration range and binding constant. J Phys Chem A 102:8063–8071, 1998.

105. KL Rundlett, DW Armstrong. Examination of the origin, variation, and proper use of expressions for the estimation of association constants by capillary electrophoresis. J Chromatogr A 721:173–186, 1996.

106. MM Garner, A Revzin. A gel electrophoresis method for quantifying the binding of proteins to specific DNA regions: application to components of the *Escherichia coli* lactose operon regulatory system. Nucleic Acids Res 9:3047–3060, 1981.

107. M Fried, DM Crothers. Equilibria and kinetics of lac repressor-operator interactions by polyacrylamide gel electrophoresis. Nucleic Acids Res 9:6505–6525, 1981.

108. WA Lim, RT Sauer, AD Lander. Analysis of DNA-protein interactions by affinity coelectrophoresis. Methods Enzymol 208:196–210, 1991.

109. C Araki. Structure of the agarose content of agar-agar. Bull Chem Soc Jpn 29:543–544, 1956.

110. MK Lee, AD Lander. Analysis of affinity and structural selectivity in the binding of proteins to glycosaminoglycans: development of a sensitive electrophoretic approach. Proc Natl Acad Sci USA 88:2768–2772, 1991.

111. K Taketa, E Ichikawa, J Sato, H Taga, H Hirai. Two-dimensional lectin affinity electrophoresis of alpha-fetoprotein: characterization of erythroagglutinating phytohemagglutinin-dependent microheterogeneity forms. Electrophoresis 10:825–829, 1989.

112. K Shimura, K Kasai. Affinophoresis of pea lectin and fava bean lectin with an anionic affinophore, bearing *p*-aminophenyl-D-mannoside as an affinity ligand. J Chromatogr 400:353–359, 1987.

113. K Shimura. Progress in affinophoresis. J Chromatogr 510:251–270, 1990.

114. K Shimura, K-I Kasai. Affinophoresis of trypsins. J Biochem Tokyo 92:1615–1622, 1982.

115. K Shimura, K Kasai. Affinophoresis: selective electrophoretic separation of proteins using specific carriers. Methods Enzymol 271:203–218, 1996.

116. NH Axelsen. Quantitative immunoelectrophoresis. New developments and applications. Scand J Immunol 4 (suppl. 2), 1975.

117. NH Axelsen. Handbook of immunoprecipitation-in-gel techniques. Scand J Immunol 17 (suppl. 10), 1983.

118. OJ Bjerrum. Electroimmunochemical Analysis of Membrane Proteins. Amsterdam: Elsevier, 1983.

119. NH Axelsen, E Bock, P Larsen, S Blirup-Jensen, PJ Svendsen, KJ Pluzek, OJ Bjerrum, TC Bøg-Hansen, J Ramlau. Immunoprecipitation in an electrical field. Scand J Immunol 17 (suppl. 10):87–96, 1983.

120. M Macheboeuf, P Rebeyrotte, J-M Dubert, M Brunerie. Microélectrophorèse sur papier avec evaporation continue du solvant (électrorhéophorèse). Bull Soc Chem Biol 35:334–345, 1953.

121. A Bussard. Description d'une technique combinant simultanément l'électrophorèse et la précipitation immunologique dans un gel: l'électrosynérèse. Biochim Biophys Acta 34:258–260, 1959.

122. N Ressler. Electrophoresis of serum protein antigens in an antibody-containing buffer. Clin Chim Acta 5:359–365, 1960.

123. S Brishammar, S Hjertén, B von Hofstein. Immunological precipitates in agarose gels. Biochim Biophys Acta 53:518–521, 1961.

124. C-B Laurell. Quantitative estimation of proteins by electrophoresis in agarose gel containing antibodies. Anal Biochem 15:45–52, 1966.

125. C-B Laurell. Quantitative estimation of proteins by electrophoresis in antibody-containing agarose gel. Protides Biol Fluids Proc Colloq 14:499–502, 1966.

126. B Nørgaard-Petersen. A highly sensitive radioimmunoelectrophoretic quantitation of human α-fetoprotein. Clin Chim Acta 48:345, 1973.

127. PJ Svendsen, C Rose. Separation of proteins using ampholine carrier ampholytes as buffer and spacer ions in an isotachophoresis system. Sci Tools 17:13–17, 1970.

128. N Ressler. Two-dimensional electrophoresis of protein antigens with an antibody containing buffer. Clin Chim Acta 5:795–800, 1960.

129. HGM Clarke, T Freeman. A quantitative immuno-electrophoresis method (Laurell electrophoresis). Protides Biol Fluids Proc Colloq 14:503–509, 1966.

130. H Løwenstein. Quantitative immunoelectrophoretic methods as a tool for the analysis and isolation of allergens. Prog Allergy 25:1–62, 1978.

131. OJ Bjerrum. Detergent-immunoelectrophoresis. General principles and methodology. In: OJ Bjerrum, ed. Electroimmunochemical Analysis of Membrane Proteins. Amsterdam: Elsevier, 1983, pp. 3–43.

132. OJ Bjerrum. Immunochemical investigation of membrane proteins. A methodological survey with emphasis placed on immunoprecipitation in gels. Biochim Biophys Acta 472:135–195, 1977.

133. OJ Bjerrum, P Lundahl. Crossed immunoelectrophoresis of human erythrocyte membrane proteins. Immunoprecipitation patterns for fresh and stored samples of membranes extensively solubilized with non-ionic detergents. Biochim Biophys Acta 342:69–80, 1974.

134. OJ Bjerrum, TC Bøg-Hansen. The immunochemical approach to the characterization of membrane proteins. Human erythrocyte membrane proteins analysed as a model system. Biochim Biophys Acta 455:66–89, 1976.

135. OJ Bjerrum, P Lundahl. Detergent-containing gels for immunological studies of solubilized erythrocyte membrane components. Scand J Immunol 2 (Suppl. 1):139–143, 1973.

136. OJ Bjerrum, I Hagen. Biomolecular characterization of membrane antigens. In: OJ Bjerrum, ed. Electroimmunochemical Analysis of Membrane Proteins. Amsterdam: Elsevier Science Publishers, 1983, pp. 77–115.

137. OJ Bjerrum, S Bhakdi. Demonstration of binding of Triton X-100 to amphiphilic proteins in crossed immunoelectrophoresis. FEBS Lett 81:151–156, 1977.

138. OJ Bjerrum. Analysis of membrane antigens by means of quantitative detergent-immunoelectrophoresis. In: A Azzi, U Brodbeck, P Zahler, eds. Membrane Proteins. A Laboratory Manual. Berlin: Springer Verlag, 1981, pp. 13–42.

139. OJ Bjerrum, J Selmer, J Hangaard, F Larsen. Isolation of human erythrocyte acetylcholinesterase using phase separation with Triton X-114 and monoclonal immunosorbent chromatography. J Appl Biochem 7:356–369, 1985.

140. WC Boyd, E Shapleigh. Specific precipitating activity of plant agglutinins (lectins). Science 119:419–419, 1954.

141.   H Debray, D Decout, G Strecker, G Spik, J Montreuil. Specificity of twelve lectins towards oligosac-
       charides and glycopeptides related to N-glycosylproteins. Eur J Biochem 117:41–55, 1981.
142.   NHH Heegaard, M Hagerup, ÅC Thomsen, PMH Heegaard. Concanavalin A crossed affinity immu-
       noelectrophoresis and image analysis for semiquantitative evaluation of microheterogeneity profiles
       of human serum transferrin from alcoholics and normal individuals. Electrophoresis 10:836–840,
       1989.
143.   T Inoue, M Yamauchi, K Ohkawa. Structural studies on sugar chains of carbohydrate-deficient
       transferrin from patients with alcoholic liver disease using lectin affinity electrophoresis. Electro-
       phoresis 20:452–457, 1999.
144.   PMH Heegaard, NHH Heegaard, TC Bøg-Hansen. Affinity electrophoresis for the characterization
       of glycoproteins—the use of lectins in combination with immunoelectrophoresis. In: J Breborowicz,
       A Mackiewicz, eds. Affinity Electrophoresis: Principles and Application. Boca Raton, FL: CRC
       Press, 1992, pp. 3–21.
145.   A Helenius, K Simons. Charge shift electrophoresis: Simple method for distinguishing between
       amphiphilic and hydrophilic proteins in detergent solution. Proc Natl Acad Sci USA 74:529–532,
       1977.
146.   S Bhakdi, B Bhakdi-Lehnen, OJ Bjerrum. Detection of amphiphilic proteins and peptides in com-
       plex mixtures. Charge-shift crossed immunoelectrophoresis and two-dimensional charge-shift elec-
       trophoresis. Biochim Biophys Acta 470:35–44, 1977.
147.   J Breborowicz, J Gan, J Klosin. Application of monoclonal antibodies for affinity electrophoresis.
       J Immunol Methods 102:101–107, 1987.
148.   A Bodenteich, LG Mitchell, CR Merril, BE Miller, HA Ghanbari, NHH Heegaard. Immunochemical
       characterization of a monoclonal antibody specific for Alzheimer's disease associated protein. J
       Neuroimmunol 41:111–116, 1992.
149.   NHH Heegaard. Interactions of IgG with specific erythrocyte membrane proteins in affinity electro-
       phoresis are highly dependent on low ionic strength conditions. Anal Biochem 208:317–322,
       1993.
150.   NHH Heegaard. Immunochemical characterization of interactions between circulating autologous
       IgG and normal erythrocyte membrane proteins. Biochim Biophys Acta 1023:239–246, 1990.
151.   NHH Heegaard, OJ Bjerrum. Immunoelectrophoretic demonstration of circulating IgG antibodies
       against detergent-solubilized membrane proteins from erythrocytes in healthy humans. In: C
       Schafer-Nielsen, ed. Electrophoresis '88. Weinheim: VCH Verlag, 1988, pp. 424–431.
152.   NHH Heegaard, FA Robey. A capillary electrophoresis-based assay for the binding of Ca^{2+} and
       phosphorylcholine to human C-reactive protein. J Immunol Methods 166:103–110, 1993.
153.   NHH Heegaard, FA Robey. Use of capillary zone electrophoresis to evaluate the binding of anionic
       carbohydrates to synthetic peptides derived from serum amyloid P component. Anal Chem 64:
       2479–2482, 1992.
154.   NHH Heegaard, PMH Heegaard, P Roepstorff, FA Robey. Ligand binding sites in human serum
       amyloid P component. Eur J Biochem 239:850–856, 1996.
155.   H Kajiwara, H Hirano, K Oono. Binding shift assay of parvalbumin, calmodulin and carbonic anhy-
       drase by high-performance capillary electrophoresis. J Biochem Biophys Methods 22:263–268,
       1991.
156.   H Kajiwara. Application of high-performance capillary electrophoresis to the analysis of conforma-
       tion and interaction of metal-binding proteins. J Chromatogr 559:345–356, 1991.
157.   Y-H Chu, GM Whitesides. Affinity capillary electrophoresis can simultaneously measure binding
       constants of multiple peptides to vancomycin. J Org Chem 57:3524–3525, 1992.
158.   Y-H Chu, LZ Avila, HA Biebuyck, GM Whitesides. Use of affinity capillary electrophoresis to
       measure binding constants of ligands to proteins. J Med Chem 35:2915–2917, 1992.
159.   Y Baba, M Tsuhako, T Sawa, M Akashi, E Yashima. Specific base recognition of oligodeoxynucleo-
       tides by capillary affinity gel electrophoresis using polyacrylamide-poly(9-vinyladenine) conjugated
       gel. Anal Chem 64:1920–1925, 1992.
160.   GE Barker, P Russo, RA Hartwick. Chiral separation of leucovorin with bovine serum albumin
       using affinity capillary electrophoresis. Anal Chem 64:3024, 1992.

161. A Guttman, N Cooke. Capillary gel affinity electrophoresis of DNA fragments. Anal Chem 63: 2038–2042, 1991.

162. JL Carpenter, P Camilleri, D Dhanak, D Goodall. A study of the binding of vancomycin to dipeptides using capillary electrophoresis. J Chem Soc Chem Commun 804–806, 1992.

163. S Honda, A Taga, K Suzuki, S Suzuki, K Kakehi. Determination of the association constant of monovalent mode protein-sugar interaction by capillary zone electrophoresis. J Chromatogr 597: 377–382, 1992.

164. SAC Wren, RC Rowe. Theoretical aspects of chiral separation in capillary electrophoresis. J Chromatogr 603:235–241, 1992.

165. Y-H Chu, LZ Avila, HA Biebuyck, GM Whitesides. Using affinity capillary electrophoresis to identify the peptide in a peptide library that binds most tightly to vancomycin. J Org Chem 58: 648–652, 1993.

166. NHH Heegaard, FA Robey. Use of capillary zone electrophoresis for the analysis of DNA-binding to a peptide derived from amyloid P component. J Liq Chromatogr 16:1923–1939, 1993.

167. NHH Heegaard, P Roepstorff. Preparative capillary electrophoresis and mass spectrometry for the identification of a putative heparin-binding site in amyloid P component. J Cap Elec 2:219–223, 1995.

168. NHH Heegaard, HD Mortensen, P Roepstorff. Demonstration of a heparin-binding site in serum amyloid P component using affinity capillary electrophoresis as an adjunct technique. J Chromatogr 717:83–90, 1995.

169. NHH Heegaard, DT Olsen, K-LP Larsen. Immuno-capillary electrophoresis for the characterization of a monoclonal antibody against DNA. J Chromatogr 744:285–294, 1996.

170. NHH Heegaard, MH Nissen, DDY Chen. Applications of on-line weak affinity interactions in free solution capillary electrophoresis. Bioseparation, 2001, in press.

171. H Nishi. Capillary electrophoresis of drugs: current status in the analysis of pharmaceuticals. Electrophoresis 20:3237–3258, 1999.

172. A Guttman, A Paulus, AS Cohen, N Grinberg, BL Karger. Use of complexing agents for selective separation in high-performance capillary electrophoresis. Chiral resolution via cyclodextrins incorporated within polyacrylamide gel columns. J Chromatogr 448:41–53, 1988.

173. N Matsubara, S Terabe. Micellar electrokinetic chromatography. Methods Enzymol 270:319–341, 1996.

174. ES Yeung. Study of single cells by using capillary electrophoresis and native fluorescence detection. J Chromatogr 830:243–262, 1999.

175. J Xian, MG Harrington, EH Davidson. DNA-protein binding assays from a single sea urchin egg: a high-sensitivity capillary electrophoresis method. Proc Natl Acad Sci USA 93:86–90, 1996.

176. H Ljungberg, S Nilsson. Protein-based capillary affinity gel elecrtophoresis for chiral separation of β-adrenergic blockers. J Liq Chromatogr 18:3685–3698, 1995.

177. S Nilsson, L Schweitz, M Petersson. Three approaches to enantiomer separation of β-adrenergic antagonists by capillary electrochromatography. Electrophoresis 18:884–890, 1997.

178. L Schweitz, LI Andersson, S Nilsson. Capillary electrochromatography with predetermined selectivity obtained through molecular imprinting. Anal Chem 69:1179–1183, 1997.

179. K Shimura, K Kasai. Determination of the affinity constants of concanavalin A for monosaccharides by fluorescence affinity probe capillary electrophoresis. Anal Biochem 227:186–194, 1995.

180. D Schmalzing, W Nashabeh. Capillary electrophoresis based immunoassays: a critical review. Electrophoresis 18:2184–2193, 1997.

181. RT Kennedy, L Tao, NM Schultz, DR Rose. Immunoassays and enzyme assays using CE. In: J Landers, ed. CRC Handbook of Capillary Electrophoresis. Boca Raton, FL: CRC Press, 1997, pp. 523–545.

182. NM Schultz, L Tao, DJ Rose, RT Kennedy. Immunoassays and enzyme assays using capillary electrophoresis. In: JP Landers, ed. Handbook of Capillary Electrophoresis. Boca Raton, FL: CRC Press, 1997, pp. 611–637.

183. JA Olivares, NT Nguyen, CR Yonker, RD Smith. On-line mass spectrometric detection for capillary zone electrophoresis. Anal Chem 59:1230–1232, 1987.

184. RD Smith, JA Olivares, NT Nguyen, HR Udseth. Capillary zone electrophoresis-mass spectrometry using an electrospray ionization interface. Anal Chem 60:436–441, 1988.

185. TJ Thompson, F Foret, P Vouros, BL Karger. Capillary electrophoresis/Electrospray ionization mass spectrometry: improvement of protein detection limits using on-column transient isotachophoretic sample preconcentration. Anal Chem 65:900–906, 1993.

186. Y-H Chu, YM Dunayevskiy, DP Kirby, P Vouros, BL Karger. Affinity capillary electrophoresis-mass spectrometry for screening combinatorial libraries. J Am Chem Soc 118:7827–7835, 1996.

187. YM Dunayevskiy, YV Lyubarskaya, YH Chu, P Vouros, BL Karger. Simultanous measurement of nineteen binding constants of peptides to vancomycin using affinity capillary electrophoresis-mass spectrometry. J Med Chem 41:1201–1204, 1998.

188. F Chen, RA Evangelista. Feasibility studies for simultaneous immunochemical multianalyte drug assay by capillary electrophoresis with laser-induced fluorescence. Clin Chem 40:1819–1822, 1994.

189. D Schmalzing, W Nashabeh, X-W Yao, R Mhatre, FE Regnier, NB Afeyan, M Fuchs. Capillary electrophoresis-based immunoassay for cortisol in serum. Anal Chem 67:606–612, 1995.

190. LB Koutny, D Schmalzing, TA Taylor, M Fuchs. Microchip electrophoretic immunoassay for serum cortisol. Anal Chem 68:18–22, 1996.

191. D Schmalzing, W Nashabeh, M Fuchs. Solution-phase immunoassay for determination of cortisol in serum by capillary electrophoresis. Clin Chem 41:1403–1406, 1995.

192. L Tao, CA Aspinwall, RT Kennedy. On-line competitive immunoassay based on capillary electrophoresis applied to monitoring insulin secretion from single islets of Langerhans. Electrophoresis 19:403–408, 1998.

193. L Tao, RT Kennedy. On-line competitive immunoassay for insulin based on capillary electrophoresis with laser-induced fluorescence detection. Anal Chem 68:3899–3906, 1996.

194. N Chiem, DJ Harrison. Microchip-based capillary electrophoresis for immunoassays: analysis of monoclonal antibodies and theophylline. Anal Chem 69:373–378, 1997.

195. M Shakuntala, C Colyer, DJ Harrison. Microchip-based capillary electrophoresis of human serum proteins. J Chromatogr A 781: 271–276, 1997.

196. N Chiem, DJ Harrison. Microchip systems for immunoassay: an integrated immunoreactor with electrophoretic separation for serum theophylline determination. Clin Chem 44:591–598, 1998.

197. NM Schultz, L Huang, RT Kennedy. Capillary electrophoresis-based immunoassay to determine insulin content and insulin seretion from single islets of langerhans. Anal Chem 67:924–929, 1995.

198. L Tao, RT Kennedy. Measurement of antibody-antigen dissociation constants using fast capillary electrophoresis with laser-induced fluorescence detection. Electrophoresis 18:112–117, 1997.

199. XC Le, JZ Xing, J Lee, SA Leadon, M Weinfeld. Inducible repair of thymine glycol detected by an ultrasensitive assay for DNA damage. Science 280:1066–1069, 1998.

200. I German, DD Buchanan, RT Kennedy. Aptamers as ligands in affinity probe capillary electrophoresis. Anal Chem 70:4540–4545, 1998.

201. RC Tim, RA Kautz, BL Karger. Ultratrace analysis of drugs in biological fluids using affinity probe capillary electrophoresis: analysis of dorzolamide with fluorescently labeled carbonic anhydrase. Electrophoresis 21:220–226, 2000.

202. NHH Heegaard. A heparin-binding peptide from human serum amyloid P component characterized by affinity capillary electrophoresis. Electrophoresis 19:442–447, 1998.

203. NHH Heegaard. Microscale characterization of the structure-activity relationship of a heparin-binding glycopeptide using affinity capillary electrophoresis and immobilized enzymes. J Chromatogr A 853:189–195, 1999.

204. AL Vergnon, YH Chu. Electrophoretic methods for studying protein-protein interactions. Methods 19:270–277, 1999.

205. Y-H Chu, CC Cheng. Affinity capillary electrophoresis in biomolecular recognition. Cell Mol Life Sci 54:663–683, 1998.

206. LZ Avila, Y-H Chu, EC Blossey, GM Whitesides. Use of affinity capillary electrophoresis to determine kinetic and equilibrium constants for binding of arylsulfonamides to bovine carbonic anhydrase. J Med Chem 36:126–133, 1993.

207. SG Penn, DM Goodall, JS Loran. Differential binding of tioconazole enantiomers to hydroxypropyl-β-cyclodextrin studied by capillary electrophoresis. J Chromatogr 636:149–152, 1993.

208. A Shibukawa, DK Lloyd, IW Wainer. Simultaneous chiral separation of leucovorin and its major metabolite 5-methyl-tetrahydrofolate by capillary electrophoresis using cyclodextrins as chiral selectors: estimation of the formation constant and mobility of the solute-cyclodextrin complexes. Chromatographia 35:419–429, 1993.

209. KL Rundlett, DW Armstrong. Effect of micelles and mixed micelles on efficiency and selectivity of antibiotic-based capillary electrophoretic enantioseparations. Anal Chem 67:2088–2095, 1995.

210. R Kuhn, S Hoffstetter-Kuhn. Capillary Electrophoresis: Principles and Practice. New York: Springer Verlag, 1993, pp. 91–93.

211. S Terabe, K Otsuka, K Ichikawa, A Tsuchiya, T Ando. Electrokinetic separations with micellar solutions and open-tubular capillaries. Anal Chem 56:111–113, 1984.

212. S Terabe, K Otsuka, T Ando. Electrokinetic chromatography with micellar solution and open-tubular capillary. Anal Chem 57:834–841, 1985.

213. H Nishi, S Terabe. Application of micellar electrokinetic chromatography to pharmaceutical analysis. Electrophoresis 11:691–701, 1990.

214. HK Kristensen, SH Hansen. Separation of polymyxins by micellar electrokinetic capillary chromatography. J Chromatogr 628:309–315, 1993.

215. KF Greve, W Nashabeh, BL Karger. Use of zwitterionic detergents for the separation of closely related peptides by capillary electrophoresis. J Chromatogr 680:15–24, 1994.

216. AS Cohen, S Terabe, JA Smith, BL Karger. High-performance capillary electrophoretic separation of bases, nucleosides, and oligonucleotides: retention manipulation via micellar solutions and metal additives. Anal Chem 59:1021–1027, 1987.

217. H Nishi, T Fukuyama, M Matsuo, S Terabe. Separation and determination of lipophilic corticosteroids and benzothiazepin analogues by micellar electrokinetic chromatography using bile salts. J Chromatogr 513:279–295, 1990.

218. E Gassmann, JE Kuo, RN Zare. Electrokinetic separation of chiral compounds. Science 230:813–814, 1985.

219. S Terabe, K Ozaki, K Otsuka, T Ando. Electrokinetic chromatography with 2-O-carboxymethyl-β-cyclodextrin as a moving "stationary" phase. J Chromatogr 332:211–217, 1985.

220. H Nishi, T Fukuyama, S Terabe. Chiral separation by cyclodextrin-modified micellar electrokinetic chromatography. J Chromatogr 553:503–516, 1991.

221. S Terabe, Y Miyashita, O Shibata, ER Barnhart, LR Alexander, DG Patterson, BL Karger, K Hosoya, N Tanaka. Separation of highly hydrophobic compounds by cyclodextrin-modified electrokinetic chromatography. J Chromatogr 516:23–31, 1990.

222. GE Barker, P Russo, RA Hartwick. Chiral separation of leucovorin with bovine serum albumin using affinity capillary electrophoresis. Anal Chem 64:3024–3028, 1992.

223. GE Barker, WJ Horvath, CW Huie, RA Hartwick. Separation of type I and III isomers of copro- and uroporphyrins using affinity capillary electrophoresis. J Liq Chromatogr 16:2089–2101, 1993.

224. Y Tanaka, K Otsuka, S Terabe. Separation of enantiomers by capillary electrophoresis-mass spectrometry employing a partial filling technique with a chiral crown ether. J Chromatogr A 875:323–330, 2000.

225. Y Tanaka, S Terabe. Separation of the enantiomers of basic drugs by affinity capillary electrophoresis using a partial filling technique and α₁-acid glycoprotein as chiral selector. Chromatographia 44:119–128, 1997.

226. AW Moore, Jr., JP Larmann, Jr., AV Lemmo, JW Jorgenson. Two-dimensional liquid chromatography-capillary electrophoresis techniques for analysis of proteins and peptides. Methods Enzymol 270:401–419, 1996.

227. DL Olson, TL Peck, AG Webb, RL Magin, JV Sweedler. High-resolution microcoil ¹H-NMR for mass-limited, nanoliter-volume samples. Science 270:1967–1970, 1995.

228. DL Olson, TL Peck, AG Webb, JV Sweedler. On-line NMR detection for capillary electrophoresis

applied to peptide analysis. In: PTP Kaumaya, RS Hodges, eds. Peptides: Chemistry, Structure and Biology. Kingswinford: Mayflower Scientific Ltd., 1996, pp. 730–731.

229. N Wu, TL Peck, AG Webb, RL Margin, JV Sweedler. Nanoliter volume sample cells for ^{1}H NMR: application to on-line detection in capillary electrophoresis. J Am Chem Soc 116:7929–7930, 1994.

230. R Consden, AH Gordon, AJP Martin. Ionophoresis in silica jelly. A method for the separation of amino-acids and peptides. Biochem J 40:33–41, 1946.

231. D Cremer, A Tiselius. Elektrophorese von eisweiss in filtrierpapier. Biochem Z 320:273–283, 1950.

232. EL Durrum. A microelectrophoretic and microionophoretic technique. J Am Chem Soc 72:2943–2948, 1950.

233. AH Gordon, B Keil, K Sebasta, O Knessl, F Sorm. Electrophoresis of proteins in agar jelly. Coll Czechoslov Chem Commun 15:1–16, 1950.

234. J Kendall, ED Crittenden. Proc 7th Congr Euro Soc Haematol 9:75, 1923.

235. N Lang. Ein verfahren zum nachweis von reaktionen vom typ der antigen-antikörperbindung. Klin Wochenschr 33:29–30, 1955.

236. F-T Chen, JC Sterberg. Characterization of proteins by capillary electrophoresis in fused-silica columns: review on serum protein analysis and application to immunoassays. Electrophoresis 15:13–21, 1994.

237. HJ Antweiler. Zur Methode der quantitativen Elektrophorese. In: HJ Antweiler, ed. Die quantitative Elektrophorese in der Medizin. Berlin: Springer Verlag, 1957, pp. 1–39.

33

Direct Enantiomeric Separations in Liquid Chromatography and Gas Chromatography

Daniel W. Armstrong

Iowa State University, Ames, Iowa

I. INTRODUCTION

The development of effective, high-efficiency enantiomeric separations is a tremendous success story. In a relatively short period of time (from the early 1980s to the mid-1990s), this field evolved from an academic novelty to a widely useful collection of related techniques that are used routinely in many branches of science and technology. This success is a tribute to the relatively few researchers who developed the field and whose work provided the impetus for the U.S. Food and Drug Administration's issuance of new guidelines for the development of stereoisomeric drugs in 1992 [1].

Prior to the early 1980s the resolution of enantiomers was considered a very difficult, time-consuming, and often intractable problem. Effective, high-efficiency separation methods for enantiomers were almost unknown or unavailable. Consequently, enantiomeric separations tended to be avoided or ignored if possible. If an enantiomeric separation had to be accomplished and if a compound had the proper structure, functional groups, properties, and/or reactivity, there were some classical, larger-scale techniques that could be tried. These included such things as fractional recrystallization of diastereomeric salts, biological digestions, or enzymatic resolutions, for example [2]. These techniques remain important preparative and process-scale methods for certain compounds today.

The possibility of resolving enantiomers by differential association with a chiral support was recognized in the early twentieth century [3]. Between 1930 and 1970 there were sporadic reports on the direct chromatographic separation of enantiomers [4–10]. The earlier papers utilized some form of liquid chromatography (LC). Often, but not always, the compounds of most concern were amino acids.

From 1970 to 1980 research involving enantioselective LC and gas chromatography (GC) increased [11–49]. There seemed to be a focused effort to develop amino acid–based chiral stationary phases for GC [24–49]. The work in LC was less focused. While much of this work is noteworthy, most methods were not easily or widely applicable. However, several relatively early papers heralded some of the advanced techniques that are used today. For example, an early series of papers demonstrated the usefulness of π-π interactions between a chiral stationary phase and racemic analyte in normal phase chromatography. Klemm and Reed first used this approach in 1960 by adsorbing TAPA (α-(2,4,5,7-tetranitro-9-fluorenylidenaminooxy) propionic acid) onto silica gel [9]. This was done using an alumina support as well [12]. These were sometimes referred to as charge-transfer stationary phases. During the 1970s Mikes and coworkers [11,13,14] and other groups [15,16] developed several bonded chiral stationary phases that contained π-electron–accepting and π-electron–donating groups.

In other early work in the 1970s, Davankov perfected chiral ligand exchange as a means to resolve amino acids and other chiral bidentate molecules [17,18]. In the same decade, Cram and coworkers synthesized a chiral crown ether that was attached to a polymeric support and used to resolve amino acids in a chromatographic mode [21–23]. Also about this time, researchers in Germany demonstrated that derivatized cellulose stationary phases could be used to separate some enantiomers [19,20]. My research group examined the theory and use of micelles and subsequently cyclodextrins in the mobile phase from the mid-1970s through 1980 [50–54]. The term pseudophase separations was coined to describe these methods [54]. Afterwards, we covalently attached cyclodextrins and their derivatives to silica gel to produce the first stable, widely useful reversed-phase mode chiral stationary phase. Still later, they became the dominant class of chiral selectors for capillary electrophoresis and gas chromatography.

After 1980 there was a veritable explosion in the development, use, understanding, and commercialization of chiral stationary phases for LC. The first two types of chiral stationary phases that became available commercially (i.e., 3,5-dinitrobenzoyl phenylglycine [55] and β-cyclodextrin [56]) have been copied, reproduced, and mimicked by companies and individuals throughout the world. Many of the chiral selectors developed for HPLC were rapidly tested and utilized in capillary electrophoresis, thereby contributing to its rapid and facile development.

The first development of a chiral stationary phases for gas chromatography (GC) was by Gil-Av and coworkers in 1966 [10]. The stationary phase consisted of acylated amino acid and dipeptide esters. Research on amino acid– or peptide-based CSPs for GC continued for over 20 years in many research groups [24–48]. However, the only commercially successful CSP of this type (Chirasil-Val) used L-valine-*tert*-butylamide coupled to a siloxane copolymer [43–45]. We wrote the following sentences in a 1988 article in *CRC Critical Reviews in Analytical Chemistry* [58]: ''The research on the separation of enantiomers by GC has decreased considerably in the last few years''; ''In this same time period, the research on CSPs in LC accelerated tremendously''; ''There is a need to develop a greater variety of chiral GC phases that will allow the resolution of some underivatized enantiomers.''

Little did we know that the biggest advances for enantioselective GC were just around the corner and that cyclodextrins would play the dominant role. König et al. found that neat, extensively derivatized cyclodextrins, with moderate-sized nonpolar groups, produced effective GC CSPs [59,60]. We devised a series of ''rules'' that allowed one to produce amorphous or liquid cyclodextrin derivatives of different polarities [61]. We also determined early on that different cyclodextrins and cyclodextrin derivatives could have opposite enantioselectivities [62]. In this same time period, Schurig et al. found that methylated cyclodextrins could be dissolved in traditional achiral GC stationary phase liquids to produce effective CSPs [63,64]. A great variety of derivatized cyclodextrin-based CSPs for GC were developed and commercialized over the next few years. The cyclodextrin-based CSPs revolutionized the GC separation of enantiomers, and they remain the dominant class of chiral selectors in this field today [59–82].

Enantiomeric separations and analyses have become increasingly important in many areas of science and technology. These include the pharmaceutical and medicinal sciences since it is known that different enantiomers can have different physiological effects and biological dispositions [83–85]. The tremendous advances in the LC analysis of enantiomers and their use provided the main impetus for the Food and Drug Administration FDA to develop and issue guidelines for the development of new stereoisomeric drugs in 1992 [1]. In the environmental arena, many chiral pesticides and herbicides have enantioselective effects and biodegradation rates [83,86,87]. Food and beverage companies are becoming increasingly interested in the analysis of enantiomers that can affect flavor, fragrance, and nutrition and can be used to monitor fermentation, age, and even the adulteration of products [88,89]. Enantiomeric separations are now the most useful and accurate way to determine the enantiomeric purity of new synthesized chiral compounds and in elucidating certain reaction mechanisms [90,91]. Stereochemical analysis can also be of importance in areas of geochemistry [92–96], geochronology [97–99], biochemistry, and in some areas of materials science [100].

In this monograph I will attempt to briefly outline the background and current state of chromatographic enantiomeric separations. This is not intended to be a comprehensive review, but rather an overview that will also give the reader a sense of the present state of the art.

II. LIQUID CHROMATOGRAPHY

One can achieve enantiomeric separations in liquid chromatography by dissolving a chiral selector in the mobile phase and then using an achiral stationary phase or by incorporating the chiral selector into the stationary phase to form a chiral stationary phase (CSP). Far and away the most common procedure has been to use CSPs. The only chiral selectors to be used extensively as chiral mobile phase additives are the cyclodextrins, which are fairly water soluble and UV transparent. The cyclodextrins have been used in this manner as frequently in TLC as HPLC [53,54,101–105]. This monograph will focus exclusively on chiral stationary phases, even though the work with cyclodextrin mobile phase additives (along with the CSPs) clearly paved the way for CE enantioseparations. The first review of chiral stationary phases for HPLC was published in 1984 [57]. A comparison of this work to subsequent reviews illustrates the tremendous progress that has been made in this field over the past 16 years [58,83,106–108].

Perhaps the easiest way to classify chiral selectors is by structural type or class (see Table 1). As can be seen, all available CSPs can be placed into one of four basic classes based on their structure and function. In a few cases a chiral selector can have characteristics of more than one class. For example, the (R)- and (S)-naphthylethylcarbamoyl derivative of β-cyclodextrin is clearly a macrocyclic chiral selector. However, in the normal phase mode it acts as a very effective π-electron/donor CSP (see below). Hybrid or miscellaneous chiral selectors can be put in a fifth class (Table 1).

A. Macrocyclic Chiral Selectors

This class of chiral selectors has had the greatest overall impact on modern analytical enantiomeric separations. Probably over 90% of all GC and CE enantiomeric separations are done with macrocyclic chiral selectors. The majority of these are done with cyclodextrins and their derivatives [58–82,90,91,109,110]. In LC, the "field" is more divided. Macrocyclic-type CSPs perform the majority of reversed-phase enantiomeric separations and polar-organic mode separations, while derivatized carbohydrate and π-complex CSPs are used in the majority of normal

Table 1 Classification of Chiral Selectors Used in LC Chiral
Stationary Phases

Macrocyclic chiral selectors
  Cyclodextrins and derivatives thereof
  Glycopeptides (macrocyclic antibiotics)
  Chiral crown ether
Polymeric chiral selectors
  Naturally occurring polymers
    Derivatized carbohydrates
    Proteins
  Synthetic polymers
π-Complex chiral selectors
  π-Electron acceptor (π-acid)
  π-Electron donor (π-base)
  Combination types (contain both π-electron and π-acceptor groups)
Ligand exchange chiral selectors
Miscellaneous and hybrid chiral selectors

---

phase separations. We were fortunate to be able to introduce two of the three main types of macrocyclic chiral selectors to modern analytical separations, i.e., cyclodextrins/derivatized cyclodextrins [53,54,56,57,83,88,111,112] and macrocyclic glycopeptides (or antibiotics) [113–119]. The third type (chiral crown ethers) was first introduced by Cram [21–23].

1. Cyclodextrins and Their Derivatives

Cyclodextrins are naturally occurring molecules that can be produced by the action of the enzyme cyclodextrin glycosyltransferase (CGT) on starch. They are cyclic oligomers of α-(1,4)-linked glucose (Fig. 1). Cyclodextrins containing 6–12 glucose units (some of which are branched, i.e., have glucose "side chain" moieties) have been identified. However, those containing 6, 7, and 8 glucose units (i.e., α-cyclodextrin, β-cyclodextrin, and γ-cyclodextrin, respec-

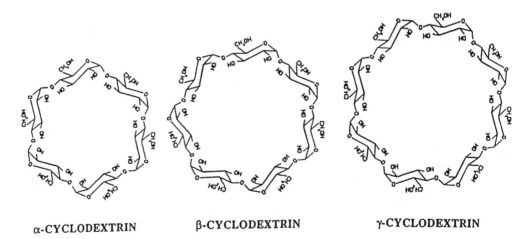

α-CYCLODEXTRIN   β-CYCLODEXTRIN   γ-CYCLODEXTRIN

Figure 1 Structures of α-cyclodextrin, β-cyclodextrin, and γ-cyclodextrin.

tively) are most common, best characterized, and available commercially. As will be discussed below, numerous derivatives have been made of these molecules. In aqueous or hydro-organic solutions, cyclodextrins are known to form host-guest complexes with a variety of molecules, ions, and solvents [120]. Since the internal cavity of the cyclodextrin tends to be more hydrophobic than either its exterior or the aqueous solvent, most nonpolar molecules (or molecules with nonpolar moieties) prefer to reside in the cyclodextrin cavity. However, the presence of appreciable quantities of nonpolar solvents could prevent the formation of an inclusion complex between a trace analyte and cyclodextrin.

Initially cyclodextrins were used as mobile phase additives in TLC to separate isomeric compounds [53,54,103–105]. Later they were immobilized to form a highly effective CSP [56,57,110–112,121–123]. The β-cyclodextrin bonded phase was the first commercially successful reversed-phase CSP. At the time of this early work, the concept of an inclusion complex and its role in retention and selectivity were foreign to most chromatographers and separation scientists. By the mid-1980s the chiral recognition mechanism of cyclodextrin in aqueous solutions was understood [56,121–124].

The utility of cyclodextrin-based CSPs was broadened in two different ways. First, several different cycodextrin derivatives were made and immobilized on a chromatographic support [111,112]. This same process of using cyclodextrin derivatives to broaden and vary enantioselectivity in LC was later used in capillary electrophoresis [109]. Figure 2 shows several of the most useful derivatized cyclodextrin chiral selectors for LC. The different moieties used to functionalize cyclodextrins can greatly alter their enantioselectivity. For example, aromatic functionalized cyclodextrins can be used in the normal phase mode as π-complex CSPs or in the reversed-phase mode where inclusion complexation dominates [111,125–128]. Completely different types of chiral molecules are resolved in each mode. Hydroxypropyl functionalized cyclodextrins (Fig. 2) may be the most useful reversed-phase CSP of this particular class of chiral stationary phases [112,129].

A somewhat different experimental approach was found to enhance the usefulness of native cyclodextrin CSPs and produce unusually enantioselectivities. In this approach, inclusion complexation is suppressed by using a non–hydrogen-bonding, polar-organic solvent (such as acetonitrile) as the main component of the mobile phase [129–135]. Although this is sometimes referred to as "the polar-organic mode," it is most closely related to normal phase separations. The acetonitrile tends to occupy the cyclodextrin cavity. It also accentuates hydrogen bonding between the hydroxyl groups on the cyclodextrin and any hydrogen-bonding groups on the chiral analyte [133–135]. A hydrogen-bonding solvent (such as methanol) can be added to decrease the retention of highly retained compounds. Very small amounts of glacial acetic acid and triethylamine are added to control the protonation of the analyte and to enhance enantioselectivity [133,134]. In this mode, the analyte is thought to reside on top of the cyclodextrin selector (somewhat like a "lid") in such a way that hydrogen bonding is maximized (Fig. 3). Chiral compounds containing two hydrogen-bonding groups (one of which should be α or β to the chiral center) and a bulky group (such as an aromatic ring) are often resolved via this approach. The Cyclobond I RN and SN CSPs also work well in this mode [127,129]. Also, this is an excellent technique for preparative separations.

There are some practical advantages to using the "polar-organic mode" versus the traditional normal phase mode (which utilizes hexane or heptane plus propanol solvent systems). First, most compounds that exist as hydrochloride salts can be dissolved and separated using the "polar-organic" mobile phase, but not with the nonpolar organic solvents used in the normal phase mode. Second, when using achiral-chiral coupled column systems, effluent from a reversed-phase column can be switched directly onto a chiral column, provided the polar-organic

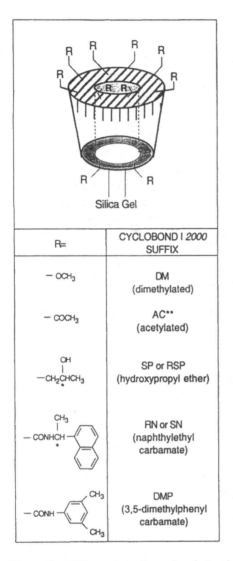

Silica Gel

R=	CYCLOBOND I *2000* SUFFIX
— OCH₃	DM (dimethylated)
— COCH₃	AC** (acetylated)
OH | —CH₂CHCH₃	SP or RSP (hydroxypropyl ether)
CH₃ | — CONHCH (naphthyl)	RN or SN (naphthylethyl carbamate)
— CONH (3,5-dimethylphenyl)	DMP (3,5-dimethylphenyl carbamate)

Figure 2 Different derivatives of cyclodextrins available as chiral stationary phases (CSPs).

type of mobile phase is used. This cannot be done when operating a CSP column with traditional normal phase solvents, which are incompatible with aqueous, buffer-containing solvents.

The polar-organic mode used in conjunction with cyclodextrin CSPs played an important theoretical and mechanistic role in cyclodextrin chemistry. It provided strong evidence for a noninclusion chiral recognition mechanism for these chiral selectors. Later, additional evidence from gas chromatography supported this new noninclusion mechanism of enantioselective separation [80].

2. Glycopeptide Antibiotics

Macrocyclic glycopeptides are the newest and fastest growing class of CSPs [87–91,113–119,136–150]. Four types of glycopeptide-based CSPs are currently available: vancomycin,

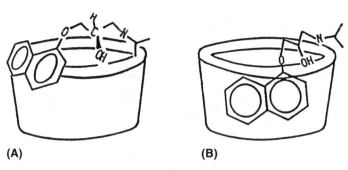

(A)   **(B)**

Figure 3 Simplified schematic illustrating two different enantioselective retention mechanisms for the native β-cyclodextrin/propranolol system. (A) The polar-organic mode where acetonitrile occupies the hydrophobic cavity and the analyte is retained via a combination of hydrogen bonding and dipolar interactions at the mouth of the cyclodextrin. Steric interactions also can contribute to chiral recognition [133–135]. (B) In the reversed-phase mode, retention is mainly due to hydrophobic inclusion complexation, while enantioselectivity also requires hydrogen bonding and steric interactions at the mouth of the cyclodextrin cavity [121–124]. (From Ref. 135.)

teicoplanism, ristocetin A, and the aglycone of teicoplanin (i.e., teicoplanin with the sugar moieties removed). These are shown in Fig. 4. The aglycone portion of all these molecules contains either three or four fused macrocyclic rings. Together they give the aglycone a basket shape that provides hydrophobic sites, hydrogen-bonding sites, dipolar sites, and π-interaction sites. The aglycone basket is semi-rigid and can flex somewhat. There is an amine moiety, a carboxylate moiety (esterified in ristocetin A), and phenolic moieties associated with each aglycone. These ionizable groups, along with an attached amino-saccharide, control the charge of these macrocycles. The aglycone portions of these molecules also contain several amide or peptide bonds (Fig. 4). One or more carbohydrate moieties is attached at various locations to each of the aglycones. Vancomycin has a single attached disaccharide while teioplanin has three monosaccharides moieties (one of which has an associated 9-carbon hydrophobic "tail" attached). Ristocetin A has a tetrasaccharide and two monosaccharides moieties associated with it (Fig. 4). The carbohydrate moieties are relatively mobile and free to change orientation compared to the aglycone portion of the molecule to which they are attached. The differences in the molecular weight and the number of stereogenic centers between the individual glycopeptides are due in large part to the number and size of the attached carbohydrate moieties.

The macrocyclic glycopeptides have very broad enantioselectivity. They can be used in all chromatographic modes (e.g., reversed-phase, normal phase, and polar-organic modes) and often have different enantioselectivities in each. The teicoplanin-based CSP (Chirobiotic T) is now the preferred means of resolving native, underivatized amino acids [137,139,140]. One of its great advantages over other CSPs (other than its high enantioselectivity for all manner of natural and unnatural amino acids) is that it requires only an unbuffered alcohol-water mobile phase. Ethanol is the preferred organic modifier, but somewhat different selectivities can be obtained with methanol or propanol, if needed [137,150]. The ristoretin A CSP (Chirobiotic R) also has somewhat different amino acid selectivities, which often complement the teicoplanin CSP.

One can obtain some very interesting separation effects by removing all carbohydrate moieties from teicoplanin (i.e., the Chirobiotic TAG column). The enantioselectivities for many amino acids and some chiral carboxylic acids are often enhanced [119]. Conversely, the enantiomeric recognition for many other compounds (e.g., neutral or cationic) was found to decrease.

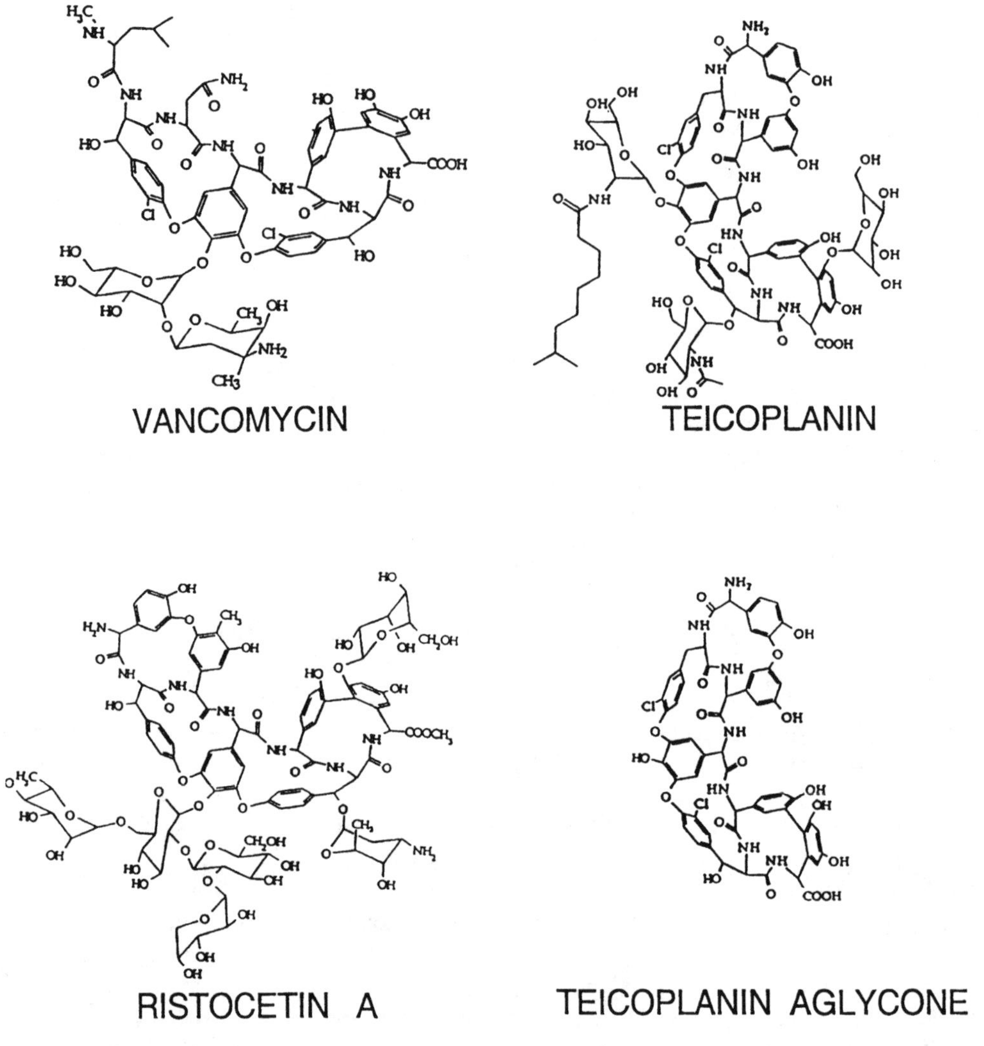

Figure 4 Structures of the four macrocyclic glycopeptides that are available as chiral stationary phases (CSPs).

This study helped to define the role of the carbohydrate moieties in chiral recognition [119]. It also provided mechanistic information regarding a specific amino acid–binding site versus other binding sites.

The "principal of complementary separations" is a very useful concept when doing method development work with glycopeptide-based CSPs [136,150]. Vancomycin, teicoplanin, and ristocetin A are similar, closely related chiral selectors. They can have similar but not identical enantioselectivities. Consequently, if a partial separation is obtained on one CSP, in many cases one can go directly to the related column and obtain a baseline separation (using identical or very similar separation conditions). This is shown for terbutaline, bromacil, and folinic acid (leucovorin) in Fig. 5.

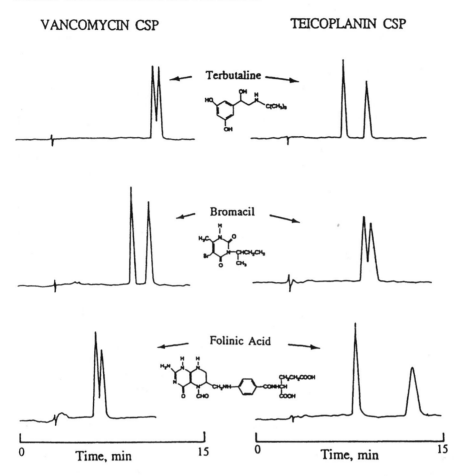

VANCOMYCIN CSP   **TEICOPLANIN CSP**

Terbutaline

Bromacil

Folinic Acid

0   Time, min   15   0   Time, min   15

Figure 5 Enantiomeric separations of terbutaline, bromacil, and folinic acid showing the complementary nature of the macrocyclic glycopeptide CSPs. If partial resolution is obtained on one column, then enhanced resolution is often obtained on the related CSP. The column dimensions were 25 cm $\times$ 0.46 cm (i.d.) and the flow rate was 1.0 mL/min in all cases. The mobile phase composition for the terbutaline separation was 55/45/0.3/0.2 (by volume) acetonitrile/methanol/acetic acid/triethylamine. The mobile phase compositions for the bromacil separation was 10/90 (by volume) tetrahydrofuran/pH 5.5, 20 mM NH_4NO_3 buffer. The mobile phase composition for the folinic acid separations were 100/1.0/0.5 (by volume) methanol/glacial acetic acid/triethylamine for the teicoplanin CSP and 50/50 (by volume) ethanol/pH 4.5 triethylammonium acetate buffer for the vancomycin CSP. (From Ref. 107.)

3. Chiral Crown Ethers

Chiral crown ether stationary phases for LC were first developed by Cram and coworkers [21–23]. Ions the size of K^+ and NH_4^+ tend to form inclusion complexes with [18]-crown-6-polyethers. Cram et al. demonstrated the chromatographic resolution of primary amine–containing chiral compounds on a CSP consisting of a chiral crown ether covalently linked to a polymeric support [21–23]. The mobile phase must be acidic so that the analyte's primary amine functional group is in the ammonium ion form. Ions that compete for inclusion in the crown ether (such as K^+) must be excluded from the mobile phase and separation solutions.

The crown ether type of chiral selector works only for primary amine–containing compounds. It will not work for secondary amines and amino acids (such as proline, hydroxyproline, etc.) or other types of compounds. The commercial version of this column (Fig. 6, p. 568) requires a $\sim 10^{-2}$ M solution of $HClO_4$ as the mobile phase [151–153]. The use of other acids either greatly diminishes or negates the separation on this column. This creates a problem when doing preparative separations, since evaporating the solvent can leave an explosive mixture of perchlorate plus organic material.

B. Polymeric Chiral Selectors

CSPs based on chiral polymers have played an important role in HPLC enantioseparations but not in GC or CE thus far. There are naturally occurring chiral polymers and synthetic chiral polymers—although the synthetic varieties have not had a significant impact on the LC separation of enantiomers to date. Proteins were the most important polymeric chiral selector in the 1980s. However, by the mid-1990s, carbohydrate-based CSPs became the most important polymeric CSPs. The use of protein-based CSPs has continued to decrease in relative importance compared to both macrocyclic CSPs and carbohydrate-based CSPs.

1. Naturally Occurring Polymers

Cellulose and amylose are among the world's most common naturally occurring chiral polymers. They are fairly poor chiral selectors in their native state. However, when their hydroxy-functional groups are derivatized (particularly with aromatic moieties via ester or carbamate linkages) and they are properly adsorbed or attached to a support, they become highly effective chiral stationary phases (Table 2). Early work on derivatized cellulose-based CSPs was done in Germany [19,20]. Subsequently, Daicel, Ltd. of Japan commercialized a series of different derivatized cellulose CSPs and, eventually, derivatized amylose. These chiral polymers are not attached covalently to the support. They are coated on wide-pore silica gel that has been silanized [154–161]. Since these derivatized carbohydrates are soluble in many solvents (e.g., chlorinated hydrocarbons, THF, etc.), some caution must be exercised when selecting the mobile phase. These CSPs are most often used in the normal phase mode (i.e., with hexane and isopropyl alcohol solvents). When these columns are not being used, they should be stored in hexane at $\sim 4°C$.

After we developed the polar organic mode for cyclodextrin and macrocyclic glycopeptide CSPs, it was found that this mode of separation also worked well with many derivatized carbohydrate CSPs [162,163]. In addition, some versions of these columns (Table 2) have been conditioned for reversed-phase use (e.g., the Chiracel OD-R and OJ-R) [160,161]. However, it should be noted that one should not use the same column in normal phase and reversed-phase modes. The configuration/orientation of the chiral polymers controls its enantioselectivity and the mobile phase can alter the configuration, often irreversibly. Hence, a column should be dedicated to either the normal phase mode or the reversed-phase mode.

The capacity of these CSPs in the normal phase mode is good, which makes them useful for preparative separations. There is overlap in the enantioselectivities of the different derivatives as well as between cellulose and amylose derivatives. Probably the most widely useful of all the derivatized carbohydrate CSPs are the 3,5-dimethylphenylcarbamate derivative of cellulose and amylose (i.e., Chiracel OD and AD, respectively). The majority of all enantioselective separations that can be accomplished on derivatized carbohydrate-based CSPs are done on one or the other of these two columns.

Proteins are well-known, naturally occurring chiral polymers. However, not all proteins are effective chiral selectors. Proteins that are good chiral selectors have a well-defined secondary

Table 2 Coated Chiral Stationary Phases Consisting of

Derivatives of Cellulose and

Amylose

Type of derivative	Commercial designation
(A) Cellulose ester	
1. R =	OJ[a]
2. R =	OB[a]
3. R =	OA
4. R =	OK
(B) Cellulose carbamate	
1. R =	OD[a]
2. R =	OC
3. R =	OG
4. R =	OF
(C) Amylose derivative	
1. R =	AD
2. R =	AS

[a] Also available as OB—H, OD—H, and OD—J for reversed-phase separations.

Table 3 Protein-Based Chiral Stationary Phases Available Commercially

Protein	Preferred analytes
Bovine serum albumin[a] (BSA)	Aromatic acids and anionic compounds
Human serum albumin[a] (HSA)	Aromatic acids and anionic compounds
α_1-Acid glycloprotein[b] (AGP)	Cyclic and 2° amines containing an aromatic group, some acids, and neutral compounds
Ovamucoid[b]	Cyclic and 2° amines containing an aromatic group, some neutral compounds
Cellobiohydrolyase (CBH)	1° amines and some 2° amines and amides

[a] The BSA and HSA columns have similar enantioselectivities.
[b] The AGP and ovamucoid columns have similar enantioselectivities.

and tertiary structure and a variety of enantioselective binding sites [164–167]. Glycoproteins have been particularly useful as CSP selectors. Protein-based CSPs are always used in the reversed-phase mode. Denaturation of the protein must be avoided. Early versions of these CSPs lacked ruggedness and longevity. Fortunately, future generations of these columns had better stability and efficiency. Table 3, summarizes proteins that have been used in commercial CSPs. There is considerable overlap in the enantioselectivites of some of these proteins, for example, CSPs made with α_1-acid glycoprotein (AGP) and ovamucoid, separate very similar types of enantiomers, particularly those that contain secondary amines and cyclic amines. In the 1980s and early 1990s the AGP column was popular because of its ability to separate a variety of chiral pharmaceutical compounds. Obviously human serum albumin (HSA) and bovine serum albumin (BSA) are very similar molecules. They have an affinity for acidic solutes. Cellobiohydrolase also prefers amine-containing compounds, but is more adept at resolving primary amines. When using a cellobiohydrolase CSP, a precolumn should be used and/or EDTA should be added to the mobile phase. This is because metals poison this protein, thereby destroying enantioselectivity [166].

Protein-based CSPs have declined in importance in the last few years for a number of reasons. Despite improvements in their stability and manufacture, they still tend to be the most labile class of CSPs. Protein-based CSPs have the least capacity of any of the chiral stationary phases and therefore are not as amenable to preparative and semipreparative separations. Also they tend to be among the more expensive CSPs. Finally, newer CSPs do many of the same separations that were accomplished previously with the protein columns. The AGP column is still the most useful column of this class [164,166]. As a class, the protein CSPs played an important role in the early development, growth, and acceptance of enantioselective separations during the 1980s [164–167].

2. Synthetic Polymers

Chiral stationary phases based on synthetic polymers were developed by Okamoto [168]. Both poly(triphenylmethylmethacrylate) and poly(diphenyl-2-pyridylmethylmethacrylate) are available as CSPs coated on wide-pore silica gel. These polymers are not made from chiral monomers, but rather are polymerized in the presence of a chiral catalyst [168]. Thus, they form either a right-hand or left-hand helical coil (depending on the catalyst used). These CSPs were very novel and interesting from an academic standpoint. However, they have never played a significant practical role in LC enantioseparations. They seem to have limited selectivity, mainly for planer, aromatic molecules. Most racemates resolved by this class of CSP are also resolved by one or more of the other CSPs mentioned in this overview.

C. π-π Interaction Chiral Selectors

The major binding interactions in these types of chiral stationary phases involves a π-π association between the chiral selector and the analyte of interest [9,11–16,55,111,127,128,169–177]. These binding sites can be π-electron donor types (π-basic) or π-electron acceptor types (π-acidic). If the CSP chiral selector is a π-electron acceptor, then the analyte should have a π-electron donor group and vice versa. Other simultaneous interactions must occur as well if an enantioselective separation is to occur. The most important of these additional attractive interactions are hydrogen bonds and dipole-dipole interactions. Steric repulsion also can be very important. The types of attractive interactions listed above are more pronounced in nonpolar solvents. The addition of polar solvents lessens π-π, hydrogen bonding, and dipolar interactions. Hence, separations on these CSP are generally accomplished in the normal phase mode (e.g., hexane and a miscible polar organic modifier). Addition of a polar organic modifier (such as isopropanol) can be used to decrease retention, increase efficiency, and modify selectivity in some cases.

As mentioned in the introduction, many of the earliest successful CSPs (between ∼ 1960 and 1980) utilized π-π interactions [9,11–16]. Later, commercialized versions were produced by Pirkle and Oi [55,169–176]. The first commercialized CSP was ionically bonded (−)-3, 5-dinitrobenzylphenylglycine (a π-electron acceptor type).

Most enantiomeric separations on π-complex CSPs require the analyte to have a complementary π-acid or π-base moiety [127,128,169–175]. Since many compounds of interest do not have these groups, an achiral derivatization step is necessary to supply the required group (interaction). Consequently, chiral stationary phases that did not require analyte derivatization and had broad selectivity (i.e., macrocyclic types and derivatized linear carbohydrate types) became more dominant. However, the newest π-complex CSPs contain both π-acid and π-base moieties (Fig. 6) and tend to be more widely applicable [176,177]. Also, a macrocyclic chiral selectors (β-cyclodextrin; see Fig. 2) has been derivatized with π-electron donor groups (i.e., Cyclobond RN and SN), making them perhaps the most useful π-basic CSPs [127,128]. Two strengths of the π-interactive CSPs is that they have good loadability in the normal phase mode and are, therefore, useful for many preparative separations. Also, both enantiomers of the chiral selectors are usually available. Therefore, one can obtain the opposite retention order, if desired.

D. Ligand Exchange Chiral Selectors

Ligand exchange chiral selectors were first developed in the early 1970s by Davankov [17,18,178–185]. This method was, perhaps, the most useful and widely utilized LC approach for the separation of enantiomers until the great "research/commercialization" expansion of the 1980s. It is the basis for the only commercially successful thin layer chromatography CSP [105,186]. Chiral ligand exchange also was used in the first capillary electrophoresis separation of enantiomers [187]. This method utilizes a chiral bidentate ligand (such as proline, hydroxyproline, penicillamine, etc.) as the chiral selector. It is usually attached to a stationary phase support, but it also can be used as a mobile phase additive. Copper(II), or another appropriate transition metal, must be added to the mobile phase. Both the chiral selector on the stationary phase and any bidentate analyte dissolved in the mobile phase can coordinate to the copper(II). In essence, the transition metal serves as a bridge between the chiral selector and the analyte of interest, orienting them and bringing them into close proximity to one another (see Fig. 7).

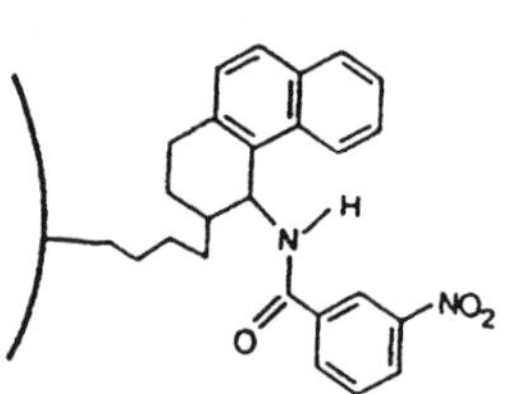

Figure 6 Examples of π-complex chiral selectors that have been immobilized on silica gel to form CSPs. They appear (from top to bottom) in chronological order of their appearance in the literature. (A) α-(2,4,5,7-Tetranitro-9-fluorenylidenaminooxy)propionic acid (a.k.a. TAPA) [11,14]; (B) 2,2′-diyl-hydrogen phosphite XL [13]; (C) dinitrobenzoyl-phenylglycine [170]; (D) 1-(α-naphthylethylamine) [172]; (E) 3,5-dinitrobenzoyl-(R)-naphthyl-glycine [175]; (F) 4-(3,5-dinitrobenzamido)tetrahydrophenanthrene [176].

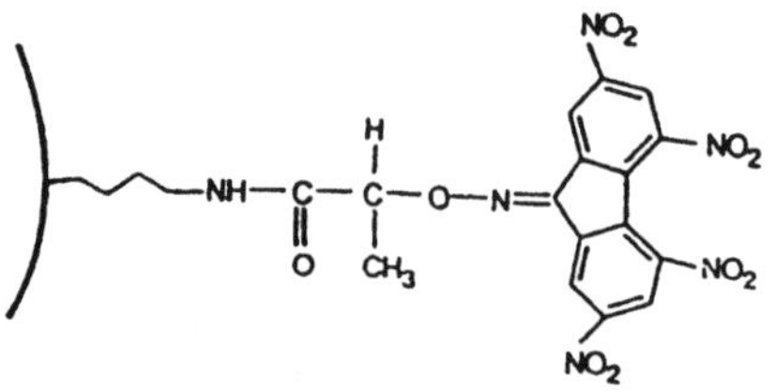

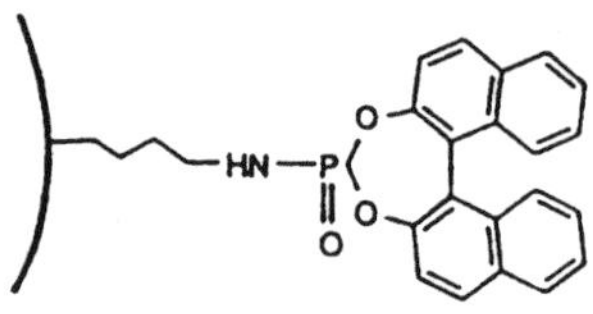

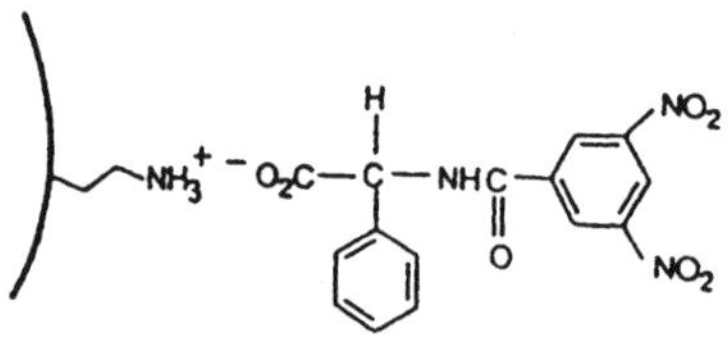

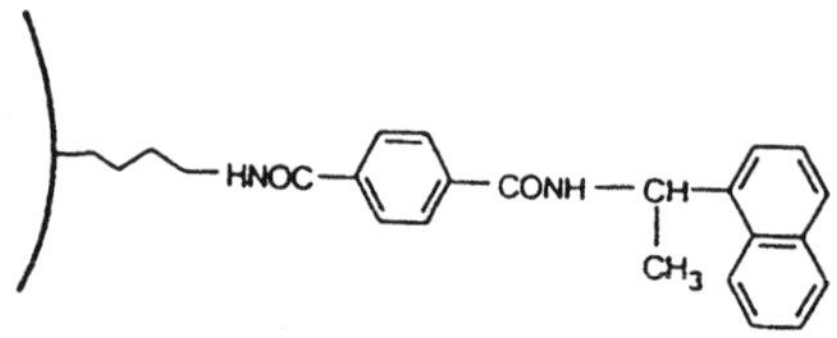

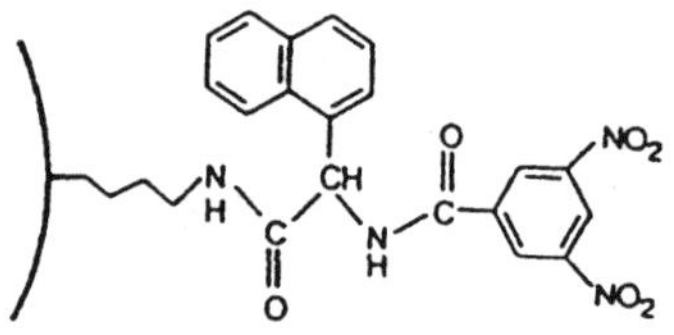

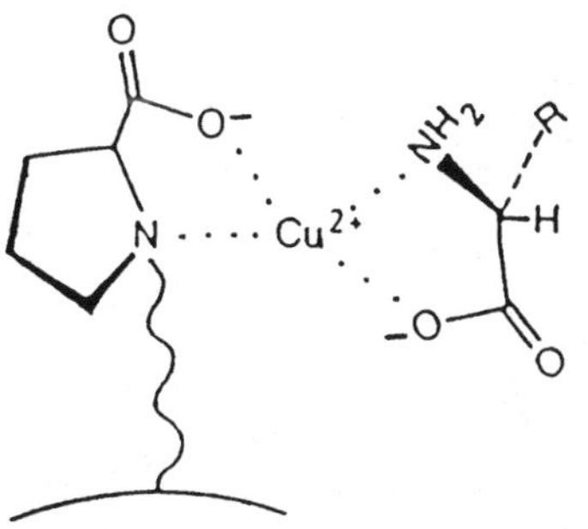

Figure 7 A chiral ligand exchange CSP where a chiral bidentate ligand is used as the chiral selector.

Ligand exchange separations are carried out with buffered, aqueous mobile phases. Some organic modifier can be added to optimize a separation. This method is limited to analytes that can act as bidentate ligands. This includes amino acids, α-hydroxy acids, diamines, etc.

The copper salt that is added to the mobile phase can give some background signal when using UV detection or an evaporative/light scattering detector. Conversely, bidendate analytes that are difficult to detect by UV absorption (e.g., nonaromatic molecules) are easily detectable at 254 nm when they form a transition metal complex with the Cu^{2+} that is in the mobile phase. When doing preparative separations, the transition metal salt must be removed from the product.

E. Miscellaneous/Hybrid Chiral Selectors

A few chiral selectors have features of two or more classes. For example, the aforementioned (R)- or (S)-naphthethylcarbamate derivative of β-cyclodextrin (i.e., Cyclobond I-RN or -SN) can separate one group of enantiomeric molecules in the reversed-phase mode via inclusion complexation within the macrocyclic ring. Conversely, in the normal phase mode it can act as a π-basic CSP and separate a totally different group of molecules via π-π surface interactions (Fig. 2). With polar organic solvents, a third group of chiral molecules can be resolved [111,125–129].

A hybrid polymeric CSP has been introduced by Allenmark et al. that uses tartaric acid as a starting material [188]. The basic monomer, N,N-dialkyl-L-tartardiamide, can be derivatized with different aromatic acid chlorides (Fig. 8). This aromatic functionalized tartardiamide is then attached and crosslinked on silica gel that was previously silanized with an unsaturated organosilane [188]. These types of CSPs are used in the normal phase mode [188].

Various alkaloids have been used as chiral selectors in LC and CE. The only ones found to be of use as a chiral stationary phases were derivatized with either a π-acid or π-basic moiety. Lindner and coworkers introduced CSPs based on derivatized cinchona alkaloids [189–192]. These chiral selectors utilize both ion exchange and π-π interactions for association and molecular recognition of chiral analytes. Their enantioselectivities tend to overlap with those of other π-π interactive CSPs.

III. GAS CHROMATOGRAPHY

Gas chromatographic CSPs can be divided into four basic groups: cyclodextrin derivatives, amino acid analogs, chiral metal-ligand complexes, and miscellaneous types. As mentioned in the introduction, the early GC-CSP work was done almost entirely on amino acid or peptide-

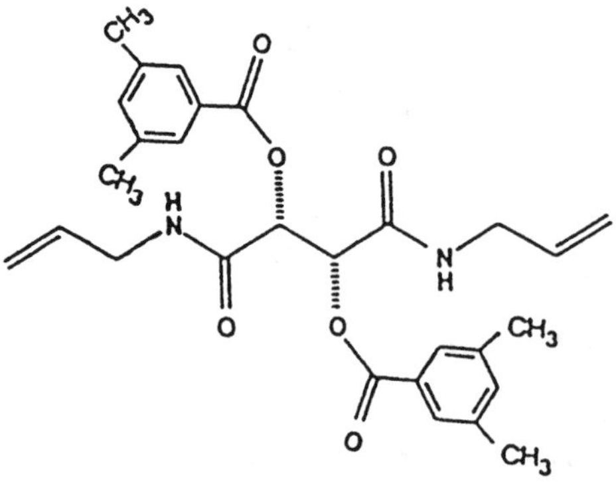

Figure 8 Structure of 3,5-dimethylphenyl esterified N,N'-diallyl-L-tartardiamide. This monomeric, chiral selector is attached and crosslinked to silica gel that was previously silanized with an unsaturated organosilane. (From Ref. 107.)

based CSPs. However, today the vast majority of all GC enantiomeric separations are done on derivatized cyclodextrin CSPs.

A. Derivatized Cyclodextrin Chiral Selectors

At about the same time that we were utilizing dissolved and bonded cyclodextrins in various forms of liquid chromatography (~1978–1981), Smolkova-Keulemansova and coworkers were experimenting with native and methylated cyclodextrins as GC stationary phases [193–200]. Since the cyclodextrins were crystalline solids and insoluble in typical achiral GC stationary phase liquids, they had to use solvents such as dimethylformamide (DMF) to dissolve the cyclodextrins and coat the stationary phase support. They found that the cyclodextrin-based stationary phases produced very interesting selectivities and that an inclusion complex mechanism was very likely involved [193–200]. Because of an efficiency problem resulting from the nature of the largely solid cyclodextrin stationary phase (the coating solvent was evaporated), only a few enantiomeric separations were reported [201]. It was not until later that König found that some amorphous derivatives of β-cyclodextrin could be used directly as stationary phase liquids [59,60]. We developed a variety of these GC CSPs as well [61,62]. Shurig and coworkers found that methylated cyclodextrins could be dissolved in appreciable quantities in conventional achiral stationary phase liquids [63,64]. These also could be used as effective GC-CSPs. Within 2 or 3 years a great variety of cyclodextrin-based GC-CSPs became available to the scientific community [59–82]. They were extremely effective in resolving a large variety of volatile chiral racemates. This included many molecules that were difficult or impossible to separate by HPLC. Molecules with very few functional groups—even chiral hydrocarbons—could be separated by cyclodextrin-based GC columns [202].

It is not widely appreciated by the separations science community that each different cyclodextrin derivative can have very different enantioselectivity [59–64,82,203–206]. The selectivity of these GC chiral selectors is controlled by the size of the cyclodextrin (α-, β-, or γ-CD; see Fig. 1) as well as the nature and location and number of the derivatizing groups. Table 4 shows a selection of different functionalized cyclodextrin CSPs and the types of compound that are most likely to be resolved on them. In the case of the CSPs made by dissolving derivatized cyclodextrins in achiral GC stationary phases, both the nature of the achiral matrix and the concentration of the dissolved cyclodextrin can affect enantioselectivity [207–212].

Table 4 Examples of Derivatized Cyclodextrins Used as GC Chiral Stationary Phases

Abbreviations	Derivative	Base type of CD used and commercial designation	Maximum operating temperature (°C)	Enantioselectivity
PH[a]	S-Hydroxypropyl ((S)-2-hydroxypropyl methyl ether)	α-CD = A-PH, β-CD = B-PH, γ-CD = G-PH	220	The β-PH has the greatest utility in this class. Separates hydrocarbons, terpenes, ethers, derivatized alcohols, amines, etc.
DA[a]	Dialkyl (2,6-di-O-pentyl)	α-CD = A-DA, β-CD = B-DA, γ-CD = G-DA	220	The B-DA has the greatest utility in this class. Separates some alkanes aromatic molecules, esters, amides, some cyclic ethers and epoxides. Often has opposite enantioselectivity of other CSPs.
TA[a]	Trifluoroacetyl (2,6-di-O-pentyl-3-trifluoroacetyl)	α-CD = A-TA, β-CD = B-TA, γ-CD = G-TA	180	The broadest selectivity GC-CSP, with the G-TA being the best of the three. Separates alcohols, ketones, acids, aldehydes, esters, amides, halogenated compounds, etc.
PN[a]	Propionyl (2,6-di-O-pentyl-3-propionyl)	γ-CD = G-PN	220	Separates epoxides, lactones, aromatic amines, stimulants (does not degrade styrene oxide)
BP[a]	Butyryl (2,6-di-O-pentyl-3-butyryl)	γ-CD = G-BP	220	Separates derivatized amino acids, amines, furans.
PM[b]	Permethyl (2,3,6-tri-O-methyl)	β-CD = B-PM	250	A broadly selective general purpose CSP. Only the G-TA separates more compounds.
DM[b]	Dimethyl (di-O-methyl)	β-CD = B-DM	250	A new CSP that may eventually prove to be more widely useful than the B-PM. Separates aliphatic, olefenic, and aromatic compounds as well as esters of nonsteroidal anti-inflammatory compounds.

[a] Neat chiral stationary phases (i.e., 100% derivatized cyclodextrins).
[b] Dissolved-type chiral stationary phase (i.e., the solid CD derivative is dissolved in an achiral polysiloxane stationary phase).

Probably the most widely useful GC-CSP is Chiraldex G-TA, which contains trifluoroace-tyldipentyl γ-cyclodextrins [82,203]. This is one of the rare examples where a γ-cyclodextrin CSP is greatly superior to its β-cyclodextrin analog. Another widely useful GC-CSP of the dissolved type is 2,6-heptakisdimethyl-β-cyclodextrin dissolved in an achiral matrix [82]. Most recently methylated cyclodextrins have been attached to siloxane polymers or incorporated into a siloxane copolymer and bonded to the wall of fused silica capillaries [73–79]. These approaches produce GC-CSPs of excellent thermal stability that are resistant to physical changes.

Cyclodextrin-based GC columns have been used in many mechanistic studies. For example, it was indicated that some compounds can be resolved on these CSPs without forming an inclusion complex [80]. Hence, there are now both HPLC and GC examples of enantiorecognition and separation by cyclodextrins that do not involve inclusion complex formation [80,133–135]. Cyclodextrin-based CSPs also have been used in a variety of thermodynamic and kinetic experiments including racemization and enantiomerization studies [80,213–222].

B. Amino Acid–Based Chiral Selectors

As mentioned in the introduction of this monograph, Gil-Av and coworkers produced the first enantioselective GC-CSP [10]. It consisted of N-trifluoroacetyl (N-TFA)-L-isoleucine lauryl ester. This CSP was used for the separation of N-TFA-α-amino acid esters. It showed excellent resolution for volatile N-TFA-α-amino acid esters (e.g., N-TFA-valine and -leucine esters of 2-butanol), but was not suitable for N-TFA-alanine esters, heptyl acetate, and α-acetoxypropionate esters of 2-butanol [10,58]. As in most of these early amino acid–based GC-CSPs, there were temperature limitations and bleeding problems. Many other research groups participated in the development and utilization of a variety of derivatized amino acid– and peptide-based CSPs (24–49). The single most effective CSP of this genre (and the only one to be successfully commercialized) was developed by Frank and coworkers. It was made by coupling L-valine-*tert*-butylamide to a copolymer of dimethylsiloxane and carboxy-alkyl-methyl-siloxane of appropriate viscosity and molecular weight [43–45,47,48]. This polymeric GC stationary phase had a higher thermal stability and a lower volatility than the monomeric stationary phases. This CSP (Fig. 8) was used to separate derivatized amino acid and some amino alcohol enantiomers, enantiomeric drugs, and metabolites [58]. This GC-CSP allowed a greater range of temperatures to be used than did previous amino acid–based stationary phases. As a result, compounds of lower volatility could be analyzed for the first time. Mass spectrometric detection was also used with this polymeric stationary phase. Even today, this remains one of the best GC-CSPs for the enantiomeric separation of trifluoroacetylated amino acid esters.

C. Metal-Ligand Complex CSPs

CSPs based on optically active complexes between chiral ligands such as chiral β-diketones and transition metal ions of nickel, manganese, cobalt, rhodium, europium, etc. were developed by Schuring and coworkers [223,234]. This approach is sometimes called complexation gas chromatography. The first CSPs of this type consisted of the appropriate metal complex dissolved in an achiral, nonpolar GC stationary phase liquid such as squalane [229]. This is the main class of GC-CSPs where π-π interactions between the analyte and selector can be of primary importance to the chiral recognition process. In later versions of this CSP, the metal-ligand complex was coupled to a siloxane polymer as in the case of Chirasil-Val [234].

D. Miscellaneous GC-CSPs

Chiral liquid stationary phases composed of di-*l*-methyl-(+)-tartrate and di-*dl*-methyl-(−)-malate were used in conjunction with glass capillary columns to separate enantiomers of amino acids, amines, and carboxylic acids [235]. Oi and coworkers also developed some amide-type stationary phases, which contained two stereogenic centers. These were used to separate a few carboxylic acid and amine enantiomers. Liquid stationary phases composed of N-(1R,3R)-trans-chrysanthemoyl-(R)-1-(α-naphthyl)ethylamine chiral selectors [236] showed better enantioselectivity than those composed of N-(1R,3R)-trans-chrysanthemoyl-lauryl-amine or N-lauroyl-(R)-1-(α-naphthyl)ethylamine, both of which contain only one asymmetric center. An O-(1R,3R)-trans-chrysanthemoyl-(S)-mandelic acid (R)-1-(α-naphthyl)-ethylamide stationary phase was synthesized and shown to have good enantioselectivity for chrysanthemic acid ester and 3-(2,2-dichlorovinyl)2,2-dimethylcyclopropanecarboxylic acid esters [237]. For the most part, the last two classes of GC-CSPs discussed in this overview have been supplanted by derivatized cyclodextrin CSPs. However, they played an important role in the history and development of GC-CSPs.

IV. EPILOGUE

Today, high-efficiency analytical enantioseparations have become so successful that it is more surprising to find racemates that cannot be resolved than ones that can. LC and GC methods are complementary to one another, with each technique able to resolve large numbers of compounds not amenable to the other. Capillary electrophoresis also has proven useful for many chiral separations [109,238–243]. It is well known that a chiral selector can sometimes produce different enantiomeric separations when dissolved in free solution versus being immobilized on a stationary phase support. Hence, CE can sometimes resolve compounds that are not separated by LC and vice versa (even when the same chiral selector is used in both procedures).

Supercritical fluid chromatography also can be used to separate enantiomers. Most of the major classes of LC-CSPs can be used with SFC. SFC seems to have the most promise for preparative separations where isolating the final product from excess solvent can be a problem. SFC has not proven as beneficial for routine analytical separations, since GC, LC, and/or CE methods usually do the same separations faster and more efficiently.

While LC has played an important role, it is the combination of all of these often complementary separation techniques that has allowed today's scientists to successfully address and solve most problems involving the resolution and analysis of enantiomeric compounds.

Enantiomeric excesses can now be determined routinely to 0.01% [90,91]. Because of the greater sensitivity and accuracy of these methods, we are finding that there are actually very few enantiomerically pure compounds in the world (either naturally or commercially).

Another important aspect of enantiomeric separations is that it has greatly enhanced our understanding of molecular recognition. CSPs also can be beneficial in the study of chromatographic theory and mechanisms. For example, one can isolate and study chiral vs. nonchiral retention mechanisms. Enantiomers can be useful probe molecules since they have the same size, shape, volume, solvent solubility, volatility, etc. Mechanistic studies involving chiral molecules and/or CSPs will undoubtedly increase in the foreseeable future.

REFERENCES

1. FDA. Chirality 4:338, 1992.
2. EL Eliel. Stereochemie der Kohlenestaffverbindungen. Weinheim: Verlag Chemie, 1966.
3. R Willsatter. Ber Dtsch Chem Ges 37:3758, 1904.
4. GM Henderson, HG Rule. Nature 141:917, 1938.
5. GM Henderson, HG Rule. Nature 142:162, 1938.

6. GM Henderson, HG Rule. J Chem Soc 1568, 1939.
7. M Kotake, T Salem, M Nakamura, S Senoh. J Am Chem Soc 73:2973, 1951.
8. CE Dalgliesh. J Chem Soc 3940, 1952.
9. LH Klemm, D Reed. J Chromatogr 3:364, 1960.
10. E Gil-Av, B Feibush, R Charles-Sigler. Tetrahedron Lett 1009, 1966.
11. F Mikes, G Boshart, E Gil-Av. J Chromatogr 122:205, 1976.
12. B Feringa, W Wynberg. Rec Trav Chim Pays-Bas 97:249, 1978.
13. F Mikes, G Boshart. J Chromatogr 149:455, 1978.
14. F Mikes, G Boshart. Chem Commun 173, 1978.
15. CH Lochmüller, JM Harris, RW Souter. J Chromatogr 71:405, 1972.
16. CH Lochmúller, RR Ryall. J Chromatogr 150:511, 1978.
17. VA Davankov, SV Rogozhin. J Chromatogr 60:280, 1971.
18. VA Davankov, AA Kurganov, AS Bocklov. In: JC Giddings, E Grushka, J Cazes, PR Brown, eds. Advances in Chromatography, Vol. 22. New York: Marcel Dekker, 1983, p. 71.
19. G Hesse R Hagel. Liebigs Ann Chem 996, 1976.
20. H Häkli, M Mintas, A Mannschreck. Chem Ber 112:2028, 1979.
21. R Helgeson, J Timko, P Moreau, S Peacock, J Mayer, DJ Cram. J Am Chem Soc 96:6762, 1974.
22. GDY Sogah, DJ Cram. J Am Chem Soc 98:3038, 1976.
23. GDY Sogah, DJ Cram. J Am Chem Soc 101:3035, 1979.
24. B Feibush, E Gil-Av. Tetrahedron 26:1361, 1970.
25. W Parr, PY Howard. J Chromatogr 71:193, 1972.
26. W Parr, PY Howard. J Chromatogr 67:227, 1972.
27. W Parr, C Yang, E Bayer, E Gil-Av. J Chromatogr Sci 8:591, 1970.
28. W Parr, PY Howard. J Chromatogr 66:141, 1972.
29. W Parr, C Yang, J Pleterski, E Bayer. J Chromatogr 50:510, 1970.
30. W Parr, PY Howard. Anal Chem 45:711, 1973.
31. WA König, W Parr, HA Lichtenstein, E Bayer, J Oro. J Chromatogr Sci 8:183, 1970.
32. W Parr, J Pleterski, C Yang, E Bayer. J Chromatogr Sci 9:141, 1971.
33. WA König, GJ Nicholson. Anal Chem 47:951, 1975.
34. WA König, K Stoelting, K Druse. Chromatographia 8:444, 1977.
35. WA König. Chromatographia 9:72, 1976.
36. JA Carbin, JE Rhoad, LB Rogers. Anal Chem 43:327, 1971.
37. W Parr, PY Howard. Angew Chem Int Ed Engl 11:529, 1971.
38. I Abe, T Kohno, S Musha. Chromatographia 11:393, 1978.
39. F Andrawes, R Brazell, W Parr, A Zlatkis. J Chromatogr 112:197, 1975.
40. N Oi, O Hiroaki, H Shimada. Bunseki Kagaku 28:125, 1979.
41. N Oi, M Horiba, H Kitahara. Bunseki Kagaku 28:607, 1979.
42. N Oi, M Horiba, H Kitahara. Bunseki Kagaku 28:482, 1979.
43. H Frank, GJ Nicholson, E Bayer. J Chromatogr Sci 15:174, 1977.
44. H Frank, GJ Nicholson, E Bayer. Angew Chem. 90:396, 1978.
45. H Frank GJ Nicholson, E Bayer. J Chromatogr 146:197, 1978.
46. MD Solomon, JA Wright. Clin Chem 23:1504, 1977.
47. H Frank, GJ Nicholson, E Bayer. J Chromatogr 167:187, 1978.
48. GJ Nicholson, H Frank, E Bayer. HRC CC 2:411, 1979.
49. T Saeed, P Sandra, M Verzek. J Chromatogr 186:611, 1979.
50. DW Armstrong, JH Fendler. Biochim Biophys Acta 478:75, 1977.
51. DW Armstrong, RQ Terrill. Anal Chem 51:2160, 1979.
52. DW Armstrong, SJ Henry. J Liq Chromatogr 3:657, 1980.
53. WL Hinze, DW Armstrong. Anal Lett 13:1093, 1980.
54. DW Armstrong. J Liq Chromatogr 3:895, 1980.
55. WH Pirkle, JM Finn, JL Schreiner, BC Hamper. J Am Chem Soc 103:3964, 1981.
56. DW Armstrong, W DeMond. J Chromatogr Sci 22:411, 1984.

57. DW Armstrong. J Liq Chromatogr 7:353, 1984.
58. DW Armstrong, SM Han. CRC Crit Rev Anal Chem 19:175, 1988.
59. WA König, S Lutz, G Weng. Angew Chem Int Ed Engl 27:979, 1988.
60. WA König, R Krebber, P Mischnick. J High Res Chromatogr 12:35, 1989.
61. DW Armstrong, W Li, C-D Chang, J Pitha. Anal Chem 62:914, 1990.
62. DW Armstrong, W Li, J Pitha. Anal Chem 62:214, 1990.
63. V Schurig, H-P Novotny. J Chromatogr 441:155, 1988.
64. V Schurig, H-P Novotny, D Schmalzing. Angew Chem 101:785, 1989.
65. WA König, S Lutz, P Mischnick-Löbbecke, B Brassat, G Wenz. J Chromatogr 447:193 1988.
66. WA König, S Lutz, G Wenz, E van der Bey. J High Resolut Chromatogr Chromatogr Commun 11:506, 1988.
67. WA König, S Lutz, P Evers, J Knabe. J Chromatogr 503:256, 1990.
68. W-Y Li, H-L Jin, DW Armstrong. J Chromatogr 509:303, 1990.
69. DW Armstrong, W Li, AM Stalcup, HV Secor, RR Izac, JI Seeman. Anal Chim Acta 234:365, 1990.
70. H-P Novotny, D Schmalzing, D Wistuba, V Schurig. J High Resolut Chromatogr 12:383, 1989.
71. JS Bradshaw, G Yi, BE Rossiter, SL Reese, P Peterson, KE Markides, ML Lee. Tetrahedron Lett 34:79, 1993.
72. G Yi, JS Bradshaw, BE Rossiter, SL Reese, P Peterson, KE Markides, ML Lee. J Org Chem 58: 2561, 1993.
73. V Schurig, D Schmalzing, U Mühleck, M Jung, M Schleimer, P Mussche, C Duvekot, JC Buyten. J High Resolut Chromatogr 13:713, 1990.
74. V Schurig, Z Juvancz, GJ Nicholson, D Schmalzing. J High Resolut Chromatogr 14:58, 1991.
75. D Schmalzing, GJ Nicholson, M Jung, V Schurig. J Microcol Sep 4:23, 1992.
76. S Mayer, V Schurig. J High Resolut Chromatogr 15:129, 1992.
77. Z Juvancz, V Schurig, K Grolimund. J High Resolut Chromatogr 16:202, 1993.
78. DW Armstrong, Y Tang, T Ward, M Nichols. Anal Chem 65:1114, 1993.
79. G Yi, JS Bradshaw, BE Rossiter, A Molik, W Li, ML Lee. J Org Chem 58:4844, 1993.
80. A Berthod, W Li, DW Armstrong. Anal Chem 64:873, 1992.
81. Y Tang, Y Zhou, DW Armstrong. J Chromatogr A 666:147, 1994.
82. Chiraldex Handbook, 5th ed. Whippany, NJ: Advanced Separation Technologies, Inc., 1999.
83. AM Krstulovic ed. Chiral Separations by HPLC. Chichester: Ellis Horwood, Ltd., 1989.
84. EJ Ariens. Clin Pharmacol Ther 42:361, 1987.
85. EJ Ariens, EW Wuis, EF Veringa. Biochem Pharmacol 37:9, 1988.
86. DW Armstrong, GL Reid III, ML Hilton, C-D Chang. Environ Pollution 79:51, 1993.
87. JM Schneiderheinze, DW Armstrong, A Berthod. Chirality 11:330, 1999.
88. DW Armstrong, C-D Chang, WY Li. J Agric Food Chem 38:1674, 1990.
89. KH Ekborg-Ott, DW Armstrong. In: S Ahuja, ed. Chiral Separations. Washington, DC: American Chemical Society, 1997, p. 201.
90. DW Armstrong, L He, T Yu, JT Lee, Y-S Liu. Tetrahedron: Asymmetry 10:37, 1999.
91. DW Armstrong, JT Lee, LW Chang. Tetrahedron: Asymmetry 9:2043, 1998.
92. G Eglington, M Calvin. Sci Am 32:167, 1967.
93. RP Philip. Chem Eng News 64:28, 1986.
94. DW Armstrong, Y Tang, J Zukowski. Anal Chem 63:2858, 1991.
95. DW Armstrong, EY Zhou, J Zukowski, B Kosmowska-Ceranowicz. Chirality 8:39, 1996.
96. A Bethod, X Wang, KH Gahm, DW Armstrong. Geochim Cosmochim Acta 62:1619, 1998.
97. PM Helfman, JL Bada. Proc Natl Acad Sci USA 72:2891, 1975.
98. PM Masters, JL Bada, JS Zigler, Jr. Nature 268:71, 1977.
99. JL Bada. In: GC Barrett, ed. Chemistry and Biochemistry of Amino Acids. New York: Chapman and Hall, 1985, p. 6.
100. K Robbie, MJ Brett, A Lakhtakia. Nature 384:616, 1996.
101. J Debowski, D Sybilska, J Jurczak. J Chromatogr 237:303, 1982.

102. T Takeuchi, H Asai, D Ishii. J Chromatogr 357:409, 1986.
103. DW Armstrong, F-Y He, SM Han. J Chromatogr 448:345, 1988.
104. DW Armstrong, JR Faulkner, Jr., SM Han. J Chromatogr 452:323, 1988.
105. SM Han, DW Armstrong. In: JC Touchstone, ed. Planar Chromatography in the Life Sciences. New York: John Wiley & Sons, 1990, p. 81.
106. DW Armstrong. Anal Chem 59:84A, 1987.
107. DW Armstrong. LC-GC Supplemental Issue, May 20:S20, 1997.
108. DW Armstrong. J Chinese Chem Soc 45:581, 1998.
109. G Vigh, AD Sokolowski. Electrophoresis 18:2305, 1997.
110. DW Armstrong, LW Chang, SSC Chang. J Chromatogr A 793:115, 1998.
111. DW Armstrong, AM Stalcup, ML Hilton, JD Duncan, JR Faulkner, Jr., S-C Chang. Anal Chem 62:1610, 1990.
112. AM Stalcup, S-C Chang, DW Armstrong, J Pita. J Chromatogr 513:181, 1990.
113. DW Armstrong, Y Tang, S Chen, Y Zhou, C Bagwill, J-R Chen. Anal Chem 66:1473, 1994.
114. DW Armstrong, KL Rundlett, J-R Chen. Chirality 6:496, 1994.
115. DW Armstrong, Y Liu, KH Ekborg-Ott. Chirality 7:474, 1995.
116. MP Gasper, A Berthod, VB Nair, DW Armstrong. Anal Chem 68:2501, 1996.
117. VB Nair, SSC Chang, DW Armstrong, YY Rawjee, DS Eggleston, JV McArdle. Chirality 8:590, 1996.
118. KH Ekborg-Ott, Y Liu, DW Armstrong. Chirality 10:434, 1998.
119. A Berthod, X Chen, JP Kullman, DW Armstrong, F Gasparrini, I D'Acquarica, C Villani, A Carotti. Anal Chem 72:1767, 2000.
120. WL Hinze. Sep Purif Methods 10:159, 1981.
121. DW Armstrong, W DeMond, BP Czech. Anal Chem 57:481, 1985.
122. DW Armstrong, TJ Ward, A Czech, BP Czech, RA Bartsch. J Org Chem 50:5556, 1985.
123. DW Armstrong, TJ Ward, RD Armstrong, TE Beesley. Science 232:1131, 1986.
124. JA Hamilton, L Chen. J Am Chem Soc 110:4379, 1988.
125. AM Stalcup, S-C Chang, DW Armstrong. J Chromatogr 540:113, 1991.
126. DW Armstrong, C-D Chang, SH Lee. J Chromatogr 539:83, 1991.
127. DW Armstrong, M Hilton, L Coffin. LC·GC 9:646, 1991.
128. A Berthod, S-C Chang, DW Armstrong. Anal Chem 64:395, 1992.
129. Cyclobond Handbook. Whippany, NJ: Advanced Separation Technology, Inc., 1999.
130. J Zukowski, M Pawlowska, DW Armstrong. J Chromatogr 623:33, 1992.
131. J Zukowski, M Pawlowska, M Nazatkina, DW Armstrong. J Chromatogr 629:169, 1993.
132. M Pawlowska, S Chen, DW Armstrong. J Chromatogr 641:257, 1993.
133. DW Armstrong, S Chen, C Chang, S Chang. J Liq Chromatogr 15:545, 1992.
134. SC Chang, GL Reid III, S Chen, CD Chang, DW Armstrong. Trends Anal Chem 12:144, 1993.
135. DW Armstrong, LW Chang, S-C Chang, X Wang, H Ibrahim, GR Reed III, TE Beesley. J Liq Chromatogr 20:3279, 1997.
136. S Chen, Y Liu, DW Armstrong, P Victory, B Martinez-Teipel. J Liq Chromatogr 18:1495, 1995.
137. A Berthod, Y Liu, C Bagwill, DW Armstrong. J Chromatogr A 731:123, 1996.
138. OP Kleidernigg, CO Kappe. Tetrahedron: Asymmetry 8:2057, 1997.
139. A Peter, G Torok, DW Armstrong. J Chromatogr A 793:283, 1998.
140. A Peter, G Torok, DW Armstrong, G Toth, D Tourwe. J Chromatogr A 828:177, 1998.
141. KB Joyce, AE Jones, RJ Scott, RA Biddlecombe, S Pleasance. Rapid Commun Mass Spec 12:1899, 1998.
142. KM Fried, P Koch, IW Wainer. Chirality 10:484, 1988.
143. HY Aboul-Enein, V Serignese. Chirality 10:358, 1998.
144. E Tesarova, A Bosakova, V Pacakov. J Chromatogr A 838:121, 1999.
145. Q Sun, SV Olesik. Anal Chem 71:2139, 1999.
146. J Lehotay, K Hrobonová, J Krupcík, J Cizmárik. Pharmazie 53:863, 1998.
147. EH Ekborg-Ott, X Wang, DW Armstrong. Michrochem J 62:26, 1999.

148. RPW Scott, TE Beesley. Analyst 124:713, 1999.
149. E Tesarova, K Zaruba, M Flieger. J Chromatogr A 844:137, 1999.
150. Chirobiotic Handbook, 3rd ed. Whippany, NJ: Advanced Separations Technologies, Inc., 1999.
151. T Shinbo, T Yamaguchi, K Nishimura, M Sugiura. J Chromatogr 405:145, 1987.
152. M Hilton, DW Armstrong. J Liq Chromatogr 14:9, 1991.
153. M Hilton, DW Armstrong. J Liq Chromatogr 14:3673, 1991.
154. Y Okamoto, M Kawashima, K Yamamoto, K Hatada. Chem Lett 739, 1984.
155. Y Okamoto, M Kawashima, K Hatada. J Chromatogr 363:173, 1986.
156. R Noyori, M Ohta, Y Hisao, M Kitamura, T Ohta, H Takaya. J Am Chem Soc 108:7117, 1986.
157. Y Okamoto, R Aburatani, K Hatano, K Hatada. J Liq Chromatogr 11:2147, 1988.
158. Y Okamoto, R Aburatani, Y Kaida, K Hatada, N Inotsume, M Nakano. Chirality 1:239, 1989.
159. M Yoshifuji, K Toyota, Y Okamoto, T Asakura. Tetrahedron Lett 31:2311, 1990.
160. H Ogoshi, K Saita, K Sakurai, T Watanabe, H Toi, Y Aoyama, Y Okamoto. Tetrahedron Lett 27: 6365, 1986.
161. R Erlandsson, R Isaksson, R Lorentzon, P Lindberg. J Chromatogr 532:305, 1990.
162. Y Tang, WL Zielinski, HM Bigott. Chirality 10:364, 1998.
163. DW Armstrong, X Wang, N Ercal. Chirality 10:587, 1998.
164. J Hermansson. J Chromatogr 269:71, 1983.
165. S Allenmark, B Bomgren, H Boren. J Chromatogr 269:63, 1983.
166. User's Guide, Chrom. Tech. Application Note, No. 13, 1994.
167. J Haginaka, C Seyama, N Kanasugi. Anal Chem 67:2579, 1995.
168. Y Okamoto, S Hondon, I Okamoto, H Yuki, S Murata, R Noyori, H Takaya. J Am Chem Soc 103: 6971, 1981.
169. WH Pirkle, JM Finn. J Org Chem 46:2935, 1981.
170. WH Pirkle, MH Hyun. J Chromatogr 322:309, 1985.
171. WH Pirkle, KC Deming, JA Burke. Chirality 3:183, 1991.
172. N Oi, M Nagase, T Doi. J Chromatogr 257:111, 1983.
173. N Oi, H Kitahara. J Chromatogr 265:117, 1983.
174. N Oi, H Kitahara. J Liq Chromatogr 9:443, 1986.
175. N Oi, H Kitahara, T Doi. Eur. patent #EP029793, 1988.
176. WH Pirkle, CJ Welch. J Liq Chromatogr 15:1947, 1992.
177. MM Maier, G Uray, OP Kleidernigg, W Lindner. Chirality 6:116, 1994.
178. SV Rogozhin, VA Davankov. Chem Commun 490, 1971.
179. VA Davankov, SV Rogozhin, AV Semechkin, TP Sachkova. J Chromatogr 82:359, 1973.
180. VA Davankov. Adv Chromatogr 18:139, 1980.
181. VA Davankov. In: JC Giddings, E Grushka, J Cazes, PR Brown, eds. Advances in Chromatography, Vol. 18. New York: Marcel Dekker, 1980, p. 139.
182. VA Davankov, AS Bochkov, AA Kurganov, P Roumeliotis, KK Unger. Chromatographia 13:677, 1980.
183. V Davankov, A Bochkov, A Kurganov, P Roumeliotis, K Unger. Chromatographia 13:677, 1980.
184. G Gübitz, W Jellenz, G Lofler, W Santi. HRC CC 2:145, 1979.
185. G Gübitz, W Jellenz, W Santi. J Chromatogr 203:377, 1981.
186. UAT Brinkman, D Kamminga. J Chromatogr 330:375, 1985.
187. E Gassmann, JE Kuo, RN Zare. Science 230:813, 1985.
188. SG Allenmark, S Anderson, P Moller, D Sanchez. Chirality 7:248, 1995.
189. W Lindner, M Laemmerhofer. Chimia 50:274, 1996.
190. M Laemmerhofer, W Lindner. J Chromatogr A 741:33, 1996.
191. M Laemmerhofer, W Lindner. J Chromatogr A 839:167, 1999.
192. V Piette, M Laemmerhofer, W Lindner. J Chrommen Chirality 11:622, 1999.
193. E Smolkova-Keulemansova, S Krysl. J Chromatogr 184:347, 1980.
194. E Smolkova-Keulemansova. J Chromatogr 251:17, 1982.
195. E Smolkova, H Kralova, S Krysl, L Feltl. J Chromatogr 257:247, 1982.

196. J Mraz, L Feltl, E Smolkova-Keulemansova. J Chromatogr 286:17, 1984.
197. T Koscielski, P Sylbilska, L Feltl, E Smolkova-Keulemansova. J Chromatogr 286:23, 1984.
198. D Sybilska, E Smolkova-Keulemansova. Inclusion Compounds 3:173, 1984.
199. E Smolkova-Keulemansova, L Feltl, S Krysl. J Inclusion Phenom 3:183, 1985.
200. E Smolkova-Keulemansova, E Neumannova, L Feltl. J Chromatogr 365:279, 1986.
201. T Koscielski, D Sylbilska, J Jurczak. J Chromatogr 280:131, 1983.
202. DW Armstrong, Y Tang, J Zukowski. Anal Chem 63:2858, 1991.
203. W Li, L Jin, DW Armstrong. J Chromatogr 509:303, 1990.
204. DW Armstrong, W Li, M Stalcup, V Secor, R Izac, JI Seeman. Anal Chim Acta 234:365, 1990.
205. DW Armstrong, HL Jin. J Chromatogr 502:154, 1990.
206. Z Jin, HL Jin. Chromatographia 38:22, 1994.
207. I Hardt, WA König. J Microcolumn Sep 5:35, 1993.
208. M Jung, V Schurig. J Microcolumn Sep 5:11, 1993.
209. C Bicchi, A D'Amato, V Manzin, A Galli, M Galli. J Chromatogr A 666:137, 1994.
210. X Zhou, H Wan, Q Qu. Fenxi Ceshi Xuebao 13:55, 1994.
211. WM Buba, K Jaques, A Venema, P Sandra. Fresenius J Anal Chem 356:679, 1995.
212. A Berthod, YE Zhou, K Le, DW Armstrong. Anal Chem 67:849, 1995.
213. V Schurig, U Leyer. Tetrahedron: Asymmetry 1:865, 1990.
214. G Weseloh, C Wolf, WA König. Angew Chem Int Ed Engl 30:74, 1991.
215. M Jung, V Schurig. J Am Chem Soc 114:192, 1992.
216. M Reist, B Testa, PA Carrupt, M Jung, V Schurig. Chiality 7:396, 1995.
217. V Schurig, A Glausch, M Fluck. Tetrahedron Asymmetry 6:2161, 1995.
218. G Weseloh, C Wolf, WA König. Chirality 8:441, 1996.
219. V Schurig, S Reich. Chirality 8:425, 1998.
220. J Oxelbark, S Allenmark. J Org Chem 64:1483, 1999.
221. G Schoetz, O Trapp, V Schurig. Anal Chem 72:2758, 2000.
222. J Krupcík, P Oswald, I Spánik, P Májek, M Bajdichová, P Sandra, DW Armstrong. In: P Sandra, AJ Rackstraw, eds. Proceedings of the 23rd International Symposium on Capillary Chromatography, 2001.
223. V Schurig. Angew Chem 89:113, 1977; Angew Chem Int Ed Engl 16:110, 1977.
224. V Schurig, W Bürkle. Angew Chem 90:132, 1978; Angew Chem Int Ed Engl 17:132, 1978.
225. V Schurig, R Weber. J Chromatogr 217:51, 1981.
226. V Schurig, W Bürkle. J Am Chem Soc 104:7573, 1982.
227. R Weber, K Hintzer, V Schurig. Naturwissenschaft 67:453, 1980.
228. B Koppenhoefer, K Hintzer, R Weber, V Schurig. Angew Chem Int Ed Engl 19:471, 1980.
229. V Schurig. In: JD Morrison, ed. Asymmetric Synthesis, Vol. 1, Academic Press, 1983, p. 59.
230. V Schurig, R Weber. J Chromatogr 289:321, 1984.
231. V Schurig, U Leyrer, R Weber. J High Res Chromatogr Chromatogr Res 8:459, 1985.
232. B Kolb. Chromatographia 9:587, 1982.
233. V Schurig, D Wistuba. Tetrahedron Lett 25:5633, 1984.
234. M Schleimer, F Fluck, V Schurig. Anal Chem 66:2893, 1994.
235. N Oi, H Kitahara, T Doi. J Chromatogr 207:252, 1981.
236. N Oi, H Hitahara, Y Inda, T Doi. J Chromatogr 213:137, 1981.
237. N Oi, H Kitahara, T Doi. J Chromatogr 254:282, 1983.
238. RMC Sutton, KL Sutton, AM Stalcup. Electrophoresis 18:2297, 1997.
239. DS Hage. Electrophoresis 18:2311, 1997.
240. P Camilleri. Electrophoresis 18:2322, 1997.
241. DW Armstrong, UB Nair. Electrophoresis 18:2331, 1997.
242. TJ Ward. Anal Chem 66:633A, 1994.
243. TJ Ward. Anal Chem 72:4521, 2000.

34

Chiral Capillary Electrophoresis

Salvatore Fanali

Istituto di Cromatografia, Consiglio Nazionale Delle Ricerche, Rome, Italy

I. HISTORY

The separation of chiral compounds is an important topic of research, especially in pharmaceutical and environmental fields where very often the two enantiomers of the drugs or agrochemicals used can exhibit different pharmacological or biological activities. Thus, in the last decade the search for new separation methods, allowing fast, cheap, and sensitive analysis, was the main topic of study of several researchers, including our group. Analytical methods so far used for the analysis of chiral compounds include high-performance liquid chromatography (HPLC), thin-layer chromatography (TLC), gas chromatography (GC), and, most recently, capillary electrophoresis (CE) [1–12].

CE is a new electromigration technique that provides analyte separation with high efficiency and high resolution in a relatively short time due to the use of high electric field and a variety of selective modes. The technique was successfully applied in different research areas such as biological, clinical, environment, pharmaceutical fields, etc. for the analysis of a wide number of different classes of compounds, including enantiomers.

Being interested in electromigration methods and pharmaceutical analysis, our group became involved in this research area, and this section will briefly illustrate this involvement in the field of separation science.

The year 1980 represents the start of research at the Institute of Chromatography (Consiglio Nazionale delle Ricerche), Monterotondo Scalo (Italy) with a part-time cooperation studying a new separation technique introduced by Prof. M. Lederer [13], namely high-performance paper electrophoresis (HPPE). Rapid separations (3–6 min) of amino acids, inorganic anions, cobalt-complexes, as well as cobalt complex enantiomers were achieved with this technique using relatively high electric fields (100–240 V/cm). The Joule heat, the main source of band broadening, was lowered by using paper strips of reduced dimensions (0.5–1 cm width $\times$ 5–10 cm length). The instrumentation was very simple—in fact, an ordinary power supply (0–1200 V) was employed, and the Joule heat was dispersed through two glass plates [14,15].

As an example of the performance of HPPE, Fig. 1 shows the separation of inorganic anions

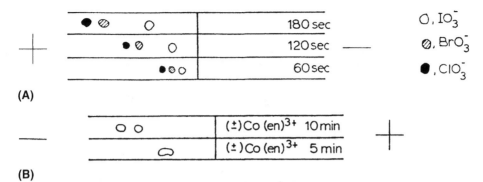

(A)

(B)

Figure 1 Electropherograms of separation by high-performance paper electrophoresis of (A) inorganic anions (B) cobalt $(en)_3^{3+}$ enantiomers. Paper strips 10 cm long and 0.5 cm wide; applied voltage (A) 700 V, electrolyte 0.1 N HCl; (B) 800 V, electrolyte 0.1 N antimony potassium (+) tartrate. (Modified from Ref. 14.)

(A) and cobalt(III) complex enantiomers (B). Unfortunately, with this technique quantitation, sensitivity, and detectability were not satisfactory, and thus in 1984 we started an interesting experiment in the field of capillary electrophoretic (CE) techniques, namely capillary isotachophoresis (ITP). Meeting with Prof. Petr Bocek and his group at Institute of Analytical Chemistry, Academy of Sciences, Brno (Czech Republic), was very fruitful for the studies carried out by our group in recent years. The 6-month stay allowed us to be introduced to ITP theory and learn

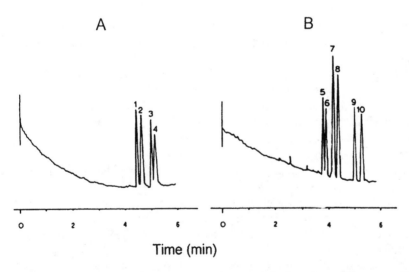

Time (min)

Figure 2 Enantiomeric separation of sympathomimetic drugs by capillary zone electrophoresis for (A) norephedrine (peaks 1,2) and ephedrine (peaks 3,4) (B) norepinephrine (peaks 5,6), epinephrine (peaks 7,8) isoproterenol (peaks 9,10). Background electrolyte, 10 mM Tris/phosphoric acid pH 2.4 and 18 mM DM-β-CD; capillary polyacrylamide coated 20 cm $\times$ 0.025 mm I.D.; applied voltage, 8 kV; injection, 6 kV, 6 s of 2×10^{-5} M for each racemic compound. (Modified from Ref. 17.)

the secrets to be used with miniaturized techniques as well as pharmaceutical analysis employing CE techniques. Our cooperation with Bocek's group continues to the present.

Since then ITP has been used by our group for the analysis of several compounds, including those of pharmaceutical interest and for the inclusion-complexation equilibria studies [16]. Finally, in 1988 we constructed, in cooperation with Dr. Frantisek Foret (Institute of Analytical Chemistry, Brno) the first capillary electrophoresis instrumentation (in our laboratory); this experience will be described later because it was the beginning of a fascinating time in the CE field. Finally, after some unsuccessful experiments using ITP for the separation of chiral compounds, we used, for the first time, capillary zone electrophoresis (CZE) employing uncharged cyclodextrin (CD) derivatives as chiral selectors [17]. These experiments were very stimulating and encouraging, and thus this first paper will be briefly discussed here.

Figure 2 shows the separation of sympathomimetic drug enantiomers by CZE employing 2,6-heptakis-di-O-methyl-β-cyclodextrin (DM-β-CD). The separation was achieved using one of the first commercially available CE instruments. It is noteworthy to mention that our interest in the use of cyclodextrins as chiral selectors in CZE was stimulated, after preliminary results, and following interesting lectures, held in our institute, by two well-recognized scientists: Prof. Danuta Sybilska (Warsaw, Poland) and Prof. Eva Smolkova-Keulemansova (Prague, Czech Republic).

II. CONTRIBUTIONS TO THE ADVANCEMENT OF SEPARATION SCIENCE

The separation of chiral compounds by CE can be dated by the work of Gassmann et al. in 1985 with the analysis of dansyl amino acid enantiomers employing a ligand exchange resolution mechanism [18]. Later, fundamental results in the field of chiral analysis by CE were obtained using either CZE [17,19,20] or ITP [21] utilizing cyclodextrins or their derivatives. Other resolution mechanisms were also investigated by us and Terabe's group employing L-tartrate or bile salts as the chiral selectors [22,23]. Since that time a wide number of publications dealing with the separation of chiral compounds have appeared and the method has been applied to the analysis of compounds of interest in the pharmaceutical, clinical, and environmental fields. This has been documented by the wide number of reviews, recently published, summarizing the state of the art of this research field [7–11,24–33].

In our first study of chiral CE, employing DM-β-CD as chiral selector for the separation of sympathomimetic drug enantiomers [17], we could not imagine the importance of cyclodextrins as additives of the background electrolyte (BGE) in CZE. However, we decided to continue with the search for new chiral selectors to be studied with electromigration methods. A recent review reported statistical data showing that CDs and their derivatives are the most popular chiral selectors employed in CE [30]. This can be explained considering the wide variety of CDs commercially available and their chemical properties, e.g., they are transparent even at low UV wavelengths, soluble in water (with some exception), and can be either uncharged or charged. In 1993 the literature of chiral CE separations studying new chiral selectors, theory, parameters influencing resolution, etc., started to grow exponentially. It was recognized that the charge of the CD strongly influenced the selectivity of the enantiomer separation. At this time our and Engelhardt's groups published a study dealing with the use of positively and negatively charged CDs, respectively [34,35], obtaining very interesting enantiomer separations.

After synthesizing monomethylamino- and dimethylamino-β-CD, we studied the effect of several parameters on chiral resolution of hydroxy acid enantiomers. Good chiral resolution of the examined compounds was achieved using polyacrylamide-coated capillaries in which the absence of the electro-osmotic flow (EOF) as well as the countercurrent movement of the chiral selectors played an important role, increasing the difference between mobility of free and complexed enantiomer [34].

We also had the opportunity to investigate by CZE the effect of a negatively charged CD, namely sulfobutyl(IV)ether-β-CD, added to BGE, on enantioselectivity and resolution of several pharmaceutical compounds. This CD is negatively charged at any pH, and thus this feature broadens its use in CE. We used SBE-β-CD for the chiral resolution of acidic, basic, and uncharged compounds [36].

III. SEPARATION SCIENCE FOR THE SOLUTION OF PRACTICAL PROBLEMS

Recently I was rereading an issue of the *Journal of Capillary Electrophoresis* in which the editor, Dr. Norberto Guzman, published a picture of me and a short bibliography. In this presentation Norberto wrote the following [37]: "The Etruscan culture flourished on the same soil upon which stands the laboratory of Dr. Salvatore Fanali. The former provided a restoration of stability to a society in turmoil; the latter created a sort of terra firma for the expansion of the study of chiral separation. . . ." From these few words it seems clear that the major recognized contribution until now in the field of separation science was the separation of enantiomers, a topic currently being studied by a great number of researchers. The chiral capillary electrophoretic method is now used for practical analysis in different fields, e.g., pharmaceutical, forensic, biomedical, environmental, etc. [38–44]. To summarize, the contribution of our group in the field of chiral separations by CE is not an easy task. In fact, our efforts were aimed at both theoretical and practical applications in this field. However, a few examples will be reported here.

The analysis of native cyclodextrins (CDs) by CE is very difficult because the analytes are not absorbing at the UV and are not charged/chargeable. We studied the analytical problem and proposed a CE separation method useful for the analysis of α-, β-, and γ-CDs. The compounds were analyzed using an absorbing background electrolyte (BGE), benzoic acid, at pH 6. At the selected conditions the CDs were moved towards the detector by the EOF and the selectivity improved by the interactions with the benzoate exhibiting mobility against the EOF. The usefulness of the studied CE method was applied to the analysis of β-CD in a pharmaceutical formulation [45]. Figure 3 shows the separation and the mechanism involved in the analysis of native CDs by CE. The above-mentioned example is indirectly involved with chiral separations because CDs are the most popular chiral selectors used in CE, and the proposed method can be useful for the purity control compounds. After this simple work, several researchers discussed the possibility of using CZE for the analysis of native or derivatized (charged or uncharged) CDs with indirect UV detection [46]. Because indirect UV detection is nonspecific, CE coupled with mass spectrometry (CE-MS) was successfully used for the characterization of CD derivative mixtures [47,48].

In cooperation with Bocek's group, we studied in our institute the theory and applicability of indirect UV detection to the analysis of inorganic cations as well as anions. In the analysis of rare earth metal ions, a selective complexing agent, namely hydroxyisobutyric acid (HIBA), was employed in order to improve the selectivity of the separation [47,48].

Glycopeptide antibiotics were also used in CE for the analysis of a wide number of enantiomers. However, the chiral selector is strongly absorbing at the wavelengths usually employed

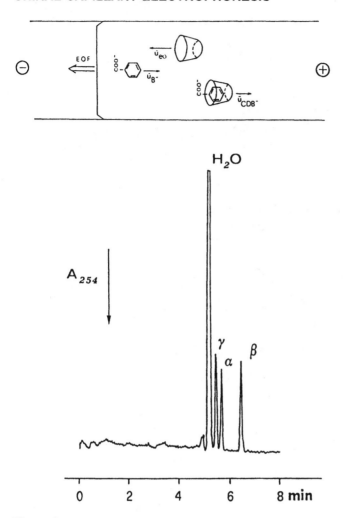

Figure 3 Electrophoretic separation of native cyclodextrins by capillary electrophoresis with indirect UV detection at 254 nm. 30 mM of benzoic acid/Tris pH 6.2 was the background electrolyte. (Modified from Ref. 45.)

in CE, and thus poor sensitivity is achieved when the chiral selector is present in the capillary and the electrode compartments. Taking in mind the above-mentioned drawbacks, we studied the chiral separation of several acidic compounds of pharmaceutical and environmental interest by using vancomycin or teicoplanin derivatives [41,49–52]. In these studies we adopted the partial filling-counter current method selecting the appropriate pH of the BGE and capillary. The method was used for the separation of acidic compounds in polyacrylamide-coated capillary in order to suppress/minimize the EOF at pH (4–7). The selection of the pH is fundamental for the success of the chiral separation because the analytes and the chiral selector must be negatively and positively charged, respectively. In such conditions, the chiral selector is moving in the opposite direction of the analytes that reach the detector in the absence of absorbing compounds. This is illustrated in Fig. 4, where the electrophoretic process can be summarized:

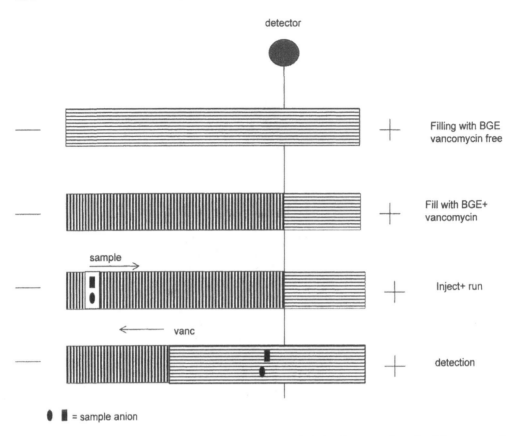

Figure 4 Scheme of the partial filling countercurrent method for the separation of acidic compound enantiomers by capillary zone electrophoresis where an absorbing chiral selector (vancomycin) is used.

(a) the capillary is filled at high pressure with the BGE (without chiral selector), (b) the BGE containing the chiral selector is injected at low pressure (only part of the capillary is filled keeping antibiotic free the detector), and (c) the sample is injected and run.

Recently we have been developing capillary electrochromatography (CEC) and CE coupled with mass spectrometry (CZE-MS and CEC-MS) for the separation and characterization of achiral and chiral compounds. At first we used commercial available RP_{18} packed capillaries, achieving interesting results for the separation of standard uncharged compounds. However, the column, after a relatively short time, no longer worked. After this experience we decided to pack in our laboratory the CEC capillaries, trying to resolve practical problems related to the preparation of such columns. We recognized that a key problem was the heterogeneity of the system. In fact the commercially available capillaries are prepared with the window just after the end of the packing material. Thus, in order to avoid bubble formation in the window, we decided to pack the whole capillary preparing the detector window in a zone where the stationary phase was present. The results were recently published, showing also the applicability of CEC to practical analysis of hydroquinone in skin toning cream. The capillary exhibited good repeatability (RSD <1%) and was used for more than 300 runs [53]. Furthermore, such a set-up can be used in two different modes, namely fast and normal, by reversing the polarity and the injection side and therefore using the shortest and the longest effective lengths.

IV. MEMORABLE MOMENTS IN SCIENTIFIC ACTIVITY

Illustrating memorable moments in the career of a separation scientist is not an easy task, because I was interested in different topics and met a wide number of colleagues and friends. However, the preparation of my thesis at the Institute of Chromatography in Rome, due to the initiation of scientific work and my youth, can be considered the most exciting time. This was, for me, the first great scientific responsibility that led me to appreciate my Indian friend, Professor S. K. Shukla, as a supervisor. The next important step was my stay at the Institute of Analytical Chemistry, where I had the opportunity to meet several friends, with whom I engaged in scientific discussion. I recall the recommendation of Professor Petr Bocek, chief of the electromigration department at that time, to remember that the referee is the person that should help the author to improve the manuscript. This was a life lesson, and I consider this advice each time that I have in my hands a paper to be examined.

Coming back from the Czech Republic, I started my research with a commercial ITP instrumentation utilizing a combined conductivity and UV detector, which was very expensive for my budget. Very often the conductivity detector did not work, and thus I decided to construct a detector following the suggestions of Professor F. Everaerts. For this purpose a block of polymeric resin with platinum wires was prepared according to the findings of previous experiments in the Czech Republic. Later the detector cell was improved by preparing the resin blocks with Araldite^R (also useful for organic solvents) and machining and finishing with platinum. Since that time, the experiments carried out by ITP were achieved with laboratory-made conductivity detector cells.

Another exciting time was during the stay of Dr. Franta Foret in our Institute (1988–1989), which for us was the beginning of CZE. In fact, we used materials and instrumentation present in the laboratory, e.g., power supply and UV detector from LKB (Bromma, Sweden) and a UV detector used for HPLC. Optical fibers were used in order to detect the separated zones in the CE runs. After several experiments the power supply no longer worked and the service was not able to repair it. Thus, we established a laboratory-made power supply (Foret's project) using old television parts that allowed us to finish the study dealing with the theory of indirect UV detection [54].

In 1990 I was invited by Professor B. J. Radola to join the editorial board of *Electrophoresis*, and since then a fruitfull cooperation has been established: I analyzed manuscripts and prepared as guest editor three special issues devoted to chiral analysis by CE. Of course, the other six journals for which I serve as referee have for me equal importance.

In 1992 I organized as chairman, the Tenth International Symposium on Capillary Electrophoresis and Isotachophoresis, ITP '92, in Rome. I remember the coworkers (Dr. C. Desiderio, A. Nardi, Mr. M. Cristalli, G. Caponecchi and A. Rosati) with which we spent time. Considering the number of participants, the quality of the communication was memorable. Even now, meeting colleagues who attended the symposium brings back memories of the time spent at the Opera Theatre and the old palace of ''Drago.''

V. CONCLUSIONS

This short report should give an idea of the scientific activity performed by our group since 1980 in the research area of ''Consiglio Nazionale delle Ricerche'' in Monterotondo Scalo, a small town 30 km from Rome where several Etruscan tombs have been discovered.

The studies of chiral analysis by capillary electrophoretic techniques have been the main topic this research, and we are planning to continue to look for new chiral selectors and different

CE modes. However, we were also involved in other studies, such as the analysis of hemoglobins [55], DNA or RNA fragments [56,57], inks [58], etc.

Thanks are due to the students who spent time in our laboratory as well as colleagues from Italy and other countries, whose names can be found in the literature. Special thanks are due to Dr. Claudia Desiderio (student, coworker, and colleague) for her skillful help since 1990 in obtaining results in the field of capillary electrophoresis.

REFERENCES

1. J Debowski, D Sybilska, J Jurczak. Resolution of some chiral mandelic acid derivatives into enantiomers by reversed-phase high-performance liquid chromatography via α- and β-cyclodextrin inclusion complexes. J Chromatogr 282:83–88, 1983.
2. G Blaschke. Chromatographic resolution of chiral drugs on polyamides and cellulose triacetate. J Liq Chromatogr 9:341–368, 1986.
3. V Carunchio, A Messina, M Sinibaldi, S Fanali. High performance ligand exchange chromatography of aminoacids on chiral stationary phases. J High Resol Chromatogr Chromatogr Comm 11:401–404, 1988.
4. NH Singh, FN Pasutto, RT Coutts, F Jamali. Gas chromatographic separation of optically active anti-inflammatory 2-arylpropionic acids using (+) or (−)-amphetamine as derivatizing agent. J Chromatogr 378:125–135, 1986.
5. F Kobor, G Schomburg. 6-Tert-butyldimethylsilyl-2,3-dimethyl-alpha-cyclodextrin, beta-cyclodextrin, and gamma-cyclodextrin, dissolved in polysiloxanes, as chiral selectors for gas chromatography—influence of selector concentration and polysiloxane matrix polarity on enantioselectivity. J High Resolut Chromatogr 16:693–699, 1993.
6. J Snopek, I Jelinek, E Smolkova-Keulemansova. Micellar, inclusion and metal-complex enantioselective pseudophases in high performance electromigration methods. J Chromatogr 452:571–590, 1988.
7. S Terabe, K Otsuka, H Nishi. Separation of enantiomers by capillary electrophoretic techniques. J Chromatogr A 666:295–319, 1994.
8. B Chankvetadze, G Endresz, G Blaschke. Charged cyclodextrin derivatives as chiral selectors in capillary electrophoresis. Chem Soc Rev 25:141–153, 1996.
9. S Fanali. Controlling enantioselectivity in chiral capillary electrophoresis with inclusion-complexation. J Chromatogr A 792:227–267, 1997.
10. ML Riekkola, SK Wiedmer, IE Valko, H Siren. Selectivity in capillary electrophoresis in the presence of micelles, chiral selectors and non-aqueous media. J Chromatogr A 792:13–35, 1997.
11. B Chankvetadze. Capillary Electrophoresis in Chiral Analysis. New York: John Wiley & Sons, 1997.
12. R Vespalec, P Bocek. Chiral separations in capillary electrophoresis. Electrophoresis 20:2579–2591, 1999.
13. M Lederer. High performance paper electrophoresis. J Chromatogr 171:403–406, 1979.
14. S Fanali, L Ossicini. High-performance paper electrophoresis. II. Comparison of separations obtained by high-voltage paper electrophoresis and high-performance paper electrophoresis. J Chromatogr 212:374, 1981.
15. S Fanali, L Ossicini. High performance paper electrophoresis. III. Influence of various parameters on electrophoretic separations. J Chromatogr 287:148–154, 1984.
16. S Fanali, M Sinibaldi. Host-guest complexation in capillary isotachophoresis. I. Alpha-, beta and gamma-cyclodextrins as complexing agents for the resolution of substituted benzoic acid isomers. J Chromatogr 442:371–377, 1988.
17. S Fanali. Separation of optical isomers by capillary zone electrophoresis based on host-guest complexation. J Chromatogr 474:441–446; 1989.
18. E Gassmann, JE Kuo, RN Zare. Electrokinetic separation of chiral compounds. Science 230:813–814, 1985.
19. A Guttman, A Paulus, AS Cohen, N Grinberg, BL Karger. Use of complexing agents for selective

separation in high performance capillary electrophoresis, chiral resolution via cyclodextrins incorporated within polyacrylamide gel column. J Chromatogr 448:41–53, 1988.

20. S Terabe. Electrokinetic chromatography: an interface between electrophoresis and chromatography. Trends Anal Chem 8:129–134, 1989.

21. J Snopek, I Jelinek, E Smolkova-Keulemansova. Use of cyclodextrins in isotachophoresis. IV. The influence of cyclodextrins on the chiral resolution of ephedrine alkaloid enantiomers. J Chromatogr 438:211–218, 1988.

22. S Fanali, L Ossicini, F Foret, P Bocek. Resolution of optical isomers by capillary zone electrophoresis. Study of enantiomeric and diastereoisomeric Co(III)-complexes with ethylenediamine and aminoacid ligands. J Microcol Sep, 1:190–194, 1989.

23. H Nishi, T Fukuyama, M Matsuo, S Terabe. Chiral separation of optical isomeric drugs using micellar electrokinetic chromatography and bile salts. J Microcol Sep 1:234–241, 1989.

24. S Fanali, M Cristalli, R Vespalec, P Bocek. Chiral separations in capillary electrophoresis. In: A Chrambach, MJ Dunn, BJ Radola, eds. Advances in Electrophoresis. Weinheim: VCH Verlagsgesellschaft GmbH, 1994, pp. 1–86.

25. M Novotny, H Soini, M Stefansson. Chiral separation. Anal Chem 66:646A–655A, 1994.

26. R Vespalec, P Bocek. Chiral separations by capillary electrophoresis: Present state of the art. Electrophoresis 15:755–762, 1994.

27. H Nishi, S Terabe. Optical resolution drugs by capillary electrophoretic techniques. J Chromatogr A 694:245–276, 1995.

28. S Fanali. Identification of chiral drug isomers by capillary electrophoresis. J Chromatogr A 735:77–121, 1996.

29. B Chankvetadze. Separation selectivity in chiral capillary electrophoresis with charged selectors. J Chromatogr A 792:269–295, 1997.

30. G Gubitz, MG Schmid. Chiral separation principles in capillary electrophoresis. J Chromatogr A 792:179–225, 1997.

31. DK Lloyd, AF Aubry, E De Lorenzi. Selectivity in capillary electrophoresis—the use of proteins. J Chromatogr A 792:349–369, 1997.

32. IS Lurie. Separation selectivity in chiral and achiral capillary electrophoresis with mixed cyclodextrins. J Chromatogr A 792:297–307, 1997.

33. H Nishi. Enantioselectivity in chiral capillary electrophoresis with polysaccharides. J Chromatogr A 792:327–347, 1997.

34. A Nardi, A Eliseev, P Bocek. S Fanali Use of charged and neutral cyclodextrins in capillary zone electrophoresis: enantiomeric resolution of some 2-hydroxy acids. J Chromatogr 638:247–253, 1993.

35. T Schmitt, H Engelhardt. Charged and uncharged cyclodextrins as chiral selectors in capillary electrophoresis. Chromatographia 37:475–481, 1993.

36. C Desiderio, S Fanali. Use of negatively charged sulfobutyl ether-beta-cyclodextrin for enantiomeric separation by capillary electrophoresis. J Chromatogr A 716:183–196, 1995.

37. NA Guzman. Editor's page. J Cap Elec 3:4A–5A, 1996.

38. C Desiderio, S Fanali, A Kupfer, W Thormann. Analysis of mephenytoin, 4-hydroxymephenytoin and 4-hydroxymephenytoin enantiomers in human urine by cyclodextrin micellar electrokinetic capillary chromatography: Simple determination of a hydroxylation polymorphism in man. Electrophoresis 15:87–93, 1994.

39. M Heuermann, G Blaschke. Simultaneous enantioselective determination and quantification of dimethindene and its metabolite n-demethyl-dimethindene in human urine using cyclodextrins as chiral additives in capillary electrophoresis. J Pharm Biomed Anal 12:753–760, 1994.

40. M Lanz, R Brenneisen, W Thormann. Enantioselective determination of 3,4-methylene-dioxymethamphetamine and two of its metabolites in human urine by cyclodextrin-modified capillary zone electrophoresis. Electrophoresis 18:1035–1043, 1997.

41. S Fanali, C Desiderio, G Schulte, S Heitmeier, D Strickmann, B Chankvetadze, G Blaschke. Chiral capillary electrophoresis-electrospray mass spectrometry coupling using vancomycin as chiral selector. J Chromatogr. A 800:69–76, 1998.

42. S Zaugg, J Caslavska, R Theurillat, W Thormann. Characterization of the stereoselective metabolism of thiopental and its metabolite pentobarbital via analysis of their enantiomers in human plasma by capillary electrophoresis. J Chromatogr A 838:237–249, 1999.

43. S Rudaz, JL Veuthey, C Desiderio, S Fanali. Simultaneous stereoselective analysis by capillary electrophoresis of tramadol enantiomers and their main phase I metabolites in urine. J Chromatogr A 846:227–237, 1999.

44. K Krause, B Chankvetadze, Y Okamoto, G Blaschke. Chiral separations in capillary high-performace liquid chromatography and nonaqueous capillary electrochromatography using helically chiral poly(diphenyl-2-pyridylmethyl methacrylate) as chiral stationary phase. Electrophoresis 20:2772–2778, 1999.

45. A Nardi, S Fanali, F Foret. Capillary zone electrophoretic separation of cyclodextrins with indirect UV photometric detection. Electrophoresis 11:774–776, 1990.

46. RJ Tait, DJ Skanchy, DP Thompson, NC Chetwyn, DA Dunshee, RA Rajewsky, VJ Stella, JF Stobaugh. Characterization of sulphoalkyl ether derivatives of β-cyclodextrin by capillary electrophoresis with indirect UV detection. J Pharm Biomed Anal 10:615 -622, 1992.

47. Y Tanaka, S Terabe. Enantiomer separation of acidic racemates by capillary electrophoresis using cationic and amphoteric beta-cyclodextrins as chiral selectors. J Chromatogr A 781:151–160, 1997.

48. Y Tanaka, Y Kishimoto, S Terabe. Analysis of charged cyclodextrin derivatives by on-line capillary electrophoresis ionspray-mass spectrometry. Anal Sci 14:383–388, 1998.

49. S Fanali, C Desiderio. Use of vancomycin as chiral selector in capillary electrophoresis. optimization and quantitation. J High Resolut Chromatogr 19:322–326, 1996.

50. S Fanali, C Desiderio, Z Aturki. Enantiomeric resolution study by capillary electrophoresis—selection of the appropriate chiral selector. J Chromatogr A 772:185–194, 1997.

51. S Fanali, Z Aturki, C Desiderio, A Bossi, PG Righetti. Use of Hepta-tyr antibiotic as chiral selector in capillary electrophoresis. Electrophoresis 19:1742–1751, 1998.

52. S Fanali, Z Aturki, C Desiderio, PG Righetti. Use of mdl63246 (hepta-tyr) antibiotic in capillary zone electrophoresis—ii. Chiral resolution of alpha-hydroxy acids. J Chromatogr A 838:223–235, 1999.

53. C Desiderio, L Ossicini, S Fanali. Analysis of hydroquinone and some of its ethers by using capillary electrochromatography. J Chromatogr A 887:489–496, 2000.

54. F Foret, S Fanali, L Ossicini, P Bocek. Indirect photometric detection in capillary zone electrophoresis. J Chromatogr 470:299–308, 1989.

55. P Ferranti, A Malorni, P Pucci, S Fanali, A Nardi, L Ossicini. Capillary zone electrophoresis and mass spectrometry for the characterization of genetic variants of human hemoglobin. Anal Biochem 194:1–8, 1991.

56. K Kleparnik, S Fanali, P Bocek. Selectivity of the separation of DNA fragments by capillary zone electrophoresis in low-melting-point agarose sol. J Chromatogr 638:283, 1993.

57. L Cellai, A Mochi Onori, C Desiderio, S Fanali. Capillary electrophoretic analysis of synthetic short-chain oligoribonucleotides. Electrophoresis 19:3160–3165, 1998.

58. S Fanali, M Schudel. Some separations of black and red water-soluble fiber tip pen inks by capillary zone electrophoresis and thin-layer chromatography. J Forensic Sci 36:1192–1197, 1991.

35

Whole Column Imaging Detection for Capillary Isoelectric Focusing

Xing-Zheng Wu

Fukui University, Fukui-shi, Japan

Jiaqi Wu

Convergent Bioscience Ltd., Toronto, Ontario, Canada

Janusz Pawliszyn

University of Waterloo, Waterloo, Ontario, Canada

I. INTRODUCTION

Isoelectric focusing (IEF) is the highest resolution technique used in biomolecule analysis [1,2]. It is traditionally performed in slab gels. As an analytical technique, gel IEF is slow, labor intensive, and generally not quantitative [3]. Since the appearance of capillary electrophoresis (CE) with high speed and quantitative properties, it was recognized that if IEF could be performed in a capillary format, it would greatly decrease IEF time, increase quantitation, make it easy to automate, and further enhance the separation resolution. In 1985, Hjerten and Zhu first reported high-resolution IEF performed in a gel-free capillary column [4]. Since then, many studies of capillary IEF (cIEF) have been reported [5–9], and CE manufacturers began developing and eventually commercialized cIEF kits and accessories offering cIEF analysis capabilities on existing CE instruments.

Despite the significant advantages of cIEF oven gel IEF, widespread acceptance as a replacement technique for gel IEF has not occurred. Several factors contribute to cIEF's slower-than-expected adoption; perhaps the most important are the performance and procedural difficulties introduced when cIEF is performed on conventional CE instruments. Conventional CE instruments are equipped with a single-point, on-column UV-Vis absorption detector that is located at one end of the capillary. Thus, most cIEF is performed in a two-step operation [4–9]. In the first step, samples such as proteins are focused (separated and concentrated) to stationary sharp zones according to their respective isoelectric points (pI) in a coated capillary, where electroosmotic flow (EOF) is eliminated by the coating. Then, as a second step, the focused protein zones are moved by electrophoretic or hydrodynamic mobilization so that they can pass the detector point to be measured and recorded. The second mobilization step introduces several problems to cIEF, such as uneven separation resolution, poor reproducibility, and increased analysis time. An alternative approach is to perform cIEF in uncoated capillaries using methyl cellulose as a buffer additive to reduce EOF to such an extent as to allow attainment of steady-state conditions, at which protein samples are focused [10–13]. The reduced EOF will continue mobilizing the focused protein zones past the detection point. Although this approach is a one-step procedure, it still has limitations associated with uneven mobilization speeds, long mobilization times for acidic proteins, and incomplete pattern detection at column locations near the capillary end.

To overcome the drawbacks of single-point detection, optical whole column detection methods based on movement of the separation capillary to the optical detection point, for example, by moving the separation capillary through the detection window of a UV absorbance [14] or fluorescence detector [15], have been proposed. Although the problems of distortion of pH gradient and uneven resolution in the focused zones moved cIEF have been improved, the mechanical movement of the capillary increases dynamic noise and analysis time. Also, it is difficult to apply to monitoring of a fast cIEF process [16]. An ideal approach to cIEF is real-time whole column imaging detection without moving any part in the system, demonstrated by Pawliszyn and Wu [17]. They and their colleagues have developed several types of real-time whole column imaging detectors for cIEF [18–24]. This approach combines gel-like IEF separation and detection with the automation, speed, and quantitation of a column-based separation technique. More recently, this technology has been commercialized and found wide application in various fields.

In this chapter, principles of the IEF and whole column imaged cIEF are discussed. Then several types of real-time whole column imaging detectors for cIEF, the commercial imaged cIEF, and its applications are briefly introduced. Finally, the current status of whole column imaged cIEF and its future trends are discussed.

II. PRINCIPLE OF REAL-TIME WHOLE COLUMN IMAGED cIEF

A. Principle of cIEF

Figure 1A illustrates the concept of IEF. For an amphoteric (zwitterionic) compound such as a protein in a buffer with a certain pH, if the pH is lower or higher than the isoelectric point (pI) of the protein, the protein molecule will be positively or negatively charged. When an electric field is applied, the positively charged protein molecule will electromigrate toward the cathode and the negative one toward the anode. If a pH gradient exists between the anode and cathode, the protein will electromigrate to a point at which pH is equal to its pI. At the pI point, the net charge of the protein is zero, thus the electromigration is stopped. This means that the protein molecules distributed in the whole pH gradient will be focused (concentrated) at the pI point to form a sharp narrow zone. However, it is usually difficult to fix a pH gradient in a bulk solution because of convection. On the other hand, the convection can be eliminated in a gel. That is why most early IEF experiments were performed in gels. A stable pH gradient is also easily formed in a coated capillary (internal diameter from $\sim$10 to $\sim$100 μm) filled with carrier ampholyte, where EOF is eliminated or reduced to near zero, in the presence of an electric field [25]. Therefore, if protein samples are mixed with the carrier ampholyte, the proteins will be focused (concentrated and separated) at different points along the capillary according to their pI values in the presence of an electric field (Fig. 1B).

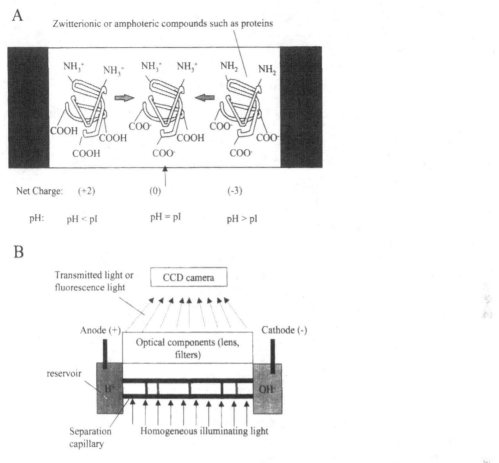

Figure 1 Illustration of the concept of IEF (A) and whole column imaging detection for cIEF (B).

B. Resolution of cIEF

Separation resolution is the most significant concern for a separation method. A theoretical consideration on the separation resolution of cIEF has been addressed, and it has been shown that the resolution has no direct relation to the length of separation capillary [16]. This means that a short column can be used in cIEF without sacrificing any separation resolution. Use of a short column has several advantages. First, the separation process over the whole column can be studied with a CCD camera imaging detector. Second, the cIEF process in a short column will be faster than that in a long column. Third, since the sample mixed with carrier ampholytes usually fills the whole column, a short column requires a smaller amount of the sample than a long column. This is important for expensive and scarce biological samples. It also simplifies instrumentation, since a relatively low direct current (DC) power supply can provide high electric field for a short column.

Based on the above theoretical consideration, the concept of short-column cIEF with whole column imaging detection is proposed.

C. Whole Column Imaging Detection of cIEF

Figure 1B illustrates the basic concept of real-time whole column imaging detection for cIEF. As stated above, the homogeneously mixed protein samples with carrier ampholytes in the coated capillary will be focused (concentrated and separated) into narrow sharp zones at different pI points in the cIEF process. This means high concentration gradients are created at the boundaries of the separated zones inside the capillary. The concentration gradients will induce refractive index gradients, which in turn generates deflection of light passed through the gradients. The deflection will change the intensity distribution of a transmitted light from the capillary, which is homogeneously illuminated along the capillary axis. Therefore, the focused sample zones can be imaged by measuring the change in intensity distribution of the transmitted light by a charged coupling device (CCD).

  If samples have optical absorption properties for light, the focused sample zones can be estimated simply by measuring the optical absorption. This is easily achieved by measuring the decrease in intensity of transmitted light. If the samples are fluorescent, fluorescence will be produced by the illuminating light with a suitable wavelength. Then the focused sample zones are also estimable by measuring the fluorescence distribution along the capillary with a sensitive CCD. Therefore, three types of whole column imaging detectors based on concentration gradient, optical absorption, and fluorescence can be established. A suitable optical arrangement, including a set of lenses and filters, is placed between the capillary and CCD to achieve the maximum detection sensitivity. Based on these considerations, real-time whole column refractive index gradient, absorption, and fluorescence imaging detectors for cIEF are developed.

III. DEVELOPMENT OF REAL-TIME WHOLE COLUMN IMAGING DETECTORS FOR cIEF

A. Refractive Index Gradient Imaging Detector

A typical instrument setup for refractive index gradient imaging detection can be constructed on the basis of either the Schileren shadowgraph method [18] or a dark field Toepler-Schliren system [19,20]. This refractive index gradient imaging detector is universal, and in real time, a detection limit of 10^{-6} M has been reached for all proteins tested under optimal conditions.

B. Fluorescence Imaging Detector

Fluorescence detection is one of the most sensitive methods for CE detection [26]. In the whole column fluorescence imaging detector for cIEF [22,27], in order to illuminate the whole column homogeneously, three introduction methods of the excitation light have been examined in detail [22,27]. A detection of 10^{-13} M or subattmole proteins was easily accomplished even without optimizing the optical detection system [27]. Because of the extremely high sensitivity, this whole column fluorescence imaging detector is expected to be the most powerful tool for cIEF of trace biomolecules in a single cell.

C. Absorption Imaging Detector

The third mode is the whole column UV-Vis absorption imaging detection [23]. Although the absorption imaging mode is less sensitive than the fluorescence one, it is more sensitive than the refractive index gradient mode and can be used for most proteins. Therefore, the absorption imaged cIEF has become the most practical mode and has been well developed. As a result, a commercial instrument, described later, is also available.

IV. COMMERCIAL IMAGED cIEF INSTRUMENT

The technique of whole column UV absorption imaged cIEF was commercialized in 1998 [3]. In this section, the structure of the first imaged cIEF instrument, *i*CE280, will be introduced and its applications will be discussed.

A. Structure of *i*CE280 Instrument

Figure 2 shows a block diagram of the *i*CE280 instrument. The separation column of the instrument is a 50 mm long, 100 μm ID, 200 μm OD silica capillary. Its outside polyimide coating is removed for whole column detection. Its inner wall is coated with fluorocarbon to substantially reduce EOF. As shown in Fig. 2, the column is packaged into a cartridge by sandwiching it between two glass slides. The two ends of the column are connected to the inlet and outlet capillaries by two sections of porous hollow-fiber membranes. The two sections of the hollow-fiber membranes isolate protein sample and carrier ampholytes within the column from external electrolytes in the two electrolyte tanks as shown in Fig. 2.

The light source of the imaging detector is a xenon lamp. The light beam from the lamp is focused onto the separation column by a bundle of optical fibers and a set of lenses. Monochromatic light is obtained by placing a 280 nm bandpass optical filter before the lamp. The whole column UV absorption image is captured by a camera, which includes an imaging lens and a CCD sensor.

The flow path of the *i*CE280 instrument is also shown in Figs. 2 and 3. The column cartridge's inlet capillary is connected by a finger-tightened nut to a two-position, 8-port PEEK

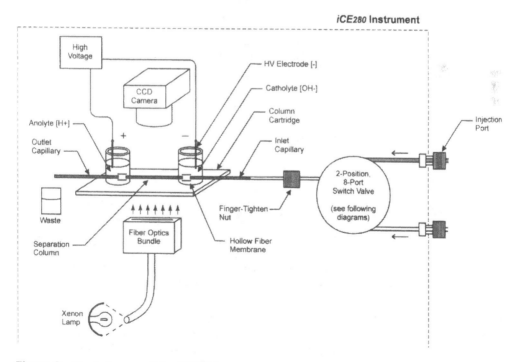

Figure 2   Block diagram of the *i*CE280 instrument.

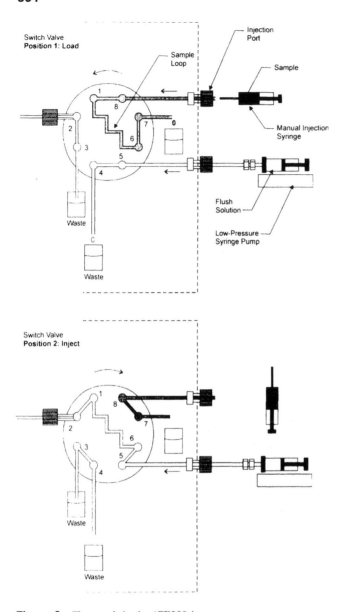

Figure 3 Flow path in the *i*CE280 instrument.

switch valve. The pressure needed for the sample introduction is provided by a low-pressure syringe pump that is connected to the switch valve and operates continuously.

B. Operation of the *i*CE280 Instrument

A column cartridge is first installed into the cartridge holder inside the instrument and connected to the switch valve. The two electrolyte tanks are filled with anolyte (usually 100 mM H_3PO_4) and catholyte (usually 40 mM NaOH). The syringe pump is filled and turned on.

The sample introduction procedure for the manual mode is similar to a conventional liquid chromatography (LC) instrument. As shown in Fig. 3, the sample loop is filled from the injection port when the switch valve is in position 1 (load position). The switch valve then rotates to position 2 (inject position) and the sample stored in the loop is pushed into the column cartridge by the syringe pump. Once the column is filled with the sample, the switch valve returns to the position 1 (load position). A 3 kV DC voltage is applied to the two electrolyte tanks to start isoelectric focusing. The focusing process usually lasts 5–7 minutes. During focusing, the process can be monitored by having the CCD camera take a picture of the whole column and display the UV absorption image of it every 30 seconds. At the end of the focusing process, the voltage is turned off and the sample and separation column are rinsed for a few seconds by the syringe pump. The instrument is then ready for the next sample. The sample introduction procedure can be automated using a LC autosampler. In the automatic mode, the sample throughput is up to 7 samples per hour.

The *i*CE280 instrument includes quantitation software for rapid batch reprocessing of electropherograms.

C. Applications of the *i*CE280 Instrument

1. *pI Determination*

Imaged cIEF provides fast and precise pI determination. Unlike in the conventional cIEF method, in imaged cIEF the pI values of proteins and peptides are determined directly from their peak positions along the separation column because all peaks are detected simultaneously by the CCD camera in the imaging detector. The conventional cIEF must use a difficult method based on retention time in order to establish pI values.

The measurement of the pI value of a protein is performed on imaged cIEF by running it with two pI makers that are mixed in the sample. Ideally, the two pI markers' peaks bracket the sample peaks. Usually, low molecular weight synthetic pI markers (such as BioMarkers, Bio-Rad, Hercules, CA) are used in the measurement. Figure 4 shows the experimental relation-

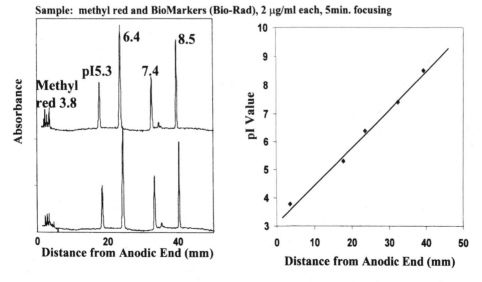

Figure 4 cIEF example of pI markers and experimental relationship between the peak position and pI.

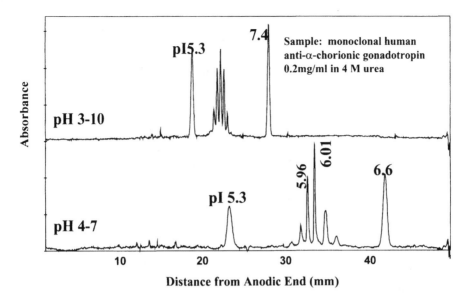

Figure 5   cIEF separation of monoclonal antibody.

ship between pI and the peak positions along the 5 cm long separation column when Pharmalyte pH 3–10 carrier ampholytes are used. The result is linear in the whole pH range.

Figure 5 shows an example of pI determination of monoclonal antibody. The pI values of its six peaks can be automatically calculated by the *i*CE280 software. Baseline resolution is achieved for two peaks having a pI difference of 0.05 pH units. The RSD of pI is less than 0.5%. Analysis times for the pI determination are routinely under 6 minutes.

2.   *Protein Identification*

Most proteins exhibit intrinsic microheterogeneity. IEF is often the separation technique of choice for characterizing the microheterogeneity. The imaged cIEF's high precision in pI determination makes it a valuable tool for quality control identification of protein products. The method for protein peak identification is the same as that for pI measurement.

3.   *Quantitation*

The imaged cIEF is a quantitative analysis method since the imaging system is based on UV absorption of the protein samples. The detector's linear range is up to 160 [3]. The RSD in the quantitation for major peaks (>20% of the total) is 5% for sample concentrations in the range of 0.1–1 mg/mL.

4.   *Analysis of Acidic Proteins and Peptides*

The whole column imaging system in the imaged cIEF detects all focused protein zones within the column simultaneously. It is an ideal tool for IEF analysis for all proteins in a wide range of pI values, especially for proteins with low pI values (acidic proteins).

Imaged cIEF can also analyze peptides as long as they have absorption at 280 nm. Figure 6 shows two runs of an acidic peptide whose pI value is 3.8. Peptides are usually difficult to be analyzed by gel IEF because they are washed away during the staining process.

Sample: Arg-Asp-Tyr[SO3H]-Thr-Gly-Trp-Nle-Asp-Phe-NH2, 5 μg/ml, 5 min.

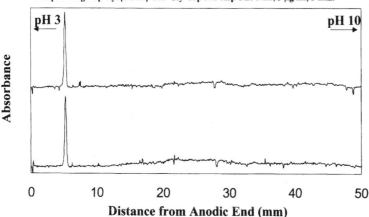

Figure 6 cIEF result of an acidic protein with a pI value of about 3.6.

V. CURRENT STATUS AND FUTURE TRENDS OF THE WHOLE COLUMN IMAGING DETECTION

A. IEF of Proteins in a Microchip

Recently there has been an increasing tendency toward miniaturization or microfabrications of a chemical or analytical system on a glass chip [28,29]. In principle, the real-time whole column imaging detectors described above are also ideal detection tools for the chip format, since the real-time dynamic information of a chemical or analytical process can be obtained. This has been demonstrated by using the whole column imaging absorption detector to monitor an IEF process of proteins performed in a microchannel fabricated in a quartz chip [30]. The microchannel fabricated by photolithography and a chemical etching process was 40 mm long, 100 μm wide, and 10 μm deep. Results of cIEF for protein myoglobin and a pI marker show that the detection limit was about 0.3 mg/mL or 24 pg for pI marker and 30 mg/mL or 2.4 ng for myoglobin in the chip.

B. On-Line Sample Preparation and Two-Dimensional Separation of HPLC-cIEF

Although both HPLC and cIEF are good separation methods with very high resolution power, the separation efficiency is still not good enough for complex biochemical samples. Therefore, it is desirable to combine them, i.e., to develop a two-dimensional (2D) HPLC-cIEF separation. Recently, the fist step toward 2D separation, i.e., on-line coupling of HPLC and cIEF with whole column imaging detection, has been shown [31]. This on-line coupling also allows on-line desalting [32] of proteins.

C. Fast cIEF of Proteins Without Carrier Ampholytes

In protein purification with an IEF process, the purified proteins have to be further separated from the carrier ampholytes. Carrier ampholytes may interact with some protein samples, reducing the sensitivity of UV detection and complicating the matrix or backgrounds when using mass spectroscopy for characterization. Accordingly, it is ideal to carry out IEF or cIEF without

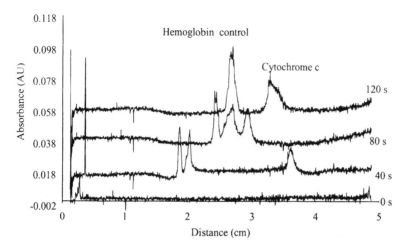

Figure 7 An example of cIEF without carrier ampholyte obtained by the whole column absorption imaging detector. The samples are devolved in water.

carrier ampholytes. Recently, a simple method for cIEF without carrier ampholytes has been demonstrated with the whole column imaging detection [33]. In either a Teflon separation capillary (200 μm ID and 6 cm in length) coated with hydroxypropylmethylcellulose (HPMC) or fused silica capillary (100 μm ID and 5 cm in length) coated with fluorocarbon, either fluorescent proteins R-phycoerythrin and GFP or nonfluorescent proteins human hemoglobin control and cytochrome c are focused and separated without the use of carrier ampholytes. The dynamic process of cIEF without carrier ampholyte is monitored with the axially illuminated fluorescence and absorption imaging detection for the fluorescent and nonfluorescent proteins, respectively. Figure 7 shows one example of the dynamic process of cIEF without carrier ampholytes. Experimental results show that cIEF without carrier ampholyte is very fast. Usually it takes less than 30 s while a cIEF with carrier ampholytes requires at least several minutes.

An alternative method to perform cIEF without carrier ampholytes is to create a temperature gradient along the capillary, since the pH of some buffers depend on temperature [34]. This has been demonstrated in a 3–4 cm long tapered capillary filled with Tris-HCl buffer containing methemoglobin (pI 7.2) and oxyhemoglobin (pI 7.0) [35]. Furthermore, a mathematical model for calculating the temperature gradient in the tapered capillary and cIEF with vertically held tapered capillary are also developed [36].

D. Theoretical Study of the Dynamics of cIEF and Its Verification

An alternative method to study cIEF is the dynamic computer simulation considering the principles of electroneutrality and conservation of mass and charge during the process of cIEF [37–39]. So far, the simulation has been successful in predicting separation dynamics, focusing on behavior of amphoteric sample components, and pH gradient formation and stability in presence of up to 15 amphoteric carrier components and up to three proteins. However, cIEF and most IEF of proteins are carried out with commercial products containing hundreds of carrier ampholyes. In order to understand the practical cIEF process of proteins theoretically, a new dynamic electrophoresis simulator that can be used with up to 150 carrier amphoteric components has been developed [40]. The predicted focusing dynamics for simple amphoteric dyes and proteins are shown to qualitatively agree with data obtained by the whole column optical imaging detector.

E. Whole Column Imaging Detection for CE

In addition to cIEF, it is easily imaginable that whole column imaging detection is also applicable to other CE modes such as capillary zone electrophoresis (CZE) with a short capillary. Whole column imaged CE has been successfully demonstrated with a 4 cm long, 200 μm ID capillary filled with 0.4% agarose gel as anticonvection medium [41]. Separation courses of proteins hemoglobin A_{1c} and myoglobin are followed consecutively by the whole column imaging absorption detector. Furthermore, it has been demonstrated that interaction between biologically active substances such as proteins, detergents, enzymes, and substrate can be studied using this technique.

In addition to whole column detection, part-column real-time fluorescence imaging has also been used for studying the dynamic process of isotachophoresis of rhodamine B [42] and CE separation of DNA fragments [43]. In these experiments, fluorescence from the middle 7 cm of a 18 cm long capillary were imaged by a CCD camera. More recently, the same group has used a 10 cm wide fiber array to couple fluorescence from the middle part of a 28 cm long capillary, where CE enantiomer separations of dansylated amino acids were performed using cyclodextrins [44]. Alternatively, a CCD video camera can also be used to image fluorescence from a capillary [45,46].

VI. CONCLUSION

Capillary isoelectric focusing–whole column imaging detection fully realizes the advantages of cIEF over slab gel IEF. The advantages are automation, high analysis speed, and quantitation. Currently, IEF in many cases is only used as a last resource for protein analysis due to its slow speed, labor intensity, and nonquantitation. By realizing the advantages of cIEF and expanding the application range of IEF, the cIEF–whole column imaging detection technique will make IEF a more routinely used protein-analysis tool.

REFERENCES

1. H Rilbe. Ann NY Acad Sci 209:11, 1973.
2. PG Righetti. Isoelectric Focusing: Theory, Methodology and Applications. Elsevier, Amsterdam, 1983.
3. J Wu, AH Watson, AR Torres. Am Biotechnol Lab 17(6):24, 1999.
4. S Hjerten M Zhu. J Chromatogr 346:265, 1985.
5. R Rodriguez-Diaz, T Wehr, M Zhu. Electrophoresis 18:2134, 1997.
6. JP Thomas. Electrophoresis 17:1195, 1996.
7. X Liu, Z Sosic, IS Krull. J Chromatogr A 735:165, 1996.
8. PG Righetti, C Gelfi, M Conti. J Chromatogr B 699:91, 1997.
9. S Hjerten, J Liao, K Yao. J Chromatogr 387:127, 1987.
10. JR Mazzeo, IS Krull. Anal Chem 63:2852, 1991.
11. JR Mazzeo, IS Krull. BioTechniques 10:638, 1991.
12. JR Mazzeo, IS Krull. J Chromatogr 606:291, 1992.
13. W Thormann, J Caslavska, S Molteni, J Chmelik. J Chromatogr 589:321, 1992.
14. T Wang, RA Hartwick. Anal Chem 64:1745, 1992.
15. CS Beale, SJ Sudmeier. Anal Chem 67:3367, 1995.
16. Q Mao, J Pawliszyn. J Biochem Biohys Methods 39:93, 1999.

17. J Wu, J Pawliszyn. Am Lab 26:48, 1994.
18. J Wu, J Pawliszyn. Anal Chem 64:224, 1992.
19. J Wu, J Pawliszyn. Anal Chem 64:2934, 1992.
20. J Wu, J Pawliszyn. Anal Chem 66:867, 1994.
21. J Wu, J Pawliszyn. Anal Chim Acta 299:337, 1995.
22. X-Z Wu, J Wu, J Pawliszyn. Electrophoresis 16:1474, 1995.
23. J Wu, J Pawliszyn. Analyst 120:1567, 1995.
24. J Wu, C Tragas, A Watson, J Pawliszyn. Anal Chim Acta 383:67, 1998.
25. S Hjerten, J Liao, K Yao. J Chromatogr 387:127, 1987.
26. TT Lee, ES Yeung. Anal Chem 64:3045, 1992.
27. T Huang, J Pawliszyn. Analyst 125:1231, 2000.
28. Micro total Analysis System '98, Proceedings of the TAS '98 Workshop, ed. F Jed Harrison and Albert van den Berg, Kluwer Academic Publishers.
29. X-Z Wu, M Suzuki, T Sawada, T Kitamori. Anal Sci 16:321, 2000.
30. Q Mao, J Pawliszyn. Analyst 124:637, 1999.
31. C Tragas, J Pawliszyn. Electrophoresis 21:227 2000.
32. J Wu, J Pawliszyn. Anal Chem 67:2010, 1995.
33. T Huang, J Pawlizyn. Anal Chem 72:4758, 2000.
34. CH Lochmuller, SJ Breiner. J Chromatogr 480:293, 1989.
35. J Pawliszyn, J Wu. J Microcol Sep 5:397 1993.
36. X-H Fang, M Adams, J Pawliszyn. Analyst 124:335, 1999.
37. M Bier, PA Palusinski, RA Saville. Science 219:1281, 1983.
38. W Thormann, RA Mosher. Adv Electrophoresis 2:45, 1988.
39. RA Mosher, DA Saville, W Thormann. The Dynamics of Electrophoresis, VCH Publishers, Weinheim, 1992.
40. Q Mao, J Pawliszyn, W Thormann. Anal Chem 72:5493, 2000.
41. A Palm, C Lindh, S Hjerten, J Pawliszyn. Electrophoresis 17:766, 1996.
42. J Johansson, DT Witte, M Larsson, S Nilsson. Anal Chem 68:2766, 1996.
43. S Nilsson, J Johansson, M Mecklenbrg, S Birnbaum, S Svanberg, K-G Wahlund, K Mosbach, A Miyabayashi, P-O Larsson. J Capillary Electrophor 2:46, 1995.
44. T Johansson, M Petersson, J Johansson, S Nilsson. Anal Chem 71:4190, 1999.
45. S Razee, A Tamura, M Khademizadeh, T Mashujima. Chem Lett 93, 1996.
46. A Tamura, K Tamura, R Saeid, T Masuhima. Anal Chem 68:4000, 1996.

36

Separation Science in Routine Clinical Analysis

Zak K. Shihabi

Wake Forest University School of Medicine, Winston-Salem, North Carolina

I. INTRODUCTION

Routine clinical laboratories focus their attention on those methods that are used very often in patient care. These methods usually have to be rugged enough to give a high degree of precision in the hands of different technicians with rapid results. On the other hand, the focus in research laboratories is on the development of new knowledge, irrespective of amount of time, difficulty, or cost involved. The role of separation science and how it has been utilized in our routine work over a span of three decades to facilitate patient care, and to enhance the image of our laboratory is discussed.

Currently, routine clinical analysis in the majority of hospitals is dominated by highly automated instruments for testing biological samples of body fluids for different compounds of clinical interest. These instruments mainly rely on different colorimetric reactions—to some extent on specific electrodes and more recently on immuno-reactions. Slowly over the last four decades, these machines have been improved and perfected so as to perform several tests simultaneously with high throughput. These tests aid the physician in patient diagnosis, treatment, and follow-up. Table 1 lists some of the common tests assayed in our hospital laboratory with the annual volume and the main clinical significance of each test. Colorimetric tests utilize specific reactions including enzymatic assays to produce a characteristic color of the analyte.

In the 1970s the introduction of enzymes as reagents for the analysis of several endpoint analytes such as glucose and cholesterol were greatly welcomed in the clinical labs. The use of enzymes as reagents has eliminated many hazardous materials such as strong acids while it provided, at the same time, a better degree of specificity and ease of automation. On the other hand, immunoassays, especially monoclonal antibodies, utilize antibodies or binding ligands to achieve a very high degree of specificity binding specifically to one analyte from a complex mixture of closely related compounds. Both the enzymatic and the immunoassay methods have been adapted successfully in the last decade for automation. Thus, in this respect, these two methods compete well for the analysis of many compounds with different chromatographic

Table 1 30 Common Tests Analyzed by Automated Instruments and Their Clinical Significance

Test	Annual volume	Analysis principle	Clinical significance
Cl	120,000	Selective electrode	Fluid balance
K	120,000	Selective electrode	Fluid balance
Na	120,000	Selective electrode	Fluid balance
Albumin	95,000	Selective electrode	Nutrition/Renal
Alkaline phosphatase	95,000	Selective electrode	Liver/Bone
Calcium	95,000	Selective electrode	Bone metabolism
Creatinine	110,000	Colorimetric	Renal disorders
Total protein	95,000	Colorimetric	Nutrition/Malignancy
Carbon dioxide	120,000	Colorimetric	Fluid balance
Cholesterol	20,000	Enzymatic	Cardiovascular marker
Glucose	110,000	Enzymatic	Diabetes
TG	20,000	Enzymatic	Cardiovascular marker
Urea nitrogen	110,000	Enzymatic	Renal disorders
B_{12}	3,100	Immunoassays	Vitamin
C-reactive protein	10,000	Immunoassays	Injury/Infection/Stress
Folic acid	900	Immunoassays	Vitamin
Troponin	27,800	Immunoassays	Cardiac injury
Carbamazepine	2,000	Immunoassays	Antiepileptic drug
Cyclosporine	3,300	Immunoassays	Immunosuppressive drug
Phenobarbital	3,700	Immunoassays	Antiepileptic drug
Procainamide	1,400	Immunoassays	Antiarrhythmic drug
Theophylline	900	Immunoassays	Antiasthmatic drug

methods, but they can quantify only one while chromatographic methods can quantify several related compounds in the same run. Irrespective of the type and size of the hospital, the tests listed in Table 1 are examples of the common diagnostic tests routinely performed.

However, teaching hospitals are faced continuously with occasional requests for many uncommon tests such as the determination of new drugs, nutrients, and some endogenous metabolites. New drugs are often introduced, and many have to be monitored because the therapeutic level is close to the toxic level. Many endogenous substances such as amino acids and nucleotides are present in the serum with many closely related compounds. In most instances these compounds do not have specific color reactions or are not amenable to specific commercial immunoassay procedures. Furthermore, the need for these tests continues to increase because of the introduction of new drugs and discovering the significance of some uncommon metabolites. In this case, separation science in the form of chromatographic or electrophoretic techniques plays an important role in the analysis of such compounds.

Because of the skills required, separation methods are employed in large teaching hospitals. The majority of the procedures are usually developed in the laboratory based on chromatographic methods and are slowly transferred for routine analysis. However, if a test is in great demand, commercial companies compete to adapt the test to automated instruments, most likely based on immunoassays. In fact, the tests growing most quickly in popularity in clinical laboratories are immunoassays. In this chapter we will discuss some of the common separation techniques used in our teaching hospital, how they were introduced during my three decades of service, and how they have been applied to improve patient care.

II. SEPARATION TECHNIQUES

Thin layer and paper chromatography were at the forefront of modern chromatographic techniques especially in the first part of the twentieth century. These methods predate the routine clinical laboratory itself. However, these techniques have very limited if any application in the modern routine clinical labs because they are qualitative in nature. On the other hand, electrophoresis, Gas chromatography (GC) and high-performance liquid chromatography (HPLC) methods did find their way to modern clinical laboratories.

A. Gas Chromatography

In the 1960s, few drugs were analyzed based on their ultraviolet spectra after solvent extraction from the blood. During this period gas chromatography (GC) using packed columns, emerged as a new method for the detection and quantification of many small molecules, especially drugs. Drug analysis, especially therapeutic drug monitoring and testing for drugs of abuse, was in its infancy. The GC became an ideal method for the analysis of these drugs in serum and urine. The tests were run on packed column 3–6 feet long after sample extraction and solvent evaporation, using the flame ionization detector. However, a great deal of knowledge and skill was necessary to operate the gas chromatograph, prepare the sample, and interpret the results. In addition, the instruments were temperamental due to the unsophisticated electronics, so the analysis of these compounds was limited to university-type institutions. As soon as I started my career at Wake Forest University School of Medicine in 1972, I introduced the GC technique for routine analysis of drugs [1]. Our analysis list included the antiepileptic and the antiasthmatic drugs. The direct result was better care not just for our hospital patients but also at nearby hospitals. At that time, clinical data had shown that the number of patients with seizures decreased if the antiepileptic drug blood level was kept within a narrow therapeutic window. Many hospitals started sending their samples for analysis to our laboratory. In many instances they transferred their patients to be admitted to our hospital.

For obvious reasons, tests in hospitals have to be fast and have to be offered on a 24-hour basis, including weekends. This, of course, placed a great demand on the laboratory to operate the GC service all the time. Unfortunately, at that time GC had several obstacles to being compatible with that type of service. Sample extraction, concentration, and derivatization were not suitable for hospital emergency work. Storing flammable gases for the GC operation was not compatible with safety practices of the hospital. Drug adsorption on the packing of the column gave nonlinear data. The column also had a low plate number. Slowly, during the mid-1970s, we, as well as many other laboratories, replaced GC with the more friendly HPLC, which started to pave a path in clinical laboratories.

GC, as a technique, has improved over the years, especially with the introduction of capillary columns, better electronics, and computerization. The capillary column yields very high resolution due to the great length of the column, with much less drug adsorption. At the present time, GC plays an important role, but only in forensic drug analysis. When GC is combined with mass spectrometry (MS), it can be considered the gold standard for confirmation of drugs of abuse.

Because of the ease and automation of the immunoassays, as mentioned earlier, these methods are now used as an initial screen for drugs of abuse in urine, while GC is used to confirm only positive samples. However, the use of the GC/MS is generally restricted to very few teaching hospitals and some commercial clinical labs.

B. High-Performance Liquid Chromatography

Early in the 1970s HPLC instruments became available. We, as well as others, attempted to explore whether HPLC offered any advantages over GC or if it could replace it for therapeutic drug testing. HPLC offered favorable conditions for the analysis of these compounds. Sample extraction and preparation was much easier and less time-consuming than with GC. Most of our sample preparation was limited only to protein removal by mixing the serum specimen with acetonitrile [2]. To further simplify the analysis and make it more suitable for emergency work, we dedicated one instrument for each test and limited it to an isocratic elution run. This eliminated the need to equilibrate the column or to change the instrument setup and conditions. Thus, all GC instruments were phased out of our lab and replaced by HPLC instruments.

Furthermore, because of the ease and speed of sample preparation, HPLC offered the opportunity to explore the development of new tests including large molecules. For example, catecholamine's metabolites, vanillylmandelic acid (VMA) and homovanillic acid (HVA), were easier to analyze by HPLC than GC, especially if an appropriate detector was used. Our lab was among the first to apply the electrochemical detector using a home-made instrument for the determination of VMA and HVA in urine after separation by HPLC [3,4]. These detectors are now commonly used. Because of their extra sensitivity, both electrochemical and fluorescence detectors greatly decrease the amount of sample injected onto the HPLC column. Based on these detectors, we found that urine or diluted serum can be injected directly without serious adverse effects on the column [2,3,5]. This procedure sped up and simplified HPLC methods greatly. For the analysis of HVA and 5-hydroxyindoleacetic acid (a metabolite of serotonin), we used electrochemical and fluorometric detectors, respectively, with direct urine injection onto the column (Fig. 1). With GC these tests required at least 2 hours for extraction, evaporation, and derivatization for analysis. Thus, after replacing GC with HPLC, our test list expanded to include many additional drugs and some endogenous compounds.

Later, commercial companies introduced specialized kits and HPLC instruments for specific tests such as glycated hemoglobin (Hb), catecholamine analysis, and drug screening. Table 2 lists the tests presently performed in our laboratory employing separation techniques, with their annual volume and clinical significance. These tests represent a very small number relative to those listed in Table 1. However, these are essential tests for the physician who is trying to diagnose specific disorders and follow up on the progress of the treatment. It is interesting to note that the drugs listed in the last part of Table 1 used to be performed by GC and HPLC; however, because their volume increased greatly over the years, the commercial companies developed automated immunoassays for their analysis. Thus, because of speed and convenience, we as well as the majority of the clinical labs replaced these procedures by automated immunoassays for the majority of the drugs.

C. Electrophoresis

Electrophoresis was one of the earliest separation techniques adopted in the clinical laboratory. As soon as paper electrophoresis became commercially available, some laboratories adopted it as a specialty procedure. The main value of this technique is in detecting disorders such as multiple myeloma and Bence Jones proteinuria, which are characterized by the presence of an

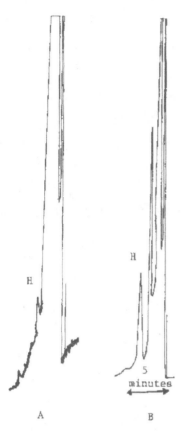

Figure 1 Comparison of the fluorescence detector to electrochemical detection (at 0.95 V glassy carbon electrode) for the analysis of 5-hydroxyindoleacetic acid. Both detectors were connected to the same column and the same pump: (A) flectrochemical detection; (B) fluorescence detection; 5HIAA, H 8.3 µg/L. (From Ref. 5.)

Table 2 Tests Analyzed in Our Institution by HPLC and CE with Current Volume

Test	Annual volume	Analysis principle	Clinical significance
Amiodarone	300	HPLC	Antiarrhythmic
Electrophoresis	1500	Agarose electrophoresis	Malignancy/Renal
Cryoglobulins	200	CE	Immune disorders
Felbamate	400	HPLC	Antiepileptic
Glycated HB	7400	HPLC	Diabetes
Homocysteine	1100	HPLC	Cardiovascular
HVA	40	HPLC	Neuroblastoma
Lamotrigine	600	HPLC	Antiepileptic
Metanephrine	100	Cation chromatography	Pheochromocytoma
VMA	63	HPLC	Pheochromocytoma

abnormal protein synthesized in large amounts (monoclonal gammapathy). These are serious disorders representing a special form of malignancy. In many instances the protein abnormalities are detectable by electropherograms before any clinical symptoms become apparent. Unfortunately, paper electrophoresis methods required a long time, about 16 hours, for completion. Thus, cellulose and agarose electrophoresis, which can accomplish the same separation in about 1–2 hours, slowly replaced paper electrophoresis. Commercially prepared agarose, due to their simplicity and good separation results, became popular, especially for separation of serum, urine, and cerebrospinal fluid (CSF) proteins. Because of its sieving effect, polyacrylamide gel electrophoresis, introduced in the 1960s, gave much better resolution for serum and urine protein analysis compared to agarose electrophoresis. However, because of its difficulty and manual nature, this technique did not gain much popularity in routine laboratories. This illustrates again the differences between research and routine work. In the routine clinical laboratory, ease of use, speed, and automation are key elements for the widespread use of any technique.

D. Capillary Electrophoresis

Capillary electrophoresis (CE) is the newest type of separation technique. It shares several features with both HPLC and gel electrophoresis. It is a more versatile technique than agarose and polyacrylamide gel electrophoresis, since it separates compounds based on several principles such as size, charge, hydrophobicity, and stereospecificity. It can be used to separate not just proteins but the majority of compounds, from small ions to large organic molecules. High resolution, speed, and low operating costs are the main features of this technique. Like HPLC, CE on-column detection allows determination of a majority of compounds without the need for staining.

We were among the first clinical laboratories to realize the potential of this technique in patient diagnosis and to adopt it in our routine work. It offered very rapid and automated separations for serum proteins without the staining step. Also, it offered simplicity of sample preparation for drug analysis [6]. We applied this technique for the analysis of cryoglobulins, a group of proteins characterized by precipitation upon cooling. This class of proteins, which can precipitate in the small blood vessels in some organs, causes cutaneous as well as renal injury. Because CE requires a very small amount of sample and is well suited for quantification, we were able to decrease the sample requirements for this test 10-fold. At the same time, the analysis was done in minutes instead of hours [7]. Also, we applied this technique for the quantification of the minor hemoglobin A_2 [8]. This test is used clinically to diagnose the condition β-thalassemia, which is characterized by mild hypochromic and microcytic anemia. This hemoglobin is usually present in the blood of normal individuals at about 3%. A high level indicates that the patient has a genetic trait where the synthesis of the β chain of the hemoglobin is reduced. Also, CE was shown to be useful for therapeutic drug monitoring [6] and polymerase chain reaction (PCR) detection. The ability to use it for DNA sequencing became very important in speeding up the completion of the Genome Project. For this reason several companies raced to offer specialized instruments with multicapillaries (array capillaries) and with laser-induced fluorescence detection for DNA sequencing.

Because the light path in the capillary is very narrow in CE, we focused our attention on improving detection by concentrating the sample directly in the capillary. The sample is injected as a large plug, about 10–20% of the capillary volume, and the velocity of the ions is altered in order to sharpen the sample zone under the influence of the electric field—"stacking." We were surprised to notice that when the sample contained high amounts of acetonitrile (about 66%) there was an unexpected high degree of sample stacking. Acetonitrile had additional advantages such as removing excess proteins in the case of the biological samples and reversing the deleterious effects of salts in the sample, which tend to deteriorate CE separation due to increased Joule's heating. We studied the mechanism further and found that the stacking is due to the low conductivity of acetonitrile, which results in high field strength. However, we were surprised to find that the presence of salts (at 1%) together with acetonitrile further enhances the degree of stacking. Unlocking the mechanism behind this type of stacking was a challenge. However, after further work it became clear that the mechanism is similar to or is a special form of isotachophoresis (ITP); we termed it "transient ITP-like." Similar to ITP, the salts in the sample act as leading ions in providing the low field strength, while the acetonitrile acts as a terminating ion in providing high field strength [9,10]. This difference in field strength is the basis for altering the ion mobility at different sites in the sample zone. The salts, having fast mobility, migrate rapidly ahead of the analytes, leaving behind an area of higher field strength. The analyte ions in the high field strength accelerate, while those behind in the low field strength slow down, leading to a sweeping action or concentration of the sample ions as the boundary of the two field strengths moves on. This stacking represents a very simple means of increasing the sensitivity 10- to 20-fold, in addition to the removal of protein interference. After the concentration step, the zone enters the separation buffer and separates into several components. Thus, the two steps of concentration and separation are accomplished electrophoretically at the same time and in the same capillary, without extra steps.

The advantages of concentration by acetonitrile in the sample solution, relative to that by isotachophoresis, are the simplicity without the need for matching the leading, terminating electrolytes, pH, and co-ions. Furthermore, the acetonitrile method can easily concentrate both the anionic and cationic compounds simultaneously [11] (Fig. 2). We applied this type of stacking to the analysis of a wide variety of compounds present in serum and other biological fluids, e.g., nitrite, nucleotides, drug analysis, and peptides [12].

Unfortunately, CE is not widely used in clinical applications. However, the technique has good potential in certain areas, such as bacterial, and viral identification based on the polymerase chain reaction, which may impact on patient diagnosis and treatment in the near future. A CE instrument with fast turnaround time and sensitive detection (laser-induced fluorescence) will make these techniques more practical and acceptable in the clinical lab. On the horizon are CE instruments that are based on microchip technology, which can speed up the analysis of DNA about 10-fold.

III. ADVANTAGES OF SEPARATION METHODS IN CLINICAL WORK

The separation techniques require skill and time while the labor cost per test is usually more expensive than that of colorimetric tests. However, these techniques offer several major advantages in clinical labs, which are not obvious or well appreciated:

1. Chromatographic methods can resolve and detect several closely related compounds simultaneously in the same run. For example, about two to three dozen compounds such as amino acids, sugars, or nucleotides can be separated and quantified in the same run. This is not possible with other techniques such as colorimetric or immunoassay methods. Screening for urinary amino acids for the detection of inborn errors of metabolism is a good example.

2. Because separation techniques resolve many compounds that can be closely related, they offer more information than one expects or anticipates. Analysis of secondary peaks might detect new or unexpected information. An example is our finding that

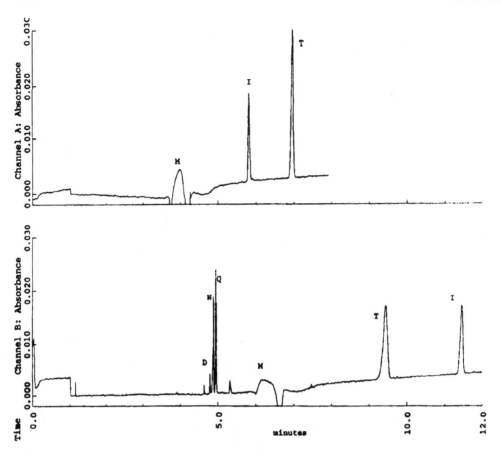

Figure 2   Stacking of cationic and anionic compounds. Effect of the separation buffer type on stacking at sample loading of 12% of the capillary volume: (top) Borate buffer 210 mM, pH 8.6; (bottom) triethanolamine 160 mM, tricine 50 mM, pH 8.6 containing 10% acetonitrile. Separation of a mixture of weakly cationic and anionic compounds in the same run: doxepin (D, 50 mg/L), N-acetylprocainamide (N, 50 mg/L), quinine (Q, 20 mg/L), theophylline (T 50 mg/L), and iothalamic acid (I, 20 mg/L) at 14 kV, 254 nm; (M = electroosmotic flow). (From Ref. 11.)

some patients with neuroblastoma, a tumor that secretes catecholamines in children, can occasionally excrete high amounts of tyramine [13]. We also found that myoglobin in serum and urine is present in two forms.

3.   Separation techniques can be used for screening for unknown compounds related to a disease. Many disorders do not have well-defined diagnostic tests. For example, we screened about 100 samples of CSF to see if any fluorescamine-reacting compounds were related to any neurological disorders. Accidentally, we noticed that two patients, one after the other, had an unknown, fast-eluting peak present in very high concentration. A quick review of the medical charts revealed that both patients had bacterial meningitis, a deadly disease if untreated. After thoughtful deduction based on the capacity factor, reaction with fluorescamine, and its relative concentration, we guessed that the compound should be taurine, an amino acid that does not participate in protein synthesis but has several other functions. Subsequent work showed that the compound

was indeed taurine, the source of which was found to be the white cells that engulf the bacteria while it sheds its taurine content into the CSF [14]. Thus, even in the absence of intact white cells, a high taurine content in the CSF indicates a high probability of meningitis [14]. Interestingly, we found taurine to be high in all the nucleated red cells, such as those of the avian species where the taurine is found to function as an osmoregulator in these cells [15].

4. Separation techniques, especially HPLC and agarose electrophoresis, can easily be used to isolate small amounts of pure compounds for further study. For example, we utilized agarose electrophoresis in isolating small amounts of the fast migrating lactate dehydrogenase isoenzyme. We used this isolated fraction for devising a propriety method to measure this isoenzyme in the sera of patients with myocardial infarction. This method was very easy to perform, so it was produced commercially.

IV. CONCLUDING REMARKS

Separation techniques, mainly GC, HPLC, and electrophoresis, including capillary electrophoresis, are essential tools in research. These methods are slightly slow to perform and require skill and patience in the routine clinical laboratory. Mostly, they are restricted to large medical centers where the expertise to use these techniques is available. In our institution, they have been very useful for patient diagnosis and follow-up for about 30 years. They offered means to measure many compounds of clinical interest which cannot be analyzed easily by other methods. It is expected that methods based on separation science will remain viable and flourish over the next few decades. Such methods enable laboratories in large medical centers to remain competitive at the forefront of technology and to distinguish themselves from laboratories in small community-type hospitals.

REFERENCES

1. ZK Shihabi. Emergency gas-chromatographic assay of phenobarbital and phenytoin and liquid-chromatographic assay of theophylline. Clin Chem 9:1630–1633, 1978.
2. ZK Shihabi. Review drug analysis with direct serum injection on the HPLC Column. J Liq Chromatogr 11:1579–1593, 1988.
3. JL Morrisey, ZK Shihabi. Assay of 4-hydroxy-3-methoxyphenylacetic (homovanillic) acid by liquid chromatography with electrochemical detection. Clin Chem 12:2045–2047, 1979.
4. JL Morrisey, ZK Shihabi. Assay of urinary 4-hdyroxy-3-methoxymandelic (vanillylmandelic) acid by liquid chromatography with electrochemical detection. Clin Chem 12:2043–2045, 1979.
5. ZK Shihabi, ME Hinsdale. Analysis of 5-hydroxyindoleacetic acid in human fluids based on anion exchange HPLC. J Liq Chromatogr 23:1903–1911, 2000.
6. ZK Shihabi. Therapeutic drug monitoring by capillary electrophoresis. In: H Shintani, J Polonsky, eds. Handbook of Capillary Electrophoresis. London: Chapman and Hall, 1997, pp. 386–408.
7. ZK Shihabi. Analysis and general classification of serum cryoglobulins by capillary zone electrophoresis. Electrophoresis 10:1607–1612, 1996.
8. ZK Shihabi, ME Hinsdale, HK Daugherty Jr. Hemoglobin A2 quantification by capillary zone electrophoresis. Electrophoresis 4:749–752, 2000.
9. ZK Shihabi. Sample stacking by acetonitrile-salt mixtures. J Capillary Electrophoresis 6:267–271, 1995.
10. ZK Shihabi. Review, stacking in capillary zone electrophoresis. J Chromatogr A 902:107–117, 2000.
11. ZK Shihabi. Stacking of weakly cationic compounds by acetonitrile for capillary electrophoresis. J Chromatogr A 817:25–30, 1998.

12. ZK Shihabi. In: S Palfrey, ed. Clinical Applications of Capillary Electrophoresis. Totowa, NJ: Humana Press, 1999, pp. 157–163.

13. J Scaro, JL Morrisey, ZK Shihabi. Urinary tyramine assay by high performance liquid chromatography. J Liq Chromatogr 3:537–543, 1980.

14. ZK Shihabi, JP White. Liquid-chromatographic measurement of taurine in cerebrospinal fluid of normal individuals and patients with meningitis. Clin Chem 8:1368–1369, 1979.

15. ZK Shihabi, HO Goodman, RP Holmes. The taurine content of avian erythrocytes and its role in osmoregulation. Comp Biochem Physiol 92:545–549, 1989.

37

Moving Microminiaturized Electrophoresis into the Clinical Laboratory

James P. Landers

University of Virginia, Charlottesville, Virginia

I. THE JOURNEY BEGINS—BIOCHEMIST MEETS CAPILLARY ELECTROPHORESIS

As a graduate student in the Department of Chemistry and Biochemistry at the University of Guelph, a 15,000 student university about an hour's drive outside of Toronto, I was familiarized with electrophoretic separation by the demands of a research project that required hundreds, perhaps thousands, of acrylamide slab gels to be run over the course of 4 years. This, of course, was in pursuit of my doctorate, which relied on the successful search for an illusive photoaffinity-labeled protein that we felt would help unravel the mysteries of gene expression. What became crystal clear from running, staining, and analyzing those innumerable gels was the revelation that electrophoretic separation was, indeed, the workhorse of the modern biochemical sciences. Perhaps more apocalyptic was the realization (brought about by a recurrent experimental schedule that required I go to the lab at 2 a.m. to shut off a gel) that there had to be a better way to do electrophoretic analysis!!! Little did I know that, roughly half a decade earlier, the Jorgenson and Everaerts groups had laid the groundwork for that apocalypse to be fulfilled.

In 1990, my translocation to Rochester, Minnesota, as a Medical Research Council of Canada postdoctoral fellow to study under Dr. Thomas Spelsberg at the Mayo Clinic coincided with the commercial release of capillary electrophoresis (CE) instruments by both Beckman and Applied Biosystems. Thomas Spelsberg, who was chairman of the Department of Biochemistry and Molecular Biology and principal investigator of a basic science research program aimed at understanding the role of steroid hormones and their receptors in controlling biological function and dysfunction, saw the value in the development of CE. What made this technology so attractive to a biochemist or molecular biologist was the ability to have instrumental control of an electrophoretic process with ultra-sensitive on-line detection capabilities. As it turns out, it was my great fortune that Tom decided I should explore the application of this technology for biomedical and clinical purposes. As biochemists, we saw the clear limitations that current technology placed on discovery in cancer biology and other biomedical arenas. It was clear that CE held the potential to begin to change this—to ratchet up technology in a manner that allowed for the science to be advanced faster and more efficiently. After exploring the application of CE to the detection of a number of biologically and clinically relevant molecules, Drs. Thomas Moyer and Henry Homburger spearheaded the establishment of the Clinical Capillary Electrophoresis Facility within the Department of Laboratory Medicine and Pathology (DLMP) at Mayo. They saw the potential value of CE, a technology that allowed for electrophoresis to be carried out in a miniaturized format with on-line detection in a format that had the instrumental control akin to high-performance liquid chromatography (HPLC). There we explored, in a collaborative manner, a number of different assays that had potential to be translated to CE.

II. CROSSING THE "CLINICAL DIAGNOSTICS" BOUNDARY

Our entrees into CE involved capillary zone electrophoresis (CZE) or free solution CE, which brought small molecules into the electrophoresis arena for the first time. This allowed for the detection of products of the reaction catalyzed by chloramphenicol acetyl transferase (CAT), an enzymatic reaction key to studying in vitro gene expression [1], and the separation of neurotransmitters by altering selectivity through borate complexation [2], all using absorbance for detection. The effectiveness of CE for small molecule analysis, not only in terms of separation efficiency but also in terms of speed, led us to develop an assay for a series of hypoglycemic drugs used to control high glucose levels in hyperglycemic patients [3]. Using relatively simple conditions, the resolution of four second- and two third-generation sulfonylurea drugs (along with an internal standard) could be accomplished in <8 minutes for urine analysis. Further exploration in this area led to some understanding of the metabolism of these drugs by exploiting diode array detection for identifying the parent compound and its metabolites [4]. This work has recently been expanded upon by others [5]. It became clear that, for many clinical applications, including this one, absorbance detection would not provide adequate sensitivity for neat sample analysis, hence, the need for a solid phase extraction procedure prior to analysis. Based on the original work of Guzman and colleagues [6], and Schwartz and Merion [7], it became clear that on-line solid phase extraction of select analytes from complex mixtures was possible. Using select silica particles immobilized in a small piece of polyethylene (PE) tubing interfaced with the separation capillary, we were able to show that hypoglycemic drugs could be extracted and concentrated on-line for subsequent electrophoretic analysis [8]. This was not only possible with small molecules but also with peptides [9,10]. This approach differed from the work of others [11], who carried out on-line SPE with solid phase membranes to accomplish "membrane preconcentration" (mPC). Interestingly, we found the mPC approach to have a capacity 40- to 80-fold lower than that obtained with silica particles in PE tubing. This may indicate that, while useful for mass spectrometry detection, mPC is of limited use for many UV detection applications.

As far as bona fide clinical applications with small molecule analysis by CZE is concerned, we developed a method to determine kidney function without the use of radioisotopes [12]. The conventional approach for this is based on subcutaneous injection of a radiolabeled marker (typically with ^{125}I) that does not undergo metabolism and is excreted intact by the kidneys, followed by detection of the marker in urine and plasma at specified times. We demonstrated that CE could mediate the measurement of GFR without the use of ^{125}I-labeled marker by detecting the nonradioactive compound directly in urine and plasma. This approach not only circumvents the obvious problems associated with using isotopically labeled compounds (cost, handling of biohazardous materials, etc.) but also reduces anxiety experienced by the patient as a result of the isotope injection. The collective advantages of this method—lower test costs due to elimination of isotopic compound purchase and disposal, decreased risk to the patient, and without loss in turnaround time—make it an attractive alternative to the current methods. This is currently on-line for routine analysis at the Mayo Clinic.

III. CE PRESENTS A NEW WORLD TO PROTEIN ANALYSIS

Historically, slab gel electrophoresis is best known for size-based analysis of proteins [13]. While it has taken some time for CE to replicate the high-resolution separations that are attainable with denaturing slab gel systems, it has been shown to be very effective for protein separations under nondenaturing conditions. This became apparent when it was demonstrated in numerous publications and by a number of different laboratories that serum protein analysis by CE was comparable and even superior to electrophoresis on acetate or agarose gels [14–18]. While this was not too surprising, the utility of CE for glycoprotein analysis was. While exploring approaches for dynamically coating fused silica capillary surfaces to reduce electroosmotic flow, we discovered the unique effect of 1,4-diaminobutane addition to borate buffer on glycoprotein separation [19]. The addition of this cationic compound to the separation buffer not only led to reduced EOF but provided unprecedented resolution of the glycoforms of ovalbumin. It turns out that this effect was not specific to 1,4-diaminobutane but was characteristic of most alkanes bearing two positive charges and having reasonable solubility [20]. These conditions were found to be exportable to the separation of transferrin glycoforms, which primarily result from varied sialic acid content [21]. Transferrin sialoform determination has clinical significance as a result of the fact that elevated levels of the lower sialoforms is indicative of chronic, excessive alcohol abuse [22]. These studies led to the development of CE conditions that allowed for the separation of transferrin sialoforms in less than 8 minutes [23]. The power of CE for transferrin sialoform analysis was realized when direct injection of serum that had undergone no pretreatment (other than dilution with ferric chloride) yielded resolution of the important sialoforms in a window devoid of other serum proteins [24–26]. While we still have not managed to fully extract the potential of this separation for clinical purposes, it is likely that others will [27].

IV. CE OFFERS SPEED TO DNA ANALYSIS

Nucleic acids are another class of clinical analytes that has been impacted by the emergence of CE. The critical role that slab gel electrophoresis, either with agarose or acrylamide, plays in the analysis of nucleic acids has contributed to it being a valuable clinical tool. Electrophoretic analysis of DNA fragments for DNA sequencing is a key issue with the Human Genome Project (HGP). CE, in the form of multicapillary systems pioneered by the Dovichi [28] and Yeung [29] groups, has provided a whole new standard for sequencing speed, which ultimately will result in expediting completion of the HGP sometime in 2001. The impact on clinical DNA analysis is not insignificant either. Qualitative assays involving PCR product analysis under low-resolution conditions or mutation detection analyses that involve high-resolution separations based on conformational differences or sequencing are key clinical diagnostic tools. After it was shown that fragments of the HIV virus could easily be detected with CE using fluorescence detection [30], the fit of CE technology for qualitative PCR analysis became clear. We exploited this capability for patient screening for herpes simplex and hepatitis C viral infections, showing that CE was faster, less labor-intensive, and less costly overall than the conventional methods

[31]. But more than this, it was clear that CE could do more in this arena than simple detection of PCR product from an exogenous source. Detection of amplified product that signaled the presence of a gene rearrangement correlating with T-cell lymphoma was possible [32], as was a number of other mutation detection schemes including single-strand conformation polymorphism (SSCP) and restriction fragment length polymorphism (RFLP) [33].

V. CHANGING PLATFORMS—MICROCHIP ELECTROPHORESIS

We soon learned what others in this field were learning, that is, that the high throughput necessary to handle the high clinical sample load of some central laboratories may be problematic with single capillary CE. It was at this point that we turned some of our attention to the future role of microfabricated chips in clinical diagnostics. By simple extrapolation of the procedures that had been developed for CE [33] and use of the high-sensitivity laser-induced fluorescence detection systems described by the Manz [34], Harrison [35], Ramsey [36], and Mathies [37] groups, the clinical potential of microchips could be shown. Using simple microchips, both T- and B-cell lymphoma could be detected in an expedient manner [38], as could herpes simplex virus prior to encephalitis complications [39]. These assays were simple, efficient, and, with the microchip, much faster than in the capillary. As with CE, the possibility for detecting mutations using the same combination of PCR and microchip electrophoresis was clear. This led us to explore SSCP [40], heteroduplex analysis (HDA) [41], and a combination of allele-specific PCR with HDA [42]. Interestingly, all of these applications, refined for detection of mutations in breast cancer susceptibility genes, were easily exported to the microchip platform, i.e., conditions for CE were effective on the microchip. As with CE, microchip electrophoresis required that the microchannel surface be passivated to reduce EOF for effective DNA separations. While the acrylamide coating devised by Hjerten [43], utilized almost universally for DNA analysis in capillaries, was adequate, we found that PVP could also function well in this capacity [44]. Not only were run numbers approaching 1000 possible without degradation of separation efficiency, but HDA analyses carried out using commercially coated capillaries were possible [45].

Our forays into microchip electrophoresis also led us to explore aspects of detection. As an alternative to absorbance detection, which is difficult to effect on a microchip, we demonstrated indirect fluorescence for detection of amino acids [46]. While this harbors some potential as an alternative detection mode for microchips, it is not applicable to all analytes, and detection sensitivity is low compared to direct fluorescence detection. Microchip detection also allowed us to dabble with optical engineering in the development of acousto-optic–based scanning for controlling the movement of a laser beam from one point to another, both spatially and temporally, on a microchip [47].

With the power of electrophoresis on a microchip apparent as early as 1990, it has now become clear that integration of functionality into microchips will be important. We have exerted efforts into this area attempting to miniaturize PCR [48–50] and DNA extraction [51] into microchips. Successful integration will not simply involve miniaturization of existing macroscale (microliter) schemes, but will require the development of new approaches for accurately heating small-volume solutions or selectively extracting analytes of interest from complex mixtures.

VI. THE JOURNEY CONTINUES

It will be interesting to see what the next decade brings with microminiaturized electrophoresis. The commercial availability of a microchip-based electrophoresis system (Agilent's Bioanalyzer) has validated that microchip electrophoresis can provide reproducible and quantitative information. However, we must be cognizant that much remains to be done. We still are relatively naive about many of the aspects of microfluidics (valving, non-EOF pumping, nanoliter volume temperature control, etc.), all of which will be critical if integrated analysis is to be successfully achieved. These issues can be solved by brute force but are more likely to be worked out effectively through collaboration with those in the engineering sector, where the foundation for microchip fabrication was laid.

REFERENCES

1. JP Landers, M Schuchard, T Sismelich, TC Spelsberg. High performance capillary electrophoretic analysis of chloramphenicol acetyl transferase (CAT) activity J Chromatogr 603:247–257, 1992.
2. JP Landers, RP Oda, MD Schuchard. Separation of boron-complexed diol compounds using high performance capillary electrophoresis. Anal Chem 64(22):2846–2851, 1992.
3. ME Nunez, JE Ferguson, D Machacek, GM Lawson, TC Spelsberg, JP Landers. Detection of hypoglycemic drugs in human urine using micellar electrokinetic chromatography. Anal Chem 67(20): 3668–3675, 1995.
4. M Roche, RP Oda, JP Landers. Capillary electrophoresis for the detection of hypoglycemic drug metabolites in human urine. Electrophoresis 18:1865–1874, 1997.
5. R Paroni, B Comuzzi, C Arcelloni, et al. Capillary electrophoresis in addition to HPLC for factitious hypoglycemia diagnosis. Clin Chem (in press).
6. (a) NA Guzman, MA Trebilcok, JP Advis. The use of a concentration step to collect urinary components separated by capillary electrophoresis and further characterization of collected analytes by mass spectrometry. J Liq Chromatogr 14:997–1015, 1991; (b) NA Guzman. Automated capillary electrophoresis apparatus. U.S. patent 5,202,010 (1993).
7. (a) ME Schwartz, M Merion. On-line sample preconcentration on a packed-inlet capillary for improving the sensitivity of capillary electrophoretic analysis of pharmaceuticals. J Chromatogr 632:209–213, 1993; (b) M Fuchs, M Merion. Apparatus for effecting capillary electrophoresis. U.S. Patent 5,246,577 (1993).
8. MA Strausbauch, SZ Xu, JE Ferguson, M Nunez, D Machacek, GM Lawson, PJ Wettstein, JP Landers. Concentration and analysis of hypoglycemic drugs using solid phase extraction-capillary electrophoresis. J Chromatogr 717:279–291, 1995.
9. M Strausbauch, JP Landers, PJ Wettstein. Mechanism of peptide separations by solid phase extraction-capillary electrophoresis (SPE-CE) at low pH. Anal Chem 68(2):306–314, 1996.
10. MA Strausbauch, BJ Madden, PJ Wettstein, JP Landers. Sensitivity enhancement and second dimensional information from the SPE-CE analysis of entire HPLC fractions. Electrophoresis 16:541–548, 1995.
11. AJ Tomlinson, NA Guzman, S Naylor. Enhancement of concentration limits of detection in CE and CE-MS: a review of on-line sample extraction, cleanup, analyte preconcentration, and microreactor technology. J Capillary Electrophor, 2(6):247–266, 1995.
12. J Bergert, R Liedtke, RP Oda, JP Landers, D Wilson. Development of a nonisotopic capillary electrophoresis-based method for measuring glomerular filtration rate. Electrophoresis 18:1827–1835, 1997.
13. PH O'Farrell. High resolution two-dimensional electrophoresis of proteins. J Biol Chem 25;250(10): 4007–4021, 1975.
14. R Clark, C Namyst-Goldberg, L Sanders, RP Oda, JA Katzmann, RA Kyle, JP Landers. Qualitative comparison of capillary electrophoresis and agarose gel electrophoresis for the analysis of serum proteins. J Chromatogr 744(1):205–214, 1996.
15. JA Katzmann, R Clark, C Namyst-Goldberg, L Sanders, RA Kyle, JP Landers. Identification of monoclonal proteins by capillary electrophoresis: quantitative comparison with acetate and agarose electrophoresis. Electrophoresis 18:1775–1780, 1997.
16. RP Oda, RJ Clark, JA Katzmann, JP Landers. Capillary electrophoresis as a clinical tool for the analysis of protein in serum and other body fluids. Electrophoresis 18:1715–1723, 1997.

17. R Clark, M Fleisher, JA Katzmann, JP Landers. Differential diagnosis of gammopathies by automated capillary electrophoresis: Analysis of serum samples problematic by agarose gel electrophoresis. Electrophoresis 19(14):2479–2484, 1998.

18. JA Katzmann, R Clark, E Sanders, JP Landers, RA Kyle. Prospective study of serum protein capillary zone electrophoresis and immunotyping of monoclonal proteins by immunosubtraction. Am J Clin Pathol 110:503–509, 1998.

19. JP Landers, RP Oda, B Madden, TC Spelsberg. High performance capillary electrophoresis of glycoproteins:the use of modifiers of endosmotic flow for analysis of microheterogeneity. Anal Biochem 205:115–124, 1992.

20. RP Oda, B Madden, TC Spelsberg, JP Landers. Alkyl quaternary amines as effective buffer additives for enhanced capillary electrophoretic separation of glycoproteins. J Chromatogr 680:85–92, 1994.

21. RP Oda, JP Landers. Effect of buffer additives on capillary electrophoretic separation of serum transferrin from different species. Electrophoresis 17(2):431–437, 1996.

22. H Stibler, O Sydow, S Van Den Berg. Quantitative estimation of abnormal microheterogeneity of serum transferrin in alcoholics. Pharmacol Biochem Behav 13:47–51, 1980.

23. RP Oda, R Prasad, RL Stout, D Coffin, WP Patton, DL Kraft, JF O'Brien, JP Landers. Capillary electrophoresis-based separation of transferrin sialoforms in patients with carbohydrate-deficient glycoprotein syndrome. Electrophoresis 18:1819–1826, 1997.

24. A Trout, MM Muza, JP Landers. Direct CE detection of CDT in serum by capillary electrophoresis using FC-coated capillaries. Electrophoresis.

25. B Giordano, MM Muza, A Trout, JP Landers. Direct CE detection of CDT in serum using dynamically-coated capillaries. J Chromatogr B Biomed Sci Appl 742(1):79–89, 2000.

26. A Trout, J Ferrance, JP Landers. Complement C3 interference in the analysis of carbohydrate-deficient transferrin in fresh serum. Anal Biochem (in press).

27. F Crivellente, G Fracasso, R Valentini, G Manetto, AP Riviera, F Tagliaro. Improved method for carbohydrate-deficient transferrin determination in human serum by capillary zone electrophoresis. J Chromatogr B Biomed Sci Appl 739(1):81–93, 2000.

28. NJ Dovichi. Advances in DNA sequencing technology. Hum Mutat 2(2):82–84, 1993.

29. N Zhang, H Tan, ES Yeung. Automated and integrated system for high-throughput DNA genotyping directly from blood. Anal Chem 15;71(6):1138–1145, 1999.

30. HE Schwartz, K Ulfelder, FJ Sunzeri, MP Busch, RG Brownlee. Analysis of DNA restriction fragments and polymerase chain reaction products towards detection of the AIDS (HIV-1) virus in blood. J Chromatogr 18;559(1-2):267–283, 1991.

31. P Pancholi, RP Oda, PS Mitchell, DA Persing, JP Landers. Clinical diagnostic detection of hepatitis C and herpes simplex viral PCR amplification products by capillary electrophoresis with laser-induced fluorescence. Mol Diagnos 2(1):27–38, 1997.

32. RP Oda, M Wick, LM Rueckert, J Lust, JP Landers. Evaluation of polymer network-capillary electrophoresis for the rapid, automated detection of T-cell gene rearrangements in lymphoproliferative disorders. Electrophoresis 17(9):1491–1498, 1996.

33. TA Felmlee, RP Oda, DA Persing, JP Landers. Capillary electrophoresis of DNA: application to clinical diagnoses. J Chromatogr 717:127–137, 1995.

34. A Manz, N Graber, HM Widmer. Miniaturized total analysis systems: a novel concept for chemical sensors. J Chromatogr 1:244–252, 1990.

35. DJ Harrison, K Fluri, K Seiler, Z Fan, CS Effenhauser, A Manz. Micromachining a minaturized capillary electrophoresis-based chemical analysis system on a chip. Science 261:895–897, 1993.

36. JM Ramsey, SC Jacobson, MR Knapp. Microfabricated chemical measurement systems. Nat Med 1(10):1093–1096, 1995.

37. J Ju, C Ruan, CW Fuller, AN Glazer, RA Mathies. Fluorescence energy transfer dye-labeled primers for DNA sequencing and analysis. Proc Natl Acad Sci USA 9;92(10):4347–4351, 1995.

38. NJ Munro, K Snow, J Kant, JP Landers. Molecular diagnostics on microfabricated electrophoretic devices: translating slab gel-based T- and B-cell lymphomoproliferative disorder assays from the capillary to the microchip. Clin Chem 45:1906–1917, 1999.

39. W Hofgaertner, AFR Hühmer, JP Landers, J Kant. Rapid diagnosis of HSV-induced encephalitis: electrophoresis of PCR products on microchips. Clin Chem 45:2120–2128, 1999.

40. H Tian, A Jaquins-Gerstl, N Munro, M Trucco, LC Brody, JP Landers. Single-strand conformation polymorphism analysis by capillary and microchip electrophoresis: a fast, simple method for detection of common mutations in BCRA1 and BCRA2. Genomics 63(2):31–39, 2000.

41. H Tian, LB Brody, JP Landers. Rapid screening of BCRA1 and BCRA2 gene mutations in using a capillary-based heteroduplex analysis method. Genome Res (in press)

42. H Tian, LB Brody, JP Landers. Rapid screening of BCRA1 and BCRA2 gene mutations using allele-specific PCR analysis method on microchips. Clin Chem

43. S Hjertep. J Chromatog 347:191–198, 1985.

44. N Munro, AFR Hühmer, JP Landers. Optimized covalent coating of glass microchips for robust and reproducible DNA analysis.

45. H Tian, LB Brody, D Mao, JP Landers. Optimizing surface coatings and polymers for capillary electrophoresis-based heteroduplex analysis. Anal Chem (submitted).

46. NJ Munro, DN Finegold, JP Landers. Indirect fluorescence detection of amino acids on electrophoretic microchips. Anal Chem 72(13):2765–2773, 2000.

47. Z Huang, NM Munro, JP Landers. Acousto-optic-based scanning of capillary and chip arrays for multichannel fluorescence detection of DNA. Anal Chem 71(23):5309–5314, 1999.

48. RP Oda, MA Strausbauch, N Borson, AFR Hühmer, S Jurrens, J Craighead, P Wettstein, B Eckloff, B Kline, JP Landers. Infrared-mediated thermocycling for ultrafast polymerase chain reaction (PCR) amplification of DNA. Anal Chem 70:4361–4368, 1998.

49. AFR Hühmer, JP Landers. Noncontact infrared heat-mediated thermocycling in capillaries: effective polymerase chain reaction (PCR) in nanoliter volumes. Anal Chem (submitted).

50. B Giordano, J Ferrance, JP Landers. Microchip-based PCR of DNA in 200 seconds. Anal Chem (submitted).

51. H Tian, AFR Hühmer, JP Landers. Evaluation of silica resins for the direct and efficient extraction of DNA from complex biological matrices in a miniaturized format. Anal Biochem 282(2):175–191, 2000.

38

A Quarter Century of Separation Science, from Paper Chromatography to Capillary Electrophoresis

Haleem J. Issaq

SAIC Frederick, NCI-Frederick Cancer Research and Development Center, Frederick, Maryland

I. INITIATION

My first encounter with separation science took place at Robert College, The American College of Istanbul, Turkey, where I was an undergraduate student. In one of the biology laboratory experiments we were asked to resolve a mixture of pigments by paper chromatography. The experiment worked and I was fascinated by it. My second encounter happened at Georgetown University, where I wrote a proposal, a requirement for the Ph.D. program, for the separation of selenium complexes by thin layer chromatography (TLC).

After graduation from Georgetown University in 1972 with a Ph.D. in analytical chemistry, where my research project dealt with the development of ultrasonic nebulization for atomic absorption [1–3], I accepted a position to head an analytical chemistry group at the newly created Frederick Cancer Research Center, where one of my main functions was to build a state-of-the-art TLC and spectroscopy laboratory. This was an excellent opportunity to expand my knowledge into different areas of analytical chemistry, especially separation science, and to use atomic absorption to study the role of metals in cancer [4]. After completing a few projects, using atomic absorption spectroscopy [5–12] our group focused our efforts on separation science, mainly chromatography. In what follows I will summarize our efforts and contributions to this area.

II. THIN LAYER CHROMATOGRAPHY

TLC was used to check the purity of bioassay compounds and other compounds of interest to the National Cancer Institute. In addition, methods were developed for the separation of aflatoxins [13–15]; separation and detection of alkylated guanines, adenines, uracils, and cytosine

[16,17]; separation, identification, and classification of antibiotics [18,19]; ion exchange TLC for the separation of amino acids in clinical samples [20]; separation of selected polycyclic aromatic hydrocarbons [21]; separation and isolation of stable photodecomposition products of benzo(*a*)pyrene [22]; separation of antitumor agents from fermentation media [23] and marine organisms [24]; and for the separation of testosterone and its metabolites from in vitro incubation mixtures [25]. Cryogenic TLC was carried out at 77°C below zero for the separation of cyclic nitrosamine conformational isomers that were not resolved at ambient temperature [26]. Also, we developed multidimensional and multimodal TLC methods for difficult separations [27,28] and studied parameters that influenced TLC separations [28–30]. For example, we developed the first on-line TLC/MS procedure [31], and off-line TLC/IR [32], TLC/atomic absorption [33]. A two-phase, reversed and normal, side-by-side TLC plate for multimodal separations was created and tested in our laboratory [34], and we then compared conventional TLC plates with Empore sheets [35]. Solvent selection in TLC has been a trial-and-error approach; we developed a statistical approach, overlapping resolution mapping [36–39], and a graphic approach for solvent selection for both TLC [40] and high-performance liquid chromatography (HPLC) [41]. It was fun to experiment with TLC separations using different shaped plates, such as a triangular plate [42]; we also developed a simple and economical antiradial development apparatus [43], for which a U.S. patent was awarded [44]. Correlation of TLC results with HPLC results was also undertaken [29,45]. After publication of a review in analytical chemistry on advances in TLC [46], I was offered a membership on the editorial board of the newly created *Journal of Liquid Chromatography*, and after 2 years I became associate editor. In this capacity, for 10 years I edited each year two issues devoted to TLC. In 1980 and 1981 I was invited to organize and chair for the Eastern Analytical Symposium in New York City a one-day session on TLC advances, after which I became a member of the Governing Board and was program chairman for three different symposia. In 1981 I was invited by Professor Tibor Devenye, of the Hungarian Academy of Sciences, to organize the First Joint American-Hungarian Symposium on Separation Science, mostly TLC, to which Eastern Bloc country scientists were invited. I was involved in organizing the next two symposia of this series, after which it was taken over by Hungarian scientists. TLC was my first serious indulgence in separation science; my next encounter was with, (HPLC).

III. HIGH-PERFORMANCE LIQUID CHROMATOGRAPHY

Using a newly introduced separation technique like HPLC has many advantages. It gives the analyst the opportunity (a) to study parameters that influence resolution, selectivity, and retention, (b) to apply it to the separation of different groups of compounds, and (c) to compare the results to other separation techniques. With the advent of commercial HPLC instrumentation in the early 1970s, it was inevitable that commercial HPLC instrumentation would become an integral part of our work. One of our first objectives was to develop scientifically sound procedures for mobile phase selection for optimum separation. The results of these efforts were two approaches: a graphical one [41] and statistical approaches to isocratic and gradient HPLC mobile phase selection [36–39,47]. We also studied the effect of column length, diameter, and volume [48–50], effect of mobile phase composition [51–57], different column packing materials [30,58–65], influence of alkyl chain length on separation, resolution, and efficiency [57,58], and effect of temperature [66,67] on the separation process. We carried out experiments to compare HPLC with TLC [29,30], with gas chromatography (GS) [68,69], and with capillary electrophoresis (CE) [70–72]. In addition, HPLC was applied to the separation of natural products with anticancer activity [73], caffeine and its metabolites [74], polychlorinated biphenyls

[75], dipeptides [76], cyclic and acyclic nitrosamines [77–80], and other compounds of interest to cancer researchers. Also, a multidimensional HPLC-CE method was developed for the peptide mapping of proteins and for the separation of complex mixtures [81,82]. Our TLC and HPLC contributions to separation science were recognized in 1987, when I was awarded The Eastern Analytical Symposium Award for Outstanding Contributions to the Field of Separation Science.

IV. CAPILLARY ELECTROPHORESIS

CE brought new excitement to separation science and a potential for new avenues for biological, biochemical, and biomedical research. Its simplicity, high efficiency, resolving power, automation, economical use, and applicability to the separation of small ions, neutral molecules, as well as large biomolecules made CE a welcome technique in our laboratory. We built our first instrument in 1988 and bought our first commercial instrument in 1989 and held the first Frederick CE Conference in 1990. Our first objective in using CE, before applying it to our work, was to understand the role of the buffer. The question we asked was how does the anion and the cation affect Joule's heating and, in turn, resolution and efficiency? What role does the organic modifier play, and how does it affect the buffer's viscosity and influence its selectivity?

As mentioned above, our first objective before applying CE to separation problems was to study and understand certain parameters that influence mobility, resolution and selectivity. We studied the role of the buffer's anion [83], cation [84], pH [85], type, and concentration [86]; the combined effect of buffer concentration and applied voltage [87]; and organic modifiers [88] and their influence on heat generation, resolution, and selectivity. The effect of pH, buffer additives, and temperature on the separation of small peptides [89], separations of heterocyclic nitrosoamino acid conformers at subambient temperature [90], and the effect of sample matrix [91] were evaluated. For a summary of these results, the reader should consult Refs. 92, 93.

The next two parallel objectives of our group were to apply micellar electrokinetic chromatography (MECC) to the separation of hydrophobic compounds in reasonable time and to develop an economical laser-induced fluorescence (LIF) system for the detection of native peptides and proteins. We first reviewed the basic considerations and current trends in MECC [94] and set out to find a practical solution to the separation of hydrophobic compounds. We realized that using a neutrally coated capillary with suppressed electroosmotic flow and reversed polarity will solve the problem of analyzing hydrophobic compounds by MECC [95–98]. We called this procedure reversed-flow MECC (RF-MECC). In RF-MECC the most hydrophobic compounds elute first. Also, the separation window of MECC is eliminated.

Detection in CE is not as sensitive as in HPLC, due to the narrow internal diameter of the capillary and on-column detection. The use of LIF is one way to increase the sensitivity of detection. Our interest in the separation and quantification of peptides and proteins in their native state and without derivatization led us to the development of a UV-LIF detection system [99]. A krypton-fluoride (KrF) pulsed UV laser, which lazes at 248 nm, formed the backbone of our detection system for proteins and peptides that contain in their structure any one of the three aromatic amino acids: phenylalanine, tryptophan, and tyrosine [99]. Also, CE with the LIF detection system was used for the separation and detection of amino acids [100–103], tryptophan and related indoles [104], acid neutral impurities in illicit heroin [105], and isomeric truxillines in illicit cocaine [106]. CE with LIF detection, using other lasers, such as He-Cd and argon ion lasers, was used for the separation and detection of DNA fragments, single-stranded DNA, and PCR products [107–109], to study the effect of temperature on the separation of DNA fragments by CE, and to compare the results to those obtained by denaturing HPLC [110].

Although we were studying CE parameters that influence mobility, resolution, efficiency, selectivity, and detection, we did not forget about the application of CE and its resolving power to solve different analytical problems, which are mainly of biological and biomedical interest [111–113]. In our laboratory CE was used, in addition to those mentioned above, for the separation and quantitation of taxol, an anticancer agent, and related taxenes from bark and needle extracts of taxus species [72,114], retinoic acid isomers [115], estrogens [116], peptides [117–119], amino acids [102], homovanillic and vanillylmandelic acids from baby urine [120], nitrate and nitrite in water and urine [121], and natural products [122]. A computer method was developed for predicting the electrophoretic mobility of peptides and for peptide mapping [123,124].

Another aspect of CE, in addition to being a powerful separation technique, is its application to the study of physicochemical properties and thermodynamic parameters. The literature has few of these studies. In 1995 we applied CE to the study of protein-drug interaction and protein-DNA interaction [125], RF-MECC for the determination of the partition coefficients for the distribution of nonpolar and moderately polar solutes between micelle and aqueous phases [95], and CE for the evaluation of monomer-dimer equilibria and stability of histidine-containing peptides [126]. A summary of our CE work for the previous decade was published in *Electrophoresis* [127]. The Frederick CE Conference, which we initiated in 1990, became an international one and attracted scientists from all over the world.

V. GAS CHROMATOGRAPHY

In addition to TLC, HPLC, and CE, GC was used in our laboratory for different routine analyses. One of the projects was the separation of polychlorinated biphenyl isomers, which total 209 isomers. The objective of the study was to find out which congener is involved in cancer as a cancer-causing agent. The study required the development of a multidimensional/multimodal GC approach in which two different selectivity columns are used [128]. The first column separates the solute mixture on the basis of vapor pressure, while the second column separates them based on the shape of the molecules (length-to-breadth ratio). The heart cut from the first capillary column (DB-1) is transferred to the head of a smectic liquid crystalline open-tubular column [128]. The liquid crystalline phases provide enhanced separation of isomeric mixtures with nearly identical vapor pressure on the basis of solute geometry [129]. The method was used for the study of the biochemical and biological effects of polychlorinated biphenyls [130–136].

VI. THE JOURNEY

I feel fortunate that my scientific career encompassed a period of great advancement in separation science. The last 25 years have been extremely rewarding, not only from a scientific point of view but from a social and human aspect. The first American-Hungarian Symposium held in Szeged, Hungary, brought me face-to-face with great scientists from the Eastern Bloc, mostly from the Soviet Union. We had friendly and beneficial scientific discussions devoid of any political talk, for fear for our Russian colleagues. After 2 years Professor Fredrick Regnier of Purdue University and I traveled to the Soviet Union, invited by the Russian Chromatography Society, to take part in a mini-symposium. We met more Soviet Union scientists, visited the Pushina Biological Research Center, saw a Bolshoi Ballet, had a nice Kosack-style barbecue and picnic, visited the Kremlin, Red Square, Moscow University, the museums, the many fantastically beautiful churches, and the Space Memorial. In 1989 I visited China as member of a joint U.S./Chinese cancer research group studying stomach cancer in Shandong Province. After finishing the study, we spent a few days in Beijing, where I visited Beijing University, the Great

Wall of China, the most beautiful Forbidden City, and Tian An Meng Square. In 1998 I was invited to South Africa to give a talk at a separation science conference. This was a very nice trip where, after the meeting had ended, I visited a gold mine and an animal reserve and established good relations with South African colleagues.

Separation science has given me the opportunity to travel to foreign countries, where I have participated in meetings and met and exchanged ideas with many great scientists from all over the world. In addition to becoming acquainted with scientists from Hungary, the Soviet Union, Poland, Romania, Czechoslovakia, China, and South Africa, as mentioned earlier, I have also met scientists from Australia, Austria, Belgium, Canada, Denmark, Egypt, England, France, Finland, Germany, Holland, Israel, Italy, Japan, Lebanon, Morocco, Saudi Arabia, Sweden, and other countries. My involvement as a member of the board and as program chairman for three symposia of The Eastern Analytical Symposium opened the door for me to meet many scientists from the United States and other countries. I truly enjoyed my journey into the halls of separation science; it was and still is a lot of fun.

ACKNOWLEDGMENT

This project has been funded in whole or in part with federal funds from the National Cancer Institute, National Institutes of Health, under Contract No. NO1-CO-56000.

By acceptance of this article, the publisher or recipient acknowledges the right of the U.S. Government to retain a nonexclusive, royalty-free license and to any copyright covering the article. The content of this publication does not necessarily reflect the views or policies of the Department of Health and Human Services, nor does mention of trade names, commercial products, or organizations imply endorsement by the U.S. Government.

REFERENCES

1. HJ Issaq, LP Morgenthaler. Utilization of ultrasonic nebulization in atomic absorption spectrometry. A study of parameters. Anal Chem 47:1661–1667, 1975.
2. HJ Issaq, LP Morgenthaler. Utilization of ultrasonic nebulization in atomic absorption spectrometry. Trace metal analysis in aqueous solutions. Anal Chem 47:1168–1169, 1975.
3. HJ Issaq, LP Morgenthaler. Utilization of ultrasonic nebulization in atomic absorption spectrometry. Trace metal analysis in samples of high salt content. Anal Chem 47:1748–1752, 1975.
4. HJ Issaq. The role of metals in tumor development and inhibition in metal ions in biological systems. In: H Sigel, ed. Carcinogenicity and Metal Ions. New York: Marcel Dekker, Inc., 1980, pp. 55–93.
5. HJ Issaq, WL Zielinski, Jr. Loss of lead from aqueous solutions during storage. Anal Chem 46:1328–1329, 1974.
6. HJ Issaq. Modification of graphite tube atomizer for flameless atomic absorption spectrometry. Anal Chem 47:2281–2283, 1975.
7. LH Chew, WR Carper, HJ Issaq. Effects of metal substitution on pig liver monoamine oxidase. Biochem Biophys Res Comm 66:217–221, 1975.
8. HJ Issaq. The determination of lead in aqueous solutions by the delves cup technique and flameless atomic absorption spectrometry. In: WH Kirchoff, ed. Methods and Standards for Environmental Measurements. Washington, DC: National Bureau of Standards Special Publication 464, 1977, pp. 497–500.
9. HJ Issaq, RM Young. Peak area vs. peak height in flameless atomic absorption measurements. Appl Spectros 21:171–172, 1977.
10. HJ Issaq. Effect of matrix on the determination of volatile metals in biological samples by flameless atomic absorption spectrometry with the graphite tube atomizer. Anal Chem 51:657–659, 1979.

11. MP Waalkes, S Rehm, KS Kasprzak, HJ Issaq. Inflammatory, proliferative and neoplastic lesions at the site of metallic identification ear tags in wistar (Crl:(WI)BR) rats. Cancer Res 47:2445, 1987.

12. ZZ Wahba, L Hernandez, HJ Issaq, MP Waalkes. Involvement of sulfhydryl metabolism in tolerance to cadmium in testicular cells. Toxicol Appl Pharm J 104:157–166, 1990.

13. HJ Issaq, EW Barr, WL Zielinski, Jr. Nondestructive distinction between aflatoxin B and ethoxyquin in thin-layer chromatography. J Chromatogr 132:115–120, 1977.

14. HJ Issaq, EW Barr, WL Zielinski, Jr. Quantitative removal of aflatoxins from thin-layer chromatographic plates. J Chromatogr Sci 13:597–598, 1975.

15. HJ Issaq, W Cutchin. A guide to thin-layer chromatographic system for the separation of aflatoxins B_1, B_2, G_1 and G_2. J Liq Chromatogr 4(6):1087–1096, 1981.

16. HJ Issaq, EW Barr, WL Zielinski, Jr. Separation of alkylated guanines, adenines, uracils and cytosines by TLC. J Chromatogr 131:265–273, 1977.

17. HJ Issaq, EW Barr. A detection reagent for adenine, guanine uracil, cytosine and their alkylated bases, nucleotides and nucleasides on TLC plates. J Chromatogr 132:121–127, 1977.

18. HJ Issaq, EW Barr, T Wei, C Meyers, A Aszalos. A TLC classification of antibiotics exhibiting antitumor properties. J Chromatogr 133:291–301, 1977.

19. HJ Issaq, Modifications of adsorbent, sample and solvent in thin-layer chromatography. J Liq Chromatogr 3:1423–1436, 1980.

20. HJ Issaq, T Devenyi. A simple ion exchange thin layer chromatography method for the separation of amino acids in clinical samples. J Liq Chromatogr 4(12):2233–2241, 1981.

21. HJ Issaq. Quantitative thin layer chromatography of polycyclic aromatic hydrocarbons. Proceeding of 2nd Biennial Symposium, Clinical and Environmental Applications of TLC. New York: J Wiley and Sons, 1982, pp. 457–462.

22. HJ Issaq, AW Andrews, GM Janini, EW Barr. Isolation of stable mutagenic photodecomposition products of Benzo[a]pyrene by thin-layer chromatography. J Liq Chromatogr 2(3):319–325, 1979.

23. HJ Issaq, HH Risser, A Aszalos. Thin-layer chromatographic separation and quantitation of the antitumor agent daunorubicin in fermentation media. J Liq Chromatography 2(4):533–538, 1979.

24. HJ Issaq, KE Seburn, P Andrews, DE Schaufelberger. Multimodal thin layer chromatographic separation of bryostatins 1 and 2 from marine organism extract employing side-by-side silica gel/C_{18} plates. J Liq Chromatogr 12:3129–3134, 1990.

25. B Shaikh, MR Hallmark, HJ Issaq, NH Risser, JC Kawalek. Use of high pressure liquid chromatography and thin-layer chromatography for the separation and detection of testosterone and its metabolites from in vitro incubation mixtures. J Liq Chromatogr 2:943–956, 1979.

26. HJ Issaq, MM Mangino, GM Singer, DJ Wilbur, NH Risser. Effect of temperature on the separation of conformational isomers of cyclic nitrosamines by thin layer chromatography. Anal Chem 51:2157–2159, 1979.

27. HJ Issaq. Multidimensional, multimodal thin-layer and high performance liquid chromatography. Chromatography 2(3):37, 1987.

28. HJ Issaq. Recent advances in multimodal thin layer chromatography. Trends Anal Chem (Trac) 9:36–40, 1990.

29. HJ Issaq, JR Klose, W Cutchin. The effect of binary solvent composition and polarity on separation in reversed phase thin layer and high performance liquid chromatography. J Liq Chromatogr 5(4):625–641, 1982.

30. HJ Issaq, JR Klose, GA Reitz. Separations in thin layer and high performance liquid chromatography using alkyl silica gel bonded phases. J Liq Chromatogr 5(6):1069–1080, 1982.

31. HJ Issaq, JA Schroer, EW Barr. A direct on-line thin-layer chromatography/mass spectrometry coupling system. Chem Inst 8:51–53, 1977.

32. HJ Issaq. A combined thin layer chromatography/microinfrared method. J Liq Chromatogr 6:1213–1220, 1983.

33. HJ Issaq, EW Barr. Combined TLC/flameless atomic absorption method for the identification of inorganic ions and organometallic complexes. Anal Chem 49:189–190, 1977.

34. HJ Issaq. Two-phase thin-layer chromatography. J Liq Chromatogr 3:841–844, 1980.

35. HJ Issaq, KE Seburn, JR Hightower. Thin layer chromatographic separations by conventional glass backed plates and empore sheets: a comparative study. J Liq Chromatogr 14(8):1511–1517, 1991.

36. HJ Issaq, J Klose, K McNitt, JE Haky, GM Muschik. A systematic statistical method of solvent selection for optimal separation in liquid chromatography. J Liq Chromatogr 4(12):2091–2120, 1981.

37. HJ Issaq. Computer assisted high performance liquid chromatography. Am Lab (Feb):41–46, 1983.

38. HJ Issaq. Statistical and graphical methods of isocratic solvent selection for optimal separation in liquid chromatography. Advances in Chromatography, Vol. 24, New York: Marcel Dekker, Inc., 1984.

39. HJ Issaq, KL McKnitt, N Goldgaber. A computer program for the selection of gradient elution in HPLC. J Liq Chromatogr 7:2535, 1984.

40. HJ Issaq, KE Seburn. Graphic presentation of binary mobile phase optimization in thin layer chromatography. J Liq Chromatogr 12:3121–3128, 1989.

41. HJ Issaq, GM Muschik, GM Janini. A graphic representation of binary mobile phase optimization in reversed phase high performance liquid chromatography. J Liq Chromatogr 6:59–269, 1983.

42. HJ Issaq. Triangular thin-layer chromatography. J Liquid Chromatogr 3:789–796, 1980.

43. HJ Issaq. A simple and economical apparatus for developing thin layer chromatography plates in the anticiruclar mode. J Liquid Chromatogr 4(8):1393–1400, 1981.

44. HJ Issaq. Antiradial chromatography device, U.S. patent 264751, Feb 7 (1984).

45. HJ Issaq, B Shaikh, N Pontzer, EW Barr. Correlation of thin-layer chromatography, high performance thin-layer chromatography, programmed multiple development results and their transfer to high performance liquid chromatography systems. J Liq Chromatogr 1:133–149, 1978.

46. HJ Issaq, EW Barr. Recent developments in thin-layer chromatography. Anal Chem 49(1):83A–96A, 1977.

47. HJ Issaq, KL McNitt. A computer program for the identification of the elution order of peaks in high performance liquid chromatography. J Liq Chromatogr 5:1771–1785, 1982.

48. HJ Issaq, RE Gourley. High performance liquid chromatography separations using short columns packed with spherical and irregular shaped ODS particles. J Liq Chromatogr 6:1375–1383, 1983.

49. HJ Issaq. HPLC separations using short columns packed with spherical ODS particles—II. Effect of mobile phase composition on resolution. J Liq Chromatogr 7:475–482, 1984.

50. HJ Issaq. High performance liquid chromatography separations using columns packed with spherical ODS particles—III. Effect of column dimensions on the resolution of a complex mixture. J Liq Chromatogr 7:883, 1984.

51. IS Lurie, AC Allen, HJ Issaq. Reversed-phase high performance liquid chromatographic separation of fentanyl homologues and analogues. I. An optimized isocratic chromatographic system utilizing absorbance ratioing. J Liq Chromatogr 7:463–473, 1984.

52. IZ Atamna, GM Muschik, HJ Issaq. The effect of alcohol chain length, concentration and polarity on separations in high performance liquid chromatography using bonded cyclodextrin columns. J Chromatogr 499:477–488, 1990.

53. HJ Issaq, GM Janini, N Schultz, L Marzo, TE Beesley. Effect of column dimensions on HPLC separations using constant volume columns. J Liq Chromatogr 11:3335–3351, 1988.

54. HJ Issaq. The mobile phase in liquid chromatography. Anal Chem Symp Series, 16 (New Approaches to LC), Budapest, Hungary, 1984, pp. 109–128.

55. HJ Issaq, GM Muschik. Ion exchange HPLC separations using a high content organic modifier mobile phase. J Liq Chromatogr 6:825–831, 1983.

56. IZ Atamna, GM Muschik, HJ Issaq. Effect of alcohol chain length, concentration and polarity on separations in high-performance liquid chromatography using bonded cyclodextrin columns. J Chromatogr 499:477–488, 1990.

57. HJ Issaq. Effect of alkyl chain length of bonded silica phases on separation, resolution and efficiency in high performance liquid chromatography. J Liq Chromatogr 4(11):1917–1931, 1981.

58. HJ Issaq, M Jaroniec. Enthalpy and entropy effects for homologous solutes in HPLC with alkyl chain bonded phases. J Liq Chromatogr 12:2067–2082, 1989.

59. CD Ridlon, HJ Issaq. Effect of column type and experimental parameters on the HPLC separation of dipeptides. J Liq Chromatogr 9:3377, 1986.

60. HJ Issaq, DW Mellini, TE Beesley. Mixed reversed phase/b-cyclodextrin packings in HPLC: single mixed support column versus two columns in series. J Liq Chromatogr 11:333, 1988.

61. HJ Issaq. The multimodal cyclodextrin bonded stationary phases for HPLC. J Liq Chromatogr 11: 2131–2146, 1988.

62. HJ Issaq, J Gutierrez. Mixed packings in liquid chromatography: II. Mixed packings vs. mixed ligands. J Liq Chromatogr 11:2851–2861, 1988.

63. IZ Atamna, GM Muschik, HJ Issaq. The effect of columndiameter on HPLC separation using constant length columns. J Liq Chromatogr 12:295–298, 1989.

64. GM Janini, HJ Issaq, GM Muschik. Electrokinetic chromatography without electroosmotic flow. J Chromatogr A 792:125–141, 1997.

65. IZ Atamna, GM Muschik, HJ Issaq. The effect of alkyl chain length and carbon loading in silica based RP-columns on the separation of basic compounds. J Liq Chromatogr 13(5):863–873, 1990.

66. HJ Issaq, SD Fox, K Lindsey, JH McConnell, DE Weiss. Effect of temperature on HPLC separations using C_1, C_4, C_8 and C_{18} alkyl chain bonded silica columns. J Liq Chromatogr 10:49, 1987.

67. HJ Issaq, ML Glennon, DE Weiss, SD Fox. Effect of temperature on retention in HPLC using a b-cyclodextrin bonded silica column. In: W Hinze, DA Armstrong, eds. Ordered Media in Chemical Separations. Washington, DC: ACS Publications, 1987, pp. 260–271.

68. HJ Issaq, GM Janini, B Poehland, R Shipe, GM Muschik. Chromatographic separation of benzo(a)-pyrene isomers on polymeric reverse phase HPLC and nematic liquid crystal GLC phases. Chromatographia 14:655–660, 1981.

69. SD Fox, GV Shaw, NE Caporaso, SE Welsh, DW Mellini, R Falk, and HJ Issaq. The determination of debrisoquine and its 4-hydroxy metabolite in urine by capillary gas chromatography and high performance liquid chromatography. J Liq Chromatogr 16(6):1315–1327, 1993.

70. HJ Issaq, GM Janini, IZ Atamna, and GM Muschik. Separations by high performance liquid chromatography and capillary zone electrophoresis; a comparative study. J Liq Chromatogr 14:817–845, 1991.

71. HJ Issaq. Comparison of chiral separations in capillary electrophoresis with other methods. Anal Instrument Technol 22(2):119–149, 1994.

72. KC Chan, AB Alvarado, MT McGuire, GM Muschik, HJ Issaq, and KM Snader. High-performance liquid chromatography and micellar electrokinetic chromatography of taxol and related taxanes from bark and needle extracts of taxus species. J Chromatogr B Biomed Appl 657:301–306, 1994.

73. KM Witherup, SA Look, MW Stasko, TG McCloud, HJ Issaq, and GM Muschik. High performance liquid chromatographic separation of taxol and related compounds from Taxus Brevifolia. J Liq Chromatogr 12:2117–2132, 1989.

74. DW Mellini, NE Caporaso, and HJ Issaq. Determination of the caffeine metabolite AFMU in human urine by column switching HPLC. J Liq Chromatogr 16(6):1419–1426, 1993.

75. HJ Issaq, J Klose, and GM Muschik. Separation of polychlorinated biphenyls (Aroclor 1254) by HPLC. J Chromatogr 302:159–166, 1984.

76. HJ Issaq. Separation of selected dipeptides by HPLC. J Liq Chromatogr 9:229–233, 1986.

77. HJ Issaq, JH McConnell, DE Weiss, DG Williams, and JE Saavedra. High performance liquid chromatography of nitrosamines: I. Cyclic nitrosamines. J Liq Chromatogr 9:1783, 1986.

78. HJ Issaq, M Glennon, DE Weiss, GN Chmurny, and JE Saavedra. High performance liquid chromatography separations of nitrosamines: II. Acylic nitrosamines. J Liq Chromatogr 9:2763, 1986.

79. HJ Issaq, D Williams, N Schultz, and JE Saavedra. HPLC separations of nitrosamines. III. Conformers of N-nitrosamino acids. J Chromatogr 452:511–518, 1988.

80. HJ Issaq, IZ Atamna, NM Schultz, GM Muschik, JE Saavedra. High performance liquid chromatography separations of nitrosamines. IV. Effect of temperature on the separation of nitrosamino conformers. J Liq Chromatogr 12:771–784, 1989.

81. HJ Issaq, KC Chan, GM Janini, and GM Muschik. A simple two-dimensional high performance

liquid chromatography/high performance capillary electrophoresis set-up for the separation of complex mixtures. Electrophoresis 20:1533–1537, 1999.

82. HJ Issaq, KC Chan, GM Janini and GM Muschik. Multidimensional multimodal instrumental separation of complex mixtures. J Liq Chrom Rel Technol 23(1):145–154, 2000.

83. IZ Atamna, CJ Metral, GM Muschik, and HJ Issaq. Factors that influence mobility, resolution and selectivity in capillary zone electrophoresis: II. The role of the buffers' cation. J Liq Chromatogr 13(13):2517–2527, 1990.

84. IZ Atamna, CJ Metral, GM Muschik, and HJ Issaq. Factors that influence mobility, resolution and selectivity in capillary zone electrophoresis: III. The role of the buffers' anion. J Liq Chromatogr 13(16):3201–3210, 1990.

85. GM Janini, KC Chan, GM Muschik, and HJ Issaq. Optimization of resolution in capillary zone electrophoresis: Effect of solute mobility and buffer pH. J Liq Chromatogr 16(17):3591–3607, 1993.

86. HJ Issaq, IZ Atamna, GM Muschik and GM Janini. The effect of electric field strength, buffer type, and concentration on separation parameters in capillary zone electrophoresis. Chromatographia 32(3/4):155–161, 1991.

87. IZ Atamna, HJ Issaq, GM Muschik, and GM Janini. Optimization of resolution in capillary zone electrophoresis: combined effect of applied voltage and buffer concentration. J Chromatogr 588: 315–320, 1991.

88. GM Janini, KC Chan, JA Barnes, GM Muschik, and HJ Issaq. Effect of organic solvents on solute migration and separation in capillary zone electrophoresis. Chromatographia 35(9–12):497–502, 1993.

89. HJ Issaq, GM Janini, IZ Atamna, GM Muschik, and J Lukszo. Capillary electrophoresis separation of small peptides: Effect of pH buffer additives and temperature. J Liq Chromatogr 1129:15, 1992.

90. GM Janini, GM Muschik, and HJ Issaq. Capillary electrophoresis separation of heterocyclic nitrosoamino acid conformers at subambient temperature. J High Resolut Chromatogr 17:753–755, 1994.

91. GM Janini, GM Muschik, and HJ Issaq. Sample matrix effects in capillary zone electrophoresis. Effect of chloride ion on nitrate and nitrite. J Cap Electrophoresis 1:116–120, 1994.

92. GM Janini, HJ Issaq. The buffer in capillary zone electrophoresis. In: NA Guzman, ed. Capillary Electrophoresis Technology. New York: Marcel Dekker, Inc., 1993, pp. 119–160.

93. HJ Issaq, GM Janini, KC Chan, and ZE Rassi. Approaches of the optimization of experimental parameters in capillary zone electrophoresis. Advances in Chromatography, Vol. 35. New York: Marcel Dekker, Inc., 1995, pp. 101–170.

94. GM Janini, and HJ Issaq. Micellar electrokinetic capillary chromatography: Basic considerations and current trends. J Liq Chromatogr 15:927–960, 1992.

95. GM Janini, GM Muschik, and HJ Issaq. Micellar electrokinetic chromatography in coated capillaries: determination of micelle/aqueous partition coefficients. J High Resolut Chromatogr 18:171–174, 1995.

96. GM Janini, GM Muschik, and HJ Issaq. Micellar electrokinetic chromatography in zero-electroosmotic flow environment. J Chromatogr B 683:29–35, 1996.

97. GM Janini, GM Muschik, and HJ Issaq. Electrokinetic chromatography in suppressed electroosmotic flow environment: use of a charged cyclodextrin for the separation of enantiomers and geometric isomers. Electrophoresis 17:1575–1583, 1996.

98. IZ Atamna, GM Muschik, and HJ Issaq. HPLC selectivities of alkyl-benzenes using C_{18} and C_{30} bonded silica gel columns under isochronal conditions. J Liq Chromatogr 12:2227–2237, 1989.

99. KC Chan, GM Janini, GM Muschik, and HJ Issaq. Pulsed UV laser-induced fluorescence detection in capillary electrophoresis. Proceedings of the 15th International Symposium on Capillary Chromatography, 16(9&10), 1993, pp. 1877–1890.

100. KC Chan, GM Janini, GM Muschik, and HJ Issaq. Laser-induced fluorescence detection of 9-fluorenylmethyl chloroformate derivatized amino acids in capillary electrophoresis. J Chromatogr 653: 93–97, 1993.

101. KC Chan, GM Janini, GM Muschik, and HJ Issaq. Micellar electrokinetic chromatography of hydroxyproline and other secondary amino acids in biological samples with laser-induced fluorescence detection. J Chromatogr 622:269–273, 1993.

102. KC Chan, GM Muschik, and HJ Issaq. Enantiomeric separation of amino acids using micellar electrokinetic chromatography after pre-column derivatization with the chiral reagent 1-(9-fluorenyl) ethyl chloroformate. Electrophoresis 16:504–509, 1995.

103. HJ Issaq, and KC Chan. Separation and detection of amino acids and their enantiomers by capillary electrophoresis: a review. Electrophoresis 16:467–480, 1995.

104. KC Chan, GM Muschik, and HJ Issaq. Separation of tryptophan and related indoles by micellar electrokinetic chromatography with KrF laser-induced fluorescence detection. J Chromatogr A 718: 203–210, 1995.

105. IS Lurie, KC Chan, TK Spratley, JF Casale, HJ Issaq. Separation and detection of acid/neutral impurities in illicit heroin via capillary electrophoresis. J Chromatogr B 669:3–13, 1995.

106. IS Lurie, PA Hays, JF Casale, JM Moore, DM Castell, KC Chan, HJ Issaq. Capillary electrophoresis analysis of isomeric truxillines and other high molecular weight impurities in illicit cocaine. Electrophoresis 19:51–56, 1998.

107. KC Chan, GM Muschik, HJ Issaq. High-speed electrophoretic separation of DNA fragments using a short capillary. J Chromatogr B 695:113–115, 1997.

108. KC Chan, GM Muschik, HJ Issaq, K Garvey, P Generlette. High-speed screening of polymerase chain reaction products by capillary electrophoresis. Anal Biochem 243:133–139, 1996.

109. HJ Issaq, KC Chan, GM Muschik. The effect of column length, applied voltage, gel type, and concentration on the capillary electrophoresis separation of DNA fragments and polymerase chain reaction products. Electrophoresis 18:1153–1158, 1997.

110. HJ Issaq, H Xu, KC Chan, MC Dean. Effect of temperature on the separation of DNA fragments by high-performance liquid chromatography and capillary electrophoresis: a comparative study. J Chromatogr B 738:243–248, 2000.

111. HJ Issaq, KC Chan, GM Muschik, GM Janini. The role of capillary electrophoresis in cancer research. Proceedings of the 16th International Conference on capillary Chromatography, Riva Del Garda, Italy, September 27–30, 1994, pp. 1835–1839.

112. HJ Issaq, KC Chan, GM Muschik, GM Janini. Applications of capillary zone electrophoresis and micellar electrokinetic chromatography in cancer research. J Liq Chromatogr 18:1273–1288, 1995.

113. HJ Issaq, GM Muschik. The role of the analytical chemist in cancer research. Am Labo (May): 23–31, 1988.

114. KC Chan, GM Muschik, HJ Issaq, KM Snader. Separation of taxol and related compounds by micellar electrokinetic chromatography. J High Resolut Chromatogr 17:51–52, 1994.

115. KC Chan, KC Lewis, JM Phang, HJ Issaq. Separation of retinoic acid isomers using micellar electrokinetic chromatography. J High Resol Chromatogr 16:558–562, 1993.

116. KC Chan, GM Muschik, HJ Issaq, RN Hoover. Separation of estrogens by micellar electrokinetic chromatography. J Chromatogr A 690:149–154, 1995.

117. GM Janini, J Lukszo, HJ Issaq. Determination of the purity of synthetic peptides by capillary electrophoresis, high performance liquid chromatography, and laser desorption mass mass spectrometry. J High Resolut Chromatogr 17:102–103, 1994.

118. H Liu, B-Y Cho, R Strong, IS Krull, S Cohen, KC Chan, HJ Issaq. Derivatization of peptides and small proteins for improved identification and detection in capillary zone electrophoresis (CZE). Anal Chim Acta 400:181–209, 1999.

119. H Yuan, GM Janini, HJ Issaq, RA Thompson, DK Ellison. Separation of closely related heptadeca-peptides by micellar electronkinetic chromatography. J Liq Chrom Rel Technol 23(1):127–143, 2000.

120. HJ Issaq, K Delviks, GM Janini, GM Muschik. Capillary zone electrophoretic separation of homovanillic and vanillylmandelic acids. J Liq Chromatogr 15:3193–3201, 1992.

121. GM Janini, KC Chan, GM Muschik, HJ Issaq. Analysis of nitrate and nitrite in water and urine by capillary zone electrophoresis. J Chromatogr B Biomed Appl 657:419–423, 1994.

122. HJ Issaq. Capillary electrophoresis of natural products: a review. Electrophoresis 18:2438–2452, 1997.

123. CJ Metral, GM Janini, GM Muschik, HJ Issaq. A computer method for predicting the electrophoretic mobility of peptides. J High Resol Chromatogr 22(7):373–378, 1999.

124. GM Janini, CJ Metral, HJ Issaq, GM Muschik. Peptide mobility and peptide mapping in capillary zone electrophoresis: experimental determination and theoretical simulation. J Chromatogr A 848: 417–433, 1999.

125. GM Janini, RJ Fisher, LE Henderson, HJ Issaq. Application of capillary zone electrophoresis for the analysis of proteins, proteins-small molecules and protein-DNA interactions. J Liq Chromatogr 18:3617–3628, 1995.

126. GM Janini, HJ Issaq, unpublished results.

127. HJ Issaq. A decade of capillary electrophoresis. Electrophoresis 21:1921–1939, 2000.

128. HJ Issaq, SD Fox, GM Muschik. The simultaneous use of solute vapor pressure and geometry in multidimensional capillary gas chromatographic separation of polychlorinated biphenyls. J Chromatogr Sc 27:172–175, 1989.

129. GM Janini, GM Muschik, HJ Issaq, RJ Laub. Neat and admixed mesomorphic polysiloxane (MEMPSIL) stationary phases for open-tubular column gas chromatography. Anal Chem 60: 1119–1124, 1988.

130. LM Anderson, LE Beebe, SD Fox, HJ Issaq, RM Kovatch. Promotion of mouse lung tumors by bioaccumulated polychlorinated aromatic hydrocarbons. Exp Lung Res 17:455–471, 1991.

131. LM Anderson, SD Fox, DE Dixon, LE Beebe, HJ Issaq. Long-term persistence of polychlorinated biphenyl congeners in blood and liver and elevation of liver aminopyrine demethylase activity after a single high dose of Aroclor 1254 to mice. Environ Toxicol Chem 10:681–690, 1991.

132. RW Nims, LE Beebe, KH Dragnev, PE Thomas, SD Fox, HJ Issaq, CR Jones, RA Lubet. Induction of hepatic CYP1A in mail F344/NCr rats by dietary exposure to Aroclor 1254: Examination of immunochemical, RNA, catalytic, and pharmacokinetic endpoints. Environ Res 59:447–466, 1992.

133. L Beebe, SD Fox, CW Riggs, SS Park, HV Gelboin, HJ Issaq, LM Anderson. Persistent effects of single dose of Aroclor 1254 on cytochromes P-450 IA1 and IIB1 in mouse lung. Toxicol Appl Pharmacol 114:16–24, 1992.

134. LM Anderson, SD Fox, CW Riggs, HJ Issaq. Selective retention of polychlorinated biphenyl congeners in lung and liver after single-dose exposure of infant mice to Aroclor 1254. J Environ Pathol Toxicol Oncol 12(1):3–16, 1993.

135. KH Dragnev, L Beebe, C Jones, P Thomas, RW Nims, SD Fox, HJ Issaq, RA Lubet. Subchronic dietary exposure to Aroclor 1254 in rats: Accumulation of PCBs in liver, blood, and adipose tissue and its relationship to induction of various hepatic drug-metablizing enzymes. Toxicol Appl Pharmacol 125:111–122, 1994.

136. LM Anderson, D Logsdon, S Ruskie, SD Fox, HJ Issaq, RM Kovatch, CM Riggs. Promotion by polychlorinated biphenyls of lung and liver tumors in mice. Carcinogenesis 15(10):2245–2248, 1994.

39

A Personal Perspective on Separation Science

Edward S. Yeung

Iowa State University, Ames, Iowa

It was an unexpected but very satisfying path for me on the way to separation science.

Ever since my first chemistry kit in eighth grade, I knew I would be doing something related to chemistry. In Hong Kong, everyone interested in science wanted to be a medical doctor. How else would you get wealth and respect at the same time? There were no electives in the high school curriculum, so I did not need to choose—that is, until twelfth grade. Many of my friends and relatives talked about studying in Canada or in the United States. By chance, I picked up a catalog from one of the small liberal arts colleges. The concept was fascinating to me. In Hong Kong I would have entered medical school right after the thirteenth grade. There would not be many more physics and chemistry courses. I wanted to keep my options open and applied to a half dozen schools in the United States. The selections were based on, not surprisingly, universities known also for their medical schools.

As acceptance letters rolled in, the decision became harder. Clearly I was not going to a place just to join their rowing team. You see, Asian students were in high demand at Ivy League schools because their physical size made them ideal as coxswains. Cornell is an interesting mix of a liberal arts college inside a large, multidisciplinary university. It is also an Ivy League school inside a state university. It was a good place to hedge my bet.

There was no official premed major at Cornell. I was told that medical schools liked hard majors such as chemistry. I fit right in. Because of the accelerated pace in my high school, I was able to place out of most freshman level courses. My first academic advisor, a physicist, was surprised that I walked into his office with a detailed plan to complete the A.B. degree in 3 years without going to summer school. There was only one problem. The chairman of the physics department did not believe in the Advanced Placement test. Even though I got a 5 (top score) on that test, he wanted me to start from the beginning. That would have shifted all my advanced courses one year later. I made a deal with him. I would enroll in the sophomore/junior physics course. If I did well in that, he would waive the freshman course retroactively. I kept my end of the bargain and so did he. Mine was the only A in the class of 20.

The chemistry department was much more flexible in terms of advanced placement. There must have been 15 of us freshmen in the second year courses, qualitative analysis and quantitative analysis. That was my introduction to analytical chemistry. It was not a topic I embraced immediately, however. Next came physical chemistry. Yes, it was physical chemistry before organic chemistry for me, despite the intent to go to medical school. It was actually the easier path, when advanced physics and calculus were still fresh in my mind.

The important event was summer undergraduate research between the second and third years. It was a nice summer ''job'' and a special entry in my resumé for medical school admission. I had to remind the chairman of the chemistry department that I had junior status in order to get on the list. Interviewing professors for a mentor was an eye-opening experience. Research is very different from course work. I even learned that Roald Hoffman was premed before he switched to chemistry. He told me it should be my choice too—and that was before he won the Nobel prize. I wanted to work more with my hands, though, so I paired up with my professor in physical chemistry lab, Richard Porter. With a quiet, scholarly manner, he instilled in me the excitement of research. Being the first person ever to observe something or to understand something is an exhilarating experience. I wrote my first paper after the summer for *Inorganic Chemistry*, even though I had no formal course work in the subject at that time.

The summer was more than just research in a laboratory. Every week the group of undergraduates got together for 2 hours of discussion. That was my introduction to group meetings. There was a memorable evening with Peter Debye. To meet someone cited in a textbook was exciting. To see someone with a Nobel prize talk with undergraduates, albeit a select group, without talking down to them or being impatient was amazing.

The right thing to do was to continue in the same research group during the academic year as part of the honors program. The project switched from NMR to the photochemistry of borazine. It was the start of my association with small molecule gas-phase photochemistry. Dick Porter insisted that I apply to graduate school. I took my GRE and applied to two graduate schools while taking more seriously my MCAT and medical school applications. Not willing to give up the excitement of research, I chose medical schools based solely on the availability of a M.D.-Ph.D. program.

Acceptances to medical school and graduate school came in the mail at about the same time. It was a difficult decision. I was about to give up the life that I had planned all along. New York University then found some quirks about the M.D.-Ph.D. fellowship program, so that the recipient must be a U.S. citizen to be eligible. On the other hand, graduate schools actually pay people to attend! All of a sudden I was heading for Berkeley.

The year in Dick Porter's group had an even bigger influence on me. Andrew Kaldor was at Berkeley before going to Cornell. He gave me the inside scoop on professors there, at least in physical chemistry. He noted that Berkeley does not usually hire their own graduates. An exception was this young professor, C. Bradley Moore, who was a student of George Pimentel, who in turn was a student of G. N. Lewis. I was urged to check him out when I got there.

Photochemistry at Berkeley meant lasers in those days. The top dogs were obviously Pimentel and Moore, in that order. When Brad Moore offered me the virgin project of formaldehyde photochemistry, the decision was easy. Actually, there was luck involved. The first student assigned that project got drafted before he was able to do anything in the laboratory. My thesis led to the first separation of isotopes by selective photodecomposition [1]. That is not the standard separation science that we normally think about. I did, however, use a mass spectrometer to do the analysis.

1972 was a tough year for academic positions. I decided to try for one at the same time I applied for postdoctoral positions. One day, Brad Moore gave me a letter announcing an opening in analytical chemistry at Iowa State University and said I might as well apply there too. Those were the days before word processors, so typing up a new application did require thought and commitment. Berkeley was not known for its analytical chemistry. My feeling was that laser spectroscopy would probably fit the description. George Pimentel was telling everyone he was hired there to teach analytical chemistry. Also, all of my teaching experience at Berkeley was in analytical chemistry courses.

The story was that Harvey Diehl did not think laser spectroscopy was analytical chemistry. Dennis Johnson had to give him a special prep talk and arranged my interview when Harvey Diehl was out of town. The physical chemists at Iowa State were obviously delighted they had found someone they could talk to. Velmer Fassel noted he had a physical chemistry degree. James Fritz had an open mind and was properly indoctrinated by his tennis partner, Gerald Small. The promise of Ames Laboratory support was all I needed to hear to turn down a postdoctoral position at Harvard to start at Iowa State. Things might have been different too if I had not bungled an interview at another university because of the flu.

My first experience with chromatography was through my partner in table tennis. Richard Chang was a student of Jim Fritz's at that time and an avid player. Together we won the annual Ames open tournament in doubles. The idea of sensitive detection in chromatography was planted. Reinforcement came when I attended a Midwest University Analytical Chemistry Conference and roomed with Jim Fritz. He thought that my proposal to use lasers for detection in chromatography had promise. The initial attempt to record Raman spectra did not go very far, as many others would confirm in subsequent years. When Michael Sepaniak showed up in my laboratory, he brought the know-how in liquid chromatography (LC) that tipped the balance. We were able to show that two-photon excited fluorescence possessed adequate sensitivity and unique selectivity for LC applications [2]. The same year, Richard Zare published on the sensitive detection of aflatoxins by using laser-induced fluorescence (LIF). The hyphen between lasers and LC had been drawn.

I was at that time fascinated by the phenomenon of polarization. The intuition was that the purity of laser beams should be beneficial to optical rotation measurements. The small size of a focused beam waist also matches nicely with LC detection volumes. Larry Steenhoek was charged with evaluating laser polarimetry. However, despite the initial purity of laser polarization, the weak link turned out to be the polarizing optics themselves. The best commercial Glan-Thompson polarizers have an extinction ratio of 1 ppm. That seemed to be the end of the road. I flipped though Larry's notebook just when we were wrapping up the project. To my surprise, some of the measured intensities appeared to be out of place. A quick check revealed that in fact one could occasionally achieve extinction ratios that were 4 orders of magnitude better than the manufacturer's specifications. The indications were that selected parts of the polarizing optics were significantly better than other parts for rejection of polarized light. The Glan-Thompson design is ideal, but the calcite material that prisms are made out of is imperfect.

LC detection materialized soon after we also learned how to put flexible, thin windows on the flow cell and how to mount each component rigidly but with fine positional adjustments [3]. It was gratifying to see a write-up in *Chemical & Engineering News* followed by a call from William Pirkle, the namesake for a class of chiral LC columns. With the help of several other coworkers, we improved the instrumentation to the point of an alpha prototype and helped several outside laboratories build their own chiral detectors. I was initiated into the world of patents, consulting, and IR-100 awards. In scientific research, there is nothing wrong with inventing something useful.

In 1982, Milos Novotny organized a U.S.-Japan workshop in Honolulu. It did not take too much to accept his invitation based just on the venue. It was a historical event not just for my career but also for separation science as a whole. The list of attendees (Fig. 1) formed a who's who of the giants in chromatography and the giants-to-be. There, the ideas of miniaturization, capillary electrophoresis (CE), lasers, and biochemistry connected with each other. The National Science Foundation and the Japan Chemical Society really got their money's worth from this conference. I did not even mind having to write a chapter in a book in return.

A trip to Phillips Petroleum research laboratories started us in a different direction. The characterization of crude oil was vital to profits in the industry. Unfortunately, the complexity of the mixture made quantitation a real challenge. As I was herded from one laboratory to another, it occurred to me that there were two unknowns in the problem—the response factor of the detector and the measured response of each sample component. That called for doing the experiment twice to set up two independent equations. The refractive index detector was an obvious system to exploit. Then budding graduate student Robert Synovec gave the concept a try on my return. The results went pretty much as expected [4]. Our era of quantitation without standards was born.

I have resisted taking on administrative duties throughout my career. It was push and shove that I was appointed Program Director in the Ames Laboratory to replace Velmer Fassel on his retirement. With that appointment came semiannual trips to Department of Energy laboratories to sort out the red tape. After two or three meetings in a row where the Human Genome Project was proposed by Charles DeLisi, I finally learned enough buzzwords to understand what the excitement was about. I had already heard about CE from James Jorgenson in Honolulu but had not until then related it to slab gel electrophoresis. There is nothing like funding to cause one to seriously focus on a problem. An afternoon interview in DeLisi's office constituted my reverse site visit on the way to research in CE.

Werner Kuhr was not a likely postdoctoral researcher for the Human Genome Project. He was an expert in electrochemistry. He could only spend one year in my laboratory because he already had a faculty position lined up after that. It was fun for the two of us to learn at the same time. Werner could not have given us a better start with two research papers and an A-page article in 12 months. The Human Genome Project was completely open-minded in those days. Even though we did not tackle DNA separations until much later, we were able to sort out many of the basic features of column surfaces and detection modes that would become useful later on.

Funding from the Human Genome Project got us started in CE. My childhood ambition to pursue a career in medicine took a different turn. As smaller and smaller capillaries were used and as more and more sensitive detection became a reality, the study of single biological cells came to mind. I was fortunate enough to convince Barry Hogan to take up the challenge. At the end of 5 years, Barry had laid all the foundation pertinent to single-cell analysis. It was the customary time to graduate. However, Barry decided to delay graduation to complete a bona fide single-cell study [5]. Not many people I know would do that. Call it the Midwest mentality. Call it personal pride. I am sure our success in characterizing a single human red blood cell would have been delayed by at least 2 years had Barry simply quit at the normal time.

The extreme degree of patience and skill required to inject single cells into the capillary was not so difficult to acquire after all. That allowed many coworkers to follow in Barry's footsteps and embark on single-cell research. We started with in vivo derivatization and soon realized that the types of species amenable to detection were limited. Indirect fluorescence is only applicable to the most abundant species. Was spectroscopy not competitive with electro-chemistry? Thomas Lee joined the group, and we started to exploit the native fluorescence of molecules. Protein fluorescence was of course well known. There is a whole lot of haemoglobin in a red blood cell. All one needed was a UV laser. We did have a monster Ar ion system that

(a)

(b)

Figure 1 (a) Photograph of participants at the U.S.-Japan Seminar on Microcolumn Separation Methods and Their Ancillary Techniques, Honolulu, Hawaii, 1982. (b) Identification of participants in photograph. 1–Nobuhiko Ishibashi; 2–Totaro Imasaka; 3–Masashi Goto; 4–Yukio Hirata; 5–James Jorgenson; 6–Frank Yang; 7–Kiyokatsu Jinno; 8–Eli Grushka; 9–Uwe Neue; 10–Bob Brownlee; 11–Wilhelm Simon; 12–Mitsuru Taguchi; 13–Akira Nakamoto; 14–Toyohide Takeuchi; 15–Edward Yeung; 16–Kiyokatsu Hibi; 17–Nobuo Suzuki; 18–Alfred Yergey; 19–Hiroo Wada; 20–Stuart Cram; 21–Jim Monthony; 22–Jack Henion; 23–Steve Bakalyar; 24–Jiroyuki Hatano; 25–Joel Harris; 26–Richard Hartwick; 27–Milos Novotny; 28–Mike McConnell; 29–Daido Ishii; 30–Shin Tsuge; 31–Ernest Daws; 32–Bjorn Josefsson; 33–Souji Rokushika; 34–Sadao Mori; 35–Hirro Sasaki; 36–Nelson Cooke.

puts out 0.5 W of deep UV light including 275 nm. Tom Lee studied a whole series of biotech products donated by various pharmaceutical companies and had great success in demonstrating the principles of native fluorescence detection in CE. Haemoglobin in single red blood cells was easily quantified [6]. The observed variability raised interesting questions such as old vs. young blood cells and the statistical implications for clinical diagnosis.

I did not pay too much attention to the catecholamines that the neuroscientists and electrochemists frequently mentioned until I saw the name 5-hydroxy-tryptamine being equated with serotonin. Surely we can make tryptamine fluoresce. H. T. Chang (very appropriate for tackling 5-HT) surveyed the fluorescence properties of the catecholamines and related compounds. While nonbiological pH conditions are needed to optimize the fluorescence efficiencies of many compounds, serotonin seems to be an ideal fluorophor. Very quickly we took advantage of our big UV laser and showed that native fluorescence can be used to follow serotonin, insulin, and adrenaline in individual cells.

Measuring the contents of individual cells reveals the heterogeneity among them. The next challenge is to follow some biological functions in individual cells. We soon realized that immobilized cells at the entrance of the capillary could be stimulated with an appropriately designed electrophoresis buffer. This is equivalent to microdialysis sampling, except that the system (cell) is smaller than the capillary diameter. The issue of calibration became a concern. The use of internal standards or external standards is not straightforward when analyzing these small samples. We decided that a two-step ratio determination would offer the most useful information. A single cell was injected, trapped, and stimulated to provide the "release" response. Afterwards, the cell was lysed to assess the residual amount of material. Therefore, the fraction released could be calculated. For the case of pancreatic β-cells, one can then distinguish those cells that possess inadequate amounts of insulin from those that fail to release insulin when stimulated [7]. The whole-cell measurement complements studies of individual exocytotic events that are based mainly on electrochemical measurements. In later experiments, we were able to simultaneously image single exocytotic events in mast cells and correlate those with pulses of serotonin electrophoretically transported through the capillary and detected by native fluorescence [8]. These represent 0.25 fl entities and contain only 2–3 amol of serotonin.

With the steady improvements in detection power of LIF in the late 1980s, many groups focused on the detection of single molecules. A large, multilabeled DNA fragment can be seen even with a standard fluorescence microscope. With LIF, the favorable absorption/fluorescence properties of the protein phycoerythrin became the key to its detection as individual molecules. At the time, Qifeng Xue had been developing the fluorescence version of enzyme-mediated microassay that originated in Fred Regnier's group. Efficient amplification via the catalytic reaction had afforded sensitive detection of biomolecules even by an absorption detector. Qifeng was able to measure lactate dehydrogenase isoforms in single white blood cells. It did not take much extrapolation to conclude that single enzyme molecules could be detected in this scheme.

While getting a large enough signal is relatively easy if the incubation time is unlimited, being sure that the signal is generated by a single molecule is much more difficult. One cannot independently obtain a single enzyme molecule and inject that into a capillary to perform an enzyme assay. We took advantage of the fact that axial diffusion inside small capillary tubes is slow enough to confine the individual molecules and their product zones so that tens of molecules can be placed in the capillary without serious overlap. A quick calculation showed that a solution of 10^{-16} M would provide such a scenario. Making up such a solution was strictly an act of faith. We neglected common sense that loss during handling or storage might not allow solutions of such low concentrations to be prepared. We drew on previous experience with lactate dehydrogenase (LDH), which has a high pI, to avoid interactions with glass or fused-silica surfaces. Contamination was less of a concern since the enzyme reaction is highly specific. The accepted method of enzyme-linked immunosorbent assay provided confidence in making up solutions at the 10^{-12} M level. Serial dilutions from that concentration did not seem so prohibitive.

We knew that LDH had a high turnover rate for generating sufficient product molecules for detection in less than an hour. Published literature pointed to dissociation of the LDH tetrameric form to the inactive monomeric form at low concentrations. We neglected that too based on the assumption that kinetics rather than equilibrium would control single-molecule reactions. To favor detection, we chose the conversion from lactate to pyruvate, which was the reverse reaction for nature's utilization of LDH. The pH was also increased to 8.5 to accelerate the reaction rate and to further prevent adsorption. The rest of the experiment was just a matter of statistics to show that on average the predicted number of molecules in the capillary was observed in the form of isolated product zones [9].

Statistics was one thing. Confirmation came from control experiments without the enzyme. However, we needed to be even more convinced that there were no artifacts. CE became the key. Since LDH and its product NADH have different electrophoretic mobilities, we rationalized that it should be possible to separate the enzyme molecule from its product zone and reincubate to produce a second product zone. That was exactly how the experiment turned out. Pairs of product zones were found that could be attributed to individual enzyme molecules. The peak areas became the most interesting results of the study. Intermolecular enzymatic activities showed six times the variance compared to intramolecular activities for the molecules. This meant that molecules had distinct catalytic characteristics even though their primary chemical structures were identical, as confirmed by bulk CE experiments. The observation led to the conclusion that enzyme molecules can exist in several persistent conformations that last for hours in solution. The experiments also constituted the first demonstration of the CE separation and manipulation of a single molecule.

Advancing a new hypothesis in science brought with it publicity as well as skepticism. We were sure enough to make the topic a central part of subsequent lectures given at various institutions and conferences. It was a freezing day in late October when I did the same at the University of Alberta. Before my seminar, Norman Dovichi showed me around his lab. To the surprise of both of us, we found that we were doing essentially the same experiments on single enzyme molecules. At the time, our work was already accepted by *Nature* pending final revisions. Dovichi's work did not appear until 18 months later. Later on, Weihong Tan in our group performed an independent experiment based on microvials rather than capillary tubes and found nearly identical results [10]. The hypothesis still has its critics, but to this date no contradictory experiments have appeared in the literature.

Receiving funding from the Human Genome Project came with it regularly scheduled workshops for all DOE grantees. The excitement level picked up substantially in the early 1990s that indeed the whole genome could be sequenced with just more speed and throughput in the traditional process. John Taylor had been developing a scheme for axial excitation for LIF detection in CE. Having seen quite a few light fixtures in shopping malls, I started to associate optical fibers as a way to distribute excitation light into multiple capillaries. It took only about a month to complete the 10-capillary experiment [11]. Around that time, Richard Mathies' group was successful in showing a 24-capillary confocal scanning arrangement. Kambara's group also published on a multicapillary sheath-flow detector. Having seen how much skill it took to insert optical fibers into capillary tubes, we modified the geometry to line excitation and CCD detection. Kyoji Ueno was the very patient postdoctoral researcher who saw the project to completion [12]. The three multiple capillary systems are now manufactured by different companies. It is fair to say that multiplexed CE was the single technological development that made the most impact on the success of the Human Genome Project as announced at an international press conference on June 26, 2000.

Multiplexed CE for DNA sequencing is not just about detection. The capillary tubes must be reusable in order to maintain a high throughput at a reasonable cost. Fairly early on, Barry Karger's group had demonstrated good separation performance in both crosslinked and non-crosslinked polyacrylamide media. Coated capillaries were employed to prevent electroosmotic flow, which goes in the opposite direction as the DNA fragments and, to a lesser extent, to prevent hydrophobic interactions at the capillary walls. These solutions and gels were generally fairly viscous to handle, especially for multiple capillaries. Longevity was also an issue. H. T. Chang began to look at alternative sieving matrices for DNA separation. He surveyed a large number of water-soluble polymers for CE sizing of DNA. It was not long before he came across the unique properties of poly(ethylene oxide) (PEO) [13], which is a variant of poly(ethylene glycol) (PEG). In a coated capillary, H. T. showed that even fragments with identical numbers of base pairs but with different base compositions could be separated from each other. He also demonstrated the concept of using a mixture of polymers with different molecular weights to optimize the separation performance over a large range of fragment sizes. The resulting matrix turned out to be much less viscous than if a single polymer solution was used.

When H. T. graduated, the project was passed on to Eliza Fung. The change in personnel often comes with a change in perspective. Eliza did not have any preconceptions about the properties of PEO. We read more about PEG in the literature. Such polymers were often utilized as a dynamic coating for capillary tubes in zone electrophoresis to reduce electroosmotic flow. Since the process of preparing coated columns was tedious, we decided to try uncoated columns for DNA sequencing. As anticipated, the separation was equally good in coated versus uncoated columns [14]. However, the performance deteriorated quickly after two to three runs, and we were very disappointed. A close examination of the electropherograms revealed that there was a gradual increase in electroosmotic flow as the walls came in contact with PEO at pH 8.3. Rinsing in between runs did not seem to provide relief. We took some time to think about why a new surface is different from a used surface. It finally became evident that the new surface was dry and only a limited fraction of the silanol groups were dissociated. On use, the high pH buffer gradually titrated these to increase the degree of ionization. Hysteresis in such ionization phenomena had been documented for silica surfaces. Eliza then used 0.1 N HCl to wash the columns in between runs. The results were better than expected. Not only could the columns be used over tens of runs, but the separation also became faster. When a new column surface was pretreated with HCl, ionization was suppressed even more effectively compared to its "dry" state. This is because moisture in the air invariably causes some ionization. The reduction in electroosmotic flow in the extreme case provided separation from 0 to 400 bp in a span of only 10 minutes. I recall receiving a fax from Eliza on a Monday night in Wurzburg (HPCE 1995) and showing the results to Norm Dovichi before my talk on Thursday afternoon. We were totally amazed.

Subsequently, we were able to perform sequencing separations at elevated temperatures, used PEO without urea, made even less viscous polymer solutions, eliminated gas bubbles that blocked capillaries, implemented pulsed-field separations, and achieved read lengths to 1000 bp. The technology was transferred to a company and reliability was enhanced with the help of automation. PEO was also successfully applied to protein separations as a dynamic coating material. At concentrations well below the entanglement limit, we found that size discrimination of kb and larger fragments was possible in CE. PEO even impacted our single-molecule studies by providing a friendly surface for experiments. It has been a long road since H. T. put some of that white powder in buffer and watched it form a gel-like solution.

As we started to think about the column surface for DNA separations, we began to ask mechanistic questions about adsorption. I was intrigued by the fact that dilute polymer solutions can separate DNA fragments. The two models that were suggested were collisions with the polymer molecules and association with one or more polymer molecules. We began to test the capillary surface as a contributing factor. Nobuo Iki did an experiment that textbooks had advised us against. He tried CE of DNA restriction digest in polymer-free buffer in a bare fused-silica capillary. Based on the slow diffusion of large DNA fragments, it was clear that the broad peak Iki recorded was more than an experimental artifact. There was some separation in such a system. The surface effect was confirmed by using smaller I.D. capillaries, and electropherograms with reasonably sharp, resolved peaks for each restriction fragment could be obtained [15]. The final piece of evidence came with parallel experiments performed on DB-wax–coated columns. There, the restriction digest sample eluted in a single sharp peak as prescribed by textbooks.

We were thinking of hydrophobic interactions all along, since that was the most common effect for protein separations. The negatively charged DNA fragments were very different from proteins and should be repelled by the surface. The DB-wax column was also more hydrophobic compared to a bare fused-silica column but showed no separation ability. Ion exchange with the surface silanol groups did not make sense either since both DNA and silanols were negatively charged. Our observations pointed to a size-to-charge selectivity. We therefore advanced the model of charge shielding at the surface to explain the results. The fused-silica surface is actually rich in positive ions compared to the bulk solution. As DNA fragments approach the surface, this layer of positive ions partially neutralizes the fragment ions to lower their electrophoretic mobilities. The smaller fragments are affected more because they can approach the surface with little steric hindrance. They are therefore swept along faster by electroosmotic flow and elute earlier. Just to be sure, we tested the effect of pressure-driven separation in the same bare fused-silica capillary. There was also size separation, but a reversed elution order was observed, in accordance with hydrodynamic chromatography.

Our interest in electrophoretic (chromatographic) surfaces extended to our single-molecule project. Nancy Xu developed a method to follow the motions of single molecules in free solution. By manipulating the excitation geometry, either the liquid layer immediately adjacent to the surface (180 nm) or a thin layer (μm) can be monitored [16]. We measured a much smaller diffusion coefficient for molecules near the surface than for molecules away from the surface. This implied that fused-silica surfaces possess electrostatic properties that can lead to chromatographic retention. More sophisticated experiments were done on proteins to compare with simple dye molecules. The average behavior of single molecules were consistent with the known pK of fused silica and the pI of the protein. What was unexpected was that immobilization on the surface for extended periods of time was not found. The proteins simply spent more time near the surface than would be predicted by simple diffusion [17]. This is contrary to the intuitive picture for chromatographic adsorption, where on-off events describe partition. Furthermore, the trapping distance for the protein molecules was in the order of 180 nm, which was much larger than predicted by double-layer theory. In fact, these trapping distances quantitatively explain the size selectivity of bare fused-silica columns in the CE of DNA fragments in our experiments described above. Seemingly unrelated findings by John Taylor on the migration behavior of carboxylated latex microspheres and measurements of reduced electroosmotic flow in electrochromatography by several other research groups also became understandable when one allowed for a large trapping distance at the surface. More quantitative determinations of these trapping distances will be needed in future experiments to allow reconciliation with electrostatic theories based on electrochemical measurements.

With the commercialization of multiplexed CE instrumentation for DNA sequencing, we turned our attention to other modes of CE. While there are many commercial CE systems available, the acceptance of CE as a standard tool like liquid chromatography has been slow. One major reason is that without a constant, externally driven flow like LC, variations in the migration times in CE are often not acceptable for validation purposes. Migration times also affect peak areas and the injected amounts. Such parameters are well known to experienced researchers in CE, but mysterious to the novice. Tom Lee was determined to seek out every single detail in CE migration and had the theoretical mindset for it. He went back to physical chemical concepts that preceded CE to realign the critical parameters. In the end, he derived a "migration index" and an "adjusted migration index" [18,19] that encompassed almost all situations in zone electrophoresis. Only the electrical current profile throughout electrokinetic injection and electrophoresis and the migration time of a neutral marker were required in the normalization procedure. Tom Lee's schemes so far have not gathered a large following, primarily because they involve some subtle ideas. The anticipated incorporation of such an algorithm into one commercial CE system should eventually bring due credit to this fine idea.

The recent buzzwords in the chemical and pharmaceutical industries are "combinatorial" and "high-throughput screening." There are even entire journals devoted to these topics. The need for chemical analysis has accelerated to the point where it has become the rate-limiting step. When absorption or fluorescence indicators can be utilized, microtiter plates in 96-well or denser formats have provided excellent support. We recognized that high-throughput separation schemes must be developed to cover the majority of analysis where the reactants, products, and contaminants are too similar to be discriminated by microtiter plate readers. The thought of having hundreds of LCs in a room is not satisfying, even if one disregards the cost of the instrumentation and the amount of solvents that have to be disposed of afterwards. Multiplexed CE seems like a natural solution. With developments in nonaqueous CE and special hydrophobic pairing reagents, almost every type of LC separation can be retooled for CE operation.

In a capillary array, reproducibility of CE runs is an even more serious problem. Each capillary has its own surface property and operational history. Temperature is also highly variable in between runs and among capillaries because much more heat needs to be dissipated. Xue Gang soon showed that renormalization based simply on two internal standards reduced the relative standard deviations of both the migration times and the peak areas by a factor of six or more [20]. This put CE on par with LC in terms of reproducibility. The use of two internal standards was not because more was better, but because two independent parameters were involved in electromigration. This scheme provided a simple solution without invoking the comprehensive corrections afforded by Tom Lee's idea.

The next hurdle in capillary array instruments that required attention was that only fluorescent species were amenable to detection. This limitation eliminates 90% of the useful applications. What was needed was a multiplexed absorption detector for CE. Actually, the diode array detector that is built into many commercial systems is already a multiplexed detector, albeit in the wavelength domain for a single capillary. Jim Jorgenson had implemented on-column multipoint detection with a diode array detector. The idea is then to turn the detector 90° relative to the capillary axis and to use different diodes for different capillaries. The initial picture seemed daunting with regard to light delivery and stray-light rejection. Having multiple ball lenses and multiple slits would create serious alignment problems. Xiaoyi Gong decided to image the capillary array directly onto the diode array without special optical elements. To our surprise, the images were clean and stray light was low. The multiplexed absorption detector for CE was much simpler and more rugged than the corresponding fluorescence detector [21].

The simple optical arrangement has a simple explanation. The capillaries are essentially cylindrical lenses. When parallel light illuminates the capillary, most of the light is focused to a point very close to the capillary and then diverges quickly. Since the detector element is at a distance many times the capillary diameter, the amount of refracted light falling on the detector becomes negligible. Only those rays of light traveling along the exact center (diameter) of the capillary can pass without being deflected and can hit the detector. These form the absorption path for each capillary tube. Stray light and crosstalk are thus minimized and absorption pathlength is maximized. The performance of the system is comparable to the best single-capillary absorption detectors.

An absorption detector will never provide the sensitivity of a fluorescence detector. However, the former is more broadly applicable. Indeed, at wavelengths below 220 nm, almost all functional groups exhibit useful absorption properties. The short absorption pathlength also means that detection of analytes is possible even in moderately opaque solvents. It is not unreasonable to consider CE as an alternative to LC in most applications. In rapid succession, we worked out protocols for PCR analysis, enzyme assay, peptide mapping of proteins, and combinatorial organic synthesis. In PCR analysis, each DNA fragment contains a large multiple of absorbing units to afford sensitive detection. In enzyme assays, one can increase the incubation time to increase the signal. For peptide mapping, the small volumes of CE injection translate to small amounts, not necessarily low concentrations, of protein being analyzed. For combinatorial synthesis, sensitivity is not an issue. In fact, one should choose less favorable absorption wavelengths to ensure a large dynamic range. With 96 capillaries, one can even optimize CE separations by using a combinatorial array of buffer systems.

The multiplexed absorption detector for CE had a fast track in terms of technology transfer. Within a year of the first experiments, the technology was licensed to a commercial manufacturer. After another 3 months, the first unit was sold. Time will tell whether it will be a commercial success, but without commercialization many fine ideas tend to become forgotten in the journals and never make an impact. It is the responsibility of scientists to reduce concepts to practice so that others can benefit from the effort.

My encounter with separation science has been truly rewarding. From small molecules to DNA, from detection to separation, from lasers back to conventional light sources, from single molecules to multiple capillaries, there was always some new challenge beyond the horizon. It all started with a chemistry kit in eighth grade.

ACKNOWLEDGMENTS

The Ames Laboratory is operated for the U.S. Department of Energy by Iowa State University under Contract No. W-7405-Eng-82. This work was supported by the Director of Science, Office of Basic Energy Sciences, Division of Chemical Sciences, and the Office of Biological and Environmental Research, and by the National Institutes of Health.

REFERENCES

1. ES Yeung, CB Moore. Isotopic separation by photopredissociation. Appl Phys Lett 21:109, 1972.
2. MJ Sepaniak, ES Yeung. Laser two-photon excited fluorescence detection for high pressure liquid chromatography. Anal Chem 49:1554, 1977.

3. ES Yeung, LE Steenhoek, SD Woodruff, JC Kuo. Detector based on optical activity for high performance liquid chromatographic detection of trace organics. Anal Chem 52:1399, 1980.
4. RE Synovec, ES Yeung. Quantitative analysis without analyte identification by refractive index detection. Anal Chem 55:1599, 1983.
5. BL Hogan, ES Yeung. Separation, detection and modulation of intracellular species at the level of a single human erythrocyte. Anal Chem 64:2841, 1992.
6. TT Lee, ES Yeung. Quantitative determination of native proteins in individual human erythrocytes by capillary zone electrophoresis with laser-induced fluorescence detection. Anal Chem 64:3045, 1992.
7. W Tong, ES Yeung. Determination of insulin in single pancreatic cells by capillary electrophoresis and laser-induced native fluorescence. J Chromatogr B 685:35, 1996.
8. H Su, ES Yeung. Study of cell degranulation with simultaneous microscope imaging and capillary electrophoresis. Appl Spectrosc 53:760, 1999.
9. Q Xue, ES Yeung. Differences in the chemical reactivity of individual molecules of an enzyme. Nature 373:681, 1995.
10. W Tan, ES Yeung. Monitoring the reactions of single enzyme molecules and single metal ions. Anal Chem 69:4242, 1997.
11. JA Taylor, ES Yeung. Multiplexed fluorescence detector for capillary electrophoresis using axial optical fiber illumination. Anal Chem 65:956, 1993.
12. K Ueno, ES Yeung. Simultaneous monitoring of DNA fragments separated by capillary electrophoresis in a multiplexed array of 100 channels. Anal Chem 66:1424, 1994.
13. H-T Chang, ES Yeung. Poly(ethylene oxide) for high resolution and high speed separation of DNA by capillary electrophoresis. J Chromatogr 669:113, 1995.
14. EN Fung, ES Yeung. High-speed DNA sequencing by using mixed poly(ethylene oxide) solutions in uncoated capillary columns. Anal Chem 67:1913, 1995.
15. N Iki, Y Kim, ES Yeung. Electrostatic and hydrodynamic separation of DNA fragments in capillary tubes. Anal Chem 68:4321, 1996.
16. X-H Xu, ES Yeung. Direct measurement of single-molecule diffusion and photodecomposition in free solution. Science 275:1106, 1997.
17. X-H Xu, ES Yeung. Long-range electrostatic trapping of single protein molecules at a liquid/solid interface. Science 281:1650, 1998.
18. TT Lee, ES Yeung. Facilitating data transfer and improving precision in capillary zone electrophoresis with migration indices. Anal Chem 63:2842, 1991.
19. TT Lee, ES Yeung. Compensating for instrumental and sampling biases accompanying electrokinetic injection in capillary zone electrophoresis. Anal Chem 64:1226, 1992.
20. G Xue, H-M Pang, ES Yeung. Multiplexed capillary zone electrophoresis and micellar electrokinetic chromatography with internal standardization. Anal Chem 71:2642, 1999.
21. X Gong, ES Yeung. Novel absorption detection approach for multiplexed capillary electrophoresis using a linear photodiode array. Anal Chem 71:4989, 1999.

40

The 1999 Belgian Dioxin Crisis: The Need to Apply State-of-the-Art Analytical Methods

Pat Sandra

Ghent University, Ghent, Belgium

Frank David

Research Institute for Chromatography, Kortrijk, Belgium

I. INTRODUCTION

Not only in Belgium but in all countries of the European Community, 1999 will be remembered as the year of the dioxin crisis. At the beginning of the crisis nobody could imagine the important role separation sciences and especially capillary gas chromatography (CGC) would play to get out of this impasse. Thousands and thousands of food samples had to be analyzed before they could be released for consumption, and this had a tremendous impact on the Belgian economy. Moreover, the toxicity of the pollutants caused the welfare of the population to be questioned, resulting in a panic situation. In Belgium, all scientists and laboratories with the knowledge to analyze these pollutants had to accept responsibility. We analyzed more than 5000 samples during a 6-month period with a minimum of instrumentation and manpower. The methodology was updated continuously, taking advantage of recent developments in sample preparation and CGC. Their implementation, however, had to be accompanied with validation studies and the analysis of certified samples. This chapter describes the different steps in the optimization procedure that resulted in a very high throughput and thus high productivity.

II. THE FACTS

In January 1999, chicken farmers observed premature death (up to 25%) and nervous disorders among chicks, combined with a high ratio of eggs failing to hatch. Initially, different hypotheses were proposed, including the possibility of too high (mortal) doses of the antibiotic salinomycine used as a growth promoter in animal feed. Two months after the start of the problem, a producer of animal feed took the initiative to send a sample of suspected animal feed and of chicken fat to a laboratory specializing in the analysis of polychlorinated dibenzodioxins (PCDDs) and

643

polychlorinated dibenzofurans (PCDFs). Both classes are referred to as "dioxins" but in fact consist of 75 dioxins and 135 furans. The results of the dioxin analyses were long in coming, most likely because nobody considered it a priority. Only at the end of April were the Belgian authorities informed that high concentrations of dioxins had been detected in the animal feed and chicken fat. Another set of samples was sent to the same laboratory for confirmation. At the end of May, the public was informed of the food contamination and measurements were taken, including the destruction of several lots of eggs, chicken meat, and related food products. At that stage, the source of the contamination was still unknown. The analyses, however, showed a 1000-fold higher concentration of PCDDs and PCDFs in the animal feed fat and in the chicken fat (1–2 ng/g fat) versus normal background values (1–5 pg/g fat) [1].

At the Research Institute for Chromatography, we could not believe that dioxins as such could contaminate food products at low ppb levels without the presence of other chlorine-containing contaminants at much higher levels. PCDDs and PCDFs are not technically chemicals, but are by-products of the synthesis of several chlorinated compounds like pentachlorophenol (PCP; an insecticide, mainly used in wood preservation), polychlorinated biphenyls (PCBs), or the chlorophenoxy acid herbicides 2,4-D and 2,4,5-T. Dioxins are also formed during badly performed combustion of chlorinated material and can therefore be present in emissions of waste incinerators. When we received the exact data of the "dioxin" analysis from a journalist, we immediately noted that the PCDFs were present in much higher concentrations than the PCDDs. Based on this, we advanced the hypothesis of a "PCB" contamination. Polychlorinated biphenyls are, in contrast to PCDDs and PCDFs, technical products that have been used intensively in electric capacitors, transformers, vacuum pumps, and even as adhesives, fire retardant, in inks, etc. Although their production has been banned for many years, they are still widely present in old electric installations. Since the disposal costs of PCB-containing materials is very high, mixing PCB oils with mineral or edible oils becomes an attractive and very profitable alternative. For many years we noticed that fats used in the production of animal feed were of low quality, containing not only polymerized lipids but also contaminants such as mineral oil. The link between dioxins, PCBs, contaminated oils, fat, animal feed, and finally chickens and eggs was very logical to us. Moreover, the hypothesis of a PCB contamination was supported by different facts. PCBs and dioxins were detected in rice in Japan in 1968 (Yusho disease) and in food in Taiwan in 1979 (Yu-Cheng disease), and, moreover, a 1972 U.S. Food and Drug Administration (FDA) survey stated that several accidents involving PCBs had contaminated animal feed and subsequently poultry and eggs intended for human consumption [2]. The relative abundance of the most toxic PCDFs (10 congeners) and PCDDs (7 congeners) in the Yusho incident (1968) and the Belgian food crisis (1999) are compared in Fig. 1, where it is obvious that the oxidation profiles of the PCBs are the same.

When one of us advanced this opinion on national radio and television, within one day we received relevant samples and could prove that ppm levels of PCBs were present in animal feed, chicken fat, and eggs. The link was made, and it was obvious that used PCB oil was the malefactor. Polychlorinated biphenyls consist of a group of 209 possible congeners ranging from mono- up to decachlorobiphenyls. Although their acute toxicity is lower than that of dioxins, several studies have shown that PCBs are linked to several negative health effects including endocrine disrupting activity, reproductive function disruption, developmental deficits in newborns and decreased intelligence in school-aged children who had in uteri exposure. Moreover, the toxicity of dioxins and organochloro pesticides increases in the presence of PCBs. PCBs accumulate in fat and build up in the food chain. Levels from 50 to 500 ppb (ng/g fat) have also been detected in human fat samples [3] and in mother's milk. It is also known that birds, fowls, and poultry are very sensitive to PCB poisoning.

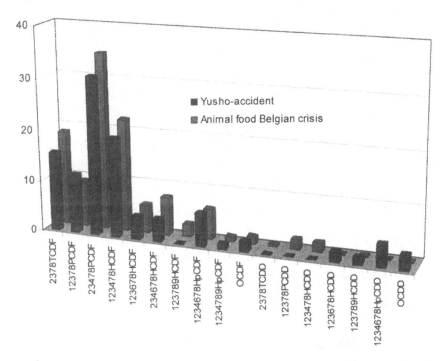

Figure 1 Relative abundance of the most toxic PCDFs and PCDDs detected in food samples during the Yusho incident (1968) and the Belgian dioxin crisis (1999).

Because PCBs are present at much higher levels than dioxins, PCB analysis is much faster and cheaper (by a factor of 10) than dioxin analysis. The analysis of PCBs was finally accepted by Belgian and later by European authorities. The European Community required for export from Belgium certificates for food products with more than 2% fat content! The norm for PCBs in fatty food (more than 2% fat content) was set at 200 ppb (200 ng/g fat) for the sum of seven PCB congeners named, according to Ballschmiter [4], PCBs 28, 52, 101, 118, 138, 153, and 180. Using the much faster PCB monitoring analysis, some 50,000 analyses were performed in 1999 in various laboratories. This made it possible not only to deliver certificates for noncontaminated food but also to localize the contaminated farms and storerooms containing contaminated food. Analysis of dioxins, on the other hand, is so time-consuming and costly that it would haven taken several years if the PCB-dioxin link had not been made. All products were released by December 1999. It is remarkable that notwithstanding good manufacturing practice (GMP), good laboratory practice (GLP), and ISO and EN norms this could happen at the end of the twentieth century!

III. METHODOLOGIES FOR THE ANALYSIS OF PCBs IN FATTY MATRICES

The analytical scheme of the official method Beltest I 014 for the analysis of PCBs in food consists of different steps: sample drying, extraction, clean-up, and CGC analysis [5]. Sample drying can be performed by freeze-drying or chemically by sodium sulfate addition. In the second step, the lipophilic contaminants are extracted from the matrix using an apolar solvent.

The extract contains the lipids, PCBs, PCDDs, PCDFs, and other apolar solutes like organo-chloro-pesticides (OCPs), polycyclic aromatic hydrocarbons (PAHs), and mineral oil. Different extraction techniques may be applied, namely, Soxhlet extraction (or the automated versions Soxtec or Soxtherm), solvent extraction using ultrasonic agitation, accelerated solvent extraction (ASE), microwave-assisted solvent extraction (MASE), and supercritical fluid extraction (SFE). All these techniques perform equally well for the extraction of fat and PCBs, as will be illustrated further. It is obvious that, especially in a crisis situation, selection should be based on sample throughput and cost. Next the PCBs are fractionated from the (co-extracted) fat matrix. For this fractionation, column chromatography on acidic silica gel and aluminium oxide is advised, although other techniques like gel permeation chromatography (GPC) and solid phase extraction (SPE) may be applied if validated. Both sample extraction and clean-up require a concentration step. Finally, the cleaned extract can be analyzed using capillary gas chromatography with elec-tron capture detection (CGC-ECD) or capillary gas chromatography with mass selective detec-tion (CGC-MS). CGC-ECD is extremely sensitive and in most cases sufficiently selective for the detection of PCBs extracted from fat. CGC-MS in the selected ion monitoring mode is somewhat less sensitive but more specific since the presence of PCBs is confirmed by the detec-tion of several ions per congener in a well-defined relative ratio. For samples positive by CGC-ECD, MS confirmation is mandatory. The limit of detection for the seven congeners is 5 ppb (5 ng/g fat per congener).

IV. SAMPLE PREPARATION METHOD FOR PCB ANALYSIS DEVELOPED AT RIC

A new and fast sample preparation technique was developed based on ultrasonic extraction followed by clean-up using matrix solid phase dispersion (MSPD) [6]. The extract is analyzed by CGC–micro ECD. Positive samples are confirmed by CGC-MS. During the crisis, high-speed CGC and the concept of retention time locking (RTL) were implemented and validated. Samples are homogenized using a blender. From fat samples (chicken or pork), a 1 g sample is weighed in a 20 mL headspace vial. From eggs, 3 g of egg yolk is taken. For animal feed samples or other meat products, a sample size corresponding to 200–500 mg of fat is taken. To the sample 2 g anhydrous sodium sulfate and 10 mL petroleum ether are added. Tetrachlo-ronaphthalene, octachloronaphthalene, or Mirex may be added as an internal standard to the sample at this stage, although external standardization can also be applied. The headspace vial is closed and placed in an ultrasonic bath at 30°C for 30 minutes. In this step, the fat and PCBs are transferred from the matrix in the petroleum ether phase. The sodium sulfate adsorbs the water present in the sample. After extraction, the sample is allowed to settle. An aliquot (typi-cally 5 mL) is transferred to a test tube, and another aliquot (2 mL) is used to determine gravimet-rically the fat content of the extract. To the test tube, 2 g of acidic silica gel (44% sulfuric acid) is added and the tube is vortexed for 10 seconds. This technique is called matrix solid phase dispersion (MSPD). In column chromatography and in SPE, the analytes are eluted through a bed and the fat is retained. In MSPD, the fat matrix is allowed to bind on the adsorbent that is mixed with the sample, while the solutes of interest stay in solution. For the fractionation of fat from PCBs, this method works very efficiently. After settlement of the adsorbent, which takes approximately 20 minutes, an aliquot of the clear solution is transferred to an autosampler vial. The final volume is not important, since the concentrations are calculated to the initial 10 mL solvent and the sample or fat weight. The whole sample preparation takes approximately one hour, and several samples can be prepared at once. One technician can handle more than 50 samples per day.

The extracts are analyzed by conventional CGC-ECD and CGC-MS. For CGC-ECD, an HP 6890 GC (Agilent Technologies, Wilmington, DE) equipped with split/splitless inlet and micro-ECD detection was used. Separations were performed on a 30 m × 0.25 mm i.d. × 0.25 μm HP-5MS column. Injection of 1 μL was done in the splitless mode at 250°C. The carrier gas was hydrogen at a constant pressure of 58 kPa. The oven was programmed from 50°C (1 min) to 150°C at 25°C/min and to 300°C at 10°C/min and hold for 2 minutes. Nitrogen at 40 mL/min was used as detector make-up gas. The detector was set at 320°C. For CGC-MS confirmation, the same conditions apply, except that helium was used as carrier gas at 1 mL/min constant flow. Detection is done in selected ion monitoring mode using two ions per congener group. The limits of detection (S/N > 3) of the micro-ECD and the MS were 0.2 pg and 0.5 pg, respectively.

Figure 2 shows the CGC-ECD chromatograms of a contaminated animal feed and egg sample we received at the beginning of June 1999. The chromatograms are very similar. The PCB profiles mainly consist of pentachloro-, hexachloro-, and heptachloro-biphenyls and correspond to the PCB profile of the Aroclor 1260 technical PCB mixture. PCB congeners 153, 138, and 180 are the predominant congeners.

Some 4000 samples, including animal feed, eggs, chicken fat, pork fat, pork meat, meat products (ham, sausages), etc. were analyzed using this methodology. The splitless liner was replaced after 100 analyses and the column after 1000 injections. On the column selected and the chromatographic conditions applied, the PCB congeners 28 and 31 are not separated, but this was not critical as both congeners were not relevant for the Aroclor 1260 pollution.

Critical in the analytical scheme is the new sample preparation method. The performance of ultrasonic extraction was in first instance compared with two other recently introduced sample

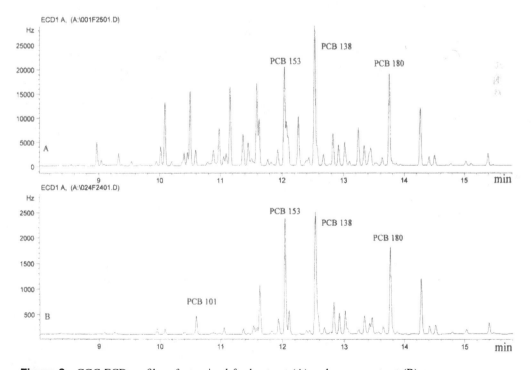

Figure 2 CGC-ECD profiles of an animal feed extract (A) and an egg extract (B).

Table 1 Comparison of PCB Concentrations in Three Contaminated Egg Samples Obtained by Ultrasonic Extraction (U), Microwave-Assisted Solvent Extraction (MASE), and Accelerated Solvent Extraction (ASE)

PCB	Egg 1 (ppb)			Egg 2 (ppb)			Egg 3 (ppb)		
	U	MASE	ASE	U	MASE	ASE	U	MASE	ASE
118	170	68	147	556	410	550	945	1023	922
153	263	309	210	1142	1051	1184	1811	2032	2042
138	240	320	234	1019	1120	1105	2031	2323	2128
180	111	166	93	696	552	686	1015	1166	1259
Sum	783	863	682	3412	3133	3525	5803	6544	6349

preparation techniques, namely, accelerated solvent extraction and microwave-assisted solvent extraction. The extraction efficiency of the three methods was evaluated with three egg samples contaminated at different levels (low, medium, and high). For ASE, a Dionex ASE 200 system (Dionex Corp., Sunnyvale, CA) was used. A 1 g sample was extracted at 100°C and 1500 psi using petroleum ether as solvent. The extraction time was 5 minutes oven heat-up time, 5 minutes static extraction, and three cycles with 60% of the extraction cell volume (22 mL). The extract was then concentrated to 10 mL. For MASE, an ETHOS SEL system (Milestone, Analis, Gent, Belgium) was applied. A 1 g sample was extracted during 20 minutes at 95°C. The extraction solvent was n-hexane (10 mL in extraction thimble, 10 mL outside thimble) using a Weflon stir bar to absorb the microwave energy in combination with the nonmicrowave absorbing solvent. After extraction, the extract was filtered and concentrated to 10 mL to obtain the same final concentration factor as the ultrasonic extraction and ASE. Clean-up and analysis was done in the same way for the three extracts of the three samples. The results based on duplicate analysis are summarized in Table 1.

For the sample with the lowest concentration (egg 1), the RSD on the PCB sums obtained by the three techniques is 12%; for the two other samples the RSDs are less than 6%. For the individual values, some small differences are noted, but in general these differences are within 10% of the average values. These results clearly demonstrate that there is no statistically significant difference between the three techniques and that equally good results are obtained. The ultrasonic method exhibit by far the highest throughput and is extremely cheap compared to ASE and MASE.

Table 2 Repeatability Tested on a Contaminated Egg and Pork Fat Sample

PCB	Egg sample		Fat sample	
	Conc (ppb)	RSD (%)	Conc (ppb)	RSD (%)
118	141	7	20	4
153	333	2	274	6
138	342	4	318	3
180	165	5	165	4
Sum	982	2	776	4

Table 3 Recovery, Linearity, and Sensitivity (S/N at 5 ppb) for Pork Fat Spiked at Six Levels (5–200 ng/g fat) with Seven PCB Congeners

PCB	Mean % recovery	Linearity	S/N at 5 ppb
28	110	0.9999	9
52	88	0.9903	8
101	91	0.9991	12
118	105	0.9996	12
153	101	0.9994	13
138	104	0.9991	11
180	105	0.9998	18

The repeatability of the method ($n = 6$) was evaluated by the analysis of a contaminated egg and fat sample (Table 2). These samples were distributed in a round robin test for Belgian laboratories during the crisis. The RSDs are all below 10% for the congeners PCB 118, 153, 138, and 180 and below 5% for the sum of the congeners.

The linearity and method sensitivity were determined by spiking a blank pork fat sample at six levels with the individual PCB congeners. The spike levels were 5, 10, 25, 50, 100, and 200 ppb (ng/g fat) per congener. The recovery was determined versus an external standard, and the linearity was measured by plotting the absolute peak areas versus the spiked concentration. The signal-to-noise ratio was also measured at the lowest spiked concentration. The results are summarized in Table 3. All recoveries are within 80–110% for the individual congeners. The linearity is better than 0.995 except for PCB 52, for which an interfering peak was observed in the CGC-ECD trace. The signal-to-noise ratio was better than 6 for all congeners at the 5 ppb level, which is the value required in the official method as LOQ.

Finally, the reproducibility and accuracy of the new method was evaluated with the determination of the PCB content in two certified reference materials of the European Community, namely, the cod liver oil sample CRM 349 and the mackerel oil sample CRM 350 (IRMM, Geel, Belgium). The analyses of these reference materials were performed by four laboratories: RIC and three laboratories to which the method was transferred and having the same CGC-ECD and CGC-MS instrumentation. The results are summarized in Table 4 for cod liver oil

Table 4 Accuracy and Reproducibility Test for the Cod Liver Oil Sample

PCB	Certified conc. (ppb)	Lab 1		Lab 2		Lab 3		Lab 4	
		ECD	MS	ECD	MS	ECD	MS	ECD	MS
28	68	61	65	73	64	104	68	75	70
52	149	126	165	141	164	144	199	159	148
101	370	333	394	385	404	296	437	356	373
118	454	390	467	421	448	479	508	397	458
153	938	975	989	886	1010	790	1030	810	1016
180	280	252	295	283	312	270	326	288	273
Sum	2259	2137	2375	2189	2402	2083	2568	2085	2338

Table 5 Accuracy and Reproducibility Test for the Mackerel Oil Sample

PCB	Certified conc. (ppb)	Lab 1		Lab 2		Lab 3		Lab 4	
		ECD	MS	ECD	MS	ECD	MS	ECD	MS
28	22.5	25	24	21	21	21	19	19	16
52	62	75	72	56	56	65	71	54	63
101	164	152	181	175	175	152	143	150	172
118	142	163	152	117	117	138	125	131	134
153	317	337	345	319	319	287	350	290	316
180	73	73	79	75	75	76	83	70	60
Sum	778.5	825	853	763	739	739	791	714	761

and in Table 5 for mackerel oil. Most values are within 80 and 110% of the certified samples. For all laboratories and for both techniques, the sum values were always between these limits.

During the interlaboratory study, we noted that the absolute retention times of the different congeners could shift between the laboratories within a window of up to 1 minute. The retention time locking (RTL) concept was therefore implemented in the CGC-ECD analysis [7]. The temperature program was changed to 70°C (2 min) to 150°C at 25°C/min, to 200°C at 3°C/min, and to 300°C (2 min) at 8°C/min, and an initial head pressure of 71 kPa hydrogen was applied. The pressure is then adjusted via the RTL software to obtain a retention time of 26.999

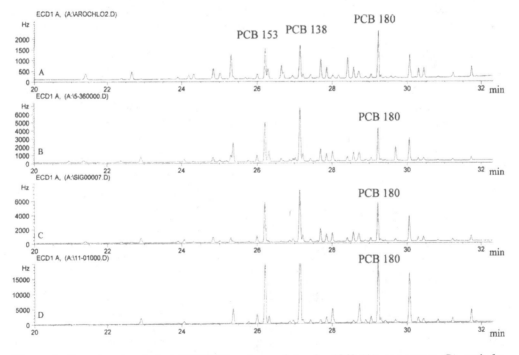

Figure 3 Retention time locked CGC-ECD analyses of Aroclor 1260 (A), egg extract (B), pork fat extract (C), and mink fat extract (D).

Table 6 Translation of Chromatographic Conditions from a Conventional to a Fast Capillary Column

30 m × 250 μm i.d. × 0.25 μm HP-5MS	10 m × 100 μm i.d. × 0.10 μm HP-5MS
71 kPa hydrogen	233 kPa hydrogen
70°C—2 min—25°C/min—150°C—3°C/min—200°C—8°C/min—300°C (2 min)	70°C—0.45 min—110°C/min—150°C—13.2°C/min—200°C—35.2°C/min—300°C (0.4 min)
Anal. time 36 min	Anal. time 8.2 min

min for p,p'-DDT. The total analysis time under these conditions increased to 36 minutes, but retention times were now very stable. Figure 3 shows the RTL-CGC-ECD profiles of Aroclor 1260, egg, pork fat, and mink fat recorded with a time interval of nearly one month. As an example, the retention time for PCB 180 varies from 29.230 to 29.234 min (mean: 29.232; s: 0.002 min; RSD% < 0.01%).

The chromatogram of the mink fat extract is interesting. The minks were fed with contaminated eggs, and although the same Aroclor 1260 profile is present, some differences are noted in the relative abundances of the PCB congeners. This can be explained by a different metabolism between the animal species. In the mink sample, the PCB concentration measured as the sum of the seven congeners was as high as 25 ppm (25 mg/kg fat). This concentration was fatal for most minks.

The relatively long analysis times under RTL conditions prompted us to evaluate fast high-resolution capillary GC for PCB analysis [8]. Presently with state-of-the-art capillary GC instrumentation, capillary columns with internal diameters of 100 μm and lengths of 10 m can be used, drastically decreasing analysis time while maintaining the resolving power compared to

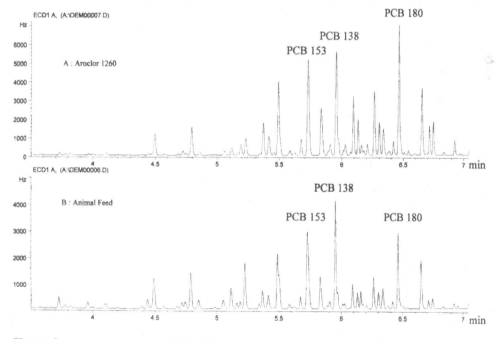

Figure 4 Fast high-resolution CGC-μECD analysis of Aroclor 1260 (A) and a contaminated egg extract (B).

a conventional capillary column. This is a prerequisite for PCB analysis, and fast CGC techniques like the use of multicapillary columns or flash GC cannot be applied because the resolution is too low to separate the specific congeners. With the help of the method translation software (MTS), chromatographic conditions of a standard analysis like the RTL-CGC-ECD profiles discussed above can be translated to a high-speed column keeping the resolution intact [8]. The translated conditions are given in Table 6.

The program predicts that the speed gain factor is 4.4, which means an analysis time of 8.2 minutes. The analyses of an Aroclor 1260 reference sample and an animal feed sample are shown in Fig. 4.

Compared to Fig. 3, PCB 180 elutes now at 6.46 minutes instead of at 29.23 minutes. The experimental speed gain factor is thus 4.5, corresponding very well with the predicted speed gain factor of 4.4. From the chromatograms, it is clear that the resolution has not been compromised compared to the analysis on the conventional columns. Important to note is that splitless injection can be applied in fast high-resolution capillary GC.

V. CONCLUSION

The contamination of Belgian food products was initially referred to as a "dioxin crisis." The main contamination source, however, was polychlorinated biphenyls. PCBs can be analyzed much faster and cheaper than PCDDs and PCDFs, and therefore the analysis of PCBs was "the" solution to ban all contaminated food. A new method, using ultrasonic extraction, followed by matrix solid phase dispersion clean-up and CGC-ECD or CGC-MS analysis, was developed and validated. Retention time locking and fast high-resolution capillary GC completed the introduction of new methodologies in the PCB screening. One technician on one instrument can easily perform 50 analyses per day on a conventional column and, with some help in sample preparation, even 100 analyses per day with fast high-resolution CGC.

REFERENCES

1. A Bernard, C Hermans, F Broeckaert, G De Poorter, A De Cock, G Houins. Nature 401:231–232, 1999.
2. *http://www.fda.gov*.
3. A Pauwels, F David, P Schepens, and P Sandra. Int J Environ Anal Chem 73(3):171–178, 1999.
4. K Ballschmiter, M Zell. Fres Z Anal Chem 302:20–31, 1980.
5. *http://beltest.fgov.be/docs_pdf/I14_NL.pdf*.
6. SA Barker. LC·GC May (suppl):S37–S40, 1998.
7. V Giarrocco, BD Quimby, MS Klee. Hewlett-Packard Application Note 228–392, December 1997.
8. F David, DR Gere, F Scanlan, P Sandra. J Chromatogr A 842:309–319, 1999.

41

Research on Separation Methods at the Max Planck Institut Between 1957 and 1995

Gerhard Schomburg*

Max Planck Institut für Kohlenforschung, Mülheim a.d. Ruhr, Germany

I. INTRODUCTION

The major topic of research of the different groups headed by Gerhard Schomburg between 1957 and 1995 was the development of methods and instrumentation in analytical separation techniques under the special aspect of miniaturization. This important feature was introduced by Golay [1] in 1958 for separations of high performance regarding efficiency, resolution, and speed. It required many years of instrumental and applicational research before the new principle of efficient gas chromatographic separations could become effective. The analysis of samples of extremely complicated composition regarding absolute and relative concentrations of target compounds (dilution), range of volatilities, and polarities of both matrices and sample components required elaborate instrumental and methodological developments.

The author applied one of the first commercial gas chromatographs, the Perkin Elmer model 154 A, in 1957 for the separation of mixtures of volatile hydrocarbons arising from the chemical research work at the Max Planck Institut under director and Nobel laureate Karl Ziegler. In 1958 the author participated in his first international symposium on chromatography at the Tropical Institute in Amsterdam. He listened there to the presentation of Marcel Golay on the theory and instrumental application of open tubular columns, which later developed into a highly effective and versatile alternative to the small particle packed columns then and even today used in practical gas chromatography (GC).

II. MINIATURIZATION IN GAS CHROMATOGRAPHY

In open tubular columns the stationary phase is contained as a homogeneous wall coating layer of high molecular weight and temperature-stable polymers of varying polarity. In the early days

---

* Retired

of this development it was shown by D. H. Desty [2,3] and others that very high separation efficiencies could be achieved in long and narrow-bore capillary columns, preferably made of glass. In modern capillary GC, the efficiency and/or the speed of separation of volatile analytes is dramatically increased compared to packed columns in capillaries with an internal diameter of <300 μm.

Capillaries 100 m long and longer could be applied because of the relatively low pressure drop in such a column type, especially when a light carrier gas such as hydrogen is used. In these bore capillaries the diffusion path lengths during the mass transfer of analytes between the gas phase of high diffusivity and the stationary phase on the walls were relatively short, resulting in very high theoretical plate numbers.

Before the important innovation by Golay could become effective in routine analysis, some principal technical problems of the new methodology had to be solved. These were, in the field of gas chromatography:

1. Capillaries were initially made from metal, then from glass [3], and finally, since 1978, from fused silica, which was introduced by Dandeneau [4]. Homogeneous and stable thin and also thicker coatings of nonvolatile and temperature-stable polymers of suitable chemical structure (polarity) on the surfaces of such capillaries were developed [5–11].

2. The very low amounts of stationary phase in highly efficient open tubular columns is related to a very low sample capacity, so that very small amounts of the target analytes had to be manipulated, vaporized, and introduced into the mobile phase for sampling, separation, and on-line detection in a monitoring mode. Samples that contain the target compound in very high dilution, i.e., as traces in a volatile matrix, had to be introduced in large volumes by special techniques in order to achieve high enough signals for quantitation. This approach led to overloading of the column by the matrix compounds. Therefore, as the natural consequence of miniaturization, sampling methods such as splitless and on-column injection [5,6] or selective sampling in MDGC systems [7] have been developed.

3. The continuous and sensitive detection of extremely small mass flows of analytes was enabled by the ''universal'' flame ionization detector (FID) [17] the electron capture detector (ECD) [18], and later especially by the gas chromatography/mass spectroscopy (GC/MS) coupling in different modes [19–21]. On-line data acquisition and handling in dedicated computers or computer systems for fast execution of qualitative and quantitative analyses of high precision and accuracy was another essential requirement for the effective application of Golay's invention in commercial equipment for routine analysis [22].

4. The application of temperature programming became the most important approach for the analytical routine in order to achieve short analysis times also for less volatile sample constituents without dramatic loss of resolution in the different groups of target components contained in samples with extremely wide range of volatilities and polarities.

5. Multidimensional separations were introduced into MDGC for preparative gas chromatography and in analytical systems mainly with involvement of interconnected capillary columns or a capillary connected to a packed column allowing intermediate trapping and monitoring the eluate flow after each column in the coupling. The valveless switching of eluate flows allows for the execution of partial analyses of complex samples at high resolution and signal-to-noise ratios [23,24]. The concept of partial analyses for

selected components or groups of components of a sample with MDGC with the aim of saving analysis time simultaneously with high performance of separation and detection seems to be in flagrant contradiction to the present trend toward "comprehensive" analysis of wide-volatility-range mixtures containing numerous compounds.

To reach the present status of performance of capillary GC separations, a number of specific instrumental developments in miniaturized GC were performed between 1960 and 1990. Some contributions to these developments in capillary GC that have been achieved in the group of G. Schomburg will be described in more detail below. They were achieved during the long period of about 25 years preceding the general acceptance of the new technology in the laboratory and especially in process control analysis. As mentioned above, the contributions concerned column technology, especially stable stationary phase coatings, sample introduction methods, automatized instrumental set-ups for multidimensional GC and for GC/MS coupling with involvement of open tubular columns, and last, but not least, on-line data handling in computerized systems.

III. MINIATURIZATION IN LIQUID CHROMATOGRAPHY

The adoption of the principle of miniaturization with the aim of optimization of separation efficiency occurred in liquid chromatography (LC) when columns for high performance with packings of chemically modified small porous particles (i.d. $< 10\mu m$) were developed [25,26]. LC separations in open tubular columns require the use of much smaller internal diameters because of the slow diffusion of the analyte molecules in the liquid phase. Therefore, in the laboratory of the author the main interest in this period of further development of high-performance liquid chromatography (HPLC) was to contribute to the chemistry of stationary phases synthesized on the basis of porous small particles with procedures of polymer coating. In this work the experiences from previous years of experimental work on the coating of surfaces in GC capillary columns were very useful. Polymer coating methods in the HPLC field were performed in comparison to the common silanization procedures for anchoring molecules of the desired "polarity" (including chirality) to porous surfaces. It was of special interest whether the modification of the surfaces within narrow pores could be done homogeneously with large polymer molecules and without much loss in porosity. This consequence of polymer coating had no serious influence on the performance of common phase systems. On the contrary, even higher efficiencies were achieved in some cases. Some separations with synthesized stationary phases for RPLC (the C_{18}-substituted polysiloxane $PMSC_{18}$, chirally substituted polysiloxanes both on silica and polybutadiene as a coating for Al_2O_3). A stationary phase for the fast ion exchange chromatography (IEC) of alkali and alkali earth cations, obtained by coating of silica by a crosslinked polybutadiene–maleic acid copolymer, was another typical result of this kind of coating method [27,28].

IV. MINIATURIZATION IN CAPILLARY ELECTROPHORESIS

The introduction of the flexible and thin-walled fused silica tubing material with narrow internal diameter ($<50\mu m$) into capillary electrophoresis (CE) by Joergenson et al. [29] in 1981 initiated another revolutionary trend in the development of highly efficient and selective analytical separation methods by miniaturization after the preliminary work of Hjertén [30] in 1967. In capillary electrophoresis separations are possible by differential electromigration and without phase transfer of the analytes vertical to the migration direction for partition as in LC. The miniaturization in capillary electrophoresis had to go further than in capillary GC in order to achieve a good dissipation of the heat generated in the strong electrical fields. This new miniaturized separation technique performed in narrow-bore capillaries encouraged the group of Gerhard Schomburg to undertake similar research activities as in GC and LC. In aqueous and nonaqueous (but dielectric) buffer media of defined pH, small and large ionic and nonionic molecules can be quickly and efficiently resolved in different modes including hybrids of CE and LC such as CEC and MECC [31]. In analogy to the coating chemistry developed and applied for GC and HPLC, surface chemistry has been one of the major topics in CE separation systems of the group since 1989 but had to be varied for the basically different separation systems of CE.

Examples of special dynamic and static coatings in separation systems of capillary zone electrophoresis (CZE), electrokinetic chromatography (EKC), and CGE include:

The modification of absolute and relative migration times (selectivities) by charged and uncharged as well as chiral buffer additives

The dynamic and permanent modification of surfaces by charged and uncharged oligomeres with the aim of manipulation of the electroosmotic flow and the suppression of undesirable analyte/wall interaction, polymer coatings by hydroxylic uncharged polymers, especially polyvinyl alcohol (PVA)

The separation of large biopolymers in CZE and CGE using gels and polymer solutions

In chromatography as well as in electrophoresis, related analogous but also different instrumental and chemical problems needed to be solved. They were consequences of miniaturization, which was adopted in CE for different reasons than in GC and LC (e.g., for the abduction of Joule heat). Although the separation mechanisms in chromatography and capillary electrophoresis are basically different, miniaturization has had a significant positive effect on the analytical performance achievable today. HPLC and HPCE have influenced each other in the development of methods and applications with regard to resolution, as can be concluded in view of the recent achievements concerning EKC and the electrochromatography (EC) modes of capillary electrophoresis, which are hybrids of LC and CE, and unite, to a certain extent, the advantages of the two separation methods In the following section some examples of instrumental developments and separations obtained with special types of surface coatings are discussed, which may give an idea of the approaches and contributions the author's group has made during more than 30 years of work.

V. SELECTED CONTRIBUTIONS TO THE DEVELOPMENT OF METHODS AND INSTRUMENTATION IN CGC, HPLC, AND CE

A. Gas Chromatography

The above statements on the development of modern gas chromatography are illustrated by separations achieved in the 1970s and early 1980s, about the time when flexible fused silica material used for open tubular columns displaced the fragile soft or borate glass as capillary material. The separation shown in Fig. 1 was obtained with a methylpolysiloxane OV-1 alkali glass column and a film thickness of about 0.1 μm. Temperatures of up to 280°C with hydrogen as carrier gas were applied, and the analysis time for these very low-volatility polyaromatic hydrocarbons of coal tar was only 42 minutes to the elution of coronene. This analysis time was quite short compared to that of several hours for a less successful separation in the best long packed columns as described by environmental analysts in the early days of analysis for the same carcinogenic benzopyrenes in engine exhaust. Later a similar influence on the perfor-

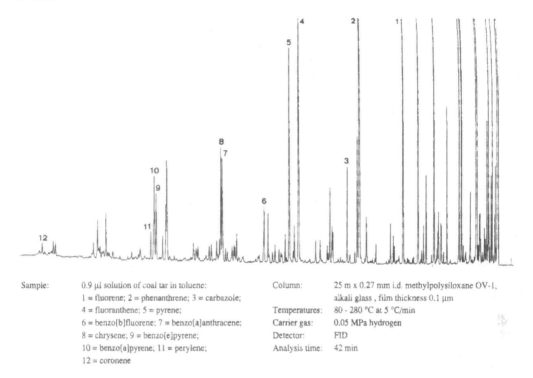

Sample:	0.9 µl solution of coal tar in toluene:	Column:	25 m x 0.27 mm i.d. methylpolysiloxane OV-1,
	1 = fluorene; 2 = phenanthrene; 3 = carbazole;		alkali glass , film thickness 0.1 µm
	4 = fluoranthene; 5 = pyrene;	Temperatures:	80 - 280 °C at 5 °C/min
	6 = benzo[b]fluorene; 7 = benzo[a]anthracene;	Carrier gas:	0.05 MPa hydrogen
	8 = chrysene; 9 = benzo[e]pyrene;	Detector:	FID
	10 = benzo[a]pyrene; 11 = perylene;	Analysis time:	42 min
	12 = coronene		

Figure 1 Separation of polyaromatic hydrocarbons of coal tar in a glass capillary column coated with the methyl polysiloxane OV-1.

mance of the separations of the numerous dioxine dibenzofurane isomers was also achieved by the introduction of capillary columns into this area of environmental ultra-trace analysis, especially with the application of GC-MS-SIM methods for specific detection. Another area of interest was the deactivation of alkali glass, borate glass, and fused silica for the tailing free separation of basic compounds. The separation of basic compounds in GC, LC, and CE systems with stationary phases based on porous silica later became the one of the fundamental problems of separation at optimum analytical performance and has been subject of several research papers from the author's group. In the area of instrumentational research, sampling techniques and multidimensional systems with involvement of capillary columns were two of the most interesting topics. Figure 2 shows typical systems for multidimensional capillary GC containing two columns of different polarity, which were coupled dead volume-free and valveless. The system in Fig. 2a is the coupling of a packed with a capillary column, whereas the system in Fig. 2b is a coupling of two capillary columns. Both systems have two detectors to monitor the eluates emerging from the first and the second columns, and could be operated automatically by computer. Intermediate trapping could be done in the inlet of the second column, and the separations in the two coupled columns could be independently executed under temperature programming. In Fig. 3, separation A (performed in the first column of the coupled system) was obtained with the polar polyethylene glycol CW 400 column (120 m of coated alkali glass capillary), while separation B was obtained with the cut taken as marked from the first separation. This cut contained all the nonpolar hydrocarbon species that cannot be resolved in the highly polar polyethylene glycol column. These hydrocarbons can be well resolved with the nonpolar methylpoly-

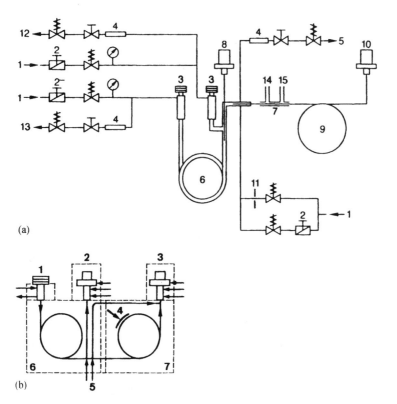

Figure 2 Instrumental system for multidimensional separations in coupled capillary columns. (a) Multi-dimensional GC in systems of coupled columns. Isothermally operated system of packed precolumn coupled to capillary column. Focusing of the transferred analytes in the temporarily cooled inlet of the capillary column. 1 = Carrier gas inlets; 2 = pressure regulators; 3 = injectors; 4 = filter columns; 5 = split; 6 = packed column; 7 = cold trap (column inlet); 8 = FID; 9 = capillary column; 10 = FID for detection of the significant analytes after main separation; 11 = blend type flow controllers for control of make-up gas flow into the coupling piece; 12 = carrier gas outlet for packed column when operated from left to right and using the left-hand injector (3); 13 = carrier gas outlet for packed column when operated from right to left using the right-hand injector (3); 14 = inlet for heating gas (nitrogen); 15 = inlet for cooling gas (nitrogen). (b) Multidimensional GC with double oven instruments. 1 = Injector; 2 = detector for monitoring the first separation; 3 = detector for monitoring the main separation; 4 = cold trap for peak focusing; 5 = make-up gas; 6 = oven 1; 7 = oven 2.

siloxane OV-1 column. The trapping within the inlet of the second column was done at −30°C. The two chromatograms illustrate the separation of two groups of compounds of very different polarity at high resolution in a single run.

This kind of multidimensional separation was first realized by Deans [23] in 1968 but in systems of coupled packed columns only. The resolution of the groups of target compounds

Figure 3 (a) Separations of a coal-derived gasoline fraction executed in the MDGC system of Fig. 2; (b) MDGC separations of TCDD's (Dioxines) for determination of 2,3,7,8-TDDD in environmental analysis.

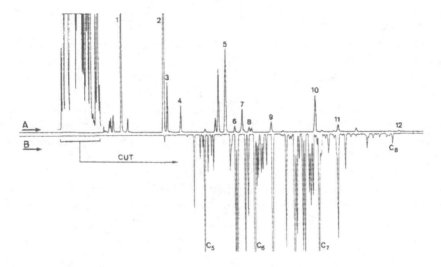

(a)

Sample: 0.2 μl. 1 = Acetone; 2 = 2-butanone; 3 = benzene; 4 = isopropyl methyl ketone; 5 = isopropanol; 6 = ethanol; 7 = toluene; 8 = propionitrile; 9 = acetonitrile; 10 = isobutanol; 11 = n-propanol; 12 = 1-butanol. Pre-column: (A) 121 m x 0.27 mm i.d. polyethylene glycol CW 400, alkali glass. Temperature: 12.6 min isothermal at 50 °C, 50 to 80 °C at 3 °C/min. Carrier gas: 0.132/0.07 MPa hydrogen; cut, 12.6 to 17.4 min. Detector: FID. Main-column: (B) 64 m x 0.27 mm i.d. methylpolysiloxane OV-1, alkali glass, 0.27 mm i.d., film thickness > 1 μm. Temperature: 17.4 min isothermal at –30 °C, –30 to 50 °C at 20 °C/min, 50 to 150 °C at 3 °C/min. Carrier gas: 0.07 MPa hydrogen. Detector: FID. Analysis time: 50 min.

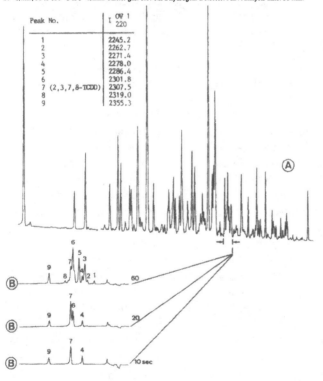

Peak No.		I OV 1 220
1		2245.2
2		2262.7
3		2271.4
4		2278.0
5		2286.4
6		2301.8
7	(2,3,7,8-TCDD)	2307.5
8		2319.0
9		2355.3

(b)

Sample: 0.6 μl solution of PCDD and PCDF
Columns: (A) 30 m x 0.27 mm i.d. cyanopropyl polysiloxane Sil - 10 C, alkali glass; coupled to
                 (B) 25 m x 0.32 mm i.d. methylpolysiloxane OV-1, fused silica,
Temperatures: (A) 60 - 160 °C ballistically, 160 - 260 °C at 3 °C/min;
                 (B) 220 °C isothermal
Carrier gas: (A) 0.115 MPa hydrogen;
                 (B) 0.050 MPa hydrogen
Detectors: ECDs
Analysis time: 35 min

was mainly reached by selectivity change at the limited separation efficiency of packed columns. The first MDGC separations in coupled capillary columns were performed in Gerhard Schomburg's group in 1971. In the systems developed by this group, the separations of selected groups of target compounds could be performed with the high efficiency of capillary columns with the additional advantage of the selectivity change as a consequence of the transfer of cuts from one column to another column of different polarity. Fig. 3b shows multidimensional separations of the isomeric methysubstituted dibenzodioxines and -furanes. From the preseparation A performed in the highly polar cyanopropyl polysiloxane Silar 10 C column, three cuts of different width in time (10 sec., 20 sec. and 60 sec.) were taken and transferred into the non-polar OV-1 methylpolysiloxane column of the MDGC system. The three chromatograms B have been achieved from these cuts. The highly toxic compound 2,3,7,8 TCDD can only be fully resolved from other much less toxic species from the narrow 10 sec. cut for undisturbed quantitation in environmental dioxine analysis, as can be seen from the three separations B.

Other areas of instrumental research work included the open-split connection of capillary columns to ion sources of MS instruments in the years 1970–1971 [20,21]. Also, since 1970 the Mülheim Computer System was developed [22] for the simultaneous on-line data handling of different methods of instrumental analysis with a Digital Equipment PDP 10 computer.

The work on the development of on-column injectors for capillary GC was published in 1977 at about the same time, that Grob et al. [14] published a study on on-column injection.

B. Liquid Chromatography

Polymer coating of porous silica and alumina for normal phase, reverse-phase, and ion exchange systems including chiral phases was mentioned in the introduction as the major topic of research in the field of HPLC [27,28].

In Fig. 4 two LC-separations A and B of strongly basic test compounds are shown which had been obtained with a coating of C_{18}-substituted polymethysiloxane on Nucleosil (a nonfired small-particle porous silica). In separation A, the highly basic 2-n-octylpyridine (4) coelutes with the nonbasic aromatic hydrocarbon n-butylbenzene (6) because of the strong interaction of the basic compound n-octyl-pyridine with the silanols, which are still present underneath the polysiloxane coating. In separation B, which had been achieved with a double coating on the same Nucleosil before the polysiloxane coating, the major portion of silanols had been precapped by trimethylsilylation. Now the interaction of the basic octylpyridine (and the other basic test compounds) is much weaker and the octylpyridine is eluted long before peak 6 of the nonpolar n-butylbenzene. Moreover, it might be noticed that the peak widths in separation B are much narrower (i.e., the separation efficiency is much higher) as without the "precapping." The figure illustrates another possibility of the "polymer coating procedure" which allows for a double layer coating also. The trimethylsilylated silica can be covered by a C_{18}-substituted polymethylsiloxane by crosslinking between the siloxane chains achieved by heating with a radical catalyst or by radiation. Chemical bonding of the final stationary phase layer to the silica surface via silanol groups as involved in the silanization method for surface modification is not required. In Fig. 5 a separation of all alkali and alkali earth cations with an ion exchange phase which was synthesized by coating of silica with polybutadiene maleic acid (PBDMA) with a similar radical crosslinking as applied for the before mentioned coating procedures of siloxane polymers [27]. The procedure of "polymer coating" allows the modification of the film thickness of the coating layer and, correspondingly and in the special case of the PBDMA phase, the variation of the concentration of the ion-exchanging groups per unit volume of the stationary phase.

The surfaces of porous alumina cannot be modified by chemical bonding in analogy of the silanization of silica because of the instability of the -Si-O-Al-bond. Therefore, coatings of

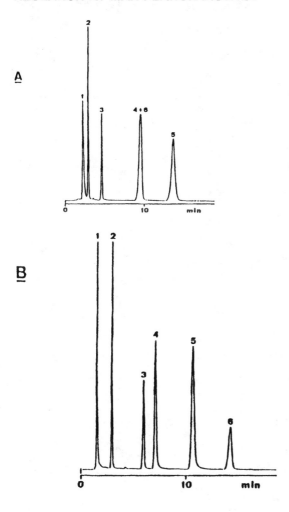

A

B

Peak identification : <u>1</u> α–picoline, <u>2</u> N,N–dimethyl–aniline, <u>3</u> ethylbenzene,
<u>4</u> 2–n–octylpyridine, <u>5</u> 2–n–nonylpyridine, <u>6</u> n–butylbenzene

Column	:	<u>A</u> 125 mm Nucleosil 100–5–$PMSC_{18}$, 4.5 mm i.d.
		<u>B</u> 125 mm Nucleosil 100–5–C_1–$PMSC_{18}$, 4.5 mm i.d.
Temperature	:	308 K
Mobile phase	:	methanol/water = 70:30 (v/v)
Flow–rate	:	1.0 ml/min
Pressure	:	<u>A</u> 11 MPa, <u>B</u> 12.3 MPa
Detection	:	UV, 254 nm

Figure 4 HPLC separations of strongly basic test compounds by a polymer ($PMSC_{18}$)–coated silica and after previous trimethyl silylation (''precapped'' silica).

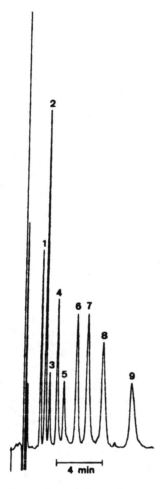

Sample   : $\underline{1}$ Li$^+$. $\underline{2}$ Na$^+$. $\underline{3}$ NH$_4$$^+$. $\underline{4}$ K$^+$. $\underline{5}$ Rb$^+$.
                : $\underline{6}$ Ca^{2+}. $\underline{7}$ Mg^{2+}. $\underline{8}$ Sr^{2+}. $\underline{9}$ Ba^{2+}

Column   :   N-7-100-PBDMA 30% (125 * 4.5 mm i.D.)
Mobile Phase   :   5 mM citric acid. 0.5 mM pyridine-2.6-dicarboxylic acid
Flow   :   1.0 ml/min
Pressure   :   2.8 MPa
Temperature   :   308 K
Detection   :   electroconductivity. 8 μS/cm full scale

Figure 5   Ion exchange HPLC separation of all alkali and alkali earth cations with a polybutadiene maleic acid (PBDMA)–coated silica.

variable film thickness of polybutadiene were generated, which were immobilized by radical catalyzed crosslinking. Such stationary reverse-phase materials exhibit very high stability even at highly alkaline pH of up to 13 and could be applied to special separations of strongly acidic compounds such as chlorophenol isomeres (Fig. 6). The author is convinced that the methodology of polymer coating could also be useful in work on stationary phases for electrochromatography.

C. Capillary Electrophoresis

The first paper of the group on CZE appeared in 1989 concerned the influence of polymer coating of capillary surfaces on the migration behavior in micellar electrokinetic capillary chromatography with sodium dodecyl sulfate (SDS) as surfactant in methylpolisiloxane as well as polyethylene glycol coatings of the fused silica surfaces [32]. Another paper in the area of

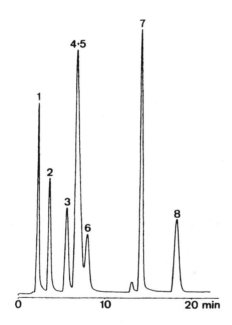

Peak identification : 1 = 4-Chlorophenol, 2 = 3,4-Dichlorophenol, 3 = 2,3,6-
Trichlorophenol, 4 = 2,4,6-Trichlorophenol, 5 = 2,3,4-
Trichlorophenol, 6 = 2,4,5-Trichlorophenol,
7 = 2,3,4,5-Tetrachlorophenol, 8 = Pentachlorophenol

Column   : 150*4.6 mm Spherisorb A 5 Y/PBD
Temperature   : 295 K
Mobile Phase   : gradient from 100 % 0.1 n Na_3PO_4 (pH 12) to 60 % 0.1 n
                  Na_3PO_4/methanol 1:2 (v/v), 6 % per minute
Flow-rate   : 0.8 ml/min
Pressure   : 3.8 MPa
Detection   : UV, 254 nm

Figure 6 Reverse-phase separation of strongly basic compounds at a pH 12 using a polybutadiene-coated porous small particle alumina as stationary phase.

capillary electrophoresis dealt with the topic of EOF suppression in polyacrylamide gel filled column and the avoidance of bubble formation during the separation of oligonucleotides. The result of this work was that the polyacrylamide coating of the surface must not be chemically bonded to the polyacrylamide gel, as proposed by authors from other groups, when bubble formation had to be avoided.

Figure 7 shows the CGE separation of poly(uridine 5′-phosphate) with more than 430 nucleotide units. During this work CGE separations with solutions of neutral hydroxylic polymers (physical gels) were investigated. These solutions of hydroxylic polymers do not require a previous coating of the capillary surfaces for suppression of the EOF and were applied to the separation of restriction fragments [33].

Meanwhile, it had been found by several authors that, unlike LC separation mechanisms, in CZE analyte/wall interaction is quite often related to a considerable decrease of efficiency in the separation of large biomolecules such as proteins and oligonucleotides and especially of small chiral species. The best way to suppress analyte/wall interaction was achieved with the

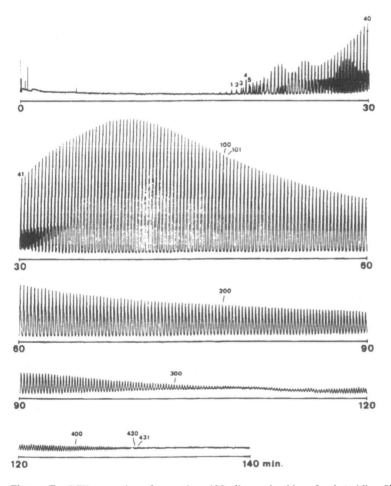

Figure 7 CGE separation of more than 430 oligonucleotides of poly(uridine 5′-phosphate).

addition of surface-modifying agents to the buffers. The EOF is also suppressed in case of the usage of neutral polar, e.g., hydroxylic oligomers. This and other methods of surface modification can be performed in the dynamic or the static mode applying chemically bonded coatings. The latter were expected to be more stable but were sophisticated enough to be reproduced. In the work of Gilges et al. [34] poly(vinylalcohol) (PVA) proved to be the most effective modifier with regard to suppression of analyte/wall interaction in a wide range of pH and also with regard to the suppression of EOF. PVA exhibits probably the weakest hydrophobic interaction with organic compounds and forms very stable coatings by strong crosslinking between the polymer chains. A procedure of stabilization of PVA coatings by thermal treatment was also described. In Fig. 8 a CZE separation of a typical test mixture of strongly basic protein indicates by the very high efficiencies observed that immobilized PVA is suited to form a coating layer on silica surfaces that is stable even at high pH and prevents the interaction of the basic molecules with the acidic silanol groups on the silica surface. Recently the reliability of the production of such PVA capillaries, which could also be successfully applied with nonaqueous buffer media, was improved by Belder et al. The modification of EOF by external electrical fields in comparison to the method of chemical surface modification was investigated by Belder [35] in his doctoral thesis.

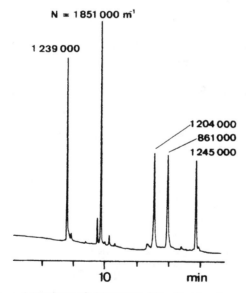

Sample:	1 cytochrome C, 2 lysozyme, 3 trypsin, 4 trypsinogen, 5 α-chymotrypsinogen
Capillary:	57 cm effective length, 70 cm total length; 75 μm i.d.
Coating	thermally immobilized PVA (M.W. 50.000, Aldrich)
Buffer:	50 mM Na-phosphate, pH 3.0
Conditions:	30 kV (429 V/cm), 69 μA; 20°C
Injection:	10 kV, 2.5 s
Detection:	UV, 214 nm

Figure 8 CZE separation of a test mixture of strongly basic proteins in a PVA-coated fused silica capillary.

REFERENCES

This list of references cannot be complete considering the excellent work of so many scientists from international research groups over so many years. The author of this text has selected only a small number of references that have influenced his ideas and intentions more than others. It is not claimed that this selection is representative from a scientific point of view.

1. MJE Golay. In: DH Desty, ed. Gas Chromatography 1958 (Amsterdam Symposium). London: Butterworths Sci. Publ., 1958, pp. 36, 139.
2. DH Desty, A. Goldup, BHF Whyman. J Inst Petroleum 45:287, 1959.
3. DH Desty, A Goldup, BHF Whyman. In: RPW Scott, ed. Gas Chromatography 1960 (Edinburgh Symposium). London: Butterworths Sci. Publisher, 1960, pp. 168–183.
4. R Dandeneau, EH Zerenner. HRC 2:351, 1979.
5. K Grob. Helv Chim Acta 48:1362, 1965.
6. G Schomburg, H. Husmann. Chromatographia 8:517–530, 1975.
7. Schomburg, R Dielmann, F Weeke, H Husmann. J Chromatogr 122:55, 1976.
8. G Schomburg, H Borwitzky. Chromatographia 12:651, 1979.
9. W Jennings. Gas Chromatography with Glass Capillary Columns. New York: Academic Press, 1978, p. 58.
10. T Welsch, W Engewald, A Engler, C Klauke. Lecture given at analytical meeting, Leipzig. Mitteilungsblatt Chem. Ges. DDR, Beiheft 9:22, 1976.
11. KD Bartle BW Wright, ML Lee. Chromatography 14:387, 1981.
12. ML Lee, BW Wright. Chromatogr 184:235, 1980.
13. K Grob, G Grob. J Chromatogr Sci 7:584–587, 1969.
14. K Grob, G Grob, K Grob, jr. HRC 2:31, 677, 1979.
15. G Schomburg, H Behlau, R Dielmann, F Weeke, H Husmann. J Chromatogr 142:87–102.
16. G Schomburg, H Husmann, F Weeke. J Chromatogr 112:205–217, 1975.
17. McWilliam, RA Dewar. Nature (London) 181:1664, 1958.
18. JE Lovelock, SR Lipsky. J Am Chem Soc 82:431, 1960.
19. RS Gohlke. Anal Chem 31:533, 1959.
20. D Henneberg, G Schomburg. Z Anal Chem 211:55–61, 1965.
21. D Henneberg, U Henrichs, G Schomburg. Chromatographia 8:449–451, 1975.
22. E Ziegler, D Henneberg, G Schomburg. Anal Chem 42:51, 1970.
23. DR Deans. Chromatographia 1:18, 1968.
24. G Schomburg, F Weeke. In: Perry, ed. Gas Chromatography. London: Institute of Petroleum, 1973, pp. 285–294.
25. JFK Huber, JARJ Hulsman. Anal Chim Acta 38:581, 1967.
26. JJ Kirkland, JJ DeStefano. J Chrom Sci 8:309, 1970.
27. P Kolla, G Schomburg, J Köhler. Chromatographia 23:465–472, 1987.
28. G Schomburg, U Bien-Vogelsang, A Deege, J Köhler. Chromatographia 19:170–179, 1984.
29. JW Jorgenson, KD Lukacs. Anal Chem 53:1298, 1981.
30. S Hjerten. Chromatogr Rev 9:122, 1967.
31. S Terabe, K Otsuka, K Ichikawa, A Tsuchiya, T Ando. Anal Chem 56:111, 1984.
32. JA Lux, H-F Yin, G Schomburg. HRC 13:145, 1990.
33. H-F Yin, JA Lux, G Schomburg. HRC 13:624–627, 1990.
34. M Gilges, MH Kleemiß, G Schomburg. Anal Chem 66:2038–2046, 1994.
35. D Belder. Thesis. University of Marburg, 1994.

42

The Secret Memories of a Member of the Blue Finger Society: A Quarter Century in Polymer Chemistry

Pier Giorgio Righetti

University of Verona, Verona, Italy

I. INTRODUCTION

In the early days of electrophoresis, there was a sure way to recognize the members associated with this secret society: shake hands. During this ceremony, you would chance a quick glance at the fingers of the chap you were meeting: blue fingers were an element of distinction, would surely point to someone furiously staining his gels with Coomassie Brilliant Blue R-250, in those days the most popular among all stains. Today this element of recognition has been lost, first because silver stains quickly took over, so you would have to look for brownish fingertips, but how could you then distinguish between a true silver staining protocol and cigarette addiction? Second, on the emotional wave of the famous James Bond movie "Goldfinger," gold staining too became a new fashion. Were this not enough, in the aseptic world of today you would not be able to sort out anybody on this basis: scientists look like Martians in modern laboratories—they wear gloves, masks, hats, sterile garments, you name it; no traces are left of their work at the end of the day. But for us pioneers, life was a real adventure: we would plunge our naked hands in basins containing gels to be stained and destained—everything involved direct contact with our fingertips. As the plot below unravels, you will recognize that we have all been intoxicated, as there were plenty of unreacted monomers floating free in these gels, readily adsorbed through our skin: since all these monomers are neurotoxins, it is surprising that the pioneers in the field did not end up in a mental hospital (do not be surprised if this chronicle seems on shaky grounds, I am still on the loose!).

My love of polymer chemistry started in the mid 1970s, when, after returning from a long period in the United States, I became intrigued by polyacrylamide gels, in those days the most popular matrix for most electrophoretic techniques, not just for proteins but also for small size DNA. It is true that, originally, during my studies at the University of Pavia, I graduated in organic chemistry; thus, it was a trip back to the very roots of my education. However, another major factor in this love for polymers was the fact that I lived very close to the mecca of Italian science: the Milan Polytechnic, where the only Nobel Laureate of Italy was living: Prof. Giulio Natta, the co-inventor, together with Ziegler, of polyethylene and polypropylene. In this account I will try to follow as closely as possible the chronology of our work in this field, which can be divided into the following main sections: polymerization kinetics, studies on photopolymerization, chemistry of novel monomers, both neutral and charged (yes, the glorious Immobiline family), and finally physicochemical studies on macroporous gels, as obtained by lateral aggregation, by thermal means, and by high levels of cross-linkers. This account will span about 25 years of my research life, more than twice as many as it took Odysseus to get back to Ithaca; too bad that it is me and not the blind poet Homer relating this story.

II. POLYMERIZATION KINETICS

In 1981, together with Cecilia Gelfi (it was, in fact, the subject of her doctorate thesis), we undertook the task of studying the efficiency of gel polymerization as a function of a number of parameters. The reason for this was the great confusion existing among the practitioners, with many possible recipes being suggested. A case in point was the proliferation of different cross-linkers: beside the standard one, N,N'-methylene bisacrylamide (Bis), others, such as N,N'-(1,2-dihydroxyethylene) bisacrylamide (DHEBA), ethylene diacrylate (EDA), N,N'-bisacrylyl-cystamine (BAC), and N,N'-diallyltartardiamide (DATD), were recommended. We made our observations directly in spectrophotometric cuvettes, either at 283 nm (disappearance of the double bond) or at 600 nm (Tyndall effect due to the turbidity of Bis, DHEBA, and BAC gels). The order of reactivity appeared to be Bis $\approx$ DHEBA > EDA $\approx$ BAC $\gg$ DATD [1]. The last cross-linker (DATD) was found to be an inhibitor of gel polymerization, leading to highly gluey and unpolymerized gels, especially at high %C values (in those days there were groups claiming that one should adopt highly cross-linked DATD gels; it was like recommending using inhibitors for firing a polymerization reaction, considering that allyl double bonds are radical sinks when admixed to highly reacting acrylic double bonds!). As a consequence of this order of reactivity, Bis and DHEBA gels, at 3–5 %C levels, were found to be polymerized within 30 minutes, whereas EDA and BAC gels, in the same %C range, required at least 3 hours. All cross-linkers, when used above 10 %C, displayed quite slow polymerization kinetics, requiring overnight reaction for good conversion of monomers into polymer chains. A second aspect explored was the effect of temperature on the final gel properties. Here too, we were spurred by recommendations of other authors that gels be polymerized immersed in coolant at 0–4°C. In fact, this last temperature was found to produce highly turbid, porous, and inelastic gels, with poor incorporation of monomers into the growing chains. At temperatures of 25°C or higher, the gels became progressively transparent, less porous, and more elastic. This phenomenon was attributed to formation of hydrogen bonds among cross-linker molecules, which would lead to chain aggregation and/or incorporation of pools of cross-linkers in selected regions (or towards the end) of the polymer, due to lower reactivities in the hydrogen-bonded state. It was thus suggested by us that polymerization at 0–4°C should be abandoned in favor of reactions at 25–30°C, since the former conditions would lead to inhomogeneous and unreproducible pore size matrices [2]. The third aspect analyzed in this series of works was the effect of catalysts on final gel structure [3]. Although most of us routinely use persulfate, it should be emphasized that, originally, in disc electrophoresis, Davis [4] recommended photopolymerization with riboflavin (later on exchanged for riboflavin-5'-phosphate) when casting the sample and spacer gels, one of the main reasons being that such a process, unlike persulfate catalysis, should be nonoxidizing and nondenaturing towards proteins. It was generally accepted that photopolymerization should proceed for 1 hour. We examined this process spectrophotometrically at 283 nm and also by titrating unreacted double bonds with permanganate. It was found that at least 8 hours of light exposure were needed to ensure 95% conversion of monomers into polymer. In addition, the viscoelastic properties of persulfate- vs. riboflavin-catalyzed gels, as measured with a dynamometer, were found to be widely different, suggesting that the final structure of these two matrices should differ considerably.

III. PHOTOPOLYMERIZATION

About a decade later we started again investigating the process of photopolymerization. We were spurred into it by a meeting I had in the fall of 1991 in Paris at the ESA (European Space Agency) headquarters. They had asked me to evaluate a project presented by two Russian scientists, Tatyana Lyubimova and Vladimir Briskman, for using photopolymerization for investigating gel polymerization in microgravity. In those days, the only access we had to low gravity was to shoot our experiments in rockets at a location in Kiruna, Finland, in the empty fields of Lapponia, where the only inhabitants were reindeer. As the rocket reached the maximum height of its parabolic flight before descent, there were about 2 minutes of microgravity to be exploited: thus, photopolymerization was the only way to initiate polymer growth. The argument was that gravity impeded gel polymerization, since it was argued that the growing polymer chains, before being frozen into space by the cross-linking events, would start to precipitate in the gelling cuvette, leading to inhomogeneous gels (more on this topic ahead). As I was not terribly excited by photopolymerization, I requested that a more thorough study be performed in order to drive the reaction to near completion, as typical of persulfate-driven catalysis. We found in fact that, upon standard 1-hour light exposure at room temperature with a 12 W neon source, only 60% conversion could be achieved [5]. But if photopolymerization was conducted at 70°C for 1 hour or in presence of a 105 W UV-A lamp, having a radiation spectrum extending also in the UV region, and in presence of 12, instead of 6, ppm of riboflavin 5′-phosphate, conversion efficiency would be augmented to approximately 95%, i.e., to the same level of incorporation guaranteed by persulfate. Oxygen, even up to a partial pressure of 900 mmHg, acted only as a retardant (i.e., it increased the lag time of reaction) but did not affect the final incorporation of monomers in photopolymerization, whereas in persulfate-driven catalysis it really acted as an inhibitor by lowering the conversion from 95% (in the absence of O_2) to only 40% at 900 mmHg pressure.

The next chain of events was a long discussion with T. Rabilloud, in Grenoble, who suggested the use of novel, even more powerful, catalysts and the fellowship awarded to Dr. Lyubimova by ESA for coming to my laboratory for a 4-month period. We thus started a new project and optimized photopolymerization in presence of a triad of novel catalysts: 100 μM methylene blue (MB) combined with a redox couple, 1 mM sodium toluenesulfinate, a reducer, and 50 μM diphenyliodonium chloride, an oxidizer (see formulas in Fig. 1). The results were spectacular: polymerization could be induced in a matter of seconds and proceeded at a terrific speed to completion within, typically, 30 minutes [6]. The viscoelastic properties of these gels were even better than those of persulfate-activated gels, suggesting a more homogeneous gel structure. The elastic modulus exhibited a maximum in correspondence of a minimum of permeability, both situated at a 5% cross-linker. A theoretical study confirmed the very high reaction kinetics of this system, together with the extraordinary conversion efficiency [7]. The "methylene blue saga" did not stop here, with the dismantling of the team (Lyubimova returning to URSS and Gelfi taking a long maternity leave). I was fascinated by the properties of this system and continued with the help of my student, Silvia Caglio. The gold nuggets we dug out were impressive:

Diphenyliodonium chloride (oxidizer) M_r 316.57

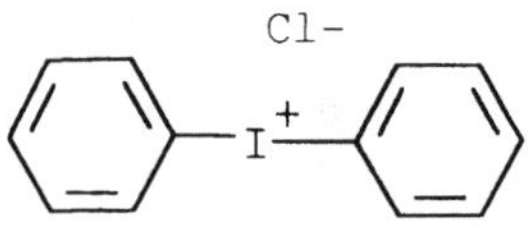

Sodium toluene sulfinate (reducer) M_r 178.18

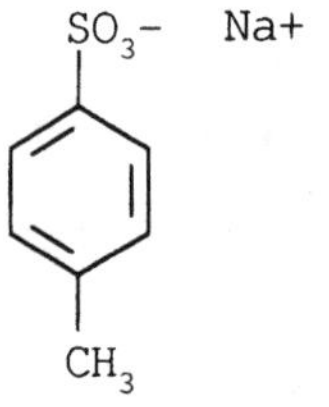

Methylene blue M_r 319.86

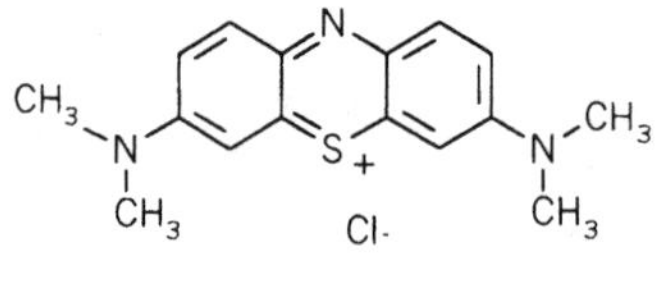

Riboflavin-5'-Phosphate M_r 514.4

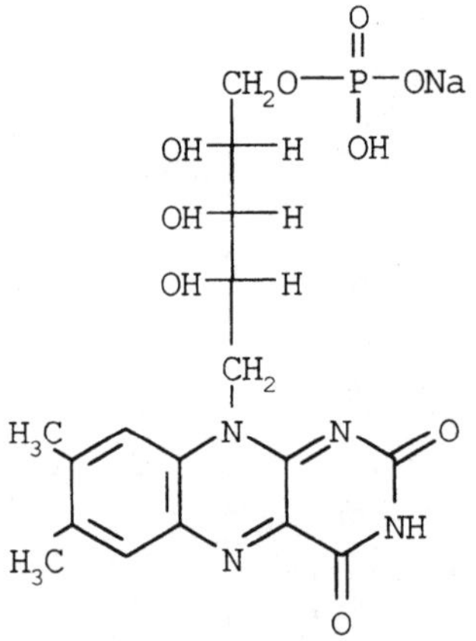

Figure 1 Formulas of the three chemicals used for photopolymerization with methylene blue and of another common photocatalyst, riboflavin-5'-phosphate.

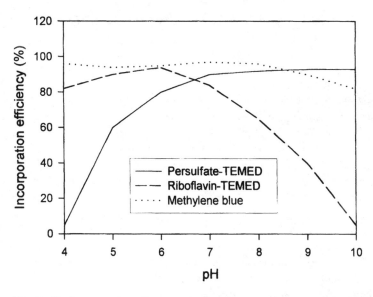

Figure 2 Incorporation efficiency of three catalyst systems: persulfate-TEMED, riboflavin-TEMED, and methylene blue-toluene sulfinate-diphenyliodonium chloride. The conversion was assessed by eluting ungrafted monomers and assessing their concentration by capillary electrophoresis. Note that, at pH 4 for persulfate and at pH 10 for riboflavin, gelation does not occur. Conversely, methylene blue ensures excellent gel formation in the entire pH 4–9 range and very good efficiency even at pH 10. (Adapted from Ref. 8.)

in a first study, we discovered another point of advantage of MB catalysis—its insensitivity to pH in the gelling solution. Persulfate-TEMED polymerization has pH optima in the pH 7–10 range; riboflavin-TEMED only in the pH 4–7 interval; ascorbic acid, ferrous sulfate, and hydrogen peroxide only at pH 4; TEMED and hydrosulfite a reasonable conversion only in the pH 4–6 interval; in contrast, MB-driven reactions work at full speed in the pH 4–9 range and still very well up to pH 10, a real record [8] (Fig. 2)! Other interesting data on matrix structure came with further studies on the MB-driven system: topological data were derived by starting polymerization at 2°C and continuing, after the gel point, at 50°C. The data suggested that, at 2°C, the nascent chains formed clusters held together by hydrogen bonds (melting point at 28°C); such clusters were subsequently "frozen" into the 3-D space as the pendant double bonds of the cross-linkers kept reacting [9]. This would lead to turbid, highly porous and inelastic gels. However, if polymerization was continued at the gel point at 50°C, clear and elastic gels would be obtained. This was additional proof of the widely held belief that turbid gels would indicate large-pore structures. Other remarkable features were then discovered: e.g., the insensitivity of MB-driven catalysis to oxygen, against a large inhibition in persulfate reactions. Conversely, whereas 8 M urea would substantially accelerate persulfate catalysis, it left the MB system undisturbed. However, when polymerization occurred in a number of hydroorganic solvents (all in a 50:50 v/v ratio—dimethyl sulfoxide, tetramethyl urea, formamide, dimethyl formamide), persulfate catalysis was severely inhibited, whereas MB photopolymerization was essentially unaffected [10]. The same applied to polymerization in detergents [11]. It thus appeared that MB catalysis was a unique process proceeding at optimum rate under the most adverse conditions and able to ensure at least 95% monomer conversion.

IV. SYNTHESIS OF *N*-SUBSTITUTED, NEUTRAL ACRYLAMIDO MONOMERS

While playing around with polymerization kinetics was quite exciting, nothing could beat, for a former organic chemist, the commotion of searching for and synthesizing novel monomers. This came in 1984, when I discovered I was living across the street from the institute of the only Italian who had ever won a Nobel prize in chemistry, Prof. G. Natta, at the Polytechnic of Milano. I thus started a collaboration with Prof. Paolo Ferrutti, who had done on his own a large amount of work on polyacrylamides. We prepared a novel couple to contrast with the classical acrylamide-Bis tandem: acryloyl-morpholine and the cross-linker bisacrylyl piperazine. At the beginning we had disastrous results, until we realized that the lack of polymerization was due to the vast amounts of inhibitors used in the synthesis and not removed during the purification. The new matrix thus prepared had some unique properties: it was fully compatible with organic solvents. Once dried, it could reswell in DMSO and dimethylformamide [12]. With the discovery of this new matrix came the adoption of a new solvent for reswelling such a matrix and for solubilizing hydrophobic proteins: sulfolane [13]. It turned out in fact that we had synthesized a rather hydrophobic matrix, which was the reason for its ability to reswell in a number of organic solvents; for proteins, the hydrophobicity was much too high, so the best results could only be obtained in a 50:50 mixture with normal acrylamide. It must have been love at first sight, this propensity for hydrophobic matrices: in 1989 we started a collaboration with AT Biochem, a small company in Malvern, Pennsylvania, which was launching a novel matrix by the name of HydroLink. We characterized it [14], and it turned out that it had the same amphiphilic character and compatibility with organic solvents as acryloyl morpholine, including 50% DMSO, 50% tetramethyl urea, 50% acetonitrile, and 50% tetrahydrofuran. Here, too, the results with proteins were not exciting, but with DNA they were outstanding [15]. It turned out that the main ingredient was dimethyl acrylamide (DMA).

My obsession with gels in these years became contagious: I had been moonlighting in my job and taking summers off at Huntsville, Alabama, at the NASA laboratories at the Red Stone Arsenal, the place from which Von Braun launched the assault on the moon. During this period we were trying to assess the maximum porosity of gels in order to see if it was truly necessary to perform free-zone electrophoresis in space rather than in earth. We measured the maximum porosity of dilute agarose gels (approximately. 500–600 nm average pore diameter, enough for viruses to swim about!) [16]; meanwhile, we also invented new, thermally reversible gels [17]. We characterized two series of such gels, with a thermal reversibility due to formation of extensive hydrogen bonds. One was formed by two polymers, 5% polyvinyl alcohol (PVA) and 4% polyethylene glycol (PEG); the other consisted of 5% PVA and 0.04% borate. These gels have extremely low melting points (16–17°C); the series made of PVA-borate can be utilized in the pH range 7–11 by progressively increasing the borate content in the pH interval 8 to 7 and concomitantly decreasing the borate levels in the pH 8–11 range. It was hypothesized that the low melting points of these gels was due to the fact of being sparsely hydrogen-bonded along the PVA chain: on the average, one-OH group out of three or four-OHs in PVA should be engaged in H-bond formation.

Our love of acrylamide monomers did not end here. In 1993 we started an extensive investigation of the properties of a series of mono- and di-substituted acrylamide monomers. We were looking for the impossible, the "marriage of Figaro" (with due permission from Mozart): trying to combine high hydrophilicity with high resistance to hydrolysis and possibly also a larger pore size than conventional polyacrylamide gels. Although we did not find the perfect monomer, we learned a few rules by examining the properties of a unique monomer: trisacryl (*N*-acryloyl-

2-amino-2-hydroxymethyl-1, 3-propane diol). This monomer had been given to me (in kilo quantities!) by an Italian friend I had met in Paris, Dr. Egisto Boschetti, chief of research at the IBF Industries in Villeneuve La Garenne, France. He used it extensively for making any sort of chromatographic beads, but there was no notion of anybody using it in electrophoresis. In previous work, we had measured its partition coefficient (P) in water/n-propanol, and it turned out to be truly unique: 0.01 [18] (considering that acrylamide has a P = 0.2, it is the most hydrophilic monomer one could ever imagine!). It turned out that, most unfortunately, this superb hydrophilicity was coupled to a ridiculous stability towards both alkaline and acid hydrolysis: this monomer degraded with first order kinetics, so it was an autocatalytic hydrolysis in pH 11 solutions. At the opposite extreme of the scale was located DMA (N,N'-dimethyl acrylamide): much too high hydrophobicity (P = 0.5), but extreme resistance to alkaline hydrolysis (approximately three orders of magnitude more stable) [19]. But what was the reason for this intrinsic instability of Trisacryl? The answer came from molecular modeling: it was found that Trisacryl was constantly forming hydrogen bonds between the -OH groups and the carbonyl of the amido group (bond distances of 0.16–0.17 nm); this activated a mechanism of "N-O acyl transfer," which led to quick degradation of the amido bond even under mild alkaline conditions [20]. Once the disease was diagnosed, the cure was just around the corner: N-acryloyl amino ethoxy ethanol (AAEE), the novel monomer combining high hydrophilicity with extreme hydrolytic stability [21]. As a free monomer, it had 10 times higher resistance to hydrolysis than plain acrylamide, but as a polymer its stability (in 0.1 N NaOH, 70°C) was 500 times higher. The stratagem: the Ω-OH group in the N-substituent was removed enough from the amido bond as to impede formation of hydrogen bonds with the amido carbonyl; if at all, the oxygen in the ethoxy moiety of the N-substituent would act as a preferential partner for H-bond formation with the Ω-OH group. "Nobody is perfect" was the British understatement uttered by the character who fell in love with the cross-dressing Jack Lemmon in the film "Some Like It Hot" and wanted to marry him even after discovering he was a man. It was however, a matter of some concern to us when we learned that our MGM (Most Glorious Monomer, as Bio Rad scientists nicknamed it) was born with a genetic defect: a mysterious and unexpected tendency to autopolymerize at random (being a liquid, it was typically stored as a 1/1 v/v water solution). Upon closer scrutiny, we discovered a unique degradation pathway, called "1-6 H-transfer," by which the C_1 (on the double bond site), by constantly ramming against the C_6 next to the ether oxygen (O_7, which in fact favors the transfer of the hydrogen atom by C_1), produced radicals that more efficiently added to the monomer facilitating autopolymerization and cross-linking. Here, too, a remedy was at hand: acryloyl amino propanol (AAP), in which the ether group on the N-substituent was removed and the chain shortened [22,23]. As a result, this monomer was even more hydrophilic than AAEE (P = 0.1 for AAP vs. P = 0.13 for AAEE) and highly resistant to hydrolysis as compared to free acrylamide (hydrolysis constant of AAP 0.008 L mol^{-1} min^{-1} vs. 0.05 L mol^{-1} min^{-1} for acrylamide, in an alkaline milieu). We could also optimize a novel synthetic route, since all the previous syntheses of monomers barely yielded 15–18% product. In this approach, an essentially pure product was obtained in a single reaction step, with a yield of >99% and an equivalent purity (>99%—not bad for a start!). The synthesis consisted in reacting acryloyl chloride at −40°C in presence of a twofold molar excess of amino propanol and in ethanol (instead of methanol) as a solvent. Other solvents, as well as the use of triethylamine for neutralizing the HCl produced (an additive adopted in all such synthetic processes), were found to give a variety of undesired byproducts. I do not want to give away well-kept industrial secrets, but I can let you imagine a number of other chemicals for which this unique synthetic route has been adopted. The AAP monomer was found to have, for instance, a unique performance in DNA separations when used as a sieving liquid polymer in capillary zone electrophoresis (CZE) [24]. As an extra bonus, poly(AAP) was found to have excellent performance also in CZE at high temperatures (60°C and higher), which are necessary for screening for DNA point mutations in temperature-programmed CZE [25]. Moreover, since we had made a whole series of Ω-hydroxyl, N-substituted acrylamides, it turned out that homo- and co-polymers of such compounds could give theoretical plate values for DNA separations in CZE in excess of 1 million. These were viscous solutions of pure 6% poly(AAP) and poly(AAB) (N-acryloyl amino butanol) as well as of their copolymers [poly(AAP-AAB)] [26].

I want to end this section on a happy note: it is known that acrylamide is a neurotoxin and possibly even (at high doses) a carcinogen; we, the early pioneers who knew nothing about it, have been so intoxicated by the free and careless handling of it as to be unfit for human consumption! Now the situation could have dramatically changed: if we assess toxicity via the ability of these monomers of alkylating free -SH groups in proteins (thus forming a cysteinyl-S-β-propionamide adduct) and evaluate this reaction by high resolution, delayed extraction MALDI-TOF-MS (matrix-assisted laser desorption ionization time of flight mass spectrometry), both in the linear and reflectron mode, this picture will change substantially: at one end of the scale

Table 1 Chemical Structure of Some N-Substituted Acrylamide Monomers

Name	Chemical formula	Mr
N,N-Dimethyl acrylamide (DMA)		99
N-Acryloyl-2-amino-2-hydroxy methyl-1,3-propane (Trisacryl)		175
N-Acryloyl morpholine		141
N-Acryloyl-1-amino-1-deoxy-D-glucitol		235
N-Acryloyl amino ethoxy ethanol		159
N-Acryloyl amino propanol		129
N-Acryloyl amino butanol		143

Table 2 Chemical Structures of Some Cross-linkers

Name	Chemical formula	Mr	Chain length
N,N′-Methylene bisacrylamide (Bis)	[structure]	154	9
Ethylene diacrylate (EDA)	[structure]	170	10
N,N′-(1,2-Dihydroxyethylene) bisacrylamide (DHEBA)	[structure]	200	10
N,N′-Diallyltartardiamide (DATD)	[structure]	228	12
N,N′,N″-Triallyl citric tramide (TACT)	[structure]	309	12–13
Polyethylene glycol diacrylate 200 (PEGDA$_{200}$)	[structure]	214	13
N,N′-Bis-acrylyl-cystamine (BAC)	[structure]	260	14
Polyethylene glycol diacrylate 400 (PEGDA$_{400}$)	[structure]	400	25
N,N′-Bis-acrylyl-piperazine (BAP)	[structure]	194	10

the big offenders remain indeed acrylamide and DMA (highly alkylating, thus highly toxic, agent), whereas at the opposed end (minimally alkylating) one can locate, yes, only AAP [27]. Tables 1 and 2 list a number of the monomers with which we worked.

V. SYNTHESIS OF *N*-SUBSTITUTED, CHARGED ACRYLAMIDO MONOMERS

We now come to another "hot spot" in our research: synthesis of the Immobiline chemicals. My group had started an intense collaboration with LKB Produkter AB in Bromma (a suburb of Stockholm, Sweden) in 1980 for developing this methodology, which had had in fact a long gestation period: it turned out that LKB had patented this idea in 1975, after a meeting I organized in Milano on isoelectric focusing and isotachophoresis in September 1974 [28]. At that meeting LKB had come under heavy criticism from the scientific audience, chastising the company for not divulging the chemical composition of Ampholine, the soluble, carrier ampholytes able to create and maintain a pH gradient in an electric field (of course, there was nothing to divulge, given that the synthetic process could generate up to 5000 different species!). Back home, LKB scientists patented a process for creating "immobilized pH gradients" (IPG) out of a few, well-characterized acrylamido weak acids and bases (i.e., nonamphoteric species with a single protolytic group of known pK and of high purity). IPGs did not lead them anywhere and were shelved with the label "the smears," since well-focused zones were not obtained. We thus joined forces and developed the technique so well that already in 1982 we could present a series of results to an international electrophoresis meeting in Athens, organized by Dimitri Stathakos [29]. We then became so excited by this technique that, in a decade, we developed perhaps up to 90% of all methodologies presently adopted in the field (analytical, preparative, 2-D maps, computer modeling, linear and nonlinear gradients) [30]. What was never clearly stated was the chemistry of these compounds, since LKB maintained a rigid silence on that, something not even observed in a Trappist monastery. Yet spies were alert on all four continents, and secret notes were circulated among the adepts. We decided to break the barrier by trying the synthesis of those phantomlike compounds in order to check our products against the commercial ones. It might be argued that today, with MS and NMR technologies, decoding the structures would be an easy task; nevertheless, in those days these techniques were scarcely available and terribly expensive. Here, too, I needed some help, since my skills as an organic chemist had rusted to a considerable extent. So I hired Marcella Chiari, who had graduated in this field in the laboratory of Prof. Enzo Santaniello at the Faculty of Medicine. Just like C. Gelfi, she had been away from the lab for a number of years, raising her daughter, but she tackled the problem with enthusiasm and dedication. The first couple of papers, on the synthesis of the acidic [31] and basic [32] Immobilines, appeared in a now defunct journal, born out of a furious litigation over the rights of *Electrophoresis*, the official journal of the International Electrophoresis Society, in 1986 at a meeting in London organized by Michael Dunn; I have thus to resurrect these articles here for my audience. So, we gave first and last names (and formulas and synthetic approaches too, of course) not only to the three acidic (pKs 3.6, 4.4, and 4.6) and to the four basic (pKs 6.2, 7.0, 8.5, and 9.3) Immobilines but, while we were at it, also added to the list a spoonful of other chemicals, too: a strong acidic titrant (pK 1.0) with an additional acidic buffer (pK 3.1) and, at the opposite extreme of the pH scale, a strong basic titrant (pK > 12), with an additional basic buffer (pK 10.3). Our appetite then grew exponentially: Why not close the gap between the pK 7.0 and pK 8.5 Immobilines with a new pK 8.05 species [33]? With a bit of magic, the two 6.2 and 7.0 pK commercial products were then transformed into 6.6 and 7.4 pK species by substituting the morpholino ring with a thio-morpholino moiety [34]. We even went as far as to substitute the commercial pK 7.0 species with a new acryloyl histamine buffer, exhibiting higher hydrophilicity and higher stability towards alkaline hydrolysis [35]. On the occasion of the synthesis of yet another "Immobiline" with a pK 6.85 value, we made an extensive study on the structure-stability relationship of all the basic species by monitoring the degradation via CZE: we could thus derive some important parameters linking the stability to alkaline hydrolysis to the flexibility of the N-substituent [36]. At the end of this long rodeo of Immobiline chemistry (we had described a total of 17 chemical compounds), we summarized our data in an article with the ironic title: The Immobiline Family, from "Vacuum" to "Plenum" Chemistry [37], with a clear allusion to the oceanic reunions of the representatives of all socialist republics in the USSR.

Table 3 Acidic and Basic Acrylamido Buffers and Titrants with Associated Formulas, Relative Molecular Masses (Mr), and pK Values

Immobiline	Formula/Mr	pK
2-Acrylamido-2-methylpropane-sulfonic acid	$CH_2\text{=}CH\text{-}CONH\text{-}C(CH_3)_2\text{-}CH_2\text{-}SO_3H$ Mr 207	1.0
N-Acryloylglycine	$CH_2\text{=}CH\text{-}CONH\text{-}CH_2\text{-}COOH$ Mr 129	3.6
4-Acrylamidobutyric acid	$CH_2\text{=}CH\text{-}CONH\text{-}(CH_2)_3\text{-}COOH$ Mr 157	4.6
2-Morpholinoethyl acrylamide	$CH_2\text{=}CH\text{-}CONH\text{-}(CH_2)_2\text{-}N\text{<morpholine, O>}$ Mr 184	6.2
2-Thiomorpholinoethyl acrylamide	$CH_2\text{=}CH\text{-}CONH\text{-}(CH_2)_2\text{-}N\text{<thiomorpholine, S>}$ Mr 200	6.6
3-Morpholinopropyl acrylamide	$CH_2\text{=}CH\text{-}CONH\text{-}(CH_2)_3\text{-}N\text{<morpholine, O>}$ Mr 198	7.0
3-Thiomorpholinopropyl acrylamide	$CH_2\text{=}CH\text{-}CONH\text{-}(CH_2)_3\text{-}N\text{<thiomorpholine, S>}$ Mr 214	7.4
N,N-Dimethylaminoethyl acrylamide	$CH_2\text{=}CH\text{-}CONH\text{-}(CH_2)_2\text{-}N(CH_3)_2$ Mr 142	8.5
N,N-Dimethylaminopropyl acrylamide	$CH_2\text{=}CH\text{-}CONH\text{-}(CH_2)_3\text{-}N(CH_3)_2$ Mr 156	9.3
N,N-Diethylaminopropyl acrylamide	$CH_2\text{=}CH\text{-}CONH\text{-}(CH_2)_3\text{-}N(C_2H_5)_2$ Mr 184	10.3

Table 3 summarizes the data on the most common Immobilines in use today. IPGs represent today the most advanced electrokinetic methodology, with a resolving power unrivaled by all other existing separation techniques, including HPLC. The impact of this technique on the field of proteomics has been tremendous, to the point at which IPGs are quickly replacing conventional IEF as a first dimension in 2-D maps. One of the latest special issues of *Electrophoresis* devoted to this topic [52] will give the readers a view of this expanding field. Figure 3 shows some recent results on 2-D maps in very alkaline pH ranges (pH 8–11): round polypeptide spots, and true equilibrium patterns, are obtained even under these extreme conditions, which represent forbidden ground for conventional IEF. These unique results have been obtained thanks to our efforts to develop this special set of Immobiline chemicals, coupled to our investigations on polyacrylamide matrices: the matrix here is made of poly(AAP), our highly hydrophilic, highly hydrolysis-resistant monomer. In conventional polyacrylamide, the pattern would have been blurred and smeared due to hydrolysis of the matrix during the focusing process. Additionally, it is of interest to know what could possibly be the reactivity of these charged

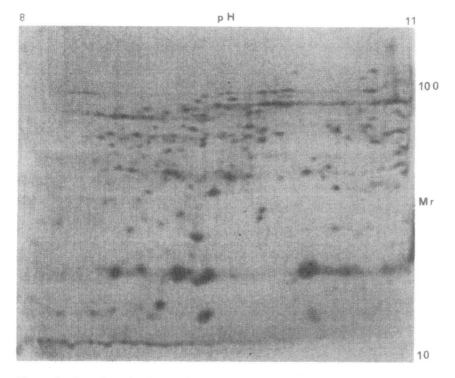

Figure 3 Two-dimensional map of a rat liver total cell lysate. First dimension: IPG pH 8–11 gradient, cast in a poly(*N*-acryloyl amino propanol) 5%T gel. Focusing for a total of 8 hours at 6000 V. Second dimension: SDS-PAGE in a 4–12%T porosity gradient polyacrylamide gel. Silver staining. The pH gradient extremes are marked on the upper side of the figure; the molecular mass (Mr) scale is marked on the right side with values in kilodaltons. (From unpublished experiments with Dr. P. K. Sinha, Berlin.)

monomers towards proteins. In an extensive investigation by MALDI-TOF, we have obtained some unique results, depicted in Fig. 4: it appears that the reactivity is almost linearly correlated with the various pK values: acidic Immobilines alkylate proteins to about 20%, neutral species to about 50%, but the more alkaline compounds alkylate to about 100% and give a multiplicity of adducts, reacting with just about any possible free -SH group available on the protein surface. The remedy to this was found at the inception of IPG methodology: extensive washing of the gels prior to use.

I could relate innumerable stories about this "Immobiline saga," but I will describe here only two additional ones for fear of annoying you to death. One is the synthesis of macroreticulate buffers or Immobiline beads [38]. We made these pearls with the purpose of controlling pH in living systems and applied them to hydroponic cultures. We grew quite successfully lettuce and endive on such beads; the results were so encouraging that I tried to sell the patent to NASA administrators to be taken on board the next manned mission to Mars. Unfortunately, although I was asking a modest price (only $1 million—peanuts compared to the total cost of the trip!), NASA did not buy this idea, which is why I still get around on a red bicycle instead of a Ferrari Testa Rossa and why the mission to Mars never took off! The last spin-off of this idea was the synthesis of soluble, isoelectric, polymeric buffers for pH control in enzymatic reactions occurring in an electric field [39]. Which brings us to the second major topic: the idea

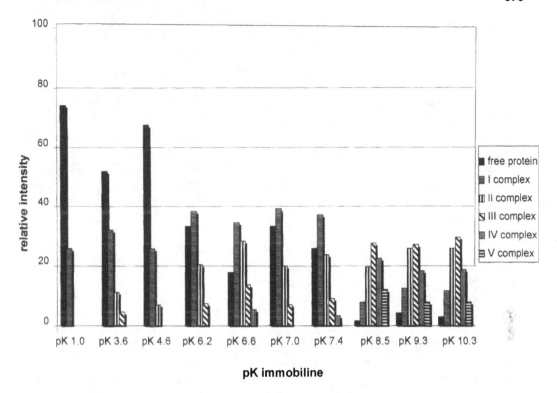

Figure 4 Measured relative intensities of free and complexed bovine α-lactalbumin as a function of Immobiline pK values. Note that the most alkaline species (pK 9.3 and 10.3) show extensive adduct formation (on the free-SH groups) with up to five Immobiline molecules (From unpublished experiments with E. Bordini and M. Hamdan, GlaxoWellcome, Verona.)

of isoelectric, Immobiline membranes to be used in preparative protein purification in multicompartment electrolyzers [40]—yes, the same ones that we now use as "immobilized enzyme reactors" [39].

VI. PHYSICOCHEMICAL STUDIES ON GEL MATRICES

In this last section, I will briefly review the sudden turns taken by our research on polymer chemistry. In 1992 a new post-doc had joined my group, Alessandra Bossi, and I decided to venture with her into some deeper work on the structure of gels. The main, still open, question was the one posed to us by ESA: How could we demonstrate that gravity was interfering with gel structure during the delicate phase of polymerization, just prior to reaching the critical point? It was not an easy question to answer, and we did not have as yet any results from those pestiferous rockets launched from Kiruna. Via Celoria at Città Studi in Milano is a street bound for glory, as we discovered one day when we went to the Department of Physics to ask Prof. Marzio Giglio (a world expert in optics) for help. He had devised an ingenious instrument for studying very small angle laser light scattering, but the problem was that our gels were optically transparent and did not produce any turbid regions able to scatter light. Nevertheless, in the basement of his institute we were able to set up a laser light observation based on shadowing: if, during polymerization, growing chains were to precipitate in the gravitational field, we could observe, by schlieren optics, refractive index gradients sculpted and frozen upon gelling of the matrix. We did, in fact, and passed many afternoons photographing such shadows by dispersing a red laser beam, focused on the gelling cuvette, on a white wall: it was like doing physical chemistry in a red light district! We had become a peripatetic team strolling down Via Celoria every afternoon, Bossi with a basin full of cuvettes and reagents and me with a tripod and my faithful Nikon F2 hanging from my neck. News spread around, and people were waiting for us around the corners asking for a portrait: I had been mistaken for the son of Ansel Adams, the famous Californian photographer, and there was no way to deny it. Our data were published under the curious title [41]: ''Is Gravity on Our Way?'' It was indeed, but we also made a fundamental mistake: we published all the relevant remedies to it, so that perfectly homogeneous gels could also be obtained at sea level. In so doing, we cut our umbilical cord: the ESA headquarters thanked us for our findings and immediately cut the funds that had supported us for a couple of years. That also ended our love affair with ESA.

Well, as the show must go on, we were not discouraged. At the same time, I had been engaged in research on highly porous gels made by growing the polyacrylamide chains in presence of a preformed polymer in solution (notably PEG of any size, typically from 1,000 to 10,000 daltons). We called these gels ''laterally aggregated,'' since we believed that the growing polyacrylamide strings were forced to form bundles of chains, held together by intrachain hydrogen bonding, by the PEG present in solution. These bundles, at the critical point, were permanently locked into the 3-D space by the intervening cross-linking events. As a result, such gels were highly turbid and macroporous, as we discovered by taking electron micrographs: the mean porosity could reach as high as 0.5 μm, i.e., a pore diameter about 2 orders of magnitude greater than that of the corresponding polyacrylamide grown in the absence of PEG [42]. Now we had the ideal sample (turbid) for measurements with the very small angle static light scattering (SALS) device assembled by Prof. Giglio. With that instrument we could in fact give evidence that the large pores were generated by a micro-segregation process caused by the competition between gelation and a phase separation of the polymers solution. The separation occurred via spinodal decomposition, and the size of the pores was determined by the stage at which the decomposition was stopped by the gelation process [43]. The large-scale structure of these laterally aggregated polyacrylamides could also be confirmed by data gathered via small-angle neutron scattering [44]. Another type of macroporous gel that could be obtained was one containing high levels of cross-linker: here, too, the SALS device could confirm the large pore structure (up to a few μm) but it also gave us unique information. Such gels had a quasi-ordered structure, thus driving us into the realm of philosophy: Is there order in the disorder (do not forget that polyacrylamides, by definition, are classified as a ''random meshwork of fibers'') [45]? Back to polymer chemistry, we devised yet another type of medium-porosity gel: a mixed-bed, polyacrylamide-agarose matrix, in which, however, the two polymers were covalently linked rather than being physically trapped into each other [46]. This extensive amount of work was summarized in a number of reviews [47–50], of which the last to be quoted [51] had the glorious title: ''Of Matrices and Men,'' in honor of the famous Steinbeck novel ''Of Mice and Men.'' (I thought at the time that this would be a smart title, but later I learned that an organic chemist had published 10 years earlier a review with the title ''Of Molecules and Men,'' which shows that somewhere, out there, there is always someone smarter than you are!)

VII.   CONCLUSIONS: AN ELECTROPHORETIC STRIPTEASE?

It is hard to know how to end a crazy review like this one. You might think that you took part in a striptease and witnessed the naked electrophoretic body (and soul) of Pier Giorgio. Well, we have a few more garments to discard before ending up fully naked, so I will suggest that you witnessed only partial nudity. But here lies a most intriguing question: Can we really have an electrophoretic striptease? Can we have a force field strong enough to disrobe us of our electrophoretic clothing (let us call it, in electrophoretic terms, our diffuse double layer)? Well, I spent half of my research time trying to answer that question, and the response is no. It would take 100,000 V/cm to reach this total stripping (called the Wien effect), and I suspect that even with the most sophisticated modern techniques we have yet to achieve such a strong force field in practice.

ACKNOWLEDGMENTS

I would like to thank many of my former (and present) collaborators who provided generous help in all the aspects of this research—in particular, Drs. C. Gelfi, A. Bossi, and M. Chiari. Among the foreign guests, I am pleased to mention Dr. E. Simò-Alfonso for his enthusiastic and relentless approach to benchwork. I am also pleased to announce that the research here described was presented to the ICES (International Council of Electrophoresis Societies) meeting, May 25–28, 1999, in Tokyo and that the Japanese Society awarded me with the most coveted Hirai Prize in honor of my friend, the late Prof. Hidematsu Hirai, former president of the society.

REFERENCES

1. C Gelfi, PG Righetti. Polymerization kinetics of polyacrylamide gels. I: Effect of different cross-linkers. Electrophoresis 2:213–219, 1981.
2. C Gelfi, PG Righetti. Polymerization kinetics of polyacrylamide gels. II: Effect of temperature. Electrophoresis 2:220–228, 1981.
3. PG Righetti, C Gelfi, A Bianchi-Bosisio. Polymerization kinetics of polyacrylamide gels. III: Effect of catalys Electrophoresis 2:291–295, 1981.
4. BJ Davis. Polyacrylamide gel electrophoresis. Ann NY Acad Sci 121:404–427, 1964.
5. C Gelfi, P De Besi, A Alloni, PG Righetti, T Lyubimova, VA Briskman. Kinetics of acrylamide photopolymerization as investigated by capillary zone electrophoresis. J Chromatogr 598:277–285, 1992.
6. T Lyubimova, S Caglio, C Gelfi, PG Righetti, T Rabilloud. Photopolymerization of polyacrylamide gels with methylene blue. Electrophoresis 14:40–50, 1993.
7. T Lyubimova, PG Righetti. On the kinetics of photopolymerization: a theoretical study. Electrophoresis 14:191–201, 1993.
8. S Caglio, PG Righetti. On the pH dependence of polymerization efficiency, as investigated by capillary zone electrophoresis. Electrophoresis 14:554–558, 1993.
9. PG Righetti, S Caglio. On the kinetics of monomer incorporation into polyacrylamide gels, as investigated by capillary zone electrophoresis. Electrophoresis 14:573–582, 1993.
10. S Caglio, PG Righetti. On the efficiency of methylene blue versus persulphate catalysis of polyacrylamide gels, as investigated by capillary zone electrophoresis. Electrophoresis 14:997–1003, 1993.
11. S Caglio, M Chiari, PG Righetti. Gel polymerization in detergents: conversion efficiency of methylene blue vs. persulphate catalysis, as investigated by capillary zone electrophoresis. Electrophoresis 15:209–214, 1994.
12. G Artoni, E Gianazza, M Zanoni, C Gelfi, MC Tanzi, C Barozzi, P Ferruti, PG Righetti. Fractionation techniques in a hydro-organic environment. I: Acryloyl-morpholine polymers as a matrix for electrophoresis in hydro-organic solvents. Anal Biochem 137:420–428, 1984.
13. G Vecchio, PG Righetti, M Zanoni, G Artoni, E Gianazza. Fractionation techniques in a hydro-organic environment. II: Sulfolane as a solvent for hydrophobic proteins. Anal Biochem 137:410–419, 1984.

14. PG Righetti, M Chiari, E Casale, C Chiesa, T Jain, R Shorr. HydroLink gel electrophoresis (HLGE). I: Matrix characterization. J Biochem Biophys Methods 19:37–49, 1989.

15. C Gelfi, A Canali, PG Righetti, P Vezzoni, C Smith, m Mellon, T Jain, R Shor. DNA-sequencing in Hydrolink matrices: extension of reading ability to >600 nucleotides. Electrophoresis 11:595–600, 1990.

16. PG Righetti, BCW Brost, RS Snyder. On the limiting pore size of hydrophilic gels for electrophoresis and isoelectric focusing. J Biochem Biophys Methods 4:347–363, 1981.

17. PG Righetti, RS Snyder. Thermally reversible gels in electrophoresis. I: matrix characterization. Theor Applied Electr 1:53–58, 1988.

18. C Gelfi, P De Besi, A Alloni, PG Righetti. Properties of novel acrylamido-monomers as investigated by capillary zone electrophoresis. J Chromatogr 608:333–341, 1992.

19. PG Righetti, M Chiari, M Nesi, S Caglio. Towards new formulations for polyacrylamide matrices, as investigated by capillary zone electrophoresis. J Chromatogr 638:165–178, 1993.

20. S Miertus, PG Righetti, M Chiari. Molecular modelling of acrylamide derivatives: the case of N-acryloylethoxyethanol versus acrylamide and trisacryl. Electrophoresis 15:1104–1111, 1994.

21. M Chiari, C Micheletti, M Nesi, M Fazio, PG Righetti. Towards new formulations for polyacrylamide matrices: N-acryloylamino ethoxyethanol, a novel monomer combining high hydrophilicity with extreme hydrolytic stability. Electrophoresis 15:177–186, 1994.

22. E Simò-Alfonso, C Gelfi, R Sebastiano, A Citterio, PG Righetti. Novel acrylamido monomers with higher hydrophilicity and improved hydrolytic stability. I: Synthetic route and product characterization. Electrophoresis 17:723–731, 1996.

23. E Simò-Alfonso, C Gelfi, R Sebastiano, A Citterio, PG Righetti. Novel acrylamido monomers with higher hydrophilicity and improved hydrolytic stability. II: Properties of N-acryloyl amino propanol. Electrophoresis 17:732–737, 1996.

24. C Gelfi, E Simò-Alfonso, R Sebastiano, A Citterio, PG Righetti. Novel acrylamido monomers with higher hydrophilicity and improved hydrolytic stability. III: DNA separations by capillary electrophoresis in poly(N-acryloyl amino propanol). Electrophoresis 17:738–743, 1996.

25. C Gelfi, M Perego, F Libbra, PG Righetti. Comparison of the behaviour of N-substituted acrylamides and celluloses on double-stranded DNA separations by capillary electrophoresis at 25° and 60°C. Electrophoresis 17:1342–1347, 1996.

26. E Simò-Alfonso, C Gelfi, M Lucisano, PG Righetti. Performance of a series of novel N-substituted acrylamides in capillary electrophoresis of DNA fragments. J Chromatogr A 756:255–262, 1996.

27. E Bordini, H Hamdan, PG Righetti. Matrix-assisted laser desorption/ionization time-of-flight mass spectrometry for monitoring alkylation of β-lactoglobulin B exposed to a series of N-substituted monomers. Rapid Commun Mass Spectrom 13:2209–2215, 1999.

28. PG Righetti, ed. Progress in Isoelectric Focusing and Isotachophoresis. Amsterdam: Elsevier, 1975, pp. 1–395.

29. D Stathakos, ed. Electrophoresis '82. Berlin: de Gruyter, 1983, pp. 61–74, 75–82, 353–361.

30. PG Righetti. Immobilized pH Gradients: Theory and Methodology, Amsterdam: Elsevier, 1990, pp. 1–400.

31. M Chiari, E Casale, E Santaniello, PG Righetti. Synthesis of buffers for generating immobilized pH gradients. I: Acidic acrylamido buffers. Theor Applied Electr 1:99–102, 1989.

32. M Chiari, E Casale, E Santaniello, PG Righetti. Synthesis of buffers for generating immobilized pH gradients. II: Basic acrylamido buffers. Theor Applied Electr 1:103–107, 1989.

33. M Chiari, L Pagani, PG Righetti, T Jain, R Shorr, T Rabilloud. Synthesis of an hydrophilic, pK 8.05 buffer for isoelectric focusing in immobilized pH gradients. J Biochem Biophys Methods 21:165–172, 1990.

34. M Chiari, PG Righetti, P Ferraboschi, T Jain, R Shorr. Synthesis of thiomorpholino buffers for isoelectric focusing in immobilized pH gradients. Electrophoresis 11:617–620, 1990.

35. M Chiari, M Giacomini, C Micheletti, PG Righetti. Synthesis of a new acrylamido buffer (acryloyl-histamine) for isoelectric focusing in Immobilized pH gradients and its analysis by capillary zone electrophoresis. J Chromatogr 558:285–295, 1991.

36. M Chiari, C Ettori, A Manzocchi, PG Righetti. Structure/stability relationship of Immobiline chemicals for isoelectric focusing as monitored by capillary zone electrophoresis. J Chromatogr 548:381–392, 1991.

37. M Chiari, PG Righetti. The Immobiline family: from 'vacuum' to 'plenum' chemistry. Electrophoresis 13:187–191, 1992.

38. PG Righetti, M Chiari, L Crippa. Macroreticulate buffers: a novel approach to pH control in living systems. J Biotechnol 17:169–176, 1991.

39. A Bossi, S Guerrera, PG Righetti. Electrically immobilized enzyme reactors: bioconversion of a charged substrate. Hydrolysis of penicillin G acylase. Biotechnol Bioeng 64:383–391, 1999.

40. P Wenger, M de Zuanni, P Javet, C Gelfi, PG Righetti. Amphoteric, isoelectric Immobiline membranes for preparative isoelectric focusing. J Biochem Biophys Methods 14:29–43, 1987.

41. PG Righetti, A Bossi, M Giglio, A Vailati, T Lyubimova, VA Briskman. Is gravity on our way? The case of polyacrylamide gel polymerization. Electrophoresis 15:1005–1013, 1994.

42. PG Righetti, S Caglio, M Saracchi, S Quaroni. Aggregated' polyacrylamide gels for electrophoresis. Electrophoresis 13:587–595, 1992.

43. D Asnaghi, M Giglio, A Bossi, PG Righetti. Large-scale microsegregation in polyacrylamide gels (spinodal gels). J Chem Phys 102:9763–9769, 1995.

44. A Deriu, F Cavatorta, D Asnaghi, A Bossi, PG Righetti. Large-scale structure of polyacrylamide gels. Physica B 234–236:271–272, 1997.

45. D Asnaghi, M Giglio, A Bossi, PG Righetti. Quasi-ordered structure in highly cross-linked poly (acrylamide) gels. Macromolecules 30:6194–6198, 1997.

46. M Chiari, L D'Alesio, R Consonni, PG Righetti. New types of large-pore polyacrylamide-agarose mixed-bed matrices for DNA electrophoresis: pore size estimation from Ferguson plots of DNA fragments. Electrophoresis 16:1337–1344, 1995.

47. PG Righetti. Macroporous gels: facts and misfacts. J Chromatogr A 698:3–17, 1995.

48. M Chiari, PG Righetti. New types of separation matrices for electrophoresis. Electrophoresis 16:1815–1829, 1995.

49. PG Righetti, C Gelfi. Electrophoresis in gel media: the state of the art. J Chromatogr B 699:63–75, 1996.

50. PG Righetti, C Gelfi. Recent advances in capillary electrophoresis of DNA fragments and PCR products in poly(N-substituted acrylamides). Anal Biochem 244:195–207, 1997.

51. PG Righetti. Of matrices and men. J Biochem Biophys Methods 19:1–20, 1989.

52. MJ Dunn, guest ed. Proteomic review. Electrophoresis 21:1037–1234, 2000.

43

Fishing for Molecules in Biological Soup: From Liquid Chromatography/Electrochemistry to Liquid Chromatography/Mass Spectrometry

Peter T. Kissinger

Purdue University, West Lafayette, Indiana

I. INTRODUCTION

The twentieth century was a lot of fun for me. I was lucky to experience the last 55% of it, by far the best half. In my mid-teens (ca. 1960) I envisioned myself as a naval aviation officer, a writer, a scientist, or an engineer. My limitations as an athlete kept me out of the Naval Academy, narrowing my choices. Mathematics was something I could do, but did not enjoy. That pretty well steered me away from engineering. Chemistry in a liberal arts setting seemed an ideal compromise. By now I've been an officer in a corporation, a pilot, a university scientist, and an engineer designing a number of instruments. I've had plenty of work published. Thus I've been lucky to achieve much of what interested me in 1960. The decade of the 1960s arguably saw the most dramatic advances in science and engineering of any decade before or since. Funding for academic and industrial research was readily available. Solid-state electronics, polymer science, and the first minicomputers transformed many fields.

II. 1960s

My first liquid chromatography and electrochemical experiments were carried out during my high school experience in 1960–1962. These were the result of an interest in electronics stimulated by my father and in organic chemistry stimulated by a neighbor. This was a time when science received a great deal of positive publicity. With some free parts from two uncles (electronics engineers), a bottle of mercury from a high school teacher, and a helical potentiometer donated by Beckman Instruments, I managed to build and use a simple polarograph for a science fair project. Since a recorder was not affordable, a current-voltage curve was plotted with a pencil, taking data from a microammeter. This took nearly an hour to complete. Including the Heyrovsky cell custom-made for me by a Fisher Scientific glass blower, I spent something like

$30 on the entire project and still have this instrument. My first chromatograph was a strip of Whatman No. 1 filter paper, and my first urine sample cost me nothing. I have now worked in these areas for 40 years! My perceived knowledge has continually declined since I was a high school student. I peaked early! Correspondingly, the number of substances detectable in urine and blood has grown dramatically.

In my third and fourth years at Union College (Schenectady, New York) I continued a polarography project for my bachelor's thesis, this time using a commercial instrument to study complexes of tellurium. E. H. Sargent made a unit the size of a refrigerator with a motor-driven potentiometer as a "sweep generator" and a clunky potentiometric recorder with a pen that seemed to either clog up or leak all over my clothes. I also worked at the General Electric R/ D Center in Schenectady, which gave me good exposure to gas chromatography and piqued my enthusiasm for industrial research. Temperature programmable F&M Scientific gas chromatographs were in every synthesis laboratory. Back then I was working on polymers to line fuel tanks for the "supersonic transport" (SST). It became quite clear that many industrial research projects never really get off the ground. There were (and are) fantastic people at GE. A Nobel prize winner, Melvin Calvin, came by to ask me what I was doing. I was, at best, a technician, but the experience at GE reinforced my enthusiasm for science like nothing else could. My mentors at GE were a substantial influence. Their beautiful R&D center convinced me that research was definitely for me. The spirits of Thomas Edison, Charles Steinmetz, and Irving Langmuir haunted that facility.

In 1967, as a first year graduate student with Charles N. Reilley (University of North Carolina), I helped a more senior student from Turkey (Dr. Attila Yildiz) by building thin-layer electrochemical cells based on transparent platinum films vapor deposited on glass. Experiments were carried out primarily for electrochemiluminescence studies in nonaqueous solvents. Our lab reeked of acetonitrile! OSHA inspectors were not available. I became interested in using thin-layer cells in flow streams to generate radical ions for ESR and optical absorbance experiments. As part of qualifying requirements at UNC, six "research proposals" were required by the analytical faculty. I submitted an idea for a thin-layer electrochemical chromatographic detector involving bands of deposited metal films. This was the usual naive graduate student proposal, to satisfy a formal academic requirement. As graduate students, we didn't like writing these proposals. It was clear to all of us that the faculty didn't like writing proposals either. In several cases we helped write their proposals as well, often over the weekend. It seemed worse than an experience at the dentist, but I learned a lot more from writing proposals than I ever have from a dentist. Involving graduate students in writing research proposals continues to strike me as a key part of their education. It sure helped me both in academia and business.

The idea of using electrochemistry as an option for LC detection is obvious and had been tried at least as early as 1946 using a dropping mercury electrode [1]. Dead volumes were huge (many mL), and the polarograph was "low performance" by today's standards. Milligrams could be detected and chromatograms lasted for hours in the 1940s. Nothing of practical significance came of this early work, although it established the concept of using electrochemistry to follow chromatographic profiles.

My 1967 proposal involved a two-dimensional stationary phase film between glass plates, which also held the electrodes (a sort of thin-layer chromatography version of LCEC). I had not yet even heard of what was soon to be called "HPLC." My LC experience consisted of the paper chromatography and TLC I tried in my home workshop in high school, separating chlorophyll and dyes, and the gas chromatography I'd enjoyed at GE.

Not long after, Prof. Reilley saw the "new LC" at DuPont, where he consulted, and at a Pittsburgh Conference. He was impressed and bubbled over with ideas, but the HPLC area did not fit the radical ion work we were doing. Reilley's group had a strong history in gas chromatography, and we had literally hundreds of stationary phases stored in closets in the hall (wasting away because by 1966 Reilley was done with GC completely and had moved in totally different directions, as he was to do several times to the amazement of many who wondered when his next GC paper would appear). I had no liquid chromatography experience as a graduate student. At the same time, great things were happening at other universities, such as in Prof. "Buck" Rogers' group here at Purdue, where graduate students and postdocs were doing early work on "bonded stationary phases."

III. 1970s AND 1980s

In the fall of 1970, I moved to the University of Kansas as a Research Associate with Prof. Ralph Adams, primarily to help keep some electrochemistry going, while Professor Adams learned neuroscience and I broadened my background in biological directions. Just as with Reilley's GC work, Adams had *totally* lost interest in organic electrochemistry in nonaqueous solvents, a field that really made his reputation. A pattern seemed to be developing. During the entire 2 years at Kansas, my "job" was to implement "in vivo voltammetry" in the living brain. All I could think of was why this wouldn't work. At night I learned how to use the newer integrated circuit op amps ("741s"), which seemed to have enormous potential for inexpensive miniaturized electrochemical instrumentation. Several of us kept our hands in the rats by day and the instruments by night. Adams wasn't inspired by this. He wanted "electrodes in the brain." We thought Adams had "electrodes on his brain." Ultimately, a pharmacologist post doc (Ben Hart) and I did the experiment to prove Adams wrong. As is now well known, Adams was right and I had the privilege of co-authoring the very first publication in this area [2]. Nevertheless, I still remained unconvinced. This also was, therefore, my last publication in this engaging field. One thing did happen. I moved from acetonitrile electrochemistry into life science problems, from solution chemistry into rat chemistry.

Prof. Adams, Mark Wightman, Jay Justice, and a number of others have now proven that in vivo electrochemistry is certainly a viable neurochemical technique when judiciously applied. The development of in vivo electrochemistry led Mark Wightman to think in broader terms about the properties of very small electrodes. As a result, the application of electrodes of small dimensions to a variety of chemical problems far exceeds their neuroscience use. This is another example of how good science pursued for one narrow purpose can lead to unpredictable applications in other areas. The lineage from rat brains to superconductor electrochemistry is shorter than many people realize. The use of carbon fiber electrodes as a capillary electrophoresis detector by the Ewing group at Penn State evolved quite logically from the in vivo work, though no one had heard of capillary electrophoresis when carbon fiber electrodes became the rage in the early 1970s.

In Adams's group, we also needed to determine catecholamines and the like in brain parts from sacrificed rats and mice. This was required for various projects in neuropharmacology. We tried "classical" fluorescence literature procedures using gravity fed ion exchange columns, we tried GC, we looked at GCMS, and we quickly (over 18 months) concluded that all existing methods were insufficient, impractical, too time-consuming, and extremely unreliable. It also became apparent that many neuroscientists were using these methods for experiments viewed as critical to our understanding of the human brain. As a physically oriented analytical chemist from the "Reilley school," it became quite evident that there was an opportunity here of real importance! A key group of scientists needed chemical data, and they didn't know how to get it.

Prof. Adams managed to find us an LDC mini-Pump and UV detector, and we explored separations of catecholamines on a DuPont pellicular cation exchange column with the UV detector, as an alternative to the gravity fed polystyrene columns we continued to use to isolate catecholamines in a fluorescence method (which required nearly 2 days to complete). Chuck Refshauge, a graduate student in Adams's lab, did the dirty work. The UV detector was fantastic for 1971, but we would see nothing at the concentrations we needed to explore. One night the whole thing crystallized. All we had to do was stick our electrodes (the miniaturized carbon paste electrodes I was fooling with) into the flow stream from the LC. A cell was already available because we were using it for batch determinations on 5 μL samples of brain tissue homogenates.

Late one night I set up the LC experiment with one of my homemade instruments and detected a few nanograms of norepinephrine and dopamine in the first hour of playing. I didn't believe it. I figured I made up the solutions wrong or that Kansas really was the Land of Oz. I didn't tell Adams. I took the whole thing apart and then tried it again. The same fantastic results appeared the next night! We had immediately beat mass spectrometers (GCMS) costing two orders of magnitude more. And no derivatives were required! I was beginning to believe this was ''real'' and developed the confidence to tell Prof. Adams (we hated to tell him something if it was to fall through, because he got so excited with a good result and so blue if we later reversed our opinion). In any event, those two nights had an enormous impact on me, and I immediately sensed that this was a breakthrough. After all, back then 1 ng was really a small amount! Today we feel ''nano is nothing to e-mail home about.''

Chuck Refshauge refined and used this first liquid chromatography/electrochemistry (LCEC) ''catecholamine analyzer,'' and the whole lab jumped in and began to work on complete assays for rat and mouse brain neurotransmitters using this simple apparatus. Soon Mark Wightman replaced me at KU as a post-doct. Mark and a generation of graduate students became enthused about neuroscience working with these simple instruments. Ivan Mefford made the first attempt to spread the idea to Urban Ungerstedt's laboratory at the Karolinska Institutet in Stockholm. Our later collaboration with Prof. Ungerstedt became an important driving force for many new developments in neuroscience. The original instrument evolved into a business segment, which has included some 25 manufacturers worldwide. Certainly a principal credit for this success is owed to Prof. Adams. He alone recognized the possibilities for continuous monitoring of transmitter substances by electrochemistry in the living brain. He fostered a laboratory atmosphere and tradition of trying the quick and dirty experiment before concluding that ''it won't work.'' In this atmosphere I began the book *Laboratory Techniques in Electroanalytical Chemistry* in 1972; the book was published in 1984. It's a long story. Recently I briefly recounted Prof. Adams's career in a special issue of *Electroanalysis* celebrating his 75th birthday [3]. Several of the earliest references are cited there.

I joined the faculty at Michigan State in 1972 and immediately gave more attention to LCEC, first continuing the neuroscience work, then exploring other applications and improving the instrumentation including the multiple electrode concepts first proposed in 1967. I was very fortunate to be able to attract a team of hardworking and very enthusiastic graduate students. At MSU we operated 7 days a week and managed to make quite rapid progress. The number of reprint requests was overwhelming, getting proposals funded seemed comparatively easy, and phone calls from instrument companies started (Waters and Princeton Applied Research in 1973). They seemed unsure of the scope of the LCEC idea. I consulted with them, but they didn't develop enough interest to act on it. I knew it was hot and decided to run with it. By late 1974 Bioanalytical Systems, Inc. (BAS) had been founded and momentum built exponentially. After all, commercial LC was growing at 50% per year.

While at Michigan State I noticed the publications of Bernard Fleet and Little in the United Kingdom [4] and those of Dennis Johnson at Iowa State [5]. Bernard revitalized the concept of using a dropping mercury electrode in LC detection and also introduced the walljet geometry for this purpose. This resulted in commercial instruments from EDT, Ltd. (United Kingdom) and from Metrohm (Switzerland). In those days Dennis and his group worked with electrodes packed with metal particles and concentrated on inorganic ion detection. It was nearly a decade later before it became clear that "triple pulse" techniques could extend the range of application of amperometric detectors. This imaginative concept applied basic anodic surface chemistry in a very clever way to detect compounds that are not usually detected by direct amperometry (such as carbohydrates and amino acids) [6]. Pulsed amperometric detectors are now available from Dionex, BAS, Princeton Applied Research, and several others.

In May 1975, virtually all of our group moved south to Purdue, the largest U.S. academic center for analytical chemistry at the time. The opportunity was simply too good to resist.

In West Lafayette, we operated BAS in several apartments. Everyone worked part-time, and most worked without any compensation. About 25% of every Purdue paycheck was used to buy electronic parts. The rest paid for rent, hamburgers, and repairs on my old Jeep. While some have said I "invented" electrochemical detection, this is certainly not true—"popularized" would be a more accurate description. Some experiments had been done at about the time I was born. It was really an obvious marriage. Most of science is this way. We build on what came before us. Thin-layer electrochemistry was born in the early 1960s for stationary solutions and modern LC was born in the late 1960s. Putting the two together was by no means a fundamental contribution. Serious attention to carbon electrodes by electrochemists, studies of electrode reactions in nonaqueous solutions, and the development of solid-state–field-effect input operational amplifiers were all quite new in the late 1960s. They all made LCEC feasible! It really was driven by the applications need in neuroscience and later proved useful in many other important areas. The early (1960s) tubular electrode voltammetry work of Blaedel's group at Wisconsin and the flow electrolysis studies of Pungor's group in Hungary could have had a large impact, but they occurred long before LC exploded into a "way of life" for many scientists. Along the way, other workers experimented with a variety of detector cell designs. They did not catch on.

Our "secret" was developing the technique in the face of an application for which there was a great need. I made this a trademark of my academic career, always striving to test chemistry ideas on important problems rather than contrived problems. The students and postdocs at Purdue took a strong interest in clinical neurochemistry, xenobiotic metabolism, and even some environmental and food chemistry. Their enthusiasm on a scientific level enabled many good things to happen. We developed the first clinical chemistry applications [7,8]. We demonstrated enzyme activity assays based on LCEC [9,10]. We provided evidence for the reaction intermediates in acetaminophen [11] and benzene [12] metabolism. We demonstrated column switching LCEC for determination of part per trillion concentrations of the carcinogen benzidine in water [13].

As an analytical chemist, I have always measured success not only by publications, but also by how many laboratories adopted our methodology to their own purposes. I suppose that meeting this psychological need drove me to start an instrument company with the motto "Helping Scientists Do Science." Several review articles from Purdue really helped us get attention and gain momentum [14–17]. Now it is fashionable for university administrators to encourage faculty to start companies that transfer technology to the community of science and create jobs for citizens. What was once viewed as angels (academics) collaborating with the devil (business) is now viewed as a strategic goal of many universities.

The first product from Bioanalytical Systems was an electrochemical detector kit purchased by the FDA San Francisco laboratory in the fall of 1974. Years later we traded them for something better because (1) we were embarrassed by what we had sold them and (2) we wanted it back for our corporate museum. This unit (LC-2) evolved into the LC-2A, many hundreds of which were eventually made in a garage. These detectors sold for an average price of about $300, and complete LC systems built around them cost $2,500–$3,000. These prices were unrealistic to support a growing business. Nevertheless, we were able to prove to the world that amperometric detectors are very useful and that we could sell lots of them. The earliest chromatographs used pellicular ion exchange resin, and chromatograms were unimpressive compared to today's results. Nevertheless, the early LCEC systems often beat LCUV by two or more orders of magnitude in detection limit and excelled in selectivity.

Over the 25 years since the LC-2, this technology has evolved significantly to include such things as multichannel detectors, processor controlled detectors, post-column reaction detectors (including reactions with bromine, photoelectrochemistry, and enzymatic reactors), chemically modified electrodes, "microbore" and "microcolumn" detectors, potential scanning detectors, triple pulse detectors, detectors for capillary electrophoresis, detectors that can be monitored over the Internet, and so on. It has been most interesting to watch these developments, to know just about all the people involved, and to do battle with the increasing number of very worthy competitors.

Since amperometric and coulometric titrations were first automated in the 1950s, no other analytical technique in Faradaic electrochemistry has attracted as many instrument manufacturers as LCEC. While I wish all the others would give up and return to BAS the monopoly we once enjoyed, it has been good to see the field gain wide acceptance. Whether in business, academics, art, or athletics, we all are made better by our competitors. They are our best teachers. Without them we would grow fat and weak. On the other hand, as Winston Churchill said, "Success is never final." It is a very fair statement to suggest that a lot of the promise of LCEC is made less relevant in the face of LC/MSMS. Naturally, I consider mass spectrometry to be gas phase electrochemistry. There are a number of parallels between the two. In fact, there are even virtues in combining them together.

IV. 1990s

The 1990s can be characterized by substantial change in bioanalytical chemistry tools and culture. I list several of these below:

1. Pharmaceutical and biotech companies began to outsource a substantial portion of their bioanalytical work, especially with respect to analytical support for all phases of clinical trials.

2. The "need for speed" accelerated dramatically as the pharmaceutical industry encountered more new molecules from various sources to build up "libraries" of compounds in numbers never before encountered. Coupled with pressure from Wall Street and patent expirations, companies began to move faster to reduce the time from discovery to market. This mandated a "fail-fast" philosophy whereby potential drugs were to be rejected from the "R&D pipeline" as quickly as possible so that more resources could be put into the more promising candidates.

3. After 20 odd years of development, LC/MS finally became a viable tool, reliable enough to operate in an environment where "quantitation" is monitored by "regulatory affairs" and terms like "method validation" and "GLP" became part of the bioanaly-

tical jargon. Coupled with the need for speed, LC/MS was rapidly adopted by drug metabolism departments. By 2000 there were several such departments at major companies that had nearly abandoned such technology as immunoassays, LCUV, LCEC, and LCF for conventional drug substances. In spite of some limitations, LC/MS became the method of choice due to the greater speed of method development (often a week or two vs. a month or 6 for other approaches). This made LC/MS the low-cost approach.

4. In association with points 2 and 3, by 2000 there was a big differential between what could be done in academics vs. industry. Academic institutions don't have the "privilege" of either paying taxes or accounting for depreciation. University research groups operated on a cash basis. As a result, graduate students did not have access to the new LC/MSMS technology and were unable to keep pace with the pressures for speed.

These trends might have been devastating to my career had BAS and Purdue ignored them. Instead, both institutions jumped on the biological mass spectrometry bandwagon even more than before. BAS was able to accelerate growth of a contract laboratory component to satisfy the new pharmaceutical services market. LC/MSMS became a major part of this initiative in both the United States and United Kingdom.

There may be some useful lessons here. It is of much greater benefit for scientists to be associated with natural phenomena (drug metabolism, neuroscience, pharmacokinetics) or a process (bioanalytical chemistry) than a tool (liquid chromatography, spectroscopy, biosensors). Tools tend to evolve more quickly, and it is important to make careful value judgments about whether a new tool is better than an old one. On the other hand, tools don't evolve *that* quickly. They can incubate for years. It took 10 years for LC to be widely accepted, 20 years for LC/MS to move from an unreliable curiosity to a real work horse for bioanalytical chemistry, and 15 years for 96-well titer plates to become the "Lego blocks" of bioanalysis. What's next? We must remain vigilant and ready for change.

V. CONCLUSION

In my view, "separation science" is an oxymoron. I view chromatography as a tool for science and not science itself. Understanding its optimization is clearly an engineering exercise of considerable value. Nevertheless, I have been far more motivated to explore applications. When my career began in the 1960s, chromatography was primarily the province of organic chemists. Gas chromatography and TLC dominated. As we closed out the twentieth century, the variety of chromatographic tools had expanded and the applications covered virtually every field of natural and commercial chemistry from fg to kg, from seconds to hours. The birth of HPLC in the late 1960s stimulated improvements in gas chromatography, TLC, preparative LC, and supercritical fluid chromatography. It opened doors for many detection schemes, including UV, IR, refractive index electrochemistry, mass spectrometry, and NMR. If we take 1970 as a rough benchmark, what could be done before that date can be described as very primitive, although the general principles of chromatography were well understood. A cabal of friends and competitors became the senior statesmen. Some were immersed in theory, some in applications, and some were primarily marketers. I suppose that everyone would have a different list. The following people would appear on many of them: Robert Brownlee, Roland Frei, Calvin Giddings, Georges Guiochon, Richard L. Henry, Csaba Horvath, J.F.K. Huber, James Jorgenson, Jack Kirkland, John Knox, Jim Little, Ronald Majors, Milos Novotny, Fred Regnier, Goran Schill, Lloyd Snyder, and Klaus Unger.

There is a separate but affiliated group that is most associated with GCMS and LCMS. Many were biochemical mass spectroscopists first, and only incidentally chromatographers. As I've grown older, it's become very clear to me that few names and few papers in science are remembered very long. Five years can be an eternity for a scientific publication. The egotistical excitement we have as youth about our latest publication is replaced by the realization that we are all part of a larger team advancing the scientific enterprise as an ecosystem rather than an egosystem. This is a very satisfying revelation to me. It suggests that our work has everlasting life.

REFERENCES

1. W Kemula. Rocz Chem 26:281, 1952.
2. PT Kissinger, JB Hart, RN Adams. Brain Res 55:209, 1973.
3. PT Kissinger. Electroanalysis 11:292, 1998.
4. B Fleet, CJ Little. J Chromatogr Sci 12:747, 1974.
5. LR Taylor, DC Johnson. Anal Chem 46:262, 1974.
6. WR LaCourse. Pulsed Electrochemical Detection in High-Performance Liquid Chromatography. New York: John Wiley & Sons, Inc., 1997.
7. PT Kissinger, RM Riggin, RL Alcorn, L Rau. Biochem Med 13:299, 1975.
8. GC Davis, RE Shoup, PT Kissinger. Anal Chem 53:156, 1981.
9. GC Davis, PT Kissinger. Anal Chem 51:1960, 1979.
10. RE Shoup, GC Davis, PT Kissinger. Anal Chem 52:483, 1980.
11. DJ Miner, PT Kissinger. Biochem Pharm 28:3285, 1979.
12. SM Lunte, PT Kissinger. Chem-Biol Interactions 47:195, 1983.
13. JR Rice, PT Kissinger. Environ Sci Tech 16:263, 1982.
14. PT Kissinger. Anal Chem 49:447A, 1977.
15. PT Kissinger, CS Bruntlett, RE Shoup. Life Sci 28:455, 1981.
16. DA Roston, PT Kissinger. Anal Chem 54:1417A, 1982.
17. PT Kissinger, WR Heineman. Laboratory Techniques in Electroanalytical Chemistry, 2nd ed. New York: Marcel Dekker, 1996.

44

Adventures in Analytical Chemistry/Biochemistry/Biotechnology

Ira S. Krull

Northeastern University, Boston, Massachusetts

I. THE BEGINNINGS

Everything begins as a child—it is all set before we even know it is being set, our interests, leanings, desires, drives, excitement, appreciations, needs, and lifetime careers. It comes from parents, environment, neighborhoods, public school, high school, college, initial employments, college advisors and instructors, and then it all gets reinforced as the career evolves into graduate school, postdoctoral training, initial permanent jobs, sabbatical leaves, and so forth. Some of us are very fortunate, we are born into families that care about education, provide a nurturing environment with books, TV, concerts, trips, more education, and the push to do better and learn more. My own Jewish upbringing in The Bronx, of all places, was filled with education, its importance for future life and work, its importance to succeed in life, its importance to earn a good living, and just the importance of continuous learning. Perhaps because we owned a small book, school supplies, and stationery store, I never went without what was essential for school, whether it was books, bookcovers, pens and pencils, slide rules, triangles, paper, book bag, and so forth. But behind it all was an understanding between child and parents that school came first, that it was the most important thing that could be done with one's childhood.

I recall a wonderful chemistry teacher in high school, Mr. Opochinsky. He was the most exciting teacher that I knew throughout grade and high schools. He loved chemistry, it was his ideal subject matter, and he was such a wonderful teacher, he just drew the students into the subject. He really made the subject so much fun, I can still see him in front of the class, gesticulating, throwing his arms around, walking back and forth in front of the class. When I showed interest in chemistry, he suggested that we do a Westinghouse Science Talent Search project in the main chemistry stockroom laboratory. I was hooked, and at such a young and tender age. I was also doing chemistry at this time at home with a friend, in our bathtub. I shall never forget the day when we succeeded in making gunpowder, and it went off just as mother walked in. What a plume of smoke, what a noise, what a smell, what a scream. Between making gunpowder at home and water repellent material in the high school laboratory, it was clear that science would be my career.

While doing the science talent search project, I was consulting with Professor E. Rochow of Harvard about how to make silicon polymers for water-repellent materials. The project was a success, and forever after, all of the glassware at Taft High School had the most hydrophobic surfaces. What an experience working with someone like Rochow, even if from a distance; he was so helpful, interested, concerned that I succeed. Of course, at that early time/stage, it was not clear which branch of chemistry it was to be: organic, inorganic, analytical, or what?

In The Bronx we had a large choice of colleges, but my preference was for City College of New York in upper Manhattan. This was the 1950s. Though I had applied to several other colleges, I really wanted to go to City. Most of my Jewish friends were going there, it was a short train ride from The Bronx, it had a very good reputation, a good chemistry department and M.S. program, and (best of all) it was totally free. I will never forget learning that I was admitted to CCNY, no restrictions, no course requirements, just show up for classes and work hard. My major was to be chemistry or chemical engineering, because at that young age I did not know which was for me. I should have known better from my high school experiences, but it was still not 100% clear to me what career differences existed between being an engineer and a scientist. So, for my first 2 years I was a ChemE major, and I took some drafting courses, engineering math courses, physics, calculus, and, of course, chemistry. After the first 2 years, I took a short break to work in industry, as a technician for a plastics firm, in order to see the differences firsthand between engineering and science/chemistry. That convinced me to do chemistry and become a scientist; it was just more exciting, fundamental, basic, intellectual, entertaining, fun, just great fun. My undergraduate honors program the last 2 years was with a truly wonderful chemistry professor, H. Meislich. I put in long hours in his labs working on reactions of N-bromosuccinimide (NBS) with hydrocarbons. We did not have a gas chromatograph (GC) and high-performance liquid chromatographs (HPLC)—all we had in those days was paper chromatography, that was our analytical and separations chemistry. We did have a Beckman DU-2 UV spectrophotometer in the department, but it was a bit early for GCs or HPLCs around teaching labs.

My earliest recollection of doing analytical chemistry was doing paper chromatography, for this was even before thin-layer chromatography (TLC). I remember the development tanks, the paper we used, preparation of the paper, iodine staining, and then trying to quantitate how much of each reaction product was formed. Quantitation was mostly guesswork, we did not have reporting integrators, we did not have densitometers, all we could do was cut and weigh or visually estimate the relative amounts of reaction products. I did manage to get honors in chemistry on graduation, but I was still a few years away from any publications.

II. AFTER COLLEGE

I was not 100% sure that I wanted to do immediate graduate studies, and so I again went to work in industry. I was fortunate to get a position at Pfizer's analytical laboratory in Brooklyn, working with a J. McGahren, a very good analytical chemist. My role was as a lab technician in the quality control (QC) division, not yet doing any real analytical research, but developing paper and thin-layer chromatography. This was real industrial chemistry, it was a first-rate analytical set-up, great laboratory facilities, great equipment and modern techniques, but again only paper and thin-layer chromatographies. We ran our own paper chromatograms for amino acid analysis, and then we made our own plates for TLC [1]. I can recall making the silica slurry, applying that to the plastic board containing the glass plates, spreading the slurry evenly, smoothing out the surface, letting the plates dry, heating them in an oven, letting them cool, and then spotting, all manually. After running, they were UV visualized, sometimes with FL if we had the right plates, sometimes with water sprays or iodine staining, but again, quantitation was difficult. However, it was a great leap beyond paper chromatography, it was much more reproducible, much better resolutions, much better quantitation (densitometry), and much easier to read and interpret.

We were really doing research and development into the application of TLC for the pharmaceutical industry, seeing if it held potential for routine qualitative analysis, quality control, and so forth. This was fun work, and I very much enjoyed getting into the laboratory and working at the bench, I was really learning a great deal. Open column liquid chromatography (LC) was just being used in the pharmaceutical industry for analytical purposes, and Pfizer was eager to pursue this approach. There were no HPLC instruments on the market in 1962–63, so everyone ran small glass or plastic columns made with large particle size packings, mainly silica gels. I remember one day packing a glass column with small particle size packings and then trying to develop a separation. I had graduated to doing more basic research, and now they let me investigate more fundamental analytical approaches, like developing LC. It was obvious that if we could pressurize the solvent-delivery system, voila, HPLC. Small particle sized packings take a very long time to elute and develop with just gravity. So, I simply put a rubber stopper on top of the glass column (not too safety conscious), applied a little pressure, and the flow rate increased. Eventually, we came up with other approaches for pressurizing the flow of mobile phase through the column, but we never approached an all-metal, stainless steel solvent delivery system, as others were already pursuing [2,3].

III. GRADUATE STUDIES

Eventually, after making some other career twists and turns, trying my hand at high school teaching in New York City, and not enjoying very much the graduate life in New Haven, I found my way to doing graduate studies, full-time, at New York University, also in the Bronx. I was not yet an analytical chemist, and organic chemistry was my first love. I had done that type of chemistry for my undergrad honors project and was really fascinated by all the possible organic compounds that could be synthesized, understanding reaction mechanisms, photochemistry, free radical chemistry; these were and are tremendously intellectually stimulating and exciting areas of chemistry. Actually, I am a closet organic chemist. Thus, my graduate and some postdoctoral studies involved synthetic organic chemistry, mechanistic organic photochemistry, Hg-sensitized vapor phase photochemistry, and so forth [4–7]. The analytical aspects of my graduate thesis were minimal, other than doing a lot of GC to identify and quantitate photoproducts. However, I got to learn about GC operations and applications and how to isolate volatile fractions from a reaction mixture by preparative GC [4]. We also invented the first GC/MS interface in about 1965–66. Using an old Varian Aerograph preparative GC instrument for fraction collection, those fractions were put on ice and brought down to Columbia University in Manhattan, where they had a Hitachi Model RMU-6D MS. The operator would then take my samples and run them on the MS, we would get the mass spectra, bring those back to The Bronx, interpret them, and voilà, we knew what we had, and we could deduce the exact structures of the photoproducts and if any hydrogen migrations had taken place during the photoreactions [4]. This was really a fantastic approach—preparative VPC fraction collection and off-line MS determinations. This was the first real interface between GC and MS, though it required a subway token in each direction to use. However, the entire experience really impressed me with what analytical chemistry could offer the organic chemist; it was essential to know about analytical approaches if one wanted to do good organic chemistry, or any chemistry, for that matter.

My graduate advisor was a D. Schuster, who had done a Ph.D. with J. Roberts at Cal Tech and a postdoc with H. Zimmerman at Wisconsin. David and I had (still do) a wonderful relationship, and we just hit it off so well that I managed to finish my thesis work in less than 3.5 years. DIS was very supportive, even after a major fire in the labs one night, and did everything possible to encourage his students to succeed. He was a wonderful role model and mentor, and he really tried to guide his students in the best possible career choices. In those days, we did not have formal graduate course work in career development, proposal preparation, ethics in science, or research/career choices. Things other than pure chemistry were learned as an aside. Things have changed, and at Northeastern we have a formal, required course for beginning graduate students in chemistry that formally instructs them in the nuts and bolts of being a scientist, succeeding in grad school and beyond, career choices, proposal writing, publication preparation, and so forth.

It was also at this time that I interacted with a William Pirkle from the University of Illinois. Bill was an organic chemist—he still is, though he does more and more analytical work, with organic requirements. He was doing some photochemistry in the 1960s, and we were working on some very similar classes of compounds [7]. We just published back-to-back, and we both got full credit for our independent work. He was and is a real gentleman, a very nice person. We have interacted ever since, for almost four decades now, and it is always great fun to see Bill at scientific meetings. Bill was starting some work on using nuclear magnetic resonance (NMR) to study the interactions of chiral substrates with enantiomeric analytes in order to determine the nature of those interactions, how NMR could be used to quantitate enantiomeric excess (ee), and how to make use of that information to develop stationary phases, using the same chiral ligands to perform chiral HPLC, which came later. This was the 1960s, but it eventually led to Bill's work in the 1970s and 1980s involving many different and very successful chiral stationary phases, perhaps the most successful work at that time in all of chiral separations. It is interesting how things coincide, accidentally overlap, and we get to interact with people in one way that leads to further interactions.

IV. AFTER GRADUATE SCHOOL

As it evolved, I had several postdocs, because I was not quite sure if it was to be an academic or industrial career. Eventually, if one takes more than a single postdoc, it suggests that they have to take the academic route. Why do two or more postdocs and then go into an industrial career, it is just not necessary or even desirable. Of course, there is no guarantee that the first or even second postdoc will lead to an academic position—one just has to keep trying, if that is what one truly wants in life. My first two postdocs were organic oriented, one involving a search for natural product anticancer drugs (NIH postdoctoral) and the other an industrial postdoc at Union Carbide Research Institute in areas of small molecule photochemistry and flash vacuum pyrolysis [5,6,8,9]. I was exposed in the natural products work to conventional, open column LC, very large columns, very large amounts of samples applied, fraction collection, TLC analysis of each fraction's components, identification of fractions isolated, structure determination by NMR (NOE), MS, IR, UV, and so forth. These two very successful postdocs were then followed by a 3-year stay at The Weizmann Institute of Science in Israel. I was very fortunate in working with some very famous chemists, such as E. Fischer, A. Yogev, A. Patchornik, M. Wilchek, E. Katchalski, E. Gil-Av, and others. This was a fantastic experience—what a wonderful, international science center, what an exciting, learning atmosphere, it was just amazing. The only problem, of course, was that it was in Israel, which may be the land of milk and honey, but there was no money. And I do mean no money, what with very high taxes, very expensive luxury items, a very high cost of living, high inflation, and very low wages. What a combination of negative economic factors. The joke used to be that if one wanted to make a small fortune in Israel, then one had to come there with a large fortune. Unfortunately, it was all true. One could do fantastic science, it was a great atmosphere to do good science, and there were really great scientists all around, from all over the world, but to live as an American in Israel was impossible. Another thing that I learned from my experiences at The Weizmann was that in business dealings, everything must be in writing, everything. If it is not in writing, then it never happens. In the United States, most of the time a professional's word is their bond, but that is not as true elsewhere. If something involves money or your career development, it just must be in writing, no exceptions. It is better to be safe than very sorry. In God, we trust; in all others, get it in writing as soon as possible.

In any event, that period of time exposed me to chiral separations by Feibush and Gil-Av, since the work was going on just down the hall [10,11]. I was able to interact with B. Feibush, the first person to actually resolve enantiomers by chromatography using GC. H. Rubinstein from the University of Lowell was on sabbatical at the Institute, working with Gil-Av and Feibush, and all of us socialized and talked chemistry. At that time (ca. 1970–71) I roomed with a new postdoc of Gil-Av's by the name of V. Schurig, who was getting involved in chiral resolutions by GC [12,13]. We would socialize, go on trips around Israel, visit Jerusalem a great deal, and all along talk chemistry. It was a wonderful life, except for the financial problems. Anyone and everyone involved in chiral resolutions eventually came through The Weizmann, and we got the chance to meet them, discuss chemistry, listen to their seminars, and so forth.

My own research involved the synthesis of strained, small-ring compounds, characterization of these materials, and some of their metalloorganic reactions [14,15]. I got to learn about photoalkylation reactions working in the group of D. Elad, which was an extension of previous photochemical interests. There was no real HPLC ongoing in Israel, but it was just beginning in Gil-Av's group, and F. Mikes was starting to do some chiral separations using chiral stationary phases in HPLC [16]. Feibush is just now a visiting scientist at Northeastern University; we have remained friends over the decades, and he pursues newer packing materials for HPLC, chiral resolutions, and organic synthesis. Then there was the really exciting work of Patchornik in the Biochemistry/Biophysics Department [17,18]. He was synthesizing polymeric reagents and then using these to modify amino acids, peptides, and extend peptide chains in solution. This was a really fascinating area, and it was obvious to me that it had immense applications in all separations areas. My 3 years at The Weizmann were ended, and soon it would be time to start paying taxes, assume Israeli citizenship, go into the Army, and remain forever. Since my financial situation was worse than when I first arrived, it was clear that to remain meant being poor for a very long time, perhaps forever. There is no honor or pride in being smart and poor. There is no pride in having to borrow funds from relatives to attend a father's funeral. It is not impossible to be wealthy or comfortable in Israel, it is just much more difficult than in the States. As a close friend used to say about living in Israel: you have to hold two full-time jobs at the very same time and then cheat the tax collectors.

V. AFTER THE POSTDOCS, THEN WHAT?

My return to the States did not coincide with an academic position, which was really what I was seeking. I was able to get into a basic research institute in Yonkers, New York, doing synthetic organic work for antimalarial drugs [19] and some more analytical work on natural products [20]. The work with natural products involved GC analysis of insect pheromones, especially for bark beetle studies. We developed methods for the separation and detection of the active pheromones from bark beetles, using an electroantennogram with antennae from the bark beetles. This permitted us to detect and isolate active components of the insect sex attractants (pheromones). These could be used to attract insect pests, and the timber industry was very interested in controlling beetle infestations. This was actually the prelude to all of the baited insect traps that later became very popular with fruit/vegetable growers and home gardeners. Isolation of the active pheromones permitted their synthesis, optically active, which led to much better control over beetle populations. This relied heavily on analytical chemistry for the isolation of pheromones and eventually to obtain chemically and optically pure materials [21]. I then wrote my very first review paper dealing with analytical chemistry–chiral resolutions. During the mid-1970s. I was also involved in the analysis for polychlorinated biphenyls (PCBs) in natural environments [22]. This was very analytically oriented, and we were responsible for developing newer HPLC methods for the analysis of these materials, found in the Hudson River. Funding for these programs soon ended, and I was fortunate to secure a research position in analytical chemistry in the Boston area. I had always had an interest in cancer research, and Thermo Electron's (TECO) Analytical Instrument Division in Waltham, Massachusetts, was involved in developing newer analytical methods for environmental carcinogens and pollutants, especially for N-nitrosamines.

VI. THERMO ELECTRON CORPORATION'S ANALYTICAL INSTRUMENT DIVISION (LATE 1970s)

The work at TECO was a combination of basic and applied R&D. This firm was heavy into making novel, selective detectors for GC and HPLC. They had developed a novel thermal energy analyzer (TEA), based on proprietary technology for nitrogen oxide gas measurements. Their detectors used a chemiluminescent reaction of the NO_x gases with ozone to generate a signal, which was then related to the levels of the original NO_x. D. Fine was the manager of this division, my direct supervisor, and a really fine engineer and manager, at times. Funding for this part of TECO came from government grants, contracts, and service work for outside firms. Because the work was externally funded, it was necessary, as at a university, to write proposals, have site visits, and visit government funding agencies. I became heavily involved in such activities. What a great experience it was. My own research involved studying in vivo formation of N-nitrosamines in laboratory animals fed diets rich in amines and nitrite. We realized a large number of publications in my 3 years at TECO [23–26].

The work was fascinating, important, needed to be done, very publishable, very presentable, and good science, by and large. We developed all types of analytical methods from a wide variety of matrices to detect low levels of N-nitroso compounds by both GC and HPLC interfacing. This was my introduction to the general area of novel detectors for chromatography. Some of the work involved animal studies, some environmental pollutant analysis, some newer HPLC protocols for a variety of commercial products that might contain N-nitroso impurities or byproducts, and it all involved modern analytical chemistry. We were not yet using MS, because the TEA was so sensitive and very selective for this class. I attended a large number of scientific meetings on the analysis and chemistry of N-nitroso compounds, I had my first real research group (three or four B.S. chemists), and I was lead author on a large number of high-quality publications and presentations.

Why do I publish so many papers, how could they be any good, there are too many of them? Some people (not all) want to believe that if you publish a lot of papers they could not be all good, perhaps only a few, perhaps none? I have always felt that one of my own goals in life was to contribute to man's knowledge and literature. What makes a scientist's life successful and fulfilling, I believe, is to do the very best science possible and to publish like crazy. One should always strive for the very best quality work. But what is wrong with having quality and quantity? And if the work can benefit mankind and lead to improved analytical methods for life-saving pharmaceuticals and biotech-derived products, all the better. Being of an academic nature, I have also believed that students must publish their own work. It serves science little if work does not get published or presented—who will ever use that work for the advancement of science and technology? I have always derived incredible satisfaction and pride from publishing high-quality work and presenting that work at meetings. There has always been, at least for me, a tremendous feeling of satisfaction and accomplishment to see our work appear in print. What is the purpose of our lives if not to contribute something to mankind that will remain after we are gone. Many say that the purpose of life, if there is any, is to have fun and to make money. Fine, do that and then leave something behind for humanity's sake, contribute something that will help other people, and advance science and health care technology.

The days at TECO were very positive, I worked very hard and learned a great deal about blending business with science and technology and how to manage a research group in the laboratory. It was all very positive and very productive. I worked long hours, but the financial rewards were not forthcoming. In industry, publishing a lot of papers is not the sine qua non, it is not the best world possible for rewards. The goal of any industry is to generate profits for management, stockholders, and employees and to grow the firm. If publications help realize those goals, fine; if not, don't publish. The goal of industry is to make money, that is always the bottom line, it is not necessarily to do the best possible science or technology. The goal is to make a better widget or service that can then be sold for the highest price and profit. If the scientist is not happy with this life, then he does not belong in industry. After almost 3 years at TECO, it was clear that I would never succeed in a corporate life. I just needed to publish and present the very best possible science. It was clearly time to move into academia, which was when an offer came from The Barnett Institute (BI) at Northeastern University (NEU) and B. Karger came into my life.

VII. A PERMANENT ACADEMIC SITUATION, AT LAST (THE 1980s)

I had always strived for a permanent, academic position, but conditions were just not right, whatever the reasons. I had always felt that I wanted a research institute and/or a university, so when Karger appeared with an opening within BI, I was there as soon as possible. He was giving me a chance to succeed, to get into a decent university through his BI, and the rest was up to me. It was a very fair situation, and most of the time it remained very fair and decent. It did not offer tenure or a tenure-track position, but rather a staff scientist, partially soft money situation, which is always less than ideal. NEU had a decent reputation in analytical chemistry, what with R. Giese and P. Vouros there, as well as some good faculty in chemistry, like K. Weiss, A. Halpern, L. Ziegler, and others. This was the late 1970s, and research in the sciences was getting a big push from the administration. My initial position did not require teaching, but I was eager to start college teaching, and so offered to teach AC courses for the forensic chemistry program. We had a very active M.S. program in forensic chemistry, with courses being taken from two departments.

I shall always be immensely grateful to BLK, for without his guidance, advice, and support (!) I would never have prospered in an academic situation and become a full-time, real faculty member of a respectable chemistry department. He was almost always supportive, guiding, advising, rarely superior or haughty. Clearly, he has been a major, major driving force behind all of separations science, and I was lucky to be able to work so long in the office next to his.

But it was always a one-way street, he was very private, very secretive, very close-to-the-chest with thoughts, ideas, advances, new science, and so forth. There was no real collaboration; even when there was collaboration, it was almost always from one direction to the other, not equally. We tried to share some graduate students, but that proved somewhat complicated and problematic. I greatly valued his scientific opinions and suggestions, but it was very difficult to get close to the man. When we eventually ended up working in very similar CE areas, it was clear that we could not remain together. The relationship soured, and I had to be on my own in chemistry from the early 1990s. Others, such as Vouros, have always been able to remain friends and colleagues with BLK, no matter what. I was just not able to master that art of relationships. BLK is truly a marvelous scientist, everyone agrees on that point. Unfortunately, as with many gifted people, especially in science and technology, his social and people skills have not evolved to the same degree of acuity and sharpness. But just to be around him, to hear him speak at colloquia and meetings, to hear his class lectures, to hear his lectures at scientific meetings, and to see his acumen at theses defenses, was just marvelous. I also got to meet a huge number of separations scientists in the earlier stages of their careers, almost all of whom then went on to become quite famous, such as Lindner, Terabe, Tanaka, Cohen, Foret, Guttman, Shimura, and many others.

Many of my initial students came to me via the Forensic Chemistry program. The M.S. in forensic chemistry also had a research component as part of the degree requirements, and so I was able to acquire several competent graduate students the first year of my stay at BI/NEU. Our initial programs involved forensic analysis, such as GC or HPLC analysis for explosives, arson residues, and drugs of abuse. It involved developing more selective and sensitive and sensitive GC/HPLC detection methods for forensic related materials [27–29]. Work on the analysis for N-nitrosoureas was our initial foray into the use of electrochemical detection (EC) in HPLC or LCEC, about 1980. This entire area grew, expanding to a large variety of analytes, usually small molecules, but we also expanded it to much larger analytes, such as peptides and proteins. We also became acquainted with a certain P. Kissinger of Purdue University and Bioanalytical Systems, Inc. (BAS). That started a very long and mutually beneficial and enjoyable relationship, leading to several grants and instrumentation donations from BAS, joint publications and presentations, and a later mini-sabbatical to BAS. Our initial work in LCEC utilized both oxidative and reductive approaches. I can recall when Pete and a graduate student, one K. Bratin, flew in from Purdue to visit my lab and find out why we could not perform reductive LCEC. It took them all of perhaps 5 minutes to decipher our problem, a short piece of Teflon tubing that was allowing oxygen to diffuse from the surrounding air into the mobile phase.

This work eventually expanded into a large number of related research projects, graduate student theses, publications, presentations, and became a major thrust of our initial work in the 1980s. It included the analysis of explosives, drugs of abuse, pharmaceuticals, and eventually biopolymers [30–33]. We pursued the use of postcolumn photolytic reactions, along with others such as Engelhardt, mainly for the analysis of organic nitro compounds, which included most of the commonly used explosives. This then also permitted us to utilize LC-hv-EC, as it became known, for a very large number of other compounds, especially drugs of abuse, pharmaceuticals, and others. These approaches allowed for the use of oxidative LCEC, rather than reductive, for a much larger variety of organic compounds, lowering detection limits, improving specificity using dual electrode ratios at various potentials, and lamp on/lamp off differences.

In the same area of analysis for explosives, we pursued GC-electron capture (ECD) and photoionization detection (PID) approaches, together with J. Driscoll at HNU Systems, Inc. [34,35]. That program involved several M.S. students, as well as some visiting scholars from China, especially Professor Xie and Mr. Ding. We pursued interfacing HPLC with ECD and

PID with support from HNU Systems, which eventually demonstrated a successful interfacing that permitted low levels of detection for PID-sensitive analytes [36]. We were also pursuing newer derivatization approaches for improved detection in GC-ECD and PID, making use of novel pentafluorophenyl reagents for a variety of volatile analytes [37].

Our work on the use of EC detection in HPLC also involved the use of photoelectrochemical methods, which irradiated the working electrode surface of the flow-through EC detector rather than post-column before the EC detection as in LC-hv-EC [38]. This work was really novel in that we were generating short-lived, free radical species from various analytes at the working electrode surface, which were detected using oxidative conditions. Very little had been done in this area for LC-selective detection, and it was different from LC-hv-EC in that intermediates in the photoreactions were being detected, rather than stable, final photoproducts. The work could have been extended to a much larger variety of analytes but for the fact that it is not always possible to continue a project once a successful graduate student has left.

The 1980s saw two other major pursuits in our research group: one was the use of element selective detection in both GC and HPLC for trace metal speciation (D. Bushee), and the other involved the use of solid phase reaction detection in HPLC. The program in GC–microwave-induced plasma (MIP) emission detection (GC-MIP) was a tandem project with interfacing of HPLC with inductively coupled plasma emission detection (ICP) [39]. It involved a collaboration with S. Smith, then Director of Research at IL. We were very interested in developing newer hyphenated techniques that would permit true, trace metal analysis and speciation, especially in environmental samples. S. Jordan was involved in the GC-MIP program, building her own interface, dual valve splitter, MIP, and then demonstrating successful applications of GC-MIP for a variety of interesting analytes and samples [40]. D. Bushee pursued studies in the development of HPLC-ICP interfacing, optimization of operational conditions, and applications to real world samples. We developed methods for improving analyte detection and specificity, such as postcolumn hydride generation in HPLC-HY-ICP, as well as sample preconcentration by liquid-liquid extraction, on-line in HPLC [41]. B. Karcher was involved in the use of postcolumn, on-line liquid-liquid extractions for improved trace metal analysis and speciation, initially using FL detection of metal chelates. Similar GC-MIP work, in other groups (Uden), eventually led to commercialization and the introduction of an element selective detector for GC. The HPLC-ICP work was then replaced by HPLC-ICP-MS in the late 1980s. This latter approach lowered detection limits for all important metal containing species, and it has become the accepted method of analysis today. Limits of detection by ICP-MS are significantly lower than by direct ICP, and this has proven a true boon to chromatographic interfacing.

A good deal of work in the trace metal detection area, especially using HPLC–direct current plasma emission (DCP), was undertaken within the Boston and then Winchester FDA laboratories (where I was a science advisor for 12 years). A large number of publications derived from the work with K. Panaro and W. Childress, involving trace metal analysis and speciation for metals of direct interest to the FDA [42,43]. M. Lookabaugh of the Boston FDA worked with us on the analysis for nitrite and nitrate using LC-hv-EC, as well as other projects in the late 1980s involving CE for insulin.

Another area that assumed significant importance in our research in the 1980s was solid phase derivatizations for separations (GC, LC, CE, CEC). We may have spent an entire decade with various students and collaborators pursuing the utility and applications of solid phase reagents, online and off-line, pre- and postcolumn, with a wide variety of detection methods [44–48]. We synthesized some newer, HPLC detector–oriented reagents, utilizing the best imaginable tags for improved UV, FL, and EC detection of nucleophilic analytes. This allowed us to make off-line or on-line derivatizations of many basic amine-type drugs, often in biofluids, prior to separation of the derivatives, without introducing any unreacted reagents into the sample or mobile phase. We even described multiple tagging with three or more derivatization reagents for the same analyte for the very same, single injection [48]. We also developed size-selective reagents that would permit a discrimination between small and large analytes, permit only the smaller analytes to enter the pores of the reagent, exclude the larger ones, and thus permit a size-selective reaction as part of the HPLC detection process [49–51]. M. Szulc was involved in this part of the polymeric reagent program, and he was perhaps the first to observe a size-selective polymeric reaction for any analyte. This work was extended to the usage of these materials for capillary electrophoresis (CE) applications [52,53].

Another major research program was started in the late 1980s, involving the interfacing of low angle laser light scattering (LALLS) with HPLC and SEC, together with multiple detectors, including UV, differential refractive index (DRI), and eventually, in the mid-1990s, viscometry (VISC) [54–57]. The LALLS work involved combining several detectors on line in SEC and then several forms of HPLC (ion-exchange, reversed-phase, hydrophobic interaction, displacement) in order to detect, quantitate, and determine the molecular weight (MW) of various proteins and antibodies. We eventually determined radius of gyration (Rg) for biopolymers, on-line in HPLC or SEC, together with MW [57]. We could thus separate aggregates of various proteins, measure their absolute MW without any external standard calibrants, and eventually determine their size and shape. This work was of unusual importance and interest for many biotechnology firms, and we then began to collaborate with several of them in order to study their unique proteins or antisense drugs. We had a wonderful collaboration with several individuals within Genentech, especially W. Hancock and W. Wu, to study several of their recombinant proteins, such as human growth hormone, for the presence of variants, aggregates, and other species. Eventually, we were able to study antibody-protein interactions, the formation of complexes, and to then determine the ratio of antibody: antigenic protein in each complex—2:1, 1: 1, and so forth [57]. This then permitted us to demonstrate the presence of antibody-antigen complexes, their stability, relative amounts present in any given sample, the nature of these complexes, and whether or not any excess, active antibody or antigen was still present in that sample [58].

Of course, as the above research was progressing well and we were realizing a large number of peer-reviewed publications and invited lectures at scientific meetings, I became a member of the Chemistry Department at NEU. That gave me regular faculty status, teaching duties, committee and administrative duties, and all of the usual responsibilities and rights of any other full-time, faculty member at a major university. NEU was becoming more and more research oriented, and recognition and promotions were being based more and more on research productivity, graduate students produced, outside funding with overhead, international recognition, national grants awarded, and so forth. Eventually I became a tenured faculty member in the early 1990s, and thus was able to remain at NEU to pursue further research areas.

VIII. THE 1990s

As with everything else in life, to every season there is a turn, everything changes, and so too with scientific research. We had done much in the areas of small molecule analysis, and we were getting somewhat bored with such approaches and studies. It was clearly time to branch out into the area of biopolymers, especially proteins, antibodies, peptides, and fusion proteins. Hence, the above-mentioned work/studies into SEC/HPLC-LALLS/DRI/UV/VISC for biopolymer characterization, an area that remains of interest, especially if it could be used to demonstrate enzyme, antibody, or antigen activity just by measuring the radius of gyration, rather than by the use of bioassays, cell-based assays, immunoassays, and so forth.

Another HPLC area that we pursued in the late 1990s involved immunorecognition in HPLC, postcolumn immunodetection in HPLC, affinity HPLC, and similar areas, again for biopolymer separations [59,60]. We developed immunoadsorbent cartridges to isolate antigens from biofluids, preconcentrate before HPLC separations, usually by reversed-phase methods, and to then detect these after the HPLC separations by conventional UV, FL, or immunodetection methods. All of this work involved working with antibodies and their antigens, isolation from biofluids, purification, immobilization, affinity recognition, and then detection. Some of this work was done in collaboration with The Upjohn Pharmaceutical Company of Kalamazoo, Michigan, using their antibodies and various antigens, such as bovine growth hormone (BGH) or BGH-releasing factor. We also started to use some CE methods for affinity recognition of BGH and its antibody, leading to involvement in affinity CE/cIEF methods [61]. This work gave us experience with and exposure to working with antibodies, their handling, stabilization, concentration, storage, analysis, activity determinations, and so forth. We were able to improve on immunochromatographic-based assays, to lower detection limits, to use ID postcolumn, and even to isolate an antigenic species on an immobilized antibody column, preconcentrate, release, tag, and then determine by HPLC or other methods.

So much had already been done in HPLC areas that we felt perhaps it was time to expand into CE areas/studies. It was now apparent that LC-MS was going to become the dominant player in all future HPLC (perhaps also CE) work. We needed to collaborate with nearby biotechnology firms that had such instrumentation, also with instrumentation firms, such as Waters, especially when they had MS divisions and instrumentation available for such collaborations. In the 1990s, we made a push in CE areas, especially interfacing of CE with MS, and those areas still dominate our thoughts.

Perhaps our very first work in these protein-based CE areas involved some studies in 1989 at the Winchester FDA laboratories with M. Lookabaugh [62]. We had already been involved in CE for small molecules, some work that Szulc had done with tagging of small analytes in the carousel vial. We now became more involved with larger molecules, mainly protein based, and started a program with J. Mazzeo in capillary isoelectric focusing (cIEF) areas and applications [63–65]. After several forays into LCEC and LC-hv-EC areas, he then focused on cIEF alone. We started in cIEF by using uncoated capillaries, based on some prior work by Hjerten and Thormann. Rather than use a two-step method, which required physical displacement of the focused proteins, we developed a method that could use a one-step approach, combining focusing with displacement, all in a single capillary without any operator intervention. Eventually, such methods were also interfaced with electrospray ionization (ESI)–MS. These cIEF conditions came together, and we had a very nice, simple, and very straightforward way to perform cIEF. There were some very nice presentations at various meetings on this work, such as a Frederick Conference, where we presented these results alongside one F. Regnier. Fred should be in heaven in the end—he is a true gentleman. Knowing that we had done very similar work, he held off showing his own results so that we could first show ours. Anyhow, we were then able to publish a number of papers, to this very day, using either the one-step or the newer, two-step method, for a wide variety of proteins, peptides, antibodies, and now even fusion proteins [66–68]. These studies continue in the areas of proteomics with S. Kazmi. We are also working with Biogen Corporation to apply these optimized cIEF methods to various antibodies and fusion proteins, eventually hoping to interface the cIEF separations with ESI-MS for characterization of protein components.

In the late 1990s we became interested in how to tag proteins and peptides fully, homogeneously, thus leading to a single, fully tagged product [69–71]. We had long realized that one of the major drawbacks in the trace analysis for proteins in all of analytical chemistry has to do with their poor UV-Vis and FL properties. Chemists had tried to devise methods of fully tagging proteins, of any size, leading to a single product, that might still have affinity properties. None of the existing literature ever demonstrated a general approach for fully tagging proteins. We considered that this was a significant area, since all of the CE of proteins suffered from unwanted protein-wall interactions, leading to tailing peaks, poor peak shape, poor resolutions, poor peak capacities, poor efficiencies, and also poor sensitivities. This was something that we really wanted to solve, because it could possibly improve all HPLC and HPCE of proteins, if a general method could be devised that would lead to a single, fully tagged product.

I recall giving a talk on some of this work at a meeting (HPLC 98) when one student in the audience asked me a very interesting question: How did you know that this approach would work? Well, of course, I did not know (!) it would work. I hoped and thought that it would work, but I never really knew for certain until we did the actual work. There is something that I used to call chemical intuition, where, based on everything that we know in chemistry and science, we can extrapolate as to whether or not a chemical reaction or approach might work. I think that is how we eventually, hopefully, know which experiments to perform and which not to try. We, hopefully, can draw on our chemical intuition or background of knowledge and use that to suggest to us (perhaps subliminally) which reactions or projects will work. In the beginning our approach did not work, but one should never be discouraged by negative results—they are what push us to find the solutions and, eventually, positive results. If your experiments are always working, then something is wrong with your entire scientific approach. One must have failures at times, or one is not doing work that is challenging enough.

We thought that the reason nobody had yet devised a way to fully tag peptides and proteins was that they had never used the correct ratio of reagent to peptide. We thought, let us start with smaller peptides, then we can graduate to much larger proteins, those are much harder. With smaller peptides, like insulin or lysozyme, if one uses only a stoichiometric ratio of reagent to tagable amino groups, one never gets a fully tagged, single product. One really has to titrate the peptide, use more and more of the reagent, until indeed only a single product results. One also has to choose the right tagging reagent, one that will eventually hydrolyze and disappear from the reaction solution, so that it does not interfere with the final HPLC or HPCE analysis. If this approach is used, then one will always get a fully tagged, single product, often active towards its antibody. That product will have ideal properties in HPCE and HPLC, but the size of the peptide is a limiting factor. The CE properties are especially enhanced, beautiful peak shapes, lovely efficiencies, low detectabilities, large peak capacities, and high plate counts. There is no longer any interaction with the bare capillary walls, because the tagged peptides are too hydrophobic, again reagent dependent.

What about larger and larger proteins? Well, the graduate student here is a very good scientist, H. J. Liu. He also spent 20 hours per week at Waters Corporation, working with S. Cohen on projects related to his thesis. We tried tagging much larger proteins in the very same way as for smaller peptides. This did not work, we always got mixtures, incompletely tagged products by HPLC and MALDI-TOFMS (matrix-assisted, laser desorption ionization–time of flight mass spectrometry). The problem must have been due to the much greater complexity of large proteins, such as bovine serum albumin (BSA) or α-chymotrypsingen A. The tertiary nature of such proteins must be hiding reactive amino groups within a sphere of protective amino acids. What was necessary was to first denature, unravel, rupture disulfide bonds and stretch out the protein so all primary amino groups would be accessible to the attacking reagent, in this case, 6-aminoquinolyl-*N*-hydroxysuccinimidyl carbamate (6-AQC). Sure enough, once we reduced and alkylated the freed thiols groups with the usual alkylating reagents and then tagged that product with an excess of 6-AQC, a single, fully tagged product resulted. This product, of course, again had fantastic CE properties, excellent peak shapes, efficiencies, limits of detection, plate counts, and so forth. They behaved as one would want proteins to behave in CZE (capillary zone electrophoresis), and they showed a single peak in all instances, whatever the CZE conditions [74]. The only problem here was that when large proteins become fully tagged with very hydrophobic tags, they do not behave well in reversed-phase HPLC, because they tend to stick to the C-18 supports and peak shapes are lost. They are single peaks, but very broad, very diffuse, unusable in HPLC modes. Solve one problem and another arises, for that is the nature of life. There is no one, universal solution to all problems in chemistry or life; one has to solve the most pressing problems first, and then either live with those that remain or find another solution to the unresolved problems.

IX. THE 2000s

Well, where do we go next? We became quite interested in capillary electrochromatography (CEC) areas in the late 1990s during a mini-sabbatical at Waters Corporation [72,73]. This could be a very exciting area of analytical chemistry and separations science. It is essential, in all of science, that we only publish results that are fully reproducible—not partly reproducible, but fully, all of the time. We are now pursuing some work in packed bed CEC, isocratic and gradient, for peptide, protein, and antibody analyses, using conventional UV/PDA detection modes. It appears that CEC may well offer some advantages over conventional HPLC. However, there are other limitations, such as sensitivity and loadability, detectability, stability of the packed beds, and so forth. These problems may well be solved in the future, and CEC could well take its place as a viable, reliable, reproducible and practical (!) analytical method. However, it may never replace conventional HPLC, which is entrenched all over the world, has a commercial market over $ 2.5 billion per year, and is used in almost all pharmaceutical, environmental, chemical, and biotech laboratories worldwide.

In our own writings so far, we feel that CEC is perhaps a niche technique, which may well find applications for certain types of analyses where it is indeed far superior to HPLC as practiced today. To expect it to replace HPLC for pharmaceutical analysis, for example, may be expecting too much. We are also quite interested in evaluating capillary LC (CLC) methods, especially for peptide mapping, protein analysis, and other biopolymer classes, with LIF and MS detection, perhaps using our now perfected, optimized, tagging approaches for such analytes (H.-J. Liu et al., unpublished) [75]. Commercial CLC-LIF/MS instrumentation is on the market, and it may well offer opportunities over conventional, analytical-scale HPLC, such as lowered limits of detection, improved peak shapes, improved sensitivity, fewer sample requirements, and so forth. Only time will tell which of these capillary methods, if any, will indeed be used on a day-to-day basis 10 or 20 years from today. MS will continue to play a larger and larger role in all of separations science—there are just too many things that it can do better or best. Hyphenated techniques are here to stay, at least for the forseeable future. Even ion-trap MS (ITMS) may not avoid an initial separation step requiring LC or CE or CEC? Multidimensional separations, involving several separation steps, such as LC-CE or LC-LC or CE-CE or CEC-CEC, together with ITMS may well provide the ultimate, today, in peak capacity realization for really complex mixtures, as found in proteomics problems.

ACKNOWLEDGMENTS

I acknowledge H. Issaq again for inviting me to write this semi-autobiographical sketch of our contributions to analytical chemistry over the past few decades. I would also like to acknowledge lots of people, but that would require another five pages of text, which the generous editor will never permit, I have already used too many pages. There are any number of graduate and undergraduate students, postdocs, and colleagues, all over the world, who have interacted with us over the past few decades. I am, of course, forever in their debt, as is the case with our numerous financial supporters and funding organizations/companies. We have been truly fortunate to have been able to collaborate with so many industrial firms over the past 20–25 years. There have been some truly great collaborations, others less so, but most of them have been very satisfying and mutually beneficial. To be able to work with industrial colleagues in helping a graduate student to succeed is indeed a worthwhile collaboration, especially when the science that comes out of it is also worthwhile and of high quality. I also acknowledge my university, which has given me the chance to succeed, to realize whatever I was capable of realizing, despite my inherent limitations (man must know his limitations, as Dirty Harry would say). We are all imperfect, flawed, just human beings, none of us is perfect, I could have done more and better work too. But I did the most that I could, with a fire burning and driving me to succeed and persevere. We all make mistakes, hopefully we learn from those mistakes and become better scientists and people in the end. We also get tired, at times lose that burning drive, slow down, it is the nature of life. The end is coming, I can feel it in my bones, as never before. We can fight this only so long, then it wins and we withdraw, just like Cyrano.

REFERENCES

1. LS Ettre, VR Meyer. LC/GC Magazine 18(7):670, 2000.
2. JJ Kirkland, ed. Modern Practice of Liquid Chromatography. New York: Wiley-Interscience Publishers, 1970.
3. LR Snyder, JJ Kirkland, Introduction to Modern Liquid Chromatography, John Wiley & Sons, New York, 1974.
4. DI Schuster, IS Krull. J Mol Photochem 1:197, 1969.
5. E Hedaya, IS Krull, RD Miller, MD Kent, PF D'Angelo, P Schissel. J Am Chem Soc 91:6880, 1969.
6. IS Krull, DR Arnold. Tet Lett 1247, 1969.
7. IS Krull, DI Schuster. Tet Lett 135, 1968.
8. SM Kupchan, TJ Giacobbe, IS Krull. Tet Lett 2859, 1970.
9. SM Kupchan, TJ Giacobbe, IS Krull, AM Thomas, MA Eakin, DC Fessler. J Org Chem 35:3539, 1970.
10. (a) E Gil-Av, B Feibush, R Charles. Gas Chromatography 1966, 6th International Symposium on GC and Related Techniques, Rome, September 20, 1966, Institute of Petroleum, London, 1967, p. 27; (b) E Gil-Av, B Feibush, R Charles-Siegler. Tet Lett 1009, 1966; (c) B Feibush, E Gil-Av. J Gas Chromatogr 5:257, 1967.
11. B Feibush, A Balan, B Altman, E Gil-Av. J Chem Soc Perkin Trans II:1230, 1979.
12. V Schurig, E Gil-Av. Israel J Chem 15:96, 1976/77.
13. V Schurig, W Burkle, A Zlatkis, C Poole. Naturwissenschaft 66:423, 1979.
14. IS Krull, A Mandelbaum. Proc Israeli Chem Soc 46, 1972.
15. IS Krull. J Organometal Chem 57:373, 1973.
16. F Mikes, G Boshart, E Gil-Av. J Chromatogr 122:205, 1976.
17. A Patchornik, MA Kraus. In: Encyclopedia of Polymer Science & Technology, Suppl. 1. New York: J. Wiley & Sons, 1976, p. 468.
18. M Kraus, A Patchornik. Chemtech 118, 1979.

19. H Gershon, R Parmegiani, VR Gianassio, IS Krull. J Pharm Sci 64:1855, 1975.
20. JAA Renwick, PR Hughes, IS Krull. Science 191:199, 1976.
21. IS Krull. Advances Chromatogr 16:175, 1978.
22. IS Krull. Residue Rev 66:185, 1977.
23. ST Fan, IS Krull, RD Ross, MH Wolf, DH Fine. In: EA Walker, M Castegnaro, L Griciute, RE Lyle, Environmental Aspects of N-nitroso Compounds. Lyon: International Agency for Research on Cancer, eds. IARC Scientific Publication No. 19, 1978, p. 3.
24. IS Krull, TY Fan, M Wolf, R Ross, DH Fine. In: G Hawk, ed. Liquid Chromatography I. Biological and Biomedical Applications of Liquid Chromatography. New York: Marcel Dekker, 1979, p. 443.
25. IS Krull, G Edwards, M Wolf, DH Fine. In: J-P Anselme, ed. N-nitrosamines. Washington, DC: American Chemical Society, 1979, p. 175.
26. Z Iqbal, SS Epstein, IS Krull, EU Goff, K Mills, DH Fine. In: EA Walker, L Griciute, M Castegnaro, M Borzsonyi, eds. N-nitroso Compounds: Analysis, Formation, and Occurrence, Lyon: International Agency for Research on Cancer, IARC Scientific Publications No. 31, 1980, p. 169.
27. IS Krull, X-D Ding, S Braverman, C Selavka, F Hochberg, LA Sternson. J Chrom Sci 21:166, 1983.
28. IS Krull, X-D Ding, C Selavka, F Hochberg. In: E Reid and ID Wilson, eds. Methodological Surveys in Biochemistry and Analysis. London: Plenum Press, 1984, p. 369.
29. IS Krull. In: I Lurie JD Wittwer, eds. HPLC in Forensic Chemistry. New York: Marcel Dekker, 1983.
30. IS Krull, C Selavka, X-D Ding, K Bratin, G Forcier. J Forensic Sci 29(2):449, 1984.
31. (a) IS Krull, X-D Ding, C Selavka, K Bratin, G Forcier. In: E Reid, ID Wilson, eds. Methodological Surveys in Biochemistry and Analysis. London: Plenum Press, 1984, p. 365; (b) L Dou, A Holmberg, IS Krull. Anal Biochem 197:377, 1991; (c) L Dou, IS Krull. J Pharm Biomed Anal 8(6):493, 1990.
32. CM Selavka, IS Krull, IS Lurie. J Chrom Sci 23(11):499, 1985.
33. IS Krull, CM Selavka, W Jacobs, C Duda. J Liquid Chrom 8(15):2845, 1985.
34. IS Krull, M Swartz, JN Driscoll. Multiple detection in gas chromatography. In: JC Giddings, E Grushka, J Cazes, PR Brown, eds. Advances in Chromatography, Vol. 24. New York: Marcel Dekker, 1984, pp. 247–316.
35. IS Krull, M Swartz, K-H Xie, JN Driscoll. In Proceedings of the International Symposium on Analysis and Detection of Explosives, FBI Academy, Quantico, Virginia, March, 1983. Washington, DC: U.S. Government Printing Office, 1984, p. 107.
36. JN Driscoll, DM Conron, P Ferioli, IS Krull, K-H Xie. J Chromatogr 302:43, 1984.
37. IS Krull, M Swartz, JN Driscoll. Anal Lett 18(B20):2619, 1985.
38. WR LaCourse, IS Krull. Trends Anal Chem 4(5):118, 1985.
39. D Bushee, IS Krull, RN Savage, SB Smith, Jr. J Liquid Chrom 5:563, 1982.
40. IS Krull, SW Jordan, S Kahl, SB Smith, Jr. J Chrom Sci 20:489, 1982.
41. BD Karcher, IS Krull. J Chrom Sci 25(10):472, 1987.
42. IS Krull, KW Panaro. Appl Spec 39(6):960, 1985.
43. IS Krull, WL Childress. In: IS Krull, ed. Trace Metal Analysis and Speciation. Amsterdam: Elsevier Science Publishers, 1991, pp. 239–288.
44. ST Colgan, IS Krull, U Neue, A Newhart, C Dorschel, C Stacey, B Bidlingmeyer. J Chromatogr 333(2):349, 1985.
45. ST Colgan, IS Krull. In: IS Krull, ed. Reaction Detection in Liquid Chromatography. New York: Marcel Dekker, 1986, pp. 227–258.
46. ST Colgan, IS Krull, C Dorschel, B Bidlingmeyer. Anal Chem 58:2366, 1986.
47. T-Y Chou, C-X Gao, N Grinberg, IS Krull. Anal Chem 61(14):1548, 1989.
48. C-X Gao, D Schmalzing, IS Krull. Biomed Chromatogr 5:23, 1991.
49. IS Krull, AJ Bourque, M Szulc, F-X Zhou, B Feibush. Proceedings of the International Congress on Analytical Sciences 91, Chiba, Japan, August 27, 1991. Anal Sci (Japan) 7:1535, 1991.
50. IS Krull, F-X Zhou, AJ Bourque, M Szulc, J Yu, R Strong. J Chromatogr B 659(1/2):19, 1994.
51. ME Szulc, P Swett, LS Krull. Biomed Chromatogr 11(3):207, 1997.
52. ME Szulc, IS Krull. J Chromatogr Rev 659(2):231, 1994.

53. IS Krull, ME Szulc, J Dai. Chemical Derivatizations for Improved Detection in Capillary Electrophoresis, Thermo Bioanalysis Corporation, San Jose, CA, 1997.
54. IS Krull, H Stuting, S Kryzsko. J Chromatogr 442:29, 1988.
55. IS Krull, R Mhatre, HH Stuting. Trends Anal Chem 8(7):260, 1989.
56. IS Krull, R Mhatre, J Cunniff. LC/GC Magazine 13:30, 1995.
57. R-L Qian, R Mhatre, IS Krull. J Chromatogr A 787:101, 1997.
58. LC Santora, IS Krull, K Grant. Anal Biochem 275:98, 1999.
59. IS Krull, B-Y Cho, R Strong, M Vanderlaan. LC/GC Magazine 15(7):620, 1997.
60. B-Y Cho, R Strong, H Zou, DH Fisher, J Nappier, IS Krull. J Chromatogr 743:181, 1996.
61. R Strong, H-J Liu, BY Cho, IS Krull. J Liquid Chrom Rel Technol 23(12):1775–1807, 2000.
62. M Lookabaugh, M Biswas, IS Krull. J Chromatogr 549:357, 1991.
63. J Mazzeo, IS Krull. BioTechniques 10(5):638, 1991.
64. J Mazzeo, IS Krull. Anal Chem 63:2852, 1991.
65. JR Mazzeo, JA Martineau, IS Krull. Methods, A Companion to Methods Enzymol 4:205, 1992.
66. B-Y Cho, R Strong, G Fate, IS Krull. J Chromatogr 697:163, 1997.
67. H Zhong, IS Krull, KC Chan, HJ Issaq. J Liquid Chromatogr & Related Techniques (submitted for publication, 2001).
68. S Kazmi, IS Krull. J Chromatogr (submitted for publication, 2001).
69. HJ Liu, RE Strong, IS Krull, SA Cohen. Anal Biochem 2001 (in press).
70. HJ Liu, B-Y Cho, R Strong, IS Krull, S Cohen, KC Chan, HJ Issaq. Anal Chim Acta 400:181, 1999.
71. IS Krull, R Strong, Z Sosic, B-Y Cho, S Beale, S Cohen. J Chromatogr B Biomed Applics 699:173, 1997.
72. R Stevenson, K Mistry, IS Krull. Am Lab 16A, 1998.
73. IS Krull, R Stevenson, K Mistry, ME Swartz. Capillary Electrochromatography (CEC) and Pressurized Flow CEC (PEC) or Electro-HPLC, An Introduction. New York: HNB Publishers, 2000.
74. HJ Liu, BY Cho, IS Krull, SA Cohen. J Chromatogr 2001 (in press).
75. HJ Liu, IS Krull, J Holyoke, SA Cohen. Anal Biochem 2001 (in press).

45

A Czech's Life in Separation Science

Zdenek Deyl

Institute of Physiology, Academy of Sciences of the Czech Republic, and Institute of Chemical Technology, Prague, Czech Republic

I. HOW I BECAME INVOLVED IN SEPARATION SCIENCE

I entered the world of separation science, chromatography, and electrophoresis in the late fifties. I was then ending my studies at the Prague Institute of Chemical Technology where the experimental work I prepared for my thesis was titled "Chromatographic Analysis of Waste Waters from a TNT-Producing Plant" (as a matter of fact this was the same plant that much later became the world-famous producer of Semtex explosives) [1]. At that time the practised separation technology was predominantly paper chromatography and indeed it flourished in Prague. Though the first monograph on chromatography that came into my hands was that of Lederer and Lederer [2], the first monograph exclusively on paper chromatography was published in Prague by Ivo Hais and Karel Macek (with whom we became close friends in the late 1960s) [3]. The reason was simple: a sufficiently large jar (frequently those used for selling pickled cucumbers which could be obtained for free from any grocery store), solvents, some reagents, and a sheet of Whatman paper (at the very beginning also regular filter paper treated by mysterious procedures) was all that what was needed to do research in this field (admittedly also some brains). Soon a number of printed bibliographies on this subject [4–7] became available to me.

II. THE AREAS OF SEPARATION SCIENCE TO WHICH WE CONTRIBUTED (DID WE REALLY?)

Our approach to separation science reflected the fact that our laboratory has been a part of the Institute of Physiology of the Czech (then Czechoslovak) Academy of Science. It has always been my belief that in such a situation one has to follow two paths in parallel: topical research, which in our case was physiology of the extracellular matrix (more precisely of collagens), and methodology, which in the early days was chromatography replaced later step-by-step by electromigration methods [8–12]. There was a common feature to all the separation procedures, whether chromatographic or electromigration—they lasted too long. It is not surprising that already in the early 1960s, when separation science was dominated (as mentioned) by planar techniques, we realized that speeding up the separation process would be of considerable importance [13]. This idea was realized by constructing a device for centrifugal chromatography which offered similar possibilities as the standard separation procedures except at a much faster rate (Fig. 1): the separations took 5–25 minutes on average, depending on the composition of the mobile phase (as compared to several hours in the standard version). Due to the economical situation in the part of Europe where we lived, nobody seemed interested in making this device commercially available. With the rise of thin layer procedures a circular sheet of Whatman

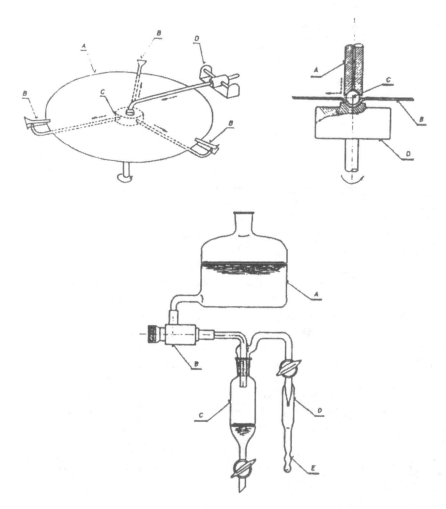

Figure 1   (Upper left) Method of fastening the chromatographic paper in the apparatus. A, Chromatographic paper; B, crocodile clips fixing the position of the paper; C, capillary distributor of the mobile phase with the ball; D, mounting of the distributor. (Upper right) Distribution of the mobile phase. A = Capillary supply tube; B = disc of chromatographic paper; C = ball of the distributor; D = bottom part of the bearing. Dotted arrow: direction of flow of the mobile phase. Solid arrow: direction of rotation. (Bottom) Scheme of indirect regulation of the movement of the mobile phase by means of mercury. A = Mercury reservoir; B = needle valve; C = reservoir of mobile phase with capillary supply of mercury; D = drop counter; E = mobile phase inlet of the chromatograph.

paper was replaced by a circular thin layer plate, and in model situations we reached run times of less than 2 minutes. This approach, however, was soon superseded by high-performance sorbents, which proved more versatile and more acceptable to the scientific community.

With respect to the subject involved, we perfected a number of chromatographic methods for the separation of parent collagen polypeptide chains, their polymers and fragments, an area in which we are still active. As an extracellular protein with an extremely slow metabolic turnover, collagen type I (present in skin, tendons, aponeuroses, and bones) was the protein of choice for depository effects (accumulation of metal ions like gold, molybdenum, lead, or zinc [14–16]) in tissues, but other collagens, namely types II and III, were studied as well. For years we dealt with nonenzymatically created cross-links (age pigments), which appear to be accumulated in connective tissue with aging [17–21]. These compounds stem from Maillard-type reactions and in some cases lead to fluorescent entities, which we succeeded in locating within the collagen molecule [22,23]. Someone may ask why we studied gold: the reason was to elucidate the mechanism of action of antirheumatic drugs possessing gold in their molecule (sodium aurothiosulfate), which turned out to be based on bound gold causing deswelling of the collagenous structure (and, perhaps, labile cross-links).

Another aspect of our work was to find out to what effect cross-linking in the extracellular matrix may be affected by feeding regimes [19–21]. Rather soon we realized that also aldehydes arising from the unsaturated fatty acid metabolism may be involved in a similar set of Maillard reactions as mentioned above (originally we were involved only in the studies about aldehydic sugars). All these studies meant a lot of peptide/protein separations. One of the puzzling observations in this respect is why the two parent α-chains of type I collagen possessing an extremely high internal homogeneity and equal in their relative molecular mass ($\sim$100,000) can be separated by standard polyacrylamide gel electrophoresis. It turned out that some secondary effects, besides sieving, must be involved (it cannot be differences in the charge, as both polypeptides are practically identical in this respect). When attempting to separate higher collagen polymers (up to 1×10^6 rel. mol. mass), we diluted our polyacrylamide gels up to 3% polyacrylamide (with some agarose added to preserve the mechanical strength for handling during detection), and it turned out that in such gels the parent α-chains comigrate while the polymers could be separated quite nicely (as a matter of fact, drawing Fergusson plots allowed distinguishing polymers of different chain composition [24]).

At a later stage, after we created our home-made capillary electrophoresis device [25], we were able to show that hydrophobic interactions between the polypeptide molecules and the inner capillary wall, particularly when separating collagen CNBr fragments, play their role in the separation process. In other words we successfully exploited a property of the proteins we had in our hands that everybody else tried to abolish, namely, adsorption of these analytes to the capillary wall. As a matter of fact, if we reduced adsorption to the inner capillary surface by appropriate inner wall modification, no or poor separations occurred [26,27]. Today we believe that such interactions can be exploited not only for proteins but for other categories of solutes provided that the capillary wall has favorable properties. We proposed, being enlighted by the advances of others achieved in electrochromatography, to call this type of separation open tubular electrochromatography. (For current status of collagen separation procedures, whether chromatographic or electromigration, see Ref. 28.) Figure 2 presents a comparison of separations obtainable in the mid-1970s to those obtainable today by both chromatographic and electromigration techniques. Figure 3 shows the potential of computerized treatment of the results.

The funniest story of our investigations is that one of our papers that is quoted quite frequently was a side product. When studying the different adducts of extracellular matrix proteins and fatty acid metabolites we regularly had two groups of laboratory rats: those fed ad libitum

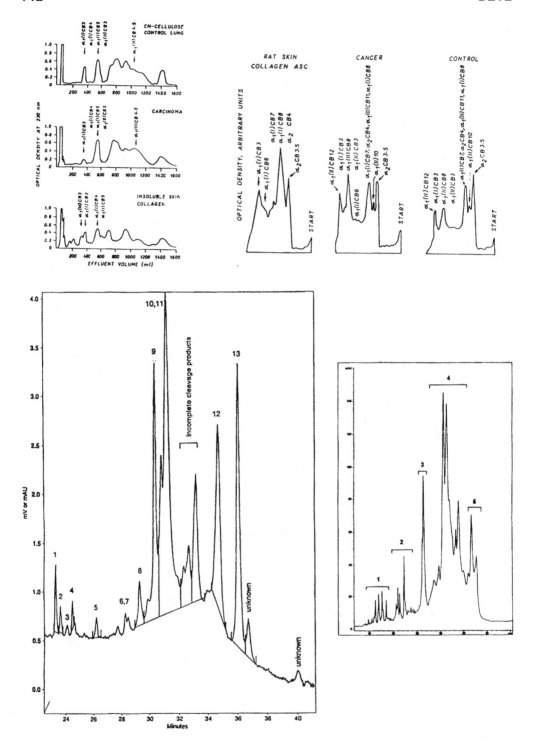

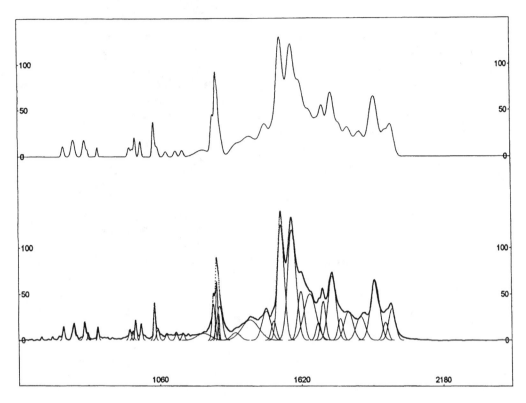

Figure 3 Separation of a natural mixture of CNBr collagen peptides obtained from rat tail tendon by capillary electrophoresis. Conditions CElect P150 capillary, 27 cm (20 cm to the detector) × 50 μm I.D., 75 mmol/L phosphate buffer containing 7.5% Pluronic (as molecular sieving exerting additive) pH 2.9, 20°C, 5 kV. The profile analyzed by the Peak Fit program. Thirty peaks can be distinguished; note that the theoretical number of peaks originating from the collagen sample should be 24; the 6 additional peaks are contaminants.

Figure 2 Comparison of different separation technologies in the 1970s with those available today. (Upper left) CM cellulose chromatogram of CNBr peptides of lung collagen. The location of peptides was done on the basis of the mobility of isolated peptide samples: 1.6 × 15 cm column was packed with CM cellulose (Whatman) CM 52. The column was operated at 42°C at a flow rate of 100 mL per hour, gradient elution (0.02–0.14 M Nacl + 0.02 M citrate buffer pH 3.6, 2000 mL total). Obtained in 1976. (Upper right) Densitometric tracing of polyacrylamide gel electrophoresis of CNBr peptides. Identification of zones based on the mobility of isolated peptides. 7% Polyacrylamide gel, stained and destained electrophoretically with Amido black. Obtained in 1976. (Lower left) Separation of collagen type 1 CNBr peptides obtained from a commercial preparation (Sigma) spiked with $\alpha_1(V)CB_1$ and $\alpha_1(III)CB_2$ by capillary electrophoresis. Background electrolyte used was Bio-Rad pH 2.5 phosphate buffer (with polymeric modifier) diluted 1: 1 with Milli-Q water. Peak identification: 1-$\alpha_1(I)CB_2$, 2-$\alpha_1(I)CB_4$, 3-$\alpha_1(V)CB_1$, 4-$\alpha_2(I)CB_1$, 5-$\alpha_1(III)CB_2$, 6-$\alpha_1(I)CB_5$, 7-$\alpha_2(I)CB_2$, 8-$\alpha_1(I)CB_3$, 9-$\alpha_1(I)CB_6$, 10-$\alpha_1(I)CB_7$, 11-$\alpha_1(I)CB_8$, 12-$\alpha_2(I)CB_4$, 13-$\alpha_2(I)CB_{3,5}$. Obtained in 1997. (Lower right) Reversed-phase chromatography of the same mixture shown in left lower panel. Peak identification: 1, $\alpha_2(I)CB_2$, $\alpha_1(I)CB_2$, 3-$\alpha_1(V)CB_1$, $\alpha_1(I))CB_5$,; 2, $\alpha_1(I)CB_4$, $\alpha_1(III)CB_3$, $\alpha_1(III)CB_6$, $\alpha_1(I)CB_3$, $\alpha_1(III)CB_4$; 3, $\alpha_1(I)CB_6$; 4, $\alpha_1(III)CB_5$, $\alpha_1(I)CB_7$, $\alpha_1(I)CB_8$, $\alpha_2(I)CB_4$, and incomplete cleavage products; 5, $\alpha_2(I)CB_{3,5}$, ($\alpha_1(III)CB_9$) $_3$. Obtained in 1999. (For experimental details, see Ref. 28.)

and those fed 50% of the amount consumed by the first group. A colleague of mine (late Dr. Stuchlikova) suggested to switch the regimes at one year of age (about half-life of a laboratory rat). The result was that the animals kept on the restricted regime for the first year got in their second half of life exceedingly fat as they developed distinct hyperfagia (many of the group weighted over 700 g, i.e. about the double of a control animal of comparable age). What, however, was most striking was that they had the best survival of all laboratory rats we have ever seen. After some reluctance we published the results. To me this means that not every type of obesity (at least in rats) is distinctly harmful [21].

III. MILESTONES IN SEPARATION SCIENCE IN THE TWENTIETH CENTURY

Perhaps everybody will agree that the progress in separation technologies over the past century has been tremendous. At the beginning chromatography was barely known and it was not until the 1930s that electromigration processes came into the game. Any evaluation of the crucial steps of the development will necessarily bear a personal imprint. Nevertheless, there are some points on which most of the separation scientists will agree. In the communication of these key steps I will limit myself to liquid phase separations as I never had the chance to be involved in gas chromatography. However, the theoretical background of both liquid phase chromatographic techniques and capillary electromigration separations was started with better understanding of processes involved in the gas phase separations.

In the 1940s and 1950s chromatographic separations were dominated by planar (one- and two-dimensional) techniques, sometimes operated in the overflow or repeated mode. Though classical silica-gel–based separations existed in parallel, they became routinely used only after the introduction of high-pressure (high-performance) sorbents of small grain sizes. They developed further into pellicular sorbents of uniformly sized and shaped particles, which helped considerably to achieve better resolution and shortening of the analysis time. A big step forward was the introduction of gel phases (soft gels), which, at their beginning, offered slow separations of macromolecular solutes with a relatively low resolution power. Nevertheless, this was a completely new way of looking at solutes so far inaccessible with other than ion exchange separations. Though, for example, Sephadex and agarose gels are still in use, particularly for preparative separations (typically in the first, enrichment step of preparation of biotechnologically engineered proteins), rigid gels withstanding high pressure (and offering better resolution) are preferred today.

A considerable step forward occurred in the late 1950s and 1960s when the first automated, ion exchange–based amino acid analyzers became available (later on modification applicable to peptide and carbohydrate analysis appeared). These analyses, however, took several hours, required precise setting of pH of the stepped gradient, and the ninhydrin-based detection exploiting detectors with segmented flow were anything but easy to use. I think that our monographs offered a good overview of liquid column chromatography and electromigration procedures of those early days [32–35].

In the 1970s analytical chemistry laboratories witnessed a booming expansion of reversed-phase techniques, which step-by-step have been applied not only for nonpolar solutes but for polar ones as well after the ion-pairing techniques became widely applied, not to mention the development of derivatization reactions. Also, the availability of reversed phases covering a wide range of polarity led to the nearly universal use of these techniques witnessed today (for a comprehensive review see Ref. 36).

Regarding electromigration techniques, the starting point was electrophoresis in free solution applied mainly to proteins. Along with the general use of flat-bed techniques, electrophoretic separations on paper, modified paper, and their layers acquired considerable popularity. Concomitantly, starch and agarose gels were used; with starch gels the problem of batch-to-batch differences of different starch preparations was never satisfactorily solved. It is to be emphasized that these separations were done for preparative rather than analytical purposes. The first main breakthrough was the introduction of polyacrylamide gels by Weber and Osborn [37] in the 1960s. The originally rod-shaped gels were relatively rapidly replaced by the flat-bed arrangement commonly used today.

It is worth mentioning that in the parallel continuous electrophoretic method, using a sheet of Whatman 3 paper for flow stabilization was developed by Grassmann and Hanning, which at a later stage (in particular for the separation of cells and cellular particles) were replaced by a thin layer of buffer placed between two planparallel glass plates [38]. Compared to polyacrylamide gel electrophoresis, these techniques (aimed at preparative purposes) have never reached the stage of general use, perhaps because the amount of the separated entities was too small and ultramicrotechniques have not been developed yet.

The next real milestone was the introduction of capillary electromigration separations. The first technique using electromigration in capillaries was isotachophoresis, introduced by Everaerts [39], another procedure that after a period of rather unsuccessful efforts for a wider application survives today mainly in the separation of low molecular weight inorganic ions.

A considerable advance in this area was the rediscovery of capillary zone electrophoresis by Jorgenson and Lukacs in the 1980s [40]. As for isotachophoresis, it is my opinion that the fact that the results of the analysis are an elevated plateau rather than a peak made it relatively less convenient than capillary zone electrophoresis, which followed. Even this, however, featured at its beginning a serious drawback: it was applicable to water-soluble, charged entities only. It was Terabe's ingenious idea [41] to introduce micellar (and later also microemulsion) [42] pseudophases that made this technique a mature, complementary approach to high-performance chromatographic separations. An alternative for water-insoluble/poorly soluble analytes (though possessing a charged moiety) is still the occasional use of nonaqueous buffers for this purpose [43]. This technique, however, yields considerable limitations. First the nonaqueous solvent used for the background electrolyte preparation has to possess a distinct dipole moment, which limits the number of solvents used to lower alcohols (mainly methanol and ethanol) and acetonitrile. Dimethyl formamide may be an alternative, but the problem is its high absorbancy in the short UV (amide bond), which precludes the use of an UV detector at wavelengths that would offer the required sensitivity.

One remarkable feature is to be noticed when reviewing the development of separation techniques over the past century. Polyacrylamide and agarose gel electrophoresis is an inevitable tool for the analysis of DNA, RNA, their fragments, PCR products, and their sequencing. Though the analytical techniques involved are routinely used today, their development mostly passed by the specialized analytical journals with, perhaps, the exception of *Analytical Biochemistry*, and the key publications appeared in the more biochemically (or nucleic acids) oriented literature [44].

IV. PERSONAL GLIMPSES

It was Karel Macek's wonderful idea to start east-west meetings on chromatography, inclusive electromigration procedures (remember that this was the time of Iron Curtain) in order to facilitate, frankly speaking, our contacts with the western world. He succeeded in this very well, and though these meetings have had several titles, they reached the international level and have been held biannually under the name Separations in BioSciences (SBS 1999 in Amsterdam being the most recent).

At the first few of these meetings there was a considerable language barrier. Though most of the attendees had a fair knowledge of German and some idea about Russian (except the Russians, who were just in the opposite situation), English was very, very rare. Consequently, the meetings were simultaneously translated into English, German, and Russian. This by itself was occasionally quite funny because you could hear remarks (in Czech) of the translators like "This idiot is too fast, I have to skip a piece" or "I cannot hear this mummy (the speaker was about 45), he talks too loudly" etc. The best of these stories is, perhaps, that of Gordon Consden, who was a prominent scientist both in planar chromatography and connective tissue proteins. At the meeting people like him (the celebrities) usually sat in the first two rows. The Russians tended to speak in their mother tongue but slowly turned to English as well. A famous theoretician from Leningrad decided to deliver his talk in English. After speaking for about 3 minutes, Gordon stood up, turned around to the audience and said: "Stop; would somebody kindly advise me on which channel this is translated into English?"

Generally speaking, it is difficult to penetrate the science world if English is not your first language. I remember that at a meeting in Texas organized by Al Zlatkis, Leslie Ettré, a Hungarian by origin, was asked to deliver the opening address, and he asked Al why he was to do it. The reply was: "If the folks here understand your accent, then they can understand anything."

In the 1970s when I was, along with Karel Macek and Jaroslav Janák, preparing the "mammoth" book on chromatography (the term used later by a referee in *Science* [32]), I was asked by the then director of the Chemistry Section at Elsevier, Marc Atkins (who, as his name suggests, is British, though he was and still is living part-time in Amsterdam), to deliver two sample pages from each contributor so that he could be sure that the messy language could be transformed into understandable English. So I asked individual contributors to supply me with these pages and, quite smartly, sent them to Amsterdam under code numbers. Because I had to supply sample pages of my own and because I was too lazy to prepare something special, I went to my secretary, took a reprint of our recent review paper published in *Journal of Chromatography Reviews* (this section ceased in the mid-1970s, if I remember correctly) and said to Mary: "Look, they want these sample pages; copy the beginning of this article and after you have written two pages send it to Marc in Amsterdam." Said and done. In a week came Marc's reaction: numbers 1–5 OK, number 6 tolerable, etc., etc., until No. 21, which was designed as absolutely not acceptable. Of course, this contribution was mine. I took the article, photocopied the first two pages (this was long before the era of xerox machines) and sent it back. I have never heard any complaints about my English since (though I am aware that it is far from being perfect and will stay so till the end of my days).

Later, when I became close with Marc and had traveled to different countries to collect papers from different symposia, we developed a game (as a matter of fact, it was Marc's idea— he kept teasing me about my improper English): at the airport in Prague where I came to meet him, "Here I am again with all my luggages, furnitures and informations"; in Sils Maria after a performance of Rhetic music in a deadly cold church, "I hope we shall see ourselves soon again" (I meant "each other"—there is no difference between these two expressions in Czech). Later the game was based on collecting improper translations of different announcements in public places. Two of these are in my view the best: in the old airport in Warsaw there was a baggage storage room which bore the message: "Luggage not claimed within 3 days will be liquefied." It is beyond my wildest dreams to see a suitcase turned into a liquid. The other message comes from the Cernomore hotel in Varna. Like any of these places, this one had on every floor instructions as to what to do in the case of fire. It ran as follows: "In case of fire take your head with you" (the real meaning was "Don't lose your head").

These stories are nearly all related to my activities with the *Journal of Chromatography*. The way I became involved starts with being late to one of the east-west meetings of Karel Macek. The place where the meeting was held had in front of the auditorium some sort of a bar or dining room where the meals were served (the place itself belonged to the then Czechoslovak Academy of Sciences, a former castle with the claim to fame that Casanova had spent a while there). The dining area was completely empty (the lectures have started about half an hour before) except for one person sitting in a corner sipping his tea. The person turned out to be Michal Lederer, the well-recognized founder and for many years the editor of *Journal of Chromatography*. "You don't care much about the lectures," he said, "You don't either " was my reaction. Ah, I have heard it so many times already, he replied (as you can see half a century later the situation has not changed much). We started talking, and it turned out that his family originated from a small town in eastern Bohemia called Náchod, that he was eager to see it, and the result was that we dropped the meeting completely, went to Náchod, stayed in two or three shabby places for refreshments, and returned shortly before midnight. When we were about to part, Michael said: "By the way, I need someone to help me with the Bibliography Section, which I would like to expand in the direction of electrophoresis. Would you like to do it?". And this is how I joined the *Journal of Chromatography* 35 years ago.

Over the years my efforts in science were a bit schizophrenic. I always believed that in order to be good in biosciences you have to possess a perfect knowledge of a technique and select a subject to which this technique could be applied. In my case, the technique was chromatography and, since the 1980s, also capillary electrophoresis, and the subject was proteins, more specifically collagen, as mentioned earlier. The proportions of the two areas altered in the course of years, and occasionally I became involved in other areas, like drug analysis and others. Coming back to collagen, there were, besides Gordon Consden, two other people I recall who had an interest in both separation science and collagen chemistry, namely the brothers Katchalskis,— Aharon and Ephraim. Aharon, who regretfully was killed during an attack of an Arabic commando on his aircraft in Athens, developed with a guy named Oplatka what they called Katchalski's perpetuum mobile. The machine was basically a wooden panel on which four wheels— two small ones and two big ones—were mounted. The wheels were joined with a string that, through four other small wheels, was directed to two jars containing a liquid. The beginning of their performance was always the same. As in a cabaret, they showed the panel from its front and back sides, emphasized that there was no engine involved, then they brought the audience to silence and usually Oplatka gave one of the wheels a push and the whole set of wheels started moving. This performance frequently preceded a festive dinner, and they let the machine run for the whole evening. The system actually works on the enthalpy of dilution, as the two liquids used were water and a highly concentrated salt solution. The string, being made from collagen, shrinks in the high-salt solution and becomes relaxed in water. Because the wheels joined were of different diameter, the momentum exerted on each was different and, consequently, all the joined wheels moved as long as the concentrations in both jars were equal (once Aharon commented on their performance by saying that this could be a way to desalinate the Dead Sea). No matter how funny (they always brought the panel covered by a piece of black cloth), the principle appeared as a Note in *Nature* some years later.

Ephraim (who later changed his name to Katzir and became in the times of Golda Meier the president of Israel) and I became acquainted during my first trip to Japan. This was at the World's Biochemistry Congress, which had, as usually, a number of satellite meetings, one of which was partly devoted to separation of collagen fractions and constituting polypeptide chains. As a celebrity in the field (this was long before his presidency), Ephraim was chairing a session during which one of the participants reported the separation of collagen fragments dissolved in 2 M sulfuric acid at elevated temperatures. The result, quite understandably, was a set of black liquids yielding smears by any applied separation method. At the end of this talk, Ephraim immediately called the next speaker without opening a discussion to the talk that had just ended. It was terribly hot, and in the evening after the session all the participants of the meeting gathered at a pool. With a great deal of respect (I was about 30 then) I approached Ephraim: "Would you mind a question, professor; why didn't you comment on that report on sulfuric acid treatment, which, in my view, was nonsense from the very beginning?" "My dear young fellow," he said, "you are not yet experienced enough in science, otherwise you would know that of the papers and contributions reported, about 98% can be criticized, and if the author is smart enough, they can be improved in one way or another; about 1% of papers are so perfect that they escape any criticism and any comment. The remaining 1% are such stupidities that talking about them is just a waste of time."

Here the story could end. However, it has an addendum. Nearly 30 years later I took part in the Bioaffinity Meeting in Kyoto, Japan. Upon entering the hotel where the meeting was to take place, I met a familiar face—Prof. Ephraim Katzir, president emeritus of Israel. I said "Hello" and remarked: "You surely don't remember me but we met some 30 years ago at the World Biochemistry Congress." "Sure, I do," was his reply. "You are the youngster astonished by dissolving collagen in sulfuric acid and chromatographing the resulting mess." We haven't seen each other since then.

V. PERSPECTIVES

From the current trends in separation science it is possible (with some degree of probability) to speculate about what breakthroughs can be expected in the near future. It is quite evident that even the most advanced separation techniques do not offer sufficient selectivity for very complex mixtures, particularly in biochemistry. Proteomics, peptide maps, nucleic acid fragments—all of them call for multidimensionality. The approach that is at hand is hyphenation, particularly the application of some of MS procedures [45]. The poor compatibility regards mainly the time needed for the separation step as compared with the time needed for obtaining a mass spectrum. On the other hand, the amount of sample elaborated in the separation step is unnecessarily high compared to that needed for a MS run. Consequently there is a need for speeding up (and perhaps miniaturization) of the separation procedures, which is likely to be separations on a chip (particularly chip-based electrophoresis). In spite of the fact that the first commercially available device for this type of separation appeared on the market this year [46], the technique and its different operational modes are far from being mature enough to achieve immediate wide applicability (most of the demonstrative examples still exploit only model mixtures). Another way of solving the problems of analyzing complex mixtures is represented by multiple hyphenation. There, however, we are facing retention (migration) time fluctuation, which is a considerable problem for automated operations. A solution proposed by Regnier [47] is based on hyphenating the so-called binar separations in which in each step only a single compound of the mixture is separated (yielding a yes-or-no answer, typically by means of biorecognition techniques).

As described in the previous section, electromigration and chromatographic techniques have developed side by side. Now we are witnessing that both of these approaches are becoming closer. The first result, already commonly used, is Terabe's micellar electrokinetic (microemulsion) chromatography [8,9] in which the separation occurs between the aqueous background electrolyte and the micellar (microemulsion) pseudophase, with the endo-osmotic flow having the effect of driving the contents of the capillary to the detector's window with electrophoretic separation of the analyte-loaded micelles in the opposite direction (in most cases). Electrochromatography [48] is an extreme of this situation: the solid phase packing is placed inside the capillary, while the applied high voltage on the capillary creates the endoosmotic flow, which essentially replaces a micropump. Whether packed columns or monolithic packing will ultimately dominate this type of separation remains to be seen. The stage in which electrochromatography is now resembles that of capillary electrophoresis in the early 1980s. The first monograph including theory, instrumentation, modes of application, and currently available applications will appear in the foreseeable future [49] (see also Ref. 50).

REFERENCES

1. Z Deyl. Chromatographic analysis of wastewaters from a TNT producing plant. Thesis Prague Institute of Chemical Technology (VSCHT), Prague 1957.
2. E Lederer, M Lederer. Chromatography, A Review of Principles and Applications. Amsterdam, 1959.
3. I Hais, K Macek, eds. Paper Chromatography, 3rd ed. (English translation), Academic Press, New York, 1964. [Papírová Chromatografie, 2nd ed. Publishing House of Czechoslovak Acad. Sci, Prague, 1959.]
4. K Macek, I Hais, eds. Stationary Phase in Paper and Thin-Layer Chromatography. Elsevier, Amsterdam, 1965.
5. K Macek, I Hais. Bibliography of Paper and Thin Layer Chromatography 1954–1956. Prague: Publishing House of the Czechoslovak Acad. Sci., 1960.
6. K Macek, I Hais, J Gasparic, J Kopecky, V Rabek. Bibliography of Paper Chromatography 1957–1960. Prague: Publishing House of the Czechoslovak Acad. Sci., 1962.
7. K Macek, I Hais, J Kopecky, J Gas/pario. Bibliography of Paper and Thin-Layer Chromatography 1961–1965. New York: Academic Press, 1967.
8. M Adam, C Dostal, Z Deyl. J Clin Chem Clin Biochem 17:495, 1979.
9. M Adam, J Musilova, Z Deyl. Clin Chim Acta 69:53, 1976.
10. E Svojtkova, Z Deyl, M Adam. J Chromatogr 84:147, 1973.
11. R Vitasek, J Coupek, K Macek, M Adam, Z Deyl. J Chromatogr 119:549, 1976.
12. O Vancikova, Z Deyl.
13. M Pavlicek, J Rosmus, Z Deyl. J Chromatogr 7:19, 1962.
14. M Adam, P Barth, Z Deyl, J Rosmus. Experientia 20:203, 1964.
15. B Bibr, Z Deyl, J Lener, M Adam. Int J Protein Peptide Res 10:190, 1977.
16. I Miksik, Z Deyl, J Herget, J Novotna, O Mestek. J Chromatogr A 852:245, 1999.
17. E Svojtkova, Z Deyl, A Smid, M Adam. Neoplasma 24:437, 1977.
18. M Adam, R Vitasek, Z Deyl, G Felsch, J Musilova, Z Olsovska. Clin Chim Acta 70:61, 1976.
19. Z Deyl, M Adam. Mechanisms Aging Development 6:25, 1977.
20. Z Deyl, M Juricova, J Rosmus, M Adam. Exp Gerontol 6:227, 1971.
21. E Stuchlikova, M Juricova-Horakova, Z Deyl. Exptl Gerontol 10:141, 1975.
22. Z Deyl, I Miksik, J Zicha. J Chromatogr A 836:161, 1999.
23. Z Deyl, I Miksik, R. Struzinsky. J Chromatogr 516:287, 1990.
24. E Svojtkova, Z Deyl, M Adam. J Chromatogr 84:147, 1973.
25. V Rohlicek, Z Deyl. J Chromatogr 487:87, 1989.
26. I Hamrnikova, I Miksik, Z Deyl, V Kasicka. J Chromatogr A 838:167, 1999.
27. I Miksik, Z Deyl. J Chromatogr A 852:325, 1999.
28. Z Deyl, I Miksik. J Chromatogr B 739:3, 2000.
29. LS Ettre. Anal Chem 43:20A, 1971.
30. LS Ettre. J Chromatogr 112:1, 1975.

31. L Zechmeister. In E Heftmann, ed. Chromatography, 2nd ed. New York: Reinhold, 1967, p. 3.
32. Z Deyl, K Macek, J Janak, eds. Liquid Column Chromatography—A Survey of Modern Techniques and Applications. Amsterdam: Elsevier, 1975.
33. Z Deyl, ed. Separation Methods. Amsterdam: Elsevier, 1984.
34. Z Deyl, FM Everaerts, Z Prusik, PJ Svendsen. Survey of Techniques and Applications, Part A, Techniques. J Chromatogr Library, Vol. 18A, Amsterdam: Elsevier, 1979.
35. Z Deyl, A Chrambach, FM Everaerts, Z Prusik. Electrophoresis, a Survey of Techniques and Applications, Part B, Applications. J Chromatogr Library, Vol. 18B. Amsterdam: Elsevier, 1983.
36. Z Deyl, I Miksik, F Tagliaro, E Tesarova. Advanced Chromatographic and Electromigration Methods in BioSciences. J Chromatogr Library, Vol. 60. Amsterdam: Elsevier, 1998.
37. K Weber, M Osborn. J Biol Chem 244:4406, 1969.
38. W Grassman, K Hannig. Naturvissenschaften 37:397, 1950.
39. FM Everaerts. In: Z Deyl, FM Everaerts, Z Prusik, PJ Svendsen, eds. Electrophoresis, Part A. Amsterdam: Elsevier, 1979, p. 193.
40. JW Jorgenson, KD Lukacs. Anal Chem 53:1981, 1298.
41. S Terabe, K Otsuka, K Ichikawa, T, Ando. Anal Chem 52:111, 1984.
42. S Terabe, K Otsuka, K Ichikawa, A Tsuchiya, T Ando. Anal Chem 56:111, 1984.
43. JL Miller, MG Khaledi, D Shea. Anal Chem 69:1223, 1997.
44. Z Deyl, J Janak, V Schwarz, I Miksik. Elsevier's Electronic Bibliography, Analytical Separations (CD-ROM). Amsterdam: Elsevier Science, 1998.
45. Z Deyl. Anal Separ News 4:7, 1999.
46. Agilent Technologies 2100 (LabChip Technology), Palo Alto, CA.
47. F Regnier, G Huang. J Chromatogr A 750:3, 1996.
48. M Dittmann, G Rozing. J Chromatogr A 744:63, 1996.
49. IS Krull, RL Stevenson, K Mistry, ME Everaerts. Capillary Electrochromatography and Pressurised Flow Capillary Electrochromatography. New York: HNB Publishing.
50. Z Deyl, F Svec, eds. Electrochromatography. Amsterdam: Elsevier, 2000, in press.

46

Stagnation and Regression Concomitant with the Advances in Electrophoretic Separation Science

Andreas Chrambach

National Institute of Child Health and Human Development,
National Institutes of Health, Bethesda, Maryland

I. INTRODUCTION

This chapter is devoted to incidences of scientific regression and stagnation encountered by the author in the course of four decades devoted to the development of electrophoretic methodology. Since those incidences reflect personal impressions, their presentation is necessarily unsystematic, nonexhaustive, and clearly subjective. Nonetheless, they may be relevant to the scientific process in this and other fields, which encompasses both learning and forgetting.

II. THE STAGNATION OF THEORETICAL AND PHYSICOCHEMICAL ELECTROPHORESIS

Modern electrophoresis surfaced with the appearance, in 1959, of a two-volume collection of chapters, edited by Milan Bier, under the title "Electrophoresis Theory, Methods and Applications," and in 1962, the distribution by the Eastman Kodak Company of a preprint under the title of "Disc Electrophoresis" comprising the theory of that method by Leonard Ornstein. The accent of these two ground-breaking publications is on electrophoretic and electrokinetic theory: The first three authors of what became known as "Bier's book" were Overbeek, Linderstrom-Lang, and Longsworth, monumental names in the history of electrophoresis and the establishment of its theory. One area of particular excitement at that time, reflected by both the "Bier book" and the "Ornstein Preprint," concerned the genesis and use in separation of "moving boundaries," i.e., concentration changes upon passage of a current in systems containing buffer discontinuities (phase interfaces). Moving boundary electrophoresis spawned an entire genealogy of theoretical developments extending into 1981, which is summarized in Ref. 1. One of the most extensive theoretical treatments of moving boundary electrophoresis, and the only one made practical through publication, albeit arcane by today's standards, of a computer program embedding it (see below), was that of T. M. Jovin [2]. Theoretical papers dealing with the fundamental unity of moving boundary electrophoresis and isoelectric focusing, and those with zone electrophoresis [3,4] followed. Thereafter, the interest in, and development of, electrophoretic theory decreased abruptly. In moving boundary electrophoresis, one of the important consequences of that stagnation is that to date, the vast majority of electrophoretic separations of native proteins proceed under a single set of conditions such as pH, buffer composition, and moving boundary displacement rate, rather than one selected from and optimized by the large number of buffer systems of both polarities and two temperatures operative across the entire pH range [5] computed by the program of Jovin. Also, the operative conditions, in exact physicochemical terms, of the most widely used gel electrophoretic moving boundary system (containing not only SDS, Tris, glycine, and chloride but also micellar SDS and complexes of buffer constituents with it) have remained largely unknown to the present day, preventing us from rationally and more effectively exploit even that single buffer system for separations.

Another incidence of stagnation in theoretical electrophoresis refers to gel electrophoresis and, in particular, the theory-based computation of molecular and gel fiber parameters as well as of optimal electrophoretic conditions on the basis of mobility data derived from several gel (or polymer) concentrations. The underlying theory for such computations, in the absence of sufficient physical data for gel structure available to date, is derived from a statistical extension of the Ogston model of gel structure [6]. It seems axiomatic that complex theory, unless embedded in user-friendly computer programs, cannot have a wide impact on the practice of electrophoretic separations, even when at times theoretical innovation in particular areas florishes. Both the moving boundary theory [5,7] and the gel electrophoretic theory [6,8], have been so embedded without, however, providing the expected massive impact on the practice of electrophoresis, which again could have activated the interest of theoreticians in the field. The failure of the available programs to have that impact is not due to their lack of user friendliness. Figure 1 depicts a representative output page providing a description of the physicochemical properties of a moving boundary system and a recipe section for setting up the particular system [5]. Corresponding buffer systems across the entire pH range, 0 and 25°C, for both orientations of the electric field have been computed [5]. Figure 2 illustrates similarly informative output with regard to optimally resolving gel concentration [9]. Other information based on the measurement of electrophoretic mobility at several gel or polymer concentrations, generated by the "PAGEPACK" [10] and "ELPHOFIT" [8] series of programs, provides recognition of family relationships among analytes and a correspondingly widened concept of molecular homogeneity, recognition of aggregation states and charge isomers, molecular weights of proteins with different values of the surface net charge (or nonideal species of SDS-proteins), values of surface charge

---

Figure 1   Representative discontinuous buffer system generated by the program of Jovin et al. [5]. (A) Physicochemical properties of the system. The trailing constituent is designated as 1, the leading constituents in phases BETA and GAMMA, respectively, as 2 and 3. The common counterion constituent is 6. Buffer phases BETA and GAMMA are those of the stacking and resolving gels as set; the corresponding operative phases formed after entrance of the moving boundary are ZETA and PI. RM is the mobility relative to that of Na$^+$. Analytes with RM between RM(1,ZETA) and RM(1,BETA) are stacked, those with RM less than RM(1,PI) are unstacked. pH, ionic strength, conductance KAPPA, and boundary displacement rate NU are shown, as is the buffer value of each phase. (B) Recipe for preparing buffers of phase composition BETA and GAMMA in polyacrylamide gels. (C) Possible subsystems with increasing values of RM(1,ZETA) and RM(1,PI). For each of those values the setting concentrations (M) of stacking gels or resolving gels are shown.

```
                    SYSTEM NUMBER
DATF · 08/26/70    COMPUTER SYSTEM NUMBER · JOV-CHP   2365
POLARITY · - (MIGRATION TOWARD ANODE)    TEMPERATURE ·   0  DEG. C.

CONSTITUENT 1 · NO.  43 , ASPARAGINE  (aspNH2)

CONSTITUENT 2 · NO.  84 , SULFATF·

CONSTITUENT 3 · NO.  99 , CHLORIDE·
CONSTITUENT 6 · NO.   9 , OR-ETHYLMORPHOLINE  (MEM)

            ALPHA(1)    ZETA(4)    BETA(2)    PI(9)    LAMBDA(8)   GAMMA(3)

C1          0.0400     0.0400                0.0624
C2                                0.0301               0.0469
C3                                                                0.1008
C6          0.0416     0.0416     0.0617    0.3212     0.3527     0.3596
THETA       1.039      1.039      2.052     5.146      7.515      3.568

PHI(1)      0.113      0.113                0.240
PHI(2)                            1.000                1.000
PHI(3)                                                            1.000
PHI(6)      0.109      0.109      0.975     0.047      0.266      0.280

RM(1)      -0.056     -0.056                -0.120
RM(2)                           -1.280               -1.280
RM(3)                                                            -1.626
RM(6)       0.066      0.066     0.595      0.028      0.162      0.171

PH          8.10       8.10       5.60      8.50       7.63       7.60

ION.STR.    0.0045     0.0045     0.0902    0.0150     0.1408     0.1008
SIGMA       0.483      0.483      10.969    1.607      17.120     21.747
KAPPA       123.       123.       2458.     393.       3663.      4831.
NU         -0.117     -0.117     -0.117    -0.075     -0.075     -0.075
BV          0.018      0.018      0.004     0.059      0.159      0.167
```

Polyacrylamide Gel Electrophoresis (PAGE) System No. 2365

	Vol. Ratio Stock Soln.	LOWER GEL (GAMMA phase)		UPPER GEL (BETA phase)		UPPER BUFFER (ALPHA phase)		LOWER BUFFER (EPSILON phase)	
		Components/100 ml Stock Solution	pH κ 25°C 1/4 dil.	Components/100 ml Stock Solution	pH κ 25°C 1/4 dil.	Components/liter	pH κ 25°C	Components/liter	pH κ 25°C
BUFFER	1	1 N HCl 40.31 ml MEM 18.87 g	7.30 8794	1 M H₂SO₄ 12.03 ml MEM 3.24 g	5.30 4959	aspNH₂ 6.00 g MEM 5.45 g	7.86 211	1 N HCl 50.0 ml MEM 8.20 g	6.29 4567
CATALYST	1	KP 60 mg RN 2 mg		KP 60 mg RN 2 mg		pH(PI) = 8.50 pH(ZETA) = 8.10 RM(1,ZETA) = 0.056 RM(2,BETA) = 1.280 RM(1,PI) = 0.120			
MONOMER SOLUTION	2	%T a) 10.0 b) 20.0 c) 40.0 %C 5		%T 6.25 %C 20		temperature = 0°C deaeration pressure = 10 mm Hg tracking dye = bromphenolblue 1 N H₂SO₄ = 0.5 N H₂SO₄			
		TD/100 ml gel (μl) a) 225 b) 100 c) 25		TD/100 ml gel (μl) 50					

Subsystem No.		OPERATIVE GEL Stacking Limit pH₀°			STACKING GEL sulfate β-MEM pH₀°			RESOLVING GEL chloride β-MEM pH₀°		
		Lower	Upper		(M)	(M)		(M)	(M)	
I	1	-0.056	-1.28	8.10	0.0301	0.0604	4.89	0.2180	0.2190	4.86
II	2	-0.081	-1.28	8.28	0.0301	0.1065	7.08	0.1503	0.2582	7.05
III	3	-0.106	-1.28	8.43	0.0301	0.1745	7.47	0.1147	0.3178	7.44
IV	4	-0.131	-1.28	8.55	0.0301	0.2688	7.73	0.0927	0.3925	7.70
V	5	-0.156	-1.28	8.65	0.0301	0.3954	7.94	0.0778	0.4819	7.91
VI	6	-0.181	-1.28	8.75	0.0301	0.5616	8.11	0.0671	0.5879	8.08
VII	7	-0.206	-1.28	8.84	0.0301	0.7777	8.27	0.0589	0.7134	8.24
VIII	8	-0.231	-1.28	8.93	0.0301	1.0575	8.41	0.0525	0.8636	8.38
IX	9	-0.256	-1.28	9.02	0.0301	1.4204	8.54	0.0474	1.0454	8.51
X	10	-0.281	-1.28	9.11	0.0301	1.8951	8.67	0.0432	1.2695	8.64
XI	11	-0.306	-1.28	9.20	0.0301	2.5244	8.80	0.0396	1.5518	8.77
XII	12	-0.331	-1.28	9.29	0.0301	3.3770	8.93	0.0366	1.9180	8.90
XIII	13	-0.356	-1.28	9.39	0.0301	4.5687	9.06	0.0341	2.4116	9.03

(A) Input File DATA06

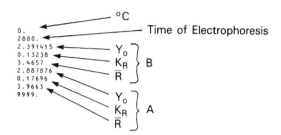

```
                          °C
0.        ◄──────────────── Time of Electrophoresis
2880.     ◄─────
2.391415  ◄─────      Y₀  ⎫
0.13238   ◄─────      Kᴿ  ⎬ B
3.4657    ◄─────      R̄   ⎭
2.887876  ◄─────
0.17696   ◄─────
3.9663    ◄─────      Y₀  ⎫
9999.     ◄─────      Kᴿ  ⎬ A
                      R̄   ⎭
```

(B) Program TOPT

Output 1

```
TEMPERATURE =   0.0      TIME = 2880.
SPECIES A: Y ZERO = 2.391415    KR =0.13238    RADIUS = 3.46570
SPECIES B: Y ZERO = 2.887876    KR =0.17696    RADIUS = 3.96630
V1 = 0.00000     V2 = 0.00000

GEL CONCENTRATION WHEN MOBILITY OF A = MOBILITY OF B    1.837677
GEL CONCENTRATION FOR MAXIMAL SEPARATION    4.665239
GEL CONCENTRATION FOR OPTIMAL RESOLUTION    7.492802  ◄────────
GEL CONCENTRATION WHEN SPECIES A UNSTACKS   2.860364
GEL CONCENTRATION WHEN SPECIES B UNSTACKS   2.602727

ANOTHER T-MAX AND T-OPT OCCURS WHEN T = ZERO

SIZE AND CHARGE SEPARATION ANTAGONISTIC; WE SUGGEST A CHANGE OF
PH, ISOELECTRIC FOCUSING, OR ISOTACHOPHORESIS UNLESS YOU ARE
DEALING WITH DIMERS OR OLIGOMERS.

GEL CONCENTRATION    SEPARATION    RESOLUTION    MOBILITY OF A    MOBILITY OF B


      0.000000        -.496461     -1.087041        2.39141          2.88788
      4.665239         0.145323     0.714342        0.57686          0.43154
      7.492802         0.107296     0.850016        0.24365          0.13635
      2.860364         0.099655     0.359430        1.00000          0.90034
      2.602727         0.081697     0.281829        1.08170          1.00000
     30.000000         0.000241     0.068598        0.00026          0.00001
```

(C)

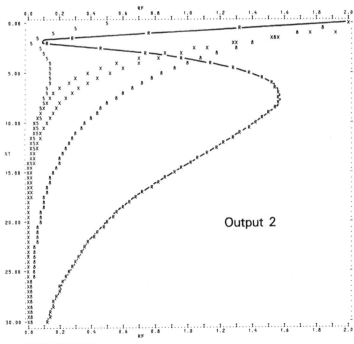

```
****LEGEND FOR GRAPH****

A--RF FOR SPECIES A
(--RF MINUS 2 STANDARD DEVIATIONS FOR SPECIES A
)--RF PLUS 2 STANDARD DEVIATIONS FOR SPECIES A
B-- RF FOR SPECIES B
<--RF MINUS 2 STANDARD DEVIATIONS FOR SPECIES B
>--RF PLUS 2 STANDARD DEVIATIONS FOR SPECIES B
S--SEPARATION
8--STANDARD DEVIATION OF DELTA RF * 4.37916
R--RESOLUTION *   1.8399
```

Output 2

(net protons/molecule), and statistically valid criteria of molecular identity or distinction. Against that available wealth of information, the present limitation of electrophoretic practice to arbitrary single gel concentrations and visual comparison of migration distances to evaluate homogeneity, identity, and molecular size appears impoverished.

III. APPARATUS

A. Analytical Apparatus

The breakthrough in analytical gel electrophoresis apparatus and procedure from a most often nonthermostated plastic box, gelation, gel staining and destaining, densitometry and evaluation of migration distances to one requiring as the sole manual operations the placing of a gel into an apparatus for computer-directed scanning of multiple lanes and applying the sample was achieved in 1994 [11]. That apparatus, in addition to its capacity for intermittent fluorescent scanning of seven gel lanes, had microgram-preparative capacity (see Sec. III. B), provided the mobility of all components during electrophoresis as well as their relative quantities (band areas), and had the capacity for software by which size and net charge of each component could be derived from mobility at any time during electrophoresis. That breakthrough from a manual to a largely automated conduct of gel electrophoresis failed to gain a sufficient market to make it economically viable. With the termination of its industrial production in 1999, the field of analytical gel electrophoresis has regressed to the previous manual procedures and the plastic-box type of apparatus.

A shocking incident of stagnation in the development of analytical gel electrophoresis apparatus concerns the two-dimensional separation of particles larger than 20 nm in diameter on agarose gels. The apparatus in question was originally designed by P. Serwer [12] and developed by D. Tietz to provide computed values of particle and gel fiber parameters and, in particular, the characterization of conjugate immunogens in terms of their sizes and net charge values [13]. This apparatus fell into oblivion, and with it a unique opportunity for the quality control in vaccine development.

B. Preparative Apparatus

I consider the function of separation science to be mainly preparative: There is no analytical substitute for being able to test directly the properties of the material separated as an electrophoretic zone. Nonetheless, preparative electrophoresis has stagnated over the last three or four decades.

Conventional microgram-preparative procedure largely consists of cutting out a stained band and eluting its contents by diffusion or, at best, by electroelution into a dialysis bag attached to the bottom of a gel. On rare occasions, electrophoresis into a cross-stream of elution buffer at the bottom of the gel is used. The first approach fails to take into account the peak distribution overlap between adjacent bands, i.e., the fact that stained bands only mark the mode of band distributions due to the insensitivity of staining. Also, diffusion from gel slices is slow and its

---

Figure 2   Representative program of the ''PAGE-PACK'' series describing the optimally resolving gel concentration between two components. (A) Input parameters. K_R and Y_0 are the slope and intercept on the (log mobility) axis of the Ferguson plot, respectively. R is the geometric mean radius of the particles. (B) Values of the optimally separating and the optimally resolving gel concentrations. (C) Graphic representation of optimal resolution (R) as a function of gel concentration.

yields are small. The second microgram-preparative method is burdened by extreme dilution unless its elution chamber features an electronic detector that allows one to elute at discontinuous flow rates. Against that background, the microgram-preparative module of Gombocz's scanning apparatus [11] discussed in Sec. I allows one to sequentially electroelute bands under computer control and with monitoring of band area as electroelution progresses [14]. With the disappearance of that apparatus from the market, microgram-preparative electrophoresis has regressed to its previous state.

Even more grave is the stagnation in the area of milligram-preparative electrophoresis. The convenient myth has it that milligrams of macromolecules are no longer needed at an age when analytical methods exist at a level many orders of magnitude less than those of a milligram. For a critical and multimethod analysis that is frequently not the case. In particular, the critical proof of homogeneity by electrophoretic reanalysis of the isolated material under various conditions of such as those of gel concentration, pH, ionic strength, and temperature requires isolation at a relatively large scale. Many assays of biological functions and many physicochemical techniques such as light scattering require relatively large samples. Nonetheless, the designs of milligram-preparative apparatus of the 1960s, based on the correct estimate of the load capacity of polyacrylamide gels of 0.1 mg/cm^2/zone, and therefore featuring gels of 20 cm^2 surface area or more (e.g., 15, 16), have only been downsized since that period. Present models with 2–3 cm^2 of gel surface are clearly microgram-, not milligram-preparative by the above-cited criteria. Moreover, the industrial production of apparatus with sufficient gel size has been discontinued for many years. Since the need for such apparatus in biochemistry continues for the reasons stated above, a restart of apparatus development at that load scale and its development (e.g., by placing a detector into the elution chamber and computer control of elution flow rate) seems inevitable at some future time.

By contrast with the regression in the area of milligram-preparative polyacrylamide gel electrophoresis, the development of preparative isolectric focusing of proteins has produced a number of new designs that formally pose no limit on the amount of isolated material [17]. However, since isoelectric focusing is inherently limited to separations based on differences in surface net charge, these methods cannot replace those of gel electrophoresis, which are responsive to differences in particle size and shape.

IV. A MULTIPLICITY OF LIKELY CAUSES FOR STAGNATION AND DECLINE

We have limited ourselves in the above discussion to two subjects—electrophoretic theory and apparatus. These two representative cases should be able to adequately illuminate the consequences of a scientific climate in which quick and dirty solutions to problems are widely preferred to those which are rigorous, methodical, quantitative, and therefore slow and laborious. That preference, again, is of course not capricious but results from a political reality in which the public support for scientific work suffices for a mere small fraction of investigators, thus engendering a desperate struggle and competition for economic survival among scientists.

Other causes of decline come into play. The decline of electrophoretic theory also appears to have had some roots in the loss of a spurt of enthusiasm similar to that engendered by a quorum, an accidental coming together of compatible scientists turning on one another in much the fashion that leads to the temporary blossoming of university departments and their decline when one or two of their leading lights depart.

In the two cited cases of programs embedding theoretical advances, however, an additional element appears to have been the spurt of youthful creative energy generated at the start of a scientific career. Jovin was a medical student when he devised his moving boundary theory and program, Rodbard a beginning post-doc when he arrived at a statistical extension of the Ogston model and conceived of its many practical corollaries. It is very likely in both cases that these first youthful bursts of creativity remained unsurpassed in the course of the subsequent careers of the authors.

Added to these accidentals, it appears nearly certain that the decline was also a consequence of changing scientific fashions by which sequentially physical chemistry, biochemistry, and immunology have been displaced by molecular biology and genetics in our time.

The failure of both the moving boundary and the gel electrophoretic programs to have a sizable impact was also due, in large part, to the fact that they preceded in time the personal computer. Thus, they were written in FORTRAN, executed on mainframe computers, and exported in form of magnetic tapes or microfiche, all of which are dinosaurs of the communication age. Belatedly, at least the gel electrophoretic programs were published in an extended form for personal computer [9] without, however, generating an appeal for a 13-year-old approach to "quantitative electrophoresis." The latest chance for the same programs, their incorporation into automated apparatus software, was also missed when the manufacturer of such apparatus discontinued its distribution (see Sec. III). The decline in the case of these programs seems in large part to consist of a missed opportunity due to missing the "right time" and to hiding the design for fear of industrial competition instead of trumpeting its virtues by means of published practical applications.

REFERENCES

1. LM Hjelmeland, A Chrambach. The impact of L. G. Longsworth on the theory of electrophoresis. Electrophoresis 3:9–17, 1982.
2. TM Jovin. Multiphasic zone electrophoresis, I, II, III. Biochemistry 12:871–890, 1973.
3. LM Hjelmeland, A Chrambach. Formation of natural pH gradients in sequential moving boundary systems with solvent counterions (I): theory. Electrophoresis 4:20–26, 1983.
4. M Bier, OA Palusinski, RA Mosher, DA Saville. Electrophoresis: mathematical modeling and computer simulation. Science 219:1281–1287.
5. TM Jovin, ML Dante, A Chrambach. Multiphasic Buffer Systems Output. Springfield, VA: National Technical Information Service, 1970.
6. D Rodbard, A Chrambach. Unified theory of gel electrophoresis and gel filtration. Proc Nat Acad Sci 65:970–977, 1970.
7. Schafer-Nielsen, PJ Svendsen. A unifying model for the ionic composition of steady-state electrophoresis systems. Anal Biochem 114:244–262, 1981.
8. D Tietz, A Chrambach. Concave Ferguson plots of DNA fragments and convex Ferguson plots of bacteriophages: evaluation of molecular and fiber properties, using desktop computers. Electrophoresis 13: 286–294, 1992.
9. D Rodbard, A Chrambach, GH Weiss. Optimization of resolution in analytical and preparative polyacrylamide gel electrophoresis. In: RC Allen and HR Maurer, eds. Electrophoresis and Isoelectric Focusing on Polyacrylamide Gel. Berlin: Walter de Gruyter, 1974, pp. 62–105.
10. A Chrambach. In: V Neuhofl, A Maelicke, eds. The Practice of Quantitative Gel Electrophoresis. Weinheim, Germany: VCH Pub., 1985, pp. 1–265.
11. E Gombocz, E Cortez. Separation, real-time migration monitoring and selective zone retrieval using a computer controlled system for automated analysis. Appl Theor Electrophoresis 4:197–209, 1994.
12. P Serwer. Two-dimensional agarose gel electrophoresis without gel manipulation. Anal Biochem 144:172–178, 1985

13. D Tietz, A Chrambach. Computer-assisted evaluation of polydisperse 2-dimensional gel patterns of polysaccharide-protein conjugate preparations with regard to size and net charge. Electrophoresis 10:667–679, 1989.

14. N Chen, A Chrambach. The preparative application of commercial automated gel electrophoresis apparatus to subcellular-sized particles: sequential isolations, fraction re-run, SDS-PAGE analysis, yield and purity. Electrophoresis 19:3096–3102, 1998.

15. T Jovin, A Chrambach, MA Naughton. An apparatus for preparative temperature regulated polyacrylamide gel electrophoresis. Anal Biochem 9:351–369, 1964.

16. G Kapadia, A Chrambach. Recovery of protein in preparative polyacrylamide gel electrophoresis. Anal Biochem 48:90–102, 1972.

17. PG Righetti, M Faupel, E Wenisch. Preparative electrophoresis with and without immobilized pH gradients. In: A Chrambach, MJ Dunn, BJ Radola, eds. Advances of Electrophoresis Vol. 5. Weinheim, Germany: VCH, 1992, GFR, pp. 159–200.

47

Lab-on-a-Chip: A Twentieth-Century Dream, a Twenty-First-Century Reality

Daniel Figeys and Jian Chen

MDS Proteomics, Inc., Toronto, Ontario, Canada

The twentieth century has seen the largest quantum leap in the development of science in human history. We have gained considerable knowledge to aid us in understanding life, nature, and beyond. All of the achievements have been due, in part, to the development of analytical technologies. These technologies provided fundamental information about the nature and the quantity essential to understand substance's functionality. The ability to rapidly investigate the nature and state of a substance opens the door to today's quality control, environmental studies, clinical diagnostic laboratories, as well as genomics and proteomics.

In the last part of the previous century analytical technology moved from highly specialized laboratories to point of care for diagnostics and field studies for pollution control and chemical warfare. Furthermore, we have also seen the burgeoning of requirements for analytical capabilities that can handle minimal amounts and volume of samples. The obvious solution to these issues was miniaturization, which offered portability, higher density of analytical test, reduction in cost, compatibility with minute amounts of sample, and speed. The driving force behind all this is a rapidly changing society in which pollution monitoring is increasingly important and chemical and biological warfare is a reality. On the bright side, the massive amounts of information now available on the genetic make-up of humans and other species require high-density analytical capability, i.e. miniaturization.

Since the 1970s, miniaturization of analytical instrumentation has gradually become a dominant trend in research and development. After the successful introduction of fused silica capillary columns for gas chromatography, researchers started to utilize fused silica capillary tubing to perform low-volume separation techniques such as liquid chromatography. Although the real improvement in column efficiency was from the introduction of small size packing material (particle size < 10 μm), researchers recognized the potential enhancement of sensitivity and handling of small sample volume that microcolumn offered. The introduction of fused silica tubing also encouraged the development of capillary electrophoresis technology. The unmatched separation efficiency of capillary electrophoresis (CE) encouraged many scientists to focus on bringing this technology into maturity in the 1980s.

Although the miniaturization of analytical instruments was evolving for all these years, it was not a paradigm shift. Such a shift occurred when scientists realized that the universal analytical process of sample preparation and analysis needed to go hand in hand for miniaturization to make sense. Surprisingly, this was accomplished from two different directions. The first direction was the development of static miniaturized system based on the microfabrication of DNA array, while the second direction was the development of microfluidic system. Harrison and Manz published in 1992 their pioneer work on microfabricated capillary channels on a planar glass chip [1]. Electrophoretic separation of a test mixture was achieved in these capillary channels.

The technology behind both the creation of static DNA array and dynamic microfluidic device is photolithography. Photolithography is the technology used for the fabrication of integrated electronic circuits and computer chips. Photolithography can create paths and control elements for electrons; it can also produce components for the control and mobilization of fluids. The integrated electrical and mechanical components fabricated with photolithography are now referred as microelectromechanical system (MEMS) devices. Like the effect that integrated circuits had on the creation of today's powerful computers, MEMS devices have the potential to integrate a roomful of analysis equipment into a compact lab-on-a-chip.

Since the pioneering work of Harrison and Manz [1,2], more and more papers have been published on various microfabrication techniques as well as applications. The number of research papers published increased exponentially (Fig. 1). More than 200 publications dedicated to microfabricated and/or microfluidic chip were published in the year 2000. Much of this effort was spawned by the growth of genomics and, recently, proteomics. Great growth in the emerging biochip technology is anticipated. One forecast predicted a sevenfold increase in the biochip market between the reported 1999 levels and 2005. During this period, biochip technology will pass the $1 billion bench mark (*thetareports.com*).

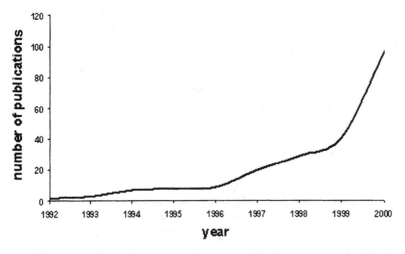

Figure 1 Number of research papers including the words microfabricated or microfluidic published from 1992 to 2000. (Data obtained by searching PubMed using key words of microfabricated or microfluidic.)

As we celebrate "a century of separation science," we will focus in this chapter on the lab-on-a-chip microfluidic chip, in particular: the chip construction techniques, separation applications, and other types of microchip devices.

I. MICROFABRICATION TECHNIQUES

A. Glass/Silicon-Based Microfluidic Systems

Most of the microfluidic devices developed so far were fabricated on glass or silicon wafers using photolithography technology. In reality the microfluidic devices were results of the intense efforts that had been made in developing the technology for integrated circuits. Photolithography was easily adapted to the fabrication of deeper pattern on silicon and glass. Harrison et al. produced their first CE chip with a well-established photolithography procedure [1]. The procedure involved masking a glass substrate with a metal layer, followed by lithographic patterning of the metal mask. The exposed glass substrate is then etched with HF-based solution to produce channels in the glass surface. The metal mask is then removed. Using this procedure, capillary channels 30 μm in width and 10 μm in depth can be obtained. A top glass/silicon substrate is then added on top of the first etched substrate (bottom plate) and bonded by heating the two substrates to their annealing temperature. The top plate had holes previously drilled through to provide access port to the channels.

A similar photolithography procedure is widely used today for integrated circuit (IC) and microchip fabrication (Fig. 2). The procedure, called reactive ion etching (RIE), can be easily

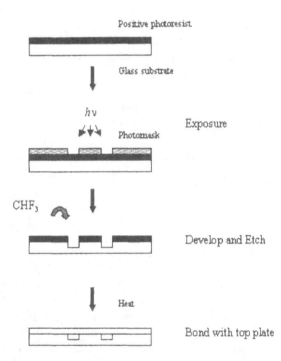

Figure 2   Illustration of reactive ion etching (RIE) photolithography procedure for creating channels on glass substrate.

adapted to produce glass and quartz microchips for separation applications [3,4]. The general procedure involves masking a glass or silica substrate with a layer of positive photoresist that breaks down upon light exposure, followed by exposure to UV light with a photomask that contains a designed pattern of capillary channels. The exposed glass or quartz wafer is then developed, leaving a channel structure on the surface of photoresist. Finally, CHF_3 gas is used to etch the glass substrate, creating a capillary channel network exactly as designed.

Glass and quartz microchips have many advantages. These chip materials provide well-defined surface properties, good electroosmotic flow properties, and excellent optical properties. Most electrophoretic or chromatographic separation applications have been reported using microfabricated planar glass or quartz wafer due to strong electroosmotic pumping on glass or quartz surfaces. However, the drawbacks of these materials are also obvious. The wet etching technique creates shallow elliptical-shaped channels with low aspect ratios, but channels with high aspect ratios (narrower and deeper) are desired. The standard photolithography procedure is well defined, yet complicated. The cost of producing such microchips is expensive, and mass production is difficult. Moreover, curtain compounds such as basic protein adsorb strongly to the surface of microfabricated glass or quartz chips.

B. Polymer-Based Microfluidic Systems

As alternatives to glass-based substrates, polymers or plastics are very attractive to be used in microchips. However, polymers have rarely been used for patterning and fabrication of small-scale objects. In particular, polymers are usually utilized in injection molding. In recent years, we have seen the development of alternative microfabrication technologies, such as "laser ablation," "imprinting and embossing," for the production of polymer-based microchips.

Among these new technologies, laser ablation is the most straightforward microfabrication technique [5]. Many commercial polymers, such as polystryrene, nitrocellulose, and polymethylmethacrylate (PMMA), can be patterned by laser ablation. The principle behind laser ablation is the absorption of pulsed laser light by the polymer substrate followed by a rapid gas phase expansion. Briefly, the polymer substrate is first covered with a chromium mask. The mask is patterned according to the desired design. The exposed regions of the polymer substrates adsorb a particular wavelength through the chromium mask. The laser light rapidly breaks chemical bonds on the polymer surface. The decomposed polymer products rapidly expand in the gas phase, leaving a photoablated cavity (Fig. 3). The depths of the channels are determined by the pulse energy and the number of laser pulses irradiating a single area on the polymer substrate. Laser ablation allows the creation of smooth, sharp, and vertical capillary channels on the surface of polymer substrates. The variations in the average wall depth are only about 0.13 μm.

The other important microfabrication techniques are imprinting and embossing. These techniques are very attractive because they provide a simple means of fabrication while keeping the cost low for small- to large-scale production. In the case of the imprinting method, small-diameter chrome wire is generally used to create channels on the surface of polymer substrate. The chrome wires are arranged to follow the layout based on architecture of the microchip. The polymer substrate is then sandwiched between two aluminum blocks and heated to its transition temperature (T_g). After the assembly is cooled, a channel network is formed on the surface of the polymer substrate by removing chrome wires [6]. The dimension of chrome wires defines the physical dimension of capillary channels. An imprinting alternative technique is to use a micromachined silicon master serving as the imprint template. The silicon master with raised three-dimensional structure can be fabricated by standard photolithographic procedures. The silicon imprint template is pressed into a polymer substrate, which is heated above its T_g. A channel network is thus formed on the surface of the polymer substrate [7,8]. In this way,

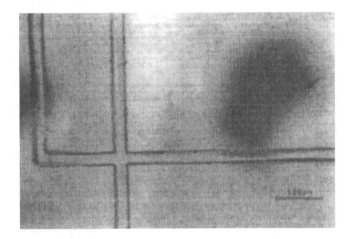

Figure 3 Photomicrograph of channel network on a glass plate. The channel width is 30 μm. (From Ref. 1.)

polymer microchips can be mass-produced at very low cost. Many polymer chips can be produced in this way. Among those, polydimethylsiloxane (PDMS) has received much attention due to its elastomeric properties. The material was used to create CE chips and other microfabricated devices, such as electrospray tips.

II. SEPARATION IN LAB-ON-A-CHIP

Separation capability is often an essential part of an analytical technique. Lab-on-a-chip technology has typically been constructed based on silicon/glass microfabrication. The process of fabrication on such a substrate and the properties of these materials initially limited the types of separation that could be performed. In particular, it became apparent that zone electrophoresis separation was the easiest one to implement on a microfabricated device.

A. Electrophoretic Separation

Fortunately, a tremendous amount of work has been devoted to the understanding of zone electrophoresis in miniaturized format, i.e., capillary electrophoresis. These results were easily applied to lab-on-a-chip separation. Capillary electrophoresis emerged in the late 1980s as a promising, highly efficient separation techniques based the movement of charged analytes in an electric field. As its name indicates, capillary electrophoresis is the separation of analytes based on a difference in mobility and performed in capillary tubing. A few hundred thousand theoretical column plates can be routinely obtained over 20–50 cm long, 50–100 μm internal diameter (i.d.) fused silica capillary. The improvement in performance is achieved by decreasing the capillary inner diameter, which improves heat dissipation and thus prevents band broadening inside the capillary tubing as well as allowing higher electrical potential to be applied.

Capillary electrophoresis was, therefore, strongly compatible with the world of microfluids. Earlier research efforts of on-chip electrophoretic separation by Harrison et al. [1,2] demonstrated that 40,000–75,000 theoretical plate numbers can be achieved on a 2.2 cm long separation channel. The width and depth of the capillary channel were 30 μm and 10 μm, respectively. A six-amino-acid mixture, all γ-fluorescein isothiocyanate (FITC)–labeled, were separated within

15 seconds. In the same report, a fast separation of a three-amino-acid mixture in a 0.75 cm long channel was achieved within 3 seconds. Although the number of theoretical plates was not as high as in capillary electrophoresis due to the geometry of the channel, the results clearly demonstrated that microfluidic systems could potentially open the doors to a new area of analytical capabilities.

Electrophoresis became an attractive technique for lab-on-a-chip as it provides electrophoretic and electroosmotic flows. Electrophoresis, as previously described, relates to the separation of charged molecule in an electric field due to differences in ionic mobility. Electroosmosis provides a bulk flow of liquid, which is induced by the charged silanol groups present on the surface of glass and silicon wafers. Therefore, applying a potential across the electrolyte-filled channel generates electroosmotic pumping, which is dependant on the presence of fixed charges on the inner channel wall, the pH of the electrolyte, ionic strength, and viscosity. In many microfabricated devices electroosmotic flow provides the sole foundation for fluid pumping and mixing. This was especially true in the earlier days of the lab-on-a-chip. By adequately applying the difference in potential between reservoirs on the device, electroosmotic pumping can be used to direct the bulk flow of solvent from one or many reservoirs, allowing mixing of different solutions.

Injection in chip-based separation techniques is typically based on electrokinetic principles, a combination of electrophoretic and electroosmosis mobility. Figure 4 shows a simplified illustration of an electrophoretic separation channel network layout on a chip. By applying a difference in potential between the sample reservoir and the waste for a predetermined time period, a sample plug can be injected from the sample reservoir. The sample plug is then separated by applying a separation potential between the buffer and the waste reservoirs. Active control of side channel potential is necessary to prevent "liquid crosstalk," caused by viscous drag induced in other channels along the separation channel.

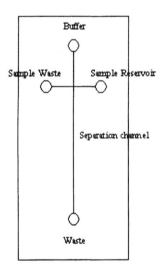

Figure 4 Simplified illustration of microfluidic chip design for the electrophoretic separation of analytes. The chip is first primed with buffer. The sample of interest is installed in the sample reservoir and mobilized towards the sample waste reservoir by applying an electric field. The cross section with the separation channel defines the injection volume. That sample plug is then separated in the separation channel by applying an electric field between the buffer reservoir and the waste reservoir.

B. High-Performance Liquid Chromatography on Chip

Capillary electrophoresis was the first separation technique to be implemented on microfluidic systems not because it is the most appropriate separation technique, but because its implementation was relatively straightforward. This was in part due to the fact that the narrow capillary channels in a microfabricated chip resemble the open capillary tubing in CE. Unfortunately, capillary electrophoresis was not at all the most popular separation technique. High performance liquid chromatography (HPLC) had dominated the robust separation of a large variety of compounds. However, the transfer of a capillary liquid chromatography (LC) method to a microfabricated separation chip was riddled with difficulties. Problems such as creating a retention frit for bed support and achieving a uniformed packing bed structure quickly appeared. Furthermore, the typical half-circle shape of the microfabricated channels required that a smaller particle size of LC stationary phase be used for the packing of the inner column. Smaller beads tend to generate higher backpressure, which can cause the devices to fail. McNeely et al. demonstrated the feasibility of directly microfabricating a network of interconnects on the chip [9]. This network of interconnects was then utilized for LC separation on chips.

Alternatively, the development of monolithic column material may provide a solution for the difficulties of constructing a LC chip [10–12]. The monolithic LC column is a polymer in nature, formed by in situ polymerization. The preparation of monolithic columns is relatively straightforward. Briefly, the column tube is filled with a mixture of monomer, activating agent, and porogen reagent. The polymerization is initiated inside the column or chip. Various pore sizes of monolithic bed can be obtained by controlling the concentration of porogen reagent. Ericson and colleagues recently reported their efforts to utilize this technique to construct monolithic capillary channels on a quartz chip [12] to explore the possibilities of creating capillary electrochromatography (CEC) and LC chips. In their experiments, a low-viscosity monomer solution was pressure-loaded into the selected channels and allowed to polymerize. Furthermore, there is no chromatographic restriction on the width of the channel or its configuration, and no retention frit is required. Both reversed- and ion-exchange phases were constructed. A mixture of alkyl phenones was separated under electrochromatographic mode, using electroosmotic pumping to drive the mobile phase across the separation channel. Furthermore, a 60-second separation of a four-protein mixture was obtained by pressure-driven anion-exchange chromatography on chip. Clearly, HPLC on a chip is still in early stage and still requires significant development. In particular, the lack of suitable valves that are compatible with HPLC solvents and pressure is a limiting factor.

In reality, substantially fewer efforts have been made so far to implement HPLC on a chip as compared to electrophoresis. Both electroosmotic pumping and pressure-driven systems are well suited for LC on a chip.

III. DETECTION SYSTEM IN LAB-ON-A-CHIP

Miniaturized detection systems have been integrated into microfabricated separation chips right from the beginning. Due to the very limited amount of sample, mostly in a nanoliter range or less, that can be injected into a separation channel, high-sensitivity detection was crucial to the performance of the lab-on-a-chip.

A. Laser-Induced Fluorescence Detection

Again, a tremendous amount of work has been done on the detection of minute amounts of sample in capillary electrophoresis. Laser-induced detection directly in capillary tubing had been shown to provide exquisite sensitivity. Because most of the separation chips were fabricated on glass or quartz-based material, which possesses great optical property, laser-induced fluorescence (LIF) became the favorite detection technique [1,2]. Furthermore, LIF detection system can be built using readily available lasers, optics, and photomultipliers, which can all be assembled in a lab for $10,000–20,000. Typically in these systems, a laser beam is either focused through an optical objective or delivered by an optical fiber to a predetermined window on a separation channel. The illuminated part of the separation channel serves as a detection window. Because the detection in situ does not require transfer of analytes to an external detection system, no extra band broadening is introduced. Typically, the laser is selected to have a wavelength corresponding the absorbance spectrum of a well-known fluorescence tag. This fluorescence tag is generally attached to the analytes of interest prior to the analysis. Once the labeled analytes pass through the illuminated window, they absorb the light and fluoresce at a higher wavelength. The fluorescence emission can be collected at different angles, focused through a spatial filter to remove scattered light and through a bandpass optical filter centered on the emission wavelength of the tag, and finally detected by a photomultiplier. Alternatively, an imaging system based on a CCD camera has been used to picture and digitize the fluorescence pattern generated in a selected area of the chip.

In addition to the common LIF technique, other detection methods are also being explored, such as electrochemical detection [13,14], holographic refractive index [15], etc. However, the sensitivity of some of these applications is limited. Efforts have also been made to introduce mass spectrometry (MS) as detection systems.

B. Mass Spectrometry

Mass spectrometry has become an indispensable tool for the pharmaceutical industry, high-throughput drug screening, proteomics, and many other scientific fields. In recent years, efforts have been focused on developing an on-chip interface to MS and different applications of chip MS. Most interfaces were fabricated to accommodate electrospray ionization (ESI), as it is the most common atmosphere pressure ionization (API) technique for both small and macromolecular analysis.

The first microfabricated device for the delivery of proteins to an ESI-MS was reported in 1997 [16]. The device consisted of a series of channels on glass for the individual delivery of samples to the MS. The device was installed in front of the MS, with the first channel aligned with the entrance of the MS. A protein sample was passed through the microfabricated device to the MS using a syringe pump attached to a reservoir. ESI was performed at the edge of the chip by applying a potential to the injection reservoir. After every analysis, the microfabricated device was translated to align the next channel to the entrance of the reservoir and the experiment was repeated with the next sample.

The development of a simple microfabricated device for the introduction of proteolytic digests to an ESI ion trap MS was reported in 1997 [17]. This device consisted of three reservoirs coupled to channels and to an ESI-MS via a transfer capillary. In this case, ESI was not generated at the edge of the chip. A transfer capillary was aligned and butt-connected at the edge of the device, with the end of the main channel through a hand-tight fitting system allowing the transfer of the effluents to capillary tubing and to a microelectrospray needle. The microfabricated devices handled and delivered a continuous low flow (100–200 nL/min) of protein digests by electroosmotic pumping to the ESI-MS/MS.

This approach was further refined to directly integrate a nanospray (electrospray at nanoliter per minute range) tip into the separation chip or to incorporate the efficient connection of a transfer capillary to a nanospray tip. The first approach was achieved by fixing a nanospray tip into a 400 μm opening drilled into the top cover plate [18]. Alternatively, a drilling technique to insert a transfer capillary at the end of a microfabricated device has also been reported [19,20]. The end result was a low dead-volume connection from the device to a transfer capillary, similar to a previously reported design. The chip is usually mounted on an XYZ stage for accurate position in front of the orifice.

The first system that integrated on-chip separation of analytes using capillary electrophoresis coupled to an ESI-MS was recently described [21]. The system consisted of three reservoirs coupled to an 11 cm long serpentine separation channel. A portion of the channels near the edge of the device was fabricated to 400 μm to accommodate a short-transfer capillary with a liquid junction ESI/MS interface. Another laboratory also reported a microfluidic CE separation system for peptides coupled to ESI-TOF-MS [18]. This design consisted of a simple, three-position microfabricated device coupled to a nanoelectrospray tip.

Other approaches have been designed to provide nanospray tip right on the surface or edge of chips. For example, Lee et al. constructed hollow needles projecting beyond the edge of a silicon wafer [22]. The same research group later fabricated a nanospray tip using an improved, simpler microfabrication process [23]. Two different types of tip, blunt and sharp tip emitters, were fabricated by applying different shapes of gold mask during a O_2 plasma etch process. The tips have performance characteristics very similar to the pulled fused silica nanospray tip. Stable ion spray can be achieved at low nanoliter per minute flow rate over a period of a few

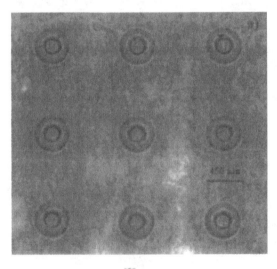

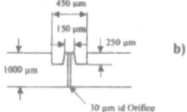

Figure 5 Microfabricated Emitter Array for the low flow electrospray ionization mass spectrometry. (a) 3 × 3 Array configuration of electrospray nozzles. (b) Schematic representation of a single nozzle. (From Ref. 24.)

hours. The microfabrication of a nanospray tip on the chip allows direct coupling of the separation channel to a MS, without introducing any extra void volume.

Researchers also attempted to fabricate nanospray nozzles on the surface of a planar glass chip or a polycarbonate substrate (Fig. 5). The construction process involves etching away surrounding material by photolithography followed by an etching process or laser etching process [24]. The advantage of this layout is its capability for possible high-throughput batch processing. Moreover, the configuration allows combining multiple layers of microfabricated chips together to form a three-dimensional structure, which can perform sample treatment, separation, and detection by mass spectrometry through nanospray tip array.

IV. LAB-ON-A-CHIP IN THE TWENTY-FIRST CENTURY

The lab-on-a-chip technology is entering a commercial phase with the recent and successful introduction by Agilent and Caliper of a lab-on-a-chip system suitable for protein, RNA, and DNA analysis. The successes of capillary electrophoresis are having a strong influence on the development of labs-on-a-chip. Capillary gel electrophoresis had an important role in the sequencing of the human genome as well as for routine DNA separation. It is interesting to see its equivalent being implemented on a chip.

Highly sensitive detection is always a challenge in chip format instrumentation. Lack of a universal detection technique can undermine the creation of a fully integrated microscale total analytical system (μTAS). However, with a suitable interface, a powerful analytical tool such as mass spectrometry may provide sensitive detection.

Over the last decade, many researchers all over the world have dedicated their efforts to creation of lab-on-a-chip. Increasing publication on the subject indicates that the miniaturizing chemical analysis system, which started in the 1970s, has entered a brand new era. Integration of sample treatment and separation technique into a single chip has been successfully achieved. The continuous advancements in technology will make ''lab-on-a-chip'' a twenty-first century reality.

REFERENCES

1. JD Harrison, A Manz, Z Fan, H Ludi, HJM Widmer. Anal Chem 64:1926–1932, 1992.
2. JD Harrison, K Fluri, K Seiler, Z Fan, CS Effenhauser, A Manz. Science 261:895–897, 1993.
3. B He, N Tait, F Regnier. Anal Chem 70:3790–3797, 1998.
4. B He, L Tan, F Regnier. Anal Chem 71:1464–1468, 1999.
5. MA Roberts, JS Rossler, P Bercier, H Girault. Anal Chem 69:2035–2042, 1997.
6. L Martynova, LE Locascio, M Galtan, GW Kramer, RG Christensen, WA MacCrehan. Anal Chem 69:4783–4789, 1997.
7. CS Effenhauser, GJM Bruin, A Paulus, M Ehrat. Anal Chem 69:3451–3457, 1997.
8. DC Duffy, JC McDonald, OJA Schueller, GM Whitesides. Anal Chem 70:4974–4984, 1998.
9. M McNeely, MK Spute, NA Tusneem, AR Oliphant. In: CH Ahn, AB Frazier, eds., Microfluidic Devices and System II. Bellingham, Washington: SPIE, 1999, pp 210–220.
10. A Maruška, C Ericson, Á Vévári, S Hjertén. J Chromatogr A 837:25, 1999.
11. C Ericson, S Hjertén. Anal Chem 71:1621–1626, 1999.
12. C Ericson, J Holm, T Ericson, S Hjertén. Anal Chem 72:81–87, 2000.
13. AT Woolley, K Lao, AN Glazer, RA Mathies. Anal Chem 70:684–688, 1998.
14. A Hilmi, JHT Luong. Anal Chem 72:4677–4682, 2000.
15. K Swinney, D Markov, DJ Bornhop. Anal Chem 72:2690–2695, 2000.

16. QF Xue, F Foret, YM Dunayevskiy, PM Zavracky, NE McGruer, BL Karger. Anal Chem 69:426–430, 1997.
17. D Figeys, Y Ning, R Aebersold. Anal Chem 69:3153–3160, 1997.
18. UM Lazar, RS Ramsey, S Sundberg, JM Ramsey. Anal Chem 71:3627–3631, 1999.
19. I Li, P Thibault, NH Bings, CD Skinner, C Wang, C Colyer, DJ Harrison. Anal Chem 71:3036–3045, 1999.
20. NH Bings, C Wang, CD Skinner, CL Colyer, P Thibault, DJ Harrison. Anal Chem 71:3292–3296, 1999.
21. B Zhang, H Liu, BL Karger, F Foret. Anal Chem 71:3258–3264, 1999.
22. A Desai, YC Tai, MT Davis, TD Lee. 1997 Inter. Conference on Solid State Sensors and Actuators (Transducers '97), Piscataway, NJ, May 1997, pp. 927–930.
23. L Licklider, X Wang, A Desai, YC Tai, TD Lee. Anal Chem 72:367–375, 2000.
24. K Tang, Y Lin, DW Matson, T Kim, RD Smith. Anal Chem 73:367–375, 2001.

Appendix: Selected Readings

The last century witnessed the publication of dozens of books dealing with all aspects of separation science; chromatography, and electrophoresis. Some discussed the theory of separation (e.g., *Dynamics of Chromatography*), others dealt with the basics of a technique (e.g., *Centrifugal Partition Chromatography*), while yet others dealt with the application of specific techniques to the separation of a closely related group of compounds (e.g., *Chiral Chromatography* or *HPLC Methods for Pharmaceutical Analysis*). Due to space limitations, a selected number of books is presented here in three different groups: classical books, published before 1965, modern books, published after 1965, and book series. Of course, a number of journals that publish articles related to separation science, which will not be listed here.

The books were selected in order to cover all areas of separation science. Some techniques are represented by more than one book. This is due to the fact that the different books present the subject matter in a different format or they were published at different times, and I felt compelled to list all of them.

A list of the volumes in a series complementary to the *Journal of Chromatography* is then presented. Each volume in the library series is an important and independent contribution in the field of chromatography and electrophoresis. The library contains no material reprinted from the journal itself.

Finally, a list of the volumes in the Marcel Dekker Chromatographic Science Series is presented.

SELECTED CLASSICAL SEPARATION SCIENCE BOOKS

Title	Author(s)/Editor(s)	Publisher	Year of publication
Chromatographie	Unknown	E. Merck AG	Unknown
The Moving Boundary Method of Studying the Electrophoresis of Proteins	A. Tiselius (thesis)	Almqvist & Wikells Boktryckeri AB	1930
Electrophoresis of Proteins and the Chemistry of Cell Surfaces	H. A. Abramson, L. S. Moyer, and M. H. Gorin	Reinhold Publishing Co.	1942
Principles and Practice of Chromatography	L. Zechmeister and L. Cholonky	Wiley	1943
Adsorption	C. L. Mantell	McGraw-Hill	1945
Chromatographic Adsorption Analysis	H. H. Strain	Interscience	1945
Vapor Adsorption	Edward Ledoux	Chemical Publishing Co.	1945
An Introduction to Chromatography	T. I. Williams	Chemical Publishing Co.	1946
Studies on Adsorption and Adsorption Analysis	Stig Gleason	Amqvist & Wiksells Boktryckeri A. B.	1946
Chromatographic Analysis	Faraday Society	Gurney and Jackson	1949
Ion Exchange: Theory and Application	F. C. Nachod, ed.	Academic Press	1949
Partition Chromatography	R. T. Williams and R. L. M. Synge	Cambridge University Press	1950
Progress in Chromatography 1938–1947	L. Zechmeister	Chapman & Hall	1950
Adsorption and Chromatography	Harold G. Cassidy	Interscience Publishers	1951
A Guide to Filter Paper and Cellulose Powder Chromatography	J. N. Balston and B. E. Talbot	H. Reeve Angel and Co.	1952
Papier Chromatographie	F. Cramer	Verlag Chemie	1952
Chromatography, A Review of Principles and Applications	Edgar Lederer and Michael Lederer	Elsevier Publishing Co.	1954
Die Papier Elektrophorese	Ch. Wunderly	Verlag H. R. Sauerlander	1954
The Elements of Chromatography	T. I. Williams	Blackie and Son Ltd.	1954
Papirova Chromatografie	I. M. Hais and K. Macek	Nakladatelstvi Ceskoslovenske Akademie Ved	1954
A Manual of Paper Chromatography and Paper Electrophoresis	R. J. Block, E. L. Durrum, and G. Zweig	Academic Press	1955
Chromatografia	Janiny Opienskiej-Blauth, Andrea Waksmundzkiego, and Marka Kanskiego	Panstwowe Wydawnictwo Naukowe	1957
Vapour Phase Chromatography	D. H. Desty, ed.	Academic Press	1957
Gas Chromatography	Howard Purnell	Wiley	1962
Ion Exchange Separations in Analytical Chemistry	O. Samuelson	Wiley	1963
Microdiffusion Analysis and Volumetric Error	Edward J. Conway	Chemical Publishing Co.	1963

SELECTED MODERN SEPARATION SCIENCE BOOKS

Title	Author(s)/Editor(s)	Publisher	Year of publication
Dynamics of Chromatography	J. Calvin Giddings	American Chemical Society	1965
Thin Layer Chromatography	E. Stahl	Springer-Verlag	1965
Principles of Adsorption Chromatography	L. R. Snyder	Marcel Dekker	1968
Separation Methods in Organic Chemistry and Biochemistry	F. J. Wolf	Academic Press	1969
An Introduction to Separation Science	B. Karger, L. Snyder, and C. Horvath	Wiley	1973
Extraction Chromatography	T. Braun and G. Ghersini, eds.	Elsevier	1975
Membranes in Separations	S-T. Hwang and K. Kammermeyer	Wiley	1975
Separation Methods in Biochemistry	C. J. O. R. Morris and P. Morris	Wiley	1976
Thin-Layer Chromatography	J. G. Kirchner	Wiley	1978
Densitometry in Thin Layer Chromatography	J. C. Touchstone and J. Sherma	Wiley	1979
Gel Chromatography	T. Kremmer and L. Boross	Wiley	1979
Introduction to Modern Liquid Chromatography	L. R. Snyder and J. J. Kirkland	Wiley	1979
Modern Size-Exclusion Liquid Chromatography	W. W. Yau, J. J. Kirkland, and D. D. Bly	Wiley	1979
Recent Developments in Chromatography and Electrophoresis	A. Fregerio and L. Renoz, eds.	Elsevier	1979
Advances in Thin Layer Chromatography	J. C. Touchstone	Wiley	1980
High Performance Liquid Chromatography	J. H. Knox	Edinburgh University Press	1980
Cell Electrophoresis in Cancer and Other Clinical Research	A. W. Preece and P. Ann Light	Elsevier	1981
Separation by Centrifugal Phenomena	H-W. Hsu	Wiley	1981
Dictionary of Chromatography	Hans-Peter Angele	Huthig	1984
Open Tubular Column GC	M. L. Lee, F. J. Yang, and K. D. Bartle	Wiley	1984
Affinity Chromatography	P. D. G. Dean, W. S. Johnson, and F. A. Middle	IRL Press	1985
Chromatographic Separations of Stereoisomers	R. W. Souter	CRC Press	1985
Modern Practice of Gas Chromatography	R. L. Grob, ed.	Wiley	1985
Techniques and Applications of Thin Layer Chromatography	J. C. Touchstone and J. Sherma	Wiley	1985
Electrophoresis	A. T. Andrews	Clarendon Press	1986

Title	Author(s)/Editor(s)	Publisher	Year of publication
Handbook of Ion Chromatography	J. Weiss and E. L. Johnson	Dionex Corp.	1986
HPLC Instrumentation and Applications	J. C. MacDonald	International Sci. Communications	1986
Optimization of Chromatographic Selectivity	Peter J. Schoenmakers	Elsevier Science Publishers	1986
Preparative Chromatography Techniques	K. Hostettmann, M. Hostettman and A. Marston	Springer-Verlag	1986
Preparative High Performance Liquid Chromatography	M. Verzele and C. Dewaele	TEC-Belgium	1986
Fundamentals of TLC	Friedrich Geiss	Huthig	1987
Introduction to Microscale HPLC	Daido Ishii, ed.	VCH	1987
New Directions in Electrophoretic Methods	J. W. Jorgenson and M. Phillps	American Chemical Society	1987
Countercurrent Chromatography: Theory and Practice	N. B. Mandava and Y. Ito	Marcel Dekker	1988
Ion Chromatography	H. Small	Plenum	1989
Affinity Electrophoresis Principles and Applications	J. Breborowicz and A. Mackiewicz	CRC Press	1991
HPLC of Proteins, Peptides and Polynucleotides	M. T. W. Hearn	VCH	1991
Chromatographia: The First 25 Years	L. S. Ettre, ed.	Vieweg	1992
Chromatography	E. Heftmann	Elsevier	1992
Introduction to Micellar Electrokinetic Chromatography	J. Vindevogel and Pat Sandra	Huthig	1992
Practice of Thin Layer Chromatography	J. C. Touchstone	Wiley	1992
Principles and Practices of Solvent Extraction	J. Rydberg, C. Musikas, and G. R. Choppin, eds.	Marcel Dekker	1992
Pulsed-Field Gel Electrophoresis	M. Burmeister and L. Ulanovsky	Humana Press	1992
Capillary Electrophoresis Technology	N. Guzman	Marcel Dekker	1993
Practical Capillary Electrophoresis	R. Weinberger	Academic Press	1993
Handbook of Derivatives for Chromatography	K. Blau and J. M. Halket	Wiley	1994
Capillary Gas Chromatography	David W. Grant	Wiley	1995
Centrifugal Partition Chromatography	A. P. Foucault	Marcel Dekker	1995
Chemical Separations with Liquid Membranes	R. A. Bartsch and J. Douglas Way	ACS Symposium Series, Vol. 642	1996
Handbook of Capillary Electrophoresis	J. P. Landers	CRC Press	1996
Handbook of Thin Layer Chromatography	J. Sherma and B. Fried	Marcel Dekker	1996
Basic Gas Chromatography	H. M. McNair and J. M. Miller	Wiley	1997

Title	Author(s)/Editor(s)	Publisher	Year of publication
Cell Separation Methods and Applications	D. Recktenwald and A. Radbruch	Marcel Dekker	1997
Chiral Separations	Satinder Ahuja	American Chemical Society	1997
Electrofocusing and Isotachophoresis	B. J. Radola and D. Graesslin, eds.	Walter de Gruyter	1997
Electrophoresis in Practice	R. Westermeier	VCH	1997
HPLC Columns, Theory, Technology and Practice	Uwe D. Neue	Wiley	1997
Introduction to Analytical Gas Chromatography	R. P. W. Scott	Marcel Dekker	1997
Practical HPLC Method Development	L. Snyder, J. Kirkland, and J. Glach	Wiley	1997
Static Headspace GC, Theory and Practice	B. Kolb and L. S. Ettre	Wiley-VCH	1997
Supercritical Fluid Chromatography with Packed Columns	K. Anton and C. Berger	Marcel Dekker	1997
Data Analysis and Signal Processing in Chromatography	A. Fellinger	Elsevier	1998
Handbook for Derivatization Reactions for HPLC	George Lunn and Louise C. Hellwig	Wiley	1998
Handbook of HPLC	E. Katz, R. Eksteen, P. Scheonmakers, and N. Miller	Marcel Dekker	1998
High Performance Capillary Electrophoresis	M. G. Khaledi, ed.	Wiley	1998
Anal. and Prep. Separation Methods of Biomolecules	Hassan Y. Aboul-Enein	Marcel Dekker	1999
Chiral Chromatography	T. E. Beesley and R. P. W. Scott	Wiley	1999
Countercurrent Chromatography	J. M. Menet and D. Thiebaut	Marcel Dekker	1999
Liquid Chromatography/Mass Spectrometry	W. M. A. Niessen	Marcel Dekker	1999
Unified Chromatography	J. F. Parcher and T. L. Chester	American Chemical Society	1999
Ion Exchange	D. Muraviev, V. Gorshkov, and A. Warshawsky	Marcel Dekker	2000
Micellar Liquid Chromatography	A. Berthold and C. G. Alvarez-Coque	Marcel Dekker	2000
Solid Phase Extraction	N. J. K. Simpson, ed.	Marcel Dekker	2000
Encyclopedia of Chromatography	J. Cazes, ed.	Marcel Dekker	2001

SELECTED SEPARATION SCIENCE SERIES

Series title	Editor	Publisher
Chromatographic Science Series	Jack Cazes	Marcel Dekker
HPLC Methods for Pharmaceutical Analysis	George Lunn	Wiley
Handbook of Chromatography	G. Zweig and J. Sherma	CRC Press
Journal of Chromatography Library	Different author/editor for each volume	Elsevier
Advances in Chromatography	J. C. Giddings, E. Grushka, P. R. Brown	Marcel Dekker
HPLC; Advances and Perspectives	Csaba Horvath	Academic Press
Advances in Electrophoresis	A. Chrambach, M. J. Dunn, and B. J. Radola, eds.	VCH
Electrophoresis Library	B. J. Radola, ed.	VCH
Chromatographic Methods	W. Bersch, H. Frank, W. G. Jennings, and P. Sandra, eds.	Huthig

JOURNAL OF CHROMATOGRAPHY LIBRARY

1. Chromatography of Antibiotics (see also Volume 26), *G. H. Wagman and M. J. Weinstein*
2. Extraction Chromatography, *edited by T. Braun and G. Ghersini*
3. Liquid Column Chromatography. A Survey of Modern Techniques and Applications, *edited by Z. Deyl, K. Macek, and J. Janák*
4. Detectors of Gas Chromatography, *J. Sevcik*
5. Instrumental Liquid Chromatography. A Practical Manual on High-Performance Liquid Chromatographic Methods (see also Volume 27), *N. A. Parris*
6. Isotachophoresis. Theory, Instrumentation and Applications, *F. M. Everaerts, J. L. Beckers, and Th. P. E. M. Verheggen*
7. Chemical Derivatization in Liquid Chromatography, *J. F. Lawrence and R. W. Frei*
8. Chromatography of Steroids, *E. Heftmann*
9. HPTLC—High Performance Thin-Layer Chromatography, *edited by A. Zlatkis and R. E. Kaiser*
10. Gas Chromatography of Polymers, *V. G. Berezkin, V. R. Alishoyev, and I. B. Nemirovskaya*
11. Liquid Chromatography Detectors, *R. P. W. Scott*
12. Affinity Chromatography (see also Volume 55), *J. Turková*
13. Instrumentation for High-Performance Liquid Chromatography, *edited by J. F. K. Huber*
14. Radiochromatography. The Chromatography and Electrophoresis of Radiolabelled Compounds, *T. R. Roberts*
15. Antibiotics. Isolation, Separation and Purification, *edited by M. J. Weinstein and G. H. Wagman*
16. Porous Silica. Its Properties and Use as Support in Column Liquid Chromatography, *K. K. Unger*
17. 75 Years of Chromatography—A Historical Dialogue, *edited by L. S. Ettre and A. Zlatkis*
18. Electrophoresis: A Survey of Techniques and Applications, Part A: Techniques, Part B: Applications, *edited by Z. Deyl*
19. Chemical Derivatization in Gas Chromatography, *J. Drozd*
20. Electron Capture. Theory and Practice in Chromatography, *edited by A. Zlatkis and C. F. Poole*
21. Environmental Problems Solving Using Gas and Liquid Chromatography, *R. L. Grob and M. A. Kaiser*
22. Chromatography. Fundamentals and Applications of Chromatographic and Electrophoretic Methods (see also Volume 51), Part A: Fundamentals, Part B: Applications, *edited by E. Heftmann*
23. Chromatography of Alkaloids, Part A: Thin-Layer Chromatography, *A. Baerheim-Svendsen and R.*

53. Hyphenated Techniques in Supercritical Fluid Chromatography and Extraction, *edited by K. Jinno*
54. Chromatography of Mycotoxins: Techniques and Applications, *edited by V. Betina*
55. Bioaffinity Chromatography, Second, completely revised edition, *J. Turková*
56. Chromatography in the Petroleum Industry, *edited by E. R. Adlard*
57. Retention and Selectivity in Liquid Chromatography: Prediction, Standardisation and Phase Comparisons, *edited by R. M. Smith*
58. Carbohydrate Analysis, *edited by Z. El Rassi*
59. Applications of Liquid Chromatography/Mass Spectrometry in Environmental Chemistry, *edited by D. Barceló*
60. Advanced Chromatographic and Electromigration Methods in BioSciences, *edited by Z. Deyl, I. Mikšík, F. Tagliaro and E. Tesařová*

MARCEL DEKKER CHROMATOGRAPHIC SCIENCE SERIES

1. Dynamics of Chromatography, *J. Calvin Giddings*
2. Gas Chromatographic Analysis of Drugs and Pesticides, *Benjamin J. Gudzinowicz*
3. Principles of Adsorption Chromatography: The Separation of Nonionic Organic Compounds, *Lloyd R. Snyder*
4. Multicomponent Chromatography: Theory of Interference, *Friedrich Helfferich and Gerhard Klein*
5. Quantitative Analysis by Gas Chromatography, *Josef Novák*
6. High-Speed Liquid Chromatography, *Peter M. Rajcsanyi and Elisabeth Rajcsanyi*
7. Fundamentals of Integrated GC-MS (in three parts), *Benjamin J. Gudzinowicz, Michael J. Gudzinowicz, and Horace F. Martin*
8. Liquid Chromatography of Polymers and Related Materials, *Jack Cazes*
9. GLC and HPLC Determination of Therapeutic Agents (in three parts), *Part 1 edited by Kiyoshi Tsuji and Walter Morozowich, Parts 2 and 3 edited by Kiyoshi Tsuji*
10. Biological/Biomedical Applications of Liquid Chromatography, *edited by Gerald L. Hawk*
11. Chromatography in Petroleum Analysis, *edited by Klaus H. Altgelt and T. H. Gouw*
12. Biological/Biomedical Applications of Liquid Chromatography II, *edited by Gerald L. Hawk*
13. Liquid Chromatography of Polymers and Related Materials II, *edited by Jack Cazes and Xavier Delamare*
14. Introduction to Analytical Gas Chromatography: History, Principles, and Practice, *John A. Perry*
15. Applications of Glass Capillary Gas Chromatography, *edited by Walter G. Jennings*
16. Steroid Analysis by HPLC: Recent Applications, *edited by Marie P. Kautsky*
17. Thin-Layer Chromatography: Techniques and Applications, *Bernard Fried and Joseph Sherma*
18. Biological/Biomedical Applications of Liquid Chromatography III, *edited by Gerald L. Hawk*
19. Liquid Chromatography of Polymers and Related Materials III, *edited by Jack Cazes*
20. Biological/Biomedical Applications of Liquid Chromatography, *edited by Gerald L. Hawk*
21. Chromatographic Separation and Extraction with Foamed Plastics and Rubbers, *G. J. Moody and J. D. R. Thomas*
22. Analytical Pyrolysis: A Comprehensive Guide, *William J. Irwin*
23. Liquid Chromatography Detectors, *edited by Thomas M. Vickrey*
24. High-Performance Liquid Chromatography in Forensic Chemistry, *edited by Ira S. Lurie and John D. Wittwer, Jr.*
25. Steric Exclusion Liquid Chromatography of Polymers, *edited by Josef Janca*
26. HPLC Analysis of Biological Compounds: A Laboratory Guide, *William S. Hancock and James T. Sparrow*
27. Affinity Chromatography: Template Chromatography of Nucleic Acids and Proteins, *Herbert Schott*
28. HPLC in Nucleic Acid Research: Methods and Applications, *edited by Phyllis R. Brown*
29. Pyrolysis and GC in Polymer Analysis, *edited by S. A. Liebman and E. J. Levy*
30. Modern Chromatographic Analysis of the Vitamins, *edited by André P. De Leenheer, Willy E. Lambert, and Marcel G. M. De Ruyter*

Index